International Reference Guide to

SPACE LAUNCH SYSTEMS

Steven J. Isakowitz

Second Edition

Updated by

Jeff Samella

In cooperation with the AIAA Space Transportation Technical Committee (STTC)

Educational Outreach Sponsors

Lockheed Martin Corporation

Thiokol Corporation

McDonnell Douglas Corporation

Published and distributed by

American Institute of Aeronautics and Astronautics

370 L'Enfant Promenade, SW

Washington, DC 20024-2518

Foreword

This edition of the guide is the second of what I hope will become a continuing effort to provide timely, accurate launch vehicle information to launch vehicle users, manufacturers, analysts, and students. The response to the first edition of this book exceeded my expectations. A recent survey was conducted of buyers of the guide to solicit their ideas for a second edition. Overall, there was overwhelming support for the guide, its extensive information, and its unquie ease of use. Based on this strong support and the number of purchases, AIAA and its Space Transportation Technical Committee enthusiatically agreed to develop this second edition.

Acknowledgments

As with the first edition, I am indebted to a number of people for their hard work and perseverance. Most importantly, I would like to thank Jeff Samella who led the team at NASA Lewis Research Center that coordinated the update for the second edition. The team consisted of individuals from the Advanced Space Analysis Office at NASA Lewis, which has a long history in the expendable launch vehicle arena and maintains an extensive database and library of launch vehicle technical information for the space agency. This edition would not have been possible without the endless hours of hard work and dedication to excellence provided by Jeff Samella.

I would also like to thank Leslie Douglas at Analex Corporation who provided invaluable editorial support; Brian Beaver at Analex Corporation who provided an extensive update to the Russia/Ukraine launch vehicle sedtions; and Karen Poniatowski and Mike Osmolovsky at NASA Headquarters for their coordination between NASA, AIAA, and industry. Below is a list of organizations and individuals who contributed technical information and support important to this second edition.

AIAA
Gayle Armstrong
Cort Durocher
Rodger Williams

Arianespace
Michelle Lyle

European Space Agency (ESA)
Jeff Lieberherr

Central Specialized Design Bureau
Gennadi Petrovich Anshakov
Dmitri Illch Kozlov

GW Aerospace, Inc.
Zuoyi Huang

Cortez III
Lorraine Feher
Irene Shaland

Indian Space Research Organization (ISRO)
S. Khrisnamurthy

The Institute of Space and Astronautical Science (ISAS)
Kayo Shoji

Khrunichev State Research and Production Space Center
Viadimir Konstantinovich Karrask

LKE International
Erik Laursen

Loral Vought Systems
Craig Vanbebber

Lockheed Martin
Susan Fields Bailey
Carol Berardino
Virnell Bruce
Don Davis
Ray Ernst
Robert White

McDonnell Douglas
Robert Carlson
Evelyn Smith

NASA Headquarters
Charles Gunn

NASA Lewis Research Center
Mark Mulac
J. Joseph Nieberding
O. Frank Spurlock
Timothy Wickenheiser

National Space Development Agency of Japan (NASDA)
Tomifumi Godai

NPO Energia
Christopher Faronetta

NPO Yuzhnoye
Yuri Alexeivich Smetanin

Orbital Sciences Corporation
Barney Gorin
Joseph Padavano
Chirstopher Schade

Plowshare Technology Limited
Alan Johnstone

PO Polyot
Alexander Ivanovich Ilyin

Shaland and Associates
Alex Shaland

STC Complex
Victor Andrushin
Yuri S. Solomanov

U.S. Department of Transportation
Damon Wells

Original Cover Art by Dennis Smith

For my wife, Monica, and children,
Matthew, Jennifer, Rachel, and Sophie.

Table of Contents

We hope the professor from Clark College [Robert H. Goddard] is only professing to be ignorant of elementary physics if he thinks that a rocket can work in a vacuum.

- *Editorial,* New York Times, *1920*

Introduction

Space is no longer the sole province of only a few elite nations. Many nations have entered the space arena as they recognize the importance of space for communications, Earth observation, planetary exploration, national security, commercial ventures, advanced technology, and national prestige. Key to attaining many of these lofty space goals is the ability to transport payloads into space with launch vehicles. Eight countries have demonstrated this ability and others are actively developing the capability. These launch vehicles not only serve the needs of their respective governments, but are often pitted against each other in a highly competitive commercial marketplace. Worldwide there are currently 19 different space launch systems that are actively being flown from 14 different geographic sites.

In an effort to summarize the proliferation of these launch programs, this reference guide has been developed. Its main purpose is to provide ready information to policymakers, planners, managers, engineers, and students who are interested in space launch systems. As noted in the title, these launch vehicles are described as "systems". Launch systems include not only the launch vehicle itself, but the whole process that makes up a successful launch: production, assembly facilities, and operations plus historical, programmatic, and organizational information. The launch systems that are included in this guide are the 19 active systems that have attained a flight record (i.e., orbital launch attempt), as well as 4 others that are expected to be launched in the near term.

To ensure quick and easy retrieval of data, this guide uses a standard format for each launch system. The standard format also helps compare various launch systems. Maximum use of a tabular format with numerical data has been emphasized. Text is used only to add information that numbers alone cannot convey.

The information for this updated edition has been collected primarily from the organizations directly involved with each launch system. Copies of the 1991 edition were sent out to each of the responsible launch system organizations for their updates and corrections. These comments were compiled, reviewed for consistency, and incorporated into the new edition. Some sections required significant additional research. In these cases user's guides and other publicly available sources were consulted to fill in the necessary data. All pricing data is labeled regarding the source. In cases where pricing data was not provided by the cognizant organizations, U.S. Department of Transportation data was used.

Significant additions have been included in this new edition. Highlights include: updated and expanded technical data for all vehicles; significantly improved data on vehicles of the former Soviet Union with data obtained directly from the vehicle designers and manufacturers; and complete launch histories through the end of 1994 for all vehicles. The H-1 (Japan) and SCOUT (U.S.) have been deactivated since the original publication and have hence been moved to the Historical Launch Systems appendix. Additionally, six new launch vehicle systems have been added in this new edition. These new systems include the J-vehicle (Japan); Rokot, Start, and Ikar (Russia/CIS); and Conestoga and Lockheed Launch Vehicle (U.S.).

While every effort has been made to verify the accuracy of the data, some information varies from source to source and discrepancies may occur. It is recommended that potential users of a launch system who require precise vehicle data contact the responsible organization noted on the cover page of each section. Additional updates to this guide are planned that will enable the inclusion of new or corrected data and improvements to format. Corrections, suggestions, and inquiries by readers can be sent to:

International Reference Guide to Space Launch Systems
AIAA
Business Developer—Books
370 L'Enfant Promenade, SW
Washington, DC 20024-2518

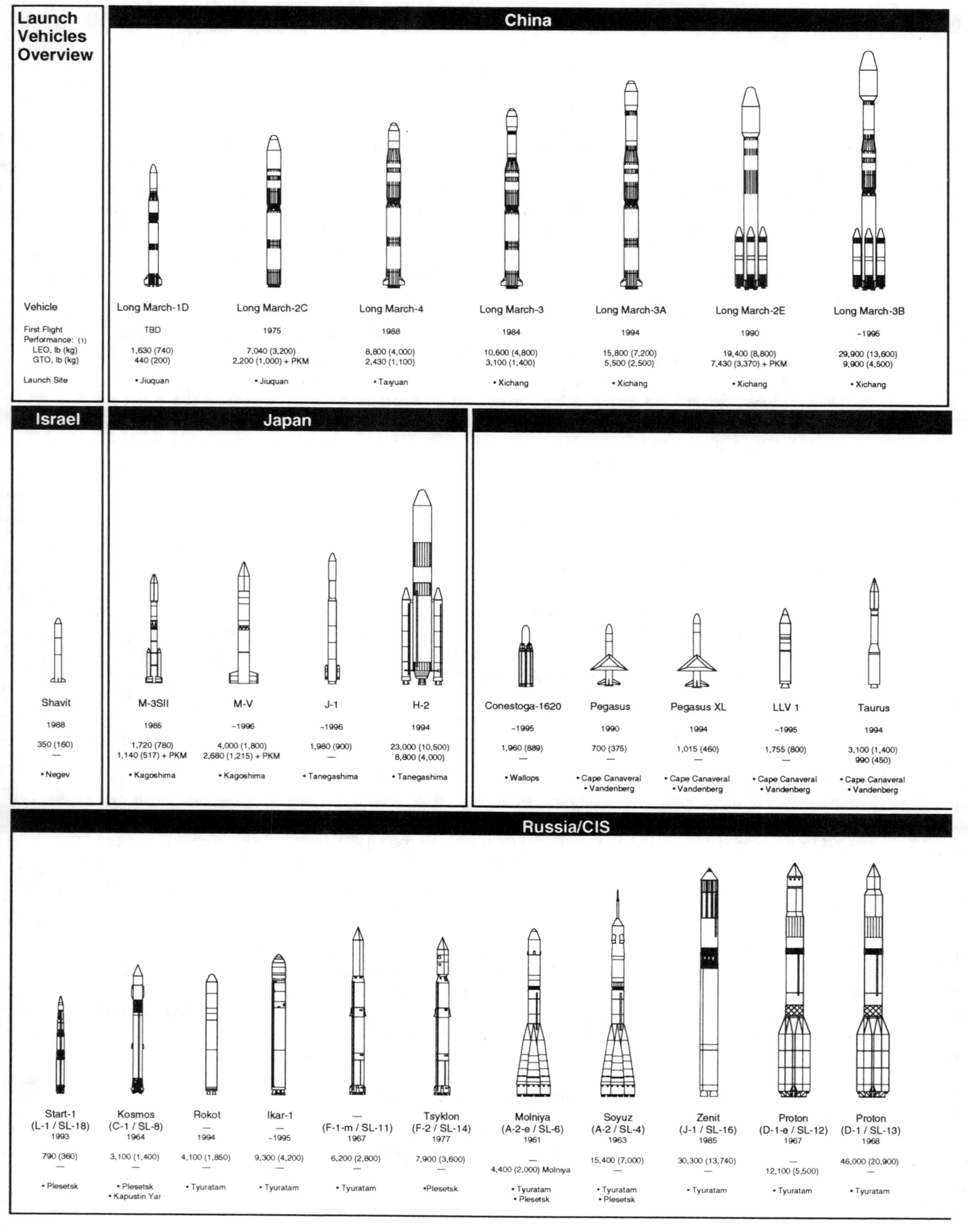

Launch Vehicles Overview
Vehicle
First Flight
Performance: (1)
LEO, lb (kg)
GTO, lb (kg)
Launch Site
China
Long March-1D
TBD
1,630 (740)
440 (200)
• Jiuquan
Long March-2C
1975
7,040 (3,200)
2,200 (1,000) + PKM
• Jiuquan
Long March-4
1988
8,800 (4,000)
2,430 (1,100)
• Taiyuan
Long March-3
1984
10,600 (4,800)
3,100 (1,400)
• Xichang
Long March-3A
1994
15,800 (7,200)
5,500 (2,500)
• Xichang
Long March-2E
1990
19,400 (8,800)
7,430 (3,370) + PKM
• Xichang
Long March-3B
~1995
29,900 (13,600)
9,900 (4,500)
• Xichang
Israel
Shavit
1988
350 (160)
—
• Negev
Japan
M-3SII
1985
1,720 (780)
1,140 (517) + PKM
• Kagoshima
M-V
~1996
4,000 (1,800)
2,680 (1,215) + PKM
• Kagoshima
J-1
~1996
1,980 (900)
—
• Tanegashima
H-2
1994
23,000 (10,500)
8,800 (4,000)
• Tanegashima
Conestoga-1620
~1995
1,960 (889)
—
• Wallops
Pegasus
1990
700 (375)
—
• Cape Canaveral
• Vandenberg
Pegasus XL
1994
1,015 (460)
—
• Cape Canaveral
• Vandenberg
LLV 1
~1995
1,755 (800)
—
• Cape Canaveral
• Vandenberg
Taurus
1994
3,100 (1,400)
990 (450)
• Cape Canaveral
• Vandenberg
Russia/CIS
Start-1
(L-1 / SL-18)
1993
790 (360)
—
• Plesetsk
Kosmos
(C-1 / SL-8)
1964
3,100 (1,400)
—
• Plesetsk
• Kapustin Yar
Rokot
—
1994
4,100 (1,850)
—
• Tyuratam
Ikar-1
—
~1995
9,300 (4,200)
—
• Tyuratam
—
(F-1-m / SL-11)
1967
6,200 (2,800)
—
• Tyuratam
Tsyklon
(F-2 / SL-14)
1977
7,900 (3,600)
—
•Plesetsk
Molniya
(A-2-e / SL-6)
1961
—
4,400 (2,000) Molniya
• Tyuratam
• Plesetsk
Soyuz
(A-2 / SL-4)
1963
15,400 (7,000)
—
• Tyuratam
• Plesetsk
Zenit
(J-1 / SL-16)
1985
30,300 (13,740)
—
• Tyuratam
Proton
(D-1-e / SL-12)
1967
—
12,100 (5,500)
• Tyuratam
Proton
(D-1 / SL-13)
1968
46,000 (20,900)
—
• Tyuratam

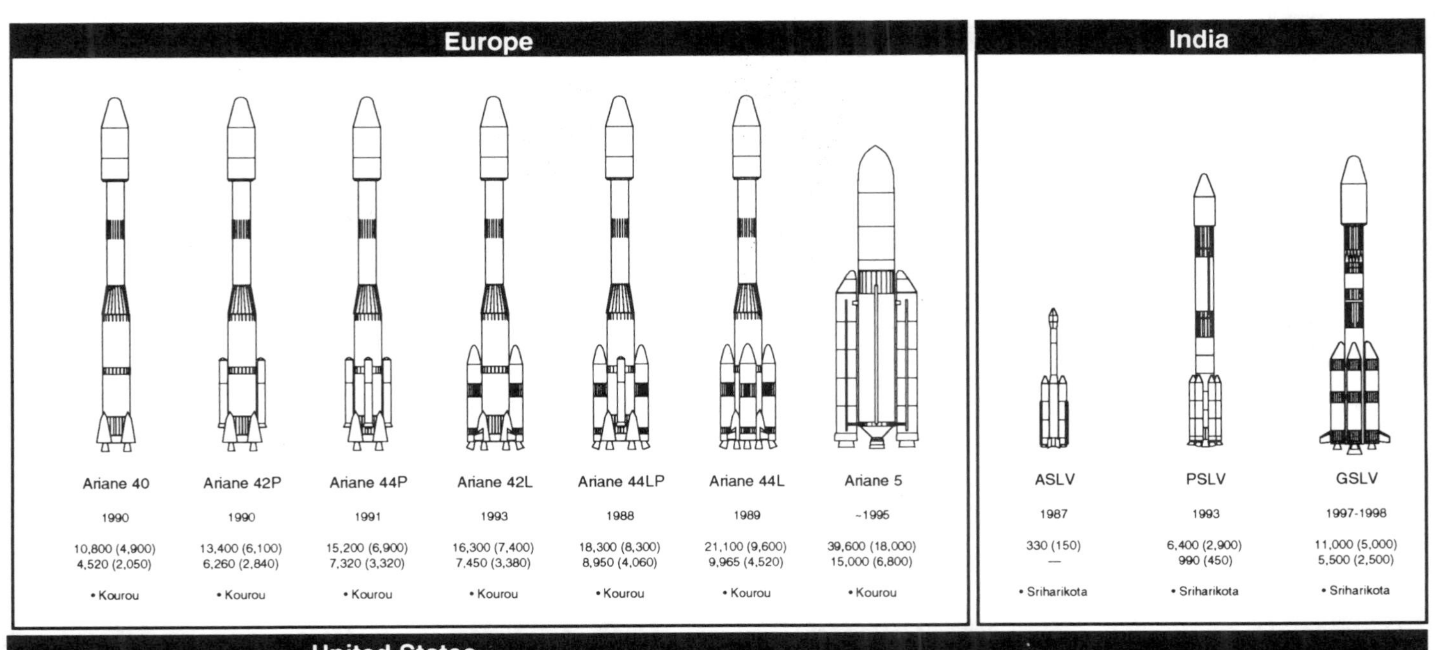

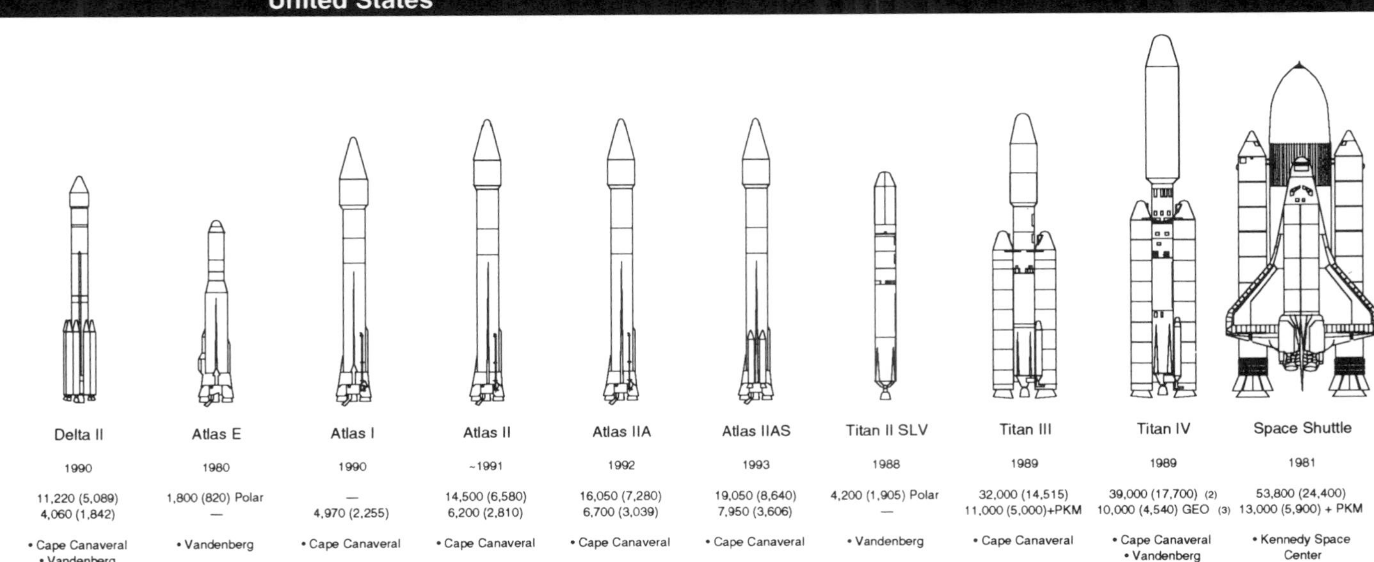

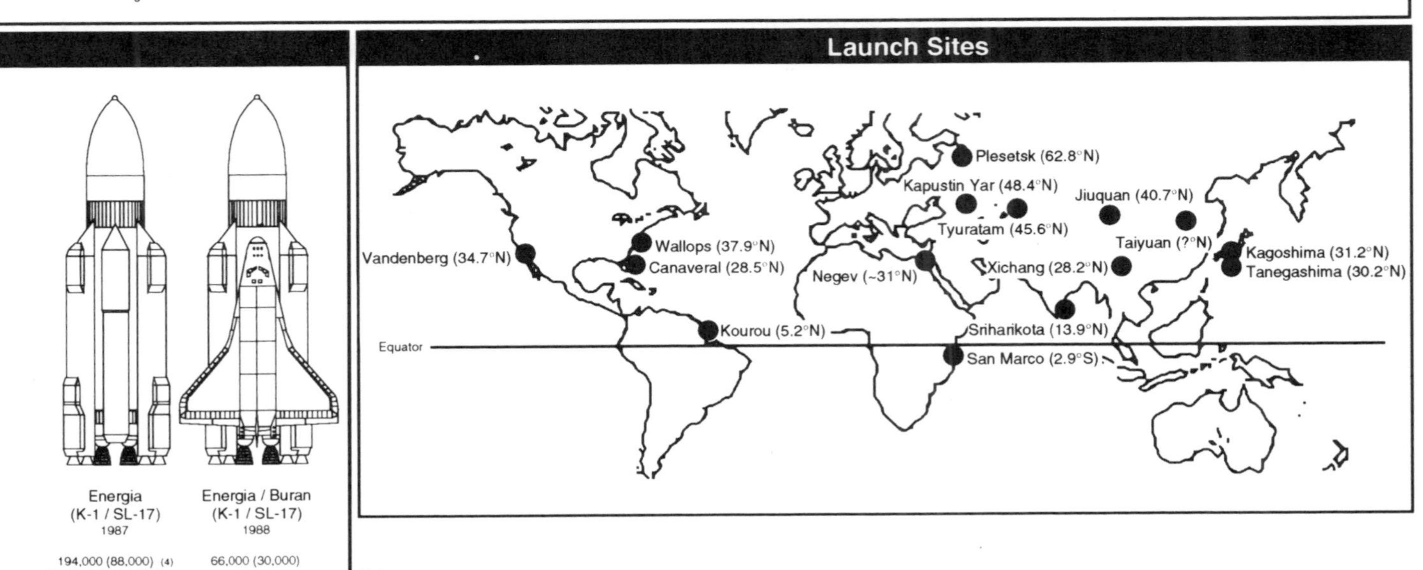

Notes:

(1) LEO - low Earth orbit
GTO - geosynchronous transfer orbit
GEO - geosynchronous orbit
PKM - perigee kick motor

(2) Plus 25% LEO with SRMU
(3) With Centaur upper stage
(4) With RCS kick stage
(5) With EUS upper stage

Format Explanation

A number of key format features are used for ease of data retrieval and convenience of use that are worth noting:

- All the launch systems are grouped by nationality
- Both nationality and vehicle names are presented in alphabetical order
- Technical data are shown both in English and metric units
- For brevity, million is often summarized as "M" and thousand as "K"
- Whenever data is unknown or uncertain a question mark "?" is shown
- Each launch system is formatted into eight standard sections described below:

(1) Cover page — Has a photograph, if available, of a launch vehicle and provides points of contact for additional information.

(2) History — Is further divided into:
- Out of Production, in Production, in Development—chronologically illustrates status of vehicles. Remaining sections of the launch system chapter focus on vehicles that are "in production" and "in development".
- Vehicle Description—provides a brief summary of each vehicle illustrated above and their technical differences.
- Historical Summary—provides a brief text history of the launch system.
- Launch Record—lists all the launches and failures when available. A failure is defined as a launch that was unable to place a payload into its intended orbit. Because criteria for failure may differ among different readers, data are provided to calculate ones own failure rate.

(3) General Description — Furnishes information of general interest to the reader. Price data are taken from publicly available sources and are only presented to give an approximate sense of launch service price. Because prices vary widely with quantity of orders and services included, the responsible launch organization should be contacted for actual prices. The reader must treat the manifest data as only a snapshot in time because they often change.

(4) Vehicle — Is further divided into:
- Overall—illustrates the entire vehicle
- Stages—provides information in a tabular form. The tables are shown by stages with each vehicle shown next to each other in comparison (e.g., Atlas I stage 1 vs. Atlas II stage 1). In the case where each member of the launch system has significant stage differences (e.g., Ariane-4 and Ariane-5), each table is dedicated to only one vehicle. Text is also provided for additional detail.
- Payload Fairing—illustrates and describes each fairing type
- Avionics—has a brief text summary
- Attitude Control System—has a brief description of attitude control mechanisms

(5) Performance — Provides performance curves for a variety of orbits

(6) Operations — Is further divided into:
- Launch Site—illustrates the geographic location of the site
- Launch Facilities—illustrates some of the key facilities such as integration facilities and launch pad
- Launch Processing—provides text description of operation process
- Flight Sequence—lists events by time that occur after launch

(7) Payload Accommodations — Provides data pertinent to the payload such as payload envelope, launch window constraints, flight environment, and injection accuracy

(8) Notes — Contains miscellaneous information such as:
- Publications—lists references used for this guide
- Acronyms—used in text
- Possible Growth—describes possible vehicle growth paths

China

Table of Contents

Long March

Launch Service Point of Contact:
China Great Wall Industry Corp.
17 Wenchang Hutong, Xidan,
P.O. Box 847
Beijing, CHINA
Phone: 8311808

U.S. Point of Contact:
Mr. Zuoyi Huang
GW Aerospace Inc.
21515 Hawthorne Blvd., Ste 1065
Torrance, CA 90503 USA
Phone: (310) 540-7706

A Long March 2E at Xichang Satellite Launch Center on Launch Complex 2 being prepared for its inaugural launch which occurred on 16 July 1990.

Long March History

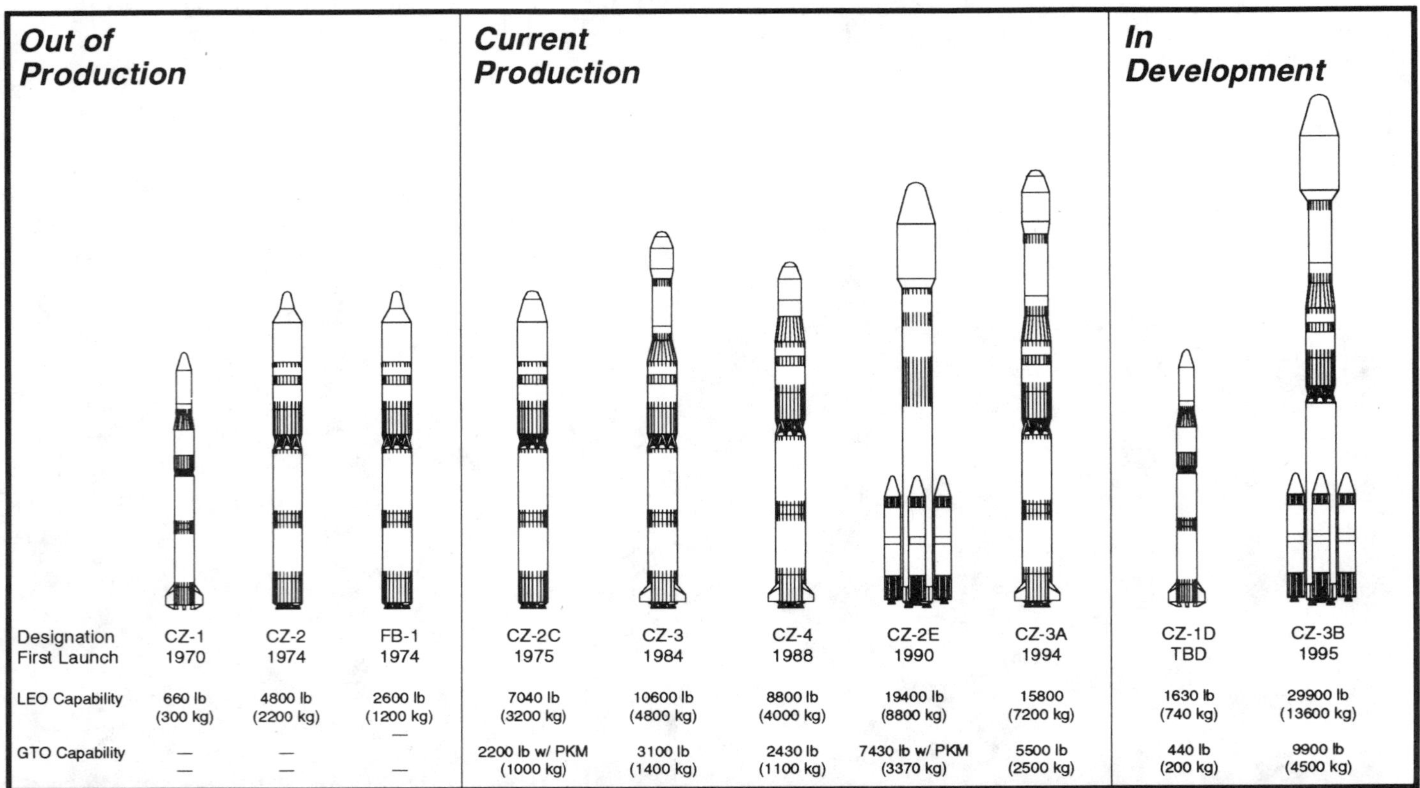

Designation	CZ-1	CZ-2	FB-1	CZ-2C	CZ-3	CZ-4	CZ-2E	CZ-3A	CZ-1D	CZ-3B
First Launch	1970	1974	1974	1975	1984	1988	1990	1994	TBD	1995
LEO Capability	660 lb (300 kg)	4800 lb (2200 kg)	2600 lb (1200 kg)	7040 lb (3200 kg)	10600 lb (4800 kg)	8800 lb (4000 kg)	19400 lb (8800 kg)	15800 (7200 kg)	1630 lb (740 kg)	29900 lb (13600 kg)
GTO Capability	— —	— —	— —	2200 lb w/ PKM (1000 kg)	3100 lb (1400 kg)	2430 lb (1100 kg)	7430 lb w/ PKM (3370 kg)	5500 lb (2500 kg)	440 lb (200 kg)	9900 lb (4500 kg)

Vehicle Description

CZ-1 Three stage vehicle derived from the CSS-2 IRBM. First two stages are nitric acid/UDMH propellant and the third stage is solid. One engine for each first and second stage, both controlled by jet vanes. 6.7 ft (2.05 m) diameter fairing.

CZ-2 Two stage vehicle derived from the CSS-4 ICBM. Both stages use N204/UDMH. Four first stage engines with gimbal control and one second stage engine with four verniers for control.

FB-1 Two stage liquid vehicle. Similar to CZ-2.

CZ-2C Same as CZ-2 except unpgraded for improved reliability and performance.

CZ-3 Same as CZ-2C except aerodynamic fins on first stage, and the addition of a LOX/LH2 four-nozzle third stage. 8.5 ft (2.6 m) and 9.8 ft (3.0 m) diameter fairings.

CZ-4 Same as CZ-2C except aerodynamic fins on first stage, stretched first and second stages and addition of a UDMH/N204 third stage. 9.5 ft (2.9 m) and 11.0 (3.35 m) diameter fairings.

CZ-2E Same as CZ-2C except stretched stages and four UDMH/N204 strap-ons for increased performance. 13.8 ft (4.2 m) diameter fairing. Chinese PKM available.

CZ-1D Same as CZ-1 except a UDMH/N204 second stage and higher orbit injection accuracy. 6.7 ft (2.05 m) diameter fairing.

CZ-3A Same as CZ-3 except stretched first two stages and a new LOX/LH2 third stage with two engines derived from the CZ-3. 11.0 ft (3.35 m) and 13.1 ft (4.0 m) diameter fairings.

CZ-3B Same as CZ-2E first stage with strap-ons, CZ-3 second stage, and CZ-3A LOX/LH2 third stage.

Historical Summary

In 1956, government policy in China laid the foundations for its astronautics industry. As a developing nation, the country recognized the need to develop space technology for its own purposes. China started its research and development on launch vehicles at the beginning of 1960s.

As a result of this effort, the first version of Long March family of launch vehicles, Long March 1, was successfully put into operation. Long March 1 (LM-1) is also designated Chang Zheng 1 (CZ-1). In April 1970, China's first satellite, Dong Fang Hong 1, was placed into the low Earth orbit (LEO) successfully on the first CZ-1. This event proved China's capability to develop its own launch technology using only Chinese materials. CZ-1 was a three-stage launch vehicle developed from the CSS-2 intermediate range ballistic missile (IRBM). The first and second stage were propelled by nitric acid/UDMH liquid engines. The third stage was powered by a solid motor. After an almost 20-year hiatus, an improved CZ-1, the CZ-1D, was made available in 1994.

CZ-1 was launched from Jiquan Satellite Launch Center (JSLC) which restricts launches to midinclination orbits due to overflight constraints. JSLC remained the sole Chinese launch site until 1984 when Xichang Satellite Launch Center (XSLC) was added for primarily geosynchronous orbit (GEO) satellites. In 1988, Taiyuan Satellite Launch Center (TSLC) was added for primarily sun synchronous orbit (SSO) missions.

In order to increase performance, the Long March 2 was developed from the CSS-4 intercontinental ballistic missile (ICBM). Long March 2 is a two stage launch vehicle propelled by storable N204/UDMH propellant. It has the capability of launching payloads seven times heavier than CZ-1 to LEO.

Long March History

(continued)

Historical Summary

(continued)

The development of CZ-2 started in 1970. In November 1974, the first Long March 2 (CZ-2) was launched, but after taking off, the vehicle lost its attitude stability, and the vehicle was automatically destroyed by the vehicle's airborne range safety device. The result of analyses showed that this failure was caused by a broken wire necessary for delivering the pitch rate gyro signal. After thorough investigation and research, the control system design was modified to improve its reliability. Quality control during production and functional checkout was improved. In the meanwhile, payload capability was increased and the modified launch vehicle was designated as CZ-2C. This vehicle became the basis for all future Long March series vehicles. In November 1975, the first CZ-2C was launched. The flight was a complete success and China's first recoverable satellite was launched into orbit.

In September 1981, China succeeded in its first multiple launch by sending three satellites into earth orbit on one launch vehicle, Feng Bao 1 (FB-1). It has been surmised that FB-1 was either a vehicle very similar to Long March 2 or a new competing vehicle. In either case, it is assumed that FB-1 was discontinued due to lower performance and/or lower reliability than Long March 2.

The Long March 3 is a three stage launch vehicle. It is mainly used for geosynchronous transfer orbit (GTO) missions. The first and second stage of CZ-3 were developed on the basis of CZ-2C. The newly developed third stage is propelled by a liquid oxygen (LOX) and liquid hydrogen (LH2) engine with restart capability. On 29 January 1984, at XSLC, the first test flight of CZ-3 was carried out. The payload was an experimental communications satellite, weighing 2050 lb (930 kg). In the actual test flight, the third stage engine thrust disappeared after restart. The mixture ratio in the gas generator of the third stage engine was abnormal, the temperature of the gas became very high, and the turbine shell was burnt out, causing the thrust to disappear. The engine was then modified and tested four times prior to the next launch. On 8 April 1984, 70 days after the failed flight of CZ-3, the second flight of CZ-3 with a modified third stage engine was carried out. It was a total success.

Long March 3A (CZ-3A) is a new launch vehicle which is being developed as an improved version of CZ-3. The first flight test of CZ-3A is expected to take place in 1992 and it will be available as a commercial launch vehicle for the international market in the middle 1990s.

Based on the technology and flight experience of CZ-2C and CZ-3, a more powerful commercial launch vehicle, Long March 2E (CZ-2E), was developed that had 3.5 times more LEO performance than CZ-2C. All the subsystems of CZ-2E are essentially the same as the subsystems of CZ-2C and the first and second stages for CZ-3, including engines, propellent feeding systems, guidance system and main structures. The CZ-2E is designed to be a vehicle for the parking orbit mission or for GTO mission if incorporated with a perigee kick motor (PKM). The first launch occurred on July 1990, 18 months after the development began. Based on the success of that first launch, China announced the development of an improved vehicle, CZ-3B, with a cryogenic upper stage for about 50% greater performance to GTO than a CZ-2E with a PKM. CZ-3B should be available by 1995.

In 1988, the first CZ-4 was launched from a new launch site, Taiyuan Satellite Launch Center (TSLC). This vehicle is used for the launch of SSO meteorological satellites and other scientific and application satellites. The vehicle is based on the CZ-2C with the addition of a storable propellant third stage using the engine of the CZ-1 second stage.

China's success in launch vehicle technology resulted in an October 1985 announcement that the Long March family would be offered commercially to customers throughout the world. The use of the recoverable satellite system was also being offered to foreign customers, providing opportunities for microgravity experimentation. Matra of France and the European Intospace consortium were the first companies to purchase this service. In 1988, Long March won commercial contracts to launch one Asiasat and two Australian Aussat satellites. In April 1990, China launched its first commercial satellite, Asiasat, aboard a CZ-3.

China's payload launch services comprise a number of interacting organizations, each of which is responsible for part of the launch services package. The China Great Wall Industry Corporation (CGWIC) is the foreign trade company responsible, under China Aerospace Corporation, for marketing and negotiating launch services, for all commercial operations with its customers, and for contract execution. CGWIC is the primary interface with customers, undertaking all coordination between the customer and the other elements of the launch services organization. CGWIC deals in the import and export of Chinese astronautics technology and products. It has a wide scope of business, one major aspect of which is launch services. CGWIC's launch vehicle partners or subcontractors in China include China Academy of Launch Vehicle Technology (CALT) and China Satellite Launch and TT&C General (CLTC), and Shanghai Bureau of Astronautics (SHBOA).

CLTC is an organization under the Commission of Science, Technology and Industry for National Defense responsible for launch operations and telemetry, tracking, and control (TT&C) and postlaunch services. CLTC runs the three launch centers, one satellite control center located in Xian and a global TT&C network.

CALT, under China Aerospace Corporation, is responsible for the development, production and testing of launch vehicles, and payload interface analysis. The Long March family is the major product of the company. It has full capability in independently developing, designing, and testing launch systems. It can undertake the complete production process from manufacturing of parts to assembly.

SHBOA is the research and production base of China Aerospace Corporation in the Shanghai area responsible for development and production of certain launch systems—first stage and second stage CZ-3 and CZ-4.

Long March History

(continued)

Launch Record

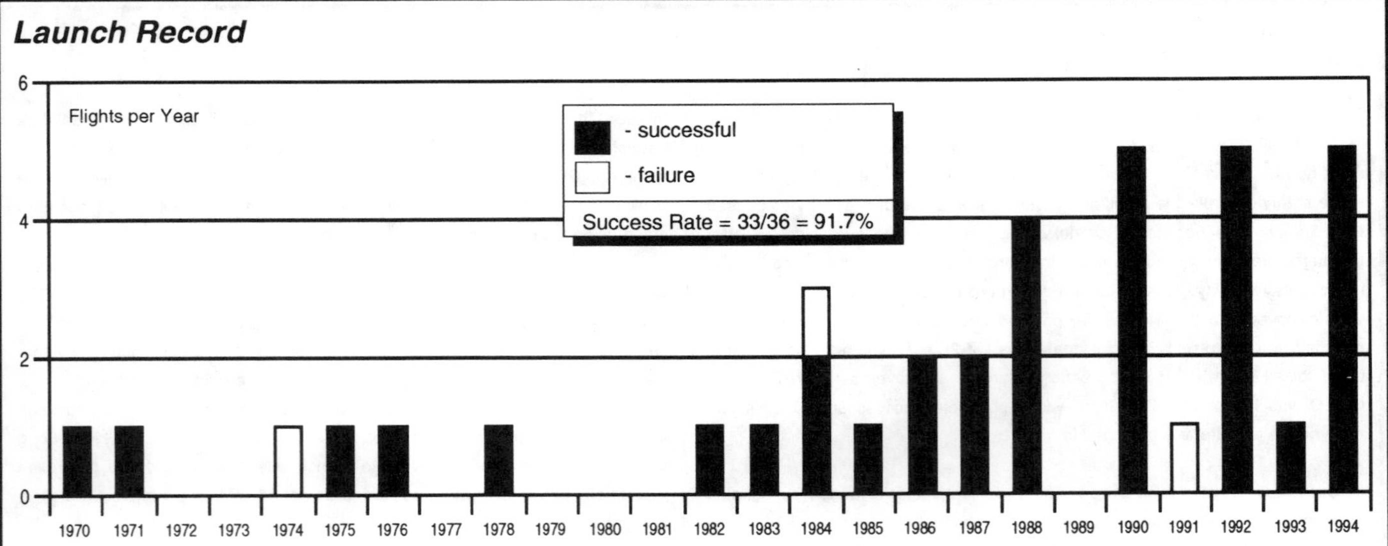

	YEAR	DATE	VEHICLE	SITE	PAYLOAD
1	1970	Apr-24	1	JSLC	Dong Fang Hong 1
2	1971	Mar-3	1	JSLC	Shi Jian 1
*	1974	Jul-12	FB-1	JSLC	SKW
		Failure			
3		Nov-5	2A	JSLC	SKW (LEO)
		Failure - lost attitude stability			
*	1975	Jun-26	FB-1	JSLC	SKW 3 (Technical Experiment)
4		Nov-26	2C	JSLC	SKW 4 (Recoverable)
*		Dec-16	FB-1	JSLC	SKW 5 (Technical Experiment)
*	1976	Aug-30	FB-1	JSLC	SKW 6 (Technical Experiment)
5		Dec-7	2C	JSLC	SKW 7 (Recoverable)
6	1978	Jan-26	2C	JSLC	SKW 8 (Recoverable)
*	1979	Jul-30	FB-1	JSLC	SKW
		Failure			
*	1981	Sep-19	FB-1	JSLC	Shi Jian 2/2A/2B (Triple Launch)
7	1982	Sep-9	2C	JSLC	(Recoverable)
8	1983	Aug-19	2C	JSLC	(Recoverable)
9	1984	Jan-29	3	XSLC	(Experimental GEO Communication)
		Failure - stage 3 failed to restart			
10		Apr-8	3	XSLC	(Experimental GEO Communication)
11		Sep-12	2C	JSLC	(Recoverable)
12	1985	Sep-18	2C	JSLC	(Recoverable)
13	1986	Feb-1	3	XSLC	STTW-1 (Operational GEO Communication)
14		Oct-6	2C	JSLC	(Recoverable)

	YEAR	DATE	VEHICLE	SITE	PAYLOAD
15	1987	Aug-5	2C	JSLC	(Recoverable)
16		Sep-9	2C	JSLC	(Recoverable)
17	1988	Mar-7	3	XSLC	STTW-2 (Operational GEO Communication)
18		Aug-5	2C	JSLC	FSW-12 (Recoverable)
19		Sep-7	4	TSLC	Feng Yung 1 (Meteorological)
20		Dec-22	3	XSLC	STTW-3 (Operational GEO Communication)
21	1990	Feb-4	3	XSLC	DFH-2A (Communication)
22		Apr-7	3	XSLC	Asiasat-1 (Commercial)
23		Jul-16	2E	XSLC	Badr-1
					"dummy" Aussat
24		Sep-3	4	TSLC	Feng Yung 2 (Meteorological)
25		Oct-5	2C	JSLC	(Recoverable)
26	1991	Dec-28	3	XSLC	STTW 5
		Failure - premature stage 3 shutdown			
27	1992	Aug-9	2D	JSLC	PRC 35 (Recoverable)
28		Aug-14	2E	XSLC	Optus B1 (Communication)
29		Oct-6	2C	JSLC	Freja / PRC 36
30		Dec-21	2E	XSLC	Optus B2 (Communication)
		Note - satellite exploded 45 sec after launch (not counted as failure)			
31	1993	Oct-8	2C	JSLC	FSW-1 F5 (Recoverable)
32	1994	Feb-8	3A	JSLC	Test + SJ 4
33		Jul-3	2D	JSLC	(Government Communication)
34		Jul-21	3	XSLC	Apstar 1
35		Aug-28	2E	XSLC	Optus B3
36		Nov-30	3A	JSLC	GTO

* FB-1 not counted as flight against Long March

JSLC (Jiuquan Satellite Launch Center) in Gansu Province
XSLC (Xichang Satellite Launch Center) in Sichuan Province
TSLC (Taiyuan Satellite Launch Center) in Shanxi Province

Long March General Description

	CZ-1D	CZ-2C	CZ-2E	CZ-3B	CZ-3	CZ-3A	CZ-4
Summary	Three stage rocket. First two stages are based on CSS-2 IRBM. Can compete with other international small launch vehicles.	Two stage liquid rocket based on CZ-2B that was developed from CSS-4 ICBM. Has been commercially available since 1985. Main role is recoverable satellites.	Based on the original CZ-2C. Represents China's largest operational launch vehicle. First launch in 1990.	Based on the original CZ-2E with the addition of the CZ-3A cryogenic third stage. Will be available in 1995.	Three stage rocket for GTO missions. The first two stages are modified from the CZ-2C and third stage is new cryogenic stage.	Growth version of the CZ-3. Can launch either one or two satellites.	Three stage rocket based on the CZ-2C with a new storable third stage. Primarily for SSO missions.
Status	In Development	Operational	Operational	In Development	Operational	In Development	Operational

Key Organizations

User - Ministry of Astronautics, commercial

Launch Service Agency - China Great Wall Industry Corp. (CGWIC)

Prime Contractor - CGWIC

Subcontractors
- Beijing Wan Yuan Industry Corp (BWYIC)
  - (launch vehicles R&D and manufacturing, mission analysis, payload interface coordination)
- China Satellite TT&C General (CLTC)
  - (launch site, launch operations, TT&C)
- Shanghai Bureau of Astronautics (SBA)
  - (production of stage 1 and 2 of CZ-3/4 launch vehicles)

Vehicle	CZ-1D	CZ-2C	CZ-2E	CZ-3B	CZ-3	CZ-3A	CZ-4
System Height	92.6 ft (28.22 m)	115 ft (35.15 m)	168 ft (51.2 m)	190 ft (57.82 m)	144 ft (43.85 m)	172 ft (52.3 m)	138 ft (42 m)
Payload Fairing Size	Diameter 6.7 ft (2.05 m) Length 13.1 ft (3.99 m)	Type A: Diameter 7.2 ft (2.2 m) Length 10.3 ft (3.144 m) Type B: Diameter 11.0 ft (3.35 m) Length 23.4 ft (7.125 m)	Diameter 13.8 ft (4.2 m) Length 39.2 ft (11.95 m)	Diameter 13.8 ft (4.2 m) Length 39.2 ft (11.95 m)	Model A: Diameter 8.5 ft (2.6 m) Length 19.2 ft (5.84 m) Model B: Diameter 9.8 ft (3.0 m) Length 23.9 ft (7.27 m)	Single: Diameter 11.0 ft (3.35 m) Length 29.2 ft (8.89 m) Dual: Diameter 13.1 ft (4.0 m) Length 39.4 ft (12.0 m)	Model A: Diameter 9.5 ft (2.9 m) Length 16.1 ft (4.91 m) Model B: Diameter 11.0 ft (3.35 m) Length 27.8 ft (8.48 m)
Gross Mass	175K lb (79.4K kg)	421K lb (191K kg)	1023K lb (464Kkg)	?? lb (?? kg)	445K lb (202K kg)	529K lb (240K kg)	549K lb (249K kg)
Planned Enhancements	None	None	None	None	None	None	Piggyback payload in the transition bay between primary payload and vehicle
Operations							
Primary Missions	LEO mid-inclination	LEO mid-inclination	LEO low-inclination & GTO with PKM	GTO	GTO	GTO	SSO
Compatible Upper Stages	Solid motor stage 3	Chinese Solid PKM	Star 63F New Chinese PKM	Cryogenic stage 3	Cryogenic stage 3	Cryogenic stage 3	Storable stage 3
First Launch	TBD	1975	1990	1995 (Projected)	1984	1994	1988

Long March General Description

(continued)

Operations (continued)	CZ-1D	CZ-2C	CZ-2E	CZ-3B	CZ-3	CZ-3A	CZ-4
Success/Flight Total	0 / 0	14 / 14	4 / 4	0 / 0	7 / 9	2 / 2	2 / 2
Launch Site	JSLC 41°N,100°E	JSLC 41°N,100°E	XSLC 28°N,102°E	XSLC 28°N,102°E	XSLC 28°N,102°E	XSLC 28°N,102°E	TSLC ??°N,??°E
Launch Azimuth	Max. inclination 57°–70°	Max. inclination 57°–70°	Max. inclination 27.5°–28.5°	Max. inclination 28°–29°	Max. inclination 29°–31.1°	Max. inclination 28°–29°	Max. inclination 96°–98°
Nominal Flight Rate	?? / year	?? / year	?? / year	?? / year	?? / year	?? / year	?? / year
Planned Enhancements	None	None	None	None	None	None	None
Performance							
108 nm (200 km) circ	At 28.5°, 300 km 1590 lb (720 kg) for spin stabilized 1630 lb (740 kg) for 3-axis At 57°, 300 km 1740 lb (790 kg) for spin stabilized 1320 lb (600 kg) for 3-axis At 70°, 300 km 1630 lb (740 kg) for spin stabilized 1210 lb (550 kg) 3-axis	At 28.5° inclination 7040 lb (3200 kg) At 41° inclination 5070 lb (2300 kg) At 90° inclination 3860 lb (1750 kg)	At 28.5° inclination 19400 lb (8800 kg)	At 28.5° inclination 29900 lb (13600kg)	At 28.5° inclination ?? lb (?? kg) At 31.1° 10600 lb (4800 kg)	At 28.5° inclination ?? lb (?? kg) At 31.1° 15800 lb (7200 kg)	At 98.5° inclination 8800 lb (4000 kg)
Geotransfer Orbit	At 28.5° inclination 440 lb (200 kg) for 3-axis	At 28.5° inclination 2,200 lb (1000 kg) with PKM from XSLC	At 28.5° inclination 7,430 lb (3,370 kg) with new Chinese PKM	At 28.5° inclination 9,900 lb (4,500 kg)	At 31.1° inclination 3,100 lb (1,400 kg)	At 31.1° inclination 5,500 lb (2,500 kg)	At 28.5° inclination 2,430 lb (1,100 kg)
Geosynchronous Orbit	220 lb (100 kg) with AKM	860 lb (390 kg) with AKM	3,300 lb (1,500 kg) with AKM	4,950 lb (2,250 kg) with AKM	1,600 lb (730 kg) with AKM	2,700 lb (1,230 kg) with AKM	1,220 lb (550 kg) with AKM
Financial Status							
Estimated Launch Price (U.S DoT estimate)	$8–12M	$15–20M	$40–50M	$60–70M	$35–40M	$35–45M	$20–30M

Manifest

1995 Qtr 1 - AsiaSat-2 on CZ-2E
Qtr 3-4 - Echostar-1 on CZ-EE
Qtr 4 - Intelsat-7A on CZ-3B
1996 Qtr 2 - IRIDIUM SV-01, 02 on CZ-2C
Qtr 3 - IRIDIUM SV-03, 04 on CZ-2C
Qtr 3 - IRIDIUM SV-05, 06 on CZ-2C
Qtr 3 - Echostar-2 on CZ-2E
1997 Qtr 2-3 - Intelsat-805 on CZ-3B
Qtr 2 - IRIDIUM on CZ-2C
Qtr 3 - IRIDIUM on CZ-2C
Qtr 3 - IRIDIUM on CZ-2C

	1995	1996	1997
CZ-1D	—	—	—
CZ-2C	—	3	3
CZ-2E	2	1	—
CZ-3	—	—	—
CZ-3A	—	—	—
CZ-3B	1	—	1
CZ-4	—	—	—

Remarks

U.S.-China trade agreement signed in 1995 allows for 11 GSO launches over 7 years.

Long March Vehicle

Overall

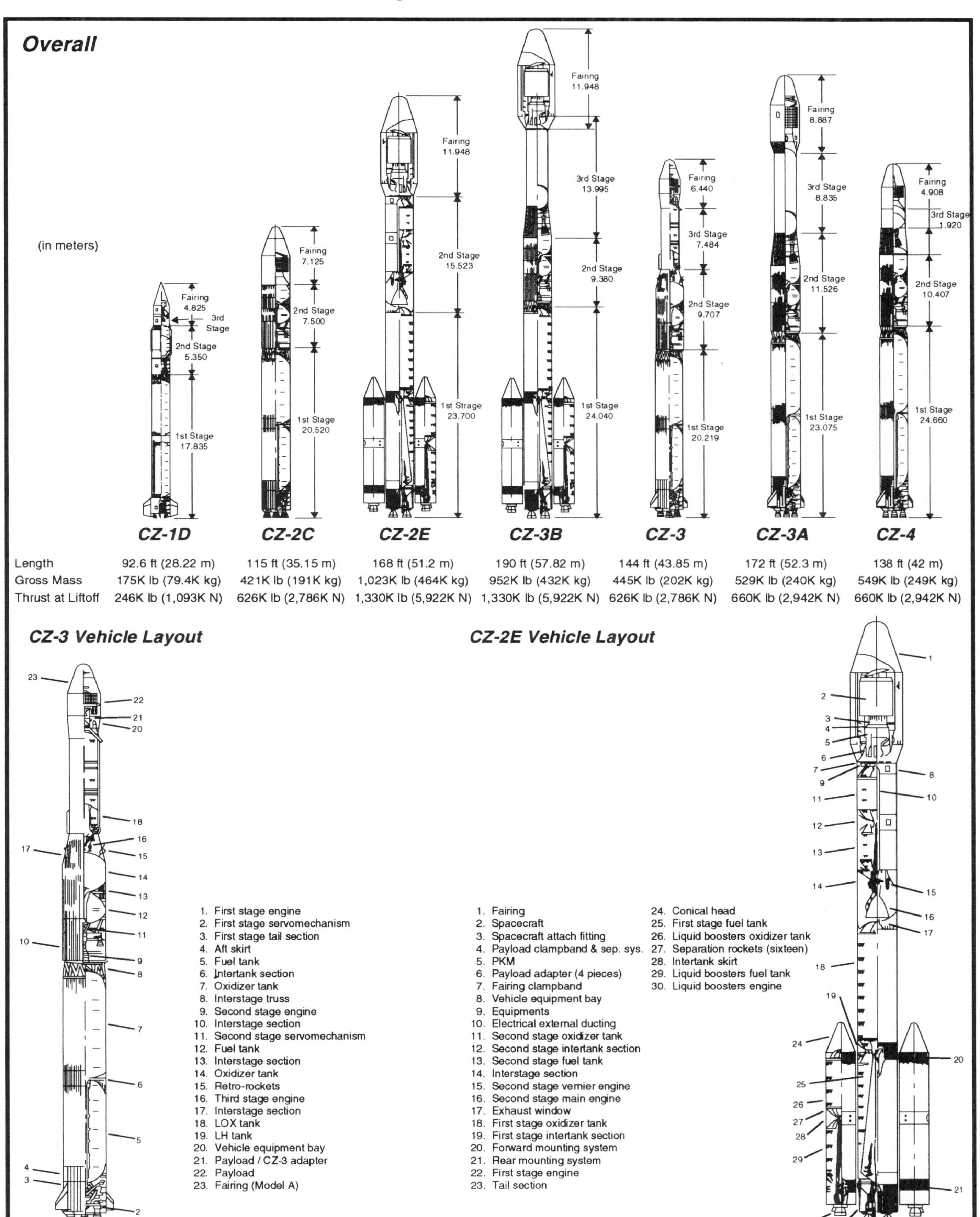

	CZ-1D	CZ-2C	CZ-2E	CZ-3B	CZ-3	CZ-3A	CZ-4
Length	92.6 ft (28.22 m)	115 ft (35.15 m)	168 ft (51.2 m)	190 ft (57.82 m)	144 ft (43.85 m)	172 ft (52.3 m)	138 ft (42 m)
Gross Mass	175K lb (79.4K kg)	421K lb (191K kg)	1,023K lb (464K kg)	952K lb (432K kg)	445K lb (202K kg)	529K lb (240K kg)	549K lb (249K kg)
Thrust at Liftoff	246K lb (1,093K N)	626K lb (2,786K N)	1,330K lb (5,922K N)	1,330K lb (5,922K N)	626K lb (2,786K N)	660K lb (2,942K N)	660K lb (2,942K N)

CZ-3 Vehicle Layout

1. First stage engine
2. First stage servomechanism
3. First stage tail section
4. Aft skirt
5. Fuel tank
6. Intertank section
7. Oxidizer tank
8. Interstage truss
9. Second stage engine
10. Interstage section
11. Second stage servomechanism
12. Fuel tank
13. Interstage section
14. Oxidizer tank
15. Retro-rockets
16. Third stage engine
17. Interstage section
18. LOX tank
19. LH tank
20. Vehicle equipment bay
21. Payload / CZ-3 adapter
22. Payload
23. Fairing (Model A)

CZ-2E Vehicle Layout

1. Fairing
2. Spacecraft
3. Spacecraft attach fitting
4. Payload clampband & sep. sys.
5. PKM
6. Payload adapter (4 pieces)
7. Fairing clampband
8. Vehicle equipment bay
9. Equipments
10. Electrical external ducting
11. Second stage oxidizer tank
12. Second stage intertank section
13. Second stage fuel tank
14. Interstage section
15. Second stage vernier engine
16. Second stage main engine
17. Exhaust window
18. First stage oxidizer tank
19. First stage intertank section
20. Forward mounting system
21. Rear mounting system
22. First stage engine
23. Tail section
24. Conical head
25. First stage fuel tank
26. Liquid boosters oxidizer tank
27. Separation rockets (sixteen)
28. Intertank skirt
29. Liquid boosters fuel tank
30. Liquid boosters engine

Long March Vehicle

(continued)

Liquid Strap-On

	CZ-2E or 3B (LB40)
Dimension:	
Length	52.5 ft (16 m)
Diameter	7.4 ft (2.25 m)
Mass:	
Propellant Mass	84K lb (38K kg) ea.
Gross Mass	90K lb (41K kg) ea.
Structure:	
Type	Skin Stringer
Material	Aluminum
Propulsion:	
Propellant	UDMH / N2O4
Average Thrust	166K lb (741K N) SL
Engine Designation	YF-20
Number of Engines	1 each strap-on
Isp	289 sec vac
Feed System	Gas Generator
Chamber Pressure	?? psia (?? bar)
Mixture Ratio (O/F)	??
Propulsion: (cont.)	
Throttling Capability	100% only
Expansion Ratio	??
Restart Capability	No
Tank Pressurization	Nitrogen bottle & self pressurization
Control-Pitch,Yaw,Roll	Fixed cant
Events:	
Nominal Burn Time	128 sec main engine
Stage Shutdown	Burn to depletion
Stage Separation	16 retro rockets

Remarks:

The liquid strap-on booster, designated LB40, has one YF-20 engine. The engine is fixed. All the subsystems of the engine are the same as that of the first stage.

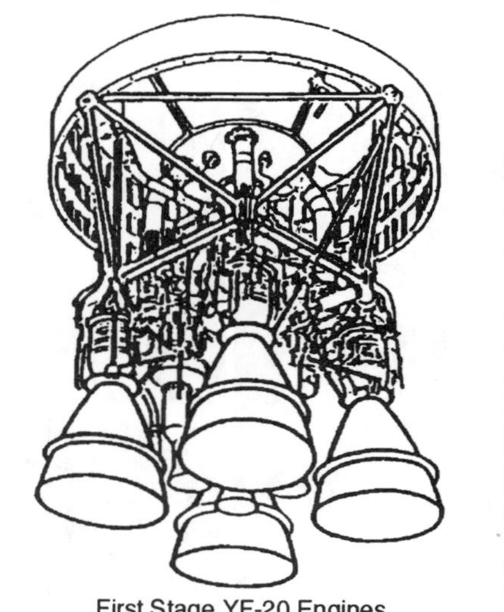

First Stage YF-20 Engines

Stage 1

	CZ-1D (L60) ?	CZ-2C (L140)	CZ-2E or 3B (L180)	CZ-3 (L140)	CZ-3A (L180)	CZ-4 (L180)
Dimension:						
Length	58.5 ft (17.84 m)	67.3 ft (20.52 m)	77.8 ft (23.70 m)	66.3 ft (20.22 m)	75.7 ft (23.08 m)	80.9 ft (24.66 m)
Diameter	7.4 ft (2.25 m)	11.0 ft (3.35 m)	11.0 ft (3.35 m)	11.0 ft (3.35 m)	11.0 ft (3.35 m)	11.0 ft (3.35 m)
Mass:						
Propellant Mass	132K lb (60K kg)	317K lb (144K kg)	412K lb (187K kg)	313K lb (142K kg)	375K lb (170K kg)	404K lb (183.2K kg)
Gross Mass	141K lb (64.1K kg)	337K lb (153K kg)	433K lb (196.5K kg)	333K lb (151K kg)	395K lb (179K kg)	425K lb (192.7K kg)
Structure:						
Type	Skin Stringer	Skin Stringer	Skin Stringer	Skin Stringer	Skin Stringer	Skin Stringer
Material	Aluminum	Aluminum	Aluminum	Aluminum	Aluminum	Aluminum
Propulsion:						
Propellant	UDMH / nitric acid	UDMH / N2O4	UDMH / N2O4	UDMH / N2O4	UDMH / N2O4	UDMH / N2O4
Average Thrust	248K lb (1101K N)SL	157K lb (697K N) SL each engine	166K lb (741K N) SL each engine	157K lb (697K N) SL each engine	165K lb (736K N) SL each engine	165K lb (736K N) SL each engine
Engine Designation	YF-2A	YF-20	YF-20	YF-20	YF-20	YF-20
Number of Engines	1 turbopump with 4 chambers	4	4	4	4	4
Isp	241 sec SL	259 sec SL	259 sec SL	259 sec SL	259 sec SL	259 sec SL
Feed System	Gas Generator	Gas Generator	Gas Generator	Gas Generator	Gas Generator	Gas Generator
Chamber Pressure	?? psia (?? bar)	?? psia (?? bar)	?? psia (?? bar)	?? psia (?? bar)	?? psia (?? bar)	?? psia (?? bar)
Mixture Ratio (O/F)	??	??	??	??	??	??
Throttling Capability	100% only	100% only	100% only	100% only	100% only	100% only
Expansion Ratio	??	??	??	10:1	??	??
Restart Capability	No	No	No	No	No	No
Tank Pressurization	??	Nitrogen bottle & self-pressurization	Nitrogen bottle & self-pressurization	Nitrogen bottle & self-pressurization	Nitrogen bottle & self-pressurization	Nitrogen bottle & self-pressurization
Control-Pitch,Yaw,Roll	Jet Vanes	Hydraulic gimbaling (±10°)	Hydraulic gimbaling (±10°)	Hydraulic gimbaling (±10°)	Hydraulic gimbaling (±10°)	Hydraulic gimbaling (±10°)
Events:						
Nominal Burn Time	500 sec ??	131 sec main engine	166 sec main engine	132 sec main engine	155 sec main engine	170 sec main engine
Stage Shutdown	Burn to depletion	Burn to depletion	Burn to depletion	Burn to depletion	Burn to depletion	Burn to depletion
Stage Separation	Stage 2 'fire-in-hole'	Stage 2 'fire-in-hole'	Stage 2 'fire-in-hole'	Stage 2 'fire-in-hole'	Stage 2 'fire-in-hole'	Stage 2 'fire-in-hole'

Remarks:

CZ-1D. The first stage is a modified intermediate range ballistic missile that is 7.4 ft (2.25 m) in diameter and uses jet vanes for control. Unlike the rest of the Long March family, this stage uses nitric acid for an oxidizer.

CZ-2, -3, and -4 series. The first stage is 11.0 ft (3.35 m) in diameter. The first stage vehicle structure consists of an interstage section, an oxidizer tank, an intertank section, a fuel tank, an aft skirt, and a tail section. The propulsion system is composed of four mutually independent YF-20 engines connected with an engine frame. Each single engine can gimbal in one direction for the attitude control of the vehicle. The pressurization system is combined nitrogen bottle pressure compensation and self-pressurization.

Long March Vehicle

(continued)

Stage 2	CZ-1D (L10) ?	CZ-2C (L35)	CZ-2E (L80) ?	CZ-3 or 3B (L35)	CZ-3A (L35)	CZ-4 (L35)
Dimension:						
Length	17.6 ft (5.35 m)	24.6 ft (7.50 m)	50.9 ft (15.52 m)	31.8 ft (9.71 m)	37.8 ft (11.53 m)	34.1 ft (10.41 m)
Diameter	7.4 ft (2.25 m)	11.0 ft (3.35 m)	11.0 ft (3.35 m)	11.0 ft (3.35 m)	11.0 ft (3.35 m)	11.0 ft (3.35 m)
Mass: (each)						
Propellant Mass	26.9K lb (12.2K kg)	77K lb (35K kg)	190K lb (86K kg)	77K lb (35K kg)	65.3K lb (29.6K kg)	78.4K lb (35.55K kg)
Gross Mass	32.7K lb (14.85K kg)	86K lb (39K kg)	202K lb (91.5K kg)	86K lb (39K kg)	74.1K lb (33.6K kg)	87.2K lb (39.55K kg)
Structure:						
Type	Skin Stringer	Skin Stringer	Skin Stringer	Skin Stringer	Skin Stringer	Skin Stringer
Material	Aluminum	Aluminum	Aluminum	Aluminum	Aluminum	Aluminum
Propulsion:						
Propellant	UDMH / N2O4	UDMH / N2O4	UDMH / N2O4	UDMH / N2O4	UDMH / N2O4	UDMH / N2O4
Average Thrust	66.1K lb (294K N)vac	162K lb (720K N)vac 10.4K lb (46.1K N) vac for 4 verniers	166K lb (740K N)vac 10.4K lb (46.1K N) vac for 4 verniers	162K lb (720K N)vac 10.4K lb (46.1K N) vac for 4 verniers	166K lb (740K N)vac 10.4K lb (46.1K N) vac for 4 verniers	162K lb (720K N)vac 10.4K lb (46.1K N) vac for 4 verniers
Engine Designation	YF-3	YF-22 main engine YF-23 verniers	YF-22 main engine YF-23 verniers	YF-22 main engine YF-23 verniers	YF-22 main engine YF-23 verniers	YF-22 main engine YF-23 verniers
Number of Engines	1 turbopump with 2 chambers	1 main engine 4 verniers	1 main engine 4 verniers	1 main engine 4 verniers	1 main engine 4 verniers	1 main engine 4 verniers
Isp	287 sec vac	296 sec vac 289 sec vac verniers	296 sec vac 289 sec vac verniers	296 sec vac 289 sec vac verniers	296 sec vac 289 sec vac verniers	295 sec vac 289 sec vac verniers
Feed System	Gas Generator	Gas Generator	Gas Generator	Gas Generator	Gas Generator	Gas Generator
Chamber Pressure	?? psia (?? bar)	?? psia (?? bar)	?? psia (?? bar)	?? psia (?? bar)	?? psia (?? bar)	?? psia (?? bar)
Mixture Ratio (O/F)	??	??	??	??	??	??
Throttling Capability	100% only	100% only	100% only	100% only	100% only	100% only
Expansion Ratio	??	??	??	25:1	??	??
Restart Capability	No	No	No	No	No	No
Tank Pressurization	??	Nitrogen bottle & self-pressurization	Nitrogen bottle & self-pressurization	Nitrogen bottle & self-pressurization	Nitrogen bottle & self-pressurization	Nitrogen bottle & self-pressurization
Control-Pitch,Yaw,Roll	Jet vanes and nitrogen gas for coast	Hydraulic gimbaling (±60°) & 4 verniers	Hydraulic gimbaling (±60°) & 4 verniers	Hydraulic gimbaling (±60°) & 4 verniers	Hydraulic gimbaling (±60°) & 4 verniers	Hydraulic gimbaling (±60°) & 4 verniers
Events:						
Nominal Burn Time	120 sec	130 sec main engine 350 sec verniers	295 sec main engine 410 sec verniers	130 sec main engine 135 sec verniers	110 sec main engine 135 sec verniers	135 sec main engine 135 sec verniers
Stage Shutdown	Burn to depletion	Command shutdown	Command shutdown	Command shutdown	Command shutdown	Command shutdown
Stage Separation	Solid retro rocket	Spring ejection	Spring ejection	Solid retro rocket	Solid retro rocket	Solid retro rocket

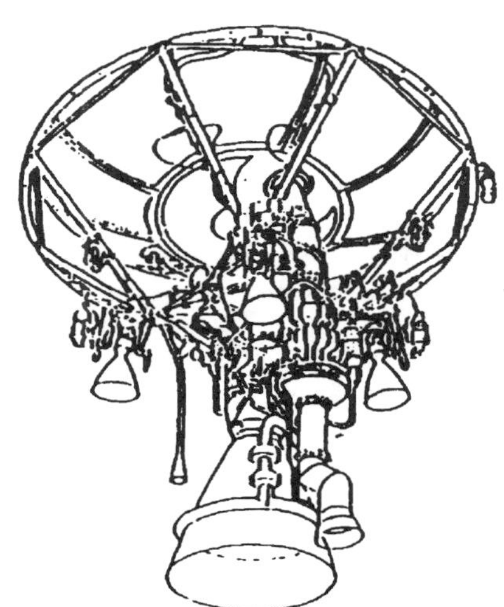

YF-22 / YF-23 Engines

Remarks:

CZ-1D. The second stage is a modified intermediate range ballistic missile that is 7.4 ft (2.25 m) in diameter and uses jet vanes for control. The UDMH / N2O4 tanks are joined by a common bulkhead. After the main engine cutoff of the second stage, CZ-1D flight is a ballistic coast phase. A nitrogen gas nozzle system is used to maintain attitude control and to minimize the ignition attitude error of the third stage.

CZ-2, -3, and -4 series. The second stage engine system consists of a YF-22 main engine and 4 vernier YF-23 engines. The second stage vehicle structure consists of an interstage section, an oxidizer tank, an intertank section and a fuel tank. The main engine is fixed at the center of a cross frame. This YF-22 engine is nearly the same as the YF-20 engine of the first stage. The YF-23 vernier engines have a common independent turbopump system. Each thrust chamber is connected with its servomechanism and can swing tangentially for attitude control of the second stage. The pressurization system of the second stage is basically the same as the first stage, combined nitrogen bottle compensation and self-pressurization.

Hot separation is used for the separation between the first and second stages. Just before separation, the second stage engine ignites, then the explosive bolts connecting the first and second stages explode when the thrust of the first stage engine reduces to a lower level after the engine shutoff and the first stage is pushed away by the second stage engine jet flow acting on the upper bottom of the first stage forward tank. There are jet flow outlets on the lower half of the interstage section.

Long March Vehicle

(continued)

Stage 3

	CZ-1D	CZ-3 (H8) ?	CZ-3A or CZ-3B (H18)	CZ-4 (L15) ?
Dimension:				
Length	7.2 ft (2.2 m)	24.6 ft (7.48 m)	30.0 ft (8.84 m)	6.3 ft (1.92 m)
Diameter	6.7 ft (2.05 m)	7.4 ft (2.25 m)	9.8 ft (3.0 m)	9.5 ft (2.9 m)
Mass:				
Propellant Mass	1,380 lb (625 kg)	18.7K lb (8.5K kg)	38.8K lb (17.6K kg)	31.2K lb (14.15K kg)
Gross Mass	1,930 lb (875 kg)	23.1K lb (10.5K kg)	45.4K lb (20.6K kg)	33.4K lb (15.15K kg)
Structure:				
Type	Monocoque	Skin Stringer ?	Skin Stringer ?	Skin Stringer ?
Material	Steel	Aluminum ?	Aluminum ?	Aluminum ?
Propulsion:				
Propellant	Solid	LOX/LH2	LOX/LH2	UDMH / N2O4
Average Thrust	?? lb (?? N) vac	9.9K lb (44.1KN) vac	17.6K lb (78.5K N) vac each engine	22.5K lb (100K N) ? vac
Engine Designation	??	YF-73	YF-75	YF-3
Number of Engines	—	1 turbopump and 4 chambers	2	1 turbopump and 2 chambers
Isp	?? sec vac	425 sec vac	440 sec vac	295 sec vac
Feed System	—	Gas Generator	Gas Generator	Gas Generator
Chamber Pressure	?? psia (?? bar)	?? psia (?? bar)	?? psia (?? bar)	?? psia (?? bar)
Mixture Ratio (O/F)	—	??	??	??
Throttling Capability	—	100% only ?	100% only ?	100% only
Expansion Ratio	??	40:1	??	??
Restart Capability	—	1 restart	1 restart ?	No
Tank Pressurization	—	Helium bottle and self-pressurization	Helium bottle and self-pressurization	??
Control-Pitch,Yaw,Roll	Spin stabilized	Gimbal 4 nozzles (±24°) Hydrazine for coast	Gimbal 2 nozzles	Hydrazine thrusters
Events:				
Nominal Burn Time	?? sec	800 sec	470 sec	135 sec
Stage Shutdown	Burn to depletion	Command shutdown	Command shutdown	Command shutdown
Stage Separation	Spring ejection	Spring ejection	Spring ejection	Spring ejection

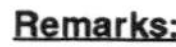

Remarks:

CZ-1D. The third stage is solid motor that is spin stabilized during its powered phase.

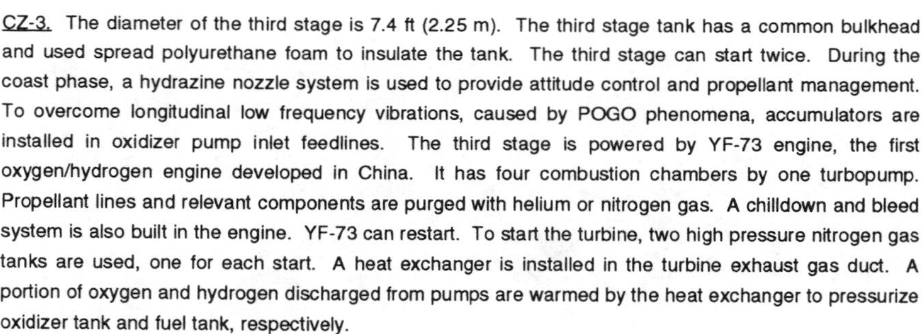

CZ-3. The diameter of the third stage is 7.4 ft (2.25 m). The third stage tank has a common bulkhead and used spread polyurethane foam to insulate the tank. The third stage can start twice. During the coast phase, a hydrazine nozzle system is used to provide attitude control and propellant management. To overcome longitudinal low frequency vibrations, caused by POGO phenomena, accumulators are installed in oxidizer pump inlet feedlines. The third stage is powered by YF-73 engine, the first oxygen/hydrogen engine developed in China. It has four combustion chambers by one turbopump. Propellant lines and relevant components are purged with helium or nitrogen gas. A chilldown and bleed system is also built in the engine. YF-73 can restart. To start the turbine, two high pressure nitrogen gas tanks are used, one for each start. A heat exchanger is installed in the turbine exhaust gas duct. A portion of oxygen and hydrogen discharged from pumps are warmed by the heat exchanger to pressurize oxidizer tank and fuel tank, respectively.

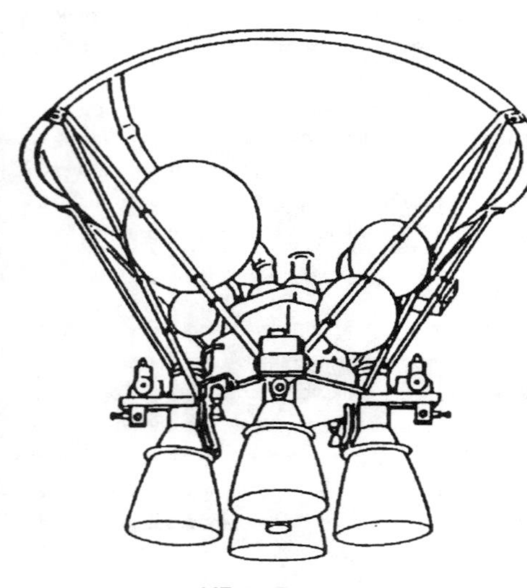

YF-73 Engine

CZ-3A and -2E/HO. The third stage structure consists of a payload adaptor, a vehicle equipment bay (VEB), and a common bulkhead tank. The propulsion system is composed of two-direction-swinging engines connected in parallel. The YF-75 is a new engine based on the YF-73. Pressurization system for LH2 tank is combined normal temperature helium bottle pressure compensation and self-pressurization, and for LOX tank is combined normal temperature helium bottle pressure compensation and cold helium bottle pressurization.

CZ-4. This newly developed third stage has a common bulkhead and uses UDMH/N204 as the propellants. The engine is the same as that used on the second stage of CZ-1D.

Long March Vehicle

(continued)

Payload Fairing

	CZ-1D (in mm)	CZ-3 (in mm) Model A	CZ-3 (in mm) Model B	CZ-3A (in mm) Single
	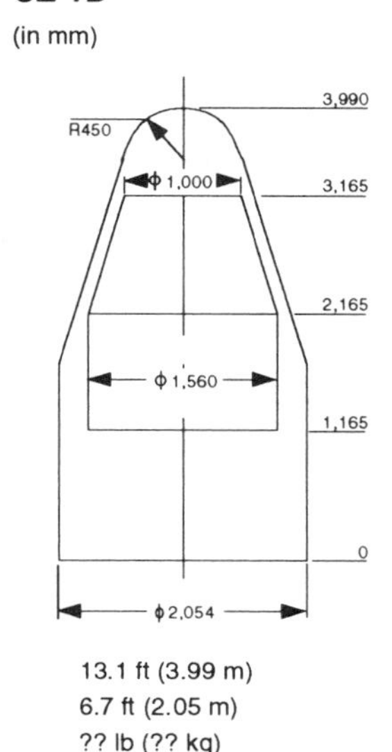	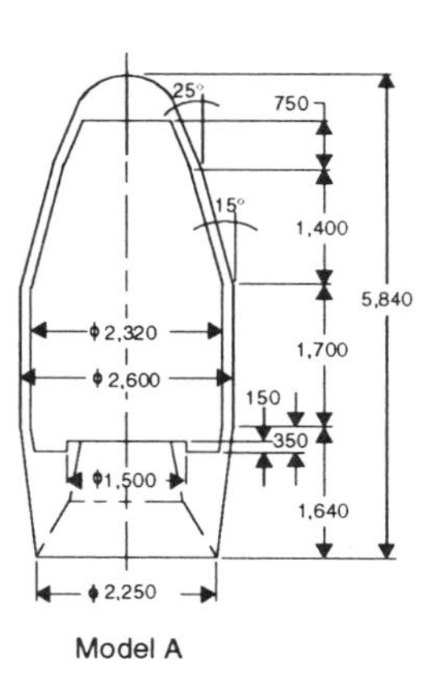	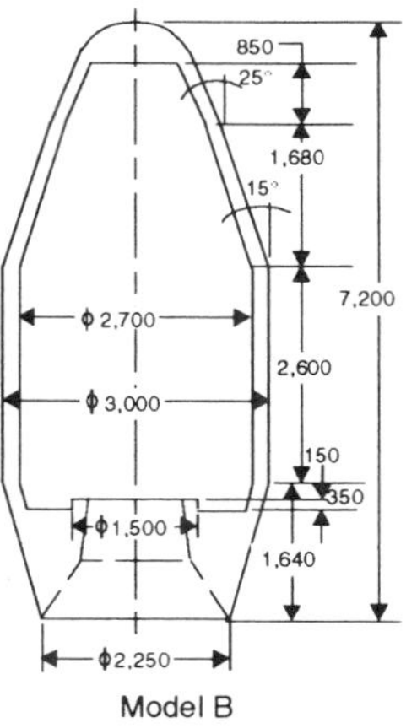	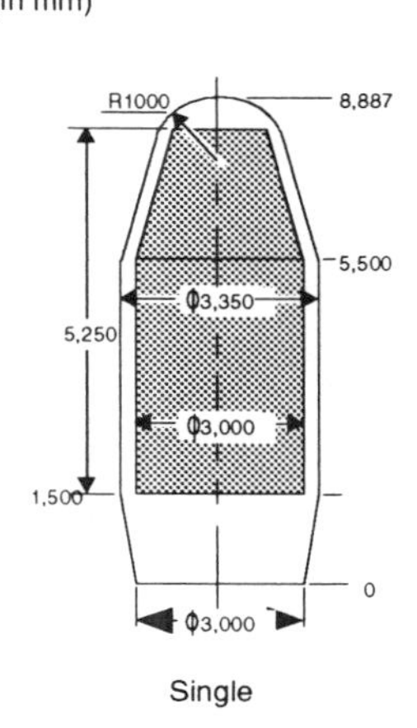
Length	13.1 ft (3.99 m)	19.2 ft (5.84 m)	23.9 ft (7.27 m)	29.2 ft (8.89 m)
Diameter	6.7 ft (2.05 m)	8.5 ft (2.6 m)	9.8 ft (3.0 m)	11.0 ft (3.35 m)
Mass	?? lb (?? kg)	?? lb (?? kg)	?? lb (?? kg)	?? lb (?? kg)
Sections	2	2	2	2
Structure	Composite sandwich	Composite sandwich	Composite sandwich	Composite sandwich
Material	Glass honeycomb	Glass honeycomb	Glass honeycomb	Glass honeycomb

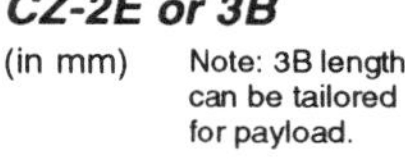

	CZ-2C (in mm) Type A	CZ-2C (in mm) Type B	CZ-2E or 3B (in mm) Note: 3B length can be tailored for payload.	CZ-4 (in mm) Model A	CZ-4 (in mm) Model B
	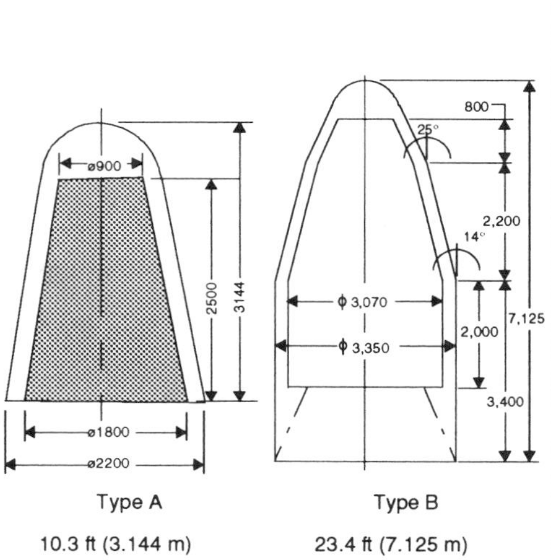		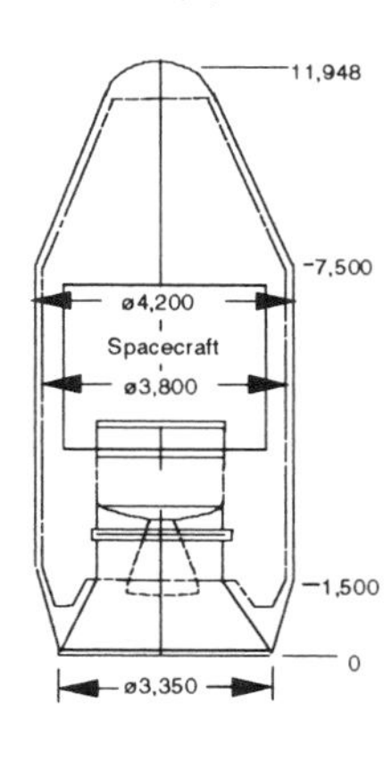	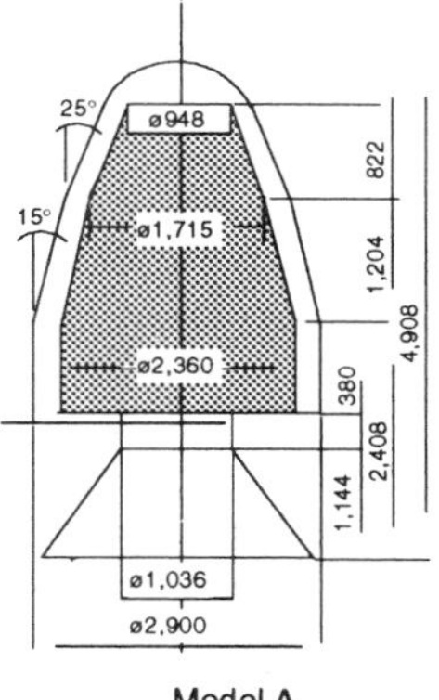	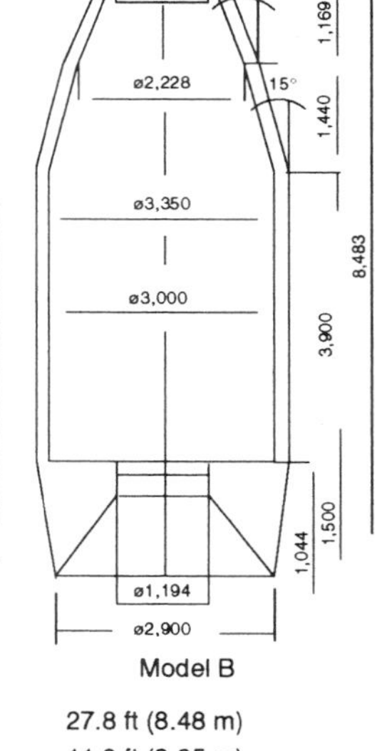
Length	10.3 ft (3.144 m)	23.4 ft (7.125 m)	39.2 ft (11.95 m)	16.1 ft (4.91 m)	27.8 ft (8.48 m)
Diameter	7.2 ft (2.2 m)	11.0 ft (3.35 m)	13.8 ft (4.2 m)	9.5 ft (2.9 m)	11.0 ft (3.35 m)
Mass	?? lb (?? kg)	?? lb (?? kg)	?? lb (?? kg)	?? lb (?? kg)	?? lb (?? kg)
Sections	2	2	2	2	2
Structure	Composite sandwich	Composite sandwich	Composite sandwich	Composite sandwich	Composite sandwich
Material	Glass honeycomb	Glass honeycomb	Glass honeycomb	Glass honeycomb	Glass honeycomb

Remarks

Payload fairing description is two half-shells structure with a longitudinal separation system. The nose dome is made of phenolic resin glass cloth. The bicone is made of glass honeycomb. The faces and the core of the honeycomb are made of barium phenolic resin and glass cloth. For most of the fairings, noncontaminative explosive cords are used for longitudinal (i.e., between two half-shells) separation. Expansion of the gas in the capsule springs open the fairing. Lateral (i.e., between fairing and the lower stage) unlocking is initiated by explosive bolts unlocking the clampband, and the clampband is pulled to the fairing side by a traction spring. For some fairing, separation is realized through the exploding bolts releasing springs. Each explosive bolt is put in a box with a baffle on the inner side to avoid contamination of the payload when the bolt explodes. The fairing is generally jettisoned above 59 mi (110 km) altitude.

Access doors may be opened on the cylindrical section to facilitate payload operations. Radio-transparent windows may be opened on the forward cone and cylindrical section at the user's request. In general, radio transparency is not less than 85%. On the launch service tower, the inside environment under the payload fairing may be air conditioned: temperature 41–59°F (5–15°C) and relative humidity not greater than 60%. Some sound deadening layers are mounted on the inner wall of the fairing so as to minimize the noise affecting the payload during flight.

Long March Vehicle

(continued)

Avionics

The main function of the control system is to make the vehicle fly along the predetermined trajectory and attitude, to cut off engines properly, and to enable the payloads to enter the predetermined orbit in accordance with various required parameters.

The control system includes three subsystems: guidance system, attitude control system, and programmed distribution system. The guidance system consists of an inertial platform and a computer which keep the vehicle flying according to the predetermined trajectory and ensure injection accuracy. The guidance is a four-axis flexible inertial platform computer scheme. For CZ-3A, it meets the requirements for dual-satellite launching. The attitude control system keeps the vehicle attitude stable during its whole flight period and makes orientation and collision avoidance maneuvers. During the powered flight phase attitude control is provided by an inertial platform, rate gyro-digital control device, servomechanism continuous control scheme, and during the inertial flight phase by a switching control scheme with small monopropellant engine. The programmed distribution system adopts a scheme that subsystems and large power equipments are supplied respectively through the main and auxiliary distributors. The system conducts timing control over all airborne systems during flight.

The telemetry system of CZ-2E is divided into three subsystems: the transmission subsystem, data acquisition subsystem and the ground receiving vehicle. The airborne system consists of sensitive elements, transform devices, transmitting devices, and recording and signal transmitting devices all over the vehicle. The vehicle flight condition is monitored by the vehicle telemetry system and ground receiving and real-time transmission equipment to provide initial injection information and range safety data.

The tracking system consists of the ground continuous wave tracking radars, single-pulse tracking radars, two airborne responders for the continuous wave radars, one responder for the single pulse radars and the antennas of these responders. The telecommanded destruction subsystem consists of the ground telecontrol radars, a safety command receiver and three linearly polarized antennas. These three antennas are installed circularly on the vehicle surface to form an all around antenna array. The above responders and receiver are all mounted inside the second stage intertank section, while the antennas on the outside surface of the same intertank section.

There are two types of destruction for range safety: radio tele-commanded destruction and airborne independent self-destruction for when the attitude angle exceed the allowable value.

Attitude Control System

CZ-1D. Jet vanes are used in first and second stages to produce the control forces. After the second stage main engine cutoff, CZ-1 flight is in a ballistic coast phase. A nitrogen gas nozzle system is used to maintain attitude control and to minimize the ignition attitude error of the third stage. The third stage is spin stabilized during its powered phase.

CZ-2, -3, and -4 series. The first stage engine is composed of four mutually independent engines connected with an engine frame. Each single engine can swing tangentially for the attitude control of the vehicle. For CZ-2E, the liquid strap-on engine is fixed.

The second stage is controlled by four YF-23 vernier engines. Each vernier engine thrust chamber is connected with its servomechanism and can swing tangentially for attitude control of the second stage. The YF-22 main engine provides pitch and yaw control.

The third stage is controlled by the YF-73 for CZ-3 and YF-75 for CZ-3A and CZ-3B. For CZ-3, each of four YF-73 combustion chambers can gimbal in a tangential direction to provide attitude control moment for the flight of the third stage. For CZ-3A and CZ-3B, each engine is composed of two-direction swinging engines connected in parallel. A reorientation control system is used for attitude control and propellant management after second-stage shutoff and payload orientation after third-stage shutoff. The hydrazine monopropellant system is pressure-fed by helium.

Long March Performance

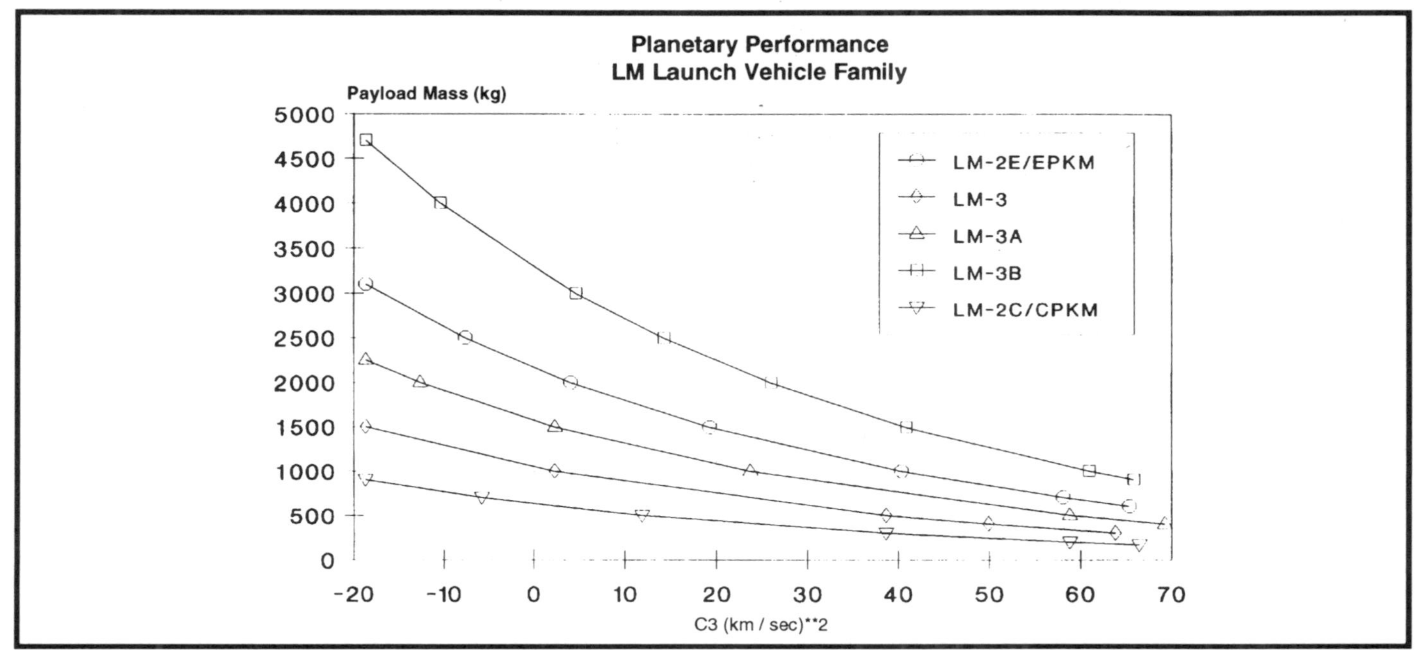

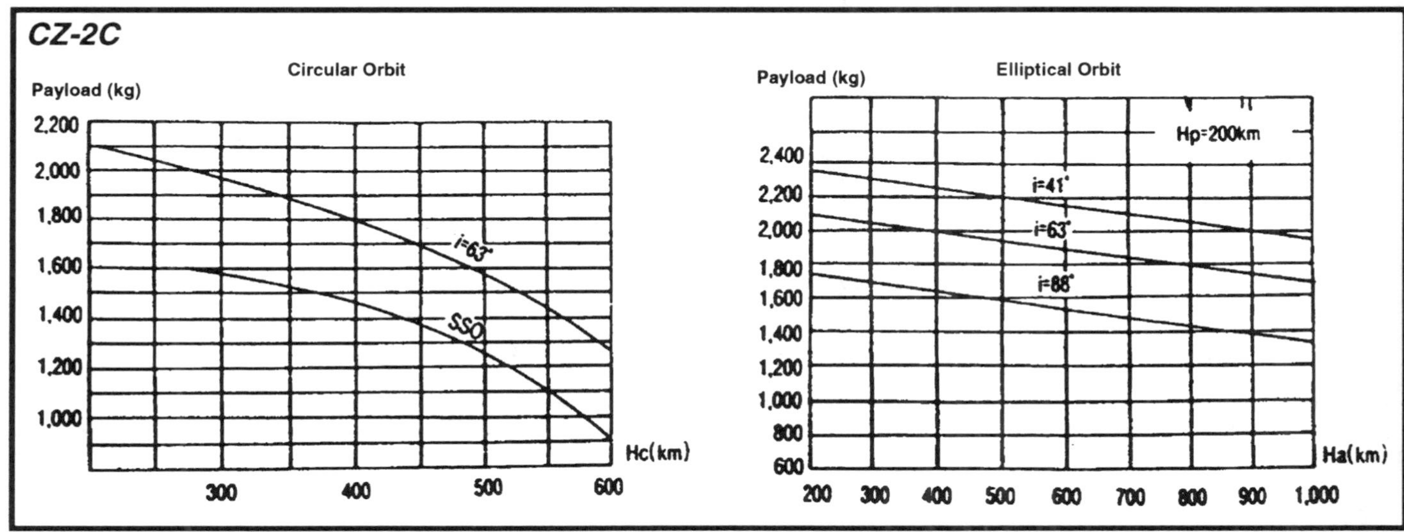

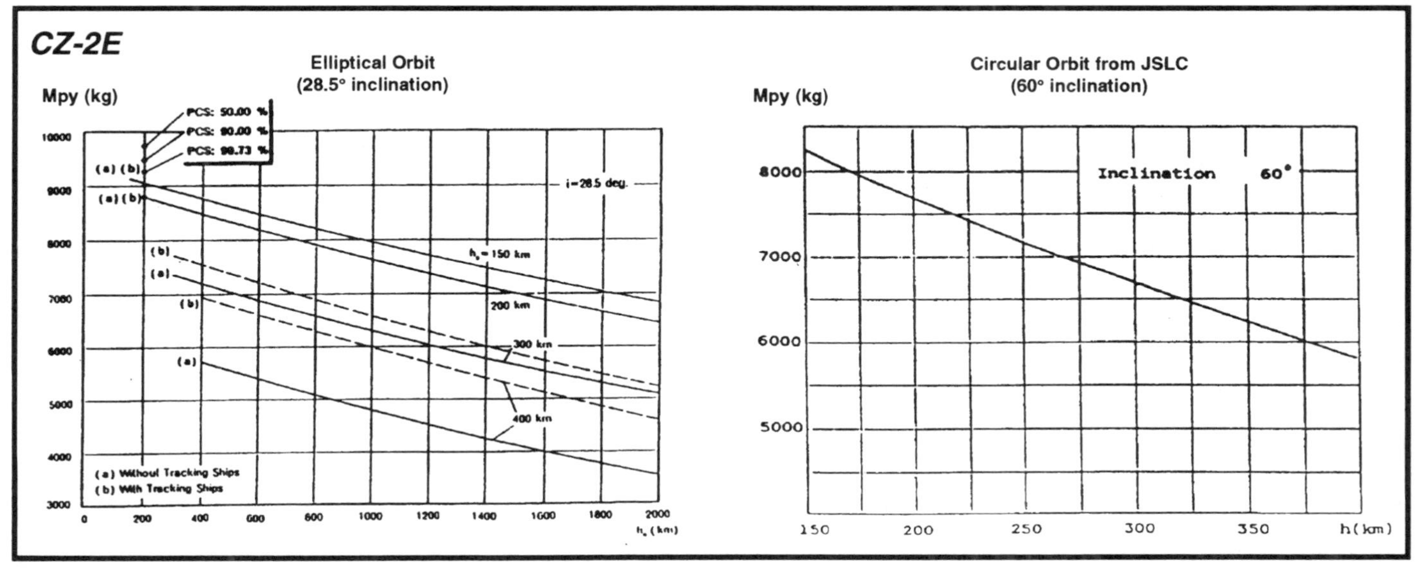

Long March Performance

(continued)

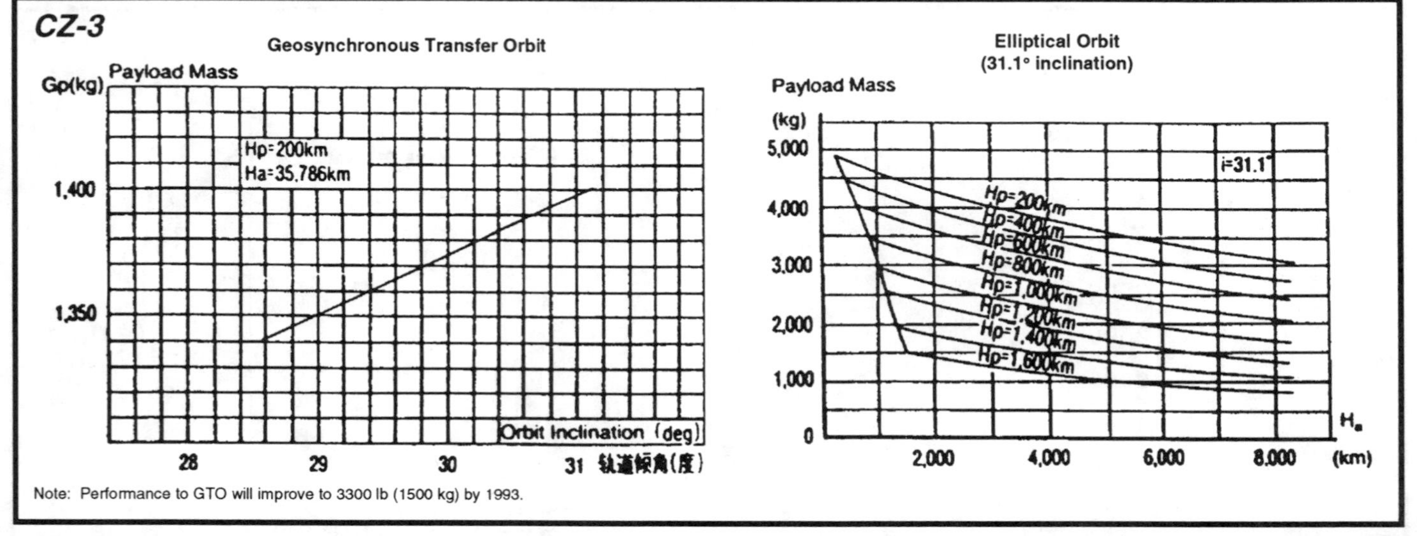

Note: Performance to GTO will improve to 3300 lb (1500 kg) by 1993.

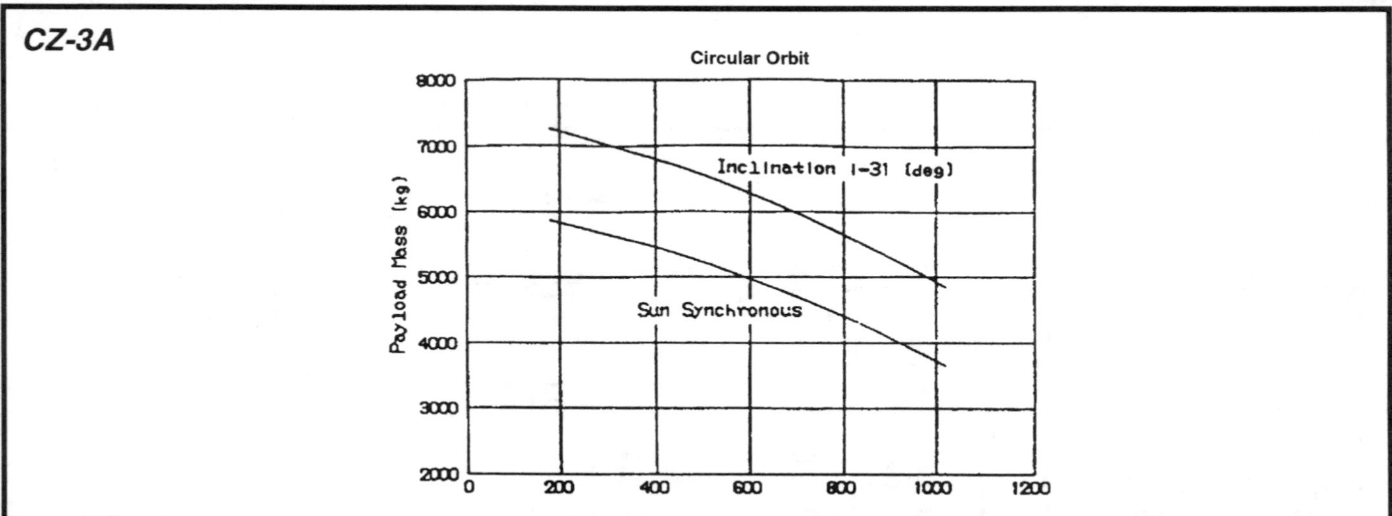

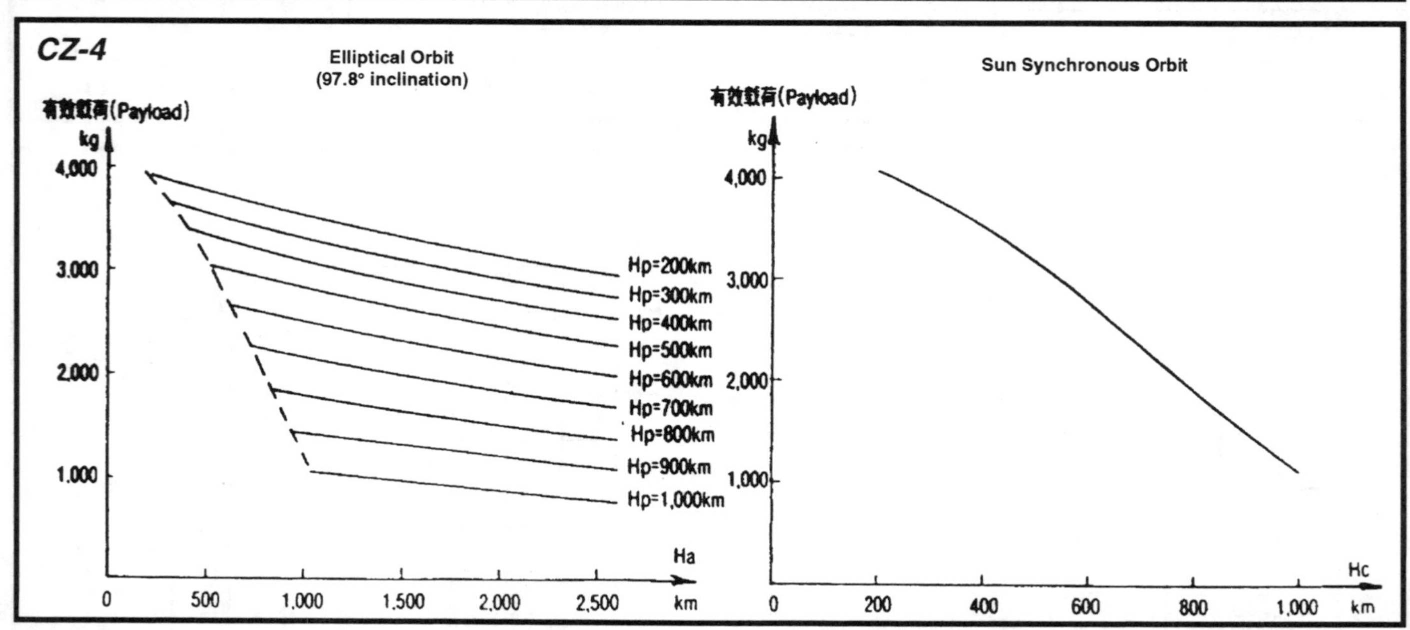

Additional performance curves not available for CZ-1D and CZ-3B

Long March Operations

Launch Site

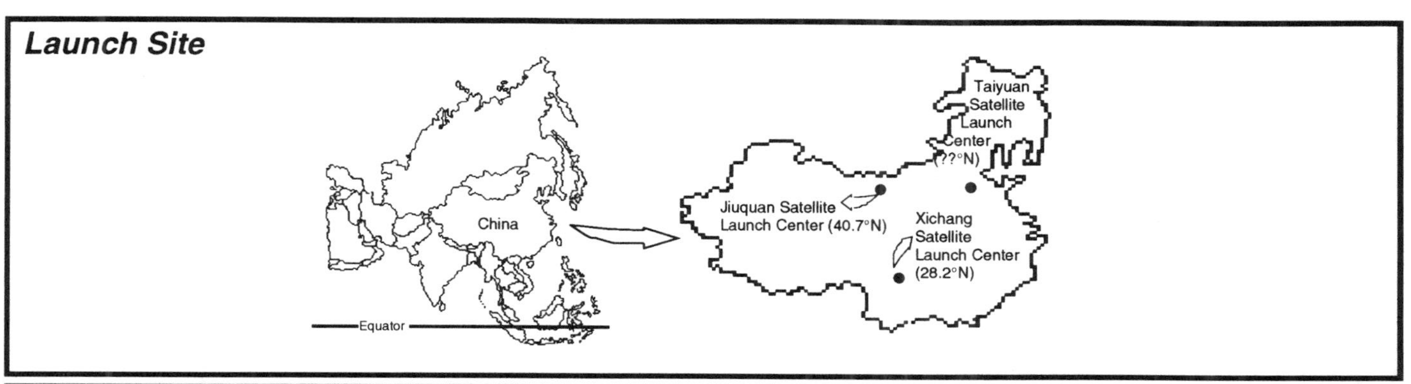

Launch Facilities

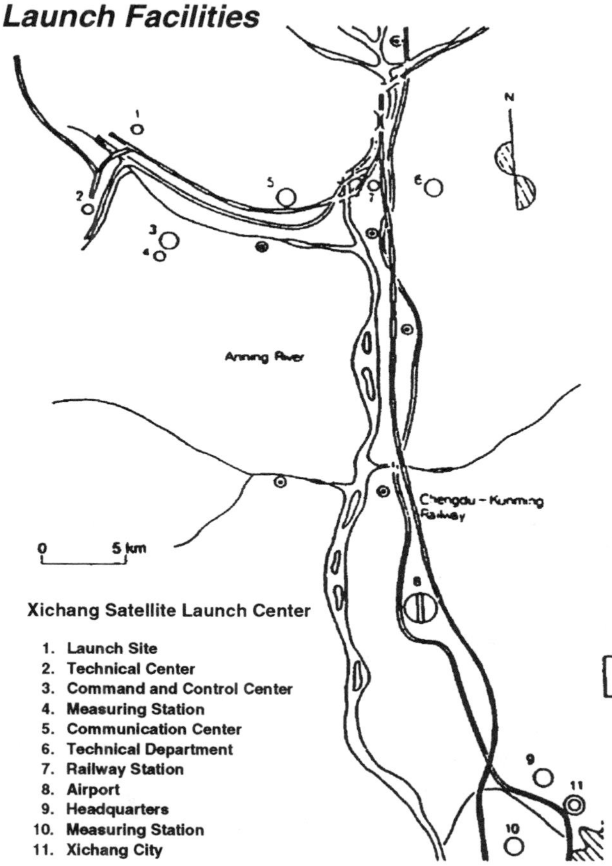

Anning River

Chengdu – Kunming Railway

0 5 km

N

Xichang Satellite Launch Center

1. Launch Site
2. Technical Center
3. Command and Control Center
4. Measuring Station
5. Communication Center
6. Technical Department
7. Railway Station
8. Airport
9. Headquarters
10. Measuring Station
11. Xichang City

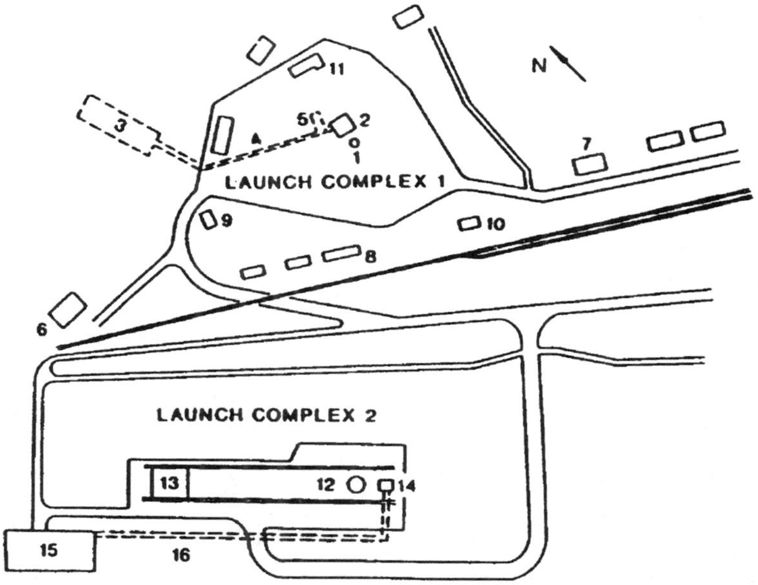

LAUNCH COMPLEX 1

LAUNCH COMPLEX 2

XSLC Launch Area

1. Launch Pad (No. 1)
2. Launch Tower (No. 1)
3. Launch Control Room
4. Cable Ditch
5. Battery Room
6. Fuel Store
7. Oxidizer Store
8. Gas Vessel Store
9. Aiming Room
10. LH Filling Room
11. LOX Filling Room
12. Launch Pad (No. 2)
13. Movable Tower
14. Launch Tower (No. 2)
15. Launch Control Room
16. Cable Ditch

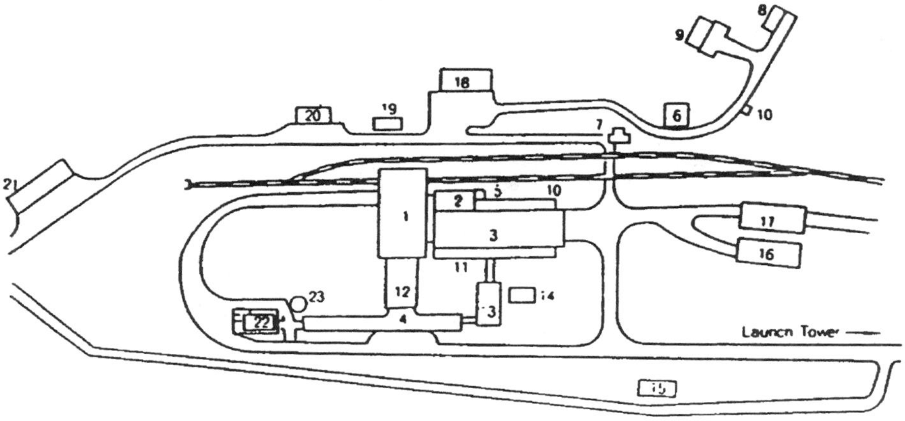

Launch Tower

XSLC Technical Center

1. Transit Hall
2. Payload Hangar BS1
3. Launch Vehicle Checkout Hangar
4. Component Checkout Hangar
5. Hydrazine Filling Hangar
6. Apogee Engine Hangar
7. Leakage Inspection Hangar
8. Pyrotechnics Store
9. Pyrotechnics Checkout Room

10-12. Working Rooms

13. Air Conditioning Equipment
14. Transformer Station
15. Boiler Room

16-21. Stores

22. Payload Hangar BS2, BS3
23. Antenna Tower

Long March Operations

(continued)

Launch Processing

China has three launch sites for the Long March family: the Xichang Satellite Launch Center (XSLC) for the CZ-3, CZ-3A, CZ-2E and CZ-3B; the Jiuquan Satellite Launch Center (JSLC) for the CZ-2C and CZ-1D; and the Taiyuan Satellite Launch Center (TSLC) for the CZ-4. In the future, as business develops, it is envisaged that the Long March vehicles will be launched from whichever site is most advantageous for the mission.

Concern has been expressed, notably in the United States, about the technical security of a spacecraft launched in China on a Long March launch vehicle. China has stated that the technical security of a foreign spacecraft is guaranteed.

Xichang Satellite Launch Center (XSLC). XSLC became operational in 1984 for launching Chinese domestic communications satellites on the CZ-3. XSLC is in a mountainous area 40 mi (64 km) northwest of Xichang City in Sichuan Province. It is approximately 5,900 ft (1,800 m) above sea level. The geographical coordinates of the launch site are 28.2°N,102.0°E.

At the site of XSLC the climate is of the subtropic type. The yearly average temperature is 61°F (16°C), with the average highest temperature at 77°F (25°C) in summer, and the average lowest temperature at 36°F (2°C) in winter. Wind speed on the launch site is very low all the year round, with a long frost-free period and clearly divided dry and rainy seasons. The rainy season is the period from June to September.

The Xichang Airport, located in the northern suburbs of Xichang City, has been upgraded to accommodate 747 and C-130 type aircraft. Xichang Airport is 8.4 mi (13.5 km) away from Xichang City and 31 mi (50 km) away from the launch site. The Chengdu-Kunming Railway and the west part of the Sichuan-Yunnan highway pass through XSLC. There is a dedicated railway branch and a dedicated highway branch leading straight to the launch site. The maximum slope of the railway is 3%. The maximum slope of the highway is 5.6%.

The Technical Center which is used for launch vehicle and payload test and checkout consists of a number of facilities such as the launch vehicle test building, payload preparation building, and hazardous processing building.

Long March stages are rail-shipped to the launch site. A vehicle will spend about five weeks at the launch vehicle checkout hangar in a horizontally mated position for checkout before being disassembled and trucked in stages about 1.3 mi (2.2 km) north to the Launch Area on a 26 ft (8 m) wide cement road. Integrated checkout of the vehicle and payload is done only at the pad, not in the Center.

The payload preparation building is for nonhazardous assembly, integration, and testing operations for spacecraft and, if required by the mission, upper stages. It is large enough to accommodate at least two spacecraft in parallel in the assembly and test hall in class 100,000 clean-room conditions. In addition, there are class 10,000 clean rooms for assembly and test operations on equipment requiring a particularly clean environment.

The hazardous processing building is for hazardous assembly operations, spacecraft propellant fueling and pressurization, solid motor integration, installation of electro-explosive devices, spin-balancing of the payload, payload and/or upper stage weighing, and hazardous testing. For the CZ-2E and CZ-3A launchers, encapsulation of the spacecraft in the launch vehicle fairing

View of Launch Complex 1 in the distance (left) and Launch Complex 2 nearby (right) with a CZ-2E on the launch pad

Long March Operations

(continued)

Launch Processing

(continued)

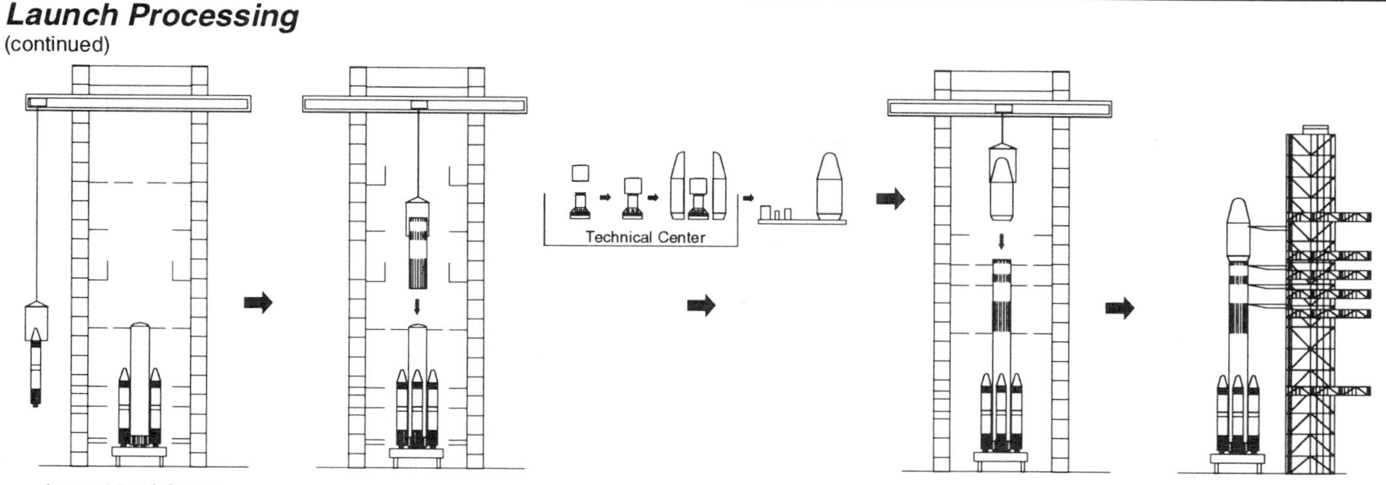

Assembly of CZ-2E

will normally take place inside the hazardous processing building. For the CZ-3, encapsulations takes place on the launch tower inside the environmental enclosure.

After leaving the facilities in the Technical Center inside an environmentally controlled payload container or inside the fairing, the spacecraft is taken to the launch area.

The Launch Area is to perform the tasks of mating, testing and checkout, direction orientation, propellant filling and launching of the Long March. At the launch area, there is a pad for the CZ-3, but two new pads about 0.5 mi (0.8 km) away are capable of handling a CZ-2E or a CZ-3A. These new pads consist of two fixed umbilical towers, and a mobile tower which can service a vehicle on either pad. It took 15 months to construct the new pad for the first launch of CZ-2E.

The 252 ft (77 m) tall CZ-3 launch tower is fixed next to the vehicle and includes a rotating column crane on top to stack stages and the payload. This is where the spacecraft is integrated with, and tested on the launch vehicle, inside an environmental enclosure at the top of the launch tower. The enclosure has a cleanliness level of class 100,000. During launch, an area of 2 sq mi (5 sq km) around the pad is evacuated and local residents are allowed to return about 10 minutes after launch.

The primary purposes of the Command and Control Center are to control operation, to receive, process and disseminate telemetry data from the launch vehicle, and to monitor and control the safety of the launch vehicle before and during flight. Downrange measuring stations under the management of XSLC are distributed in Xichang and Yibin of Sichuan Province and Guiyang of Guizhou Province.

There is a network of TT&C (Telemetry, Tracking, and Control) stations, including two tracking ships, which are used for TT&C operations during the liftoff and flight. XSLC has a weather station which is connected to the national weather network. Other facilities include a chemical analysis laboratory, workshop facilities and comprehensive medical facilities at the launch site, and a hospital at Xichang City.

Jiuquan Satellite Launch Center (JSLC). JSLC, previously known in the West as Shuang Cheng Tzu, is China's original launch site. To date, most of China's satellites have been launched there. About 1,000 mi (1,600 km) west of Beijing, to the north at the tail of the Great Wall on the edge of the Gobi Desert, this base is located 40.7° N 100° E, in Gansu Province in northwestern China. It is approximately 3,300 ft (1,000 m) above sea level. Historically, orbital missions are launched southeast to avoid overflying Mongolia and the Soviet Union and enter LEO in a narrow band between 56.9° and 69.9° inclination. Although XSLC is now used for GEO launches, JSLC continues to be used for recoverable reconnaissance and Earth resources satellites.

The airport is south of the launch site. A rail system connects the airport to the Technical Center and the launch area. The Technical Center provides facilities for the assembly, integration and testing of payloads for the CZ-2C, and from 1991, the CZ-1D. The current facilities are less sophisticated than those at XSLC, but are still adequate. Other facilities at JSLC are similar to those at XSLC.

Taiyuan Satellite Launch Center (TSLC). TSLC is based amid rugged terrrain in Shanxi Province 270 nm (500 km) southwest of Beijing. The site is used to launch CZ-4 vehicles southward into polar orbit. It was used for the first time for a Long March launch in September 1988 for China's first weather satellite. This was the first launch of a CZ-4. Few details have been released about this launch site.

Start Activity (days)	End Activity (days)	Duration (days)	Activity
T-11	T-9	2	Launch vehicle erection and mating
T-11	T-9	2	Checkout preparation
T-9	T-7	2	Launch vehicle checkout
T-7	T-6	1	Payload integration and fairing installation
T-6	T-4	2	Overall checkout
T-4	T-3	1	Prelaunch rehearsal
T-3	T-1	2	Filling preparation and filling
T-1	T-0	1	Countdown sequence

Sequencing and Duration of CZ-2E Launch Processing

Time Sequence	Operation
T-5 hrs	Vertical adjustment, aiming
T-4 hrs	Charging and prelaunch functional checkout
T-2 hrs	Status checkout and moving back the working platform
T-60 min	Prelaunch charging checkout and entering into one-hour launch sequence
T-40 min	Pressurizing and removing gas pipes
T-20 min	Switching on onboard telemetry system
T-1 min	Withdrawal of swing cable rod,automatic ignition
T-0 sec	Liftoff

CZ-2E Final Countdown

Long March Operations
(continued)

Flight Sequence

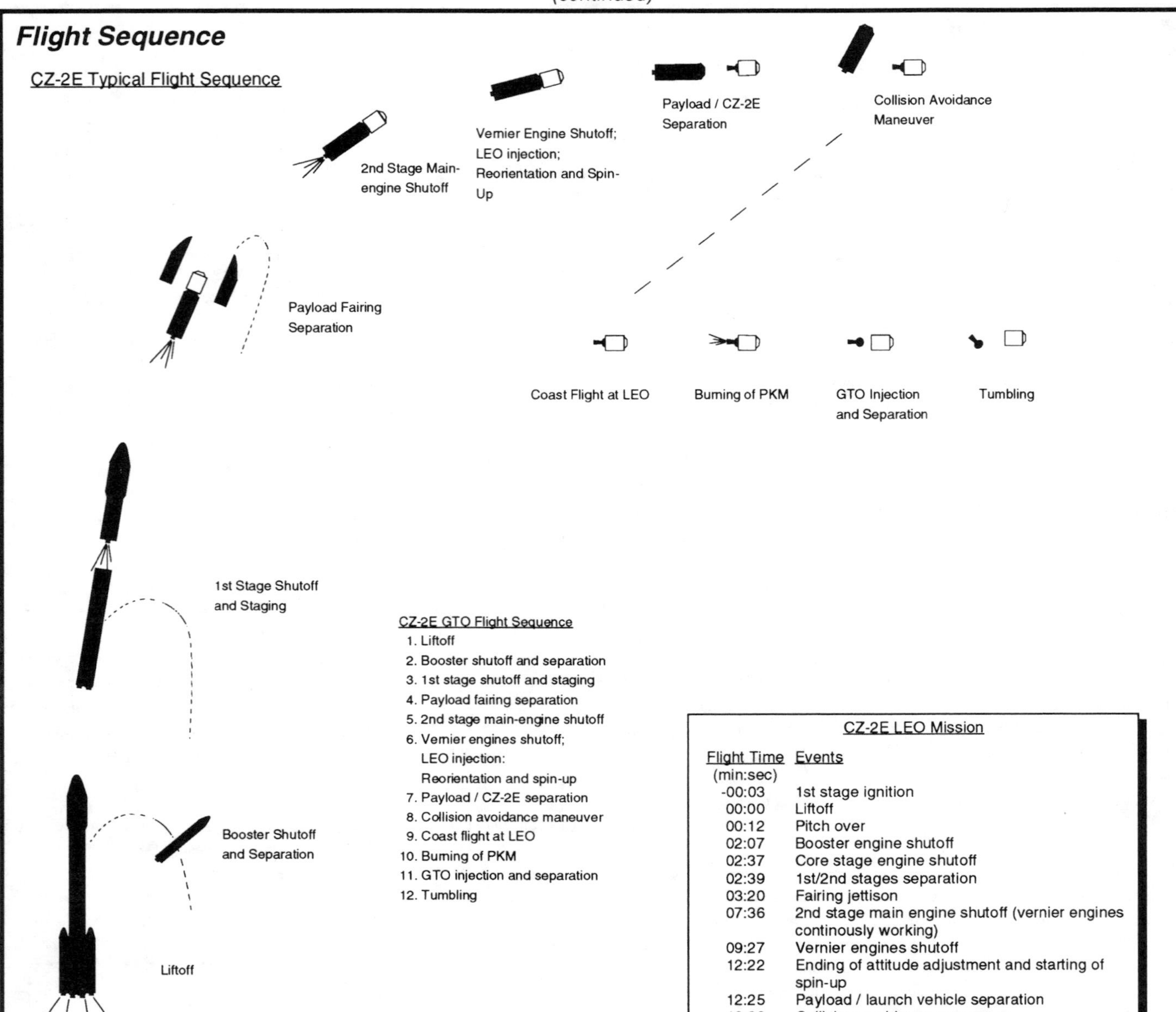

CZ-2E GTO Flight Sequence
1. Liftoff
2. Booster shutoff and separation
3. 1st stage shutoff and staging
4. Payload fairing separation
5. 2nd stage main-engine shutoff
6. Vernier engines shutoff;
   LEO injection:
   Reorientation and spin-up
7. Payload / CZ-2E separation
8. Collision avoidance maneuver
9. Coast flight at LEO
10. Burning of PKM
11. GTO injection and separation
12. Tumbling

CZ-2E LEO Mission

Flight Time (min:sec)	Events
-00:03	1st stage ignition
00:00	Liftoff
00:12	Pitch over
02:07	Booster engine shutoff
02:37	Core stage engine shutoff
02:39	1st/2nd stages separation
03:20	Fairing jettison
07:36	2nd stage main engine shutoff (vernier engines continously working)
09:27	Vernier engines shutoff
12:22	Ending of attitude adjustment and starting of spin-up
12:25	Payload / launch vehicle separation
12:28	Collision avoidance maneuver

CZ-2C Typical Flight Sequence

CZ-2C LEO Mission

Flight Time (min:sec)	Events
-00:03	1st stage ignition
00:00	Liftoff
00:08	Beginning of sequence pitch over
00:44	Flight Mach 1
02:07	Precommand of 1st stage engine shutoff
02:08	Maincommand of 1st stage engine shutoff
03:25	Jettison of fairing
03:56	Precommand of 2nd stage main engine shutoff
03:57	Maincommand of 2nd stage main engine shutoff
07:55	Command of 2nd stage vernier engines shutoff
07:58	Main satellite / launch vehicle separation

CZ-3 Typical Flight Sequence

CZ-3 GTO Mission

Flight Time (min:sec)	Events
-00:03	1st stage ignition
00:00	Liftoff
02:07	1st stage shutoff
02:08	1st/2nd stage separation
04:15	2nd stage main engine shutoff
04:19	Fairing jettison
04:22	2nd stage vernier engine shutoff
04:23	2nd/3rd stage separation
11:29	3rd stage first shutoff
15:36	3rd stage restart
20:54	3rd stage second shutoff
21:02	Velocity adjustment
21:32	Spin-up
21:33	Spacecraft separation

Long March Payload Accommodations

	CZ-2E	Others
Payload Compartment		
Maximum Payload Diameter x Maximum Cylinder Length x Maximum Cone Length	CZ-1D: 61.4 in (1560 mm) x 39.4 in (1000 mm) x 39.4 in (1000 mm) CZ-2C: Type A - 70.9 in (1800 mm) x (no cylinder) x 98.4 in (2500 mm) Type B - 120.9 (3070 mm) x 133.9 in (3400 mm) x 118.1 in (3000 mm) CZ-2E: 149.6 in (3800 mm) x 236.2 in (6000 mm) x 135.2 in (3435 mm) CZ-3: Model A - 91.3 in (2320 mm) x 80.7 in (2050 mm) x 85.1 in (2161 mm) Model B - 107.1 in (2720 mm) x 116.1 in (2950 mm) x 99.6 in (2530 mm) CZ-3A: Single - 118.1 in (3000 mm) x 157.5 in (4000 mm) x 49.2 in (1250 mm) Dual - 141.7 in (3600 mm) x ?? x ?? CZ-4: Model A - 92.9 in (2360 mm) x 34.8 in (884 mm) x 79.8 in (2026 mm) Model B - 118.1 in (3000 mm) x 153.5 in (3900 mm) x 102.7 in (2609 mm)	
Payload Adapter		
Interface Diameter	47.1 in (1,197 mm) 63.1 in (1,604 mm) 64.8 in (1,645 mm)	CZ-2C: 36.9 in (937 mm) 47.0 in (1,194 mm) 58.9 in (1,497 mm) CZ-3A: 47.0 in (1,194 mm)
Payload Integration		
Nominal Mission Schedule Begins	T-?? months	
Launch Window		
Latest Countdown Hold Not Requiring Recycling	T-?? min	
On-Pad Storage Capability	?? hours for a fueled vehicle	
Latest Access to Payload	T-?? hours or T-?? hours through access doors	
Environment		
Maximum Load Factors	+5.8 g axial, ±0.8 g lateral at stage 1/2 separation	CZ-3: +6.1 g axial, ±0.8 g lateral at stage 1 shutoff CZ-3A: +5.4 g axial at , ±0.8 g lateral at stage 1 shutoff, ±1.0 g lateral at max. dynamic pressure
Minimum Lateral / Longitudinal Payload Frequency	10 Hz / 26 Hz	
Maximum Overall Acoustic Level	142 dB (full octave) 137 dB with blankets	CZ-3A: 142 dB (full octave)
Maximum Flight Shock	2,000 g from 1,500–5,000 hz	CZ-3A: 2,000 g from 1,500–5,000 hz
Maximum Dynamic Pressure on Fairing	?? lb/ft^2 (?? N/m^2)	CZ-3A: 532 lb/ft^2 (25.5K N/m^2) CZ-2C: 982 lb/ft^2 (47K N/m^2)
Maximum Pressure Change in Fairing	?? psi/s (?? KPa/s)	
Cleanliness Level in Fairing (Prior to Launch)	Class 10,000	
Payload Delivery		
Standard Orbit and Accuracy (1 sigma)	Perigee: 108 nm (200 km) ±1.1 nm (2 km) Apogee: 216 nm (400 km) Inclination: 28.5° ±0.05°	CZ-3: Perigee: 108 nm (200 km) ±0.6 nm (1.1 km) Apogee: 19,323 nm (35,786 km) ±51 nm (94 km) Inclination: 31.1° ±0.07°
Attitude Accuracy (3 sigma)	For nonspin - Pitch and Yaw: ±0.5 deg, ±0.2 deg/sec Roll: ±0.2 deg, ±0.15 deg/sec	CZ-2C: Pitch and Yaw: ±0.467 deg, ±0.117 deg/sec Roll: ±0.167 deg, ±0.083 deg/sec
Nominal Payload Separation Rate	1.6 ft/s (0.5 m/s)	CZ-3: 2.8 ft/s (0.85 m/s)
Deployment Rotation Rate Available	2 to 5 rpm with spin rockets for L/V-S/C Higher with spin table	
Loiter Duration in Orbit	?? hours	
Maneuvers (Thermal / Collision / Telemetry)	Yes	

Long March Notes

Publications

User's Guide

A Briefing of Launch Vehicle of Long March 2 Family, Beijing Institute of Astronautical Systems Engineering, April 1988.

LM-2E User Guide, Issue 1, China Great Wall Industry Corporation (CGWIC), February 1988.

Long March 3A, Beijing Institute of Astronautical Systems Engineering.

Technical Publications

"Briefing of CALT", China Academy of Launch Vehicle Technology, June 1990.

"Long March-3 Briefing", China Great Wall Industry Corporation, June 1990.

"Long March-2E Briefing", China Great Wall Industry Corporation, June 1990.

Space Log, TRW, 1957-1989.

"Oxygen / Hydrogen Rocket Engine for CZ-3", Wang Zhiren & Gu Mingchu, China Ministry of Aero Space Industry, IAF-89-299, 40th Congress of the International Astronautical Federation, Malaga Spain, October 7-13, 1989.

"China Clarifies Long March", Stephen Archer, Vega Space Systems Engineering, Space, May 1989.

"Long March Launch Vehicle Family - Current Status and Future Development", H. Zuwei & R. Xinmin, China Ministry of Astronautics, IAF-87-183, 38th Congress of the International Astronautical Federation, Brighton, United Kingdom, October 10-17, 1987.

"China Great Wall Industry Corp.", China Great Wall Industry Corporation (Brochure).

"Long March Launch System", China Great Wall Industry Corporation (Brochure).

"Space in China", China Great Wall Industry Corporation (Brochure).

"China Satellite and TT&C General", China Satellite and TT&C General (Brochure).

"China Xian Satellite Control Center", China Satellite and TT&C General (Brochure).

"China Xichang Satellite Launch Center", China Satellite and TT&C General (Brochure).

Acronyms

AKM - apogee kick motor
CALT - China Academy of Launch Vehicle Technology
CGWIC - China Great Wall Industry Corporation
CLTC - China Satellite Launch & TT&C General
CZ - Chang Zheng
FB - Feng Bao (Storm Booster)
GEO - geosynchronous orbit
GTO - geosynchronous transfer orbit
H18 - cryogenic stage (18 metric tons propellant)
ICBM - intercontinental ballistic missile
IRBM - intermediate range ballistic missile
JSLC - Jiuquan Satellite Launch Center
L140 - liquid stage (140 metric tons propellant)
LEO - low Earth orbit
LH2 - liquid hydrogen
LM - Long March
LOX - liquid oxygen
N2O4 - nitrogen tetroxide
PKM - perigee kick motor
SHBOA - Shanghai Bureau of Astronautics
SSO - sun synchronous orbit
TSLC - Taiyuan Satellite Launch Center
TT&C - telemetry, tracking and control
UDMH - unsymmetrical dimethylhydrazine
VEB - vehicle equipment bay
XSLC - Xichang Satellite Launch Center

Other Notes - Long March Growth Possibilities

China is studying an even larger class vehicle than the CZ-3B named the Long March-X series of rockets. The Long March-X vehicle would be fueled by LOX/kerosene propellant and would be capable of lofting 66,000 lb (30,000 kg) into LEO. Development cost of the 180 ft (55 m) tall Long March-X is estimated at $400 million.

Europe

Table of Contents

Page

Government Point of Contact:
European Space Agency (ESA)
8-12, rue Mario Nikis
F-75738 Paris Cedex 15, FRANCE
Phone: 33 (1) 42.73.76.54

Centre National d'Etudes Spatiales (CNES)
French Space Agency
2, place Maurice Quentin
F-75039 Paris Cedex 01, FRANCE
Phone: 33 (1) 45.08.75.00

Industry Point of Contact:
Arianespace
Boulevard de l'Europe, BP177
F-91006 Evry Cedex, FRANCE
Phone: 33 (1) 60.87.60.00
or
1747 Pennsylvania Ave NW Suite 875
Washington, DC 22006, USA
Phone: (202) 728-9075

The successful launch on 15 June 1988, of the first Ariane-4 marked the conclusion of its six year development program aimed at providing Europe with a launcher that will meet foreseeable market demand in the 1990s.

Ariane History

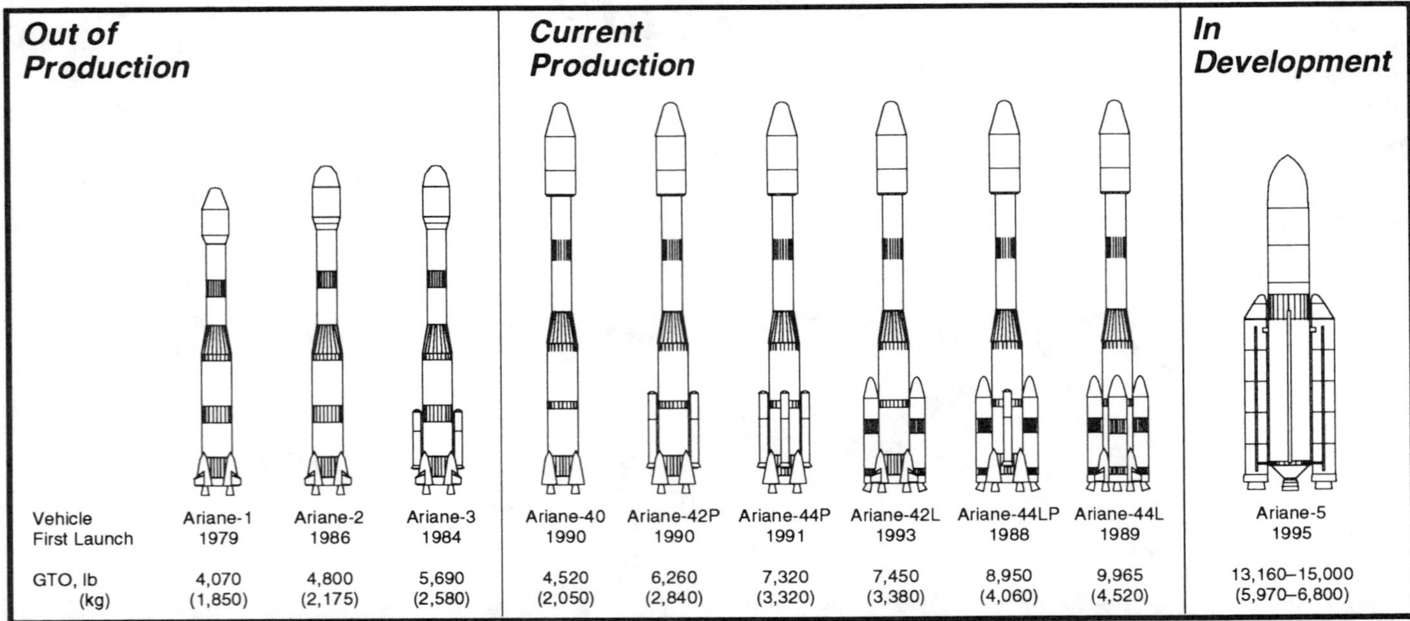

Vehicle	Ariane-1	Ariane-2	Ariane-3	Ariane-40	Ariane-42P	Ariane-44P	Ariane-42L	Ariane-44LP	Ariane-44L	Ariane-5
First Launch	1979	1986	1984	1990	1990	1991	1993	1988	1989	1995
GTO, lb	4,070	4,800	5,690	4,520	6,260	7,320	7,450	8,950	9,965	13,160–15,000
(kg)	(1,850)	(2,175)	(2,580)	(2,050)	(2,840)	(3,320)	(3,380)	(4,060)	(4,520)	(5,970–6,800)

Vehicle Description

Ariane-1 — Stage 1 and 2 storable propellant, stage 3 cryogenic propellant, 10.5 ft (3.2 m) diameter fairing.

Ariane-2 — Same as Ariane-1 except increased thrust for stage 1 and 2 engines, stretched stage 3 for 25% more propellant, 4 sec specific impulse increase in stage 3 engine, increased volume in fairing via modified forward conic section (bicone), and Sylda structure available for 2 payloads under fairing.

Ariane-3 — Same as Ariane-2 except two solid strap-on boosters are added.

Ariane-4 — Same as Ariane-3 except stretched and strengthened stage 1 for 61% more propellant, new water tank and new propulsion bay layout; strengthened stage 2 and 3; new vehicle equipment bay; new onboard computer and new laser gyro backup; longer and wider (13 ft or 4 m) fairing; Spelda structure for two payloads; and a mix of boosters either solid strap-ons (30% more propellant than Ariane-3 solids) or liquid strap-ons.

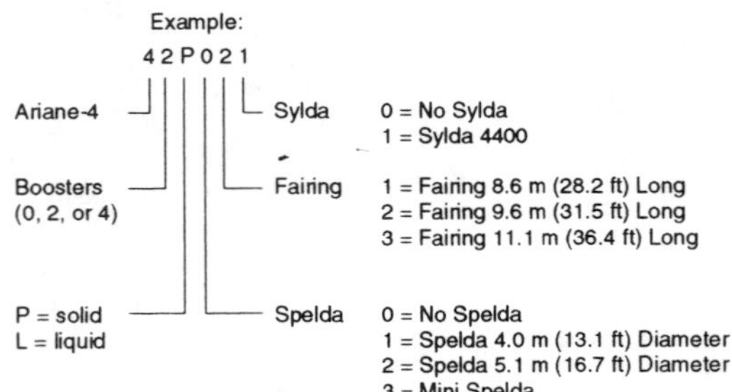

Ariane-5 — Lower composite, which is mission independent, consisting of two large solid strap-ons, and cryogenic propellant core, and upper composite comprised of a final stage, Vehicle Equipment Bay, 17.7 ft (5.4 m) diameter fairing and, if required, a Speltra structure for two or three payloads.

Historical Summary

Ariane is Europe's second attempt to develop its own launcher, following the unsuccessful Europa project, which was cancelled in April 1973 following a series of launch failures. Even before the official cancellation of Europa, the French space agency, CNES, proposed its replacement launcher, code-named L3S (the French acronym for third-generation substitution launcher), as well as a detailed financial plan. France guaranteed it would finance more than half of the program itself and would take responsibility for cost overruns up to 15% of program cost. When the Europa program was finally abandoned, France began to rally its European neighbors behind the Ariane launch system (in Greek mythology, Ariadne, whose translation to French is Ariane, is the daughter of Minos who gave Theseus the thread with which to find his way out of the Minotaur's labyrinth). The substitution launcher was given a go-ahead for development in July 1973.

There are three key organizations involved in the development, operation and management of Ariane—European Space Agency (ESA), Centre National d'Etudes Spatiales (CNES), and Arianespace.

ESA, a multinational space organization, was set up in December 1973. It was formed out of, and took over the rights and obligations of, two earlier European space organizations: the European Space Research Organization (ESRO) and the European Launcher Development Organization (ELDO). The purpose of the agency is to provide for and to promote, for exclusively peaceful purposes, cooperation among European states in space research and technology and their space applications. Its member states are: Austria, Belgium, Denmark, France, Germany, Ireland, Italy, Netherlands, Norway, Spain, Sweden, Switzerland, and the United Kingdom; Finland is an associate member, and Canada is a cooperating state. ESA's role with Ariane is direction of the development program and financing of the facilities construction (launch pads and payload preparation facilities).

CNES was created in 1962 to promote the development of French space activities. It developed the Diamant series of launch vehicles. The first launch took place in 1965 from Hammaguir and the last in 1975 from Kourou. Its role on Ariane is management of the development program, construction of the space center facilities, coordination of operations, and operation of the launch base and payload preparation facility.

Ariane History

(continued)

Historical Summary

(continued)

Arianespace was set up in March 1980. Under terms of an intergovernmental agreement, ESA member states transferred production, marketing, and launch responsibilities for the operational Ariane and its uprated versions to Arianespace. The lead responsibility for Ariane's overall development is routed through ESA to CNES. Once declared operational, the vehicles are turned over to Arianespace for commercial exploitation. Participating in the creation of Arianespace were 36 of Europe's key aerospace and avionics manufacturers, 13 major European banks, and CNES. Shareholding distribution among the three top nations is France 59%, Germany 20%, and Belgium 4%.

A three way coordination for launching the Ariane is conducted at the Guiana Space Center (CSG) in Kourou, French Guiana. CSG was set up by the French Government in April 1965 and built by CNES. It became operational in April 1968 with the launch of a Veronique sounding rocket. Following the Diamant and Europa programs, CSG was selected for Ariane launch operations based on proximity to the equator, wide opening on the ocean allowing all inclination missions, no hurricanes or earthquakes, and low population density. Currently, CSG is operated for ESA by CNES. CNES is also responsible for expanding the launch facilities to meet growing mission demands. Operation and maintenance of the launch facilities are the duty of Arianespace, and CNES is reimbursed by Arianespace for the personnel, facilities, and materials used to support launch operations.

Ariane-1 was defined in 1973 and was intended to achieve a geosynchronous transfer orbit (GTO) lift capacity for satellite masses of up to 4,070 lb (1850 kg). The development and qualification of Ariane covered a period of 8.5 years, from mid-1973 through 1981. The first three years were devoted to qualifying stage components (engines, structures, and equipment). Later, systems trials were conducted with dynamic and electric mockups of the launch vehicle and with the propulsion systems at the stage level. Construction of the SIL (Site d'Integration des Lanceurs—Launch Vehicle Assembly Building) at Aerospatiale's Les Mureaux Center was completed in 1976. The second phase of development consisted of qualification at stage level, and the construction of the launch facilities in French Guiana. The first launch took place in December 1979 at launch site ELA-1 (Ensemble de lancement Ariane No. 1). The development cost for Ariane-1 was about 2,000 MAU (EC86).

As soon as development was complete on Ariane-1, it became apparent that if Ariane was to remain competitive it would have to be able to launch two PAM-D class payloads simultaneously. So in July 1980, the Ariane-2/3 program was started. The launcher was derived from Ariane-1, incorporating a series of modifications which increased lift capacity to GTO by 50%. To allow double payload launches, a new structure was developed—the Sylda (Systeme de lancement double Ariane). The Ariane-2 version was identical to Ariane-3, but without the two solid strap-on boosters. The development cost for Ariane-2/3 was about 144 MAU (EC86).

The decision to develop Ariane-4 was taken on by ESA in January 1982, when the need became evident for a more powerful and more flexible launch vehicle (compared to Ariane-2/3) to match the trend in the payload market. The major program objectives were to: (a) achieve a significant increase in launch capability (payload mass and volume); (b) maintain the capability of multiple launches; (c) create a range of mission-adaptable configurations; and (d) improve the flexibility of launch operations.

The increase in performance, by up to 90%, over Ariane-3 has been achieved by increasing the first-stage propellant mass and by attaching newly developed, powerful solid or liquid boosters. The diameter of the payload fairing has also been increased to accommodate bigger payloads. A new supporting structure, the Spelda (Structure porteuse externe pour lancements doubles Ariane) allows multiple payload launches. Fairings of various heights and different sizes of Spelda allow adaptation of the launcher to the volume of the payloads to be launched. Six different versions of Ariane-4 are offered by varying the number of liquid and/or solid strap-ons. A new launch complex, ELA-2, has been built to reduce the interval between launches to one month. This allows both a higher launch rate and greater flexibility in launch scheduling.

The majority of the Ariane-4 development work was completed by 1986; manufacture of the flight hardware for the demonstration flight started in 1985. All launcher hardware except the third stage was ready by November 1987. The demonstration flight had to be postponed because of development difficulties on the third-stage engine. The launch campaign finally started in December 1987, before being interrupted and then resumed, leading up to the flawless first launch on 15 June 1988. The development cost for Ariane-4 was about 476 MAU (EC86). Highest contributors to Ariane-4 development were France 62%, Germany 18%, and Italy 5%.

Ariane-5, the successor to Ariane-4, is a completely new design. The Ariane-5 development was decided on in 1988 after a preparatory program initiated in 1984. The objectives of the Ariane-5 development program are: (a) deliver into GTO one or more satellites with a total mass of 15,000 lb (6,800 kg) for single-launch configuration, or 13,000 lb (5,900 kg) in the dual-launch configuration; (b) inject the automatic transfer vehicle with a payload at least 26,000 lb (12,000 kg) or the crew transfer vehicle for rendezvous with the international space station; (c) provide a volume for payloads represented by a cylinder of 15 ft (4.57 m) diameter; (d) meet reliability target of 0.98 for the launcher's total mission; (e) comply with safety target of 0.999 with respect to manned flights, and (f) satisfy a cost goal of using Ariane-5, for a dual launch into GTO, that will be at least 10% lower than that of using an Ariane-44L, assuming eight launches a year including four into GTO (or 45% reduction in the cost-per-pound compared with Ariane-44L).

Building of a simplified launch zone, ELA-3, will enable 10 launches per year. Ariane-5 development plan includes two qualification flights scheduled in 1995 and 1996. The development cost for Ariane-5 will be about 4,114 MAU (EC86) that includes 3,070 MAU for launcher development (216 for system work, 819 for P230 motor, 964 for HM60 engine, 537 for upper section, 234 for man-rating), 212 MAU for management, 566 MAU for ground infrastructure, and 266 MAU for in-flight qualification. Highest contributors to Ariane-5 development are France 45%, Germany 22%, and Italy 15%.

Ariane History

(continued)

Launch Record

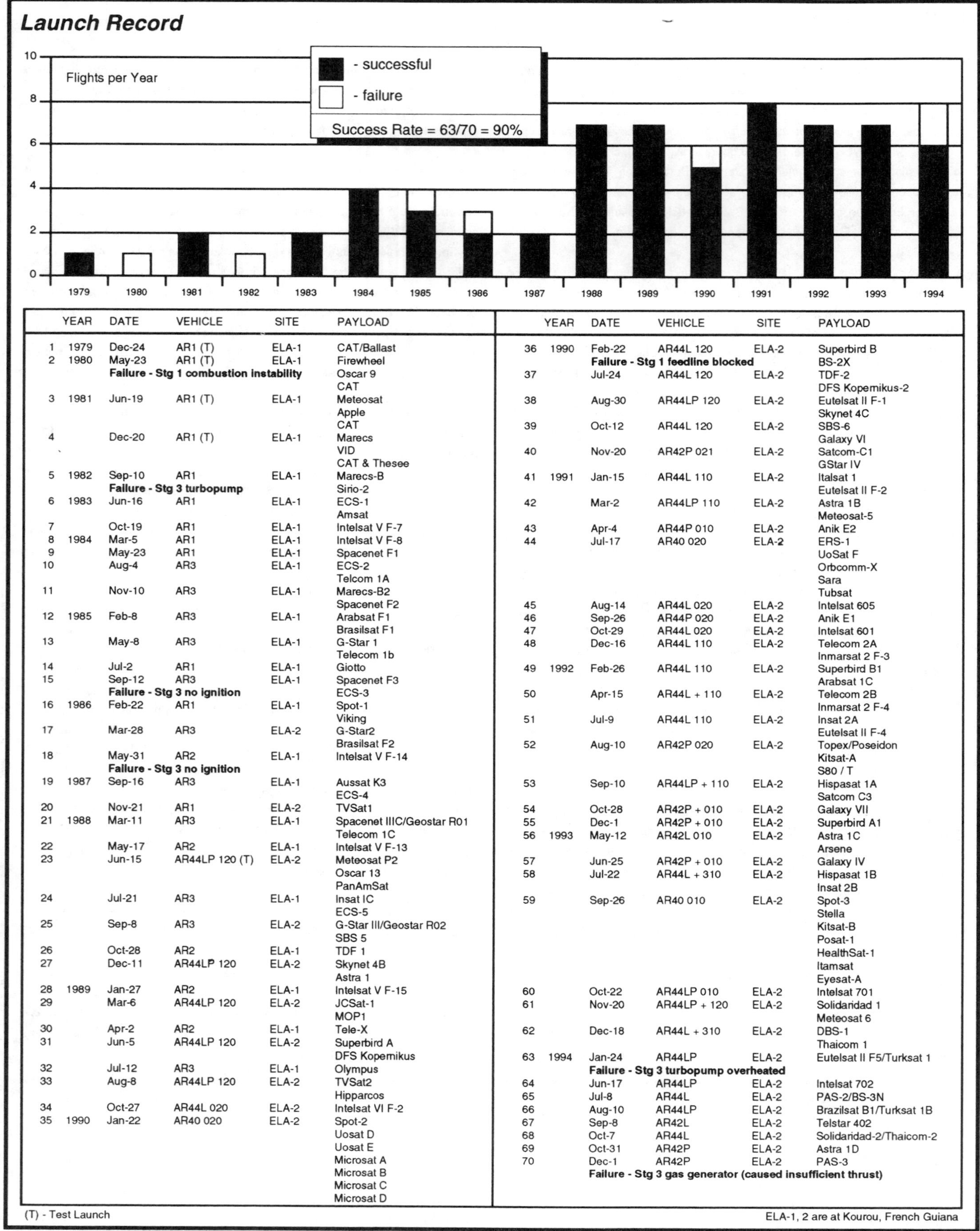

	YEAR	DATE	VEHICLE	SITE	PAYLOAD
1	1979	Dec-24	AR1 (T)	ELA-1	CAT/Ballast
2	1980	May-23	AR1 (T)	ELA-1	Firewheel Oscar 9 CAT
		Failure - Stg 1 combustion instability			
3	1981	Jun-19	AR1 (T)	ELA-1	Meteosat Apple CAT
4		Dec-20	AR1 (T)	ELA-1	Marecs VID CAT & Thesee
5	1982	Sep-10	AR1	ELA-1	Marecs-B Sirio-2
		Failure - Stg 3 turbopump			
6	1983	Jun-16	AR1	ELA-1	ECS-1 Amsat
7		Oct-19	AR1	ELA-1	Intelsat V F-7
8	1984	Mar-5	AR1	ELA-1	Intelsat V F-8
9		May-23	AR1	ELA-1	Spacenet F1
10		Aug-4	AR3	ELA-1	ECS-2 Telcom 1A
11		Nov-10	AR3	ELA-1	Marecs-B2 Spacenet F2
12	1985	Feb-8	AR3	ELA-1	Arabsat F1 Brasilsat F1
13		May-8	AR3	ELA-1	G-Star 1 Telecom 1b
14		Jul-2	AR1	ELA-1	Giotto
15		Sep-12	AR3	ELA-1	Spacenet F3 ECS-3
		Failure - Stg 3 no ignition			
16	1986	Feb-22	AR1	ELA-1	Spot-1 Viking
17		Mar-28	AR3	ELA-2	G-Star2 Brasilsat F2
18		May-31	AR2	ELA-1	Intelsat V F-14
		Failure - Stg 3 no ignition			
19	1987	Sep-16	AR3	ELA-1	Aussat K3 ECS-4
20		Nov-21	AR1	ELA-2	TVSat1
21	1988	Mar-11	AR3	ELA-1	Spacenet IIIC/Geostar R01 Telecom 1C
22		May-17	AR2	ELA-1	Intelsat V F-13
23		Jun-15	AR44LP 120 (T)	ELA-2	Meteosat P2 Oscar 13 PanAmSat
24		Jul-21	AR3	ELA-1	Insat IC ECS-5
25		Sep-8	AR3	ELA-2	G-Star III/Geostar R02 SBS 5
26		Oct-28	AR2	ELA-1	TDF 1
27		Dec-11	AR44LP 120	ELA-2	Skynet 4B Astra 1
28	1989	Jan-27	AR2	ELA-1	Intelsat V F-15
29		Mar-6	AR44LP 120	ELA-2	JCSat-1 MOP1
30		Apr-2	AR2	ELA-1	Tele-X
31		Jun-5	AR44LP 120	ELA-2	Superbird A DFS Kopernikus
32		Jul-12	AR3	ELA-1	Olympus
33		Aug-8	AR44LP 120	ELA-2	TVSat2 Hipparcos
34		Oct-27	AR44L 020	ELA-2	Intelsat VI F-2
35	1990	Jan-22	AR40 020	ELA-2	Spot-2 Uosat D Uosat E Microsat A Microsat B Microsat C Microsat D
36	1990	Feb-22	AR44L 120	ELA-2	Superbird B BS-2X
		Failure - Stg 1 feedline blocked			
37		Jul-24	AR44L 120	ELA-2	TDF-2 DFS Kopernikus-2
38		Aug-30	AR44LP 120	ELA-2	Eutelsat II F-1 Skynet 4C
39		Oct-12	AR44L 120	ELA-2	SBS-6 Galaxy VI
40		Nov-20	AR42P 021	ELA-2	Satcom-C1 GStar IV
41	1991	Jan-15	AR44L 110	ELA-2	Italsat 1 Eutelsat II F-2
42		Mar-2	AR44LP 110	ELA-2	Astra 1B Meteosat-5
43		Apr-4	AR44P 010	ELA-2	Anik E2
44		Jul-17	AR40 020	ELA-2	ERS-1 UoSat F Orbcomm-X Sara Tubsat
45		Aug-14	AR44L 020	ELA-2	Intelsat 605
46		Sep-26	AR44P 020	ELA-2	Anik E1
47		Oct-29	AR44L 020	ELA-2	Intelsat 601
48		Dec-16	AR44L 110	ELA-2	Telecom 2A Inmarsat 2 F-3
49	1992	Feb-26	AR44L 110	ELA-2	Superbird B1 Arabsat 1C
50		Apr-15	AR44L + 110	ELA-2	Telecom 2B Inmarsat 2 F-4
51		Jul-9	AR44L 110	ELA-2	Insat 2A Eutelsat II F-4
52		Aug-10	AR42P 020	ELA-2	Topex/Poseidon Kitsat-A S80 / T
53		Sep-10	AR44LP + 110	ELA-2	Hispasat 1A Satcom C3
54		Oct-28	AR42P + 010	ELA-2	Galaxy VII
55		Dec-1	AR42P + 010	ELA-2	Superbird A1
56	1993	May-12	AR42L 010	ELA-2	Astra 1C Arsene
57		Jun-25	AR42P + 010	ELA-2	Galaxy IV
58		Jul-22	AR44L + 310	ELA-2	Hispasat 1B Insat 2B
59		Sep-26	AR40 010	ELA-2	Spot-3 Stella Kitsat-B Posat-1 HealthSat-1 Itamsat Eyesat-A
60		Oct-22	AR44LP 010	ELA-2	Intelsat 701
61		Nov-20	AR44LP + 120	ELA-2	Solidaridad 1 Meteosat 6
62		Dec-18	AR44L + 310	ELA-2	DBS-1 Thaicom 1
63	1994	Jan-24	AR44LP	ELA-2	Eutelsat II F5/Turksat 1
		Failure - Stg 3 turbopump overheated			
64		Jun-17	AR44LP	ELA-2	Intelsat 702
65		Jul-8	AR44L	ELA-2	PAS-2/BS-3N
66		Aug-10	AR44LP	ELA-2	Brazilsat B1/Turksat 1B
67		Sep-8	AR42L	ELA-2	Telstar 402
68		Oct-7	AR44L	ELA-2	Solidaridad-2/Thaicom-2
69		Oct-31	AR42P	ELA-2	Astra 1D
70		Dec-1	AR42P	ELA-2	PAS-3
		Failure - Stg 3 gas generator (caused insufficient thrust)			

(T) - Test Launch

ELA-1, 2 are at Kourou, French Guiana

Ariane General Description

	Ariane-4	Ariane-5
		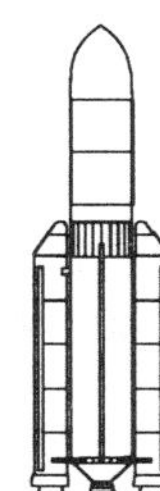
Summary	Development work for Ariane-4 began in 1982, and the first launch occurred in 1988. The wide range of masses that can be orbited, plus the flexibility of double launch, permits the Ariane-4 to cover the needs of the commercial marketplace. The system has been optimized for launches into GTO with Kourou providing about a 15% mass advantage over Kennedy Space Center.	Development work for Ariane-5 began in 1988, and the first launch is projected for 1995. The two prime requirements for Ariane-5 are to improve reliability and lower cost in order to remain competitive in the commercial marketplace. Ariane-5 will be capable of launching one, two or three payloads, as well as elements of the manned space transportation system.
Status	Operational	In Development
Key Organizations	User - ESA, commercial Launch Service Agency - Arianespace Prime Contractor - CNES Major Contractors Aerospatiale (Ariane System, Overall Studies, Stage 1 & 3, Sylda) MBB-Erno (Stage 2, Liquid Strap-on) SEP (Stage 1, 2, 3 and Liquid Strap-on Propulsion) SNIA-BPD (Solid Strap-on) Air Liquide (Stage 3 Tanks) Matra (Vehicle Equipment Bay) British Aerospace (Spelda) Contraves (Payload Fairing) Ferranti (Gimbal Gryo) SFENA (Ring Laser Gyro)	User - ESA, commercial Launch Service Agency - Arianespace Prime Contractor - CNES Major Contractors Aerospatiale (Ariane System, Overall Studies, Stage H155 & P230) DASA-Erno (Stage L9) SEP (H155 Propulsion) Europropulsion (Stage P230 Propulsion) Regulus (Stage P230 Propellant) Matra (Vehicle Equipment Bay) Dornier (Speltra) Contraves (Payload Fairing)
Vehicle		
System Height	Up to 192 ft (58.4 m)	Up to 177 ft (54.05 m)
Payload Fairing Size	Diameter 13.1 ft (4.0 m) for fairing and Spelda. Height for Short fairing - 28.2 ft (8.6 m), Long fairing - 31.5 ft (9.6 m), Extra-Long fairing 36.5 ft (11.12 m), Short Spelda - 9.2 ft (2.8 m), Long Spelda 12.5 ft (3.8 m).	Diameter 17.7 ft (5.4 m) for fairing and Speltra. Height for Short fairing - 37.9 ft (11.55 m), Long fairing - 55.8 ft (17 m), Speltra 18.5 ft (5.65 m).
Gross Mass	AR40 - 529,000 lb (240,000 kg) AR42P- 747,000 lb (339,000 kg) AR44P - 789,000 lb (358,000 kg) AR42L - 882,000 lb (400,000 kg) AR44LP - 925,000 lb (420,000 kg) AR44L - 1,040,000 lb (470,000 kg)	Ariane-5 / GTO Double Launch - 1,570,000 lb (710,000 kg)
Planned Enhancements	Plan to stretch the H10 Stage 3 by 13 in. (32 cm) for 700 lb (300 kg) more propellant. This plus operational experience and some dry mass reductions is projected to add 110 lb (50 kg) to GTO for AR40, 42P, 42L and 440 lb (200 kg) for AR44LP, 44P, 44L. A shortened Spelda, named SDS or Spelda Dedicated Satellite, (7.9 ft or 2.4 m height) is being developed to accommodate small payloads from 880 to 1760 lb (400 to 800 kg). For piggyback payloads less than 110 lb (50 kg) there is a circular built platform, named ASAP or Ariane Structure for Auxiliary Payloads.	Options being studied are detailed in the Ariane Notes.

Ariane General Description

(continued)

	Ariane-4	Ariane-5
Operations		
Primary Missions	GTO, Polar	LEO, GTO, Polar
Compatible Upper Stages	H10+	L9
First Launch	1988	Projected 1995
Success / Flight Total	AR40 - 3 / 3 AR42P - 6 / 7 AR44P - 2 / 2 AR42L - 2 / 2 AR44LP - 12 / 13 AR44L - 14 / 15	0 / 0
Launch Site	ELA-2 - (5.2°N, 52.8°W)	ELA-3 - (5.2°N, 52.8°W)
Launch Azimuth	0°–108° (max. inclination is 5.2°-100.5°)	0°–108° (max. inclination is 5.2°-100.5°)
Nominal Flight Rate	Up to 10 / year	Up to 10 / year (possibility of 15 / year)
Planned Enhancements	None	None
Performance	Data below must include mass of Spelda 860 or 970 lb (390 or 440 kg) or Sylda 420 lb (190 kg) and adapter 60 - 110 lb (27 - 51 kg)	
100 nm (185 km) circ, 5.2°	LEO data below is currently being refined: AR40 - 10,800 lb (4,900 kg) AR42P- 13,400 lb (6,100 kg) AR44P - 15,200 lb (6,900 kg) AR42L - 16,300 lb (7,400 kg) AR44LP - 18,300 lb (8,300 kg) AR44L - 21,100 lb (9,600 kg)	For automatic missions: 39,600 lb (18,000 kg) at 300 nm (550 km) circ, 28.5°
100 nm (185 km) circ, 90°	AR40 - 8,580 lb (3,900 kg) AR42P- 10,600 lb (4,800 kg) AR44P - 12,100 lb (5,500 kg) AR42L - 13,000 lb (5,900 kg) AR44LP - 14,500 lb (6,600 kg) AR44L - 16,900 lb (7,700 kg)	26,400 lb (12,000 kg) at 430 nm (800 km) circ, 98.6°
Geotransfer Orbit, 7°	AR40 - 4,520 lb (2,050 kg) AR42P- 6,260 lb (2,840 kg) AR44P - 7,320 lb (3,320 kg) AR42L - 7,450 lb (3,380 kg) AR44LP - 8,950 lb (4,060 kg) AR44L - 9,965 lb (4,520 kg)	Single payload - 15,000 lb (6,800 kg) Dual payload - 13,160 lb (5,970 kg)
Geosynchronous Orbit	Requires spacecraft kick motor	Require spacecraft kick motor
Financial Status		
Estimated Launch Price	Dedicated launch: AR40 - $45–60M (U.S. DoT estimate) AR42P- $60–75M (U.S. DoT estimate) AR44P - $80–95M* (U.S. DoT estimate) AR42L - $75–85M (U.S. DoT estimate) AR44LP - $80–95M* (U.S. DoT estimate) AR44L - $90–110M (U.S. DoT estimate) **Launch price for launch on AR44P and AR44LP have historically been the same*	Dedicated launch: $120M for dual launch (Arianespace)
Manifest	Site / 1995 / 1996 ELA-2 / 13 / 6	Site / 1995 / 1996 ELA-3 / 1 / 2
Remarks	—	There will be two test launches in 1995 and 1996.

Ariane Vehicle

Overall

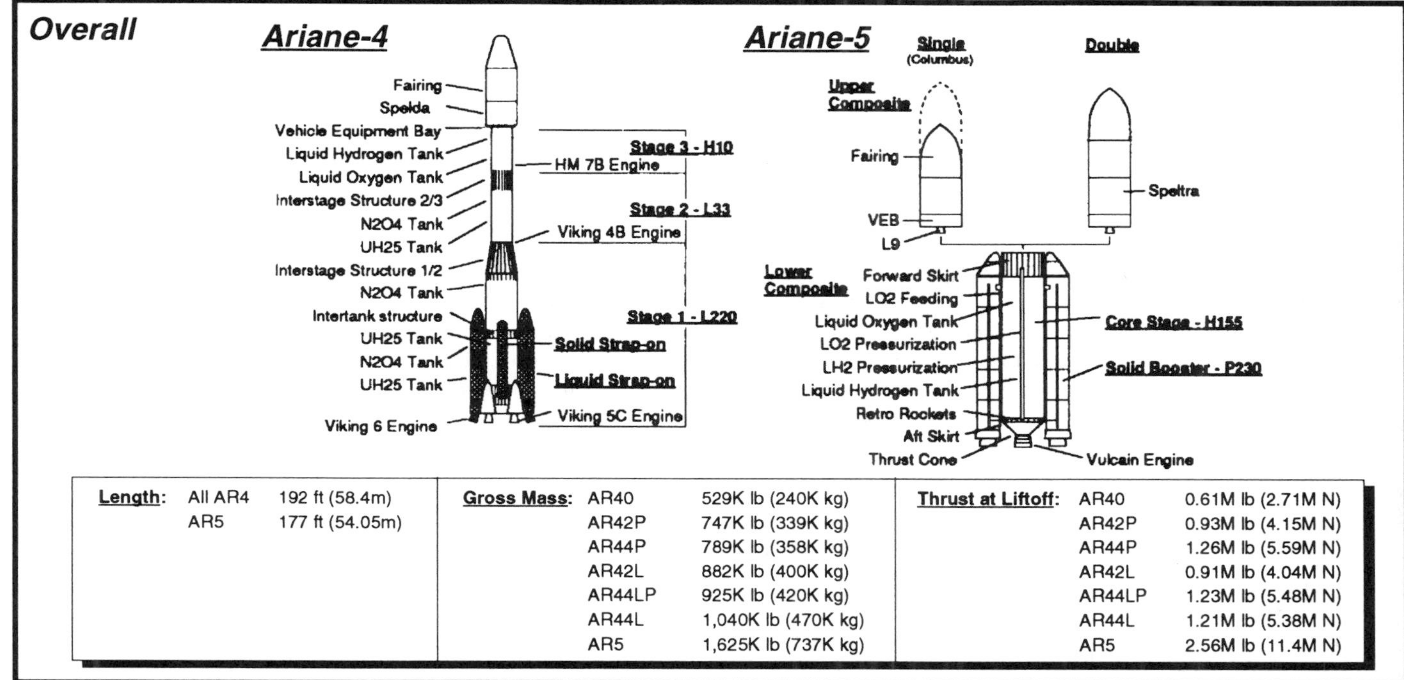

Length:			Gross Mass:		Thrust at Liftoff:	
All AR4	192 ft (58.4m)		AR40	529K lb (240K kg)	AR40	0.61M lb (2.71M N)
AR5	177 ft (54.05m)		AR42P	747K lb (339K kg)	AR42P	0.93M lb (4.15M N)
			AR44P	789K lb (358K kg)	AR44P	1.26M lb (5.59M N)
			AR42L	882K lb (400K kg)	AR42L	0.91M lb (4.04M N)
			AR44LP	925K lb (420K kg)	AR44LP	1.23M lb (5.48M N)
			AR44L	1,040K lb (470K kg)	AR44L	1.21M lb (5.38M N)
			AR5	1,625K lb (737K kg)	AR5	2.56M lb (11.4M N)

Ariane-4 Stages

	Solid Strap-On (P9.5 or PAP)	Liquid Strap-On (L40 or PAL)	Stage 1 (L220)	Stage 2 (L33)	Stage 3 (H10)
Dimension:					
Length	37.7 ft (11.5 m)	61.0 ft (18.6 m)	83.3 ft (25.4 m)	38.1 ft (11.6 m)	32.5 ft (9.9 m)
Diameter	3.51 ft (1.07 m)	7.12 ft (2.17 m)	12.5 ft (3.8 m)	8.5 ft (2.6 m)	8.5 ft (2.6 m)
Mass: (each)					
Propellant Mass	20.9K lb (9.5K kg)	86K lb (39K kg)	514K lb (233K kg)	77.6K lb (35.2K kg)	23.8K lb (10.8K kg)
Gross Mass	27.8K lb (12.6K kg)	95.9K lb (43.5K kg)	553K lb (251K kg)	84.9K lb (38.5K kg)	26.7K lb (12.1K kg)
Structure:					
Type	Monocoque	Semimonocoque	Semimonocoque	Semimonocoque	Semimonocoque
Material	Steel	Steel	Steel	Aluminum	Aluminum
Propulsion:					
Propellant	CTPB	N2O4 / UH25	N2O4 / UH25	N2O4 / UH25	LOX / LH2
Average Thrust (each)	162K lb (720K N) SL	150K lb (667K N) SL 167K lb (737K N) vac	152K lb (676.9K N) SL 171K lb (758.5K N) vac	177K lb (785K N) vac	14.1K lb (62.7K N) vac
Engine Designation	—	Viking 6	Viking 5C	Viking 4B	HM7B
Number of Engines	0-4 PAPs (1 segment ea.)	0-4 PALs (1 engine ea.)	4	1	1
Isp	241 sec SL	248 sec SL 278 sec vac	248.5 sec SL 278.4 sec vac	293.5 sec vac	444.2 sec vac
Feed System	—	Gas Generator	Gas Generator	Gas Generator	Gas Generator
Chamber Pressure	??	848 psia (58.5 bar)	848 psia (58.5 bar)	848 psia (58.5 bar)	508 psia (35 bar)
Mixture Ratio (O/F)	—	1.7	1.7	1.7	4.77
Throttling Capability	—	100% only	100% only	100% only	100% only
Expansion Ratio	??	??	10.48:1	30.8:1	62.5:1
Restart Capability	No	No	No	No	No
Tank Pressurization	—	Pressurized helium	Gas generator exhaust	Pressurized Helium	Fuel-GH2, Ox-cold GHe
Control-Pitch, Yaw -Roll	Fixed 12°, controlled by stage 1	Fixed 10°, controlled by stage 1	Hydraulic gimbaling (4 nozzles) (±6°)	Hydaulic gimbaling(±3°) 2 hot gas thrusters	Hydaulic gimbaling(±3°) Gaseous hydrogen
Events:					
Nominal Burn Time	34 sec	140 sec	205 sec	126 sec	725 sec
Stage Shutdown	Burn to depletion	Burn to depletion	Half thrust decay	Predetermined velocity	Predetermined velocity
Stage Separation	Springs	Retro-rockets	8 Retro-rockets	Retro-Rockets	Attitude thrusters

Ariane Vehicle

(continued)

Ariane-4 Stages

(continued)

Remarks:

The stage 1 (designated L220) consists of: (a) two identical cylindrical steel propellant tanks connected by an intertank skirt of the same diameter; (b) a cylindrical steel water tank located in the intertank skirt, containing maximum of 14,700 lb (6,700 kg) of water, with thermal protection on the lower hemispherical bulkhead; (c) a conical interstage skirt connecting the first and second stages; (d) a forward skirt on which the eight first stage retro-rockets are mounted and to which the interstage skirt is connected; and (e) a cylindrical thrust frame, the upper part of which is connected to the UH25 tanks and on whose lower part the four Viking-V engines are mounted. The water tank is completely new and has been specially developed and fully qualified under the Ariane-4 program. The other parts of the stage have been strengthened to cope with Ariane-4 loads. The capacity of the propellant tanks has been increased to extend the burn-time from 135 sec to about 205 sec. The stage 1 propulsion system consists of four Viking-V engines. Each engine is an independent assembly supplied with propellant and water via its own valves. The propellants used are UH25 (a mixture of 75% unsymmetrical dimethylhydrazine and 25% hydrazine hydrate) as fuel and N2O4 (nitrogen tetroxide) as the oxidizer. Propellant is off-loaded for the lower performing Ariane-4's (e.g., AR40). During the propulsion phase, the first stage consumes about 2,200 lb (1,000 kg) of propellant per second. Each Viking engine has a gas generator supplied with propellant and water for cooling the gases. The gases feed the turbine driving the propellant and water pumps and also serve to pressurize the tanks. The throats of the Viking engines, made of Sephen (carbon-based composite), have been strengthened for the longer burn time. The engine turbopump bearings have also been modified for the same reason.

The stage 2 (known as L33) has one Viking-IV engine. It carries 'storable' propellants in aluminum alloy tanks and 1,230 lb (560 kg) of water. The tanks form a cylindrical structure with hemispherical bulkheads divided into two chambers by an intermediate bulkhead, also hemispherical, and with its concave face upwards. Both tank compartments are pressurized by helium gas stored in spherical bottles at a pressure of 4,350 psi (300 bar), at ambient temperature. During the waiting period on the launch pad before liftoff, the stage 2 tanks are protected by a thermal shield, the temperature of which is controlled with cool air, to limit heat exchange between the fuels and the ambient environment. The shield is jettisoned at liftoff.

The stage 3 (called H10) has one HM-7B engine. The two tanks holding the cryogenic propellants, i.e., liquid oxygen and liquid hydrogen, are made from an aluminum alloy, and have a common intermediate bulkhead (double vacuum loaded skin). They are pressurized during flight by gaseous hydrogen (the hydrogen tank) and by cold helium (the oxygen tank). Externally, the tanks are coated with thermal insulation to avoid rapid heating of the cryogenic propellants.

Ariane-4 can use a combination of either liquid and/or solid strap-on boosters. The liquid strap-ons (called L40) are a completely new element for Ariane-4. They are essentially a Viking-VI engine with two identical separate steel tanks, an intertank skirt, a forward skirt and a nose cone. Water for the Viking engine is supplied from the first stage's central tank. After burnout, the strap-ons are released by pyrotechnic cutting devices and jettisoned by small rockets. The solid strap-ons (called P9.5) are derived from the Ariane-3 solids by increasing propellant mass 30%. Each strap-on is jettisoned after burn-out by a system of four strong springs.

The Ariane-4 core stages are separated by explosive cords fitted into the rear skirts of the second and third stages. The stages are moved away from each other by retro-rockets mounted on the lower stage. Acceleration rockets fitted to the upper stage allow a small acceleration to be maintained to ensure homogeneous propellant flow to the engine during ignition. Separation of the first and second stages is initiated by the onboard computer when the inertial guidance platform detects half-thrust decay in the first stage (due to depletion of one of the propellants). Separation of the second and third stages is initiated by the onboard computer when the increase in velocity due to the second stage thrust has reached a predetermined value.

Ariane Vehicle

(continued)

Ariane-5 Stages

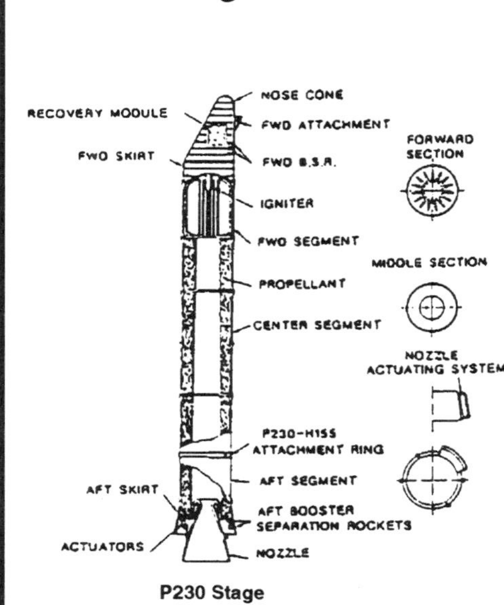

P230 Stage

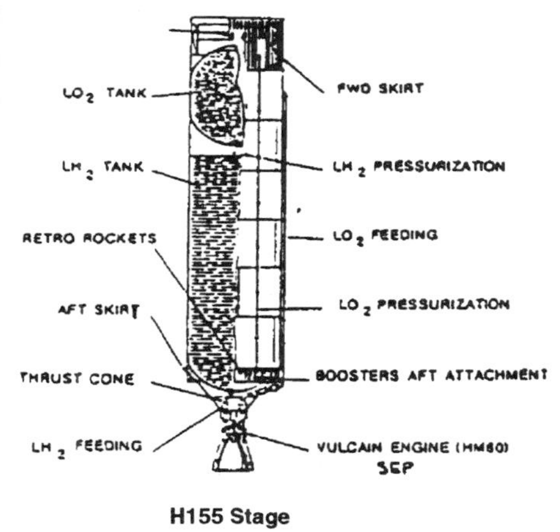

H155 Stage

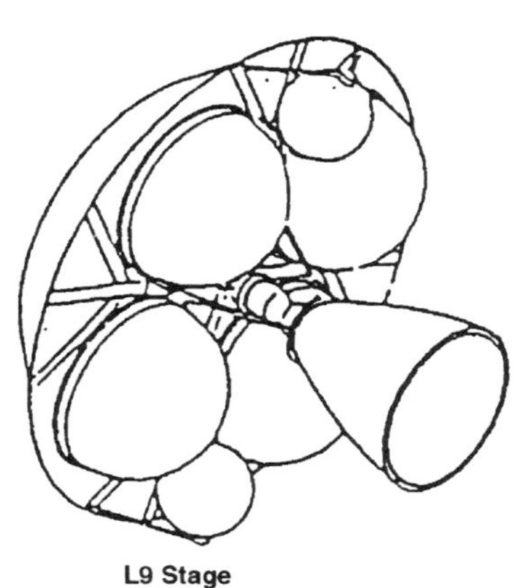

L9 Stage

	Solid Booster (P230)	**Core Stage (H155)**	**Upper Stage* (L9)**
Dimension:			
Length	102 ft (31.2 m)	95 ft (29 m)	14.8 ft (4.5 m)
Diameter	10 ft (3.04 m)	17.7 ft (5.4 m)	17.7 ft (5.4 m)
Mass: (each)			
Propellant Mass	506K lb (230K kg)	342K lb (155K kg)	21.4K lb (9.7K kg)
Gross Mass	583K lb (265K kg)	375K lb (170K kg)	24.0K lb (10.9K kg)
Structure:			
Type	Monocoque	Semi-Monocoque	Monocoque
Material	Steel	Aluminum	Aluminum and composite sandwich
Propulsion:			
Propellant	HTPB	LH2 / LOX	N2O4 / MMH
Average Thrust (each)	1.43M lb (6.36M N) SL	180K lb (800K N) SL 250K lb (1,120K N) vac	6.14K lb (27.5K N) vac
Engine Designation	—	Vulcain (HM60)	L9
Number of Engines	2 motors (3 segment ea.)	1	1
Isp	?? sec SL 273 sec vac	?? sec SL 430 sec vac	324 sec vac
Feed System	—	Gas Generator	Pressure fed
Chamber Pressure	930 psia (64 bar)	1,450 psia (100 bar)	145 psia (10 bar)
Mixture Ratio (O / F)	—	5.3	2.05
Throttling Capability	—	100% only	None
Expansion Ratio	9.7:1	45:1	83:1
Restart Capability	No	No	Multiple
Tank Pressurization	—	Helium for LO_2 tank GH_2 for LH_2 tank	Pressurized helium
Control-Pitch, Yaw, Roll	Hydraulic gimbaling(±6°)	Hydraulic gimbaling(±6°) VEB hot gas thrusters	Electrical gimbaling(±6°) VEB hot gas thrusters
Events:			
Nominal Burn Time	123 sec	590 sec	800 sec
Stage Shutdown	Burn to depletion	Command shutdown	Command shutdown
Stage Separation	Retro-rockets	Pyrotechnic actuator	VEB ACS

**VEB not included*

Remarks:

Lower Composite. The lower composite, which is mission independent, comprises a main cryogenic stage ignited on the ground, powered by an HM60 (Vulcain) engine, and two large solid-propellant boosters.

The core stage (known as H155) has one Vulcain engine. It carries cryogenic propellants in aluminum alloy tanks. The tanks are separated by a common dome bulkhead. During one period on the launch pad before liftoff, the stage is protected by a thermal shield.

The solid boosters (called P230) are strapped to the side of the H155. Each solid booster stage incorporates a recovery system permitting systematic recovery for inspection. The thrust profile is tailored to reduce thrust during maximum dynamic pressure so that the HM60 is not required to throttle.

Upper Composite. For automatic missions the upper composite comprised: (a) an upper stage used for low Earth or sun synchronous orbit missions and for putting a payload into GTO; (b) a vehicle equipment bay (VEB); and (c) an upper section that consists of a short or a long fairing, and for dual launches, the Speltra (Structure porteuse externe lancements triples Ariane) structure and the short fairing. The upper stage (designated L9), a pressure-fed propulsion stage, delivers between 330 and 4,900 ft/s (100 and 1,500 m/s) velocity increment depending on the mission. It is located atop the VEB and carries the VEB with it when released from the core stage. The stage consists of four propellant tanks (55.5 in. (1,410 mm) diameter), helium bottles, actuators, and a regeneratively cooled gimballed engine. The stage has a loaded stand-by duration of 60 days before launch.

The lower composite flight is virtually identical for all missions. It begins with ignition of the HM60 Vulcain engine on the ground; once the proper functioning of this has been checked, the command is given to ignite the P230 solid boosters and this leads to liftoff. At P230 burnout the boosters are jettisoned and the H155 core stage continues powered flight on its own. For a GTO mission, the L9 is ignited immediately after H155 separation. For LEO missions, at the end of the first phase, the L9 together with the VEB and payloads is placed in its final orbit.

Ariane Vehicle

(continued)

Payload Fairing

Ariane-4

in inches (mm)

Type 01

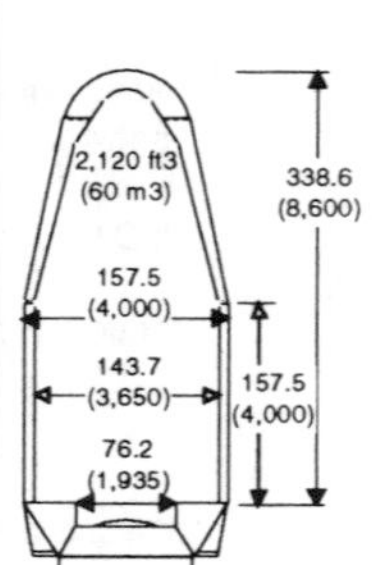

No Spelda

Type 02

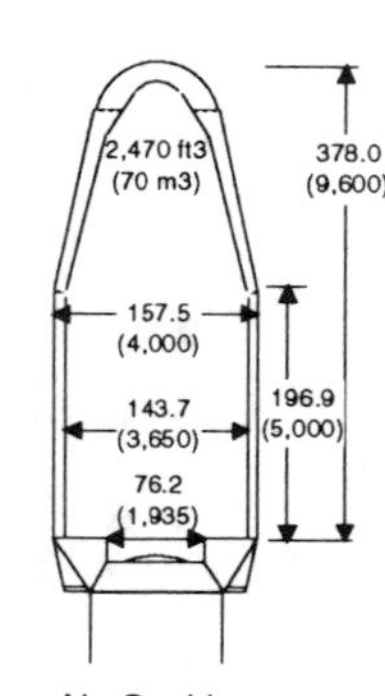

No Spelda

Type 03

Fairing available on special request only

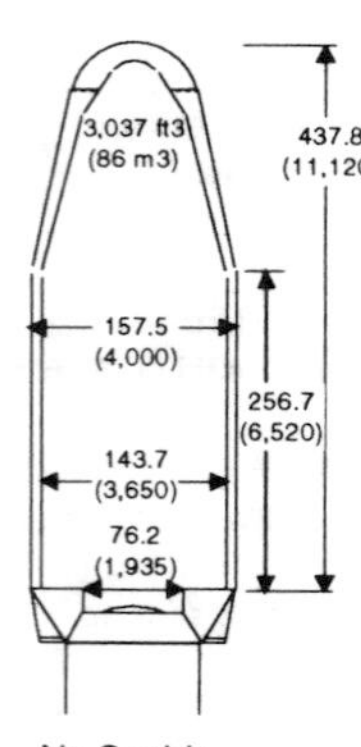

No Spelda

Type 11

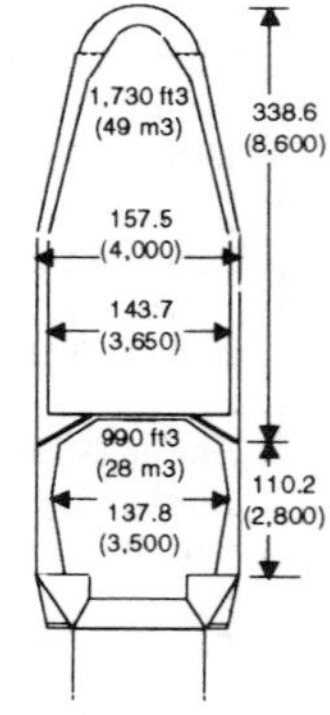

Short Spelda

Type 12

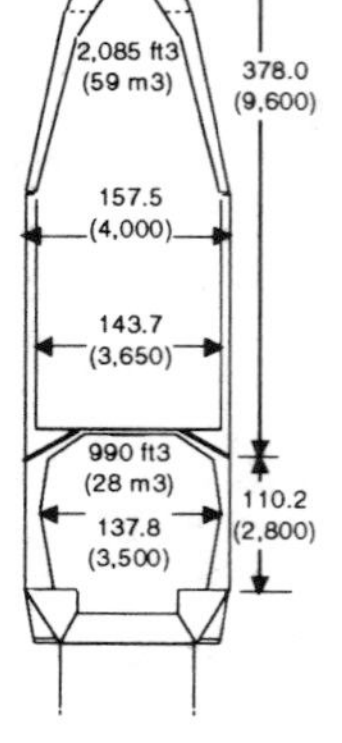

Short Spelda

Type 13

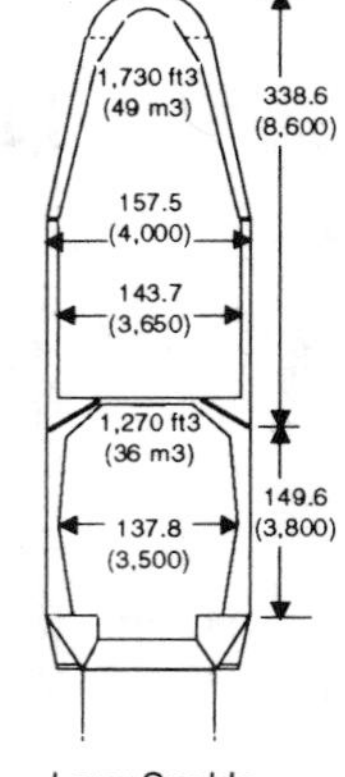

Long Spelda

Length	Fairing - 28.2 ft (8.6 m), 31.5 ft (9.6 m), 36.5 ft (11.12 m)
	Spelda - 9.2 ft (2.8 m), 12.5 ft (3.8 m)
Diameter	13.1 ft (4.0 m)
Mass	Fairing - 1,675 lb (760 kg), 1,800 lb (815 kg), 1,960 lb (890 kg)
	Spelda - 917 lb (417 kg), 992 lb (450 kg)
Sections	2
Structure	Composite sandwich
Material	Aluminum honeycomb / graphite epoxy

Remarks:

The fairing is made of carbon-fiber reinforced plastics in order to keep its mass low. At separation, which is normally initiated at an altitude of about 59 nm (110 km), the clamp band that holds the fairing to the vehicle is released. A pyrotechnic cord then cuts the fairing into two vertical halves and pushes them apart. The Ariane-4 Spelda supports the fairing, encloses the bottom spacecraft, and supports the upper spacecraft. Separation is achieved by cutting the structure along a horizontal plane and springs jettison the top Spelda. NOTE: An "egg-shaped" Ariane-3 Sylda support structure (9.2 ft (2.8 m) diameter, 14.4 ft (4.4 m) height, 440 lb (200 kg) mass) can be placed inside a fairing.

Ariane-5

in inches (mm)

Dual Launches:
- Short Fairing
- Speltra

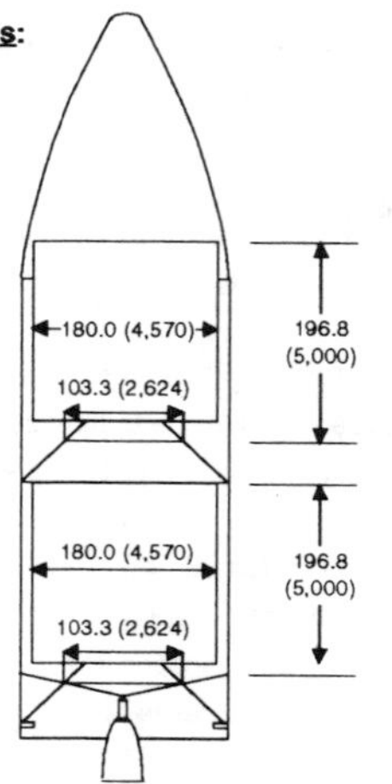

Single Launches:
- Long Fairing
- Short Fairing

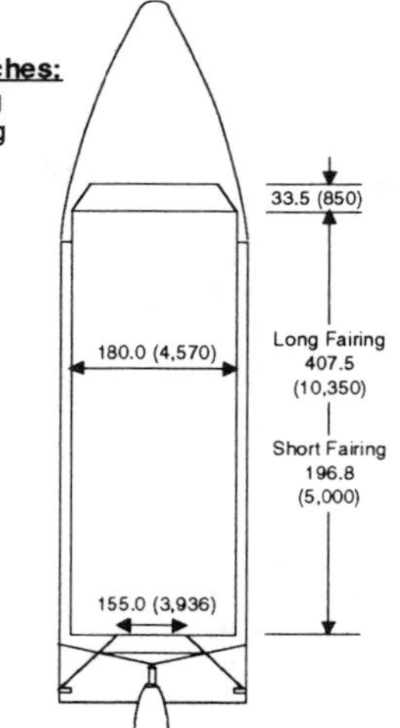

Length	Fairing - 41.7 ft (12.7 m), 55.8 ft (17.0 m)
	Speltra - 23.0 ft (7.0 m)
Diameter	17.7 ft (5.4 m)
Mass	Fairing - 4,400 lb (2,000 kg), 5,950 lb (2,700 kg)
	Speltra - 1,980 lb (900 kg)
Sections	2
Structure	Composite sandwich
Material	Aluminum honeycomb / graphite epoxy

Remarks:

The fairing structure is made up of sandwiched panels with an aluminum honeycomb core between CFRP skin. It is a two half shell structure with a longitudinal separation system similar to that used on Ariane-4. The horizontal separation is located in the lower flange and is based on a pyrotechnic "cordeau." The long fairing is made of two parts, the upper one being similar to the short fairing. The Speltra is an external support structure permitting double launches. It is made up of a cylinder interfacing with the VEB and the fairing, and a cone carrying in its upper section a frame connecting it with the payload. The structure is made of sandwiched panels with an aluminum honeycomb core between carbon-fiber resin composite skins. The lower flange of the Speltra contains a pyrotechnics separation system.

Ariane Vehicle

(continued)

Avionics

Ariane-4. The VEB, which carries most of the launcher's electrical equipment, has been entirely reconfigured from Ariane-2/3 to cope with the higher loads introduced by the larger and heavier payloads, Spelda, and fairing. Its structure is fabricated in four parts from a honeycomb material with carbon fiber facing. This new configuration has two considerable advantages compared with Ariane-2/3: first, it allows separation of the equipment-carrying platform from the load-carrying structure, thus allowing completely independent integration of the payloads with the Spelda and the fairing in the clean rooms of the payload-preparation buildings, after the equipment platform has already been installed on the launcher; and, secondly, it allows easy access to the launch-vehicle equipment for checkout or any necessary interventions after the Spelda/fairing payload cluster has been mounted on the launcher.

The Ariane-4 guidance systems consists of two platforms: a classical inertial platform (like that on Ariane-2/3) used in normal circumstances, and a gyro laser platform to which the onboard computer switches if the main platform should malfunction. A digital flight-control systems on Ariane-4 replaces the analog control system used on Ariane-2/3. The tracking system, with two redundant radar transponders; the fully redundant destruct system, which can receive a destruct command from the ground (the only intervention possible from the ground); and the telemetry system are practically unchanged compared with Ariane-2/3.

Ariane-5. The Ariane-5 guidance system consists of two ring laser gyros located in the VEB.

Attitude Control System

Ariane-4. The four stage 1 engine nozzles pivot in the plane tangent to the thrust frame, providing pitch, yaw, and roll control. Aerodynamic stability during atmospheric flight is improved by four 21 sq-ft (2 sq-m) fins. The stage 2 Viking IV engine can be swivelled about two axes to allow yaw and pitch control. Roll control is provided by two tangential thrusters (11 lb (50 N) of thrust), supplied with hot gas from the stage 2 gas generator. The stage 3 engine, designated HM7B, is linked to the conical thrust frame through a gimbal joint, allowing swivelling of the engine for pitch and yaw attitude control. Gaseous-hydrogen thrusters provide rotational momentum for roll control. These thrusters, together with additional hydrogen thrusters, ensure attitude control of the stage and the attached payload about all three axes after engine cutoff.

Ariane-5. The solid boosters are equipped with a nozzle activation system (GAT) enabling gimballing in pitch and yaw. The core stage is equipped with an engine activation system (GAM) enabling swivelling of the Vulcain engine for pitch and yaw control and a cold gas system for roll control. A reaction control system located in the VEB and using N_2O_4/MMH propellant is used for roll control for the H155 and L9. In orbit, this system is used as an attitude control system before payload separation.

Ariane Performance

Ariane-4

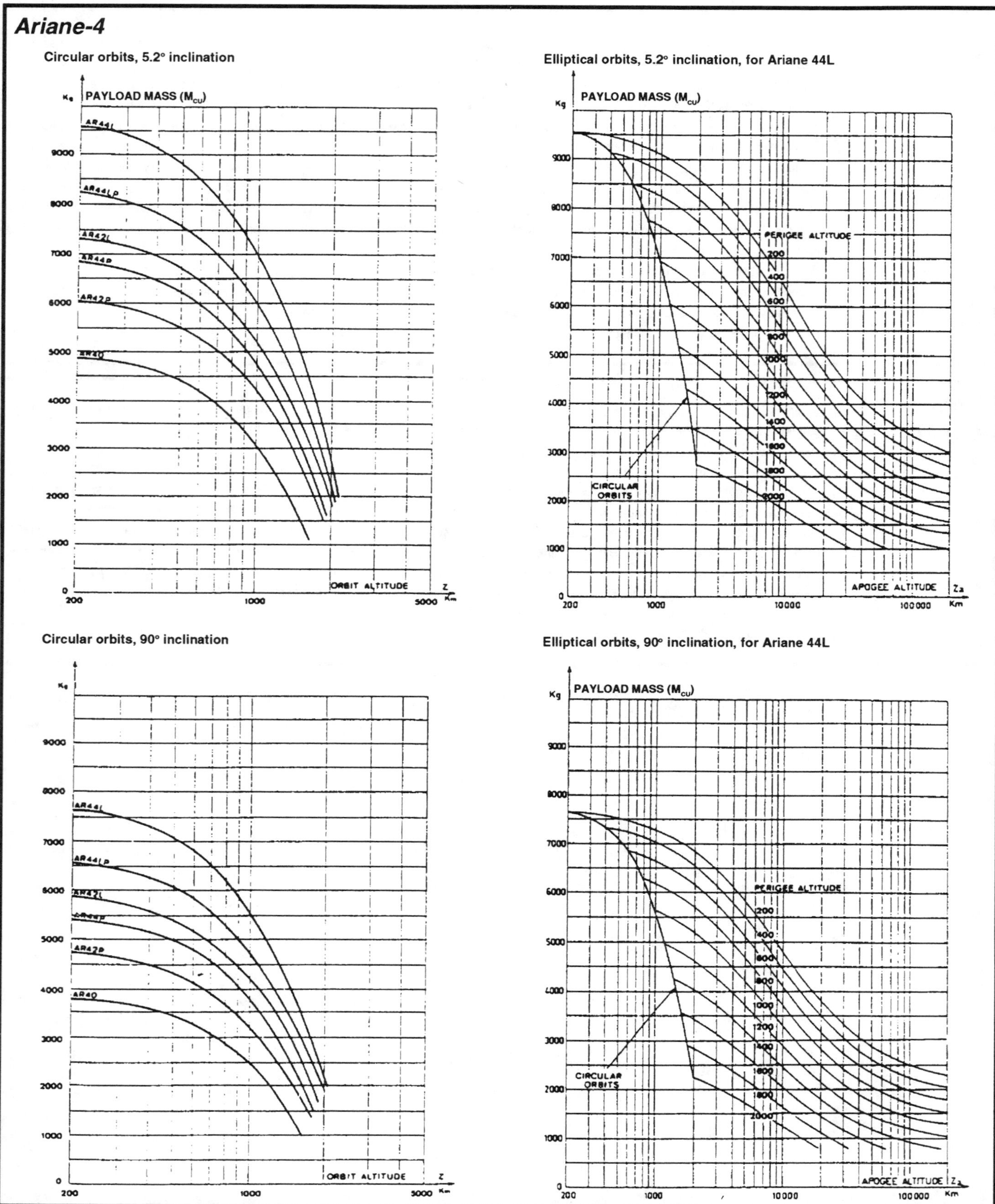

Ariane-5

Performance Curves Not Available.

Ariane Operations

Launch Site

Surinam
Kourou, French Guiana (5.2°N)
Equator
Brazil

ELA 2
North
5° 13' 56"
52° 46' 32"
1
2
3
4
5
RN 1
KOUROU
Town
Harbour
Radar
Telemetry
IR Kinetheodolite
0 1 2 3 4 5 km

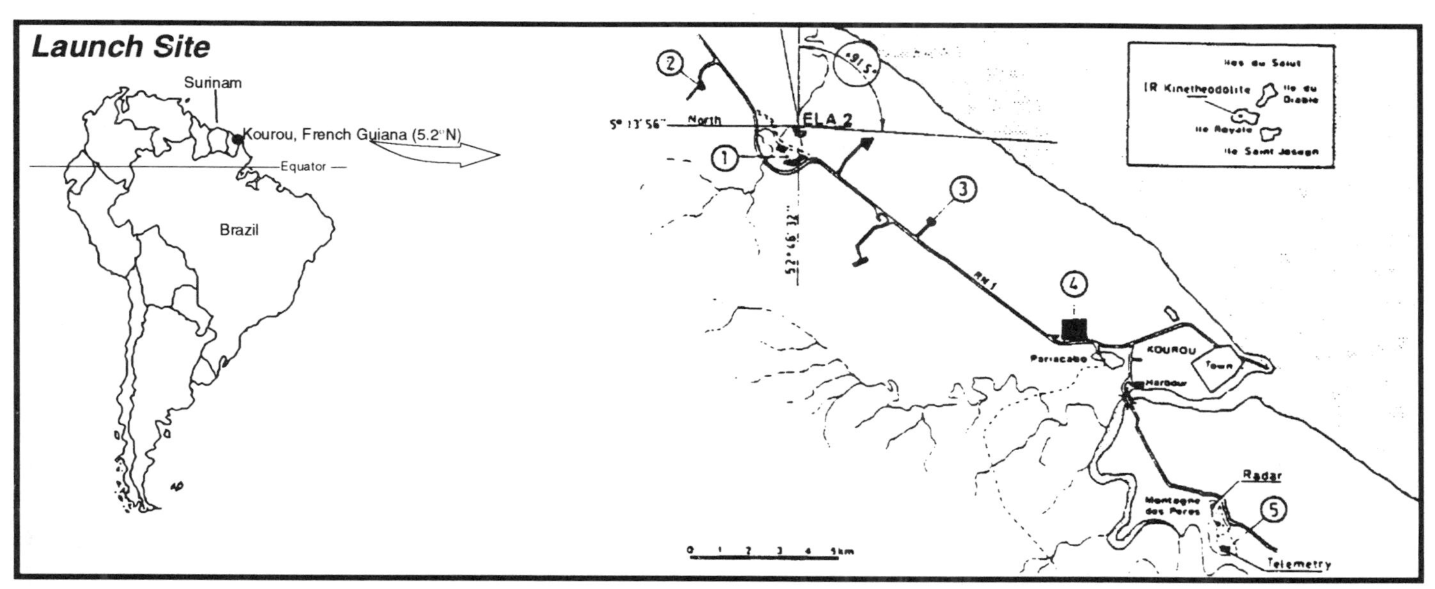

Launch Facilities

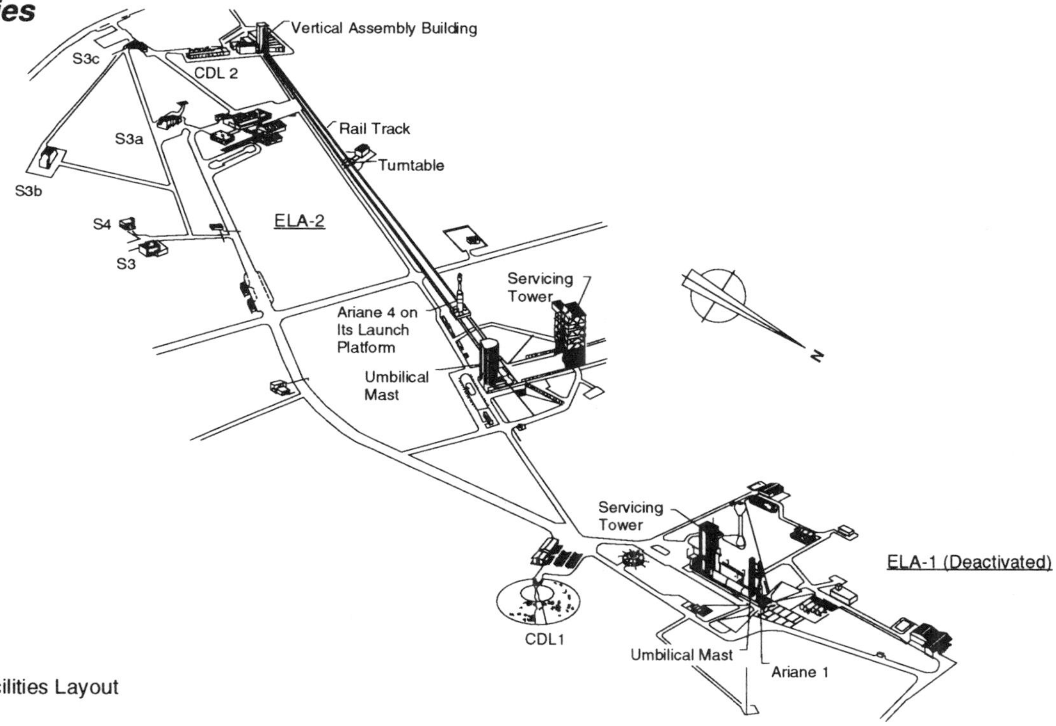

Vertical Assembly Building
S3c
CDL 2
Rail Track
S3a
Turntable
S3b
S4
ELA-2
S3
Servicing Tower
Ariane 4 on Its Launch Platform
Umbilical Mast
N
Servicing Tower
ELA-1 (Deactivated)
CDL1
Umbilical Mast
Ariane 1

ELA-1 and ELA-2 Facilities Layout

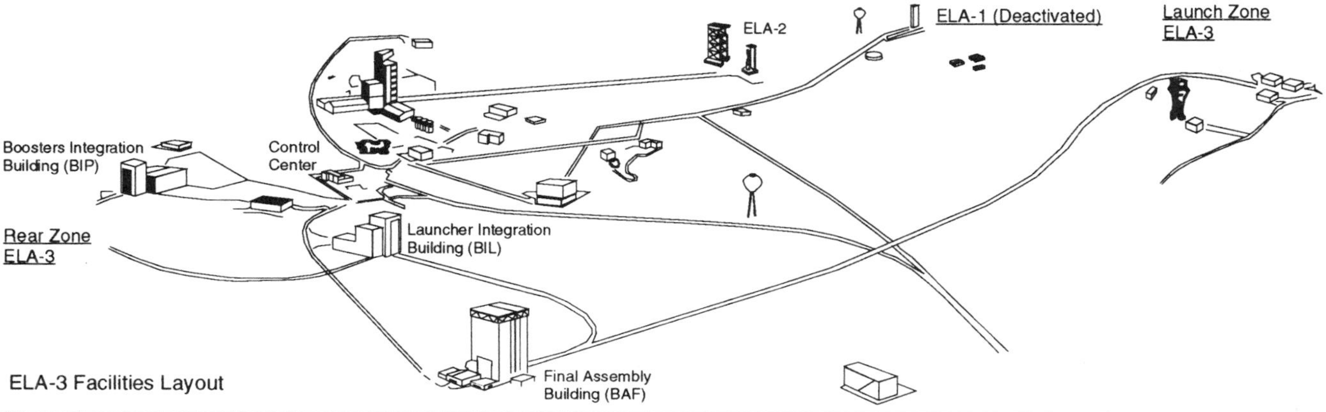

ELA-2
ELA-1 (Deactivated)
Launch Zone ELA-3
Boosters Integration Building (BIP)
Control Center
Rear Zone ELA-3
Launcher Integration Building (BIL)
Final Assembly Building (BAF)

ELA-3 Facilities Layout

Ariane Operations

(continued)

Launch Processing

Ariane launches are made from the Ariane launch site (ELA) at the Guiana Space Center (CSG) near Kourou. All facilities used for a launch campaign sequence (i.e.; Ariane launch complex, Payload preparation complex, CSG facilities including the down-range stations) form the Ariane launch base.

Guiana Space Center (CSG). CSG was set up by the French Government in April 1965 and built by the CNES. It became operational in April 1968 with the launch of a Veronique sounding rocket. Following the Diamant program and the Europa project, the CSG is now being used for Ariane launch operations. The first Ariane flight took place on 24 December 1979.

The CSG facilities are spread over a 11 mi (18 km) strip of the Atlantic coast, between Kourou and Sinnamary. Close to the equator (latitude 5.23° N), the CSG is geographically well located, with a launch sector over the Atlantic Ocean extending from North to East (- 10.5° to + 93.5° launch azimuth), which makes it well suited for launching satellites into geostationary orbit. The CSG is responsible not only for supplying the overall logistic support during launch activities, but also operating the tracking and telemetry networks (including the processing of all launch vehicle telemetry data) and for the safety of personnel and the protection of the facilities. Owned by the ESA, the launch sites are operated by Arianespace, which also maintains the launch facilities.

ELA-1, which was originally used for Europa 2, was used for launching Ariane-1, -2, and -3 before being deactivated. ELA-2 is used for Ariane-3 and -4, and ELA-3 for Ariane-5. For the eastwards launches, the CSG radar, telemetry, and telecommand stations are completed by the three down-range stations located at Natal, Brazil, on Ascension Island, and near Libreville, Gabon, in order to continuously receive data on the launcher's trajectory and behavior in flight.

Payload processing occurs at the Payload Preparation Complex (EPCU), which includes all facilities placed at the disposal of Arianespace customers for the preparation of their satellites, from their arrival in Guiana up to the actual mounting of the payload on the Ariane launcher. Designed for the preparation of satellites in a single or dual launch configuration, the EPCU consists of a number of geographically dispersed buildings: (a) buildings S1A and S1B are located in the CSG Technical Center, and provide clean-room facilities for satellite preparation; (b) buildings S2 and S4, near the two ELA's are designed for solid kick-motor preparation and X-ray operations; (c) building S3A and S3B near the two ELA's are assigned for satellite propellant filling operations and final integration, assembly of the satellites on a Sylda or Spelda dual launch system, and satellite encapsulation into the Ariane-4 nose-fairing; and (d) building S3C is located close to S3A and S3B, and is used for monitoring and control of hazardous operations conducted in the latter. Owned by ESA, the EPCU is operated by the CSG for the benefit of Arianespace customers.

Ariane-4. The Ariane-4 launcher is manufactured by approximately 50 European industrial companies, providing employment for up to 10,000 people. Eight firms play the role of technical main contractors, responsible for a major technical domain or a major element of the launch vehicle. The main contractors receive contracts directly from Arianespace and they subcontract a significant part of the work throughout Europe. The industrial manufacturing facilities are presently capable of producing up to eight Ariane-4s per year in a nominal production cycle, and up to 10 vehicles per year by shift working.

Ariane-4 is prepared at and launched from ELA-2. This new facility allows the minimum interval between two launches to be reduced to one month (two months on ELA-1) thus providing considerably greater flexibility in launch scheduling. While the concept of the ELA-1 complex called for the vehicle erection and assembly directly on the pad, the ELA-2 vehicle assembly and checkout takes place in a remote Vehicle Assembly Building (VAB). The vehicle is then moved, on its launch table, to the launch pad. While this vehicle is undergoing final preparations on the pad, the next one can already be erected in the VAB on a second mobile launch table.

A typical Ariane-4 launch campaign starts about nine weeks prior to the launch, with the transport, first on the River Seine from Les Mureaux to Le Havre and then by boat to Cayenne, French Guiana, of all the vehicle hardware stored in special containers. Propellants (except liquid oxygen which is produced in Kourou) are also loaded on board at Le Havre. Some 10 days later the vessel arrives at the port of Cayenne, from where the launcher and propellants are transfered by road to the launch site, some 9.3 mi (15 km) west of Kourou. In the VAB, the launch vehicle is erected on the mobile launch table and liquid strap-ons, if required, are attached. Mating and checkout in the VAB take about four weeks. The vehicle, on one of its two mobile tables, is pulled by truck for about 50 minutes along a rail track 0.6 mi (1 km) long to the pad. At the pad, solid strap-ons, if required, are attached (need 1 day to attach each solid).

In parallel, the payload, which has been flown to Cayenne, is prepared at the EPCU. At building S3A/B, payloads are encapsulated making a transport container superfluous and reducing the time needed for payload preparation in the launch tower. The fully integrated upper composite consisting of the VEB structure, the payload, and the fairing in the case of a single payload; or the VEB structure, the Spelda, the payloads, and the fairing in the case of a dual launch is transported to the launch pad and installed on top of the vehicle five working days before launch.

The launch countdown, covering mainly the filling of the stages with propellants, takes about 38 hrs distributed over three days. At about 10 and 5 hrs prior to launch, balloons are released for winds aloft measurements. The lightning, whose season is mostly July through August, keepout zone is 5 mi (8 km). During the last 6 min before launch initiation, the ground checkout system verifies the proper functioning of the vehicle. It also separates the propellant transfer arms from the third stage 5 sec before the end of the sequence, and finally commands ignition of the first four first stage engines and of the liquid propellant boosters.

After ignition, the ground checkout system monitors the functional parameters of the ignited engines. If their status is correct, the command to open the jaws holding the launcher is given at the same time as that to ignite the solid propellant boosters. The typical flight time for the Ariane-4 launcher, up to third-stage engine shutdown, is 17 to 18 minutes. The ELA-2 launch complex is also capable of supporting 10 launches per year.

Before Launch (hr:min:sec)

Start Activity	End Activty	Activity
10:15:00	~8:45:00	Spacecraft functional checks
--	7:15:00	Tower withdrawal decision
6:15:00	5:00:00	Tower withdrawal
6:00:00	4:25:00	Stage 3 engine lines flushing
4:25:00	3:55:00	Margin
4:00:00	0:45:00	Stage 3 filling and pressurization
0:20:00	0:08:00	Margin
0:06:00	--	Start automatic sequence
0:03:30	--	Payload ready for launch
0:00:09	--	Inertial platform release
0:00:05	--	Cryogenic arms opening
0:00:00	+0:00:03	Ignite liquid engines and check status
+0:00:03	--	Ignite solid motors and release vehicle 0.4 sec later

In Case of Launch Hold Between:

0:06:00	0:05:00	Reset clock to 6 min (at T-9 sec, last payload requested hold) If miss window, postpone 24–48 hrs (3 hrs to return Tower)
0:05:00	+0:00:03	Postpone launch minimum 10 days (33 hrs to return Tower)

Ariane-4 Final Countdown Phase

Ariane Operations

(continued)

Launch Processing

(continued)

Ariane-5. The Ariane-5 launch site has four main specifications: (1) launch 10 Ariane-5s per year; (2) increase the reliability to be able to launch manned flights; (3) minimize the investment costs; (4) reduce the operations and maintenance costs. To fulfill these requirements, the following characteristics were chosen: (a) All the preparation and check out phases of the launcher are done in parallel, thus, reducing the total time of a campaign to six weeks. The payload will be checked and mated in the fairings before being installed on the launcher so the time is minimal between the time when the payload is ready and the launch. (b) No set-back in the configuration of the launcher during all the different steps of its preparation. All the connections between the launcher and the ground are checked one time. (c) One simple launch pad is used where only the cryogenic filling up and the launch take place. This means minimal investments and maintenance costs and the simple pad design minimizes the risk and reduces the time to resume the launches after an accident. (d) Each building of the complex will be optimized to only one main operations phase: i.e., mating and checking of the solids, erection and checking of the cryogenic stage, etc.

These design characteristics will be reflected in the new launch complex, ELA-3, being built for the Ariane-5 launcher close to the ELA-1 and ELA-2 launch complexes at Kourou. ELA-3 consists essentially of two zones, the Preparation Zone and the Launch Zone.

The Preparation Zone consists of four main buildings—Boosters Integration Building (BIP), Launcher Integration Building (BIL), Final Assembly Building (BAF), and Preparation and Launch Control Center (CDL 3). The separation of these buildings is dictated by safety constraints imposed by the use of solid-propellant boosters.

The Boosters Integration Building (BIP) is where the P230s are assembled and checked out. The center and forward booster segments are built in a plant located near the BIP. The forward segment is produced in Italy. Regulus, a joint venture with SNEP and BPD, will produce the propellant locally. In the BIP, the segments are mated in two integration rooms on two special devices which are used to transfer the booster. The boosters are fully prepared and checked before leaving.

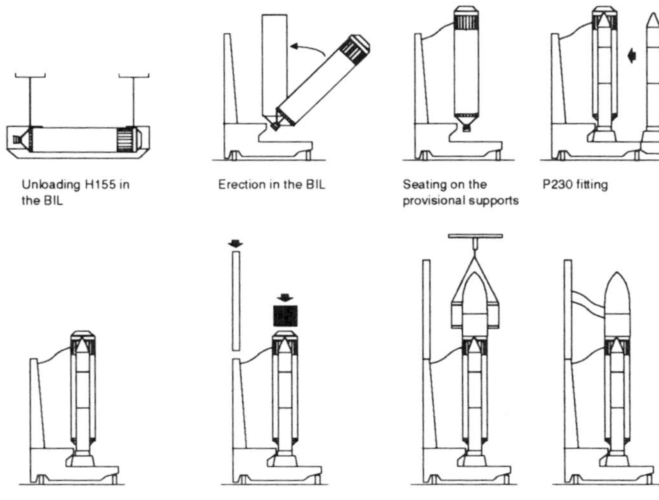

Illustration of Launch Vehicle Integration

The Launcher Integration Building (BIL) is where all the assembly and checkout operations are carried out on the main body of the launcher (H155 and L9) and where the P230 boosters are mated with it. The cryogenic stage, shipped from Europe, is erected and mated on the launch table with the L9 stage and the VEB. When the core has been checked, the two boosters are fitted. The vehicle is now prepared for mating with the payload.

The Final Assembly Building (BAF) is where the payload composite is assembled and erected, the fairing is assembled, the L9 tanks are filled, and the final electrical checkout is done. The building has two main parts: an upper part that is a clean room where the payload is mated, and a lower part which protects the launcher and the table. This building represents a major change compared with the ELA-2 process in that it replaces the mobile gantry and its function, namely final check on the launcher, and mating of the payloads. After integration and test, the launcher is transferred on a twin railway track from the Preparation Zone to the Launch Zone in a fully-integrated state.

The Launch Control Center (CDL3) is where the checkout and command equipment allow the activation of two launchers at the same time, and where operations are controlled and monitored down to the moment of launch.

The Launch Zone consists of the launch pad. This area includes only deflectors and exhaust ducts, a small building protecting the electrical and fluid connections, the cryogenic propellant storage, and a water tower to suppress noise. This simple design ensures a sustained launch rate and, in case of a pad accident, minimal standdown. Simplifying the Launch Zone means that the launcher preparation and payload integration are done in the Preparation Zone at a safe distance from the Launch Zone. From 1994, the Launch Zone will be used for running development and qualification tests on the Vulcain engine in conjunction, initially, with heavy tanks and then with the H155 qualification stage. This will permit a saving on dedicated test facilities in Europe and will enable a valuable contribution to be made to the qualification of ELA-3. Similarly, some ELA-3 facilities intended for booster preparation are to be used for tests on the P230s on a special stand site 2 mi (3 km) away from ELA-3 and linked to it by a rail track.

The launcher tracking, safety and telemetry reception facilities will be adapted for Ariane-5 launches. In particular, further downrange facilities to be used or set up for the various Ariane-5 trajectories (GTO, LEO, or SSO) are envisaged in addition to the existing downrange stations.

ELA-3 is also comprised of a production plant for liquid oxygen and liquid nitrogen and another for liquid hydrogen.

Phase	Start Activity (days)	End Activity (days)	Duration (days)	Activity
Phase 1	T-61.5	T-17.5	44	Preparation of the 2 Solid Rocket Boosters in the BIF
Phase 2	T-27.5	T-17.5	10	Preparation of Central Body in the BIL plus checks
Phase 3	T-17.5	T-14.5	3	Transfer of Solid Rocket Boosters from the BIP to the BIL and mating with Central Body
	T-14.5	T-10.5	4	Verification of Launcher in the BIL without payloads
Phase 4	T-18.5	T-10.5	8	Preparation of payloads and nose fairing in the BAF
	T-10.5	T-9.5	1	Transfer of Launcher from BIL to BAF
Phase 5	T-9.5	T-4.5	5	Assembly of payloads and fairing on the launcher, plus checks
	T-4.5	T-1.5	3	Preparation for countdown
Phase 6	T-1.5	T-0.5	1	Transfer of launcher to launch pad
	T-0.5	T-0.0	0.5	Countdown and launch

Sequencing and Duration of Ariane-5 Launch Processing.

Ariane Operations

(continued)

Flight Sequence

Ariane-4 Typical Flight Sequence

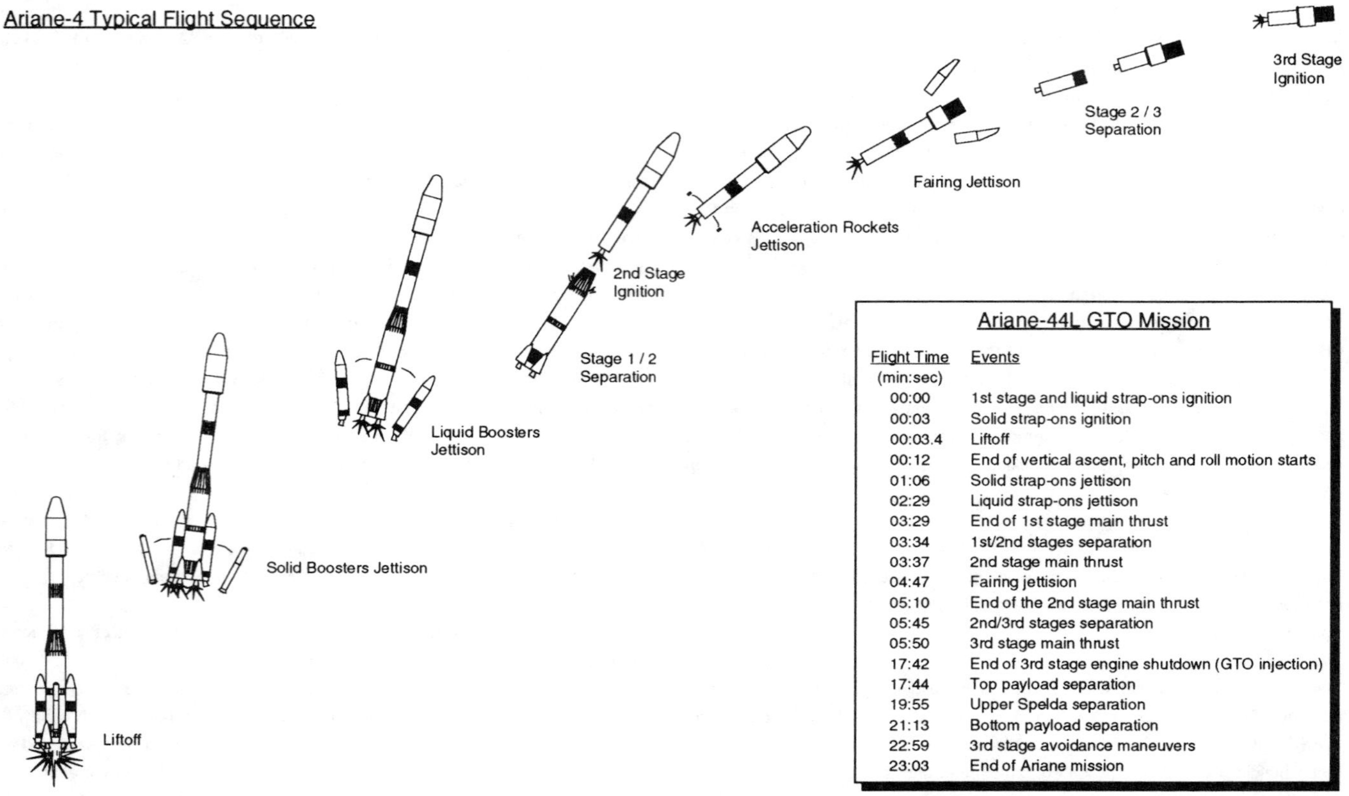

Ariane-44L GTO Mission

Flight Time (min:sec)	Events
00:00	1st stage and liquid strap-ons ignition
00:03	Solid strap-ons ignition
00:03.4	Liftoff
00:12	End of vertical ascent, pitch and roll motion starts
01:06	Solid strap-ons jettison
02:29	Liquid strap-ons jettison
03:29	End of 1st stage main thrust
03:34	1st/2nd stages separation
03:37	2nd stage main thrust
04:47	Fairing jettision
05:10	End of the 2nd stage main thrust
05:45	2nd/3rd stages separation
05:50	3rd stage main thrust
17:42	End of 3rd stage engine shutdown (GTO injection)
17:44	Top payload separation
19:55	Upper Spelda separation
21:13	Bottom payload separation
22:59	3rd stage avoidance maneuvers
23:03	End of Ariane mission

Ariane-5 Typical Flight Sequence

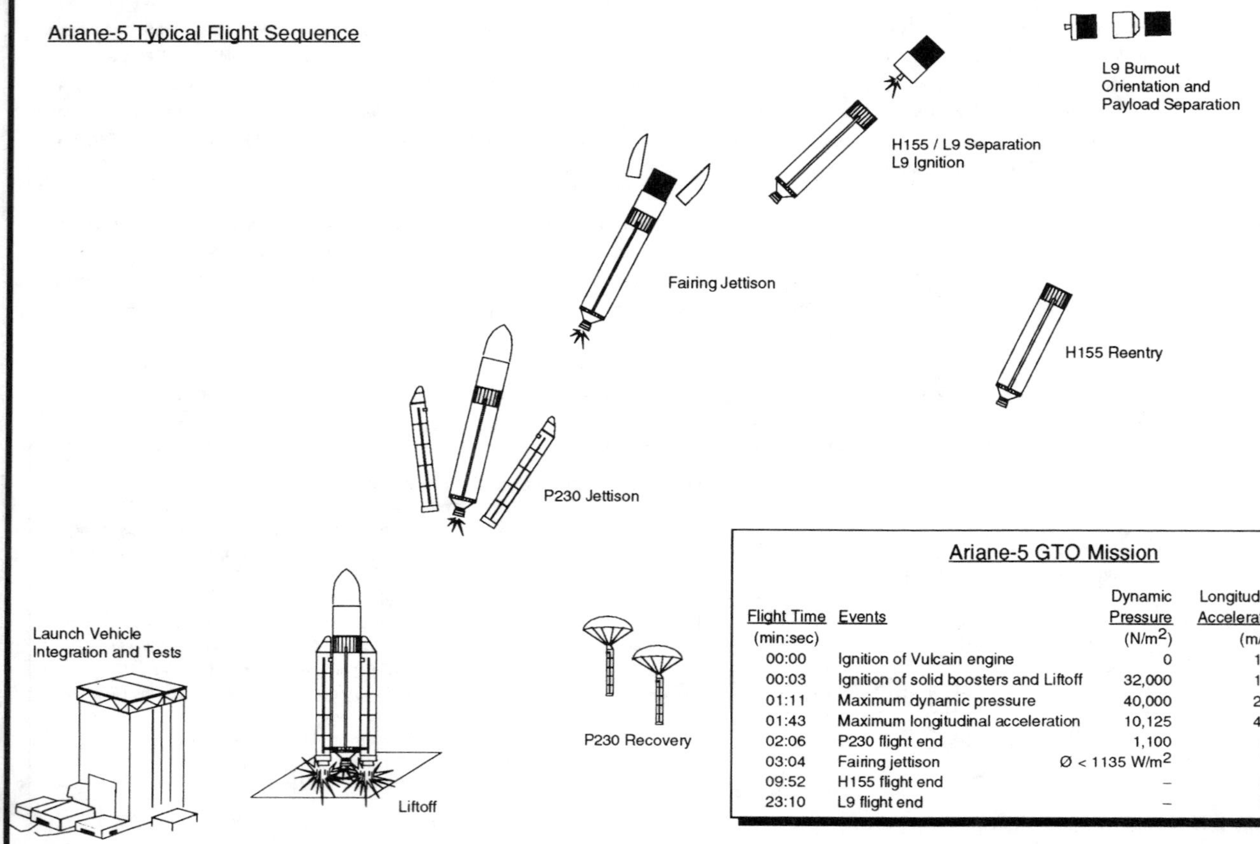

Ariane-5 GTO Mission

Flight Time (min:sec)	Events	Dynamic Pressure (N/m^2)	Longitudinal Acceleration (m/s^2)	Altitude (km)
00:00	Ignition of Vulcain engine	0	16.6	0
00:03	Ignition of solid boosters and Liftoff	32,000	17.6	8
01:11	Maximum dynamic pressure	40,000	24.5	15
01:43	Maximum longitudinal acceleration	10,125	43.8	35
02:06	P230 flight end	1,100	6.6	56
03:04	Fairing jettison	Ø < 1135 W/m^2	7.8	106
09:52	H155 flight end	–	0.5	141
23:10	L9 flight end	–	2.8	1077

Ariane Payload Accommodations

	Ariane-4	Ariane-5
Payload Compartment Maximum Payload Diameter	Fairing - 143.7 in (3650 mm) Spelda - varies from 143.7 ft (3650 mm) to 125.6 ft (3190 mm)	180 in (4570 mm)
Maximum Cylinder Length NOTE: The "egg-shaped" Sylda is 110 in (2800 mm) maximum diameter and 173 in (4400 mm) height.	Single payload (assume 1920 mm interface): Short fairing - 155 in (3940 mm) Long fairing - 194 in (4940 mm) Extra long fairing - 254 in (6460 mm) Dual payload (assume 1920 mm interface): Short fairing - 107 in (2720 mm) Long fairing - 146 in (3720 mm) Short Spelda - 109 in (2760 mm) Long Spelda - 150 in (3810 mm)	Single payload (assume 2624 mm interface): Long fairing - 472 in (12000 mm) Multi payload (assume 2624 mm interface): Short fairing - 177 in (4500 mm) Speltra - 177 in (4500 mm)
Maximum Cone Length	Fairing - 168 in (4255 mm) Short Spelda - 34.6 in (880 mm) Long Spelda - 32.7 in (830 mm)	Fairing - ?? Speltra - 4.6 ft (1.4 m)
Payload Adapter Interface Diameter	36.9 in (937 mm) 47.0 in (1194 mm) 58.9 in (1497 mm) 65.6 in (1666 mm)	75.6 in (1920 mm) for payload adapter 105.3 in (2624 mm) for L9 155.0 in (3636 mm) for VEB
Payload Integration Nominal Mission Schedule Begins	T-29 months	T-?? months
Launch Window Latest Countdown Hold Not Requiring Recycling	T-6 min (launch postponed after T-5 sec requiring minimum 10 days until next launch attempt)	T-?? min
On-Pad Storage Capability	2 ? days fueled stage 2, ?? hrs fueled stage 3	?? hrs for fueled core
Latest Access to Payload	T-8 days (single) to T-9 days (dual) or T-2 days through access doors	T-2 days through access doors
Environment Maximum Load Factors	+4.5 g axial, ±0.2 g lateral	+4.5 g axial, ±?? g lateral
Minimum Lateral/Longitudinal Payload Frequency	10 Hz / 31 Hz	?? Hz / ?? Hz
Maximum Overall Acoustic Level	142 dB (one third octave)	144 ?? dB (one third octave)
Maximum Flight Shock	2,000 g from 1,500–4,000 hz	?? g from ?? hz
Maximum Dynamic Pressure on Fairing	?? lb/ft2 (?? N/m2)	835 lb/ft2 (40,000 N/m2)
Maximum Pressure Change in Fairing	0.5 psi/s (3.2 KPa/s)	?? psi/s (?? KPa/s)
Cleanliness Level in Fairing (Prior to Launch)	Class 10,000	Class 10,000
Payload Delivery Standard Orbit and Accuracy (3 sigma)	GTO - 108 nm (200 km) is ±0.5 nm (1.0 km) by 19323 nm (35786 km) is ±28 nm (52 km) at 7.0° is ±0.018°	LEO - 300 nm (500 km) circ is ±2.2 nm (4 km) at 28.6° is ±0.06° GTO - 150 nm (280 km) is ±0.5 nm (1.0 km) by 19,323 nm (35,786 km) is ±54 nm (100 km) at 10.0° is ±0.03°
Attitude Accuracy (3 sigma)	All axes: ±3 deg; ±?? deg/sec	All axes: ±?? deg; ±?? deg/sec
Nominal Payload Separation Rate	1.6 ft/sec (0.5 m/s)	1.6 ft/sec (0.5 m/s)
Deployment Rotation Rate Available	0–5 rpm	0–5 rpm
Loiter Duration in Orbit	?? hrs	?? hrs
Maneuvers (Thermal / Collision / Telemetry)	Yes	Yes

Ariane Notes

Publications

User's Guide

ARIANE-4 User's Manual, Arianespace, Issue No. 1, Rev. 7, July 1993.

Technical Publications

"Ariane Launch Record: V62", 1994.

Launch Vehicle Catalogue, ESA, December 1993.

"Development of the Ariane 5 Upper Stage", H. Holsten, MBB/ERNO, IAF-90-156, 41st Congress of the International Astronautical Federation, Dresden, Germany, October 6-12, 1990.

"ATV - Ariane Transfer Vehicle Status", C. Bonnal, F. Theiller, Aerospatiale, and D. Salt, British Aerospace, IAF-90-167, 41st Congress of the International Astronautical Federation, Dresden, Germany, October 6-12, 1990.

"Ariane 5 Propulsion Requirements and Constraints for a Cost Effective Unmanned (Commercial)/Manned (HERMES) Compatible Advanced Launcher", R. Rault, Aerospatiale, AIAA-90-2699, AIAA 26th Joint Propulsion Conference, Orlando, FL, July 16-18, 1990.

"Ariane 4 Performance and Promise", D.A. Heydon & E.H. Weinrich, Arianespace, AIAA-90-2716, AIAA 26th Joint Propulsion Conference, Orlando, FL, July 16-18, 1990.

"Ariane 5: A Mature Development", C. Johnson, CNES, J.F. Lieberherr, ESA, IAF-89-201, 40th Congress of the International Astronautical Federation, Malaga, Spain, October 7-12, 1989.

"Ariane 5 Transfer Vehicle", C. Bonnal, P. Eymar, Aerospatiale, IAF-89-211, 40th Congress of the International Astronautical Federation, Malaga, Spain, October 7-12, 1989.

"ARIANE 4 and ARIANE 5 Launch Systems", P. Rasse, Arianespace, presentation at Cape Canaveral, October 2, 1989.

"Flight Performance of the ARIANE 4", D.A. Heydon, Arianespace, AIAA-89-2742, AIAA 25th Joint Propulsion Conference, Monterey, CA, July 10-12, 1989.

"ARIANE 5", Arianespace, April 1988.

Acronyms

ACS - attitude control system
AR 40, etc. - Ariane-40, etc.
BAF - Final Assembly Building
BIL - Launcher Integration Building
BIP - Boosters Integration Building
CDL 3 - Launch Control Center
CNES - Centre National d'Etudes Spatiales (French Space Agency)
CSG - Guiana Space Center
ELA - Ensemble de lancement Ariane (Ariane Launch Site)
EC - European Community
EPCU - Payload Preparation Complex
ESA - European Space Agency
GTO - geosynchronous transfer orbit
H155 - Cryogenic stage (155 metric tons propellant)
HTPB - hydroxy terminated polybutadiene
L9 - Storable stage (9 metric tons propellant)
LEO - low Earth orbit
LH2 - liquid hydrogen
LOX - liquid oxygen
MAU - Millions Accounting Units (In 1986 equal $0.73, in 1991 equal $1.22)
MMH - monomethyl hydrazine
P230 - Solid booster (237 metric tons propellant)
Spelda - Structure porteuse externe pour lancements doubles Ariane
Speltra - Structure porteuse externe pour lancements triples Ariane
SSO - sun synchronous orbit
Sylda - Systeme de lancements double Ariane
VEB - Vehicle Equipment Bay

Ariane Notes

(continued)

Other Notes - Ariane Growth Possibilities

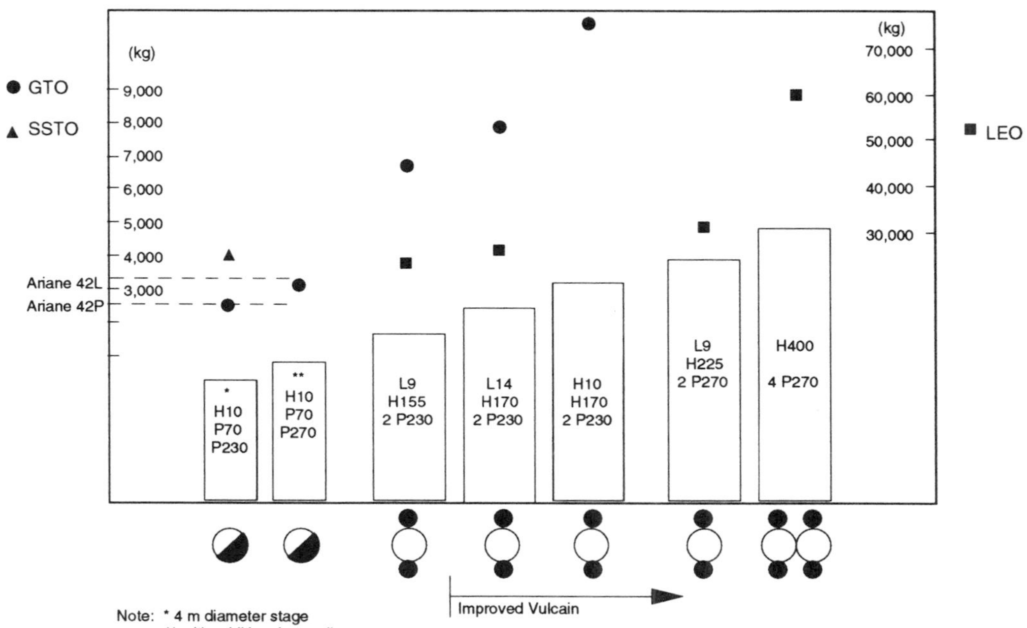

Exactly as Ariane-1 proved to be the forebear of a large family of vehicles, it is expected that Ariane-5 will play a similar role. Conservative measures were taken, whenever possible at low cost, to meet the requirement for growth potential, either in the Ariane-5 launcher design or in ground infrastructures and test facilities. Below is a review of various potential evolutions at the stage level and then what possible family of vehicles could be derived. More advanced concepts featuring vehicle recovery and/or air breathing propulsion are also being studied.

Improvement to Solid Booster. There is also a development potential of the P230 by adding one short segment to increase propellant mass up to 594K lb (270K kg). It is also possible to derive a much smaller motor of 154K lb (70K kg) propellant that would keep the same metallic structures, propellant composition thermal protection materials and pyrotechnics. New developments would concern the nozzle assembly, propellant grain definition and a light casing.

Improvements to Core Stage. A much heavier cryogenic stage can be derived from the basic hardware. Propellant mass will vary from about 450 to 495K lb (205 to 225K kg) according to the mixture ratio that will be chosen for the improved Vulcain. The main features would be: LOX and LH2 tanks with separate bulkheads; intertank skirt derived from present forward skirt; booster thrust introduction at intertank level; cluster of two Vulcain engines. These engines would have to be integrated with a completely new thrust frame, but practically all equipment would be kept unaltered.

Cryogenic Upper Stage. The modifications of the H10 (upper stage of Ariane 4) to adapt it to Ariane 5 have already been studied but the development was postponed because no mission really required it. The redesign mainly concerns the LH2 tank that must be of a larger diameter. The LOX tank is almost unchanged, as are the thrust frame and the propulsion system subassembly.

Improvement to L9. It is possible to increase the L9 propellant mass up to 17 times and at the same time to modify the engine thrust at 35 km.

Ariane Transfer Vehicle (ATV). The ATV has been studied to be the "smart end" of a vehicle which bridges the Ariane-5 launcher and an orbital element to be serviced. The reference mission is logistics resupply of Space Station. After an attached period of up to six months at the station, the vehicle can remove a cargo of station waste products. This expendable stage could use L9 components and gross mass will be about 11,000 lb (5,000 kg).

Ariane-5 Derivatives. A large variety of launch vehicles can be created by combining the basic stages that would be available: solid propellant boosters (P230, P70, P270); main cryogenic stage (H155, H170, H225); and upper stage (H10, L9). Some combinations are adapted to heavy payloads in LEO and others to GTO missions.

P230-270/P70/H10 - This "small" vehicle requires the development of the specific P70. It can be rather easily developed if market evolution and commercial competition requires it. It could be launched from ELA-2 pad with an adapted launch table.

2P230/H155/L9 - reference Ariane-5

2P230/H170/L14 - improved Vulcain and increased propellant mass and use L14 stage. Performance improvement is 1400 lb into GTO.

2P230/H170/H10 - provides a large increase of GTO performance. This vehicle could also launch planetary missions.

2P270/H205-225/L9 - puts 66,000 lb (30,000 kg) in LEO 59 x 250 nm, 28.5°(110 x 463 km).

4P270/H400 - derived by clustering two of the above mentioned versions. Very few missions are projected for this vehicle and it would not be cost effective to develop very large (about 25 ft (7.5m) diameter) tanks skirts and thrust frame structures.

India

Table of Contents

SLV-Series

Government Point of Contact:
Indian Space Research
Organization (ISRO)
Launch Vehicle Program Office
Antariksha Bhavan
New BEL Road
Bangolore 560094 INDIA
Phone: 080-3334474

U.S. Point of Contact:
Counselor of ISRO
Embassy of India
2107 Massachusetts Avenue NW
Washington, DC 20008, USA
Phone: (202) 939-7075

The Polar Satellite Launch Vehicle (PSLV) is shown launched from Sriharikota, India.

SLV-Series History

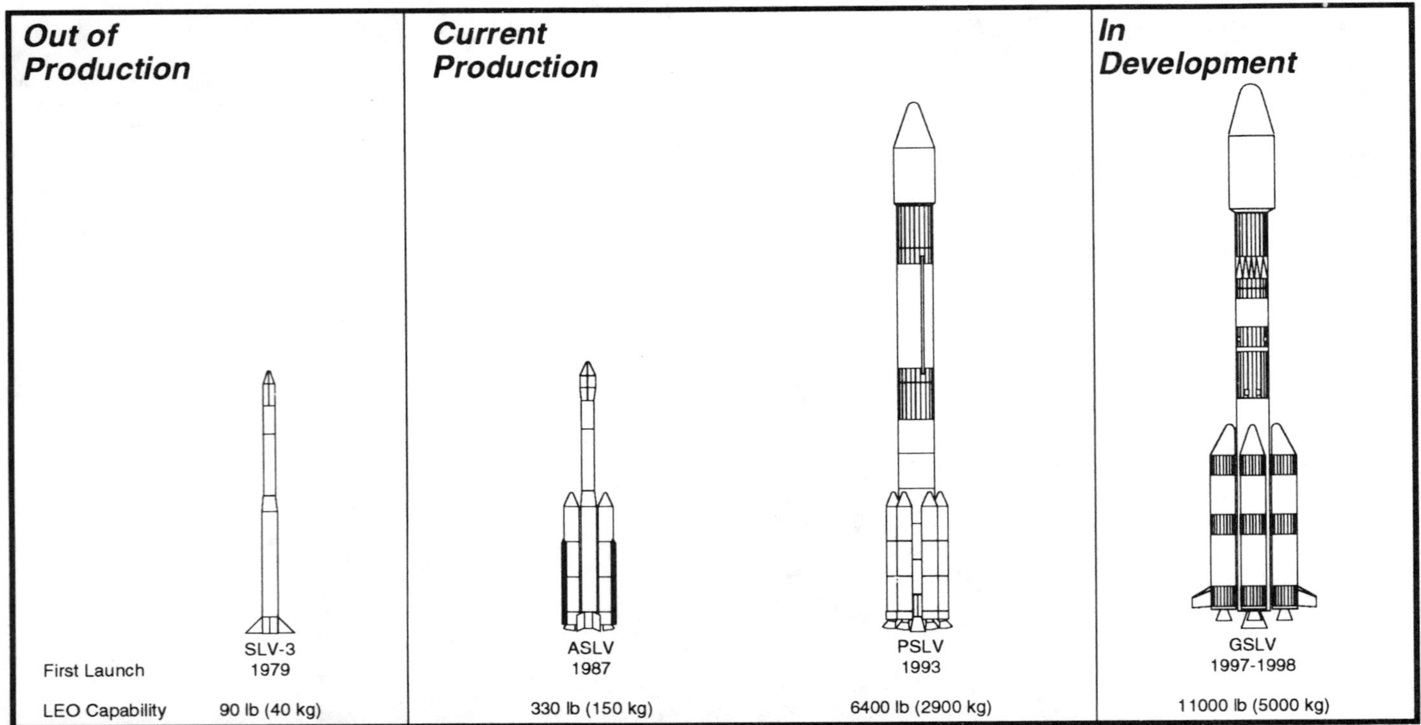

Vehicle Description

SLV-3 Satellite Launch Vehicle (SLV) is a four stage, solid-propellant vehicle based on earlier sounding rockets. Payload fairing diameter is 2.6 ft (0.8 m).

ASLV Augmented Satellite Launch Vehicle (ASLV) is an upgraded version of the SLV-3 with the first stage motor used as two strap-ons. Payload fairing diameter is 3.3 ft (1 m).

PSLV Polar Satellite Launch Vehicle (PSLV) has six solid strap-ons similar to the ASLV, a solid-propellant first and third stage, and liquid second and fourth stages. Payload fairing diameter is 10.5 ft (3.2 m).

GSLV Geostationary Satellite Launch Vehicle (GSLV) is derived from PSLV by replacing the six solid strap-ons of PSLV with four liquid strap-ons similar to the second stage of PSLV. A cryogenic upper stage will replace the last two stages of PSLV.

Historical Summary

The main goal of the Indian space program is to provide operational space services to the nation. The focus is on satellite communications and remote sensing including weather. These services are sought to be achieved through indigenous development of operational spacecraft and launch vehicles with emphasis on self-reliance.

Based on the desire to achieve self-reliance in launch vehicle capability, the Indian Space Research Organization (ISRO) initiated a launch program in the early 1970s. In the late 1980s, ISRO spent nearly 40% of its approximate $300 million space budget on launch vehicles. The launch vehicle program is based on the development of four generations of launchers, each one utilizing the pedigree of the previous generation.

The first-generation launcher, Satellite Launch Vehicle (SLV-3), began in 1973 after the success of the suborbital sounding rocket program. The SLV-3 was built in order to orbit Rohini-class satellites, which weigh about 90 lb (40 kg). The first flight of SLV-3 took place in 1979 and the last in 1983. The first test flight failed in 10 August 1979. It was determined that a valve in the second-stage system malfunctioned.

The second flight test occurred on 18 July 1980. The vehicle successfully launched the 77 lb (35 kg) Rohini-1 satellite. The main purpose of the Rohini-1 was to monitor the performance of the SLV-3 during launch.

On 31 May 1981, the third flight test of the SLV-3 launched the 84 lb (38 kg) Rohini-2 satellite. The satellite was intended to stay in orbit for approximately 90 days but it reentered on 8 June 1981, after only 9 days, due to an improper orbit. Instrumentation aboard the spacecraft included a spinscan, solid state imaging system with a planned resolution of about 0.6 mi (0.9 km).

The fourth and last launch of the SLV-3 occurred on 17 April 1983. This launch successfully placed the Rohini-3, a 91 lb (41.5 kg) remote sensing satellite, into orbit. The purpose of the mission was to verify the operation of the solid state imaging system and it sent high-quality pictures of the areas scanned by it.

The second-generation launcher, Augmented Satellite Launch Vehicle (ASLV), is an upgraded version of SLV-3. It was developed to orbit stretched Rohini-class satellites (SROSS) weighing about 330 lb (150 kg) to low Earth orbit (LEO). SROSS satellites are designed for selected scientific, technological, and remote sensing missions. The ASLV vehicle uses the basic SLV-3 core plus two strap-on boosters for initial thrust. The strap-ons use the same motor as the first-stage core. The first two launches of the

SLV-Series History

(continued)

Historical Summary

(continued)

ASLV occurred in 1987 and 1988, both ending in failure. The first flight terminated when the core stage engine failed to ignite after separation of the boosters. This mishap is attributed to either a loose connection or a random malfunction of the igniter safe/arm.

The second vehicle failed shortly after the strap-on boosters shut off. A redesign of the vehicle was undertaken, particurlary related to the transition between the strap-ons and first stage.

The third development flight of ASLV on May 20, 1992, successfully injected SROSS "C", carrying two scientific experiments. This flight validated all the corrective actions in the design of ASLV.

This was followed by the launch of ASLV-D4 on May 4, 1994. ASLV-D4 successfully injected the SROSS C2 satellite in the intended orbit. The satellite is functioning well, giving valuable technical data through its two experiments Retarding Potential Analyser and Gamma Ray Burst Detector.

The third-generation launcher, Polar Satellite Launch Vehicle (PSLV), is derived from ASLV and aims to deliver remote sensing IRS satellites weighing 2,200 lb (1,000 kg) to polar sun synchronous orbit. The PSLV is expected to be the workhorse for the Indian space program for remote-sensing missions. The core of the vehicle is a 2.8 m diameter solid motor with and six solid strap-ons similar to the ASLV. The first and third stages will be solid stages and the second and fourth will be liquid stages. The technology of the second stage liquid engine, Vikas, was acquired from France and is based on the Viking-IV engine used for the Ariane program. The first developmental launch of PSLV (PSLV-D1) took place on September 20, 1993. Although nearly all the systems performed to expectation, PSLV-D1 could not place IRS-1E satellite into orbit due to a software implementation error.

The second developmental launch, PSLV-D2, took place on October 15, 1994. PSLV-D2 successfully placed 1,769 lbs (804 kg) IRS-P2 remote sensing satellite into an 488 mi x 533 mi (802 km x 875 km) polar sun synchrounous orbit.

The fourth-generation launcher, and most ambitious, is the Geostationary Satellite Launch Vehicle (GSLV). Its primary mission will be to deliver communication satellites to geosynchronous transfer orbit (GTO). The vehicle will use the basic propulsion systems from PSLV for the first two stages. In addition, four liquid strap-ons derived from the PSLV second stage will augment the booster. The third and final stage will be a cryogenic stage. GSLV will be capable of 5,500 lb (2,500 kg) to GTO. First development launch is expected by 1997–1998.

All ISRO launches are conducted from Sriharikota Range (SHAR) Center, on the southeastern coast of India.

Organizationally, the Space Commission lays down the national policies in all matters concerning space, while the Department of Space (DOS) executes those policies through the ISRO. Both DOS and ISRO are headquartered in Bangalore.

R&D laboratories of ISRO are spread over the country: the Vikram Sarabhai Space Center (VSSC) at Trivandrum; SHAR Center at Sriharikota; ISRO Satellite Center (ISAC) at Bangalore; Space Applications Center (SAC) at Ahmedeabad; Liquid Propulsion Systems Center (LPSC) with its laboratories and test facilities at Bangalore, Trivandrum, and Mahendragiri; ISRO Telemetry, Tracking, and Command (ISTRAC) network of stations around the country; the ISRO Range Complex (IREX) with Ranges in Sriharikota, Trivandrum, and Balasore, the Development and Educational Communication Unit (DECU) at Ahmedabad, and ISRO Inertial Systems Unit (IISU) at Trivandrum.

Launch Record

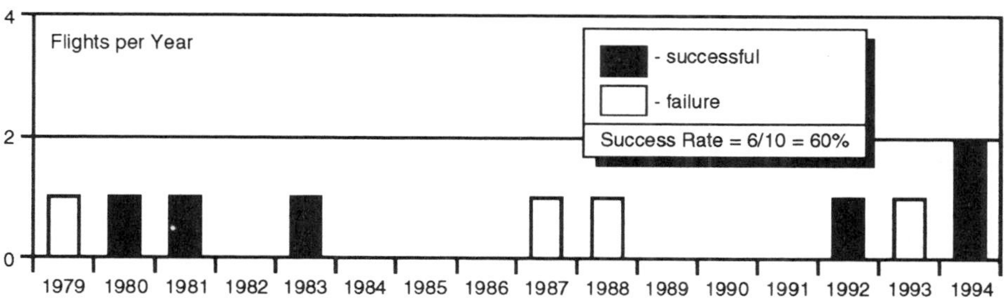

	YEAR	DATE	VEHICLE	SITE	PAYLOAD
1	1979	Aug-10	SLV-3	SHAR Center	Rohini
		Failure - stg 2 control system valve malfunction			
2	1980	Jul-18	SLV-3	SHAR Center	Rohini-1
3	1981	May-31	SLV-3	SHAR Center	Rohini-2
4	1983	Apr-17	SLV-3	SHAR Center	Rohini-3
5	1987	Mar-24	ASLV	SHAR Center	SROSS (Stretched Rohini)
		Failure - stg 1 did not ignite			
6	1988	Jul-13	ASLV	SHAR Center	SROSS (Stretched Rohini)
		Failure - transition between strap-on and first stage			
7	1992	May-20	ASLV	SHAR Center	SROSS-C
8	1993	Sep-20	PSLV	SHAR Center	IRS-1E
		Failure - software implementation error			
9	1994	May-4	ASLV	SHAR Center	SROSS-C2
10		Oct-15	PSLV	SHAR Center	IRS-P2

SHAR Center is located at Sriharikota, India

SLV-Series General Description

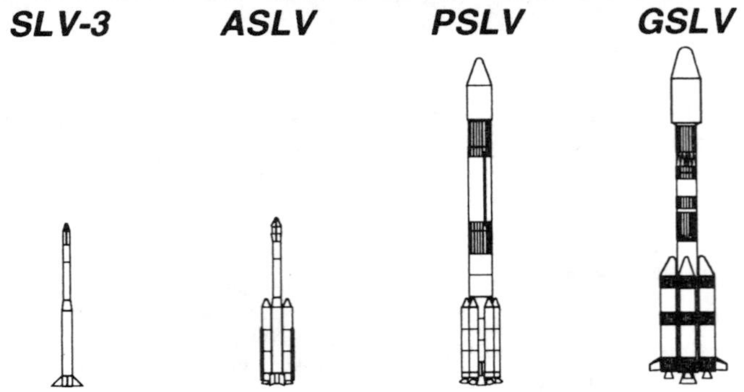

SLV-3 **ASLV** **PSLV** **GSLV**

Summary

Based on their desire to achieve self-reliance in space launch capability, India initiated development of four basic vehicles. The first, SLV-3, was a four-stage solid-propellant vehicle flown from 1979 to 1983. The second vehicle, ASLV, is basically an SLV-3 with two strap-ons. It failed on both of its two initial launch attempts, but has been redesigned and successfully tested twice. The third vehicle, the much larger PSLV, is for the launch of polar satellites. The first and third stages are solid propellant and the second ard fourth are liquid. The fourth vehicle, GSLV, will be a 5,500 lb (2,500 kg)-class vehicle for GTO missions. The GSLV has four liquid strap-ons, a solid first stage, a storable liquid second stage, and a cryogenic third stage.

Status

SLV-3 - out of production
ASLV - operational
PSLV - one development flight (successful)
GSLV - under development

Key Organizations

User & Manufacturer -
Indian Space Research Organization (ISRO)

Other Organizations -
Nearly 150 industries, national laboratories, and academic institutes

Vehicle

System Height
SLV-3 - 74.5 ft (22.7 m)
ASLV - 77.1 ft (23.5 m)
PSLV - 145 ft (44.2 m)
GSLV - 165 ft (50 m)

Payload Fairing Size
SLV-3 - 2.6 m (0.8 m) diameter
ASLV - 3.3 ft (1.0 m) diameter, 9.8 ft (3 m) length
PSLV - 10.5 ft (3.2 m) diameter, 27 ft (8.3 m) length
GSLV - 11.2 ft (3.4 m) diameter

Gross Mass
SLV-3 - 37,300 lb (16,900 kg)
ASLV - 90,400 lb (41,000 kg)
PSLV - 623,900 lb (283,000 kg)
GSLV - about 880,000 lb (400,000 kg)

Planned Enhancements
PSLV - up to 3,900 lb (1,300 kg) sun synchronous orbit
GSLV - up to 6,600 lb (3,000 kg) GTO

Operations

Primary Missions
SLV-3 - LEO
ASLV - LEO
PSLV - LEO, polar
GSLV - LEO, GTO, polar

Compatible Upper Stages
Fourth stage included

First Launch
SLV-3 - 1979
ASLV - 1987
PSLV - 1993
GSLV - 1997 (projected)

Success / Flight Total
SLV-3 - 3/4, ASLV - 2/4, PSLV - 1/2, GSLV - 0/0

Launch Site
SHAR Center at Sriharikota (13.9°N, 80.4°E)

Launch Azimuth
140° (max inclination range is 18°– 50°)

Nominal Flight Rate
1– 2 / yr

Planned Enhancements
—

Performance

216 nm (400 km) circ, 43°
SLV-3 - 90 lb (40 kg)
ASLV - 330 lb (150 kg)
PSLV - 6,400 lb (2,900 kg)
GSLV - 11,000 lb (5,000 kg)

486 nm (900 km) circ, 99°
PSLV - 2,200 lb (1,000 kg)

Geotransfer Orbit, 18°
PSLV - 990 lb (450 kg)
GSLV - 5,500 lb (2,500 kg)

Geosynchronous Orbit
N/A

Financial Status

Estimated Launch Price
Unknown

Manifest

	1995	1996	1997	1998
ASLV	—	—	—	—
PSLV	1	1	1	1
GSLV	—	—	1	—

Remarks

Launching PSLV from another launch site that has no launch azimuth restrictions would increase PSLV polar capability by 60%.

SLV-Series Vehicle

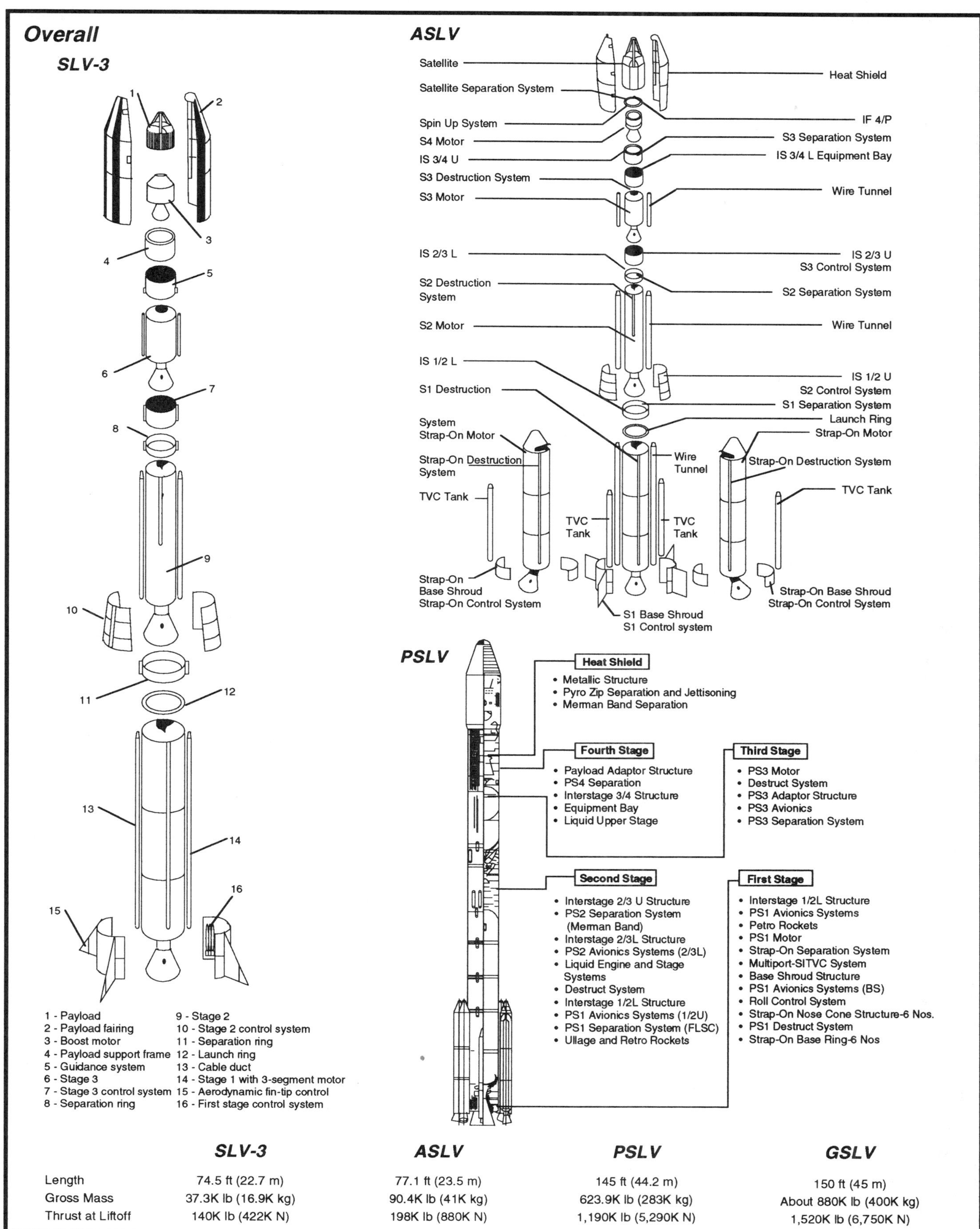

	SLV-3	ASLV	PSLV	GSLV
Length	74.5 ft (22.7 m)	77.1 ft (23.5 m)	145 ft (44.2 m)	150 ft (45 m)
Gross Mass	37.3K lb (16.9K kg)	90.4K lb (41K kg)	623.9K lb (283K kg)	About 880K lb (400K kg)
Thrust at Liftoff	140K lb (422K N)	198K lb (880K N)	1,190K lb (5,290K N)	1,520K lb (6,750K N)

SLV-Series Vehicle

(continued)

SLV-3 Stages

	Stage 1	Stage 2	Stage 3	Stage 4
Dimension:				
Length	33 ft (10 m)	21 ft (6.4 m)	7.5 ft (2.3 m)	4.9 ft (1.5 m)
Diameter	3.3 ft (1 m)	2.6 ft (0.80 m)	2.67 ft (0.815 m)	2.15 ft (0.657 m)
Mass: (each)				
Propellant Mass	19,100 lb (8,660 kg)	6,940 lb (3,150 kg)	2,340 lb (1,060 kg)	578 lb (262 kg)
Gross Mass	23,800 lb (10,800 kg)	10,800 lb (4,900 kg)	3,300 lb (1,500 kg)	795 lb (360 kg)
Structure:				
Type	Monocoque	Monocoque	Monocoque	Monocoque
Case Material	Steel	Steel	Steel	Steel
Propulsion:				
Propellant	PBAN	PBAN	HEF-20	HEF-20
Average Thrust (each)	95K lb (422K N) SL	60K lb (267K N) vac	20.4K lb (90.7K N) vac	6.03K lb (26.83K N) vac
Number of Motors	1	1	1	1
Number of Segments	3	1	1	1
Isp	253 sec vac	266.9 sec vac	276.9 sec vac	282.9 sec vac
Chamber Pressure	640 psia (44.1 bar)	555 psia (38.3 bar)	640 psia (44.1 bar)	426 psia (29.4 bar)
Expansion Ratio	6.7:1	14.2:1	25.7:1	30.5:1
Control - Pitch, Yaw, Roll	SITVC & movable fin tips	RCS	RCS	Spin stabilized
Events:				
Nominal Burn Time	49 sec	39.9 sec	45 sec	33 sec
Stage Shutdown	Burn to depletion	Burn to depletion	Burn to depletion	Burn to depletion
Stage Separation	Pyrotechnic charge	Pyrotechnic charge	Pyrotechnic charge	Spring ejection

Remarks:

The SLV-3 is a four-stage solid-propellant vehicle. Its major subsystems are four solid-propellant rocket motors to provide propulsive energy; the interstages connecting the forward skirt of one stage with the rear skirt of the next stage and housing control guidance electronics and pyro-subsystems, inertial guidance and control systems to steer the vehicle along a predetermined trajectory; and a heat shield to protect the fourth stage and the satellite from the aerodynamic heating during initial flight through atmosphere. The vehicle carries an instrumentation package to measure its performance and to monitor the flight events.

ASLV Stages

	Stage 0 (AS0)	Stage 1 (AS1)	Stage 2 (AS2)	Stage 3 (AS3)	Stage 4 (AS4)
Dimension:					
Length	36 ft (11 m)	33 ft (10 m)	20.8 ft (6.35 m)	8.0 ft (2.44 m)	4.6 ft (1.4 m)
Diameter	3.3 ft (1 m)	3.3 ft (1 m)	2.6 ft (0.8 m)	2.67 ft (0.815 m)	2.15 ft (0.655 m)
Mass: (each)					
Propellant Mass	19,040 lb (8,637 kg)	19,600 lb (8,900 kg)	7,050 lb (3,200 kg)	2,340 lb (1,060 kg)	700 lb (317 kg)
Gross Mass	25,575 lb (11,600 kg)	26,000 lb (11,800 kg)	9,700 lb (4,400 kg)	3,770 lb (1,710 kg)	1,130 lb (512 kg)
Structure:					
Type	Monocoque	Monocoque	Monocoque	Monocoque	Monocoque
Case Material	Steel	Steel	Steel	Graphite Epoxy	Graphite Epoxy
Propulsion:					
Propellant	HTPB	HTPB	HTPB	HEF-20	HEF-20
Average Thrust (each)	95K lb (422K N) SL	158K lb (702K N)vac	68.3K lb (304K N)vac	20.4K lb (90.7K N) vac	7.9K lb (35K N) vac
Number of Motors	2	1	1	1	1
Number of Segments	3	1	1	1	1
Isp	253 sec vac	259 sec vac	275.9 sec vac	277 sec vac	281 sec vac
Chamber Pressure	640 psia (44.1 bar)	640 psia (44.1 bar)	555 psia (38.3 bar)	640 psia (44.1 bar)	426 psia (29.4 bar)
Expansion Ratio	6.5:1	6.7:1	14.2:1	25.7:1	28.6:1
Control - Pitch,Yaw, Roll	SITVC Cold gas RCS	SITVC RCS & ACS	Bipropellant RCS	Monopropellant RCS	Spin stabilized
Events:					
Nominal Burn Time	49 sec	45 sec	36 sec	45 sec	33.4 sec
Stage Shutdown	Burn to depletion	Burn to depletion	Burn to depletion	Burn to depletion	Burn to depletion
Stage Separation	Pyrotechnic charge	Pyrotechnic charge	Pyrotechnic charge	Pyrotechnic charge	Spring ejection

Remarks:

ASLV is an augmented version of the SLV-3 vehicle. All the five stages of ASLV use solid propellants. The zero stage consists of a pair of booster motors strapped to the core vehicle to provide initial thrust. The four stages of the core vehicle are similar to the configuration of the SLV-3. Strap-on motor nozzles were redesigned to obtain a fixed 9 degree cant. To obtain reproducible and minimal dispersion in the strap-on motors, the mixed propellant slurry was equally distributed among the two pair motors. The propellant system of the first and second stage was improved to provide higher energetics by improving the solid loading. While retaining the same grain configuration, the aluminum content in the propellant was changed from 12% to 18%. This yielded an improvement in specific impulse of 9 sec. ASLV uses navigational computers and a closed-loop control guidance system.

SLV-Series Vehicle

(continued)

PSLV Stages	Strap-Ons (PSOM or S9)	Stage 1 (PS1 or S125)	Stage 2 (PS2 or L37.5)	Stage 3 (PS3 or S7)	Stage 4 (PS4 or L2)
Dimension:					
Length	33 ft (10 m)	66.6 ft (20.3 m)	37.7 ft (11.5 m)	11.5 ft (3.5 m)	8.5 ft (2.6 m)
Diameter	3.3 ft (1 m)	9.2 ft (2.8 m)	9.2 ft (2.8 m)	6.6 ft (2 m)	4.4 ft (1.34 m)
Mass: (each)					
Propellant Mass	19,700 lb (8,920 kg)	284,400 lb (129,000 kg)	82,700 lb (37,500 kg)	15,900 lb (7,200 kg)	4,400 lb (2,000 kg)
Gross Mass	24,100 lb (10,930 kg)	353,200 lb (160,200 kg)	94,800 lb (43,000 kg)	18,300 lb (8,300 kg)	6,400 lb (2,920 kg)
Structure:					
Type	Monocoque	Monocoque	Semimonocoque	Monocoque	Monocoque
Material	Steel	Steel	Aluminum	Kevlar	Titanium
Propulsion:					
Propellant	HTPB	HTPB	UDMH / N2O4	HTPB	MMH / N2O4
Average Thrust (each)	95K lb (422K N) SL ea.	1,090K lb (4,860K N) vac 806K lb (3,587K N) SL	163K lb (725K N) vac	73.9K lb (328.7K N) vac	1.7K lb (7.5K N) vac ea.
Engine Designation	—	—	Vikas (based on Viking 4)	—	LVS
Number of Engines	6 motors (3 segments ea.)	1 motor (5 segments)	1	1 motor (1 segment)	2
Isp	262 sec vac	264 sec vac	293 sec vac	291 sec vac	307 sec vac
Feed System	—	—	Gas Generator	—	Pressure-fed
Chamber Pressure	640 psia (44.1 bar)	853 psia (58.8 bar)	763 psia (52.6 bar)	876 psia (60.4 bar)	123 psia (8.5 bar)
Mixture Ratio (O/F)	—	—	1.87	—	1.4
Throttling Capability	—	—	No	—	No
Expansion Ratio	6.6:1	8:1	31:1	53:1	60:1
Restart Capability	No	No	No	No	Yes
Tank Pressurization	—	—	Helium pressurization	—	Pressurized helium
Control-Pitch,Yaw, Roll	Fixed cant, stg 1 control SITVC on two strap-ons	SITVC RCT, ACS	Hydraulic gimbaling (±4°) 2 hot-gas thrusters	Gimbaling nozzle (±2°) Stage 4 RCS	EMA 2 nozzles (±3°) Gimbaling + RCS Coast
Events:					
Nominal Burn Time	49 sec	97 sec	150 sec	75.2 sec	420 sec
Stage Shutdown	Burn to depletion	Burn to depletion	Predetermined velocity	Burn to depletion	Predetermined velocity
Stage Separation	Pyrotechnic charge	Retro-rockets	Retro-rockets	Retro-rockets	Spring ejection & RCS

Remarks:

The PSLV has four stages, the first and third stages having solid motors and second and fourth stages powered by liquid engines. The thrust of the first stage is augmented by six strap-on solid motors (PSOM) that are similar to the ASLV motors. To maximize the payload, the ignition sequence of PSOM's is designed such that two PSOMs are ignited along with the first stage at liftoff and other four PSOMs are ignited 29 seconds later.

The first stage (S125) of the PSLV is a five segment solid rocket motor of HTPB propellant and a composite nozzle. Each segment is 9.2 x 11 ft (2.8 x 3.4 m) long and joined by 144 pins. The central segments are interchangeable. The pitch and yaw control of the PSLV during the thrust phase of the solid motor is achieved by injection of aqueous solution of strontium perchlorate in the nozzle divergent at 35% of the length of the nozzle from throat to exit. The injectant (strontium perchlorate) and pressurant (nitrogen) are stored in two aluminum tanks strapped to the solid rocket motor.

The second stage (L37.5) is powered by a turbo-pump fed, film-radiation cooled engine. The engine used by ISRO, designated Vikas, is based on the Viking-IV engine technology of SEP, France. The Vikas differs from the Viking mainly in its larger propellant load and longer burn time. The propellants, nitrogen tetroxide (N204) and unsymmetrical dimethyl hydrazine (UDMH), are fed to the combustion chamber by pumps mounted on a single shaft rotating at 9,400 rpm. The turbine is driven by hot gases produced in the gas generator by the combustion of UDMH and N204 and cooled by water spray. The engine has gimballing capability in two planes and a radiatively cooled nozzle. The propellant tank is of a common bulkhead construction with slosh suppressors. The thrust frame and the structural elements connecting this stage to other stages of the launch vehicle are basically of sheet stringer construction. Apart from the engine, this stage has a number of subsystems which include a pressurization system to ensure minimum specified pressures in the tanks during flight, command system to provide start and stop command to the engine and reference pressures for actuating regulators and valves, fill and drain system for loading the required quantity of propellants, pogo corrector system to suppress low frequency longitudinal feedline oscillations during flight, gimbal control system for attitude control of the stage, and roll control systems with thrusters which use hot-gas bled from gas generator to develop 67 lb (300 N) thrust.

The third stage (S7) is a solid motor that is located inside an aluminum skin stringer skirt and is equipped with a flex bearing seal thrust vector control system. The motor design provides an option for off-loading propellant.

The fourth stage (L2) is used as the terminal stage of the PSLV to provide guided injection. This high performance stage is powered by two identical pressure fed engines using monomethyl hydrazine (MMH) and nitrogen tetroxide (N204) as propellants. The propellants are stored in the two compartments of a titanium alloy tank separated by common bulkhead. The tank is designed for an operating pressure of 282 psia (19.5 bar). The tank has Propellant Acquisition System (PAS) with catch tanks to ensure propellant supply for starting the engine under adverse 'g' conditions. The PAS is of surface tension type with an outer perforated cylinder provided with appropriate mesh.

The control power plants, separation systems, instrumentation, and stage electronics are housed in the interstages. The Vehicle Equipment Bay (VEB) is placed on the fourth stage alongside the propellant tank and houses the navigation, guidance, and control systems.

SLV-Series Vehicle

(continued)

GSLV Stages	Stage 0 (L-40) (GS0)	Stage 1 (S-125) (GS1)	Stage 2 (L-375) (GS2)	Stage 3 (CS) (GS3)
Dimension:				
Length	65 ft (19.7 m)	66 ft (20 m)	38 ft (11.5 m)	28.6 ft (8.72 m)
Diameter	6.9 ft (2.1 m)	9.2 ft (2.8 m)	9.2 ft (2.8 m)	9.2 ft (2.8 m)
Mass: (each)				
Propellant Mass	4 x 88,200 lb (4 x 40,000 kg)	284,400 lb (129,000 kg)	82,700 lb (37,500 kg)	27,600 lb (12,500 kg)
Gross Mass	4 x 100,300 lb (4 x 45,500 kg)	344,000 lb (156,000 kg)	95,000 lb (43,000 kg)	32,200 lb (14,600 kg)
Structure:				
Type	Skin stringer ?	Monocoque	Skin stringer ?	Skin stringer ?
Material	Aluminum	Steel	Aluminum	Aluminum ?
Propulsion:				
Propellant	UDMH / N2O4	HTPB	UDMH / N2O4	LOX / LH2
Average Thrust (each)	165K lb (735K N) vac ?	1090K lb (4860K N) vac	165K lb (725K N) vac	17K lb (75K N) vac ?
Engine Designation	Vikas	—	Vikas	—
Number of Engines	1 / booster (canted)	1 motor (5 segments)	1	1
Isp	280 sec vac	264 sec vac	293 sec vac	450 sec vac
Feed System	Gas Generator	—	Gas Generator	Gas Generator
Chamber Pressure	763 psia (52.6 bar)	853 psia (58.8 bar)	763 psia (52.6 bar)	?? psia (?? bar)
Mixture Ratio (O/F)	1.7	—	1.87	??
Throttling Capability	No	—	No	Yes
Expansion Ratio	13.9:1	8:1	31:1	198:1
Restart Capability	No	No	No	Yes
Tank Pressurization	Helium	—	Helium pressurization	??
Control-Pitch, Yaw,	Gimbaling (±5°)	SITVC	Hydraulic gimbaling (±4°)	Gimbaled engine
Roll		RCT	2 hot gas thrusters	and cold gas
Events:				
Nominal Burn Time	158 sec	93 sec	150 sec	800 sec
Stage Shutdown	Burn to depletion	Burn to depletion	Predetermined velocity	Predetermined velocity
Stage Separation	Retro-rockets	Retro-rockets	Retro-rockets	Spring ejection & RCS

Remarks:

GSLV is basically derived from PSLV by replacing the six solid strap-ons with four liquid strap-ons derived from the PSLV second stage and replacing the upper two stages by a single cryogenic stage. Current plans are to use a Russian cryogenic stage and eventually replace it with a domestically developed cryogenic engine.

SLV-Series Vehicle

(continued)

Payload Fairing

	SLV-3	ASLV	PSLV	GSLV
Length	7.2 ft (2.2 m)	10.5 ft (3.2 m)	27 ft (8.3 m)	25.6 ft (7.8 m)
Diameter	2.6 ft (0.8 m)	3.3 ft (1.0 m)	10.5 ft (3.2 m)	11.2 ft (3.4 m)
Mass	?? lb (?? kg)	330.7 lb (150 kg)	2,425 lb (1,100 kg)	2,760 lb (1,250 kg)
Sections	2	2	2	2
Structure	Skin stringer	Skin stringer	Skin stringer	Skin stringer ?
Material	Phenolic glass	Aluminum alloy	Aluminum alloy	Aluminum alloy
Remarks	The payload fairings are two half-shell structures with a longitudinal separation system.			

Avionics

ASLV, PSLV, and GSLV vehicles use an inertial guidance system. A redundant Strapdown Inertial Navigation System (RESINS) using Dry Tuned Gyros (DTG) and Servo Accelerometers (SA) is used for PSLV. RESINS is designed with three DTGs in a skewed configuration and four servo accelerometers positioned along the orthogonal axes, with a redundant unit along the thrust axis. A Navigation Processor (NGP) carries out strap-down navigation computation. Navigation software includes preprocessing of sensor signals (fixed drift/bias/scale factor compensation); Failure Detection and Isolation (FDI) logic; transformation from body coordinates to inertial coordinates; and generation of attitude, velocity, and position information.

SLV-3 used an inertial measurement system.

Attitude Control System

SLV-3. Stage 1 is controlled by secondary injection thrust vector control (SITVC) and movable fin tips, stage 2 by a reaction control system (RCS) using inhibited red fuming nitric acid (IRFNA) and hydrazine propellant, stage 3 by a monopropellant RCS, and stage 4 by spin stabilization.

ASLV. Stage 0 control is provided by SITVC for pitch and yaw and cold RCS for roll. Stage 1 and 3 use a monopropellant RCS for roll control and stage 2 a bipropellant. Stage 4 is spin stabilized.

PSLV. Stage 1 of the vehicle is controlled by an omni-axis multiport Secondary Injection Thrust Vector Control (SITVC) system in the pitch and yaw axes. Strontium perchlorate, used as the injectatant, is stored in two tanks each of 27.6 in (700 mm) diameter, strapped on either side of the booster, in the pitch plane. Twenty four injection valves are mounted on the Stage 1 nozzle, through which the strontium perchlorate is injected. The valves are actuated by electromechanical actuators. The power plant is capable of producing 4.5K lb (20 KN) side force. The roll control of Stage 1 is achieved by means of two bipropellant thrusters that swivel of 1,400 lb (6,400 N) each, mounted below the SITVC tanks. Nitrogen textroxide and monomethyl hydrazine are used as propellants. The thrusters burn continuously during Stage 1 flight. Electromechanical actuators operating in proportional mode deflect the thrusters in opposite directions to produce the required roll control moments. Additional roll control is provided by SITVC in two strap-ons. The pitch and yaw control of Stage 2 is provided by gimballing the engine using an electrohydraulic actuator system. The actuators are mounted at an angle of 45 degrees to the pitch and yaw planes, and depending on the errors, the engine is deflected in proportional mode in the required direction. The roll control is accomplished by means of two hot-gas on-off thrusters mounted tangentially and back to back, on the interstage. The hot gas drawn from the gas generator is expanded through the hot-gas thrusters to produce the roll control moment.

A flex nozzle system is used to control Stage 3 in the pitch and yaw axes. The nozzle is connected to the motor by means of a flex bearing. Electromechanical actuators deflect it in proportional mode during the thrusting phase. Roll control is provided by the Reaction Control System (RCS) of Stage 4. During the coast phase flight, following Stage 3 burn, the Stage 4 RCS provides control in all three axes.

During the Stage 4 thrusting phase, three-axis control is achieved by gimballing the engines. Each engine is supported with the stage structure through a universal gimbal mount for enabling two plane gimballing with maximum deflection of +3 degrees in each axis. The RCS, consisting of two modules with three thrusters in each module, provides the attitude control during the coast phase. Each thruster produces a thrust of 11 lb (50 N) and uses MMH/N204 propellants drawn from the main propellant tank.

GSLV. Unknown.

SLV-Series Performance

PSLV

Launches from SHAR Center at 43° Inclination

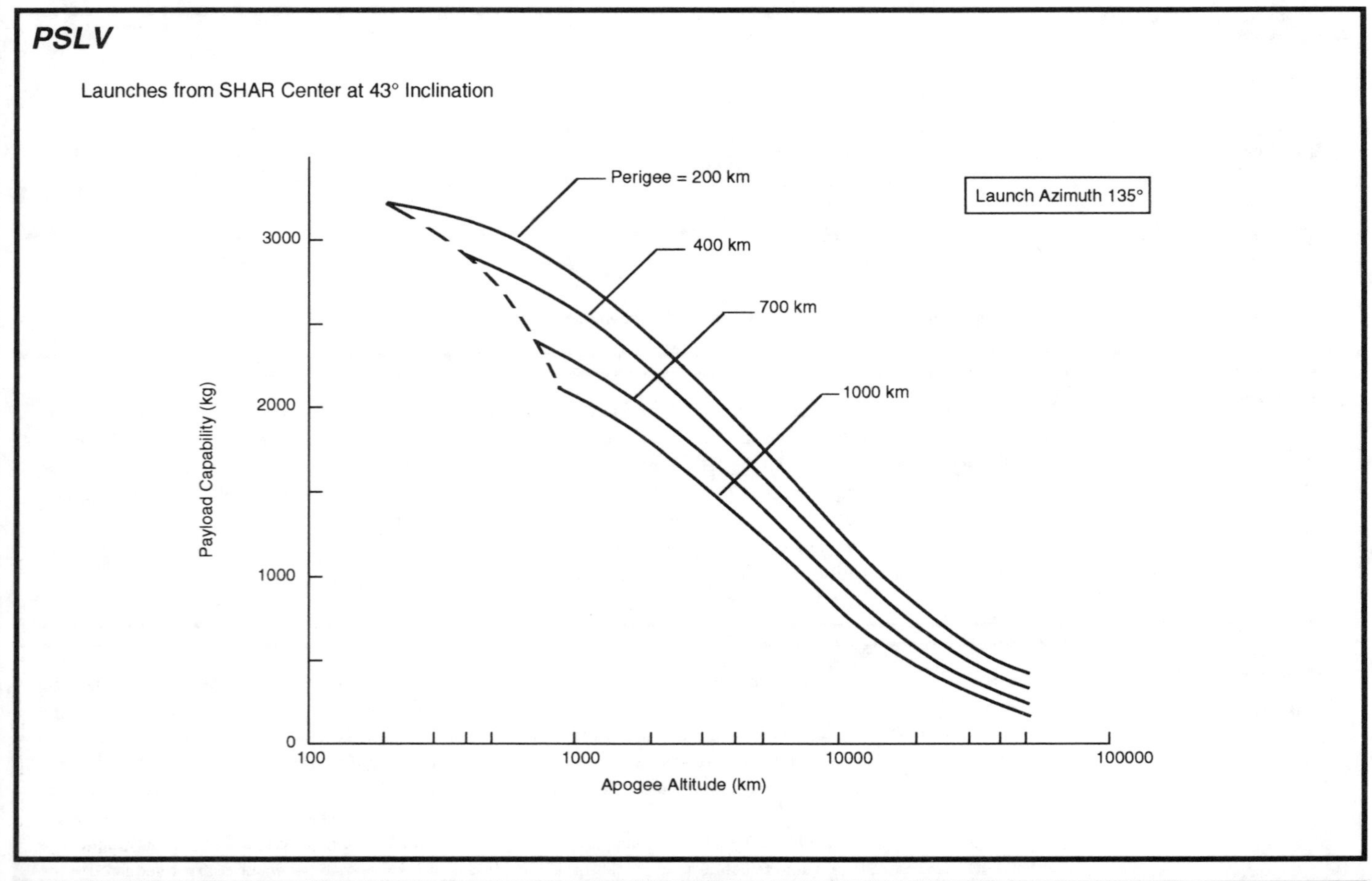

SLV-3, ASLV, GSLV

Performance Curves Not Available.

SLV-Series Operations

Launch Site

India
Equator

Bangalore
Space Commission
Department of Space
ISRO Headquarters
Trivandrum
Vikram Sarabhai Space Center
Liquid Propulsion Systems Center
ISRO Inertial Systems Unit
Balasore
Meteorological Rocket Station
Sriharikota
SHAR Center (13.9°N)
Mahendragiri
Liquid Propulsion Test Facilities

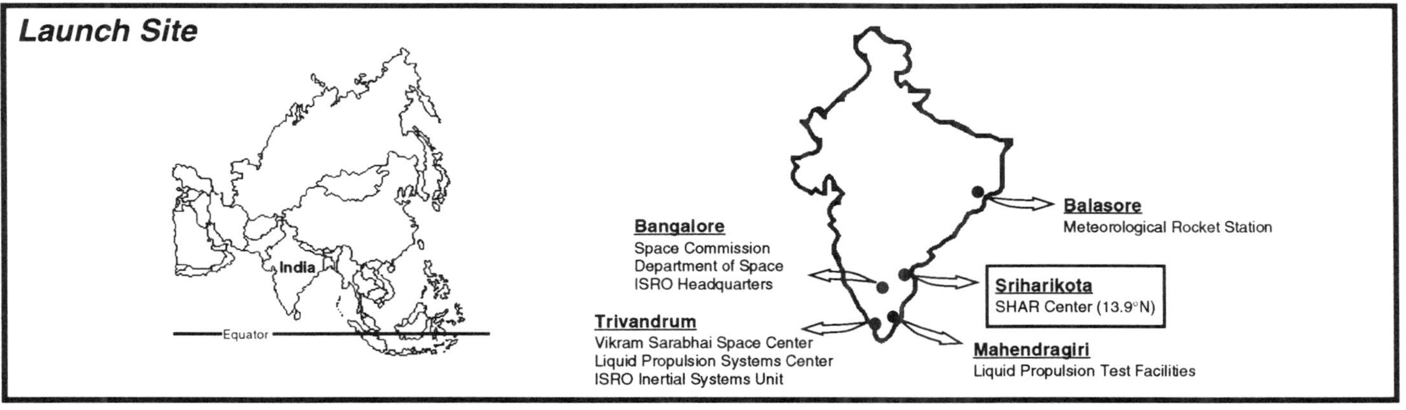

Launch Facilities

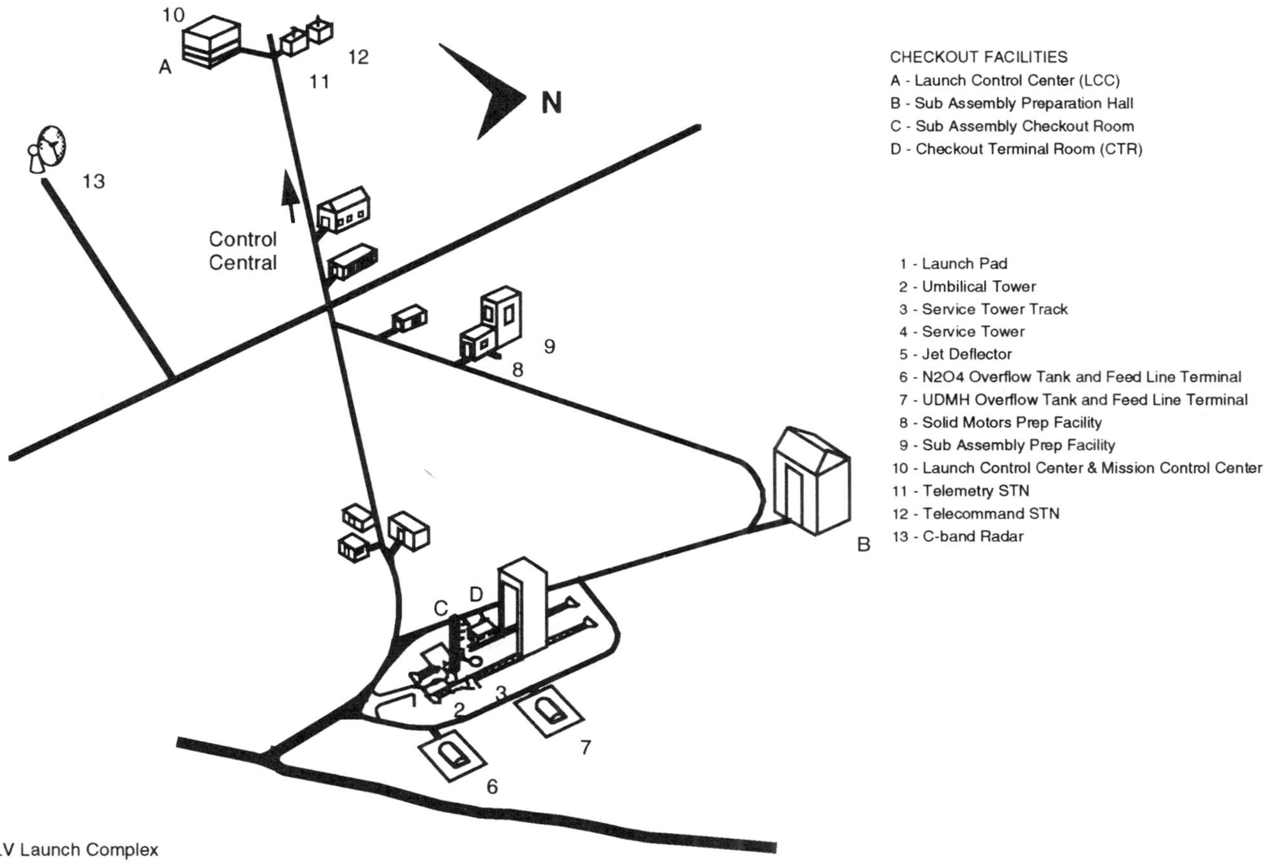

N
Control Central
A
B
C
D
CHECKOUT FACILITIES
A - Launch Control Center (LCC)
B - Sub Assembly Preparation Hall
C - Sub Assembly Checkout Room
D - Checkout Terminal Room (CTR)
1 - Launch Pad
2 - Umbilical Tower
3 - Service Tower Track
4 - Service Tower
5 - Jet Deflector
6 - N2O4 Overflow Tank and Feed Line Terminal
7 - UDMH Overflow Tank and Feed Line Terminal
8 - Solid Motors Prep Facility
9 - Sub Assembly Prep Facility
10 - Launch Control Center & Mission Control Center
11 - Telemetry STN
12 - Telecommand STN
13 - C-band Radar

PSLV Launch Complex

Launch Processing

SHAR Center. Sriharikota Range (SHAR) Center is located in southeast India in the Bay of Bengal, and encompasses the Sriharikota Island on the east coast of Andhra Pradesh. It is operated by ISRO. In addition to the SHAR launch facilities the complex includes the Thumba Equatorial Rocket Launching Station and the Balasore Rocket Launching Station much further north. The SHAR Center supports 140 Yanadi families who live on the island. India's first successful SLV-3 launch was carried out here on 18 July 1980.

The location of SHAR Center on the southeastern coast of India imposes severe launch window and range safety constraints on launches into polar orbits. A north or south polar launch from Sriharikota is impossible since heavily populated areas (India and Sri Lanka) lie in those directions. Therefore, launch azimuth is limited to 140°, requiring polar-orbit missions to be launched in a southwesterly direction, followed by an energy-intensive 55° yaw maneuver. Upgrades at the Balasore Meteorological Rocket Station, near Calcutta, has been studied as a new site that will eliminate most of these constraints and increase the PSLV's polar sun synchronous orbit capacity to a 99° inclination from 2,200 lb (1,000 kg) to 3,500 lb (1,600 kg).

SLV-Series Operations
(continued)

Launch Processing
(continued)

SLV Processing. The SLV-3 is processed by horizontal transfer of the vehicle to the pad and vertical erection on the pad. The ASLV is similar except the two strap-on boosters are integrated on the pad. Launch pad facilities include vertical assembly and integration tower. The PSLV will be integrated on the pad. A mobile structure has been built to provide for the vehicle vertical assembly and integration. Facilities for GSLV are still to be determined.

Mission Support. The ISRO Telemetry, Tracking, and Telecommand network (ISTRAC) comprises stations at SHAR, Trivandrum, Ahmadabad, Bangalore, and Lucknow. SHAR station has telemetry (TM) receivers, a dual radar tracking facility (TR), as well as a telecommand facility (TC) which is required for vehicle destruction in case of malfunctioning of the onboard system that could result in an unacceptable deviation in the flight path of the vehicle. The stations at Trivandrum, Bangalore, Ahmedabad, and Lucknow are equipped with S-band TM and ranging facilities.

For PSLV, the flight is monitored from SHAR ground stations from liftoff until the end of the long coast phase after third stage separation when loss of signal occurs. The vehicle is tracked in S-band from Trivandrum ground station from about T-130 sec. until 100 sec after ignition of fourth stage, thus providing overlapping coverage. The entire fourth stage flight is monitored by the Down Range Station (DRSN) at Mauritius. Further, the tracking of the fourth stage stage and the spacecraft until injection and of the separated fourth stage with VEB for about 200 sec enables DRSN to provide look angle predictions from the S-band telemetry stations for tracking the spacecraft from foreign ground stations and also from SHAR and other Indian Stations in the 6/7th orbit. Preliminary Orbit Determination (POD) is also carried out during this phase for declaring the orbit, thus completing the mission.

Flight Sequence

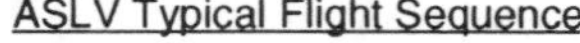

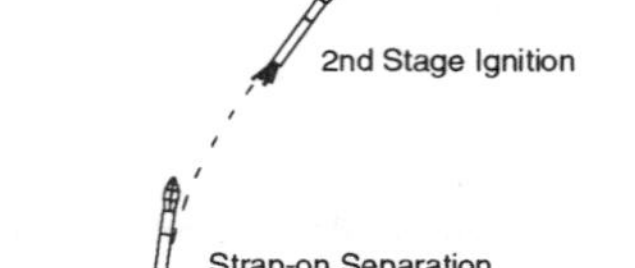

SLV-3 LEO Mission

Flight Time (min:sec)	Events
00:00	1st stage ignition
01:12	1st stage separation
03:10	2nd stage separation
06:31	3rd stage separation
12:06	4th stage separation

ASLV LEO Mission

Flight Time (min:sec)	Events
00:00	Strap-on booster ignition and liftoff
00:49.5	1st stage ignition
00:52	Strap-on booster separation
01:44.5	2nd stage ignition
02:37.5	3rd stage ignition
03:24.5	3rd stage burnout
08:48	4th stage ignition
09:22.5	Injection of satellite

PSLV Typical Flight Sequence

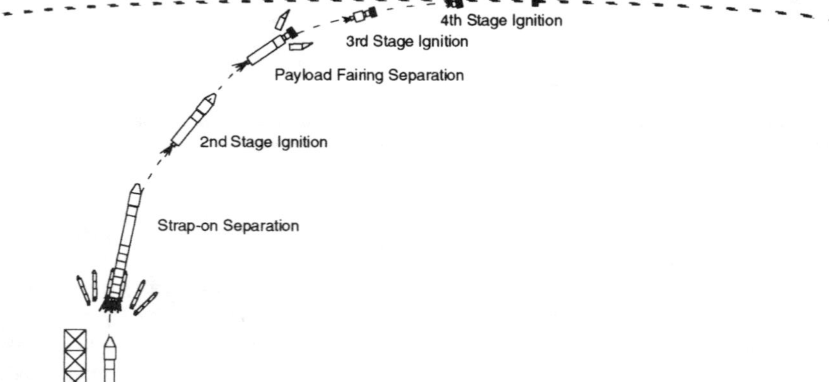

PSLV Polar Sun-Synchronous Mission

Flight Time (min:sec)	Events
00:00	Two strap-on motors and 1st stage ignition
00:29	Four remaining strap-on motors ignition
00:52	First set of strap-on motors burnout
01:17	Second set of strap-on motors burnout
01:33	1st stage burnout and separation 2nd stage ignition
02:10.2	Payload fairing separation
04:06.2	2nd stage burnout and separation
04:10	3rd stage ignition
05:42.2	3rd stage burnout
11:10	4th stage ignition
17:38	4th stage burnout Satellite injection

GSLV Flight Sequence Not Available

SLV-Series Payload Accommodations

	ASLV	PSLV
Payload Compartment		
Maximum Payload Diameter	31.5 in (800 mm)	114.2 in (2,900 mm)
Maximum Cylinder Length	?? in (?? mm)	107.9 in (2,740 mm)
Maximum Cone Length	?? in (?? mm)	126.0 in (3,200 mm)
Payload Adapter		
Interface Diameter	?? in (?? mm)	?? in (?? mm)
Payload Integration		
Nominal Mission Schedule Begins	T-?? months	T-?? months
Launch Window		
Latest Countdown Hold Not Requiring Recycling	T-?? min	T-?? min
On-Pad Storage Capability	?? hours for a fueled vehicle	?? hours for a fueled vehicle
Latest Access to Payload	T-?? hours or T-?? hours through access doors	T-?? hours or T-?? hours through access doors
Environment		
Maximum Load Factors	+9.5 g axial, ±?? g lateral at stage 2 separation	+?? g axial, ±?? g lateral
Minimum Lateral / Longitudinal Payload Frequency	?? Hz / ?? Hz	?? Hz / ?? Hz
Maximum Overall Acoustic Level	?? dB (full octave)	?? dB (full octave)
Maximum Flight Shock	?? g from ??–?? hz	?? g from ??–?? hz
Maximum Dynamic Pressure on Fairing	?? lb/ft^2 (?? N/m^2)	1,670 lb/ft^2 (80,000 N/m^2)
Maximum Pressure Change in Fairing	?? psi/s (?? KPa/s)	?? psi/s (?? KPa/s)
Cleanliness Level in Fairing (Prior to Launch)	Class ??	Class ??
Payload Delivery		
Standard Orbit and Accuracy (3 sigma)	±?? nmi (?? km), ±?? deg	Polar orbit: ±7.3 nmi (13.5 km), ±0.08 deg
Attitude Accuracy (3 sigma)	All axes: ±?? deg, ±?? deg/sec	All axes: ±?? deg, ±?? deg/sec
Nominal Payload Separation Rate	?? ft/s (?? m/s)	?? ft/s (?? m/s)
Deployment Rotation Rate Available	?? rpm	?? rpm
Loiter Duration in Orbit	?? hrs	?? hrs
Maneuvers (Thermal / Collision / Telemetry)	??	??

SLV-Series Notes

Publications

Technical Publications

"PSLV-D1 Mission", U. R. Rao, S. C. Gupta, G. Madhavan Nair and D. Narayana Moorthi, ISRO Headquarters, Current Science, Vol. 65, No. 7, 10 October 1993.

"Mission Studies to Reduce Aerodynamic Loads for the Polar Satellite Launch Vehicle", K. Sivan, Sudhakara K. Rao, Vikram Sarabhai Space Center, IAF-90-158, 41st International Astronautical Federation, Dresden, GDR, October 6-12, 1990.

"Some Recent Developments in Liquid Propulsion Systems", A. E. Muthunayagam, ISRO Liquid Propulsion Systems Center, IAF-90-241, 41st International Astronautical Federation, Dresden, GDR, October 6-12, 1990.

"Indian Space Program Budget", Dr. A. Bhatra, ISRO Representative at the Indian Embassy, presentation to the AIAA International Activities Committee, 1989.

"India's Mixed Bag of Launcher's", Hormuz P. Mama, Stephane Chenard, Space Markets, pg. 255-258, 4 qtr 1989.

"Compendium of Small Class ELV Capabilities, Costs and Constraints", Karen S. Poniatowski, NASA Headquarters, Washington, D.C., 1989.

TRW Space Log, 1957-1987, 1988, 1989 TRW, El Segundo, CA.

Launch Vehicle Catalogue, European Space Agency, December 1989.

"ISRO's Solid Rocket Motors" R. Nagappa, M.R. Kurup, A.E. Muthunayagam, ISRO, IAF-88-232, 39th Congress of the International Astronautical Federation, Bangalore, India, October, 8-15, 1988.

"The Polar Satellite Launch Vehicle and Mission for the Indian Remote Sensing Spacecraft", M. Anjaneyulu, K. Sudhakara Rao, S. Srinivasan, Vikram Sarabhai Space Centre, IAF-88-168, 39th Congress of the International Astronautical Federation, Bangalore, India, October, 8-15, 1988.

Jane's Spaceflight Directory 1987, Jane's Publishing Inc., New York, NY.

Foreign Space Launch Vehicle Performance Document, Battelle Columbus Laboratories, December 30, 1983.

Acronyms

AP - ammonium perchlorate
ASLV - Augmented Satellite Launch Vehicle
DOS - Department of Space
EMA - electromechanical actuators
GSLV - Geostationary Satellite Launch Vehicle
GTO - geosynchronous transfer orbit
HEF - ??
HTPB - hydroxy terminated polybutadiene
IRFNA - inhibited red fuming nitric acid
ISRO - Indian Space Research Organization
LEO - low Earth orbit
LH2 - liquid hydrogen
LOX - liquid oxygen
MMH - monomethyl hydrazine
N/A - not applicable
N2O4 - nitrogen tetroxide
PAS - Propellant Acquisition System
PBAN - polybutadiene acrylonitrile acrylic acid
PSLV - Polar Satellite Launch Vehicle
PSOM - PSLV Strap-On Motors
RCS - reaction control system
SHAR - Sriharikota Range
SITVC - secondary injection thrust vector control
SLV - Satellite Launch Vehicle
SROSS - Stretched Rohini Satellite System
UDMH - unsymmetric dimethylhydrazine
VEB - vehicle equipment bay
VSSC - Vikram Sarabhai Space Center

Other Notes—SLV-Series Growth Possibilities

PSLV was designed for future growth with the addition of modules and minor modifications to the systems. The first stage, now comprised of five segments can be augmented by one more segment, thus uprating the thrust of the booster. Instead of having six small strap-on motors, two large strap-on motors can be attached to the core as zero stage to the vehicle. The second stage propellant loading can be augmented by using a larger tank, which improves the payload capability substantially. The upper stages can be replaced by a cryogenic stage yielding a substantial increase in the payload. It is this capability of PSLV and its propulsion modules that will enable ISRO to develop the GSLV.

Vehicle Configuration	Payload in GTO 108 x 19,400 nm (200 x 36,000 km)
2S150 + S150 + L37.5 + S7 *	5,400 lb (2,450 kg)
2S150 + S150 + L55 + S7 *	5,600 lb (2,550 kg)
(8S9 + S125) + L37.5 + C14	4,000 lb (1,810 kg)
(8S9 + S150) + L37.5 + C14	4,960 lb (2,250 kg)
2S150 + S150 + L37.5 + C14	11,000 lb (5,000 kg)
2S150 + S150 + L55 + C25	12,800 lb (5,800 kg)

Note: S, L, and C stand for solid, liquid, cryogenic stages, respectively. The number before the letter indicates the number of motors. The number after the letter indicates propellant mass in metric tons.

Israel

Table of Contents

Shavit

Government Point of Contact:
Israel Space Agency
23 Arania St., P.O. Box 7182
Tel Aviv 61070 ISRAEL
Phone : 972-3-260226

Industry Point of Contact:
Israel Aircraft Industries
Electronics Division
P.O. Box 45
Beer Yakou 70350 ISRAEL
Phone : 972-8-272333

Second launch of the Offeq spacecraft on the Shavit launch vehicle on 3 March 1990.

Historical Summary

The Shavit launch vehicle was first launched on 19 September 1988, allowing Israel to become the eighth country to attain a domestic space launch capability. The Offeq-1 payload, which translates to Horizon, was a small technology demonstration satellite designed to give industry experience in developing space hardware. The payload's mission is to gather information on space environment conditions and the Earth's magnetic field. The payload, weighing 340 lb (155 kg), was launched into a 81 nm (250 km) by 620 nm (1150 km) orbit at a 142.9 degree inclination. On 2 April 1990, Offeq-2, a similar payload which weighed 350 lb (160 kg), was launched into a 113 nm (210 km) by 810 nm (1500 km) orbit at a inclination of 143 degrees.

Shavit, which translates to Comet, is a three-stage solid-propellant vehicle. It is a modified version of the Jericho II intermediate range ballistic missile. The vehicle is launched westward from the Negev desert into a retrograde orbit to prevent pieces of the booster from falling on Arab territory. The Shavit is the responsibility of the Israel Space Agency and is built and operated by Israel Aircraft Industries.

Launch Record

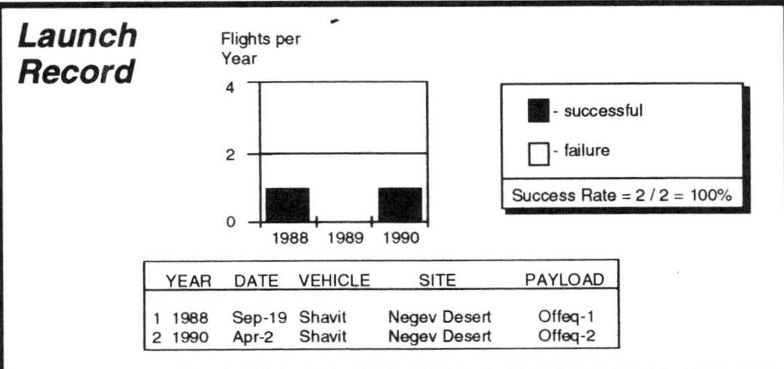

	YEAR	DATE	VEHICLE	SITE	PAYLOAD
1	1988	Sep-19	Shavit	Negev Desert	Offeq-1
2	1990	Apr-2	Shavit	Negev Desert	Offeq-2

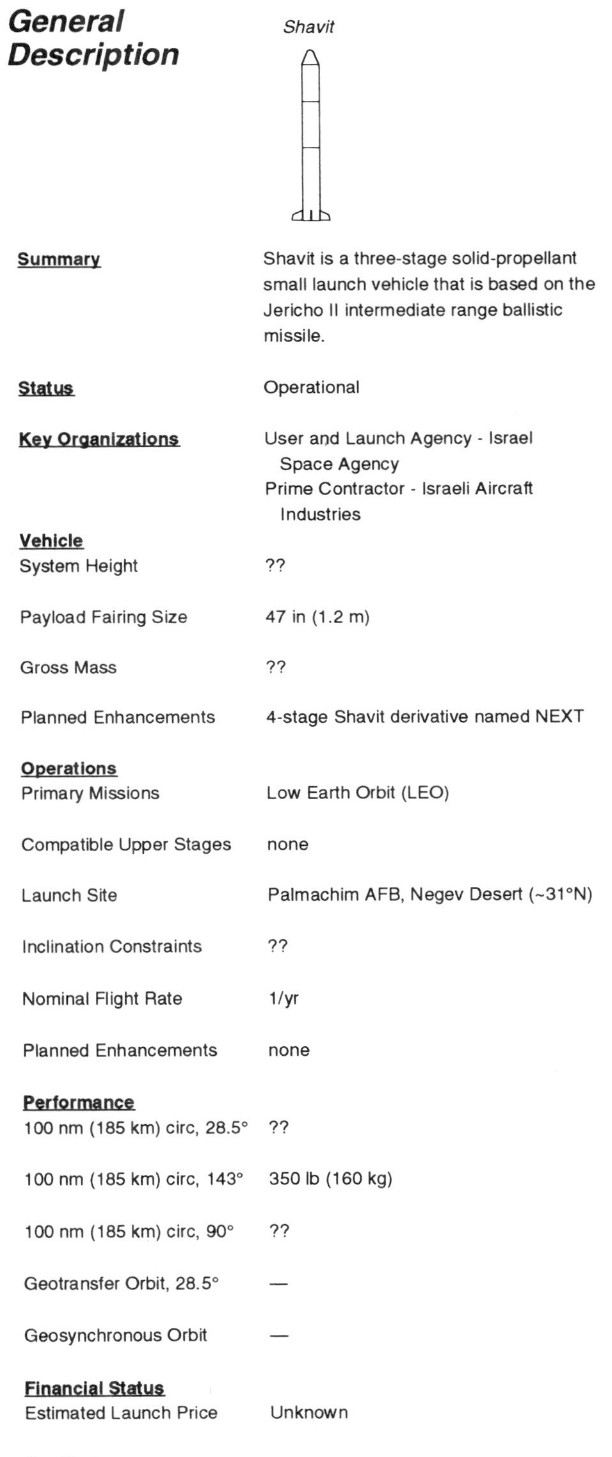

General Description

Summary	Shavit is a three-stage solid-propellant small launch vehicle that is based on the Jericho II intermediate range ballistic missile.
Status	Operational
Key Organizations	User and Launch Agency - Israel Space Agency Prime Contractor - Israeli Aircraft Industries
Vehicle	
System Height	??
Payload Fairing Size	47 in (1.2 m)
Gross Mass	??
Planned Enhancements	4-stage Shavit derivative named NEXT
Operations	
Primary Missions	Low Earth Orbit (LEO)
Compatible Upper Stages	none
Launch Site	Palmachim AFB, Negev Desert (~31°N)
Inclination Constraints	??
Nominal Flight Rate	1/yr
Planned Enhancements	none
Performance	
100 nm (185 km) circ, 28.5°	??
100 nm (185 km) circ, 143°	350 lb (160 kg)
100 nm (185 km) circ, 90°	??
Geotransfer Orbit, 28.5°	—
Geosynchronous Orbit	—
Financial Status	
Estimated Launch Price	Unknown
Manifest	—
Remarks	—

Shavit

Operations

Palmachim AFB
Negev Desert
About 31°N

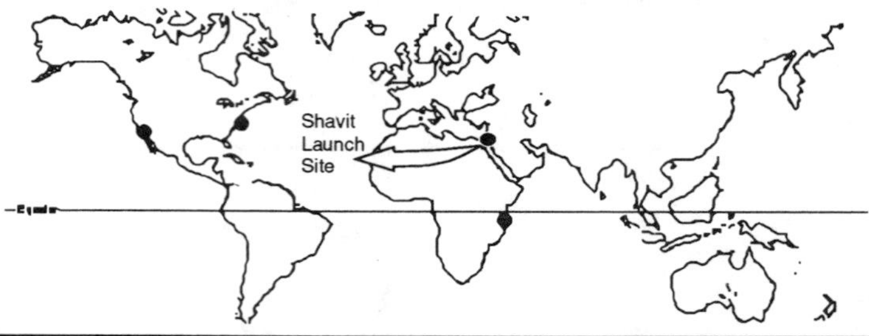

Publications

"Israel Orbits Offeq-2 Spacecraft", Aviation Week & Space Technology, pg. 20, April 9, 1990.

"Israel's Emerging Ability Demonstrated With Offeq-2 Launch", Space News, pg. 3, 21, April 9-16, 1990.

"Israeli Lightsat Is Only First Step", Space Business News, pg. 7-8, October 3, 1988.

"Israeli Satellite Launch Sparks Concern About Middle East Missile Buildup", Aviation Week & Space Technology, pg. 21, September 26, 1988.

TRW Space Log 1988, TRW.

Other Notes—Shavit Growth Possibilities

Israel Aircraft Industries (IAI) is proposing the NEXT launch system to provide a low-cost launch service capability for small satellites. The NEXT launch system is an adaptation of the existing IAI developed Shavit launch vehicle system. A substantial amount of hardware and software of NEXT has been flight proven in the Shavit program.

The four-stage NEXT vehicle is composed of three propulsion stages using solid rocket motors and one upper propulsion module using a bipropellant thruster. Two identical solid rocket motors (with exception of the expansion ratio of the nozzles) are proposed for the first two stages.

The three rocket-motor casings are wound from graphite fiber in an epoxy resin matrix and are an essential part of the launch vehicle structure. The interstages are built of aluminum skin and stiffeners. The bipropellant module structure is based on graphite epoxy composite. The different stages are joined by external clamps and the flight staging is carried out by activating pyrotechnic bolts.

The avionics system is composed of two identical and independent systems in order to achieve redundancy; the avionic units of each of the avionic systems are interconnected by a MIL-STD-1553B dual redundant communication bus. Every avionic system has its own pyrotechnical system, which is redundant by itself.

The navigation system includes one low-cost strapdown platform (SDP), an Inertial Navigation Unit (INU), and a global positioning system (GPS) receiver. The SDP provides the guidance and control system with the data needed for navigation and control. The gimbaled platform (INU) is located in stage-II and used for initial alignment and for accurate navigation during the two first stages. The GPS unit is located in stage-4 and takes over the INU role after stage two jettison.

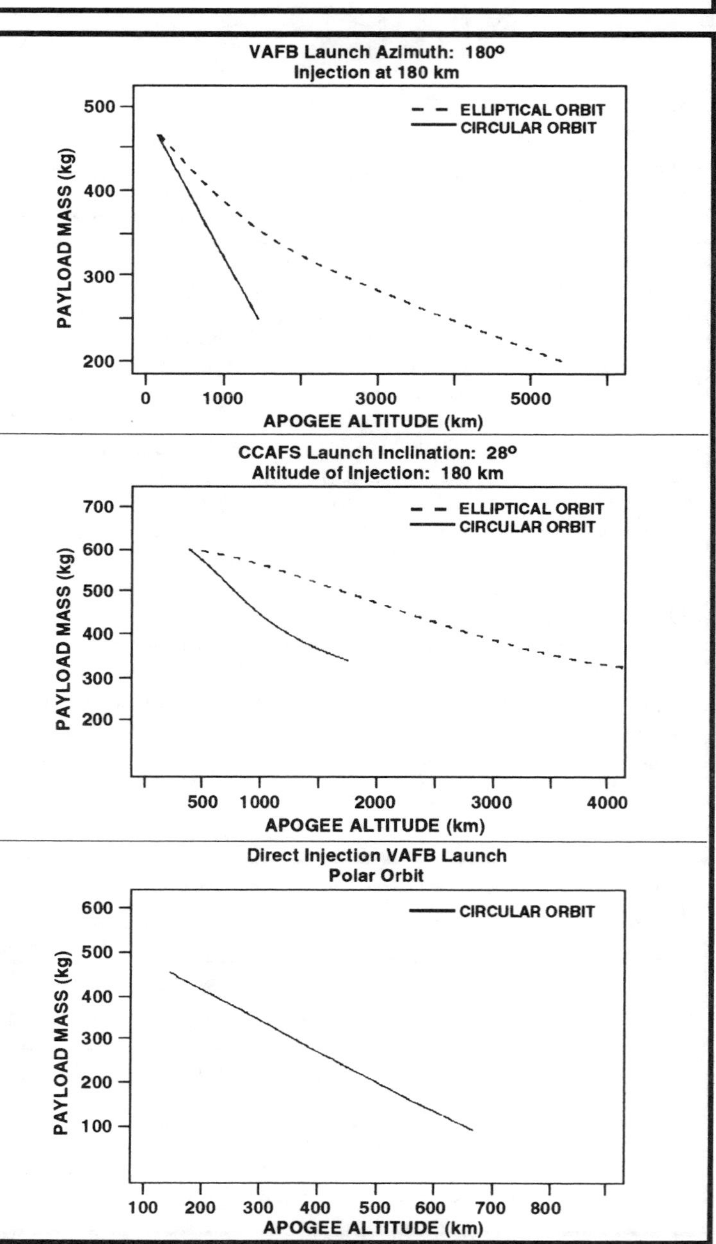

Japan

Table of Contents

H-Vehicle

Government Point of Contact:
National Space Development
Agency of Japan (NASDA)
World Trade Center Building
2-4-1, Hamamatsu-cho,
Minato-ku, Tokyo 105-60, JAPAN
Phone: (3) 5470-4111
Fax: (3) 3433-0796
or
NASDA Washington Office
1301 K Street, NW, Suite 560-E
Washington, DC 20005, U.S.A.
Phone: (202) 333-6844
Fax: (202) 333-6845

Industry Point of Contact:
Rocket System Corporation
Hamamatsucho Central Bldg. 4F
1-29-6 Hamamatsu-cho,
Minato-ku, Tokyo 105, JAPAN
Phone: (3) 212-3111
Fax: (3) 5470-7950

Mitsubishi Heavy Industries Ltd.
Space Systems Department
2-5-1 Marunouchi,
Chiyoda-ku, Tokyo 100, JAPAN
Phone: (3) 3212-3111
Fax: (3) 3212-9869

The first flight of H-2 was conducted successfully on 4 February 1994, launching the Vehicle Evaluation Payload (VEP) and the Orbital Reentry Experiment (OREX) from the new Yoshinobu launch site in Tanegashima Space Center.

H-Vehicle History

Out of Production

Current Production

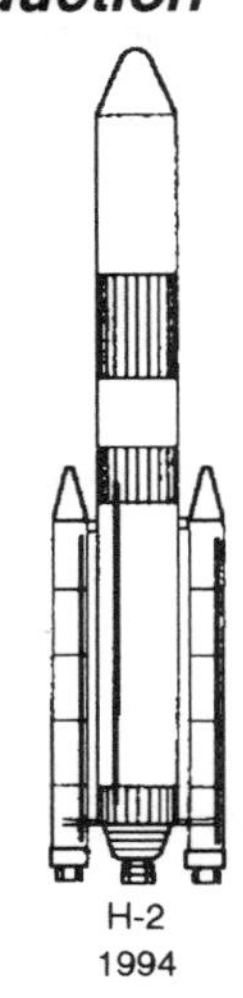

Designation	N-1	N-2	H-1	H-2
First Launch	1975	1981	1986	1994
Total Length	107 ft (32.6 m)	116 ft (35.4 m)	132 ft (40.3 m)	161 ft (49 m)
Core Diameter	8.0 ft (2.44 m)	8.0 ft (2.44 m)	8.0 ft (2.44 m)	13.1 ft (4.0 m)
Total Weight	199,000 lb (90,400 kg)	297,000 lb (135,000 kg)	308,000 lb (140,000 kg)	582,000 lb (264,000 kg)
LEO Payload	2,600 lb (1,200 kg)	4,400 lb (2,000 kg)	7,000 lb (3,200 kg)	23,000 lb (10,500 kg)
GTO Payload	800 lb (360 kg)	1,600 lb (730 kg)	2,400 lb (1,100 kg)	8,800 lb (4,000 kg)
GEO Payload	290 lb (130 kg)	770 lb (350 kg)	1,200 lb (550 kg)	4,800 lb (2,200 kg)

Vehicle Description

N-1 Derived from a version of the Thor-Delta launcher—three solid Castor II strap-ons, LOX/RJ-1 first stage, NTO/A-50 second stage, radio guidance, solid spinning upper stage, and 5.4 ft (1.65 m) diameter fairing.

N-2 Same as N-1 except nine solid Castor II strap-ons, first stage tank extended, second stage engine improved, inertial guidance, and 8.0 ft (2.44 m) diameter fairing.

H-1 Same as N-2 except new LOX/LH2 second stage and engine, higher mass fraction third stage, and improved inertial guidance.

H-2 New vehicle fully developed with Japanese technology—two large solid strap-ons, LOX/LH2 first stage, a LOX/LH2 derived H-1 second stage, inertial guidance with ring laser gyros, and 13.4 ft (4.07 m) diameter fairing.

Historical Summary

In 1955, Hideo Itokawa, at the University of Tokyo, assembled a team that produced the "pencil rocket", a small sounding rocket for collecting information about the atmosphere. In 1964, the Japanese government started taking more notice of Itokawa's work with rockets when satellite images of the Tokyo Olympics were broadcast by the United States. The Science and Technology Agency then created the National Space Development Center to look in to what practical benefits might come from space. In 1969, the Center became the National Space Development Agency of Japan (NASDA).

That same year the Japanese government negotiated an agreement with the United States government for the transfer of Delta launch vehicle technology and assistance in development of the N-vehicle. This resulted in development of the N-1 vehicle by NASDA, intended to launch much larger spacecraft than the M-vehicle, which was developed by the Japanese Institute for Space and Astronautical Science (ISAS).

NASDA differs from ISAS in that it is responsible for space application missions such as communication satellites whereas ISAS is responsible for space research and technology such as astronomical satellites. The Space Activities Commission (SAC), which reports to the prime minister, is responsible for establishing space policy and coordinating NASDA and ISAS efforts. NASDA receives about 70% of the Japanese space budget.

The Japanese were licensed by the U.S. Department of State to build the entire N-vehicle in Japan. Japan was restricted from commercially offering the vehicle. For reasons of cost and convenience, they purchased many items in the United States rather than building these components themselves. McDonnell Douglas Corporation, the U.S. builder of the Delta launch vehicle, provided the bulk of the assistance on the N-vehicle (overall design, production, and launch operations). Mitsubishi Heavy Industries was prime contractor for NASDA on the N-vehicle program. Mitsubishi produced the Delta booster under license from McDonnell Douglas and the first-stage engine (MB-3) under license from Rocketdyne. Through the technical assistance of Rocketdyne, they developed the second stage engine (LE-3), and through McDonnell Douglas, they developed the second stage structure and nose fairing. Delta tankage (McDonnell Douglas), Castor II strap-ons (Thiokol), solid propellant third stage (Thiokol), control systems (Honeywell) and system analysis (TRW) were purchased from the U.S.

From 1975 to 1982, seven N-1 vehicles were successfully flown. The three stage N-1 was capable of delivering 290 lb (130 kg) to geostationary orbit (GEO). Because the N-1 performance was not sufficient for operation of commercial communication systems or other applications, the N-2 was developed. The N-2 used nine strap-ons instead of three, and improved the first, second, and third stages and the inertial guidance

H-Vehicle History

(continued)

Historical Summary

(continued)

system. From 1981 to 1987, eight N-2 vehicles were successfully launched. All 15 N-vehicle launches were flown from the Osaki launch site at the Tanegashima Space Center on the southern tip of Japan.

Although the N-2 increased performance to GEO by 2.7 times the capability of N-1, application satellites required even larger launch vehicles. The research and development of a larger launcher, which was in progress since 1977, led to the development of the H-vehicle program. The H-vehicles significantly enhance Japan's autonomous capability in the design and use of launch vehicles.

The H-1 vehicle, developed as a successor of the N-series rockets, employed a new domestically-developed cryogenic second stage, inertial guidance system and third stage solid motor, while the first-stage, strap-on boosters and fairing remained the same as the N-2 vehicle, that is manufactured by license. A three-stage H-1 rocket could launch a 1,200 lb (550 kg) payload into GEO. A two-stage H-1 was available for lower orbits. Two successful test flights of the H-1 were flown, a two-stage version on August 13, 1986, and a three-stage version on August 27, 1987. A total of nine H-1 launches were conducted through 1992.

Based on the experience gained through the H-1 development program, NASDA has developed the H-2 launch vehicle entirely with Japanese technology. The H-2 rocket is designed to serve as NASDA's main workhorse in the 1990s to meet the demand for larger satellite launches at lower cost, and still maintain a high degree of reliability. The H-2 vehicle consists of cryogenic first and second stages and a pair of solid rocket boosters. The first stage propulsion engine (LE-7) was designed so that the experience gained in the H-1 second stage engine (LE-5) could be utilized and extended. The solid rocket boosters (SRBs) were selected because of the experience with solid rocket motors developed in Japan. The H-2 second stage was chosen to be a modified version of the H-1 second stage. A strap-down inertial guidance system with ring laser gyros is applied for the guidance and control of the H-2 while tuned platform gyros were used for the H-1. The H-2 has a 13 ft (4 m) diameter payload fairing that can accommodate either one or two payloads. A 16 ft (5 m) Space Shuttle-class fairing is also available. The payload capability is 4,800 lb (2,200 kg) into GEO or four times the capability of H-1.

The H-2 development program was approved by the SAC in 1984, based on the results of trade-off studies by NASDA. Having verified the feasibility of the vehicle design with the systems study and component tests, NASDA began the development in 1985, and established the vehicle baseline configuration at the Preliminary Design Review in May 1987, three months prior to the first H-1 launch. In order to verify their analyses on the launch environment and SRB separation, NASDA launched a series of one-quarter scale, suborbital test rockets (TR-1) in 1988 and 1989. At the Critical Design Review in July 1990, the H-2 design was set. Due to engineering development problems associated with the LE-7 first stage cryogenic engine, it was decided at the review to reduce the thrust of the LE-7 to 90% of the original design target and to delay the first launch from 1992 to 1994. Fortunately, however, the LE-7 performance loss was offset by higher performance values of the SRB and second stage ascertained by test results.

The first H-2 launch occurred on February 4, 1994 from the Tanegashima Space Center. Two payloads were flown on this flight. The Vehicle Evaluation Payload (VEP) which measured the rocket's performance and the Orbital Re-entry Vehicle (OREX) an aerodynamic shell body designed to collect atmospheric reentry data. The flight was termed a "100 percent success".

The H-2 launches are conducted from Tanegashima Space Center from newly designed launch facilities, Yoshinobu Launch Complex, on the Osaki Range. After three test flights, the operational H-2 rocket will launch various low Earth orbit (LEO) and GEO missions. Numerous concepts are currently being studied to evolve the H-2 into a larger vehicle. The H-2 is expected to launch an unmanned H-2 Orbiting Plane (HOPE) by the late 1990s.

Launch Record

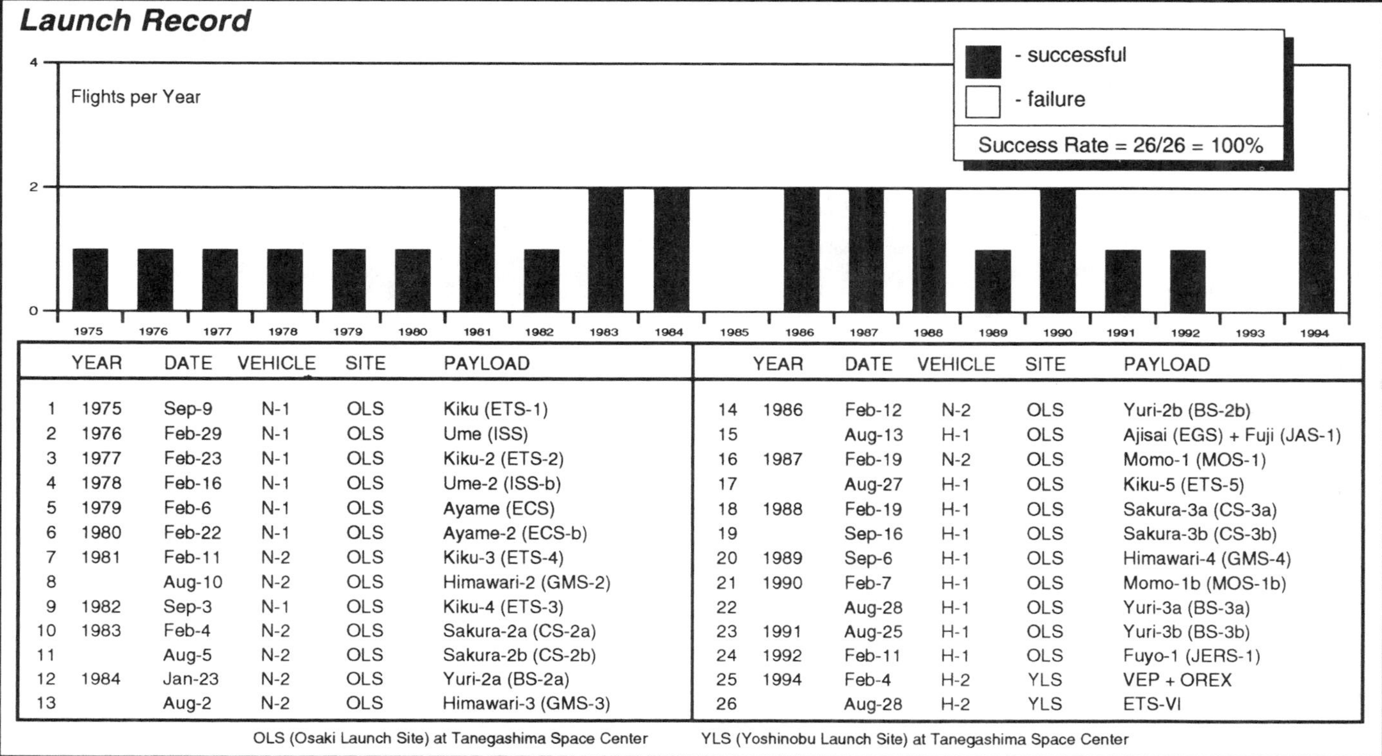

	YEAR	DATE	VEHICLE	SITE	PAYLOAD
1	1975	Sep-9	N-1	OLS	Kiku (ETS-1)
2	1976	Feb-29	N-1	OLS	Ume (ISS)
3	1977	Feb-23	N-1	OLS	Kiku-2 (ETS-2)
4	1978	Feb-16	N-1	OLS	Ume-2 (ISS-b)
5	1979	Feb-6	N-1	OLS	Ayame (ECS)
6	1980	Feb-22	N-1	OLS	Ayame-2 (ECS-b)
7	1981	Feb-11	N-2	OLS	Kiku-3 (ETS-4)
8		Aug-10	N-2	OLS	Himawari-2 (GMS-2)
9	1982	Sep-3	N-1	OLS	Kiku-4 (ETS-3)
10	1983	Feb-4	N-2	OLS	Sakura-2a (CS-2a)
11		Aug-5	N-2	OLS	Sakura-2b (CS-2b)
12	1984	Jan-23	N-2	OLS	Yuri-2a (BS-2a)
13		Aug-2	N-2	OLS	Himawari-3 (GMS-3)

	YEAR	DATE	VEHICLE	SITE	PAYLOAD
14	1986	Feb-12	N-2	OLS	Yuri-2b (BS-2b)
15		Aug-13	H-1	OLS	Ajisai (EGS) + Fuji (JAS-1)
16	1987	Feb-19	N-2	OLS	Momo-1 (MOS-1)
17		Aug-27	H-1	OLS	Kiku-5 (ETS-5)
18	1988	Feb-19	H-1	OLS	Sakura-3a (CS-3a)
19		Sep-16	H-1	OLS	Sakura-3b (CS-3b)
20	1989	Sep-6	H-1	OLS	Himawari-4 (GMS-4)
21	1990	Feb-7	H-1	OLS	Momo-1b (MOS-1b)
22		Aug-28	H-1	OLS	Yuri-3a (BS-3a)
23	1991	Aug-25	H-1	OLS	Yuri-3b (BS-3b)
24	1992	Feb-11	H-1	OLS	Fuyo-1 (JERS-1)
25	1994	Feb-4	H-2	YLS	VEP + OREX
26		Aug-28	H-2	YLS	ETS-VI

OLS (Osaki Launch Site) at Tanegashima Space Center   YLS (Yoshinobu Launch Site) at Tanegashima Space Center

H-Vehicle General Description

H2

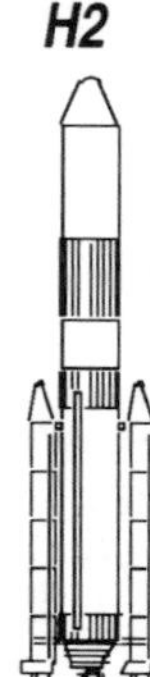

Summary

The H-2 vehicle has been developed fully with Japanese domestic technology. The H-2 consists of cryogenic first and second stages and a pair of solid rocket boosters, and is capable of four times more payload to GEO than the H-1. The first flight was conducted on February 4, 1994. A newly formed Japanese consortium, Rocket System Corporation, which is attempting to lower the cost of H-2, is expected to market the H-2 on a commercial basis.

Status

Operational

Key Organizations

User - NASDA

Launch Service Agency - NASDA

Prime Contractor - Rocket System Corporation

Other Major Contractors
- Mitsubishi Heavy Industries (System Integration, LE-7 & LE-5A Engines, Stages 1 & 2, Vehicle Assembly)
- Kawasaki Heavy Industries (Payload Fairing)
- Ishikawajima - Harima Heavy Industries (LE-7 & LE-5A Engine Turbopumps, RCS)
- Nissan Motor Company (Solid Rocket Boosters)
- NEC Corporation (Onboard Equipment)

Vehicle

System Height: 164 ft (50 m)

Payload Fairing Size:

Single Launch:
13.4 ft (4.07 m) diameter by 39 ft (12 m) height
or
16.7 ft (5.1 m) diameter by 39 ft (12 m) height

Dual Launch:
13.4 ft (4.07 m) diameter by 47.5 ft (14.5 m) height
or
13.4 ft (4.07 m) and 16.7 ft (5.1 m) diameter by 46.2 ft (14.1 m) height

Gross Mass: 573,000 lb (260,000 kg)

Planned Enhancements: Launch HOPE atop an enhanced H-2 in late 1990s (see H-vehicle Notes for possible growth options).

Operations

Primary Missions: LEO, GTO, Polar

Compatible Upper Stages: None

First Launch: 1994

Success / Flight Total: 1 / 1

Launch Site: Yoshinobu Launch Site (30.2°N, 130.6°E)
At Tanegashima Space Center

Launch Azimuth: 85°–135° (max, inclination is 28.5°–100°)

Operations (cont)

Nominal Flight Rate: 2 / yr

Planned Enhancements: Up to 4 / yr if the capability of facilities is enhanced.

Performance

100 nm (185 km) circ, 30°: 23,000 lb (10,500 kg)

100 nm (185 km) circ, 90°: 14,500 lb (6,600 kg)

Geotransfer Orbit, 28°: 8,800 lb (4,000 kg)

Geosynchronous Orbit: 4,800 lb (2,200 kg) - require spacecraft propulsion systems

Financial Status

Estimated Launch Price: $14–17 billion Yen (~$150–190M) (NASDA)

Orders: Payload/Agency:
ETS-6/NASDA in Summer 1994
Space Flyer Unit & GMS-5/NASDA in Winter 1995
ADEOS/NASDA in Winter 1996
COMETS/NASDA in Winter 1997
ETS-7 & TRMM/NASDA in Summer 1997

Manifest

1995	1996	1997	1998
1	1	2	1

Remarks

NASDA decided to improve H-2 to increase reliability and reduce the cost. The new version will enter the production phase in 1998.

H-Vehicle Vehicle

H-2 Overall

Length: 164 ft (50 m)
Gross Mass: 573,000 lb (260,000 kg)
Thrust at Lift-off: 890,000 lb (3,959,200 N)

Vehicle

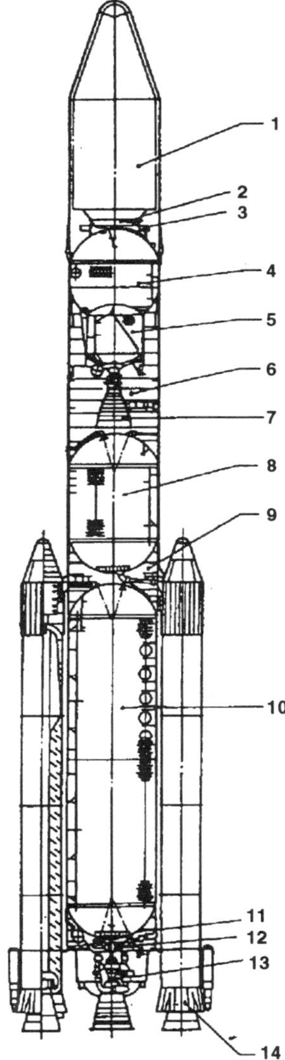

1. Payload Fairing
2. Payload Attachment Fittings
3. Onboard Electronics
4. Second Stage LH_2 Tank
5. Second Stage LOX Tank
6. Interstage Section
7. Second Stage Engine (LE-5A)
8. First Stage LOX Tank
9. Center Body Section
10. First Stage LH_2 Tank
11. First Stage Engine Section
12. Auxiliary Engines (2)
13. First Stage Engine (LE-7)
14. Solid Rocket Boosters (SRBs) (2)

Stage 1

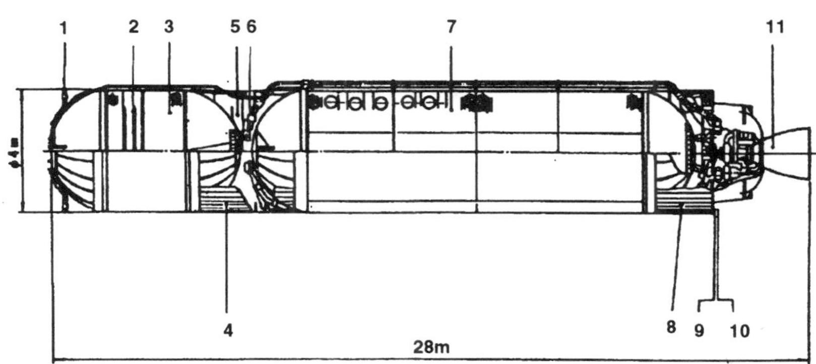

1. GOX Vent Port
2. Anti-Slosh Baffle
3. LOX Tank
4. First Stage Center Body Section
5. Electrical Equipment
6. GH_2 Vent Port
7. LH_2 Tank
8. First Stage Engine Section
9. Umbilical Connector
10. LOX/LH_2 Filling Ports
11. First Stage Main Engine (LE-7)

Stage 2

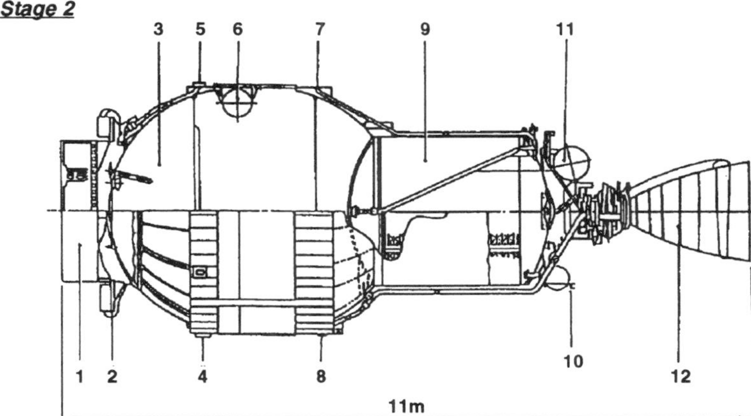

1. Payload Attachment Fitting
2. Electrical Equipment
3. LH_2 Tank
4. GH_2 Vent Port
5. Umbilical Connector
6. Cryogenic He Bottle
7. LOX Filling Port
8. LH_2 Filling Port
9. LOX Tank
10. Reaction Control Module
11. Room-Temperature He Bottle
12. Second Stage Engine (LE-5A)

Solid Rocket Booster

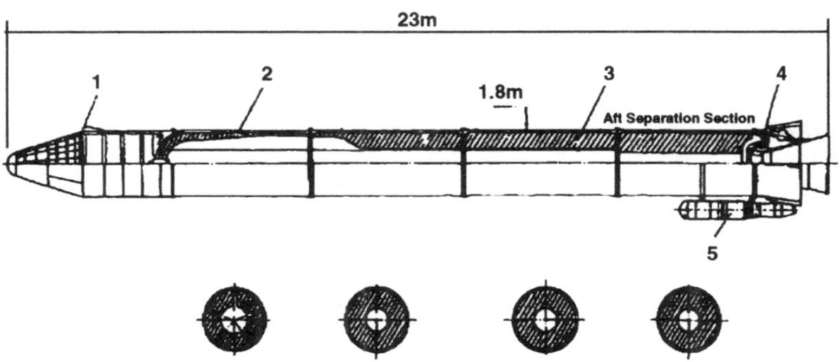

1. Forward Separation Section
2. Motor Casing
3. HTPB / AP / A1 Propellant
4. Movable Nozzle
5. External Hydraulic Tank

H-Vehicle Vehicle

(continued)

H-2 Stages

	Solid Rocket Booster	Stage 1	Stage 2
Dimension:			
Length	76.8 ft (23 m)	95 ft (29 m)	35.8 ft (11 m)
Diameter	5.94 ft (1.8 m)	13.1 ft (4.0 m)	13.1 ft (4.0 m)
Mass: (each)			
Propellant Mass	131K lb (59 ton)	190K lb (86 ton)	37K lb (17 ton)
Gross Mass	155K lb (70 ton)	216K lb (98 ton)	43K lb (20 ton)
Structure:			
Type	Monocoque	Isogrid	Isogrid
Material	Steel	Al-alloy	Al-alloy
Propulsion:			
Propellant	HTPB	LH2 / LOX	LH2 / LOX
Average Thrust (each)	350K lb (1560K N) SL	189K lb (843K N) SL 243K lb (1080K N) vac	27K lb (122K N) vac
Engine Designation	—	LE-7	LE-5A
Number of Engines	2 SRBs	1	1
Isp	273 sec vac	445 sec vac	452 sec vac
Feed System	—	Pump-fed (Staged Combustion)	Pump-fed (Hydrogen Bleed)
Chamber Pressure	812 psia (56 bar)	2090 psia (144 bar)	555 psia (38 bar)
Mixture Ratio (O/F)	—	6.0	5.0
Throttling Capability	—	100% only	100% only
Expansion Ratio	10:1	52:1	130:1
Restart Capability	No	No	Multiple
Tank Pressurization	—	Fuel-GH2, Ox-GHe	Fuel-GH2, Ox-GHe
Control-Pitch,Yaw,	Hydraulic gimbaling(±5°)	Hydraulic gimbaling(±7°)	Hydraulic gimbal(±3.5°) RCS for coast
Roll		Auxiliary engines for coast (after SRB separation)	RCS
Events:			
Nominal Burn Time	94 sec	346 sec	609 sec (max.)
Stage Shutdown	Burn to depletion	Burn to depletion	Command shutdown or Burn to depletion
Stage Separation	8 separation motors	Spring ejection	RCS

Remarks:

The H-2 is a two-stage vehicle capable of delivering a 2-ton class payload into geostationary orbit. The H-2 is equipped with a high-performance rocket engine using liquid oxygen and liquid hydrogen. The thrust of the first stage is also supplemented by solid rocket boosters.

The basic H-2 launch vehicle is designed to deliver a single 2-ton satellite into geostationary orbit. Changing its configuration allows the H-2 to be used for launching low- or medium-altitude satellites, multiple satellites, or planetary probes.

The first stage consists of two propellant tanks, the engine section, the center body section and the interstage section. The propellant tanks consist of an 18 m long LH2 tank with a diameter of 4 m and an 8 m long LOX tank with a diameter of 4 m. Both tanks are aluminum alloy cylinders with their inner surface machined into isogrids. Each end of the cylinder is jointed to a welded dome bulkhead. To prevent propellant from evaporating, foamed-resin insulant is applied to the outer surface of the tank. The first stage engine, called LE-7 employs a staged-combustion cycle and was developed based on technology of the H-1 second stage engine, LE-5. Compared with the gas generator cycle used in the LE-5, the staged-combustion cycle produces combustion gases of higher temperature and pressure. Therefore, the development of the LE-7 required a number of technical problems to be solved. Nevertheless, the LE-7 has the advantage of dramatically increased launch performance.

The second stage consists of the propellant tanks and the on-board equipment section. The tanks are separated by a common bulkhead that was developed based on technology for the second stage propellant tank of the H-1. The common bulkhead integrates a 4.4 m long LH2 tank with a diameter of 4 m and a 3.2 m long LOX tank with a diameter of 2.4 m into a single unit. Both tanks are aluminium alloy cylinders with their inner surface machined into isogrids. Foamed resin insulant is applied to the outer surface. The second stage engine, called LE-5A, employs a hydrogen belled cycle. LE-5A is an improved version of LE-5 and yields increased performance and reliability with a simpler configuration. The first burn of the second stage places the H-2 in a low earth parking orbit and then the second burn delivers the payload into a geostationary transfer orbit.

The SRBs use composite solid propellant and generate an average thrust of 159 tons at sea level. They have pivoting nozzles to control attitude and are separated and jettisoned during the first stage flight. An adaptor at the forward section of each SRB has a device for separation from the first stage, consisting of electronics, four small solid motors for separation and a self-destruction device. The two SRB aft skirts support the entire weight of the H-2 on the launch pad. Like the forward section, the aft skirt has four separation motors. The nozzles are connected to the motor casings by flexible rubber joints and can be tilted up to 5 degrees to control attitude.

To meet the growing needs to launch larger satellites, the H-2 has large payload fairings with diameters of 4 m and 5 m.

The H-2 uses an inertial guidance and control system. The H-2 system, however, is based on a strapped-down system, instead of the platform system used in the H-1.

H-Vehicle Vehicle

(continued)

Payload Fairing

in inches (mm)

Model 4S

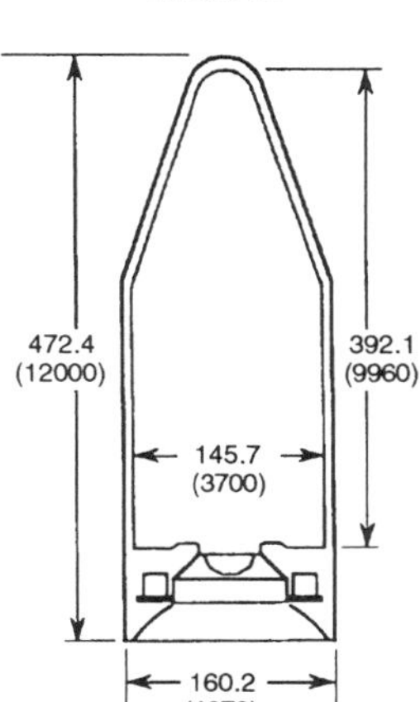

Model 5S

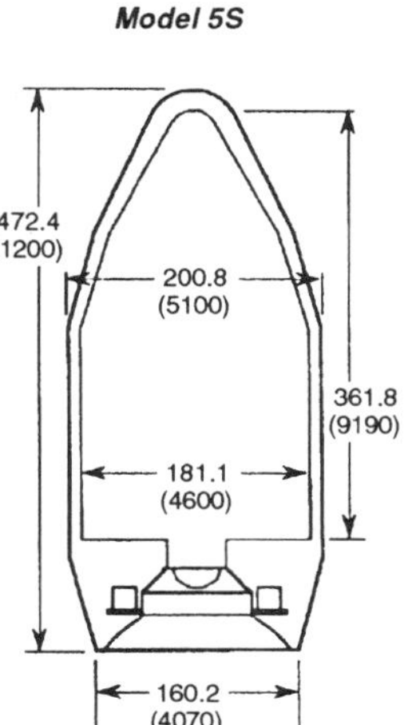

Model 4/4D

570.9 (14500)
324.0 (8230)
145.7 (3700)
141.3 (3590)
145.7 (3700)
160.2 (4070)

Model 5/4D

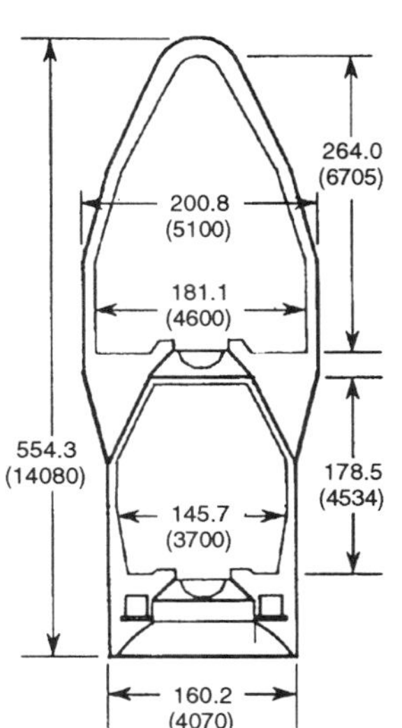

Length	Single - 39 ft (12 m) Dual - 49 ft (15 m) max.
Diameter	13.4 ft (4.07 m) or 16.4 ft (5.0 m)
Mass	Approx. 3,100 lb (1,200 kg)
Sections	2 (max.)
Structure	Aluminum or composite sandwich
Material	Aluminum honeycomb / graphite epoxy

Remarks

The payload fairing is made up of a cone and cylinder. The nose is made of aluminum alloy. Other parts are made of aluminum alloy honeycomb sandwich shell structures to minimize vehicle weight. With the separation system, the payload fairing is divided into symmetrical half shells. To prevent heat from atmospheric friction during flight, silicon insulant is applied to the nose cone. To reduce acoustic vibration, if necessary, acoustic blankets can be installed. The fairing is separated and jettisoned from the vehicle once there is no longer need to protect the spacecraft (at around an altitude of 100 km). The frangible bolts that connect the bottom of the fairing to the second stage and that connect the two half shells of the fairing are released by pyrotechnics. The fairing is forced to be opened, like a clam shell, with springs installed on the bottom of the fairing.

Avionics

The guidance and control system includes navigation, attitude control, and sequence control functions. The navigation function determines position, velocity, and attitude by verifying data from the inertial sensors. The guidance function gives information such as the thrust vector of the engines and engine cutoff timings from the navigation data. The attitude control function corrects attitude errors around the pitch, yaw, and roll axes, respectively. Sequence control generates event signals such as separation.

Unlike the platform system of the H-1, the inertial measuring unit of the H-2 has a strapped-down system to measure data such as acceleration and angular velocity of the vehicle. Acceleration sensors and gyros are strapped down onto the fixed box. Instead of the mechanical gyro in the H-1, the H-2 uses a newly developed ring-laser gyro that provides high precision over an extensive range of measurement requirements.

The lateral acceleration measurement system is equipped with an acceleration sensor, which provides acceleration data used to reduce aerodynamic force to the vehicle.

A 16-bit computer with an inertial guidance program is installed in the H-2. Based on data from the inertial sensors and other equipment, the computer calculates navigation, guidance, and control parameters and then sends control signals to the control devices through the data interface unit.

Attitude Control System

During SRB burn, attitude control of the vehicle is accomplished by means of gimballing both the LE-7 and SRB nozzles. The nozzle of each SRB is connected to the motor casing by a flexible rubber joint and can be tilted up to 5 degrees by hydraulic actuators which are powered with blow-down hydraulic pressure. The LE-7 can be tilted up to 7 degrees by hydraulic pressure supplied through a hydraulic pump driven by hydrogen gas which is tapped off the LE-7 to pressurize the LH2 tank.

After SRB burn-out, pitch and yaw control is provided by gimbaling the LE-7 and roll control by the auxiliary engine system. A single auxiliary engine has a thrust of 0.17 ton in vacuum. It uses a mixture of combustion gas from the LE-7 preburner and hydrogen gas from the LE-7 nozzle cooling. Each of the two engine modules are installed 180 degrees apart on the sides of the engine section of the first stage.

During the second stage powered flight, pitch and yaw control is provided by gimbaling the LE-5A and roll control by the reaction control system (RCS). After LE-5A cutoff, the RCS is used for three axis attitude control. The RCS consists of monopropellant hydrazine thrusters with thrust of 18 N and 50 N in vacuum.

H-Vehicle Performance

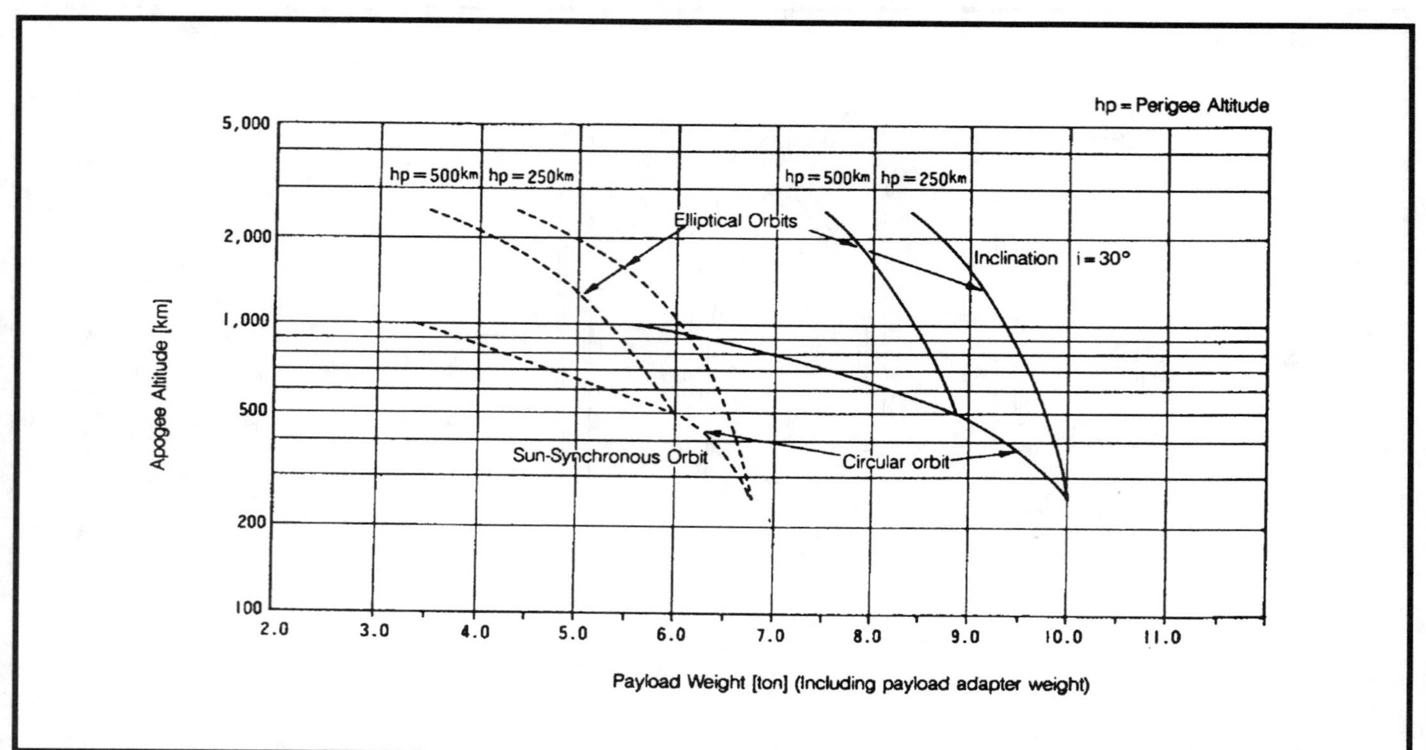

H-Vehicle Operations

Launch Site

Tanegashima, Japan (30.2°N)
Equator

Tsukuba Space Center, NASDA
NASDA Headquarters
The Institute of Space and Astronautical Science (ISAS) (Space Data Analysis Center (SDAC))
Earth Observation Center NASDA
Kakuda Propulsion Center, NASDA
(Usuda Deep Space Center, ISAS)
Katsuura Tracking & Data Acquisition Station, NASDA
(Kagoshima Space Center, ISAS (31° 15' N 131° 05' E))
Tanegashima Space Center, NASDA (30° 24' N 130° 58' E)

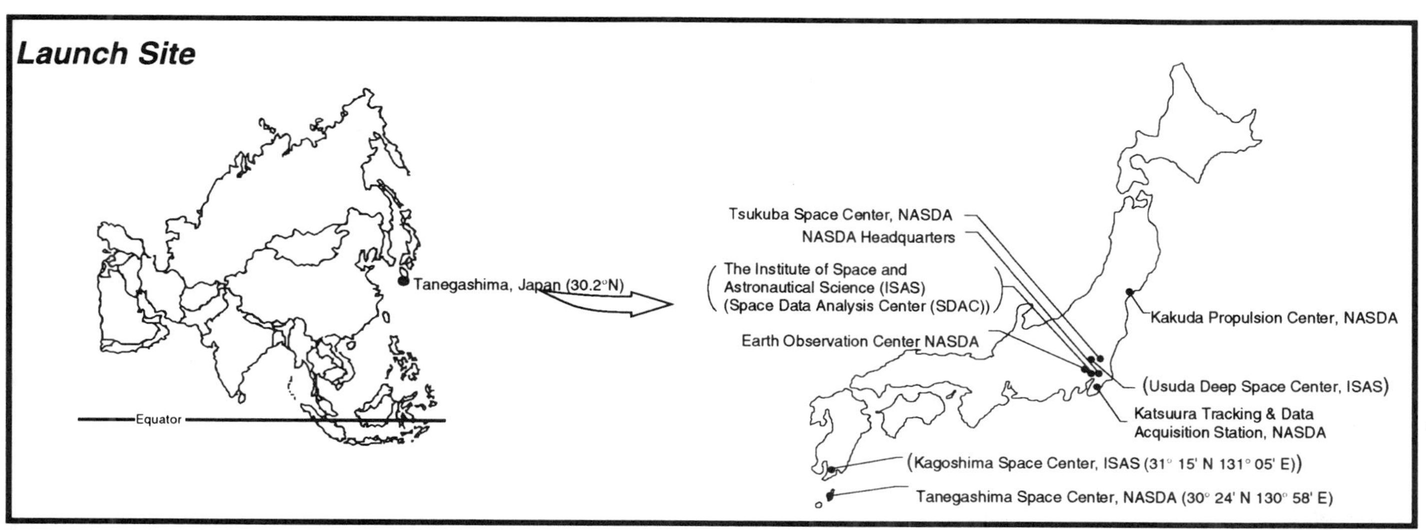

Launch Facilities

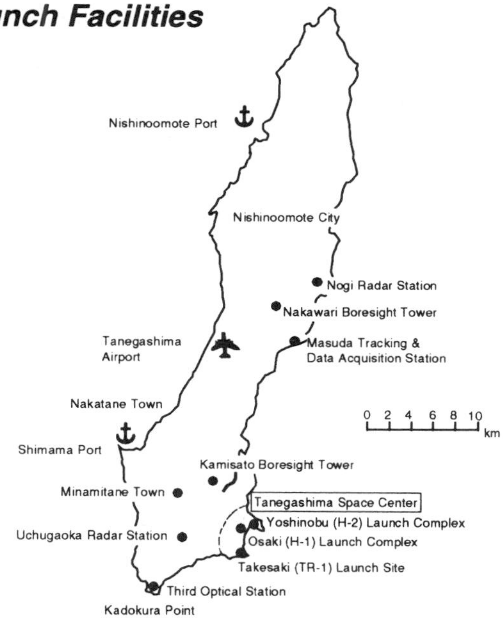

Nishinoomote Port
Nishinoomote City
Nogi Radar Station
Nakawari Boresight Tower
Tanegashima Airport
Masuda Tracking & Data Acquisition Station
Nakatane Town
0 2 4 6 8 10 km
Shimama Port
Kamisato Boresight Tower
Minamitane Town
Tanegashima Space Center
Yoshinobu (H-2) Launch Complex
Uchugaoka Radar Station
Osaki (H-1) Launch Complex
Takesaki (TR-1) Launch Site
Third Optical Station
Kadokura Point

Location of NASDA Facilities on Tanegashima Island

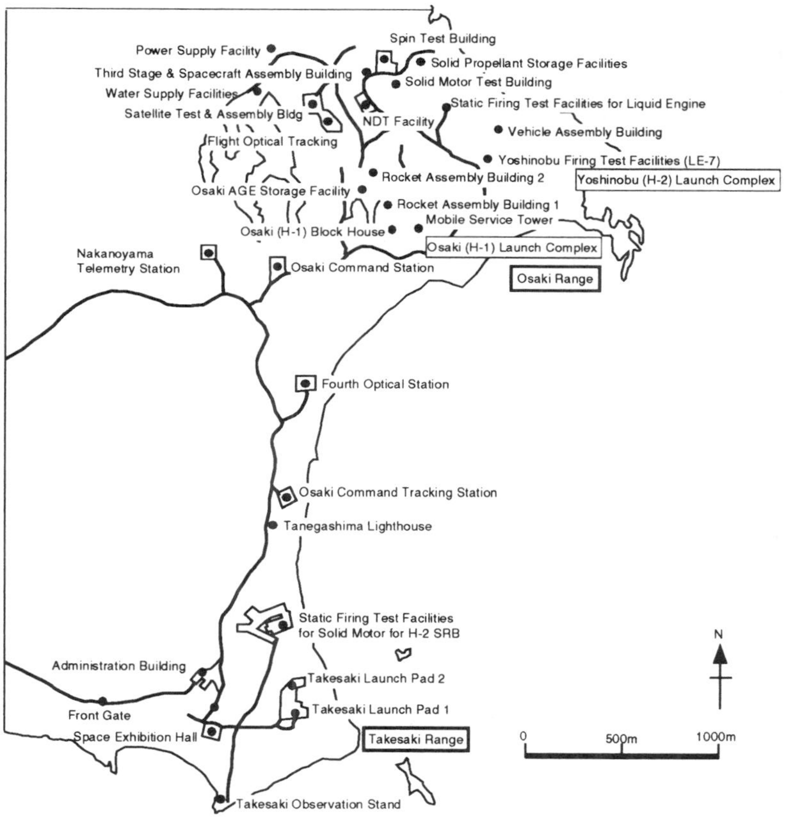

Spin Test Building
Power Supply Facility
Solid Propellant Storage Facilities
Third Stage & Spacecraft Assembly Building
Solid Motor Test Building
Water Supply Facilities
Static Firing Test Facilities for Liquid Engine
Satellite Test & Assembly Bldg
NDT Facility
Vehicle Assembly Building
Flight Optical Tracking
Yoshinobu Firing Test Facilities (LE-7)
Rocket Assembly Building 2
Yoshinobu (H-2) Launch Complex
Osaki AGE Storage Facility
Rocket Assembly Building 1
Mobile Service Tower
Osaki (H-1) Block House
Osaki (H-1) Launch Complex
Nakanoyama Telemetry Station
Osaki Command Station
Osaki Range
Fourth Optical Station
Osaki Command Tracking Station
Tanegashima Lighthouse
Static Firing Test Facilities for Solid Motor for H-2 SRB
N
Administration Building
Takesaki Launch Pad 2
Takesaki Launch Pad 1
Front Gate
Space Exhibition Hall
Takesaki Range
0 500m 1000m
Takesaki Observation Stand

Location of Facilities at Tanegashima Space Center

INSTRUMENTATION BUILDING
VEHICLE ASSEMBLY BUILDING
LH STORAGE
BURN POND
LE-7 FIRING TEST FACILITY
HIGH PRESSURE GAS STORAGE
PAD SERVICE TOWER
LOX STORAGE

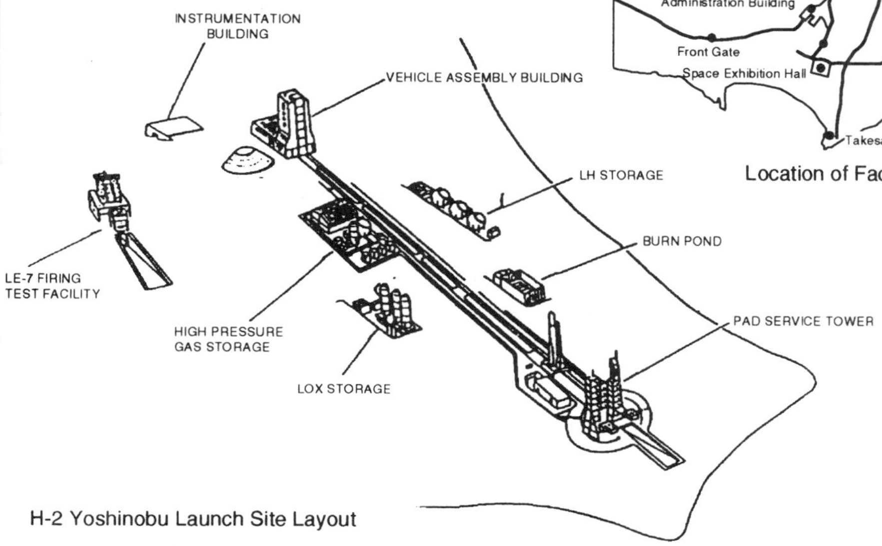

H-2 Yoshinobu Launch Site Layout

H-Vehicle Operations

(continued)

Launch Processing

Tanegashima Space Center. The Tanegashima Space Center is located on the southeast of Tanegashima Island, Kagoshima. Tanegashima Island, which is 36 mi (58 km) in diameter with a population of about 43,000, is located 50 mi (80 km) off the southern coast of Kyushu, the southern most island in the main Japanese chain. The facilities include the Takesaki Range for small rockets and the Osaki Range for the H-2 launch vehicles. The Center also includes the Masuda Tracking and Data Acquisition Station, the Nogi and Uchugaoka radar stations, and three optical tracking stations. The Center occupies approximately 3.3 sq mi (8.6 sq km) of land, where combustion test facilities on the ground for liquid and solid rocket engines are installed. It is the largest launch site in Japan. The major tasks of the Center are to check, assemble and launch rockets, and to perform tracking and control after launch. It plays a major role of applications satellite launching and combustion tests for solid rocket motors and liquid rocket engines. Due to fishermen's objections to the noise and hazards associated with the launches over their fishing grounds, launches have been restricted to two launch periods of each year—January 15 to February 28 and August 1 to September 15.

Operations. The Yoshinobu Launch Complex was newly constructed at the Osaki Range, starting in 1985, for the launch of H-2 rockets. This launch site has been designed so that parallel operations can be done in both the Vehicle Assembly Building (VAB) and the Pad Service Tower (PST). The major facilities in the complex are the VAB, the Mobile Launcher (ML), the PST, the Block house and propellants and high-pressure gas storage. Adjacent to those facilities, a firing test stand was constructed to conduct tests of the LE-7. Also, the computer tomography (CT) facility is used for nondestructive inspections of Solid Rocket Boosters (SRBs).

In the 217 ft (66 m) high VAB, the two SRBs are assembled horizontally on the ML. Each SRB is secured by four explosive bolts to a hold-down post mounted on the ML. Once the two SRBs are erected, the cryogenic core stages are attached to the SRBs. The SRBs bear the entire weight of the vehicle as it sits on the ML. Major umbilicals to the H-2 are routed through two tail service masts mounted on the ML. About 1.5 months is spent in the VAB before rolling out to the pad. Currently, only one H-2 can be processed at a time. Another ML is under construction to be capable of launching two H-2s within a relatively short period. Only basic check-out of the vehicle is done in the VAB. Most of electrical and propulsion system checkouts are done at the pad. After check-out in the VAB, the H-2 is carried by the ML along a 1,640 ft (500 m) rail to the oceanside launch pad.

The PST is a 246-ft (75 m) tower at the pad, which can open to both sides to house the vehicle on the ML for preflight check-outs.

Payloads are encapsulated in the payload fairing at the Fairing and Satellite Assembly Building. After encapsulation, the fairing and payloads are transported to the pad, hoisted to the top of the PST and mated on the second stage. The final operations include vehicle system check-out, pyrotechnic system arming, propellant loading, and terminal countdown.

The total launch processing period is about three months, but expected to be shorter in the future.

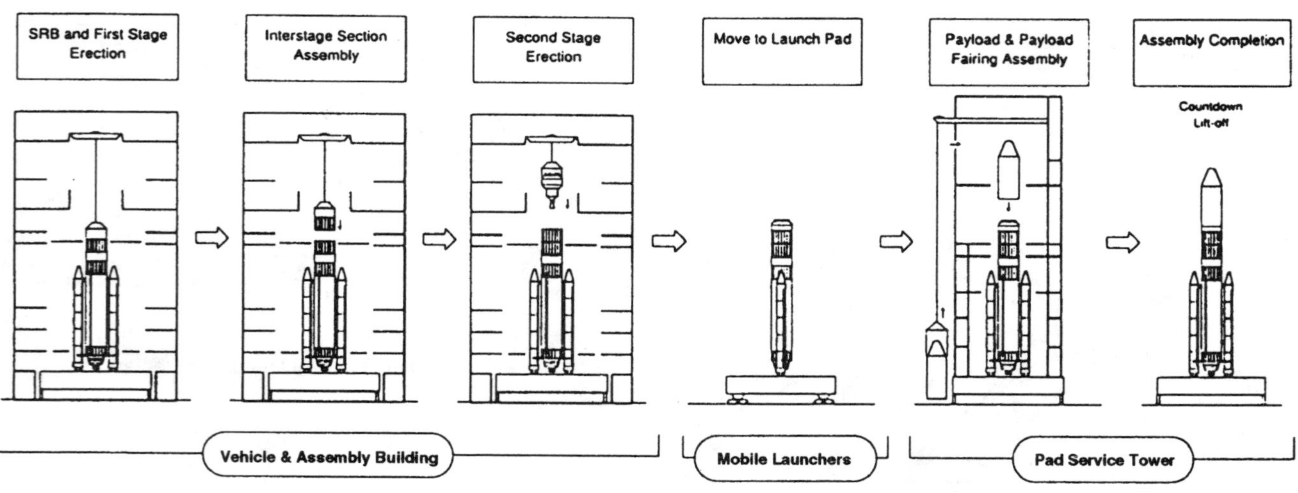

H-2 Launch Vehicle Processing Flow

H-Vehicle Operations
(continued)

Flight Sequence

Flight Sequence (Launch to Geostationary Transfer Orbit)

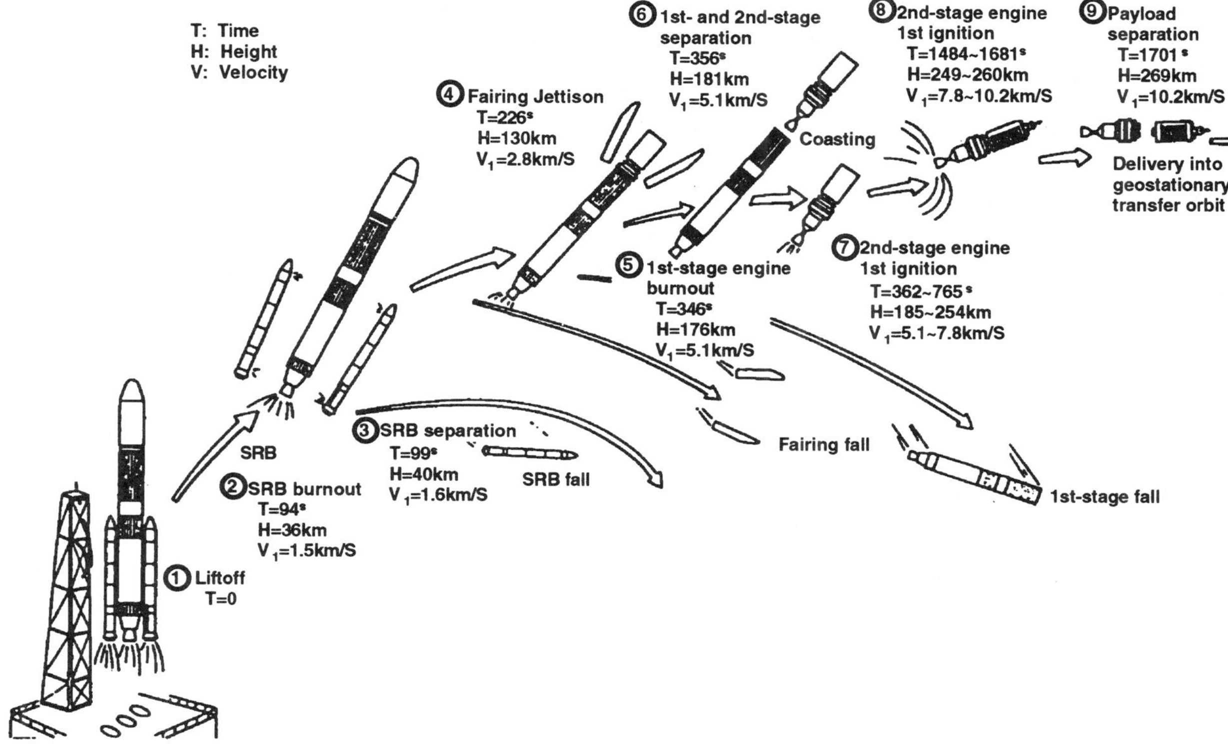

Phase	Time (sec)	Height (km)	Inertial Velocity (km/s)	Note
Ignition (LE-7, 2 SRBs)	0 (reference)	0	0.4	Rotational velocity of the Earth
SRB Burnout	94	36	1.5	
SRB Separation	99	40	1.6	
Fairing Jettison	226	130	2.8	
1st-Stage Main Engine Burnout	346	176	5.1	
1st-Stage Separation	356	181	5.1	
2nd-Stage Engine 1st Ignition	362	185	5.1	
2nd-Stage Engine 1st Cutoff	765	254	7.8	Delivery into parking orbit
2nd-Stage Engine 2nd Ignition	1,484	249	7.8	Above the equator
2nd-Stage Engine 2nd Cutoff	1,681	260	10.2	Delivery into transfer orbit
Satellite Separation	1,701	269	10.2	

H-Vehicle Payload Accommodations

Payload Compartment	
Maximum Payload Diameter	147.7 in (3,700 mm) for 4-m fairing 181.1 in (4,600 mm) for 5-m fairing
Maximum Cylinder Length	137.8 to 196.9 in (3,500–5,000 mm)
Maximum Cone Length	173.2 in (4,400 mm)
Payload Adapter	
Interface Diameter	92.9 in (2,360 mm)
Payload Integration	
Nominal Mission Schedule Begins	T-36 months
Launch Window	
Latest Countdown Hold Not Requiring Recycling	T-7 min (next opportunity in 1 hr)
On Pad Storage Capability	?? hours for a fueled vehicle
Latest Access to Payload	T-10 hours through access doors
Environment	
Maximum Load Factors	+5.0 g axial at MECO ±2 g lateral at lift-off
Minimum Lateral / Longitudinal Payload Frequency	10 Hz / 30 Hz
Maximum Overall Acoustic Level	141 dB (full octave)
Maximum Flight Shock	2,000 g from 800–2,500 Hz
Maximum Dynamic Pressure on Fairing	1,220 lb/ft2 (58,420 N/m2)
Maximum Pressure Change in Fairing	0.34 psi/s (2.3 KPa/s)
Cleanliness Level in Fairing (Prior to Launch)	Class 10,000
Payload Delivery	
Standard Orbit and Accuracy (3 sigma)	LEO - 342 nm (450 km) circ ± 11 nm 20 km) at 28.5° ±0.03° GTO (Apogee Altitude) - 19,560 nm (36,226 km) ±135 nm (250 km) at 28.5° ±0.03°
Attitude Accuracy (3 sigma)	All axes: min ±0.3 deg, ±0.5 deg/sec
Nominal Payload Separation Rate	As required
Deployment Rotation Rate Available	0 to 5 rpm without spin table 50 rpm with spin table
Loiter Duration in Orbit	?? hrs
Maneuvers (Thermal / Collision / Telemetry)	Yes

Publications

Technical Publications

"National Space Development Agency", NASDA (brochure), 1994.

"H-2 Rocket", NASDA (brochure), 1994.

"H-2 Rocket" (press kit), February 1994.

"Development Status of H-2 Rocket Cryogenic Propulsion System", A. Konno, NASDA, et al., IAF-91-263, 42nd Congress of the International Astronautical Federation, Montreal, Canada, October 5-11, 1991.

"Evolution in HOPE Concept and Flight Experimental Plan", H. Miyabe, et al., NASDA, IAF-90-162, 41st Congress of the International Astronautical Federation, Dresden, Germany, October 6-12, 1990.

"Japanese Launch Vehicle Propulsion - Status and Direction", Y. Yamada, NASDA, et al., AIAA-91-263, AIAA/SAE/ASME 27th Joint Propulsion Conference, Sacramento, CA, June 24-26, 1991.

"Development Status of the LE-7 Engine", R. Nagao, NASDA, et al., IAF-93-470, 44th Congress of the International Astronautical Federation, Graz, Austria, October 16-22, 1993.

"LE-5/LE-5A Engine", Mitsubishi Heavy Industries Ltd. (brochure), 1990.

"Idle Mode Operation of LE-5A Engine", Y. Kakuma, Mitsubishi Heavy Industries Ltd., et al., IAF-89-297, 40th Congress of the International Astronautical Federation, Maraga, Spain, October 7-13, 1989.

Acronyms

GEO - geosynchronous orbit
GTO - geosynchronous transfer orbit
HOPE - H-2 Orbiting Plane
IMU - inertial measurement unit
ISAS - Institute for Space and Astronautical Science
LAMU - Lateral Acceleration Measurement Unit
LEO - low Earth orbit
LH2 - liquid hydrogen
LOX - liquid oxygen
ML - Mobile Launcher
NASDA - National Space Development Agency of Japan
OLS - Osaki Launch Site
OREX - Orbital Reentry Experiment
PST - Pad Service Tower
RCS - reaction control system
SAC - Space Activities Commission
SRB - solid rocket booster
VAB - Vehicle Assembly Building
VEP - Vehicle Evaluation Payload

H-Vehicle Notes

(continued)

Other Notes—H-2 Vehicle Growth Possibilities

OREX
Mass: 865 kg

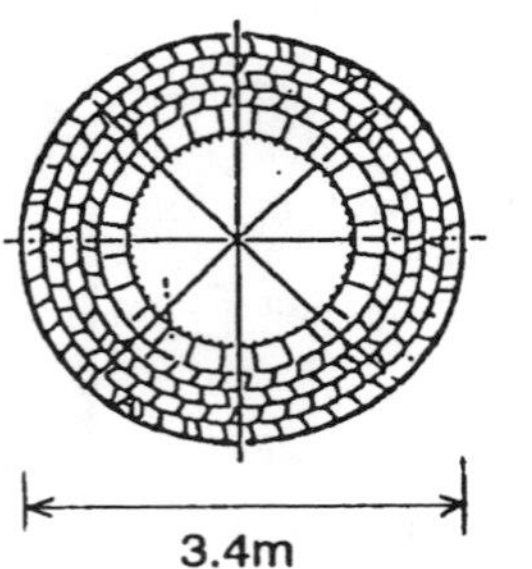

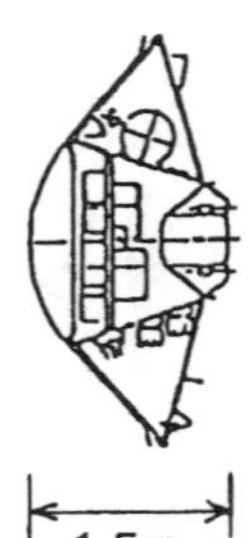

HYFLEX
Mass: 1,040 kg

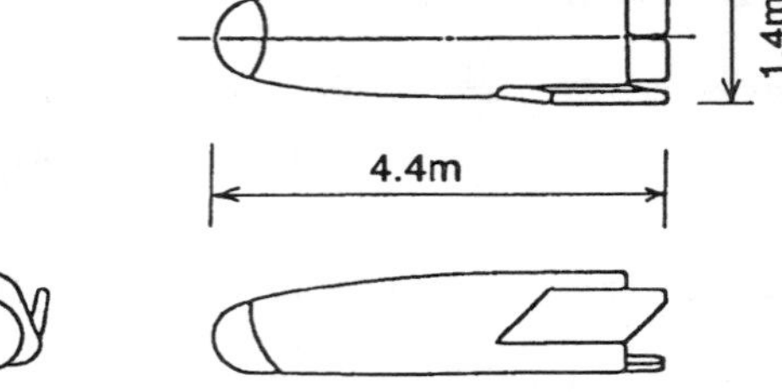

ALFLEX
Mass: 760 kg

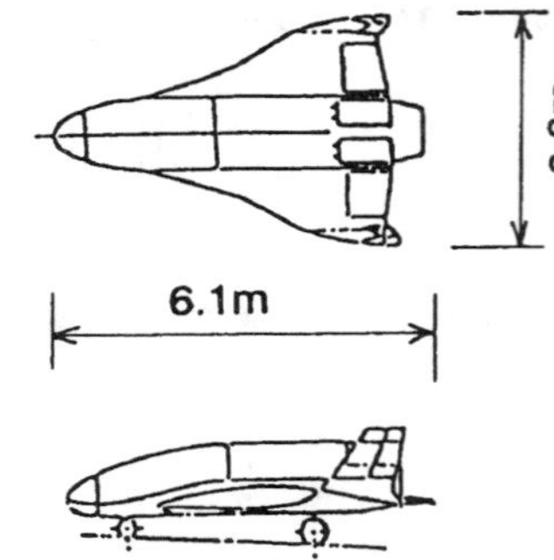

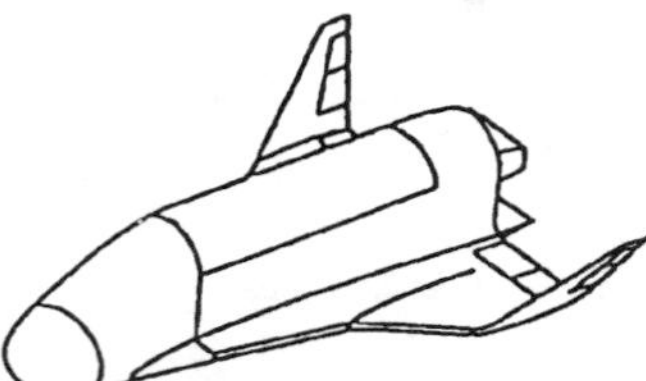

HOPE-X
Mass: 8,500 kg

LAUNCH CONFIGURATION OF HOPE-X

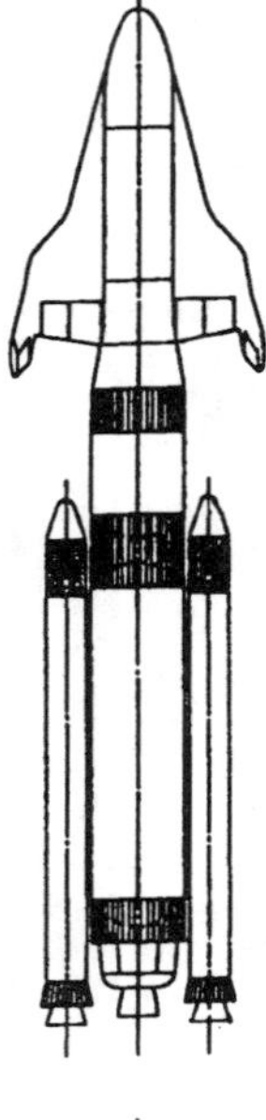

HOPE. HOPE (H-2 Orbiting Plane) is a reusable space transportation system which is an unmanned, winged vehicle launched on top of H-2 (or H-2 derivative rocket) and returns horizontally to Earth. NASDA is now under early development phase of HOPE.

The purpose of HOPE is:
- to carry cargo to/from Japanese Experimental Module (JEM), which will be attached to the International Space Station (ISS).
- to perform autonomous scientific and engineering experiments on-orbit.
- to provide on-orbit services (checks, repairs, resupply, parts replacement, recovery, etc.)

The requirements for HOPE are:
- to be able to carry payloads to/from ISS and platforms:
  - Payload: 3–5 tons
  - Mission Time: 4–7 days
  - Reliability: 2 fail safe for ISS Proximity Operations
- to land horizontally on a specified runway:
  - Runway Length: 3,000–3,500 m
  - Cross Range Capability: 1,500 km or more
- to be reusable - 20 flights per vehicle.

Development program for operational HOPE is grouped into four major flight experiments of research vehicles:
- Orbital Reentry Experiment (OREX)
  An aerodynamic shell body launched by H-2 Test Flight #1 (TF#1) to collect the atmospheric reentry data for HOPE.
- Hypersonic Flight Experiment (HYFLEX)
  An aerodynamic winged body launched by J-1 TF#1 to collect the hypersonic flight data for HOPE.
- Automatic Landing Flight Experiment (ALFLEX)
  A 37% HOPE scale model being carried by and separated from a helicopter to establish the unpowered automatic landing technology for HOPE.
- Reentry Flight Experiment (HOPE-X)
  A full-size simplified HOPE launched by H-2 rocket without its second stage, carrying out the suborbital flight test to integrate and verify the results of OREX, HYFLEX and ALFLEX.

Following are the major profiles of HOPE-X experiment:

Launch Site:	Tanegashima Space Center
Landing Site:	TBD (2,000 m class runway)
Vehicle Weight:	8.5 tons
Vehicle Separation:	Altitude 100 km, Velocity 7.5 km/sec
Flight Duration:	30 minutes

HOPE-X has the same dimensions as HOPE and is intended to evolve into the operational HOPE by modifying the system, mainly incorporating the additional components for its operational use.

H-Vehicle Notes

(continued)

H-2 Derivative Rocket. The H-2 Derivative Rocket, which will have a capability of launching 15–20-ton class payloads to Low Earth Orbit (LEO), is under system study phase.

The purpose of the H-2 Derivative Rocket is:

- to reinforce H-2 capability to launch larger payloads like HOPE and ISS missions.
- to reduce payload launch costs by introducing a multi-payload launch system.
- to cope with Lunar/Mars observation or exploration missions.

The basic concept of the H-2 Derivative Rocket is:

- to be used in parallel with H-2 as a family member of H-2 launch system.
- to use the same core stage as H-2, increase payload capacity by upgrading booster capability.
- to minimize required time for development by applying or improving H-2 technology and hardware as much as possible.

Three major candidate configurations for H-2 Derivative Rocket are under consideration. Following are the characteristics of 20-ton LEO class H-2 Derivative Rockets:

- Solid-Propellant Booster Augmented Configuration
  Six large solid boosters which are to be newly developed based on M-V rocket technology replace the two H-2 SRBs. Four SRBs ignite at liftoff and the remaining two after burn-out the first of four. Reinforcement of the core stage structure and development of LE-7 engine altitude ignition technology are required.
- Hydrogen Booster Augmented Configuration
  Two hydrogen boosters, which are the same as the first stage of the H-2 core rocket, are added to the basic H-2 rocket. Cost reduction effect will be maximized in this configuration by the mass production of H-2 first stage hardware. Cross-feed technology for the LE-7 engine, instead of altitude ignition which means the core stage LE-7 burns from lift-off using booster tank propellant, is required.
- Hydrocarbon Booster Augmented Configuration
  Two hydrocarbon boosters, which are to be newly developed, replace the two H-2 SRBs. Conventional kerosene is considered as its propellant to utilize existing engine technologies. The payload capability will be maximized in this configuration without exceeding the required launch pad safety distance because of the higher safety margin of hydrocarbon fuel. LE-7 engine altitude ignition capability is also required.

Payload capabilities can be adjusted by reducing the number of attached boosters. For example, LEO 15-ton class Hydrogen Booster Augmented Configuration is achieved by adding one hydrogen booster instead of two. Geostationary Orbit (GEO) transportation capability is achieved by adding H-2 second stage. LEO 20-ton class capability corresponds to GEO 4-ton class capability with the H-2 second stage.

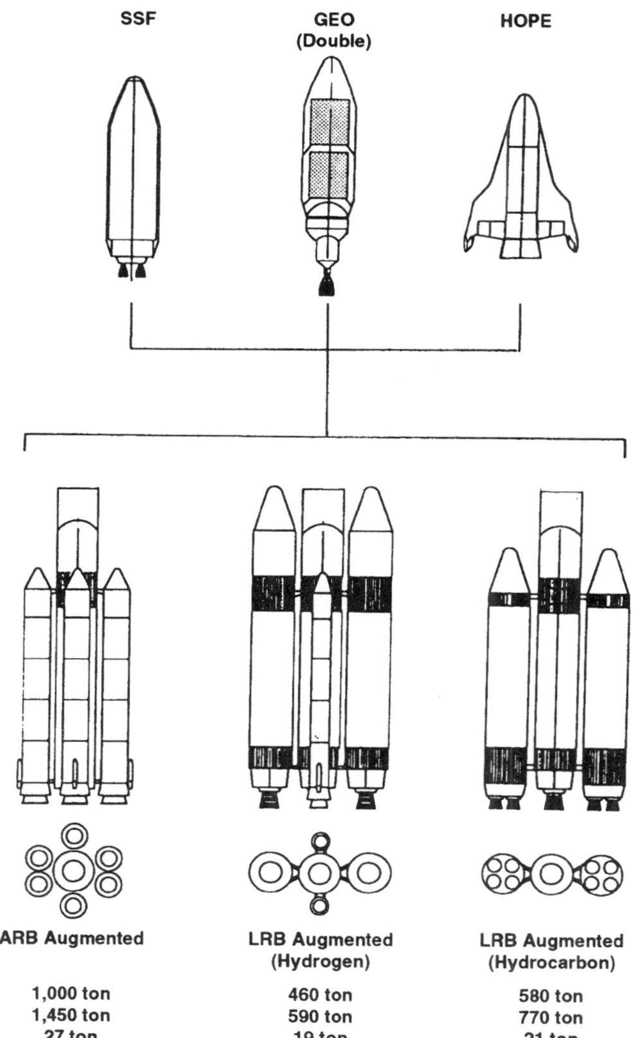

	ARB Augmented	LRB Augmented (Hydrogen)	LRB Augmented (Hydrocarbon)
Liftoff Mass:	**1,000 ton**	**460 ton**	**580 ton**
Liftoff Thrust:	**1,450 ton**	**590 ton**	**770 ton**
Payload (LEO):	**27 ton**	**19 ton**	**21 ton**

J-Vehicle

Government Point of Contact:
National Space Development Agency of Japan (NASDA)
World Trade Center Building
2-4-1, Hamamatsu-cho,
Minato-ku, Tokyo 105, JAPAN
Phone: (3) 5470-4111
or
NASDA Washington Office
1301 K Street, NW, Suite 560E
Washington, DC 2005, USA
Phone: (202) 333-6844

Industry Point of Contact:
Nissan Motor Company, Ltd.
3-5-1 Momoi,
Suginami-ku, Tokyo 167, JAPAN
Phone: (3) 3390-1111

The J-1 Launch Vehicle is the combination of the Solid Rocket Booster (SRB) of the H-2 developed by NASDA and the upper stages of the M-3SII developed by the Institute of Space and Astronautical Science (ISAS).

J-Vehicle History

In Development

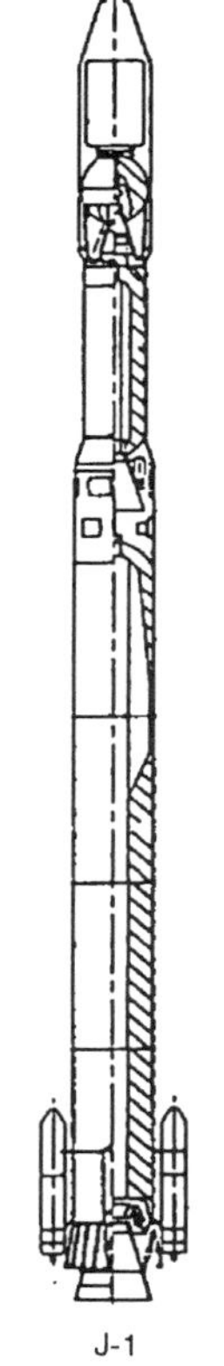

Designation	J-1
First Launch	1996
Number of Stages	3
Total Length	108.6 ft (33.1 m)
Diameter	5.9 ft (1.8 m)
Total Weight	196,000 lb (89,000 kg)
Payload to 250 km, 30°	1,800 lb (830 kg) (minimum)

Vehicle Description

Combination of the H-2 Solid Rocket Booster (SRB) and the M-3SII upper stages (second and third stages with the payload fairing).

Historical Summary

In Japan, the two key national space organizations have been developing upgraded versions of their current launch vehicles. The National Space Development Agency of Japan (NASDA), which is responsible for applications missions such as communications satellites, has completed development and flown the latest in the line of H-Vehicles, the H-2. The Institute of Space and Astronautical Science (ISAS), which is responsible for space research and technology missions, is currently developing a new M-Vehicle, the M-V. In order to fill the need for a new class of satellite launcher, a smaller vehicle is being developed by NASDA using existing solid rocket technologies. The basic concept behind the development of this new vehicle, named J-1, is to combine the H-2 Solid Rocket Booster (developed by NASDA) with the M-3SII upper stages and payload fairing (developed by ISAS).

In 1991, a conceptual study of the J-1 was conducted which showed the feasibility of this new rocket as an H-2 derived vehicle. At this time, development of the H-2 Solid Rocket Booster was almost complete, and the M-3SII Rocket had undergone several successful flights. In 1992, NASDA went forward to the system design level for the J-1. In 1993, the J-1 preliminary design was conducted and most of the subsystem designs were fixed. In 1994, the critical design was underway and some development testing had begun. Only a few years are needed for development of the J-1, because the new items are very limited and most of them can be developed by using a protoflight model, which implies that the hardware for the development tests can be refurbished and used for the test flight model.

Launch Record

The first J-1 flight is scheduled for 1996. This flight will carry the Hypersonic Flight Experiment (HYFLEX), an aerodynamic winged body designed to collect hypersonic data for the H-2 Orbiting Plane (HOPE).

J-Vehicle General Description

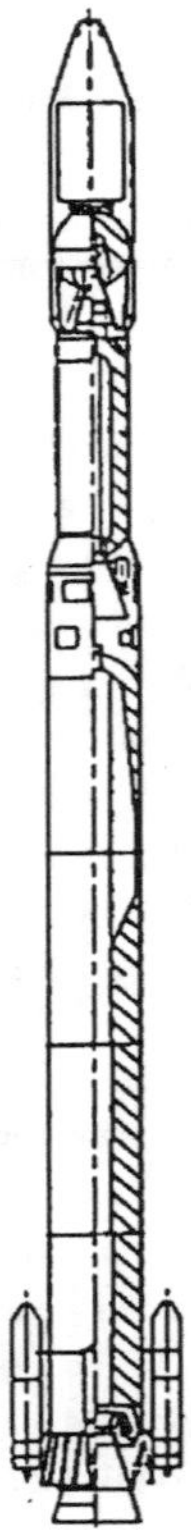

Summary

The J-1 is a three-stage solid-propellant launch vehicle developed with Japanese domestic technology. The first launch (test flight in two-stage configuration) is scheduled for 1996.

Status

In Development

Key Organizations

User - NASDA

Launch Service Agency - NASDA

Prime Contractor - None

Other Major Contractors
- Nissan Motor Co. (Solid Motor, Structure, and Fairing Production)
- Mitsubishi Heavy Industries (1-2 Interstage, Control Software)
- Ishikawajima-Harima Heavy Industries (Vernier Engine Module)
- NEC Corp., Mitsubishi Precision, Fujitsu and Matsushita Communication Industrial Co. (Onboard Electronics)

Vehicle

System Height — 108.6 ft (33.1 m)

Payload Fairing Size — 5.4 ft (1.65 m) diameter by 22.5 ft (6.85 m) height

Gross Mass — 196,000 lb (89,000 kg)

Planned Enhancements — Large fairing and modification of upper stages (concept study)

Operations

Primary Missions — Circular or elliptical Earth orbit

Compatible Upper Stages — No definite plan

First Launch — 1996

Success / Flight Total — No Flight

Launch Site — Osaki Launch Site (30.4°N, 130.97°E)
At Tanegashima Space Center

Launch Azimuth — 85°–125°

Nominal Flight Rate — 2 / yr

Planned Enhancements — No definite plan

Performance

100 nm (185 km) circ., 30° — 1,980 lb (900 kg)

100 nm (185 km) circ., 90° — No capability

Geotransfer Orbit, 30° — No capability with three-stage configuration

Geosynchronous Orbit — No capability with three-stage configuration

Financial Status

Estimated Launch Price — $4.8 billion Yen ($50–55M) (NASDA)

Orders: Payload / Agency — HYFLEX / NASDA in 1996
OICETS / NASDA in 1998

Manifest

1996	1997	1998	1999	2000	2001
1	—	1	?	?	?

Remarks

Further payloads are being planned

J-Vehicle Vehicle

Overall

Length: 108.6 ft (33.1 m)
Gross Mass: 196,000 lb (89,000 kg)
Thrust at Liftoff: 350,700 lbf (1,560 kN)

(all dimensions in mm)

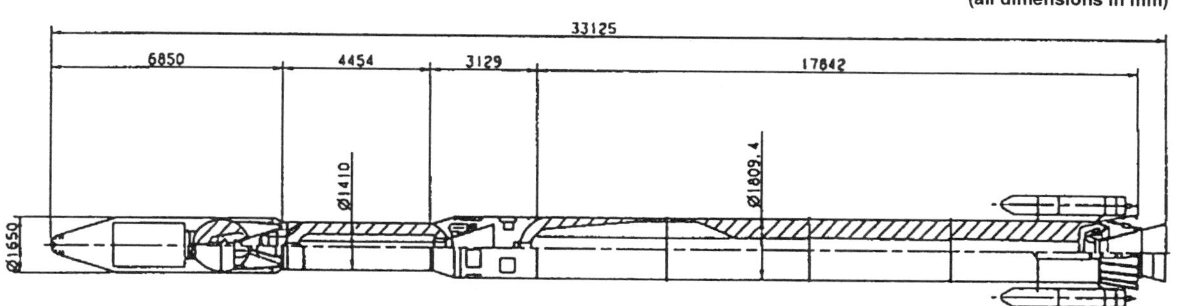

Stages

	Stage 1	Stage 2	Stage 3
Dimension:			
Length	68.9 ft (21.0 m)	20.7 ft (6.3 m)	8.9 ft (2.7 m)
Diameter	5.9 ft (1.8 m)	4.6 ft (1.4 m)	4.9 ft (1.5 m)
Mass: (each)			
Propellant Mass	130,500 lb (59,200 kg)	22,900 lb (10,400 kg)	7,300 lb (3,300 kg)
Gross Mass	156,700 lb (71,100 kg)	28,600 lb (12,800 kg)	7,900 lb (3,600 kg)
Structure:			
Type	Monocoque	Monocoque	Monocoque
Material	Steel	Steel	Titanium (6A1-4V)
Propulsion:			
Propellant	HTPB	HTPB	HTPB
Average Thrust (each)	350K lb (1,560K N) SL	118K lb (524K N) vac	29.7K lb (132K N) vac
Engine Designation	—	M-23	M-3B
Number of Motors	1	1	1
Number of Segments	4	1	1
Isp	273 sec vac	282 sec vac	294 sec vac
Chamber Pressure	812 psia (58 bar)	? bar	? bar
Expansion Ratio	10.0	23.24	51.84
Control-Pitch,Yaw,	MNTVC	LITVC	Spin stabilized
Roll	Vernier engines	Hydrazine side jets	—
Events:			
Nominal Burn Time	94 sec	73 sec	87 sec
Stage Shutdown	Burn to depletion	Burn to depletion	Burn to depletion
Stage Separation	Separation nuts / springs	Separation nuts / springs	V-clamp / springs

Remarks:

The basic concept of the J-1 Launch Vehicle is a combination of the H-2 Solid Rocket Booster (SRB) developed by NASDA, and the upper stages (i.e., the second and third stages with the payload fairing) of the M-3SII Rocket developed by ISAS.

The J-1 first stage motor is identical to the SRB of the H-2 Launch Vehicle. It was flight-proven with the successful H-2 test flight in February 4, 1994. The second stage motor (M-23), the third stage motor (M-3B) and the payload fairing are exactly the same as those used in the M-3SII Rocket since 1985, with seven successful flights.

The J-1 adopts a radio guidance method similar to the M-3SII, and an attitude correction command can be transmitted from the ground station if necessary. The ignition timing of the third stage motor can also be corrected by the guidance command.

J-Vehicle Vehicle
(continued)

Payload Fairing

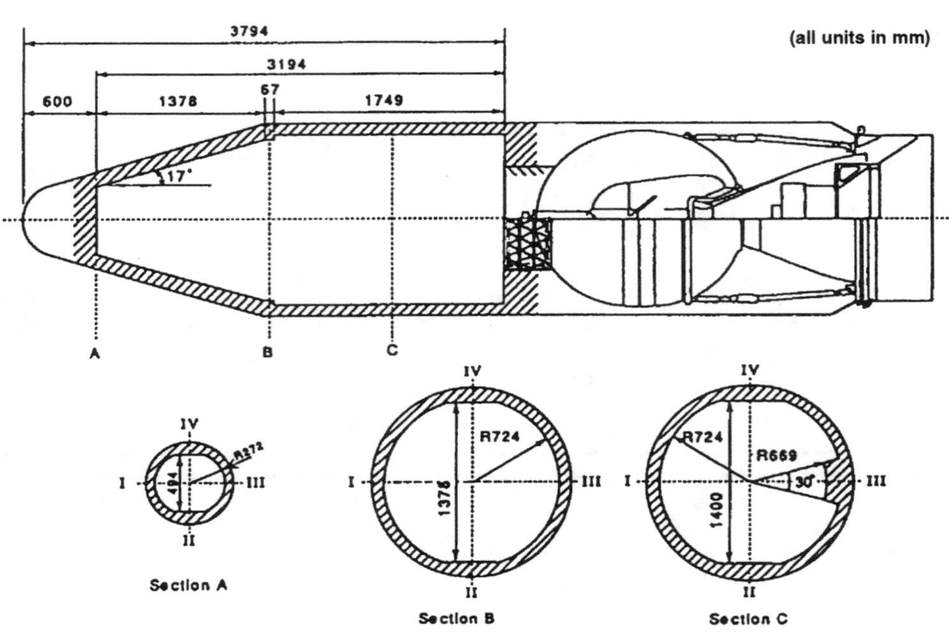

Length	22.47 ft (6.85 m)
Diameter	5.4 ft (1.65 m)
Mass	1,150 lb (520 kg)
Sections	2
Structure	Composite sandwich
Material	Honeycomb sandwich with Glass Fiber Reinforced Plastic (GFRP) face plate
Remarks	The payload fairing is identical to that of the M-3SII. Two pieces of the payload fairing are clamped to the M-23 motor by V-clamp bands. Ignition of the pyrotechnics initiates the fairing jettison. The pyrotechnic devices provide sufficient relative velocity to each fairing piece even under 5g axial acceleration.

Avionics

The instrumentation is basically similar to that of the M-3SII, though some modification is required because of the difference in the radio frequency regulation and the flight safety requirement.

The onboard instruments include electronics for radar tracking (C band), command receiving, attitude reference, control computing, and telemetry monitoring.

Attitude Control System

The attitude control devices for the first stage are the Movable Nozzle Thrust Vector Control (MNTVC) for pitch and yaw during the motor burn and two bipropellant External Vernier Engine (EVE) modules for roll during burn and for 3-axis during coasting. The second stage has the Liquid Injection Thrust Vector Control (LITVC) and the Side Jet (SJ) thrusters (a hydrazine reaction control system with 16 thrusters). The LITVC and SJ thrusters are the same as the M-3SII second stage. The third stage is spin stabilized, which is spun up by the SJ thrusters before separation of the the second and third stages.

J-Vehicle Performance

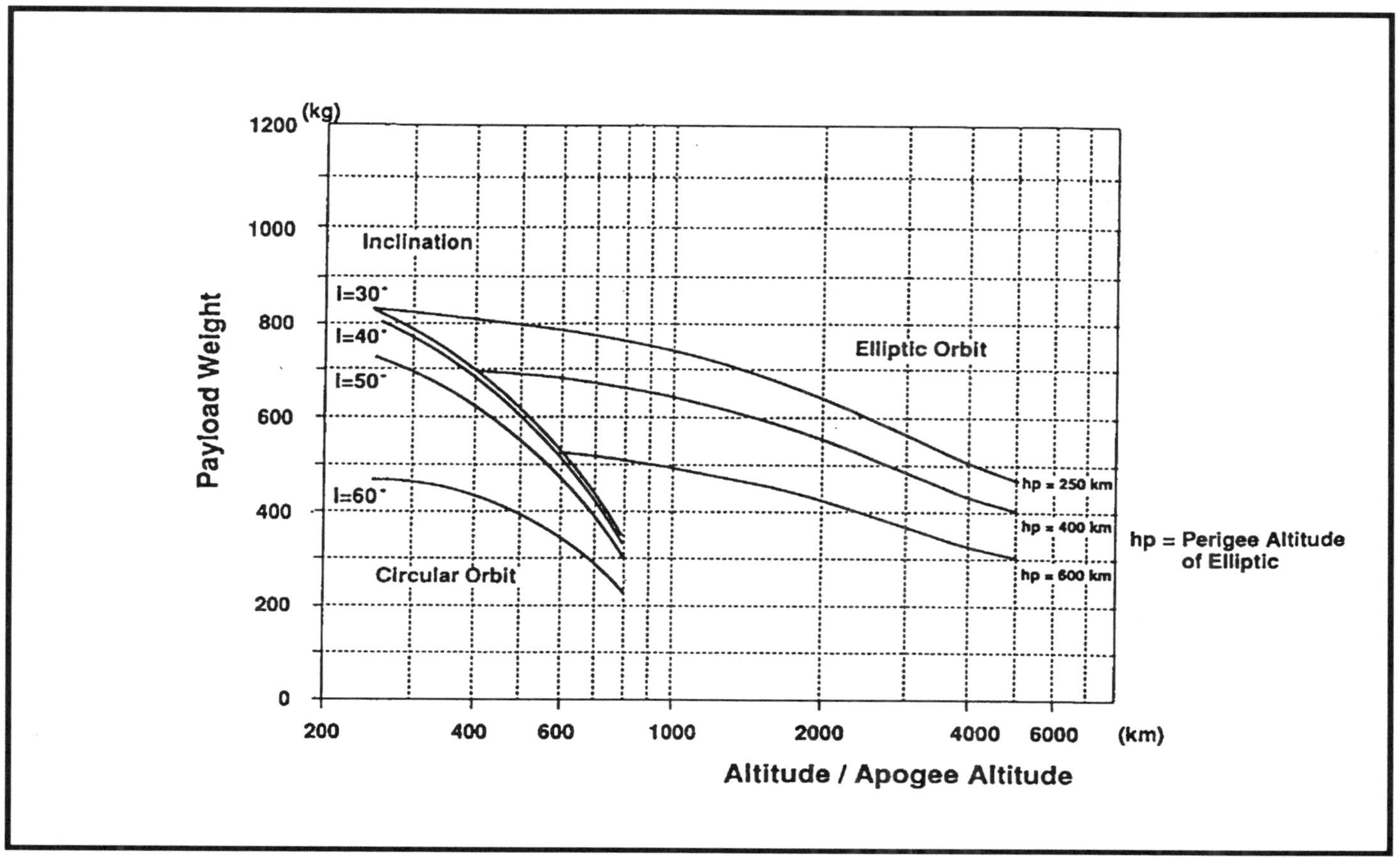

J-Vehicle Operations

Launch Site

The Tanegashima Space Center (TNSC) is located in the Tanegashima Island. The location of the launch pad is indicated as 30° 24'N, 130° 58'E (or 30.40° N, 130.97° E).

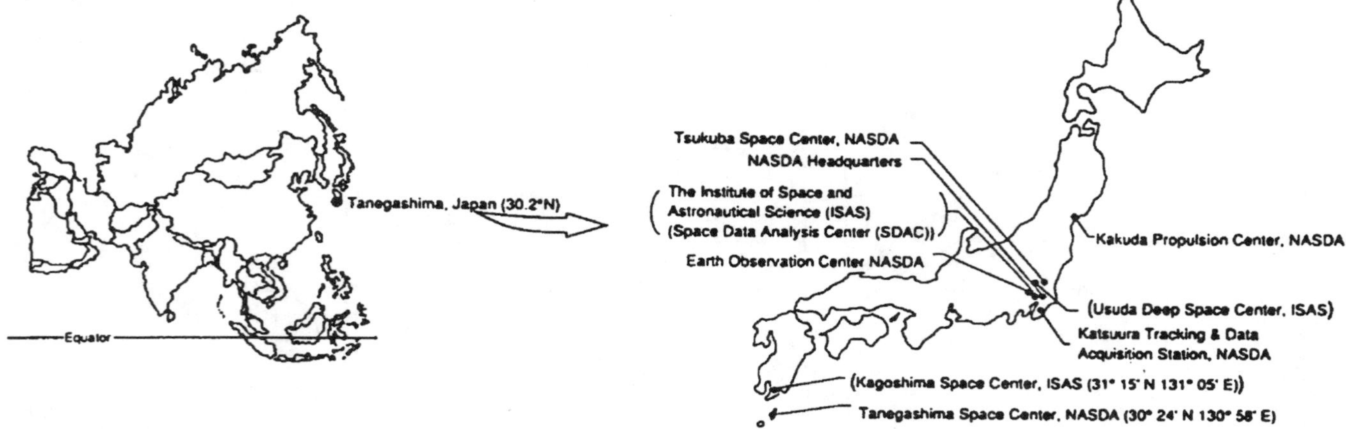

Launch Facilities

The launch facilities to be used for J-1 are those originally used for the N-1, N-2, and H-1 Launch Vehicles in the Tanegashima Space Center (TNSC). As the H-1 Launch Vehicle Program ended in 1992, the H-1 facilities in TNSC, including the launch pad with the mobile service tower and the block house, were to be dismantled with no future utilization plans. With the start of the J-1 program, NASDA decided to maintain those facilities and modify them for the J-1. By doing so, the J-1 program has the minimum cost for ground facilities.

On-site facilities include the Nakanoyama Telemetry Station, the Masuda Tracking and Data Acquisition Station, the Nogi Radar Station, and the Uchuugaoka Radar Station.

An illustration showing the location of the old H-1 launch facilities can be seen in the H-vehicle section.

Launch Processing

The J-1 vehicles are transported to Tanegashima in elements such as the first stage motor segments, the second stage motor, the third stage motor, the payload fairing, the 1-2 interstage, the second stage instrument installation section, and so on. They are transported by ship to the Tanegashima Island and by tracks on the island to TNSC.

The vehicle is integrated vertically on the launch pad. The fourth segment of the first stage motor, mated with the aft skirt, comes first and the following segments are added sequentially. The 1-2 interstage (including the first stage instrumention), the second stage motor, and the second stage instrument installation section are then added to the stack. The third stage motor is integrated with the 2-3 interstage structure and the spacecraft in the assembly building, and carried on top of the second stage. The payload fairing is finally fixed in the service tower.

Though the assembly work mentioned above needs about four weeks, some marginal weeks may be added to avoid interference with the H-2 launch operation.

Flight Sequence

Typical Flight Sequence
(tentative 500 km mission)

Time (min:sec)	Events
-00:05	1st stage EVE start
00:00	1st stage ignition, MNTVC start
01:34	1st stage burnout
01:50	1st-2nd separation
01:52	2nd stage ignition, LITVC and SJ start
03:01	2nd stage burnout
03:20	Payload fairing jettison
06:25	3rd stage spin-up start
07:16	2nd-3rd separation
07:18	3rd stage ignition
08:40	3rd stage burnout
09:40	3rd spacecraft separation

J-Vehicle Payload Accommodations

Payload Compartment	
Maximum Payload Diameter	55.1 in (1400 mm)
Maximum Cylinder Length	68.9 in (1749 mm)
Maximum Cone Length	54.2 in (1378 mm)
Payload Adapter	
Interface Diameter	?? in (?? mm)
Payload Integration	
Nominal Mission Schedule Begins	T-12 months
Launch Window	
Latest Countdown Hold Not Requiring Recycling	T-?? min
On-Pad Storage Capability	?? hrs for a fueled vehicle
Latest Access to Payload	T-?? hrs or T-?? hrs through access doors
Environment	
Maximum Load Factors	±?? g axial, ±?? g lateral
Minimum Lateral / Longitudinal Payload Frequency	?? Hz / ? Hz
Maximum Overall Acoustic Level	?? dB (full octave)
Maximum Flight Shock	?? g from ??-?? hz
Maximum Dynamic Pressure on Fairing	?? lb/ft^2 (?? N/m^2)
Maximum Pressure Change in Fairing	?? psi/s (?? KPa/s)
Cleanliness Level in Fairing (Prior to Launch)	Class ??
Payload Delivery	
Standard Orbit and Accuracy (3 sigma)	Altitude ±?? nm (?? km) Inclination ±?? deg
Attitude Accuracy (3 sigma)	All axes: ±?? deg, ±?? deg/sec
Nominal Payload Separation Rate	?? ft/s (?? m/s)
Deployment Rotation Rate Available	?? rpm
Loiter Duration in Orbit	?? hrs
Maneuvers (Thermal / Collision / Telemetry)	??

J-Vehicle Notes

Publications

Technical Publications

"Japan's New Solid Rocket Booster and its Derivative Launcher", M. Miyazawa, Y. Fukushima, H. Miyaba, NASDA and T. Asai, Nissan Motor, IAF-92-0629, 43rd Congress of IAF, Washington, D.C., 1992.

"System Concept of J-1 Launch Vehicle", N. Saki, M. Miwada, H. Miyaba, NASDA and J. Onoda, ISAS, 5th ISCOPS, Shanghai, 1993.

Acronyms

GFRP - glass fiber reinforced plastics
EVE - External Vernier Engine
HTPB - hydroxyl terminated polybutadiene
HYFLEX - Hypersonic Flight Experiment
ISAS - Institute for Space and Astronautical Science
LITVC - Liquid Injection Thrust Vector Control
MNTVC - Movable Nozzle Thrust Vector Control
NASDA - National Space Development Agency of Japan
OICETS - Optical Inter-Orbit Communications Engineering Test Satellite
SJ - side jet
SRB - solid rocket booster
TNSC - Tanegashima Space Center

Other Notes—J-1 Launch Vehicle Growth Possibilities

The J-1 Launch Vehicle is a combination of existing stages from other vehicles, and thus is not an optimized vehicle. One of the main problems with the vehicle is that the size of the payload fairing is small compared with the payload mass capability. Therefore, the satellite designers have requested larger volume (especially the diameter) for the payload envelope. To improve this situation, NASDA is going to develop the large fairing with a diameter of about 2.1 m, which will provide about 6 m^3 of additional volume in the payload cylindrical section. The payload capability will go down because of the weight and drag increase. This performance loss may be compensated by adding strap-on boosters or by modifying the upper stages.

Another point is the payload injection accuracy. As the J-1 is an all-solid-propellant rocket, it is not possible to control the value of velocity at payload injection. A simple survey shows that some expected payloads require a high degree of injection accuracy: these payloads include docking experiments, Earth observation satellites, and recovery capsules. NASDA may need a liquid propellant upper stage in the future. There is no definite plan for a J-1 growth vehicle at this time.

M-Vehicle

Government Point of Contact:
The Institute of Space and
Astronautical Science
3-1-1 Yoshinodai
Sagmihara, Kanagawa 229
JAPAN
Phone: (0427) 513911

Industry Point of Contact:
Nissan Motor Company, Ltd.
5-1, 3-Chome, Momoi
Suginami-ku Tokyo 167
JAPAN
Phone: (3) 390-1111

The M-3SII is the most powerful version of the M-family of solid propellant launch vehicles developed by The Institute of Space and Astronautical Science (ISAS) and is mainly used to launch ISAS spacecraft for science and technology.

M-Vehicle History

	Out of Production				Current Production	In Development
Designation	L-4S	M-4S	M-3C	M-3H / M-3S	M-3SII	M-V
First Launch	1970	1970	1974	1977 / 1980	1985	1996
No. of Stages	4	4	3	3	3	3
Total Length	54.1 ft (16.5 m)	77.4 ft (23.6 m)	66.3 ft (20.2 m)	78.1 ft (23.8 m)	91.1 ft (27.8 m)	101 ft (30.8 m)
Diameter	2.41 ft (0.735 m)	4.63 ft (1.41 m)	4.63 ft (1.41 m)	4.63 ft (1.41 m)	4.63 ft (1.41 m)	8.2 ft (2.5 m)
Total Weight	20,700 lb (9,400 kg)	96,100 lb (43,600 kg)	91,700 lb (41,600 kg)	107,000 lb (48,700 kg)	136,000 lb (61,700 kg)	303,000 lb (137,500 kg)
Payload to 100 nm (185 km), 31°	57 lb (26 kg)	400 lb (180 kg)	430 lb (195 kg)	640 lb (290 kg)	1,720 lb (780 kg)	4,000 lb (1,800 kg)

Vehicle Description

L-4S This Lambda rocket was first to orbit a payload. Four-stage solid-propellant, with first three stages unguided and fourth stage with attitude control.

M-4S First member of the M-family. Four stage solid propellant vehicle utilizing fins and spinning for attitude stabilization. Guidance provided by radio guidance.

M-3C Same as M-4S except stage 2 was improved, stage 3 motor was replaced by enlarged M-4S stage 4, and stage 2 used liquid injection thrust vector control (LITVC) and hydrazine side jets for roll control.

M-3H Same as M-3C except longer stage 1, fairing was lengthened, and option existed for a stage 4.

M-3S Same as M-3H except added stage 1 LITVC and small motors for roll control (SMRC).

M-3SII Same as M-3S except enlarged strap-on boosters with steerable nozzles, lengthened stage 2, enlarged stage 3, and wider and longer fairing.

M-V New development with larger diameter stage motors and fairing.

Historical Summary

In 1966, the Japanese National Space Activities Council (SAC) proposed a long-range program concerning the launch and use of satellites. It urged that Japan develop the technology required to launch its own scientific and experimental applications satellites. Early attempts to orbit a spacecraft used a modified sounding rocket known as the L-4S. After four failures, the Japanese successfully orbited the 52 lb (24 kg) Ohsumi satellite (named after the launch site location) on 11 February 1970, using the L-4S vehicle. Although the satellite carried little more than accelerometers and thermometers, the launch represented Japan's first successful orbital mission.

The success of the final L-4S launch gave the Japanese the confidence to proceed with their efforts. To orbit a larger payload, a new vehicle designed specifically as a satellite launcher was needed. The Institute of Space and Aeronautical Science at the University of Tokyo was given responsibility for the design, development and operation of this new launcher designated M (or Mu). The first vehicle in the M-family of launchers was the M-4S. The M-4S was defined as an all-solid-propellant vehicle which made use of aerodynamic and spin-stabilization control techniques. Responsibility for vehicle production and motor fabrication was assigned to the Nissan Motor Co., Ltd. Work was subcontracted by Nissan to other aerospace corporations in Japan. The first attempt to launch the complete vehicle was made in September 1970. The launch was unsuccessful due to failure of the fourth stage to ignite. The M-4S later successfully orbited three satellites of 150–165 lb (67–75 kg) mass.

To expand launch capabilities, the M-3C was designed as a second generation launcher. The three-stage M-3C retained the first stage of the M-4S but incorporated an improved second stage and a new third stage. The second-stage motor used secondary fluid injection for thrust vector control in response to an on-board autopilot. The M-3C was successfully tested on 16 February 1974. There have been three other launches of the M-3C. Two were successful in orbiting satellites of 190–220 lb (85–100 kg) mass, and the other launch was a failure due to a malfunction of a second-stage thrust-vector control system.

An improved version of the the M-3C was designated the M-3H. This vehicle successfully launched a test satellite on 19 February 1977. Two other launches were successful, carrying satellites of 200–290 lb (90–130 kg) mass. The second and third stages of this vehicle remained the same as on the M-3C, while the length of the first stage core motor was increased by one third. The capability of the M-3H is 50% larger than that of the M-3C.

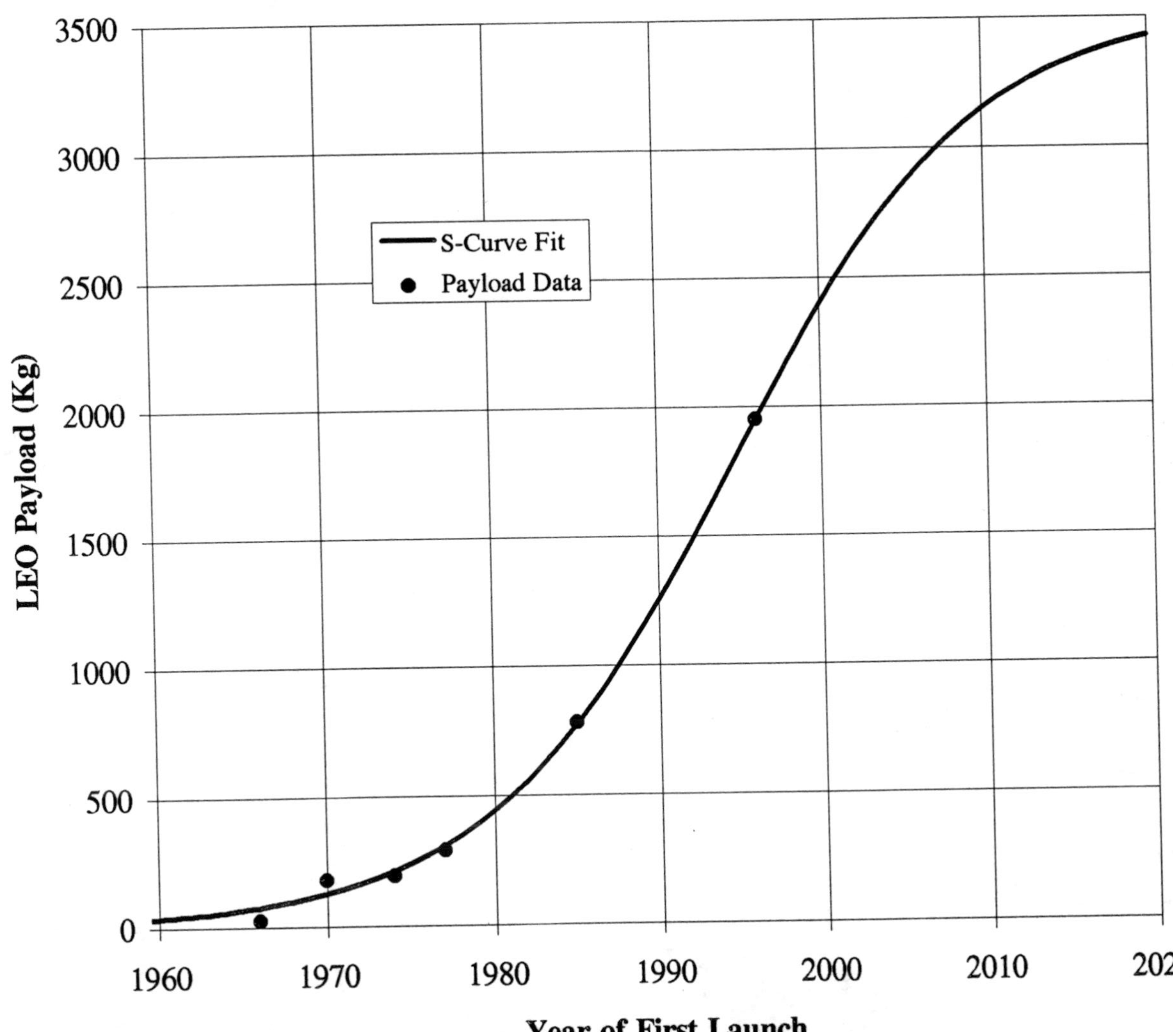
Japanese L/M Launch Vehicle Family
S-Curve Fit
Payload Data
LEO Payload (Kg)
0
500
1000
1500
2000
2500
3000
3500
1960
1970
1980
1990
2000
2010
2020
Year of First Launch

M-Vehicle History

(continued)

Historical Summary

(continued)

The next version of the M-series was the M-3S. This launch vehicle was similar to M-3H, except that the first stage had guidance and control capability. A 400 lb (181 kg) satellite was successfully orbited during the first flight on 17 February 1980. Three more successful launches followed through 1984.

The Institute of Space and Astronautical Science (ISAS) was newly established on 14 April 1981, by reorganizing the Institute of Space and Aeronautical Science at the University of Tokyo (the former ISAS). ISAS is one of the National Interuniversity Research Institutes belonging to the Ministry of Education, Science, and Culture of Japan. ISAS is in charge of research and development of scientific satellites and their launch vehicles (the M-vehicles), while the National Space Development Agency (NASDA) is in charge of development of application satellites and their launch vehicles (the N-, J-, and H-vehicles). The administration of space activities in Japan is coordinated by the Space Activities Commission (SAC) under the Prime Minister's Office. ISAS receives about 13% of Japan's space budget or, in 1989, approximately 21 billion Yen ($145 million).

In 1981, ISAS began the research and development of the M-3SII, built primarily for the Halley's Comet mission, the first Japanese interplanetary flight. This launch vehicle is a three-staged solid rocket with two strap-on boosters and a kick stage that is optionally added for high-energy missions such as interplanetary flights. In contrast with the step-by-step improvements, as were usual in the evolution of the M-family, significant enhancement of the launch capability was required this time. Only the first stage was inherited from the predecessor, M-3S. The size and thrust of the second and third stages were increased. The eight small strap-on motors of the M-3S were replaced by two much larger strap-on boosters. These modifications enabled the launch of a 1,720 lb (780 kg) payload into low Earth orbit or 375 lb (170 kg) into heliocentric orbit.

Among the other major items to be developed were the interstage structures and separation mechanisms, the nose fairing and the attitude control system. A movable nozzle system (MNTVC) was introduced to the strap-on booster, while the secondary liquid injection thrust vector control systems (LITVC) had been commonly used in the M-family. The control system was digitalized to increase control logic design flexibility and to reduce power and weight needs. The first two M-3SII launches, with an optional fourth stage, injected the Japanese first interplanetary probes, Sakigake and Suisei, in 1985 toward Halley's Comet. M-3SII has 2.7 times the performance of the M-3S.

In 1989, ISAS was given approval by the SAC to embark on the development of a new solid propellant launch vehicle to replace the M-3SII currently in use. This follows an amendment to Japan's official space policy. Previously, the Institute was limited to the development of the 4.6 ft (1.4m) diameter M-series of rockets to avoid competition with the NASDA's N- and H-series launchers and to prevent double investment. The new rocket, designated M-V, will have a 8.2 ft (2.5 m) diameter, an 80% increase. Increase in demand from foreign organizations for the launch of large scientific satellites by ISAS has lead to the decision to develop the new launch vehicle.

The M-V is a three-stage solid-propellant rocket, with an optional kick stage for high-energy missions. Due to range safety restrictions, the vehicle is limited to 240,000 lb (110,000 kg) propellant. As a result, the vehicle will be capable of 4,000 lb (1,815 kg) to low Earth orbit or 660-880 lb (300-400 kg) for planetary missions. To achieve this capability all the stages are to be newly developed, that is, no direct heritage from the current M-3SII. Sizing effects and design simplifications such as removal of the strap-on boosters will minimize recurring cost, even though new technology will be introduced. At the launch site, only the launch tower used for the M-3SII will be modified. A sea level test stand will be constructed at Noshiro Testing Center (ISAS's static firing test facility for the developments of both solid and cryogenic engines) for the first stage motor firing test. First launch is 1996.

All M-rocket launches are conducted at the Kagoshima Space Center (KSC) located on the southern tip of Kyushu, the southern most point in Japan's major island chain. Construction of a site for launching of sounding rockets began in February 1962, with extensions for satellite launches by M-vehicles completed by 1966. Although launch dates are greatly restricted because of fishing rights, Kagoshima has remained available to ISAS for launching scientific satellites.

Launch Record

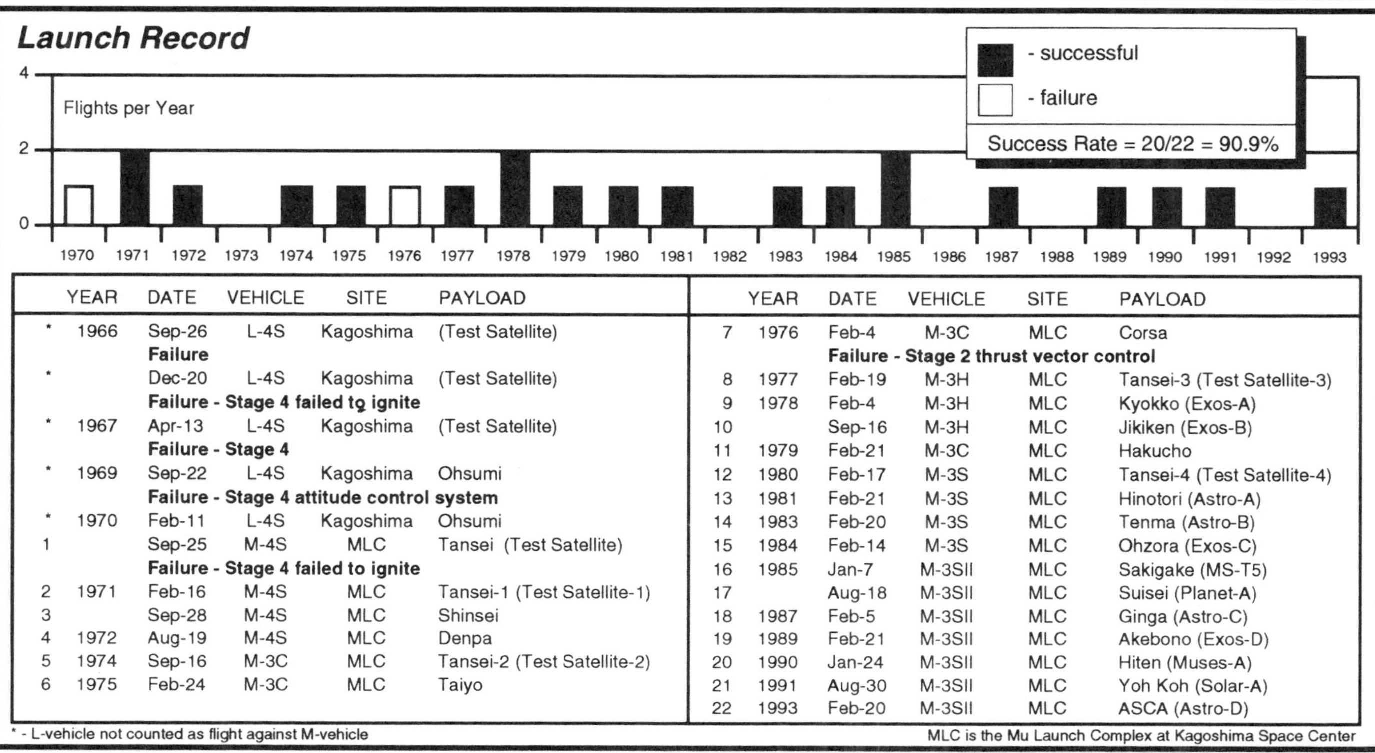

	YEAR	DATE	VEHICLE	SITE	PAYLOAD
*	1966	Sep-26	L-4S	Kagoshima	(Test Satellite)
		Failure			
*		Dec-20	L-4S	Kagoshima	(Test Satellite)
		Failure - Stage 4 failed to ignite			
*	1967	Apr-13	L-4S	Kagoshima	(Test Satellite)
		Failure - Stage 4			
*	1969	Sep-22	L-4S	Kagoshima	Ohsumi
		Failure - Stage 4 attitude control system			
*	1970	Feb-11	L-4S	Kagoshima	Ohsumi
1		Sep-25	M-4S	MLC	Tansei (Test Satellite)
		Failure - Stage 4 failed to ignite			
2	1971	Feb-16	M-4S	MLC	Tansei-1 (Test Satellite-1)
3		Sep-28	M-4S	MLC	Shinsei
4	1972	Aug-19	M-4S	MLC	Denpa
5	1974	Sep-16	M-3C	MLC	Tansei-2 (Test Satellite-2)
6	1975	Feb-24	M-3C	MLC	Taiyo
7	1976	Feb-4	M-3C	MLC	Corsa
		Failure - Stage 2 thrust vector control			
8	1977	Feb-19	M-3H	MLC	Tansei-3 (Test Satellite-3)
9	1978	Feb-4	M-3H	MLC	Kyokko (Exos-A)
10		Sep-16	M-3H	MLC	Jikiken (Exos-B)
11	1979	Feb-21	M-3C	MLC	Hakucho
12	1980	Feb-17	M-3S	MLC	Tansei-4 (Test Satellite-4)
13	1981	Feb-21	M-3S	MLC	Hinotori (Astro-A)
14	1983	Feb-20	M-3S	MLC	Tenma (Astro-B)
15	1984	Feb-14	M-3S	MLC	Ohzora (Exos-C)
16	1985	Jan-7	M-3SII	MLC	Sakigake (MS-T5)
17		Aug-18	M-3SII	MLC	Suisei (Planet-A)
18	1987	Feb-5	M-3SII	MLC	Ginga (Astro-C)
19	1989	Feb-21	M-3SII	MLC	Akebono (Exos-D)
20	1990	Jan-24	M-3SII	MLC	Hiten (Muses-A)
21	1991	Aug-30	M-3SII	MLC	Yoh Koh (Solar-A)
22	1993	Feb-20	M-3SII	MLC	ASCA (Astro-D)

* - L-vehicle not counted as flight against M-vehicle

MLC is the Mu Launch Complex at Kagoshima Space Center

M-Vehicle General Description

M-3SII **M-V**

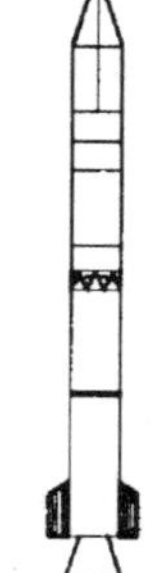

Summary

The M-3SII is an improved version of the M-3S. It is now the main launch vehicle for scientific satellites in Japan. The M-3SII has about 2.7 times the capability of the M-3S because of improvements such as more powerful second and third stage motors and two strap-on boosters. In addition, newly developed movable nozzles allow for thrust direction control of the strap-on boosters. Its first launch was 1985. M-V is the sixth generation in the family of M-vehicles. ISAS was given approval in 1989 for its development. Its diameter will be 80% larger and will be have 2.4 times more performance than the M-3SII. Initial launch is expected in 1996.

Status

M-3SII - Operational
M-V - In Development

Key Organizations

User - ISAS

Launch Service Agency - ISAS

Prime Contractor - Nissan (Vehicle Integration, Operations, Propulsion Motor Production)

Vehicle

System Height: 91.1 ft (27.78 m) for M-3SII
101 ft (30.8 m) for M-V

Payload Fairing Size: 5.41 ft (1.65 m) diameter by 22.5 ft (6.85 m) height for M-3SII
8.2 ft (2.5 m) diameter by 28.9 ft (8.8 m) height for M-V

Gross Mass: 136,000 lb (61,700 kg) for M-3SII
303,000 lb (137,500 kg) for M-V

Planned Enhancements: None

Operations

Primary Missions: Circular / elliptical Earth orbits and interplanetary missions

Compatible Upper Stages: KM-P, KM-D, KM-M for M-3SII
KM-V for M-V

First Launch: 1985 for M-3SII, projected 1996 for M-V

Success / Flight Total: 7 / 7 for M-3SII, 0 / 0 for M-V

Launch Site: KSC - MLC (31.2°N, 131.1°E)

Launch Azimuth: MLC - 90°–160° (max. inclination is 31°–85°)

Nominal Flight Rate: 1 / yr

Planned Enhancements: None

Performance

100 nm (185 km) circ, 31°: 1,720 lb (780 kg) for M-3SII
1,940 lb (880 kg) for M-3SII with optional stage 4
4,000 lb (1,800 kg) for M-V

100 nm (185 km) circ, 90°: 1,300 lb (590 kg) for M-3SII
1,500 lb (680 kg) for M-3SII with optional stage 4
2,860 lb (1,300 kg) for M-V

Geotransfer Orbit, 31°: 1,140 lb (517 kg) for M-3SII with optional stage 4
2,680 lb (1,215 kg) for M-V with optional stage 4

Geosynchronous Orbit: 460 lb (209 kg) for M-3SII with optional stage 4 plus apogee kick motor
1,080 lb (490 kg) for M-V with optional stage 4 plus apogee kick motor

Financial Status

Estimated Launch Price: M-3SII: $25–30M (U.S. DoT estimate)
M-V: $55–65M (ISAS)

Orders: Payload / Agency: Muses-B / ISAS in 1996
Lunar-A / ISAS in 1997
Planet-B / ISAS in 1998
Astro-E / ISAS in 1999

Manifest

Site	1995	1996	1997	1998	1999
M-3SII	—	—	—	1	1
M-V	1	1	1	—	—

Remarks

Cost of the M-3SII could be cut by 20% if several, about three to five, are ordered simultaneously and design simplifications were introduced. M-3SII is the last of the M-vehicles limited by Japanese space policy to 4.8 ft (1.4 m) motor diameter. M-V will be 8.2 ft (2.5 m) in diameter. M-V total development cost is expected to be about 24.2B Yen ($167M) consisting of 5.5B Yen ($38M) for first vehicle, 13.1B Yen ($90M) for development and 5.6B Yen ($39M) for ground facility requirements.

M-Vehicle Vehicle

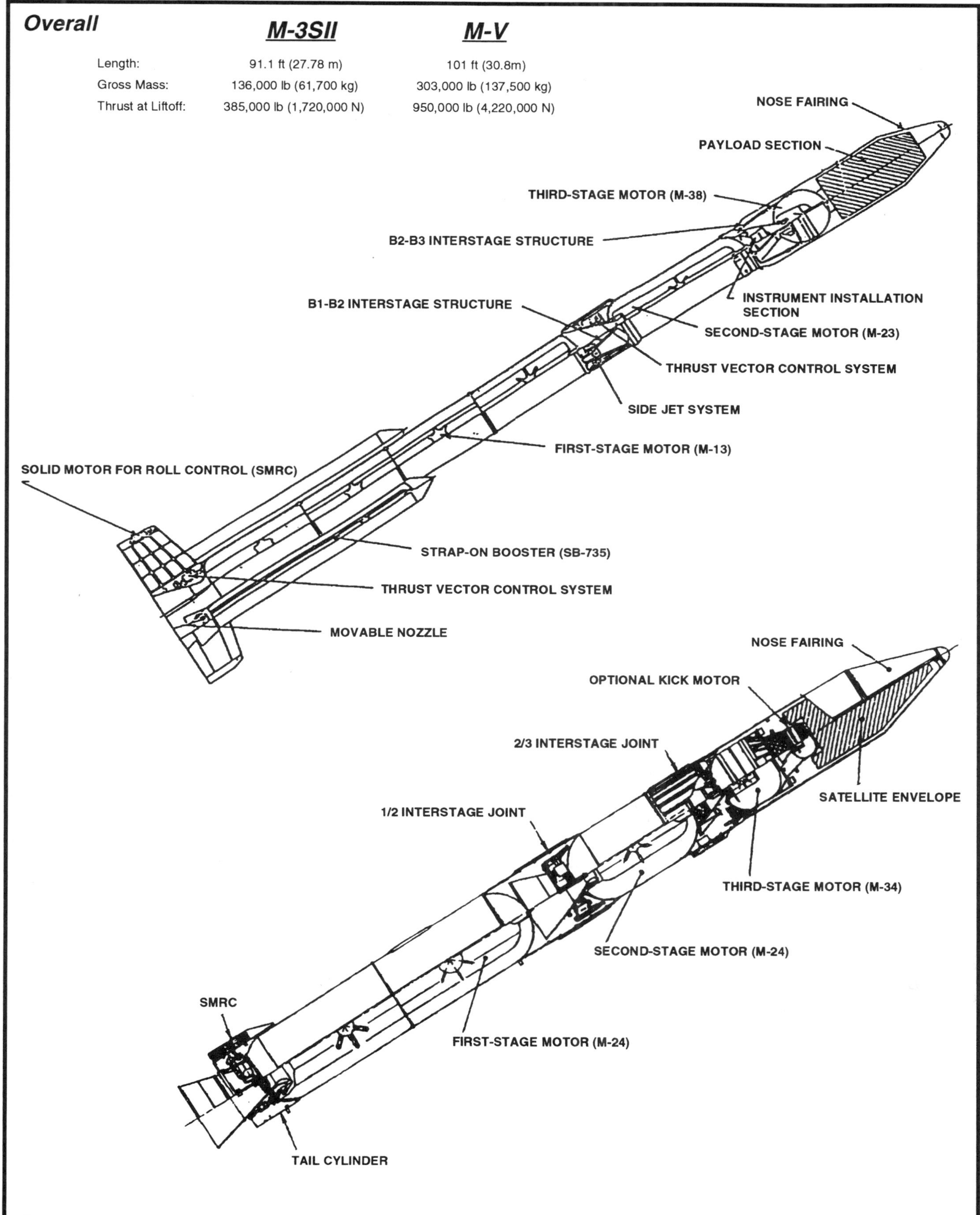

M-Vehicle Vehicle

(continued)

M-3SII Stages	Strap-on Booster (SB-735)	Stage 1 (M-13)	Stage 2 (M-23)	Stage 3 (M-38)	Stage 4 - Optional (KM-P)	(KM-D)	(KM-M)
Dimension:							
Length	27.1 ft (8.25 m)	48.4 ft (14.75 m)	20.3 ft (6.19 m)	8.79 ft (2.68 m)	3.92 ft	3.22 ft	4.30 ft
Diameter	2.41 ft (0.735 m)	4.63 ft (1.41 m)	4.63 ft (1.41 m)	4.91 ft (1.495 m)	2.59 ft	2.07 ft	2.63 ft
Mass: (each)							
Propellant Mass	8,800 lb (4000 kg)	59,700 lb (27,100 kg)	22,900 lb (10400 kg)	7,230 lb (3280 kg)	923 lb	617 lb	1,113 lb
Gross Mass	11,000 lb (5000 kg)	76,500 lb (34,700 kg)	28,900 lb (13100 kg)	7,910 lb (3590 kg)	1,016 lb	670 lb	1,213 lb
Structure:							
Type	Monocoque	Monocoque	Monocoque	Monocoque		Monocoque	
Case Material	Steel	Steel	Steel	Titanium (6Al-4V)	Titanium	Titanium	Composite
Propulsion:							
Propellant	CTPB	CTPB	HTPB	HTPB	HTPB	HTPB	HTPB
Average Thrust (each)	73.5K lb (327K N) vac 63.8K lb (284K N) SL	287K lb (127.5K N) vac 258K lb (114.7K N) SL	118K lb (524K N) vac 85.2K lb (379K N) SL	29.8K lb (132K N) vac	7.0K lb	3.9K lb vac	7.2K lb
Number of Motors	2	1	1	1	1	1	1
Number of Segments	1	4	1	1	1	1	1
Isp	266 sec vac	266 sec vac	282 sec vac	294 sec vac	286 s vac	294 s vac	293 s vac
Chamber Pressure	?? psia (?? bar)	?? psia (?? bar)	?? psia (?? bar)	?? psia (?? bar)	?? psia	?? psia	?? psia
Expansion Ratio	9.13	7.80	23.24	51.84	46.13	??	64.97
Control-Pitch, Yaw,	MNTVC	LITVC	LITVC	Spin stabilized		Spin stabilized	
Roll		SMRC	Hydrazine side jets			–	
Events:							
Nominal Burn Time	38 sec	70 sec	73 sec	87 sec	44 sec	38 sec	63 sec
Stage Shutdown	Burn to depletion	Burn to depletion	Burn to depletion	Burn to depletion		Burn to depletion	
Stage Separation	Sep. nuts / struts	Sep. nuts / spring	Sep. nuts / spring	V-clamp / springs		V-clamp / springs	

Remarks:

The first stage is composed of an M-13 motor, two SB-735 strap-on boosters, four fins and their shrouds, and 1st-2nd interstage joint. The M-13 rocket motor is made of four segments. The jettison of the strap-on boosters is initiated by the ignition of the pyrotechnics (separation nuts). The 1st -2nd stage-separation is also initiated by the ignition of the separation nuts. Each of the six petals of the 1st-2nd interstage joint rotates around the bottom hinges in order to assure separation. Spring force provides nominal 3 ft/s (1 m/s) relative velocity. Secondary liquid injection thrust vector control (LITVC) devices are installed at the nozzle of M-13 motor, which control the pitch and the yaw attitudes. The roll attitude is controlled by the movable nozzle device of the strap-on boosters, as well as by the small Solid Motors for Roll Control (SMRC) installed at the tips of the fins.

The second stage is composed of an M-23 motor, a nose fairing, a 2nd-3rd interstage joint an an instrument installation section. The 2nd-3rd interstage joint is V-clamp type. At the separation, six petals rotate around the bottom hinge just as the 1st-2nd interstage joint. The instrument installation section houses flight control systems, telemetry, range safety systems, tracking, and power systems. The pitch and yaw attitudes are controlled by the LITVC device of the M-23 motor. The roll attitude is controlled by the hydrazine Side Jet (SJ) device installed at the nozzle of M-23 motor. The third stage is composed of spherical M-3B motor and the payload adapter. The payload adapter is a V-clamp type interstage joint.

The first stage, strap-on boosters, and second stage use a maraging steel case and a reinforced plastic nozzle. The third stage uses a spherical motor case, fabricated by welding titanium alloy cups with a highly submerged nozzle that is bell-shaped and fabricated from reinforced plastics. All stages have a seven-point-star propellant grain, internal burning configuration. Their grain is a cast-in-case aluminized composite propellant with a CTPB binder for the strap-on boosters and first stage and HTPB for the second and third stages. Their igniter is a flame producing rocket motor type with dual squib initiators.

M-Vehicle Vehicle

(continued)

M-V Stages

	Stage 1 (M-14)	Stage 2 (M-24)	Stage 3 (M-34)	Stage 4 - Optional
Dimension:				
Length	45.3 ft (13.8 m)	22.3 ft (6.8 m)	11.8 ft (3.6 m)	4.9 ft (1.5 m)
Diameter	8.2 ft (2.5 m)	8.2 ft (2.5 m)	7.2 ft (2.2 m)	3.9 ft (1.2 m)
Mass: (each)				
Propellant Mass	157,600 lb (71,490 kg)	68,500 lb (31,060 kg)	22,000 lb (10,000 kg)	2,890 lb (1,312 kg)
Gross Mass	184,200 lb (83,560 kg)	76,000 lb (34,470 kg)	24,300 lb (11,000 kg)	3,150 lb (1,430 kg)
Structure:				
Type	Monocoque	Monocoque	Monocoque	Monocoque
Case Material	Steel	Steel	CFRP Filament Wound	CFRP Filament Wound
Propulsion:				
Propellant	HTPB	HTPB	HTPB	HTPB
Average Thrust (each)	850K lb (3,780K N) vac	280K lb (1,240 N) vac	65.2K lb (290K N) vac	11.7K lb (52K N) vac
Number of Motors	1	1	1	1
Number of Segments	2	1	1	1
Isp	276 sec vac	288 sec vac	301 sec vac	298 sec vac
Chamber Pressure	853 psia (59 bar)	853 psia (59 bar)	853 psia (59 bar)	710 psia (49 bar)
Expansion Ratio	11	31	96	89
Control-Pitch, Yaw,	MNTVC (±5°)	LITVC	MNTVC	MNTVC
Roll	SMRC	SMSJ	Hydrazine SJ	Hydrazine SJ
Events:				
Nominal Burn Time	46 sec	71 sec	102 sec	73 sec
Stage Shutdown	Burn to depletion	Burn to depletion	Burn to depletion	Burn to depletion
Stage Separation	FLSC / FIH	V-clamp/Spring	V-clamp/Spring	V-clamp/Spring

Remarks:

The first stage is composed of an M-14 motor, a tail shroud and 1st-2nd interstage joint. The M-14 motor is made of two segments. The 1st-2nd interstage separation is a fire-in-the-hole (FIH) type. Stage separation is initiated by the ignition of Flexible Linear Shaped Charges (FLSCs) which cut the interstage structure. Movable Nozzle Thrust Vector Control (MNTVC) device is installed at the nozzle of M-14 motor, which control the pitch and yaw attitudes. The nozzle employs a flexible bearing to achieve the required thrust vector deflection of 5 degrees. The roll attitude is controlled by the small Solid Motors for Roll Control (SMRC) installed around the tail shroud.

The second stage is composed of an M-24 motor, a nose fairing, a 2nd-3rd interstage joint and an instrument installation section. The 2nd-3rd interstage joint is V-clamp type. The M-24 motor nozzle is of a submerged, slightly contoured bell-conical design with an Extendible and Expandable Exit Cone (E/EEC) system. The E/EEC system allows a large nozzle expansion ratio despite interstage restrictions. The E/EEC is to be deployed just after motor ignition. The pitch and yaw attitudes are controlled by the secondary Liquid Injection Thrust Vector Control (LITVC) devices of the M-24 motor. The roll attitude is controlled by the SMRC device, installed at the 2nd-3rd interstage section. The pitch, yaw, and roll attitudes after the M-24 motor burnt out are controlled by the Solid Motor Side Jet (SMSJ) device installed around the M-24 motor nozzle.

The third stage is composed of a semispherical M-34 motor, an instrument installation section, and a payload adapter. The M-34 motor nozzle is of a movable, slightly contoured bell-conical design with a simple collapsible cup type EEC system. The MNTVC device controls the pitch and yaw attitudes. Hydrazine Side Jets (SJ) device is installed around the M-34 motor nozzle, which controls the roll attitude during the motor firing and the pitch, yaw, and roll attitudes after motor burnout.

The first stage and second stage use maraging steel cases and reinforced plastics nozzles. The third stage uses a carbon fiber epoxy Filament-Wound (FW) case and a reinforced plastic nozzle. All stages have a star perforated propellant grain, internal burning configuration. Their grains are cast-in-case aluminized composite propellants with an HTPB binder. Their igniters are flame producing rocket motor type, with dual squib initiators.

M-Vehicle Vehicle

(continued)

Payload Fairing

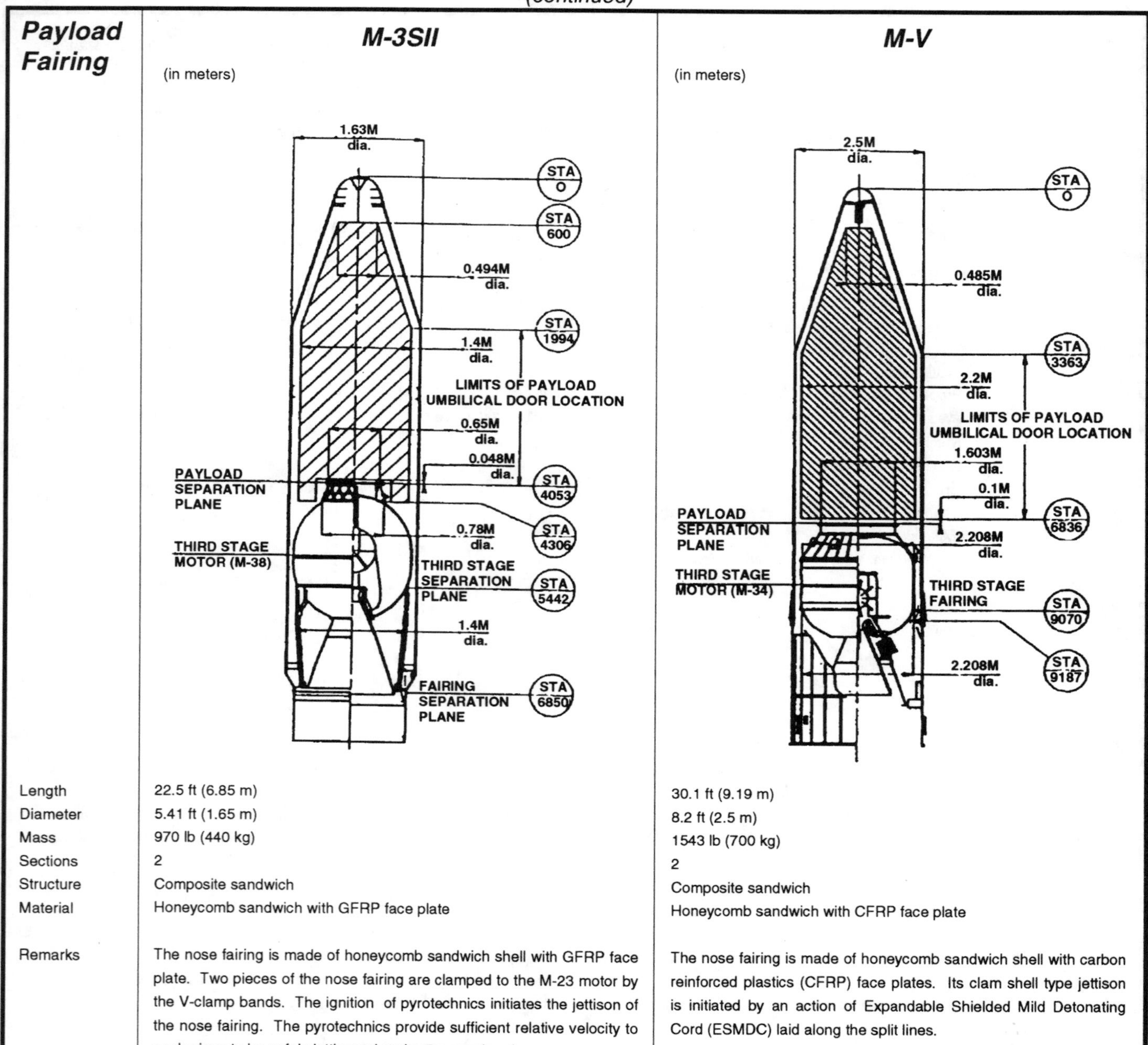

	M-3SII	M-V
Length	22.5 ft (6.85 m)	30.1 ft (9.19 m)
Diameter	5.41 ft (1.65 m)	8.2 ft (2.5 m)
Mass	970 lb (440 kg)	1543 lb (700 kg)
Sections	2	2
Structure	Composite sandwich	Composite sandwich
Material	Honeycomb sandwich with GFRP face plate	Honeycomb sandwich with CFRP face plate
Remarks	The nose fairing is made of honeycomb sandwich shell with GFRP face plate. Two pieces of the nose fairing are clamped to the M-23 motor by the V-clamp bands. The ignition of pyrotechnics initiates the jettison of the nose fairing. The pyrotechnics provide sufficient relative velocity to each piece to be safely jettisoned under 5 g acceleration.	The nose fairing is made of honeycomb sandwich shell with carbon reinforced plastics (CFRP) face plates. Its clam shell type jettison is initiated by an action of Expandable Shielded Mild Detonating Cord (ESMDC) laid along the split lines.

M-Vehicle Vehicle

(continued)

Avionics

M-3SII. The instrumentation system for the M-3SII vehicle includes vehicle-borne equipment required to provide ground monitoring of subsystems during prelaunch checkout and countdown, telemetry monitoring of the flight performance of the first two stages, radar beacon tracking, and command destruct electronics for receiving and decoding of command destruct information. The radar transponder systems employed on the M-3SII vehicle are C-Band and L-Band. The C-Band transponder system is also available for commanding 20 control items of guidance and control systems. The radar tracking data are generally available up to the ignition time of the third stage.

The guidance and control systems provide an attitude reference and the resultant control torques necessary for stabilizing a vehicle as well as for carrying out precise orbit injection. Attitude reference unit on the second stage instrument installation section consists of 3 RIG's (rate-integrating-gyros) for pitch, yaw, and roll axes mounted on a spin-stabilized platform and a roll rate gyro. Three rate gyros on the first stage provide damping signals for first stage pitch, yaw, and roll control.

Attitude determination and control calculations are carried out by onboard microprocessor system which is comprised of three CPUs. The 16 bit CPUs with bipolar type are adopted. For the first stage control, bending of up to fifth mode is considered and a state feed-back control algorithm is used.

M-V. The guidance and control systems provide an attitude reference, orbit determination, and the resultant control torques necessary for stabilizing a vehicle as well as for carrying out precise orbit injection. The system includes Inertial Navigation System (INS) which comprised strap-down Fiber Optical Gyros (FOGs), accelerometers and computers on the third stage. Three rate gyros on the first stage provide damping signals for the first stage pitch, yaw, and roll control. The attitude and orbit determination and control calculations are carried out by an onboard microprocessing system. The primary guidance for orbit injection is done from the ground by a tracking radar and a real-time computer. As a backup an onboard INS is used.

The instrumentation system for the M-V vehicle is similar to that of M-3SII series: (1) The guidance and control system, S-Band telemetry system and C-Band radar transponder are on the third stage; (2) The VHF telemetry system is on the first and second stages; (3) The L-Band radar transponder in on the second stage; (4) The C-Band transponder system is also used for commanding of guidance and control; (5) The radio guidance is primarily used and the Inertial Navigation System (INS) onboard the vehicle is used as a backup.

Attitude Control System

M-3SII. The Solid Motor for Roll Control (SMRC) device controls the first stage roll angle. Four SMRC devices are on the tips of aerodynamic's fins. The Movable Nozzle Thrust Vector Control (MNTVC) device on the strap-on boosters is also used for controlling the first stage roll angle. The thrust vectors of the strap-on boosters are deflected to give a roll torque.

The LITVC devices on the first stage and second stage are used for pitch and yaw control during the burn phase of each stage motor. The control thrust of the TVC is obtained by injecting sodium perchloride gas normal to the main thrust, resulting in a small deflection of the thrust direction. The hydrazine Side Jet (SJ) is used for roll axis control during the second stage motor burn. It is also used for 3-axis control after the second stage motor has burnt out. The SJ system is composed of four modules, each having four thrusters.

M-V. The Movable Nozzle Thrust Vector Control (MNTVC) device is used for controlling the 1st stage pitch and yaw angles. The Solid Motor for Roll Control (SMRC) device controls the first stage roll angle.

The Liquid Injection Thrust Vector Control (LITVC) device is used for the second stage pitch and yaw control and the SMRC device is used for roll control while the main motor is burning. The Solid Motor Side Jet (SMSJ) is used for pitch, yaw, and roll control after the motor burnt out.

The MNTVC is used for the third stage pitch and yaw control while the motor is burning. The Side Jet (SJ) system is used for roll control during the third stage motor burn and is also used for pitch, yaw, and roll control after the motor has burnt out. Sodium perchloride is used for LITVC and hydrazine is used for SJ.

M-Vehicle Performance

M-3SII

Circular Orbits (Also with KM-M Optional 4th Stage),
31° Inclination

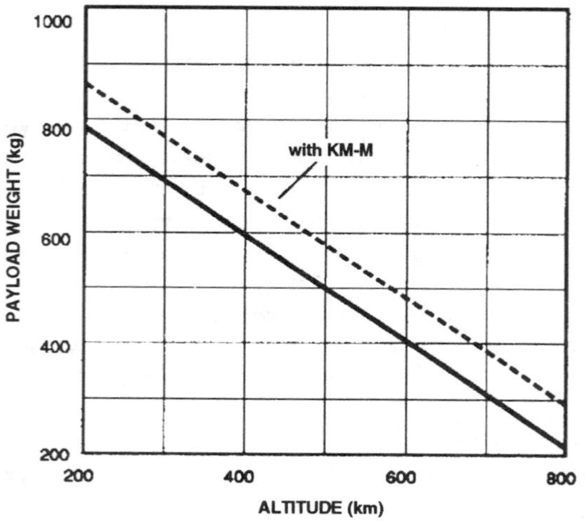

PAYLOAD WEIGHT (kg)
1000
800
600
400
200
200
400
600
800
ALTITUDE (km)
with KM-M

Circular Orbits at Different Inclinations
(Also with KM-M Optional 4th Stage)

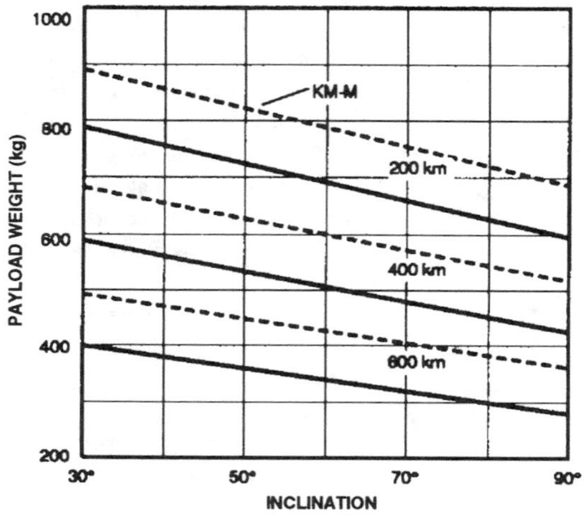

PAYLOAD WEIGHT (kg)
1000
800
600
400
200
30°
50°
70°
90°
INCLINATION
KM-M
200 km
400 km
600 km

Elliptical Orbits,
31° Inclination

PAYLOAD MASS (kg)
800
700
600
500
400
300
200
100
1000
10000
100000
APOGEE ALTITUDE (km)
INCLINATION = 31°
PERIGEE = 200 km
300 km
400 km
500 km
600 km
700 km
CIRCULAR
ORBIT

M-V

Elliptical Orbits,
(31° inclination)

PAYLOAD (kg)
2,000
1,600
1,200
800
400
100
1,000
10,000
100,000
APOGEE ALTITUDE (km)
PERIGEE ALTITUDE
150 km
250 km
500 km
700 km

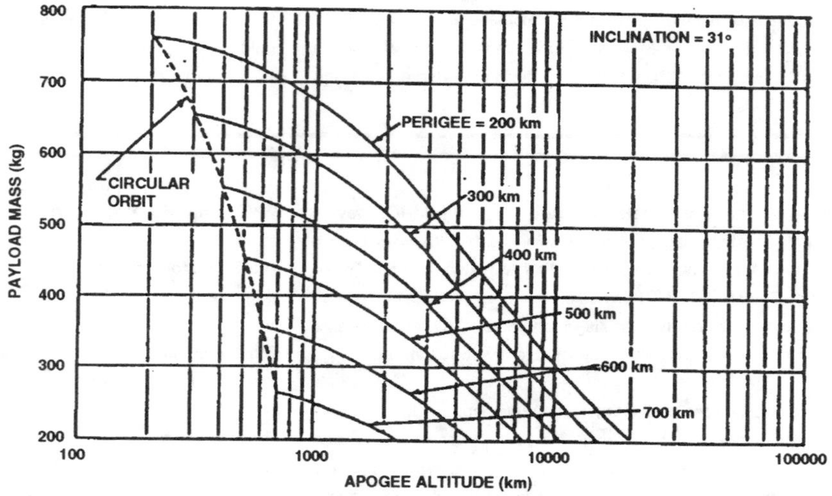

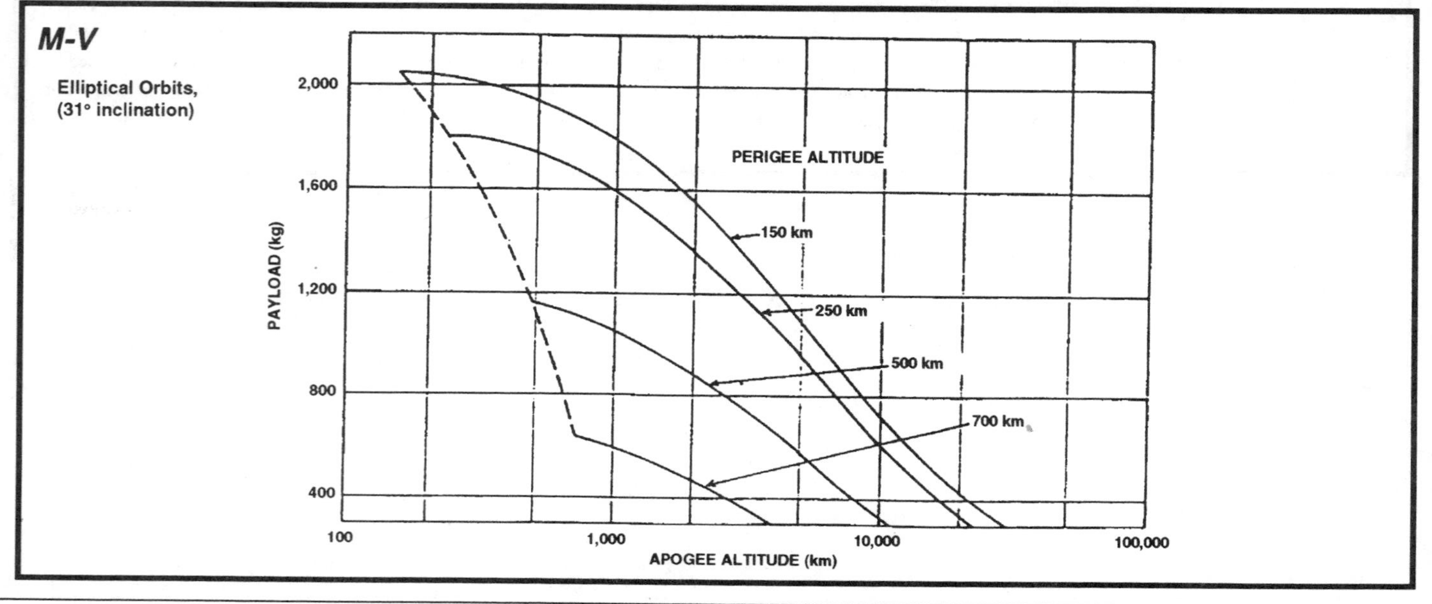

M-Vehicle Operations

Launch Site

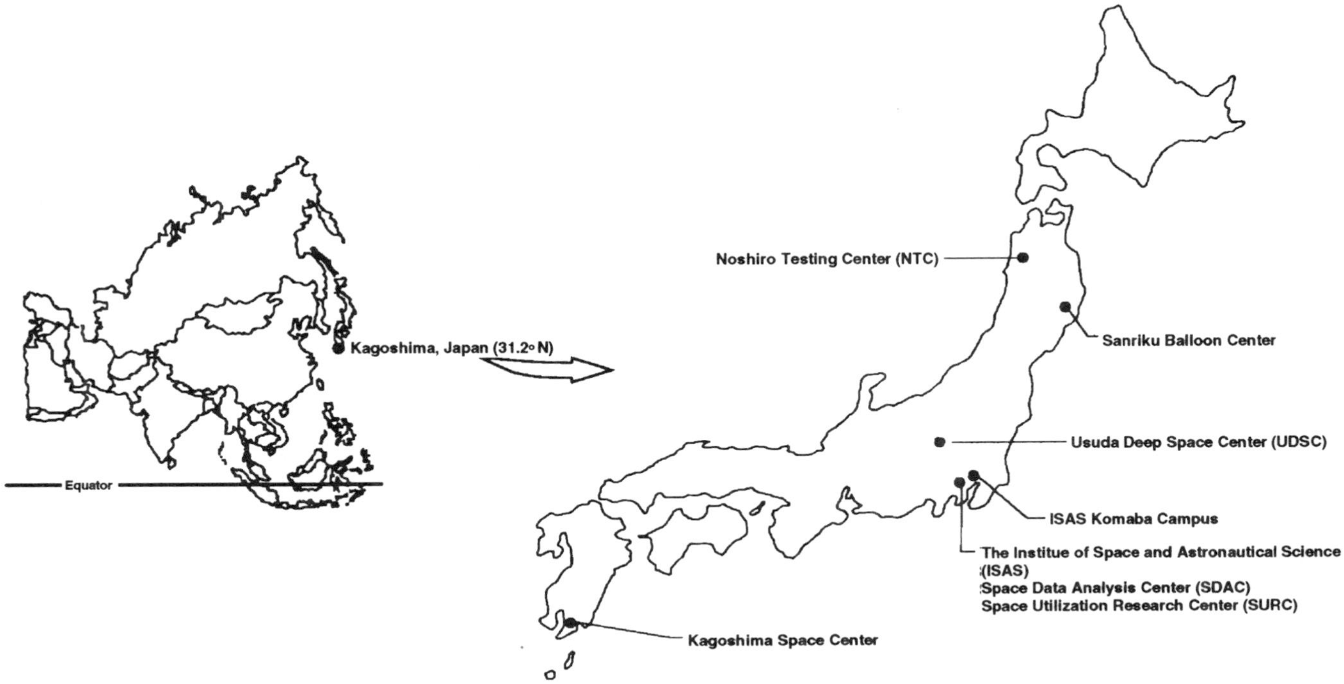

Launch Facilities

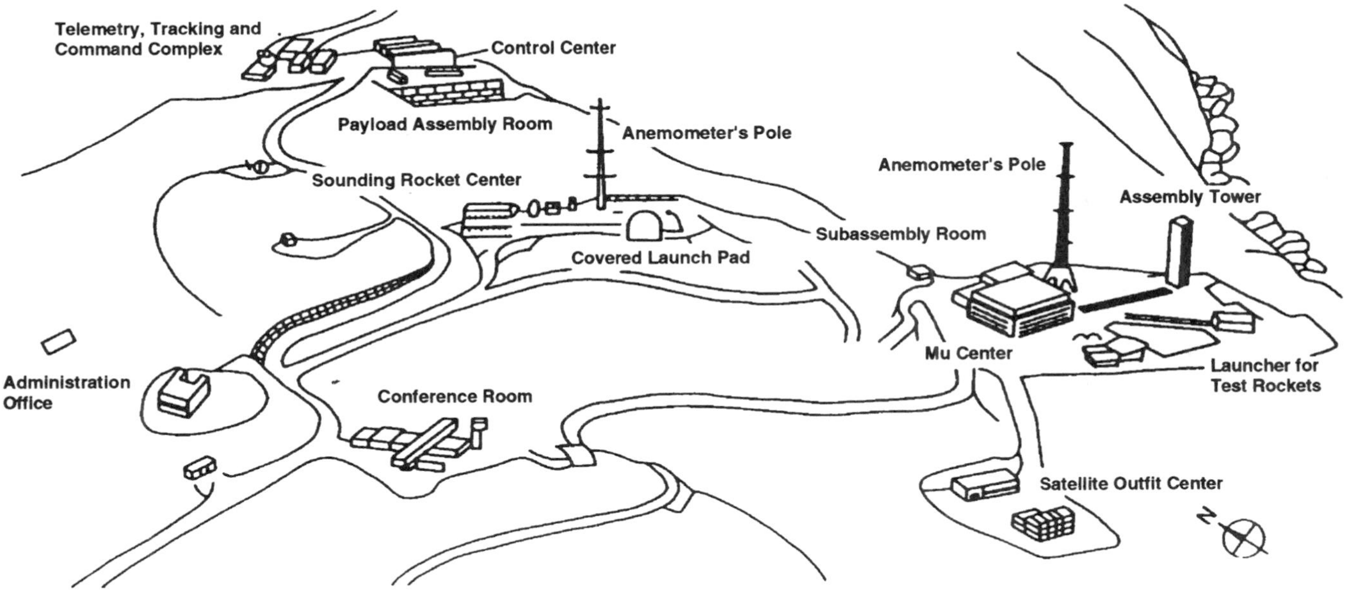

M-Vehicle Operations

(continued)

Launch Processing

The Kagoshima Space Center (KSC) is located in Uchinoura-cho, which lies on the east coast of Ohsumi Peninsula, Kagoshima Prefecture. Starting in February 1962, the first stage of construction was completed with the Lambda and other smaller rocket launch facilities. In October 1966, the launch facilities for M rockets were built.

The land around KSC is mostly hilly. Various facilities, such as those for launching rockets; telemetry, tracking, and command stations for rockets and satellites; and optical observation posts are placed in areas prepared by flattening the tops of several hills, and connected to each other with roads. Buildings in the Mu Center have the total floor space of 137,000 sq-ft (12,750 sq-m).

Since 1962, when the temporary use of KSC began, a total of 293 rockets (Mu: 19, Lambda: 25, Kappa: 116, S and test rockets: 133) have been launched through February 1987. Objections by local fishermen to the noise associated with launches over their fishing areas have resulted in space launches being restricted to roughly two months a year (generally February and September).

The facilities used for M launch preparation include the Satellite Preparation Building (SPB), Mu Assembly Building (MAB) and Mu Launch Complex (MLC) at the Mu Center, KSC.

M-3SII. Vehicle elements such as stage motors, fins, and nose fairing are transported separately to the launch site by trucks. The most massive piece is the second stage motor, weighing about 13 tons, since the four-segmented first stage is transported segment by segment. Vehicle integration of these elements is carried out inside the assembly tower, building the launcher vertically. Segments of the first stage are assembled first, then the strap-on boosters, the second stage motor, and so on. The spacecraft is integrated with the third stage and covered by the nose fairing in advance at the MAB and then transported to the tower. At launch the launcher swings out of the tower to the launch position with the vehicle on it.

It is usual at KSC to integrate the vehicle on site two months before launch using a dummy spacecraft. This takes about three weeks. Another two weeks are spent just prior to launch to bring the spacecraft aboard the launch vehicle and check out the whole system, launch vehicle and spacecraft. Though the above procedure provides margin in the schedule, these two pieces might be merged into one if the spacecraft could be prepared in time. Then, the total launch operation would take four weeks with 150 to 250 personnel involved.

M-V. The vehicle is expected to be processed through a procedure similar to M-3SII.

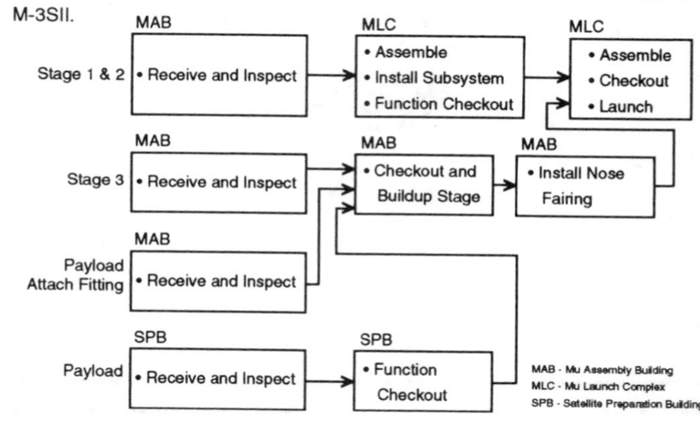

Typical Launch site M-3SII Operation Flow

Flight Sequence

M-3SII Typical Flight Sequence

The M-3SII first stage is launched with an elevation angle greater than 75 deg. and follows a zero lift trajectory based on an averaged wind profile of the season. An inclined launch method is employed to augment range safety. The first stage TVC system starts to operate at 6 sec after liftoff. At 40 sec after liftoff, the strap-on boosters are separated. Prior to booster separation, the roll control actuator is switched over from the MNTVC to the SMRC. During the last phase before second stage ignition, the preprogrammed attitude profile of the second stage, which approximates a gravity turn trajectory, is biased by radio command so that the nominal injection condition is achieved based on the actual first stage trajectory. This assumes that the third stage and kick stage are fired equivalently at the apogee of the second stage trajectory. After the second stage burnout, the nose fairing is jettisoned at 69 nm (128 km). The third stage is aligned to the injection attitude and is given a spin of 2 Hz by the side jet system of the second stage. The ignition time and the attitude of the third stage (and the kick stage) are modified again by radio command based on a ground measured second stage trajectory.

Flight Time (min:sec)	Events
00:00	1st stage and strap-on boosters ignition
00:05	MNTVC start
00:06	1st stage LITVC start
00:35	MNTVC stop
00:36	SMRC start
00:40	Strap-on boosters separation
00:55	1st stage LITVC stop
01:20	SMRC stop
01:24	1st stage separation
01:26	2nd stage ignition, 2nd stage LITVC & roll SJ start
02:21	2nd stage LITVC stop
02:22	Pitch and yaw SJ start
02:35	Nose fairing jettison
03:15	3rd stage spin up start
04:06	2nd stage separation
04:08	3rd stage ignition
06:00	3rd stage separation

M-V Typical Flight Sequence

Flight Time (min:sec)	Events
00:00	Ignition, MNTVC start
01:10	1st stage separation, 2nd stage ignition LITVC & SMSJ start
02:57	Nose fairing jettison
03:00	2nd stage separation
03:25	3rd stage ignition
05:50	3rd stage separation

M-Vehicle Payload Accommodations

	M-3SII	M-V
Payload Compartment		
Maximum Payload Diameter	55.1 in (1,400 mm)	86.6 in (2,200 mm)
Maximum Cylinder Length	81.1 in (2,059 mm)	136.7 in (3,473 mm)
Maximum Cone Length	54.9 in (1,394 mm)	100.9 in (2,563 mm)
Payload Adapter		
Interface Diameter	25.6 in (650 mm)	63.1 in (1603 mm)
Payload Integration		
Nominal Mission Schedule Begins	T-30 months	T-30 months
Launch Window		
Latest Countdown Hold Not Requiring Recycling	T-120 min	T-120 min
On-Pad Storage Capability	Indefinite	Indefinite
Latest Access to Payload	T-1.5 hrs	T-?? hrs
Environment		
Maximum Load Factors	+7.0 g axial, ±5.0 g lateral	+5.0 g axial, ±7.0 g lateral
Minimum Lateral / Longitudinal Payload Frequency	20 Hz / 20 Hz	15 Hz / 15 Hz
Maximum Overall Acoustic Level	143.3 dB	148.8 dB
Maximum Flight Shock	7.32 g	10 g
Maximum Dynamic Pressure on Fairing	3,070 lb/ft2 (147,000 N/m2)	?? lb/ft2 (142,000 N/m2)
Maximum Pressure Change in Fairing	0.7 psi/s (4.9 KPa/s)	0.4 psi/s (2.9 KPa/s)
Cleanliness Level in Fairing (Prior to Launch)	Class 100,000	Class 100,000
Payload Delivery		
Standard Orbit and Accuracy (3 sigma)	±5 nmi (±10 km) ±0.01° inclination	±2.7 nmi (±5 km) ±0.01° inclination
Attitude Accuracy (3 sigma)	All axes: ±1 deg	All axes: ±1 deg
Nominal Payload Separation Rate	6.6 ft/s (2 m/s)	66 ft/s (2 m/s)
Deployment Rotation Rate Available	?? to ?? rpm	?? to ?? rpm
Loiter Duration in Orbit	?? hrs	?? hrs
Maneuvers (Thermal / Collision / Telemetry)	Yes	Yes

M-Vehicle Notes

Publications

User's Guide

M-3SII User's Guide, The Institute of Space and Astronautical Science, March 1984.

Technical Publications

"An Air-Launch Vehicle as a Derivative of the Japanese M-V Rocket", R. Akiba, H. Matsuo, M. Kohno, The Institute of Space and Astronautical Science, T. Makino, Nissan, Y. Nagao, Fuji Heavy Industries, IAF-90-178, 41st Congress of the International Astronautical Federation, Dresden , Germany, October 6-12, 1990.

"Cost Estimation of the M-V Launch Vehicle", R. Akiba, H. Hinada, H. Matsuo, The Institute of Space and Astronautical Science, 1990.

"ISAS' New Satellite Launcher M-V", R. Akiba, H. Matsuo, M. Kohno, ISAS, 3rd PISSTA, Los Angeles, CA, November 1989.

"ISAS's New Launch Vehicle [M-3S-II] for Scientific Use", R. Akiba, H. Matsuo, Y. Matogawa, ISAS, T. Hosomura, Nissan, 1989.

"The Institute of Space and Astronautical Science", Japanese Ministry of Education, Science and Culture (brochure), 1989.

"Compendium of Small Class ELV Capabilities, Costs and Constraints", Karen S. Poniatowski, NASA Headquarters, Washington, D.C., 1989.

Acronyms

CFRP - carbon reinforced plastics
CPU - computer processing unit
E/EEC - extendible and expandable exit cone
FIH - fire in the hole
FLSC - flexible linear shaped charges
HTPB - hydroxy terminated polybutadiene
INS - inertial navigation system
ISAS - The Institute of Space & Astronautical Science
KSC - Kagoshima Space Center
LITVC - liquid injection thrust vector control
MAB - Mu Assembly Building
MLC - Mu Launch Complex
MNTVC - movable nozzle thrust vector control
NASDA - National Space Development Agency
SJ - side jet
SMRC - solid motors for roll control
SMSJ - solid motors for side jet
SOB - strap-on booster
SPB - Satellite Preparation Building
TVC - thrust vector control

Other Notes—M-Vehicle Growth Possibilities

ISAS has examined a concept of an air-launch system as a derivative of the M-V rocket. As a preparatory work, ascent phase trajectory optimization has been carried out for various initial conditions, corresponding to the ground, balloon and airplane launch modes. Based on the flight path analysis, a winged three-stage horizontal air-launch configuration was selected. The gross mass of the air-launch rocket is 114,300 lb (51,850 kg), and the payload mass into 135 nm (250 km) is 2,800 lb (1,270 kg). The payload ratio reaches 2.5% versus 1.5% of the original M-V rocket.

First Stage. The first stage of the air-launch rocket is designed on the basis of the second stage of the M-V rocket. The original M-V second stage motor has a LITVC system that is replaced by aerodynamic control actuators. The main wing of the air-launch rocket is installed at just aft of the separation point between the first and second stages. The wing span is 21.6 ft (6.6 m) and its area is 167 ft^2 (15.5 m^2). The wing and tails are composed of CFRP material, and the additional inert mass of the aerodynamic surface is estimated up to 2,200 lb (1,000 kg). The aerodynamic surface can produce 0.2g lift force with an attack angle of 20 degree. This lift force is utilized for the separation from the carrier and the pull-up maneuver after the ignition.

Second Stage. The second stage of the air-launch rocket is exactly the third stage of the M-V rocket. The original M-V third stage motor is installed in the payload fairing of the M-V rocket. However, the second stage of the air-launch rocket flies in the atmosphere, therefore, the fairing is attached to the second stage during its powered ascent. During the second stage burn, the attitude is controlled by an MNTVC system, which is used in the original stage. After the second stage burnout, the payload fairing is separated, and then the attitude for the third stage ignition and spin-up maneuver are carried out by a reaction control system (RCS) installed on the second stage.

Third Stage. The third stage of the M-3SII is used for the third stage motor of the air-launch rocket. The M-3B stage does not have any attitude control system, therefore, the third stage flies with spin stabilization. The attitude orientation and spin-up maneuver are carried out by the second stage RCS.

Air-Launch Rocket

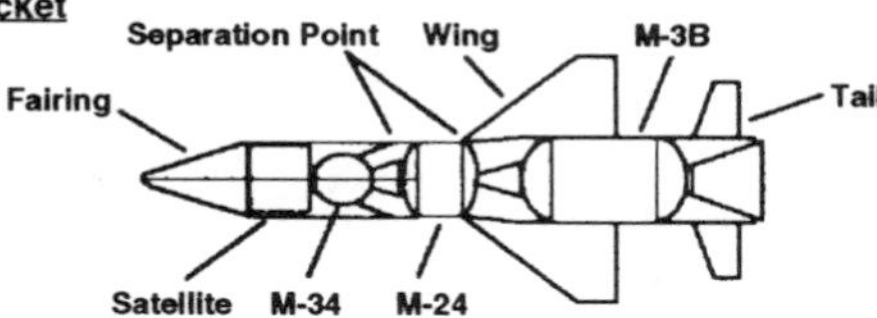

Payload. The satellite is installed on the payload attach fitting of the third stage. A cylindrical envelope of 6.6 ft (2.0 m) diameter and 5.6 ft (1.7 m) length is available in the fairing that is 7.2 ft (2.2 m) in diameter. The payload envelope allows launch of a low density 15.5 lb/ft3 (250 kg/m3) satellite.

Carrier Aircraft. The liftoff mass of the air-launch rocket can be carried by large airplanes such as Boeing 747, C-5, and Antonov 225. Based on availability and procurement cost, the B-747 seems suitable. The air-launch rocket is loaded upon the B-747 fuselage since it cannot be flown from its wing. With the air-launch rocket, the B-747 can continue flying for 3,560 nm (6,600 km) without refueling. This flight range of the carrier allows the launch of the rocket at the equator area. The carrier can fly to near equator from a Japanese airfield, and fly back to Japan after launch, even when the carrier must carry back the rocket to Japan. This brings an operational flexibility, which is probably the most important feature of the air launch. In a normal flight operation, the B-747 with the air-launch rocket can climb up to an altitude of 6.1 nm (11.4 km) at a speed of Mach 0.82. At this flight condition, the air-launch rocket can separate safely from the B-747 using its lift force.

Cost. The air-launch rocket is estimated to cost 4.2B Yen ($29M) per vehicle plus 0.3B Yen ($2M) for launch operations and 0.15B Yen ($1M) for B-747 maintenance. The cost per unit payload mass is 20% higher than the M-V due primarily to the M-V first stage being relatively inexpensive and the need for aerodynamic surfaces but it is 40% cheaper than the M-3S-II.

In conclusion, this concept will relax the launch constraints which are strictly imposed now in Japan mainly due to the heavy sea traffic along the Japanese coast line and will provide an economical scheme to cover the payload spectrum in its medium to small size range.

Russia / Ukraine

Table of Contents

Table of Contents

Launch Service Point of Contact:
Rocket Space Corporation Energia
Lenin Street 4a
Kaliningrad, Moscow area
The Russian Federation, CIS 141070
Fax: (7095) 187-98-77

USA Marketing Representative:
NPO Energia Ltd.
631 South Washington Street
Alexandria, Virginia 22314, USA
Phone: (703) 836-1999
Fax: (703) 836-1995

Photo Courtesy of Space Commerce Corp.

The Buran orbiter, which flew its first mission atop the Energia heavy lift rocket on 15 November 1988, is being mated to Energia in the assembly facility at the Baikonur Cosmodrome.

Energia History

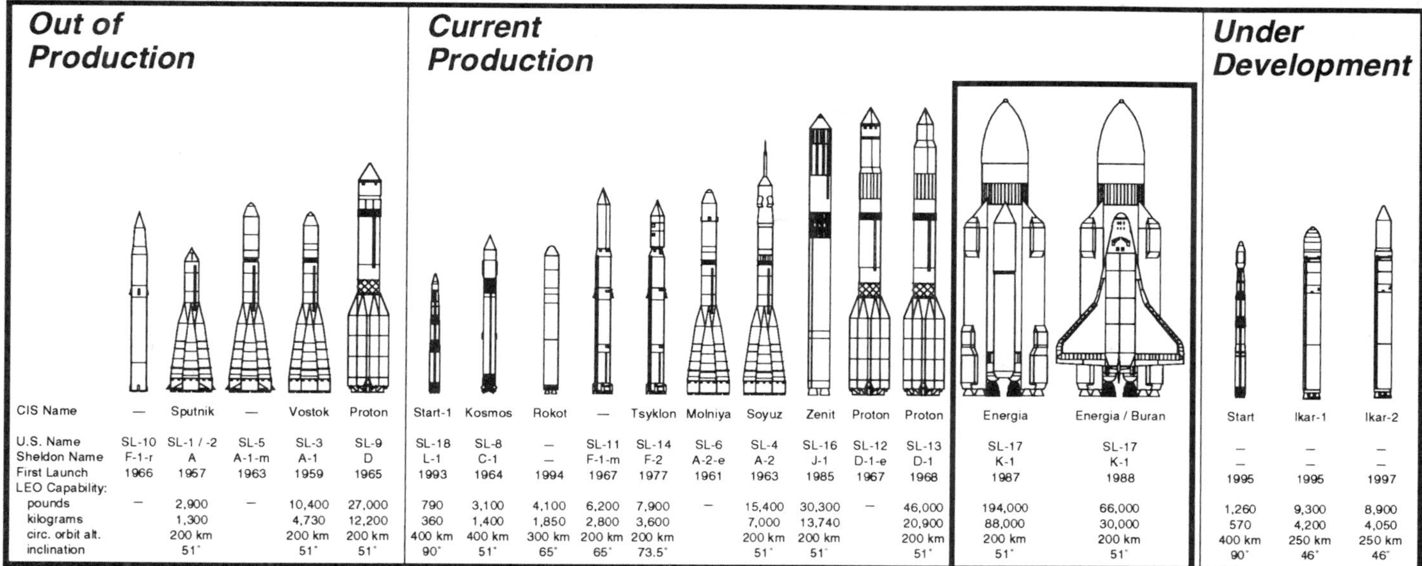

Out of Production

CIS Name	—	Sputnik	—	Vostok	Proton
U.S. Name	SL-10	SL-1 / -2	SL-5	SL-3	SL-9
Sheldon Name	F-1-r	A	A-1-m	A-1	D
First Launch	1966	1957	1963	1959	1965
LEO Capability:					
pounds	—	2,900	—	10,400	27,000
kilograms		1,300		4,730	12,200
circ. orbit alt.		200 km		200 km	200 km
inclination		51˚		51˚	51˚

Current Production

CIS Name	Start-1	Kosmos	Rokot	—	Tsyklon	Molniya	Soyuz	Zenit	Proton	Proton	Energia	Energia / Buran
U.S. Name	SL-18	SL-8	—	SL-11	SL-14	SL-6	SL-4	SL-16	SL-12	SL-13	SL-17	SL-17
Sheldon Name	L-1	C-1	—	F-1-m	F-2	A-2-e	A-2	J-1	D-1-e	D-1	K-1	K-1
First Launch	1993	1964	1994	1967	1977	1961	1963	1985	1967	1968	1987	1988
LEO Capability:												
pounds	790	3,100	4,100	6,200	7,900	—	15,400	30,300	—	46,000	194,000	66,000
kilograms	360	1,400	1,850	2,800	3,600		7,000	13,740		20,900	88,000	30,000
circ. orbit alt.	400 km	400 km	300 km	200 km	200 km		200 km	200 km		200 km	200 km	200 km
inclination	90˚	51˚	65˚	65˚	73.5˚		51˚	51˚		51˚	51˚	51˚

Under Development

CIS Name	Start	Ikar-1	Ikar-2
U.S. Name	—	—	—
Sheldon Name	—	—	—
First Launch	1995	1995	1997
LEO Capability:			
pounds	1,260	9,300	8,900
kilograms	570	4,200	4,050
circ. orbit alt.	400 km	250 km	250 km
inclination	90˚	46˚	46˚

Vehicle Description

CIS vehicles are identified either by their CIS, U.S., or Sheldon name. In the Soviet Union, it was standard practice to name a launch vehicle after its original payload (e.g., Kosmos, Proton). The U.S. names (developed by the U.S. Department of Defense) are alphanumeric designations based roughly on chronological appearance. The Sheldon names, a most commonly used system that was published by Dr. Charles Sheldon of the U.S. Library of Congress in 1968, emphasize the basic families of launch vehicles with special indicators for variants within a family.

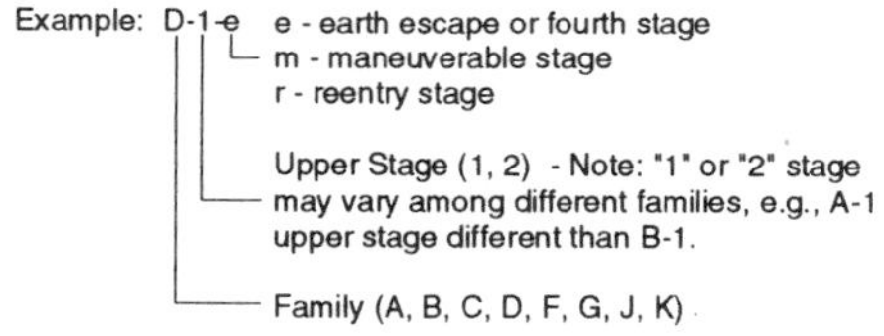

Energia (K-1/SL-17) LOX/LH2 cryogenic core vehicle with four LOX/kerosene liquid strap-ons based on Zenit first stage. Payloads are located in a sidemounted carrier. The Buran space shuttle can also be attached for manned launches. An LOX/LH2 Energia upper stage (EUS) and LOX/kerosene retro & correction stage (RCS) are being developed for high energy and low energy orbit changes, respectively.

Launch Record

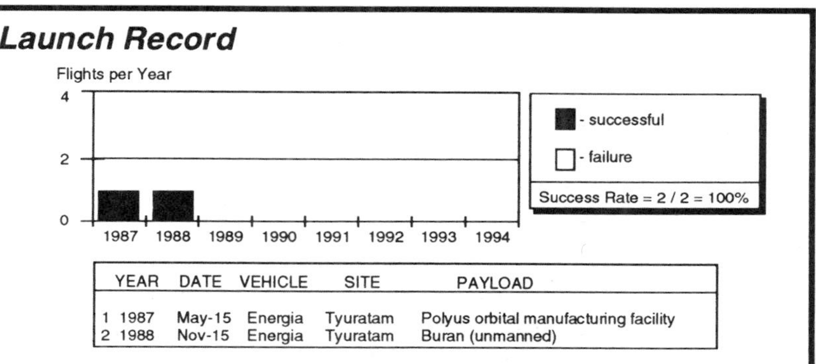

	YEAR	DATE	VEHICLE	SITE	PAYLOAD
1	1987	May-15	Energia	Tyuratam	Polyus orbital manufacturing facility
2	1988	Nov-15	Energia	Tyuratam	Buran (unmanned)

Historical Summary

The history of Energia (translates to "Energy") and Buran (translates to "Snowstorm") dates back to the unsuccessful Soviet moon program to be flown on the N-1 (SL-15, G-1-e) heavy lift vehicle. The behemoth N-1 was 370 ft (113 m) tall, 56 ft (17 m) in diameter at the base, and employed 30 engines in the first stage for a total thrust of about 10.1 million pounds (45 million newtons). After four consecutive unsuccessful launch attempts from 1969 to 1972, the program was cancelled and a new heavy lift launch vehicle program took its place—Energia/Buran. Energia was to be a modular heavy-lift launch vehicle capable of launching a variety of payloads, including the Buran orbiter.

As the Energia/Buran program started in the 1970s, the design of the Energia/Buran system was placed under the jurisdiction of the Energia Scientific Production Organization (NPO Energia—now called Rocket Space Corporation Energia), whose direct predecessor was an enterprise headed by Sergei Korolev, the designer of the first Sputniks and Vostok as well as the N-1. This enterprise was subordinate to the Ministry of General Machine Building.

Energia History

(continued)

Historical Summary (continued)

The head designer of this concern, Valentin Petrovich Glushko, took the overall direction of the project. He was one of the pioneers of rocket techniques in the former Soviet Union. Beginning in the late 1920s, he created the first Soviet rocket engines and then, headed the organization studying and designing powerful engines for all Soviet space rockets. He had also been a strong opponent of the N-1 vehicle design arguing for a more energetic propellant like that of Energia. He was the head of the Energia organization from 1974 until his death in 1989. Yuri Pavlovich Semenov is the current head of Rocket Space Corporation Energia.

The principal designer of the giant rocket Energia was Boris Ivanovich Gubanov. The Energia was designed with four strap-on liquid oxygen / kerosene propellant boosters around a large diameter core that is capable of delivering 194,000 lb (88,000 kg) to low Earth orbit (LEO) with a kick stage. The core uses four liquid oxygen / liquid hydrogen propellant engines, the first such cryogenic engines of its kind in the CIS fleet. Payloads can be carried in either a side-mounted unmanned cargo carrier or the Buran orbiter. Energia can be launched from three different pads at the Baikonur Cosmodrome (i.e., Tyuratam). At liftoff, the 198 ft (60 m) tall, 66 ft (20 m) wide Energia generates 7.8 million pounds (35 million newtons) of thrust from a total of 20 thrust chambers burning simultaneously from the strap-on and core stages.

The Buran orbiter was developed to transport crews and supplies to space stations, as well as for deployment, maintenance, and recovery of satellite and other payloads. It is capable of flying in a manual or automatic mode (i.e., without human piloting). Buran was designed so that main propulsion was located on Energia, with only a modest orbit maneuvering stage to achieve orbit and deorbit located on Buran. Energia/Buran is capable of delivering 66,000 lb (30,000 kg) to LEO. The overall chief designer of the Buran system was Semenov. He was also responsible for the design of manned spacecraft and stations (Soyuz, Mir, Progress, etc). Another organization, NPO Molniya, designed the Buran orbiter airframe under the direction of General Director Gleb Yevgenyevich Lozino-Lozinsky. The guidance system of the Buran was developed by the Scientific and Production Association for Automation and Instrument Engineering (NPO AP). KB Obscheye Mashinostroyenya, under general designer V. P. Barmin, developed the launch system ground infrastructure.

The strap-on booster engines of the rocket (RD-170) were designed at NPO Energomash under the direction of Glushko. The RD-0120 core cryogenic engines were designed at what is now KB Khimavtomatiki under the direction of chief designer A. D. Konopatov.

The introduction of Energia and Buran at Baikonur Cosmodrome required extensive infrastructure development and construction. The work began in 1978 with the construction of a landing strip for the shuttle. It was tested more than 60 times using a remote controlled Tu-154 aircraft, with the crew taking over only at the last moment. It is this landing strip that the world's largest aircraft, the AN-225 Mriya, uses to bring in the various parts of the Energia/Buran system. At the same time, the construction of large support buildings and the three launch pads was in progress in the area. In 1982, the first elements of Energia were delivered to Baikonur.

In 1985, the Zenit launch vehicle was successful on its maiden launch. This was significant for Energia since its four strap-ons are based on the Zenit first stage. Numerous successful launches of Zenit and a successful ground-test program ultimately led to the decision to proceed with the first Energia launch.

In 1987, Energia was ready for launch. General Secretary Mikhail Gorbachev visited Leninsk and the Baikonur Cosmodrome from 11–13 May 1987. The report of his visit stated that the Baikonur Cosmodrome was preparing for a launch of a new all-purpose carrier rocket. News of the actual launch was delayed until the following day and gave the time of launch from the Baikonur Cosmodrome as 1730 GMT on May 15. The launch was conducted from a static test stand that was pressed into service for the first test launch. The announcement stated that the first stage landed in Soviet territory as planned, and that the core stage delivered the test payload to the calculated position and at that time separated from the payload and fell into the preplanned area of the Pacific Ocean. However, a malfunction of the Polyus microgravity manufacturing platform's propulsion system resulted in the payload failing to reach orbit. Although the mission was something less than a success, the Energia functioned as planned and the launch announcement claimed that the "the aims and objectives of the first launching have been fully met."

Although the Buran program was publicly denied and decried as economically unjustified, it was no longer secret by the spring of 1983 due to the cosmonaut Igor Volk, who announced during the Paris Air Show that the Soviets were well into the process of building a space shuttle, the dimensions of which were approximately those of the United States Shuttle. The most complicated problems in developing the Buran were its thermal protection system for the aluminum structure, carbon-carbonic structural material, and an automatic landing system for a wide range of weather conditions. Valuable data on heat resistant reentry materials and transonic aerodynamics was achieved through four flights of the lifting-body shaped BOR-4 subscale space plane during 1982–1984 on the Kosmos launch vehicle. Experience with Buran's handling capability during the final atmospheric flight portion was obtained by a flight crew in an experimental orbiter. This orbiter was equipped with jet engines which allowed it to take off from a normal airport. Overall, 18 of 24 flights performed were fully automatic.

The first launch test of the Buran was scheduled on 29 October 1988. During prelaunch procedures, however, about one minute before the launch, the automatic launch control system commanded "automatic termination of the flight" because of a separation delay of an instrument unit from the body of the launch vehicle. The investigation and redesign took about two weeks. The Energia was launched with the Buran orbiter as payload on 15 November 1988. The Buran orbiter returned to Earth after two orbits and landed automatically.

Since Energia does not place its payload directly into LEO (the core remains sub-orbital), additional stages are needed to move into high Earth orbits or escape trajectories. Two new auxiliary stages are currently under development to fill this role. These are the Retro and Correction Stage (RCS) and the Energia Upper Stage (EUS). The stages will be used separately or together depending on the mission and will fit inside a universal cargo container 138 ft (42 m) long and 22 ft (6.7 m) in diameter. Given the Energia's current situation, the exact development status of these stages is not known.

Currently, the Energia and Buran programs are in a "mothballed" status, with three Energias and two Burans in storage at Baikonur. The industrial base required to manufacture additional Energias has been preserved; with appropriate funding, Energia production could presumably resume two years from go-ahead. The ability to build additional Burans, however, has been lost for all practical purposes.

A scaled down version of the Energia, designated Energia-M, is under active development (see *Energia Notes*). This launch vehicle has a shortened core with a single RD-0120 engine, two standard Energia strap-ons, and an in-line payload fairing. The first test flight of the Energia-M is planned for 1997. Various upper stages are also under development for the Energia-M.

Energia General Description

Energia **Energia / Buran**

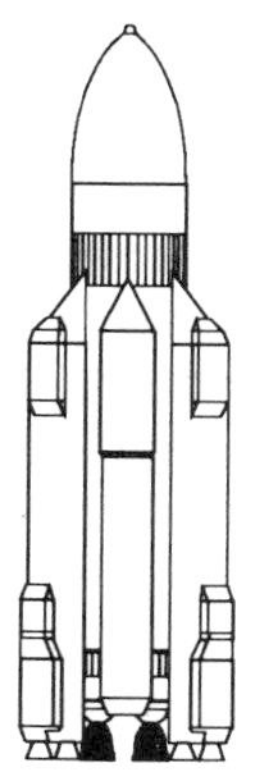

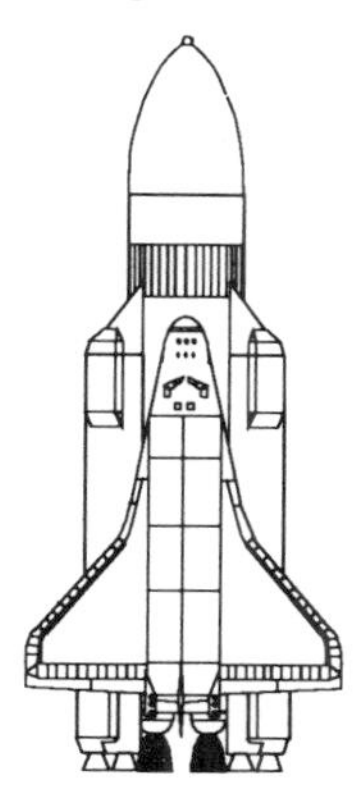

Summary

The Energia, introduced in 1987, is five times more powerful than any other CIS booster. The four strap-on boosters, same as the Zenit vehicle first stage, burn liquid oxygen and kerosene. The large central core contains liquid oxygen/liquid hydrogen burning engines. Energia's design is modular and allows for the attachment of the Buran Space Shuttle or cargo modules. In development are a cryogenic Energia upper stage (EUS) and a retro & correction stage (RCS) for orbit changes. The Buran orbiter, flown only once as unmanned, is able to carry up to 10 cosmonauts for seven days as well as deploy and return cargo.

Status

Energia - Operational/Suspended
Buran - Test/Suspended
Upper stages - Development (?)

Key Organizations

Users - Russian Ministry of Defense, Russian Academy of Sciences, commercial

Launch Service Agency - Rocket Space Corporation Energia

Key Organizations

- Rocket Space Corporation Energia (Design & Development)
- NPO Molniya (Buran orbiter)
- NPO Yuzhnoye (strap-on boosters)
- NPO Energomash (strap-on engines)
- KB Khimavtomatiki (core engines, Buran OMS engines)

Vehicle

System Height	Up to 198 ft (60.3 m)
Cargo Carrier Size	First flight cargo - 13 ft (4 m) diameter by 125 ft (38 m) height New cargo carrier - 22 ft (6.7 m) diameter by 138 ft (42 m) height Buran cargo bay - 15.4 ft (4.7 m) diameter by 59 ft (18 m) height
Gross Mass	5.3M lb (2.4M kg)
Planned Enhancements	Parachute recovery and reuse of strap-ons. Possible wing recovery of core. Varying number of strap-ons from two to eight. Possible in-line forward mounted fairing 26 ft (8 m) in diameter. Scaled down version (Energia-M) is under active development (See *Energia Notes*).

Operations

Primary Missions	LEO, GEO, interplanetary
Compatible Upper Stages	EUS, RCS
First Launch	1987 for Energia and 1988 for unmanned Buran
Success / Flight Total	2 / 2
Launch Site	Baikonur Cosmodrome (45.6°N, 63.4°E) near Tyuratam. One Energia pad and two Energia/Buran pads.
Launch Azimuth	??°–??° (max. inclination is 50.7°- 110°)
Nominal Flight Rate	1 / yr planned; no flights since 1988
Planned Enhancements	None

Performance

Suborbital	231,000 lb (105,000 kg) for Energia 205,000 lb (93,000 kg) for Energia + new cargo carrier
108 nm (200 km) circ, 51.6°	194,000 lb (88,000 kg) for Energia + new cargo carrier + RCS 66,000 lb (30,000 kg) for Energia/Buran
108 nm (200 km) circ, 98°	176,000 lb (80,000 kg) for Energia 35,000 lb (16,000 kg) for Energia/Buran
Geotransfer Orbit, 28.5°	—
Geosynchronous Orbit	48,500 lb (22,000 kg) for Energia / EUS

Financial Status

Estimated Launch Price	Unknown

Manifest (proposed)

1996–2000	High Capacity Communications Satellites
1996–2000	Unmanned Flight of Buran Orbiter
2000+	International Manned Missions to Moon and Mars
2000+	Disposal of Nuclear Waste in Space

Remarks

Proposed Energia manifest provided by NPO Energia, LTD.

Energia Vehicle

Overall

Length	198 ft (60.3 m)
Gross Mass	5.3M lb (2.4M kg)
Thrust at Liftoff	7.83M lb (34.8M N)

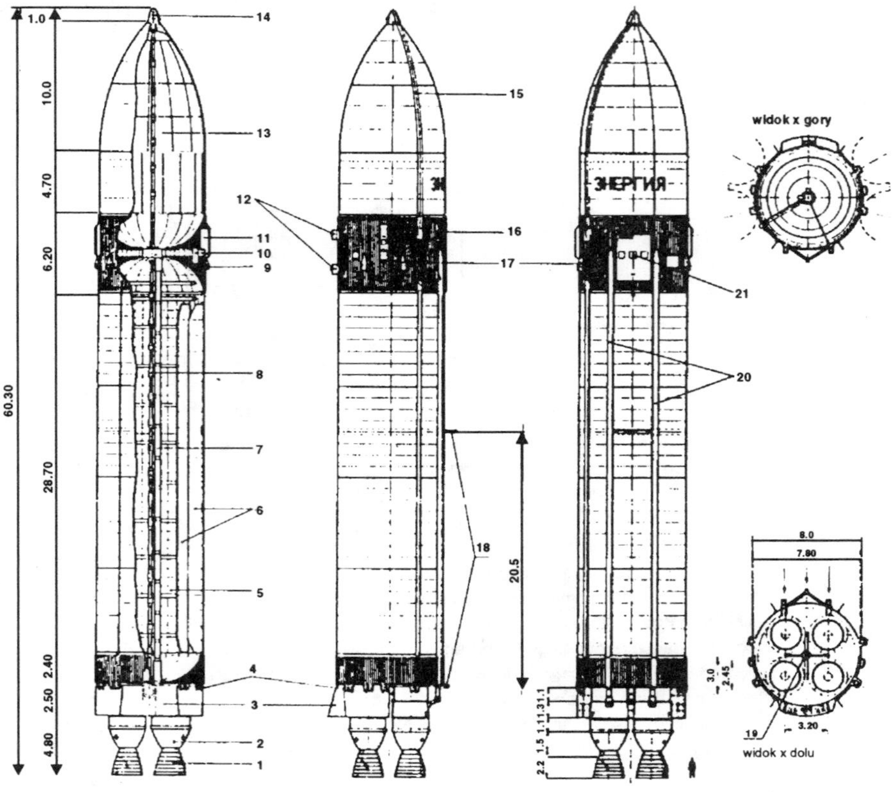

Sustainer Core Stage

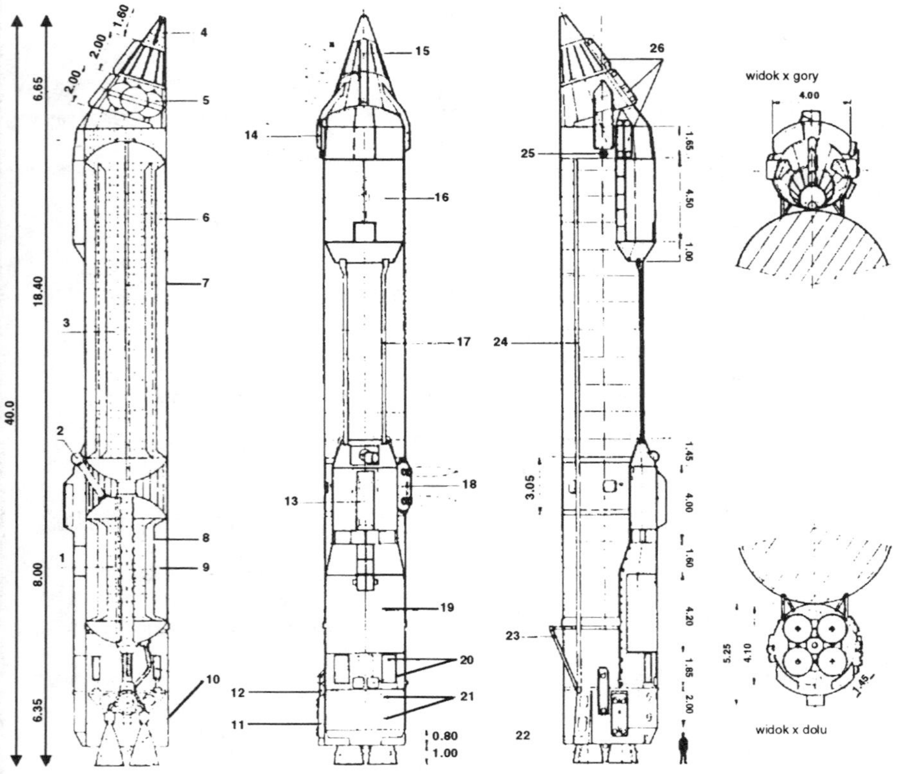

Strap-On Booster Stage

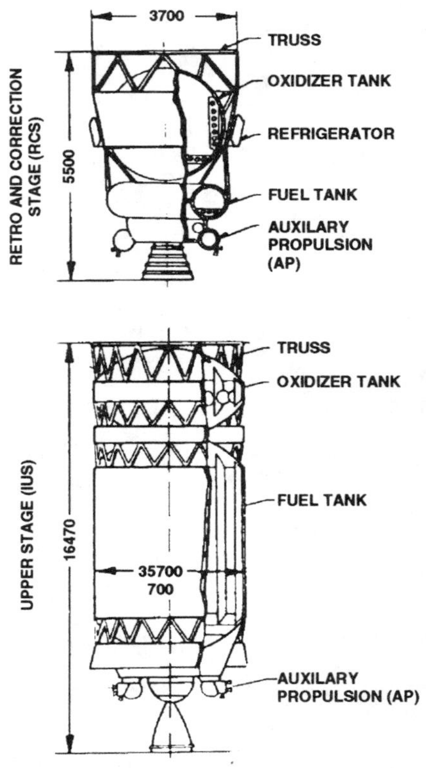

Upper Stages

Energia Vehicle
(continued)

Energia Stages

	Stage 1 (Strap-ons)	Stage 2 (Core)	EUS - optional	RCS - optional
Dimension:				
Length	131 ft (40 m)	197 ft (60 m)	54 ft (16.47 m)	18 ft (5.5 m)
Diameter	12.8 ft (3.9 m)	26 ft (8.0 m)	18.7 ft (5.7 m)	12 ft (3.7 m)
Mass: (each)				
Propellant Mass	705K lb (320K kg)	1810K lb (820K kg)	154K lb (70K kg)	33 lb (15K kg)
Gross Mass	783K lb (355K kg)	1995K lb (905K kg)	170K lb (77K kg)	37K lb (17K kg)
Structure:				
Type	Wafer	Skin-Stringer	Skin-Stringer ?	Monocoque
Material	Aluminum	Aluminum	Aluminum	Aluminum
Propulsion:				
Propellant	LOX / kerosene	LOX / LH2	LOX / LH2	LOX / kerosene
Average Thrust (each)	16,32K lb (7,268K N) SL 17,77K lb (7,906K N) vac	326K lb (1451K N) SL 441K lb (1962K N) vac	326K lb (1451K N) SL 441K lb (1962K N) vac	19.1K lb (85K N) vac
Engine Designation	RD-170	RD-0120	RD-0120	1158DM
Number of Engines	1 turbopump + 4 chambers	4	1	1
Isp	309 sec SL 337 sec vac	354 sec SL 452.5 sec vac	490 sec vac	351.8 sec vac
Feed System	Staged combustion	Staged combustion	Staged combustion	Staged combustion
Chamber Pressure	3556 psia (245 bar)	3000 psia (206 bar)	3000 psia (206 bar)	1124 psia (77.5 bar)
Mixture Ratio (O/F)	2.43	6.35	6.35	2.6
Throttling Capability	49–102%	75–100%	75–100%	100% only
Expansion Ratio	26:1	85:1	85:1	189:1
Restart Capability	No	No	No	7 restarts
Tank Pressurization	Stored pressurized gas	Gas Generator ?	Gas Generator ?	Stored pressurized gas
Control-Pitch,Yaw,Roll	Gimbal nozzles (±5°)	Gimbal nozzles (±8.6°)	Gimbal nozzles (±8.6°)	ACS
Events:				
Nominal Burn Time	145 sec	480 sec	?? sec	680 sec
Stage Shutdown	Burn to depletion	Command shutdown	Predetermined velocity	Predetermined velocity
Stage Separation	Retro-rockets	Retro-rockets	APE	APE/Spring ejection

Remarks:

The Energia consists of a central core with four liquid hydrogen/liquid oxygen engines, and nominally four strap-on boosters (based on the first stage of the Zenit launcher), each with a single four-chamber liquid oxygen/kerosene engine. All core and strap-on engines are ignited at liftoff; engine throttling is used to manage axial acceleration and dynamic pressure. Following propellant depletion and jettison of the strap-ons, the core continues to burn until the desired shutdown state is reached. The Energia is unique among expendable launch vehicles in that it does not deliver its payload directly to orbit. In order to facilitate a predictable and safe reentry of the expended core, the core engines are shut down, and the core jettisoned, before orbital velocity is reached. The final velocity increment required to achieve orbit must be provided by either the payload itself or an upper stage. Two upper stages are currently under development: a large cryogenic stage (EUS) for high energy missions (including geosynchronous) and a smaller stage (RCS) based on Proton stage 4 (Block DM) for low Earth orbit insertion of large payloads.

The Energia vehicle design was driven by a number of operational requirements. One requirement was its ability to launch both large unmanned payloads and the Buran orbiter. This resulted in a launch vehicle payload with near-orbital capability of 230,000 lb (105,000 kg) and a side-mounting payload arrangement. Containing all the propulsion elements in Energia made possible independent development of the Energia and orbiter. Another important requirement was the ability to launch a range of payloads from 22,000–440,000 lb (10,000–200,000 kg). Thus, the Energia is modular in design, with the number of strap-ons variable from two to eight. A third requirement was the ability for an omni-azimuthal launch along with depositing of spent hardware in safe land zones. This was satisfied by a two stage configuration with impact of the first stage relatively near the launch site (about 250 mi or 400 km downrange). Such a decision made possible a low first stage jettison velocity (4000 mph or 1.8 km/sec) and, hence, possible recovery of these stages. The strap-ons have been designed for 10 uses.

Another requirement was the need for improved reliability. To provide this, the Energia vehicle was based on the principle of redundancy so a failure of one system does not affect the success of the launch. Energia can sustain a shutdown of one strap-on stage engine or one of the four main core stage engines in midflight and still reach Earth orbit. Even a second failure in the same system does not affect flight safety. This principle is accomplished by redundant vital systems and units. Reliability is confirmed by multilevel autonomous tests of systems and units, as well as the complex checkouts and tests of the vehicle as a whole by both cold and hot firings. If an emergency return to launch site is required (for Buran missions), dumping of oxygen from a defective strap-on stage is planned. Fuel dumping begins at an altitude not less than 650 ft (200 m). In case of emergency during the first seconds of liftoff, there is sufficient thrust to weight to divert the vehicle to a safe distance.

In addition to design parameter selection to deal with probable abnormal situations, emergency protective devices on the first and second stage engines continuously monitor their stage, record failures and cut off a failed engine without its disintegration. To monitor and suppress fire onboard the launch vehicle, special fire- and explosion-prevention devices sense temperature and analyze gas composition; neutral gas reserves are used to ventilate and fill the affected compartments.

Energia Vehicle

(continued)

Orbiter

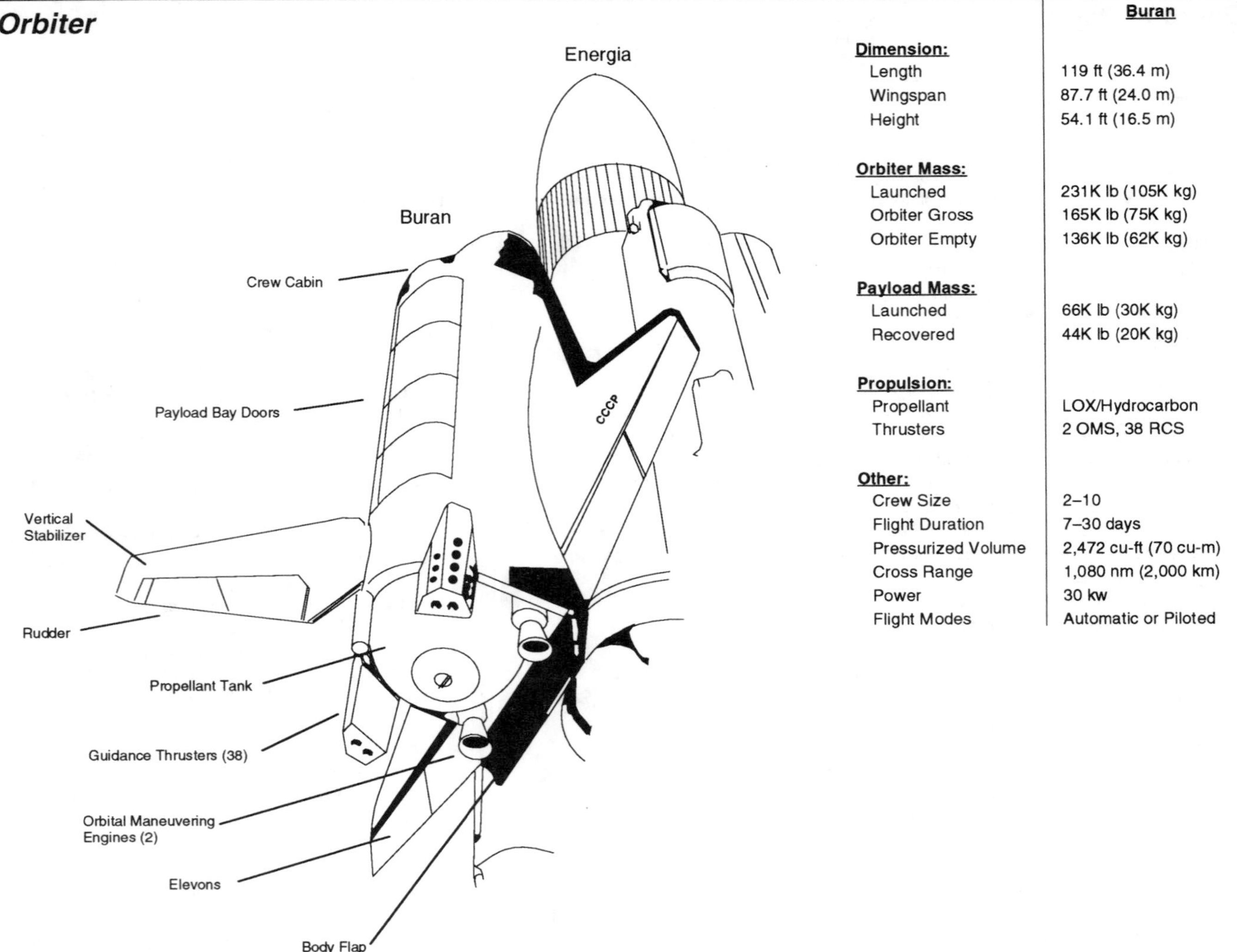

	Buran
Dimension:	
Length	119 ft (36.4 m)
Wingspan	87.7 ft (24.0 m)
Height	54.1 ft (16.5 m)
Orbiter Mass:	
Launched	231K lb (105K kg)
Orbiter Gross	165K lb (75K kg)
Orbiter Empty	136K lb (62K kg)
Payload Mass:	
Launched	66K lb (30K kg)
Recovered	44K lb (20K kg)
Propulsion:	
Propellant	LOX/Hydrocarbon
Thrusters	2 OMS, 38 RCS
Other:	
Crew Size	2–10
Flight Duration	7–30 days
Pressurized Volume	2,472 cu-ft (70 cu-m)
Cross Range	1,080 nm (2,000 km)
Power	30 kw
Flight Modes	Automatic or Piloted

Remarks:

The Buran orbiter is very similar to the U.S. Space Shuttle orbiter in appearance and intended function. Although the Buran has not flown since its initial unmanned test flight in 1988 and no firm future launch dates have been announced, two orbiters are stored in flight condition at the Baikonur Cosmodrome. The design lifetime of each orbiter is 100 flights. The Buran has three sections: (1) a pressurized forward section, (2) a payload bay, and (3) a tail section. The forward section consists of three compartments which can be hermetically sealed from each other. They are the command section, the equipment section, and the living quarters. An air lock can be installed in the living quarters. All the manual controls are located in the command section. The Buran can accommodate a flight crew of from two to four and up to six additional mission specialists. The unpressurized fuselage is composed of a combination of composite materials and alloys of beryllium, nickel-titanium, niobium, and boraluminum, instead of traditional aluminum alloys. During reentry, Buran can fly in three different control modes: (1) an entirely automatic mode, where the Buran reenters and lands entirely without human intervention; (2) a classic manually piloted mode; and (3) a controlled autopilot mode, during which the Buran flies autonomously, but provisions are made for pilot intervention and/or override at any time.

In space, a "unified engine system" is used that consists of two orbital maneuvering engines to attain orbit or deorbit conditions and 38 guidance thrusters for attitude control. Upon atmospheric reentry, control authority is transfered from the unified engine system to the Buran's aerodynamic control surfaces: The elevons (pitch and roll controls for delta wing aircraft), vertical stabilizer and rudder (yaw control), air braking system, and body flap. The elevons are located at the wings' trailing edges and are used to control pitch and roll. The body flap, located under the rear of the fuselage, is used for aerodynamic trim purposes, as well as to assist the elevons in pitch control. The air brake consists of two flaps located at the trailing edge of the vertical stabilizer.

Energia Vehicle

(continued)

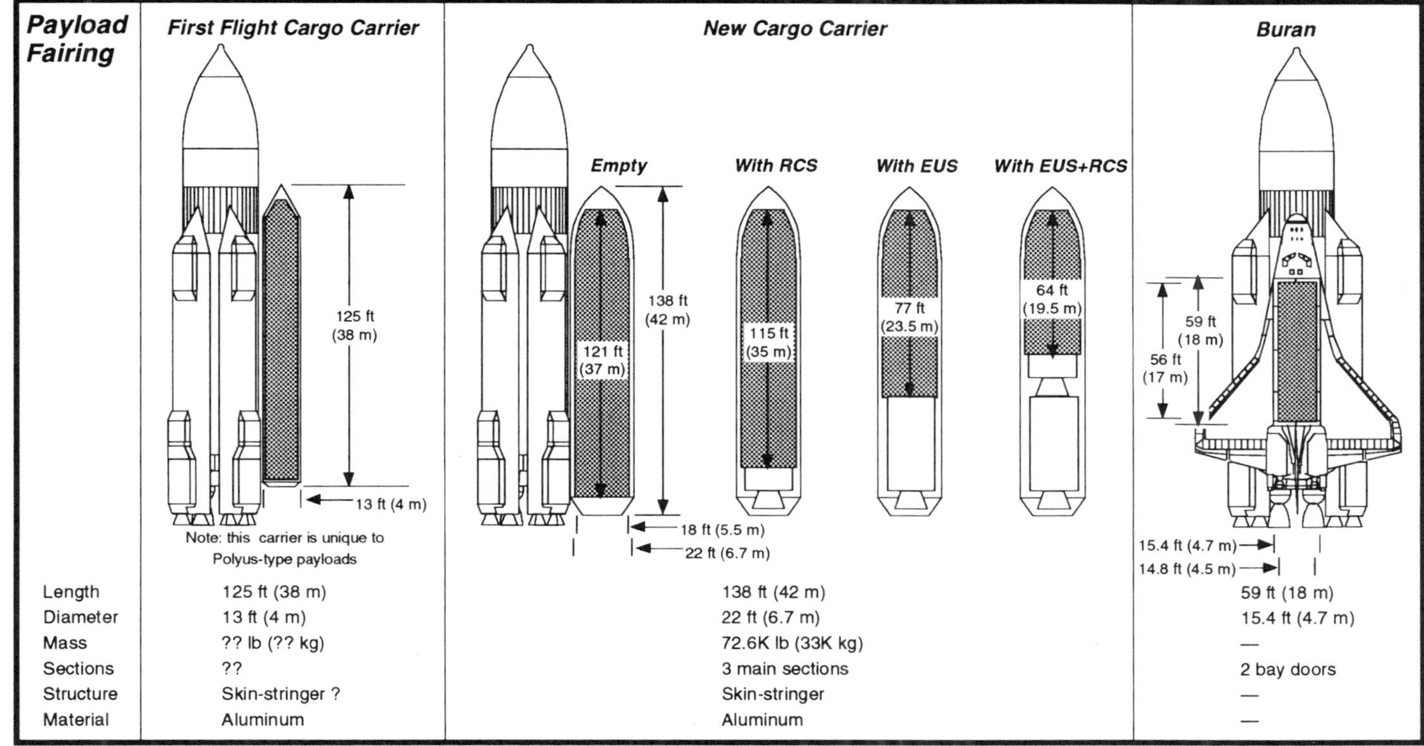

Payload Fairing	***First Flight Cargo Carrier***	***New Cargo Carrier***	***Buran***
Length	125 ft (38 m)	138 ft (42 m)	59 ft (18 m)
Diameter	13 ft (4 m)	22 ft (6.7 m)	15.4 ft (4.7 m)
Mass	?? lb (?? kg)	72.6K lb (33K kg)	—
Sections	??	3 main sections	2 bay doors
Structure	Skin-stringer ?	Skin-stringer	—
Material	Aluminum	Aluminum	—

Avionics

Energia. No data available.

Buran. The basic element of the guidance system consists of three gyrostabilized pads, which have three accelerometers to measure acceleration during the function of the engines or during atmospheric braking. Nine angular velocity sensors, three on each of the three axes, interact with the spacecraft structure. During certain flight periods, accelerometers fastened to the spacecraft structure are also used.

The gyrostabilized pad allows the acclerometers which are affixed to it, to maintain a permanent orientation in relation to space. Normally, a gimbal system with 3 degrees of freedom is utilized for this, although, such a scheme places limitations on one of the rotation axes. This is the reason that a gimbal system with 4 degrees of freedom was created for Buran, ensuring a satisfactory orientation regardless of the spacecraft's attitude. This means that instead of the usual three angles whose values allow movement control, there are now four with, consequently, more complex algorithms.

All data from the sensors are fed into the onboard computer complex. This complex consists of four identical computers, which operate on a parallel basis throughout the flight, all solving the same problem. The system continues to function even if two of the computers go down simultaneously.

The computer complex transmits its commands to the unified engine system (for space flight) and to various controls for atmospheric flight. The computer watches over the 50 onboard systems, detecting the anomalies, automatically disconnecting malfunctioning equipment, and connecting reserve equipment.

Attitude Control System

Energia. The strap-on booster and core stage provide control by independent gimbaling engines. For the Energia Upper Stage (EUS) and Retro and Correction Stage (RCS) attitude control is provided by an auxiliary propulsion engine (APE) that contains two sets of small fixed thrusters for three axis control.

Buran. It has a "unified engine system" of engines for orientation, space maneuvering, and to provide sufficient braking impulse for reentry. This same system is also used at the beginning of the mission to supply sufficient impulse for orbit insertion since the last stage of Energia does not quite reach orbital velocity. The system comprised two principal rocket engines of 6,600 lb of thrust each, referred to as "orbital maneuvering engines", and 38 guidance thrusters with relatively low thrust (660 lb (2935 N) primary, 44 lb (195 N) secondary). The guidance thrusters are used for stabilizing the orbiter, as orientation and maneuvering require only slight thrust. They also serve for stabilization during the initial reentry phase. The orbiter can successfully reenter with only one orbital maneuvering engine functional.

The orbital maneuvering engines and guidance thrusters are fed from common tanks of a hydrocarbon-oxygen propellant which have a maximum capacity of 31,000 lb (14,000 kg). With 18,000 lb (8,000 kg) of propellant, it is possible to place a 66,000 lb (30,000 kg) payload at an altitude of 108 nm (200 km). With 31,000 lb (14,000 kg) of propellant, a payload of 53,000 lb (24,000 kg) can attain an altitude of 243 nm (450 km). If necessary, it is possible to install extra tanks in the payload bay for an additional 31,000 lb (14,000 kg) of propellant, in order to reach an altitude of 430–540 nm (800–1,000 km).

Energia Performance

Performance Characteristics of Energia:

Suborbital, 50.7°	231K lb (105K kg) including upper stage and cargo carrier; 205K lb (93K kg) not including cargo carrier
108 nm (200 km) circular, 50.7°	194K lb (88K kg) not including cargo carrier, RCS, and circularization propellant
220 nm (407 km) circular, 51.6° International Space Station	183K lb (83K kg) not including cargo carrier, RCS, and consumed propellant
108 nm (200 km) circular, 98°	176K lb (80K kg) including upper stage and cargo carrier

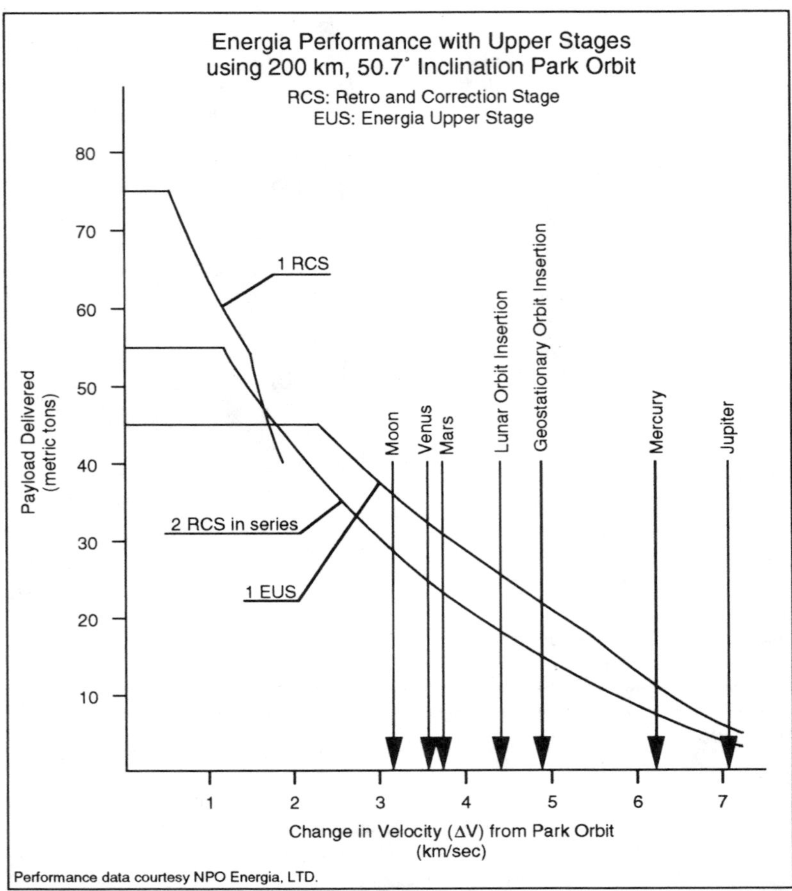

Performance data courtesy NPO Energia, LTD.

Energia / Buran Performance for Different Orbits (Seven Day Mission)

Inclination	Circular Altitude		
	108 nmi (200 km)	243 nmi (450 km)	540 nmi (1000 km)
51.6 deg	66K lb (30K kg)	53K lb (24K kg)	15.4K lb (7K kg)
65 deg	55K lb (25K kg)	42K lb (19K kg)	6.6K lb (3K kg)
97 deg	35K lb (16K kg)	22K lb (10K kg)	2.2K lb (1K kg)

Performance Characteristics of Energia-M (see *Energia Growth Possibilities*):

Orbit	Upper Stage			
	Integral Spacecraft Propulsion	RCS	Growth RCS	LOX/LH2 Stage
108 nm (200 km) circular, 50.7°	75.0K lb (34.0K kg)	66.1K lb (30.0K kg)	—	—
324 nm (600 km) circular, 97°	54.0K lb (24.5K kg)	45.2K lb (20.5K kg)	—	—
20200 x 92 nm (37500 x 170 km) Molniya orbit, 63°	—	22.5K lb (10.2K kg)	23.1K lb (10.5K kg)	27.6K lb (12.5K kg)
Geostationary orbit	—	6.6K lb (3.0K kg)	9.9K lb (4.5 K kg)	15.4K lb (7.0K kg)
Moon	—	—	19.8K lb (9.0K kg)	26.5K lb (12.0K kg)
Venus/Mars	—	—	15.4K lb (7.0K kg)	20.9K lb (9.5K kg)

Notes:
Physical characteristics and development status of growth RCS and LOX/LH2 stage are unknown.

LOX/LH2 stage for Energia-M is **not** EUS.

Like Energia, Energia-M delivers payload to a suborbital state. Additional propulsion (either integral to spacecraft or as an upper stage) is required.

Energia Operations

Launch Site

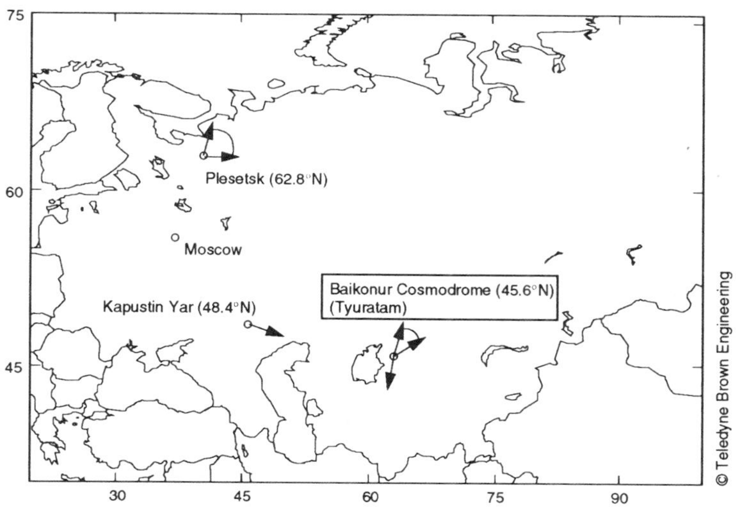

Launch Facilities

Energia on the Launch Pad

Energia / Buran on the Launch Pad

Energia Operations

(continued)

Launch Processing

Energia is launched from the Baikonur Cosmodrome located just east of the Aral Sea in Kazakhstan, north of Afghanistan. This site is usually referred to in the West as Tyuratam, a reference to a nearby railway station. Tyuratam is roughly analogous to the U.S. launch facilities at Cape Canaveral. All manned space flights originate from Tyuratam, as do all geosynchronous, lunar, and planetary launches. It is the only facility capable of launching the Proton, Zenit, and Energia/Buran. All other Soviet launch vehicles can be launched from Tyuratam with the exception of the small Kosmos booster.

Since the breakup of the Soviet Union, Russia has been in the difficult situation of having its largest space launch facility located in a foreign country (i.e., Kazakhstan). Despite political turmoil, the Baikonur Cosmosdrome has remained in operation virtually without interruption. In the summer of 1994, Russia and Kazakhstan signed a long-term agreement whereby Russia would lease Baikonur from Kazakhstan, with day-to-day operations of the cosmodrome remaining under the control of the Russian Space Forces.

The Energia strap-on boosters are transported by railway and the core stage and orbiter by the AN-225 heavy lift aircraft from the manufacturing facility to the cosmodrome. At the Space Vehicle Assembly Building (MIK), the Energia elements are assembled horizontally. The assembly building, which is 200 ft (60 m) high and 800 ft (240 m) on each side, is similar in appearance to the Vertical Assembly Building (VAB) at Kennedy Space Center, though less than half its height. Each vehicle is tested before it is launched and the engines are fired on experimental launch platforms. In the case of an Energia/Buran mission, the Buran is mated horizontally atop the Energia launcher. Once integration is complete, the large vehicle is carried by a massive transporter which is pulled by four diesel locomotives on two sets of rails 60 ft (18 m) from one another. This allows them to pull the Energia to one of three pads. Once at the pad, a hydraulic lift tilts the entire stack upward where it is positioned on the concrete pad.

There are three pads available for Energia. The experimental pad No. 1 is the one which was utilized on 15 May 1987 for the first launch of the Energia, and is located the farthest from the technical area. The other two pads, No. 2 (used on 15 November 1988) and No. 3 intended for the Buran shuttle, are the nearest ones. They are centrally placed and located in an area originally used for the the Soviet N-1 moon program. The distance between the two Energia/Buran pads and the Energia-only pad is 1.9 mi (3 km), whereas the Buran landing strip is 7.5 mi (12 km) away.

Pad No. 1 is placed over an enormous flame deflector 131 ft (40 m) deep. A service gantry, 210 ft (64 m) in height, is located on the side of the base on which the rocket is mounted. A mobile service gantry is moved on rails prior to launching. The rails, which lead to the stand, are mounted on supports which span the deflector pit. There are also four lighting towers, with a total of 670 projectors, allowing work to proceed throughtout the night, and two gigantic lightning rod towers, each 738 ft (225 m) in height. Cryogenic fuel, oxygen and hydrogen, is kept in spherical reservoirs 1.2 mi (2 km) away and is brought to the launch pad through pipelines. Before launch, prior to hydrogen fueling, all personnel are evacuated within a radius of 9 mi (15 km). To prevent the hydrogen from mixing with air (risk of explosion), one of the umbilical masts remains in place until launch time.

The launch bunker is located 330 ft (100 m) from the rocket. The launch preparations are conducted through an automated command system using three dual-processing computers (the system is able to function with only two of the six computers in operation). The launch control center consists of three consoles with color screens: the first gives information on the command systems, the second shows the rocket condition, and the third is held in reserve. Each operator only has to do three to four operations during the countdown (in case of unexpected situations, the necessary instructions appear on the screen). Any last minute difficult decisions are made by the chief designer and by the chief in charge of testing at the cosmodrome.

Pad No. 2 is placed on a cylinder, 66 ft (20 m) in diameter, at the bottom of which there are three flame deflection pits 75 ft (23 m) deep, placed at 120 degrees from one another, and end with two windows each. Two fixed service towers are located on each side of the stand on which the rocket is erected. A 330 ft (100 m) tall mobile service tower pivots around its own axle to access the shuttle. This tower is probably the one that was used in the moon program to launch the N-1. One tower supports an orbiter access arm for cosmonauts to enter the vehicle through its side hatch. The Buran pad does not use a rotating service structure to enclose the orbiter as does the U.S. shuttle. In case of emergency on the pad, the Buran crew can be evacuated by means of two tubes, 10–13 ft (3–4 m) diameter, which end up in a bunker comprised of sixteen protected areas. One tube has an elevator on rails, the other, a toboggan. It takes 15 seconds for the crew to reach safety.

Pad No. 3 is identical to pad No. 2 except that the pivoting service tower is considerably shorter (about 200 ft (60 m) tall). There are also two other pads near the Energia/Buran pads that may be developed for Energia-only launches.

Photo Courtesy of Space Commerce Corp.

Horizontal Transportation of Energia / Buran to the Pad

Energia Operations
(continued)

Flight Sequence

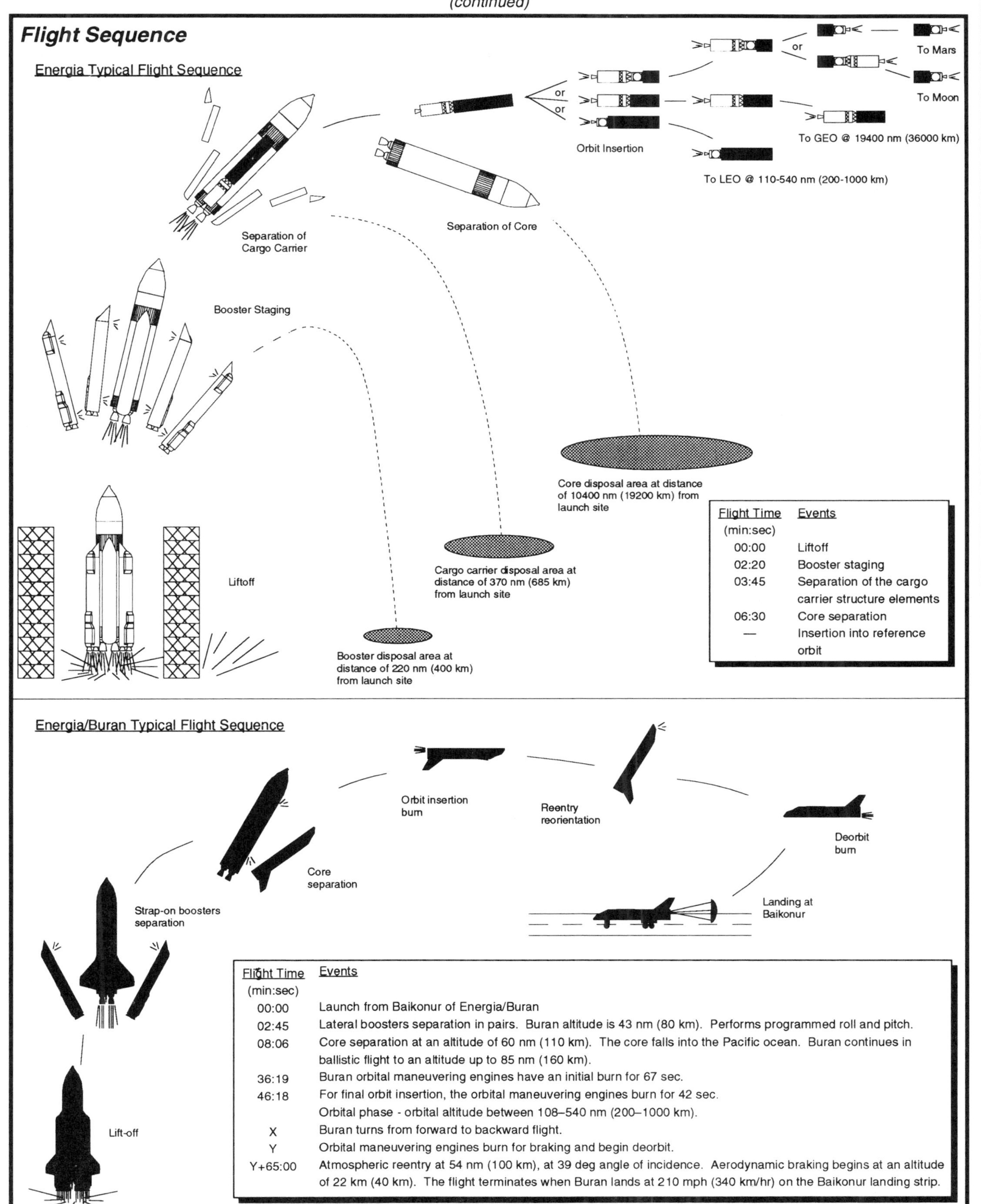

Flight Time (min:sec)	Events
00:00	Liftoff
02:20	Booster staging
03:45	Separation of the cargo carrier structure elements
06:30	Core separation
—	Insertion into reference orbit

Flight Time (min:sec)	Events
00:00	Launch from Baikonur of Energia/Buran
02:45	Lateral boosters separation in pairs. Buran altitude is 43 nm (80 km). Performs programmed roll and pitch.
08:06	Core separation at an altitude of 60 nm (110 km). The core falls into the Pacific ocean. Buran continues in ballistic flight to an altitude up to 85 nm (160 km).
36:19	Buran orbital maneuvering engines have an initial burn for 67 sec.
46:18	For final orbit insertion, the orbital maneuvering engines burn for 42 sec. Orbital phase - orbital altitude between 108–540 nm (200–1000 km).
X	Buran turns from forward to backward flight.
Y	Orbital maneuvering engines burn for braking and begin deorbit.
Y+65:00	Atmospheric reentry at 54 nm (100 km), at 39 deg angle of incidence. Aerodynamic braking begins at an altitude of 22 km (40 km). The flight terminates when Buran lands at 210 mph (340 km/hr) on the Baikonur landing strip.

Energia Payload Accommodations

Payload Compartment	
Maximum Payload Diameter	New Cargo Carrier - 22 ft (5.5 m) Buran Cargo Bay - 14.8 ft (4.5 m)
Maximum Length	New Cargo Carrier empty - 121 ft (37 m) New Cargo Carrier plus RCS - 115 ft (35 m) New Cargo Carrier plus EUS - 77 ft (23.5 m) New Cargo Carrier plus EUS plus RCS - 64 ft (19.5 m) Buran Cargo Bay - 56 ft (17 m)
Payload Adapter	
Interface Diameter	?? in (?? mm)
Payload Integration	
Nominal Mission Schedule Begins	T-?? months
Launch Window	
Latest Countdown Hold Not Requiring Recycling	T-?? min
On-Pad Storage Capability	?? hrs for a fueled vehicle
Latest Access to Payload	T-?? hrs or T-?? hrs through access doors
Environment	
Maximum Load Factors	+?? g axial, ±?? g lateral (minimum acceleration up to 3 g's)
Minimum Lateral/Longitudinal Payload Frequency	?? Hz / ?? Hz
Maximum Overall Acoustic Level	?? dB (full octave)
Maximum Flight Shock	?? g from ??-?? hz
Maximum Dynamic Pressure on Fairing	?? lb/ft^2 (?? N/m^2)
Maximum Pressure Change in Fairing	?? psi/s (?? KPa/s)
Cleanliness Level in Fairing (Prior to Launch)	Class ??
Payload Delivery	
Standard Orbit and Accuracy (3 sigma)	±?? nm altitude ±?? deg inclination
Attitude Accuracy (3 sigma)	All axes: ±?? deg, ±?? deg/sec
Nominal Payload Separation Rate	?? ft/s (?? m/s)
Deployment Rotation Rate Available	?? rpm
Loiter Duration in Orbit	4 days for EUS, 1–2 yrs for RCS
Maneuvers (Thermal / Collision / Telemetry)	??

Publications

Technical Publications

"Soviets Developing Smaller Version of Energia Heavy-Lift Booster", Aviation Week and Space Technology, pg. 79, December 24, 1990.

"Energia/Buran", W. Wirin, Space Commerce Corp., AIAA-90-3818, AIAA Space Programs and Technologies Conference, Huntsville, AL, September 25-28, 1990.

Space Transportation Propulsion USSR Launcher Technology, Rockwell International, BC-90-71, June 1990.

"Space Ways" B.I. Gubanov, NPO Energia, from Space Studies Institute, 1990.

"Energia-A New Versatile Rocket - Space Transportation System", B.I. Gubanov, Glavkosmos, IAF-89-202, 40th Congress of the International Astronautical Federation, Malaga, Spain, October 7-12, 1989.

"Soviet Commercial Launch Vehicles", A. Dula, Space Commerce Corp., B. Gubanov, NPO Energia, Y. Smetanin, Glavkosmos, AIAA-89-2743, AIAA 25th Joint Propulsion Conference, Monterey, CA, July 10-12, 1989.

TRW Space Log, 1957-1991,1992,1993 TRW, El Segundo, CA.

The Soviet Year in Space 1986, 1988, 1989, Nicholas Johnson, Teledyne Brown Engineering, Colorado Springs, CO.

Launch Vehicle Catalogue, European Space Agency, December 1989.

"Perestroika and Glasnost in the Soviet Space Program", Space Policy, November 1989.

"USSR RD-170 Liquid Rocket Engine", NASA Johnson Space Center, August 25, 1989.

"Versatile Rocket Space Transportation System Energia", Scientific-Industrial Enterprise, 1989.

"Opportunities from Soviet Space Industry, A Commercial User's Guide", A. Dula, Space Commerce Corp., A.I. Dunayev, Glavkosmos, Paris Air Show 1989.

"Dzis I Jutro Systems Energia", Astronautika NR 6, (616), 1989.

"The Energiya-Buran System Complex", from the "Journal Orbite-Organe du Cosmos Club de France", translated from French by N. Timacheff, 1989.

Soviet Space Programs: 1981-1987, United States Senate, Committee on Commerce, Science, and Transportation - Part 1 - May 1988.

Jane's Spaceflight Directory 1987, Jane's Publishing Inc., New York, NY.

Jane's Spaceflight Directory 1994-95, Jane's Publishing Inc., New York, NY.

"Soviet Launch Vehicle Designations", Ralph Gibbons, Spaceflight, pg. 54-60,80, February 19, 1977.

Space Directory of Russia, February 1993 Edition, Euroconsult/Sevig Press, Paris, France.

The Russian Space Directory 1994, European Space Report, Munich, Germany.

"Energia-M", promotional brouchure, NPO Energia.

Acronyms

APE - auxiliary propulsion engines
CIS - Commonwealth of Independent States
EUS - Energia Upper Stage
GEO - geosynchronous orbit
GTO - geosynchronous transfer orbit
KB - Design Bureau
LEO - low Earth orbit
LH2 - liquid hydrogen
LOX - liquid oxygen
MIK - Space Vehicle Assembly Building
NPO - Scientific Production Organization
OMS - orbital maneuvering system
RCS - retro and correction stage
USSR - Union of Soviet Socialist Republics

Energia Notes

(continued)

Other Notes—Energia Growth Possibilities

1 - Energia with Four Strap-Ons (Baseline)
2 - Energia with Six Strap-Ons (Not acknowledged by Energia NPO)
3 - Energia with Eight Strap-Ons and In-Line Cargo

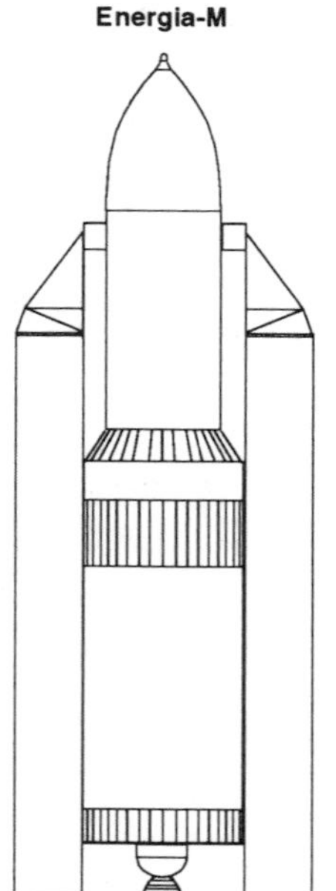

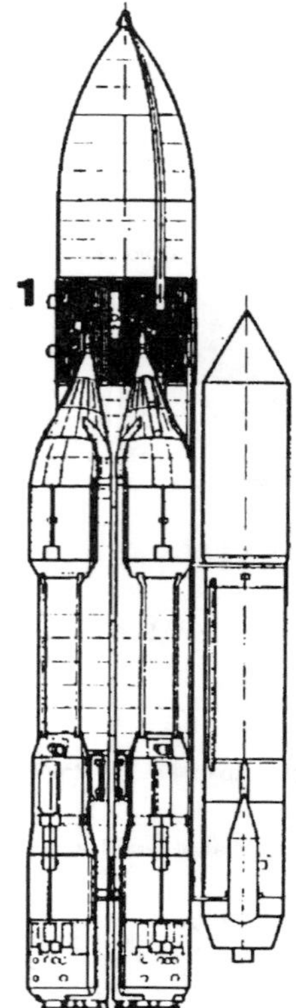

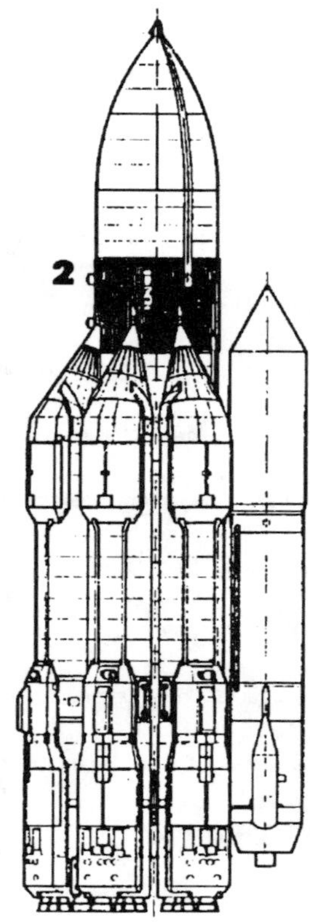

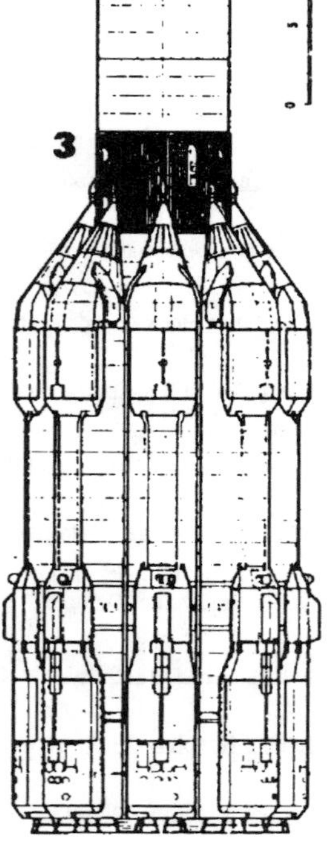

Approximate near-orbital performance for Energia with Varying Number of Boosters

Inclination	Number of Strap-on Boosters				
	2 (Energia-M)	2	4	6	8
	One Core Engine	Four Core Engines			
50.7 deg	75K lb (34K kg)	140K lb (65K kg)	231K lb (105K lb)	330K lb (150K kg)	440K lb (200K kg)
65 deg	—	130K lb (60K kg)	210K lb (95K lb)	—	410K lb (185K kg)
98 deg	—	110K lb (50K kg)	176K lb (80K lb)	—	365K lb (165K kg)

Other possible modifications to the Energia that are being examined provide greater payload flexibility, improved efficiency, and eventual reusability. The modular concept of the Energia permits development of a scaled down launch vehicle. For larger payloads, eight modified strap-ons and a modified core with increased fuel reserves can form a launch vehicle for payloads of up to 440,000 lb (200,000 kg).

For smaller payloads, two strap-ons and a scaled-down core can be used to assemble a rocket with payload capacity of about 90,000 lb (40,000 kg). A test version of this smaller vehicle, designated Energia-M, is expected to be launched by 1997. Energia-M is 154 ft (47 m) tall and generates 3.68M lb (16.4M N) of liftoff thrust. The payload fairing is placed in-line with the core.

Despite the difficult economic conditions within the CIS, development work still continues on the Energia-M, although at a much slower pace. To offset the loss of government funding for the Energia-M program, a joint stock company called ENERM was founded in June 1992. To date ENERM has raised some international capital for the Energia-M development program. However, the Energia-M program is still in need of additional capital to complete the development and launch testing of the vehicle.

Ikar

Launch Services Point of Contact:
NPO Yuzhnoye
Krivorozhskaya Street 3
Dniepropetrovsk, 8
Ukraine 320008
Phone: (70562) 42-00-22
Fax: (70562) 92-50-41
Telex: 143547 BRON

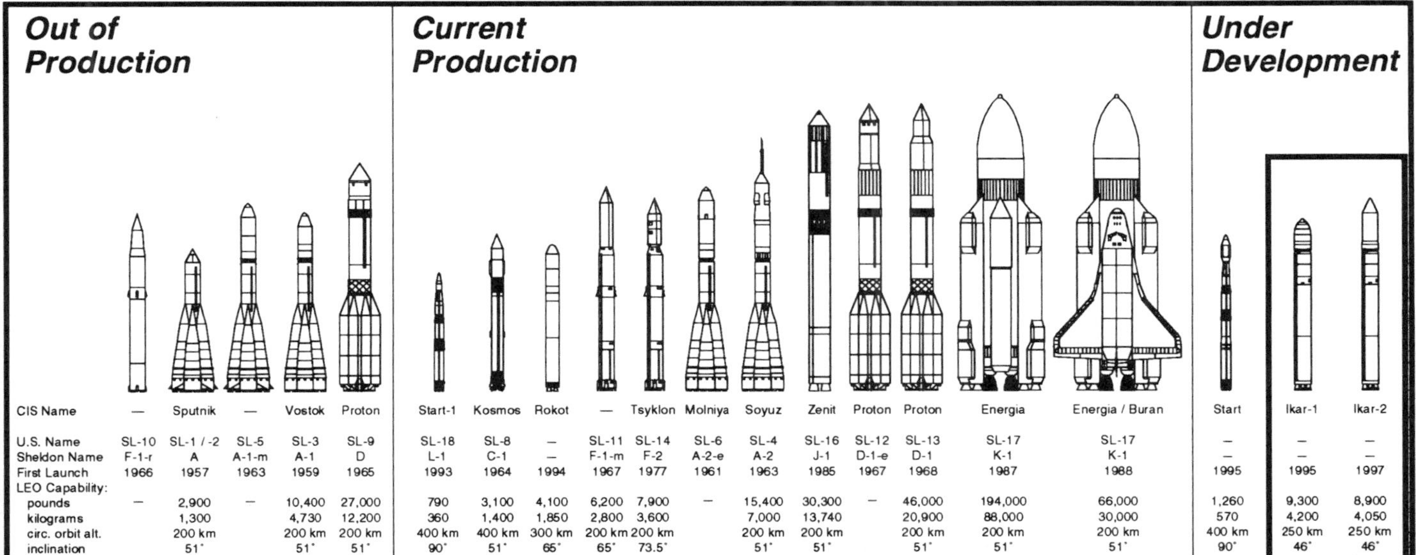

	Out of Production					Current Production					
CIS Name	—	Sputnik	—	Vostok	Proton	Start-1	Kosmos	Rokot	—	Tsyklon	Molniya
U.S. Name	SL-10	SL-1 / -2	SL-5	SL-3	SL-9	SL-18	SL-8	–	SL-11	SL-14	SL-6
Sheldon Name	F-1-r	A	A-1-m	A-1	D	L-1	C-1	–	F-1-m	F-2	A-2-e
First Launch	1966	1957	1963	1959	1965	1993	1964	1994	1967	1977	1961
LEO Capability:											
pounds	—	2,900	—	10,400	27,000	790	3,100	4,100	6,200	7,900	—
kilograms		1,300		4,730	12,200	360	1,400	1,850	2,800	3,600	
circ. orbit alt.		200 km		200 km	200 km	400 km	400 km	300 km	200 km	200 km	
inclination		51°		51°	51°	90°	51°	65°	65°	73.5°	

	Current Production						Under Development		
CIS Name	Soyuz	Zenit	Proton	Proton	Energia	Energia / Buran	Start	Ikar-1	Ikar-2
U.S. Name	SL-4	SL-16	SL-12	SL-13	SL-17	SL-17	–	–	–
Sheldon Name	A-2	J-1	D-1-e	D-1	K-1	K-1	–	–	–
First Launch	1963	1985	1967	1968	1987	1988	1995	1995	1997
LEO Capability:									
pounds	15,400	30,300	—	46,000	194,000	66,000	1,260	9,300	8,900
kilograms	7,000	13,740		20,900	88,000	30,000	570	4,200	4,050
circ. orbit alt.	200 km	200 km		200 km	200 km	200 km	400 km	250 km	250 km
inclination	51°	51°		51°	51°	51°	90°	46°	46°

Vehicle Description

CIS vehicles are identified either by their CIS, U.S., or Sheldon name. In the Soviet Union it was standard practice to name a launch vehicle after its original payload (e.g., Kosmos, Proton). The U.S. names (developed by the U.S. Department of Defense) are alphanumeric designations based roughly on chronological appearance. The Sheldon names, a most commonly used system that was published by Dr. Charles Sheldon of the U.S. Library of Congress in 1968, emphasize the basic families of launch vehicles with special indicators for variants within a family. To date, the Ikar vehicles have been assigned neither U.S. nor Sheldon names. NPO Yuzhnoye also refers to the Ikar-1 and -2 as the RS-20K1 and RS-20K2, respectively.

Ikar-1 Based on the SS-18 "Satan" ICBM (Russian designation RS-20), it is composed of three stages (2 boost stages plus a small insertion stage) utilizing storable propellants.

Ikar-2 Stages 1 and 2 are identical to those of Ikar-1. Tsyklon stage 3 (designated "S5M") replaces insertion stage; Tsyklon-based payload fairing replaces smaller Ikar-1 payload fairing.

Historical Summary

The Ikar (Icarus) launch vehicle family is based on the SS-18 "Satan" (Russian designation RS-20) heavy ICBM developed by what is now NPO Yuzhnoye and first deployed in 1975. The SS-18 is silo-launched and utilizes storable UDMH and N_2O_4 liquid propellants. The SS-18 underwent two major upgrades, in 1980 and 1988.

In accordance with the terms of the Strategic Arms Reduction Treaty 2 (START 2) of 1993, all 308 SS-18 ICBMs are being decommissioned and removed from their silos in Russia and Kazakhstan. Instead of simply destroying these deactivated missiles, their manufacturer, NPO Yuzhnoye, is refurbishing them into orbital launch vehicles of roughly the same capability as the Tsyklon. The Ikar-1 is essentially identical to the SS-18, with the missile's post-boost vehicle serving as a small orbital insertion stage. The small size of this third stage limits the Ikar-1 to delivering payloads to relatively low altitudes. The Ikar-2 vehicle will employ the third stage of the Tsyklon vehicle (S5M), which, through its larger propellant capacity and multiple restart capability, will provide increased performance to higher circular and elliptical orbits (1,000+ km altitude). Ikar-2 will also use a payload fairing based on that of the Tsyklon .

No known orbital missions have been flown to date, although the SS-18 has demonstrated a success rate of 97% over 188 test flights. An orbital demonstration flight of the Ikar-1 is planned for 1995; current projections call for Ikar-2 to be operational by 1997.

Launch Record

No orbital launches of an Ikar launch vehicle have occurred to date. The SS-18 missile system, upon which the Ikar is based, has achieved a 97% success rate in 188 test flights. The proposed third stage of the Ikar-2 (S5M) has flown successfully over 100 times as the third stage of the highly reliable Tsyklon launch vehicle.

Ikar General Description

Ikar-1 | Ikar-2

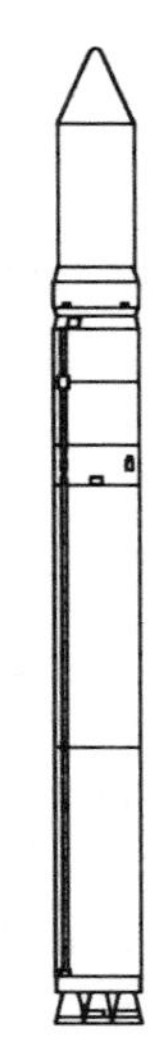
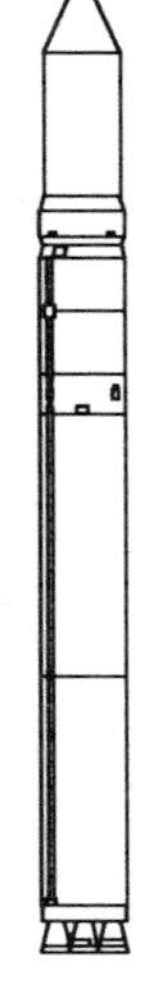

Summary The Ikar launch vehicles are based on refurbished SS-18 heavy ICBMs. Both the Ikar-1 and -2 are three-stage vehicles utilizing storable liquid propellants. The Ikar-1 is essentially identical to the SS-18; its undersized stage 3 restricts the Ikar-1 to relatively low altitude orbits. The Ikar-2 replaces this small injection stage (based on the SS-18 post-boost stage) with the more capable Tsyklon stage 3, and adds a longer payload fairing.

Status Under development

Key Organizations Users - Ukrainian Space Agency, NPO Yuzhnoye, commercial

Key Organizations - NPO Yuzhnoye (design bureau)

Vehicle

System Height	112.5 ft (34.3 m) for Ikar-1 128.6 ft (39.2 m) for Ikar-2
Payload Fairing Size	Ikar-1 - 9.8 ft (3.0 m) diameter by 17.2 ft (5.25 m) height Ikar-2 - 8.9 ft (2.7 m) diameter by 31.3 ft (9.54 m) height
Gross Mass	464,700 lb (210,800 kg) for Ikar-1 439,400 lb (212,900 kg) for Ikar-2
Planned Enhancements	None

Operations

Primary Missions	LEO
Compatible Upper Stages	SS-18 post-boost stage for Ikar-1; S5M (Tsyklon stage 3) for Ikar-2
First Launch	1995 (Ikar-1), 1997 (Ikar-2)
Success / Flight Total	0/0 (97% success rate for SS-18 missile test launches)
Launch Site	Baikonur Cosmodrome (46.0°N, 63.4°E)
Launch Azimuth	35°, 82°, 177.5°, 193° Corresponding to inclinations of 65°, 46°, 87°, 98°
Nominal Flight Rate	Max. 8–9 per pad annually; 2 pads (converted missile silos)
Planned Enhancements	None

Performance

135 nm (250 km) Circ. Orbit

Inclination: 46°	9,300 lb (4,200 kg) for Ikar-1 8,900 lb (4,050 kg) for Ikar-2
65°	6,600 lb (3,000 kg) for Ikar-1 8,400 lb (3,800 kg) for Ikar-2
87°	6,200 lb (2,800 kg) for Ikar-1 7,900 lb (3,600 kg) for Ikar-2
98°	5,800 lb (2,650 kg) for Ikar-1 7,600 lb (3,450 kg) for Ikar-2
Geotransfer Orbit, 51.6°	—
Geosynchronous Orbit	—

Financial Status

Estimated Launch Price — $6–10 million (NPO Yuzhnoye)

Manifest Ikar-1 demonstration flight in 1995

Remarks —

Ikar Vehicle

Overall

Ikar-1

Length	112.5 ft (34.3 m)
Gross Mass	464.7K lb (210.8K kg)
Thrust at Liftoff	? K lb (? kN)

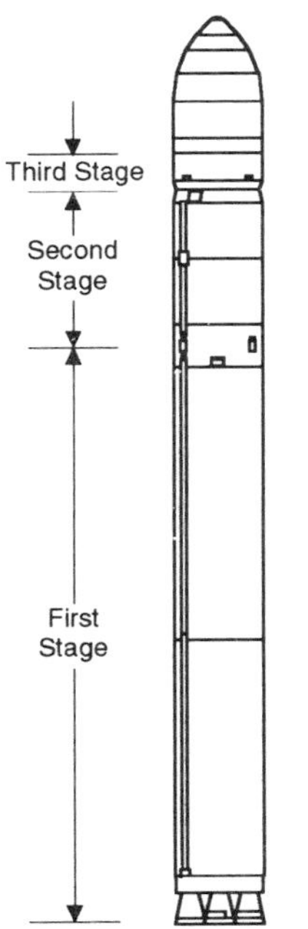

Ikar-2

Length	128.6 ft (39.2 m)
Gross Mass	469.4K lb (212.9K kg)
Thrust at Liftoff	? K lb (? kN)

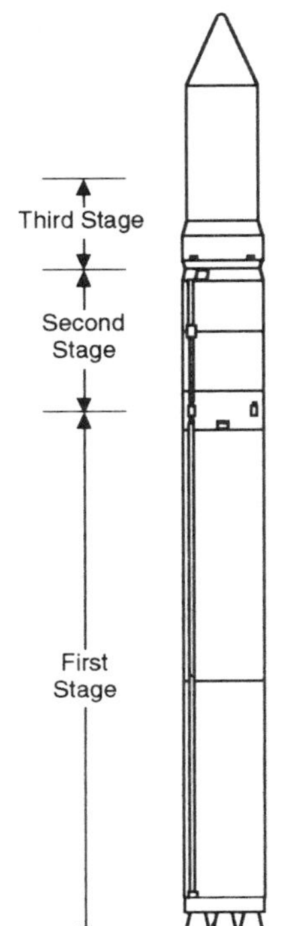

Ikar-1 / Ikar-2 Stages

Remarks:

Very few technical details are available for the Ikar-1 and Ikar-2 vehicles. The Ikar-1 is essentially identical to the silo-launched SS-18 (Russian designation RS-20) ICBM, with the missile warhead replaced by a spacecraft payload. All three stages employ storable N2O4 and UDMH propellants. Stage 1 is powered by four thrust chambers; although it has been reported in Western sources that this stage uses the RD-251 engine, official NPO Yuzhnoye information indicates that the RD-251 is used by the Tsyklon vehicle. The stage 1 control method (gimbaled engines, verniers, etc.) is not known. Virtually no information about the stage 2 engine configuration is available. Western sources list the stage 2 engine as the RD-252, while NPO Yuzhnoye states that the RD-252 is the Tsyklon stage 2 engine. Information concerning the number of chambers and the flight control method is not available. The limited information that is available on Ikar stages 1 and 2 makes it appear unlikely that Ikar and Tsyklon (and thus the SS-18 and SS-9 systems upon which these launchers are based) share the same stage 1 and 2 propulsion systems. Stage 3 of the Ikar-1 is apparently the SS-18's post-boost vehicle. Its propulsion system configuration is believed to consist of a single main chamber and four verniers. The principal diameter of the Ikar-1 is 3.0 m.

Stages 1 and 2 of the Ikar-2 are identical to those of Ikar-1. In order to increase performance to higher altitude orbits, NPO Yuzhnoye is planning to replace Ikar-1's small stage 3 with the larger S5M stage, currently employed as stage 3 of the Tsyklon. This stage also burns N2O4 and UDMH propellants. As with Tsyklon, the 2.2 m diameter, 2.7 m long S5M stage will be enclosed within the Ikar-2 payload fairing. The characteristics of this stage are defined in the Tsyklon section of this document.

Payload Fairings

Dimensions in inches (mm)

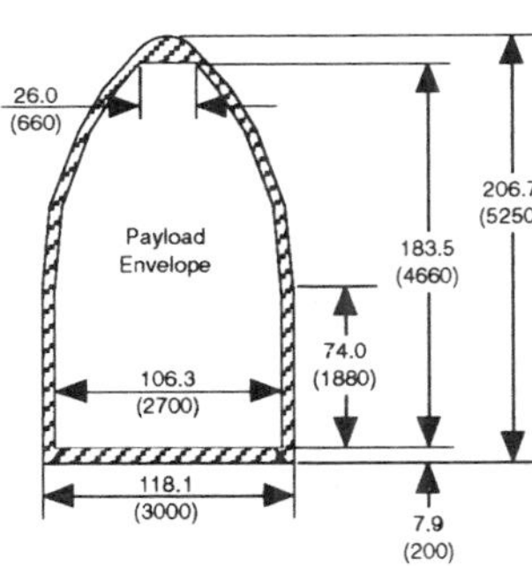

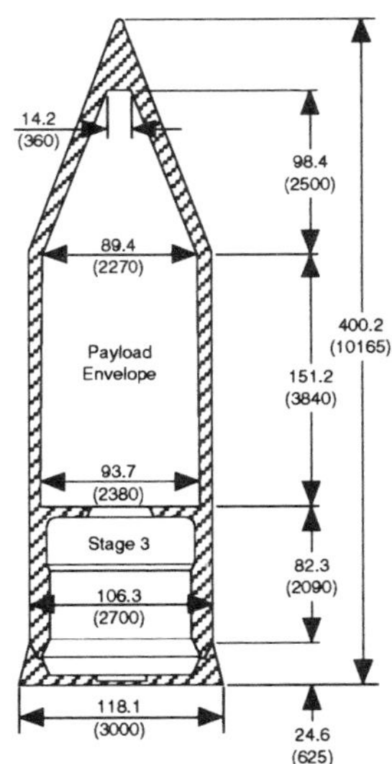

	Ikar-1	Ikar-2
Length	17.2 ft (5.25 m)	33.4 ft (10.16 m)
Diameter	9.8 ft (3.0 m)	8.9 ft (2.7 m)
Mass	?? lb (?? kg)	?? lb (?? kg)
Sections	?	2 ?
Structure	Skin-stringer ?	Skin-stringer ?
Material	Aluminum ?	Aluminum
Remarks	--	Based on Tsyklon fairing.

Avionics

Ikar-1 uses the standard SS-18 computer-controlled inertial guidance system. Ikar-2 will use the SS-18 control system for stages 1 and 2, while stage 3 will be controlled by its own autonomous flight control system.

Attitude Control System

Information on stage 1 and 2 attitude control systems is not currently available. Ikar-1 stage 3 is believed to use four verniers for attitude control and steering. The S5M stage proposed for use with Ikar-2 employs a system of 8 turbine exhaust nozzles for control during powered flight and an 8-thruster RCS system during coast phases.

Ikar Performance

Performance Characteristics of Ikar-1 and -2 Launched from Baikonur Cosmodrome (courtesy NPO Yuzhnoye)

Ikar-1 Performance to Circular Orbits at Various Inclinations

Payload Mass (kg)

Circular Orbit Altitude (km)

46˚ 65˚ 87˚ 98˚

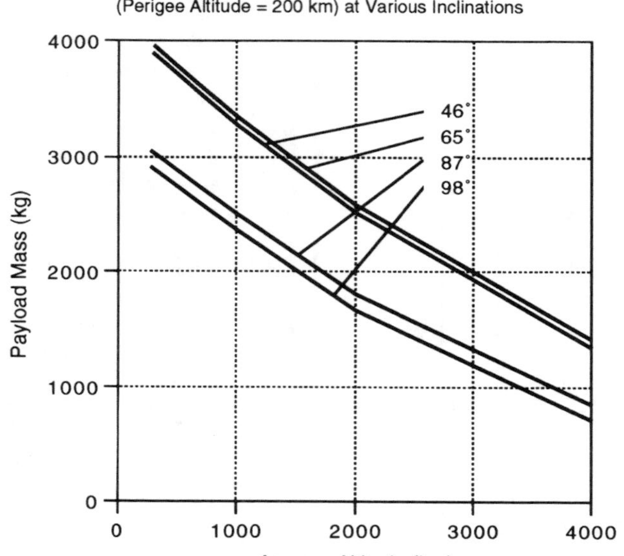

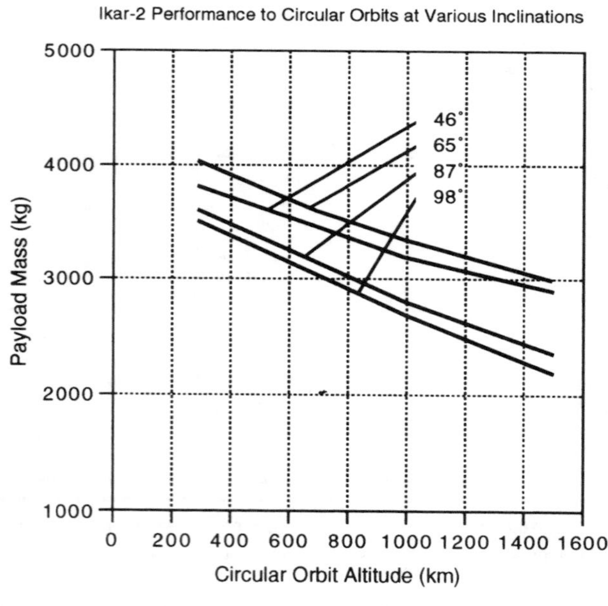

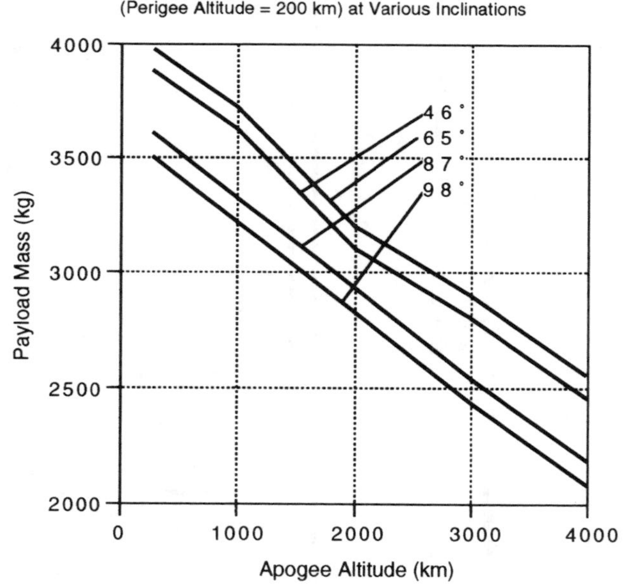

Ikar Operations

Launch Site

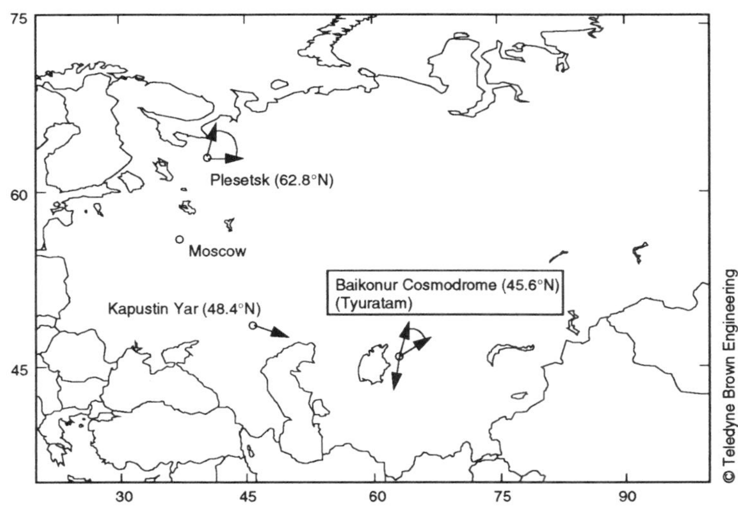

Launch Facilities

No illustrations available.

Launch Processing

Ikar will be launched from two SS-18 silos at the Baikonur Cosmodrome. It is presumed that, like the SS-18, Ikar will be transported in, and launched from inside, a protective canister. It is unknown how this unique launch approach will affect payload processing, pad access, and environmental control.

Flight Sequence

No data available.

Ikar Payload Accommodations

Payload Compartment	
Maximum Payload Diameter	Ikar-1 - 106.3 in (2,700 mm) Ikar-2 - 93.7 in (2,380 mm)
Maximum Cylinder and Cone Length	Ikar-1 - 74.0 in (1,880 mm) for cylinder, 109.4 in (2,780 mm) for cone Ikar-2 - 151.2 in (3,840 mm) for cylinder, 98.4 in (2,500 mm) for cone
Payload Adapter	
Interface Diameter	Variable, according to customer requirements
Payload Integration	
Nominal Mission Schedule Begins	T- 9 to 12 months
Launch Window	
Latest Countdown Hold Not Requiring Recycling	T-?? min
On-Pad Storage Capability	Limited only by spacecraft requirements
Latest Access to Payload	T- 2.5 to 3.0 hrs
Environment	
Maximum Load Factors	+8.3 g axial (Ikar-1), +6.7 g axial (Ikar-2), ±0.36 g lateral (both)
Minimum Lateral/Longitudinal Payload Frequency	25 Hz longitudinal, 15 Hz lateral
Maximum Overall Acoustic Level	140 dB, 20 to 2,000 Hz
Maximum Flight Shock	1,000 g at 1,000 Hz
Maximum Dynamic Pressure on Fairing	1,400 lb/ft^2 (7,000 kgf/m^2) for Ikar-1, 1,700 lb/ft^2 (8,500 kgf/m^2) for Ikar-2
Maximum Pressure Change in Fairing	?? psi/s (?? kPa/s)
Cleanliness Level in Fairing (Prior to Launch)	Class 100,000
Payload Delivery	
Standard Orbit and Accuracy (3 sigma)	Ikar-1: Orbit: 108 nm (200 km) / 324 nm (600 km) Radius: ±1.9 nm (3.5 km) / ±3.0 nm (5.5 km) Period: ±2.5 sec / ±4.0 sec Inclination: ±2.0 min / ±2.5 min Ikar-2: Orbit: 324 nm (600 km) / 513 nm (600 km) Radius: ±11 nm (20 km) / ±13 nm (25 km) Period: ±10 sec / ±15 sec Inclination: ±10 min / ±12 min
Attitude Accuracy (3 sigma)	All axes: ±2 deg, ±1 deg/sec After separation: ±3 deg, ±2 deg/sec
Nominal Payload Separation Rate	?? ft/s (?? m/s)
Deployment Rotation Rate Available	?? rpm
Loiter Duration in Orbit	?? hrs
Maneuvers (Thermal / Collision / Telemetry)	No ?

Ikar Notes

Publications

Technical Publications

"Main Characteristics of the SS-18K Space Launch Vehicle Based Upon a Military Rocket",Y. Smetanin and O. Drobakhin, NPO Yuzhnoye, Ukraine. ESA SR-362, March 1994.

Jane's Space Directory 1994-95, Jane's Information Group Inc., Alexandria, VA.

Jane's Strategic Weapons Systems, Jane's Information Group Inc., Alexandria, VA, 1994.

Space Directory of Russia, February 1993 Edition, Euroconsult/Sevig Press, Paris, France.

The Russian Space Directory 1994, European Space Report, Munich, Germany.

Acronyms

CIS - Commonwealth of Independent States
GEO - geosynchronous orbit
GTO - geosynchronous transfer orbit
LEO - low Earth orbit
N2O4 - nitrogen tetroxide
NPO - Scientific Production Organization
UDMH - unsymmetrical dimethylhydrazine
USSR - Union of Soviet Socialist Republics

Launch Service Point of Contact:
PO Polyot
226 B. Khmelnitskovo St.
Omsk 644021
The Russian Federation
Phone: (3812) 57 92 00
Fax: (3812) 57 92 00

Marketing Representative:
Plowshare Technology, Ltd.
P.O. Box 9
Horsham, West Sussex RH13 5YN
United Kingdom
Phone: (44) 403 210 494
Fax: (44) 403 210 494

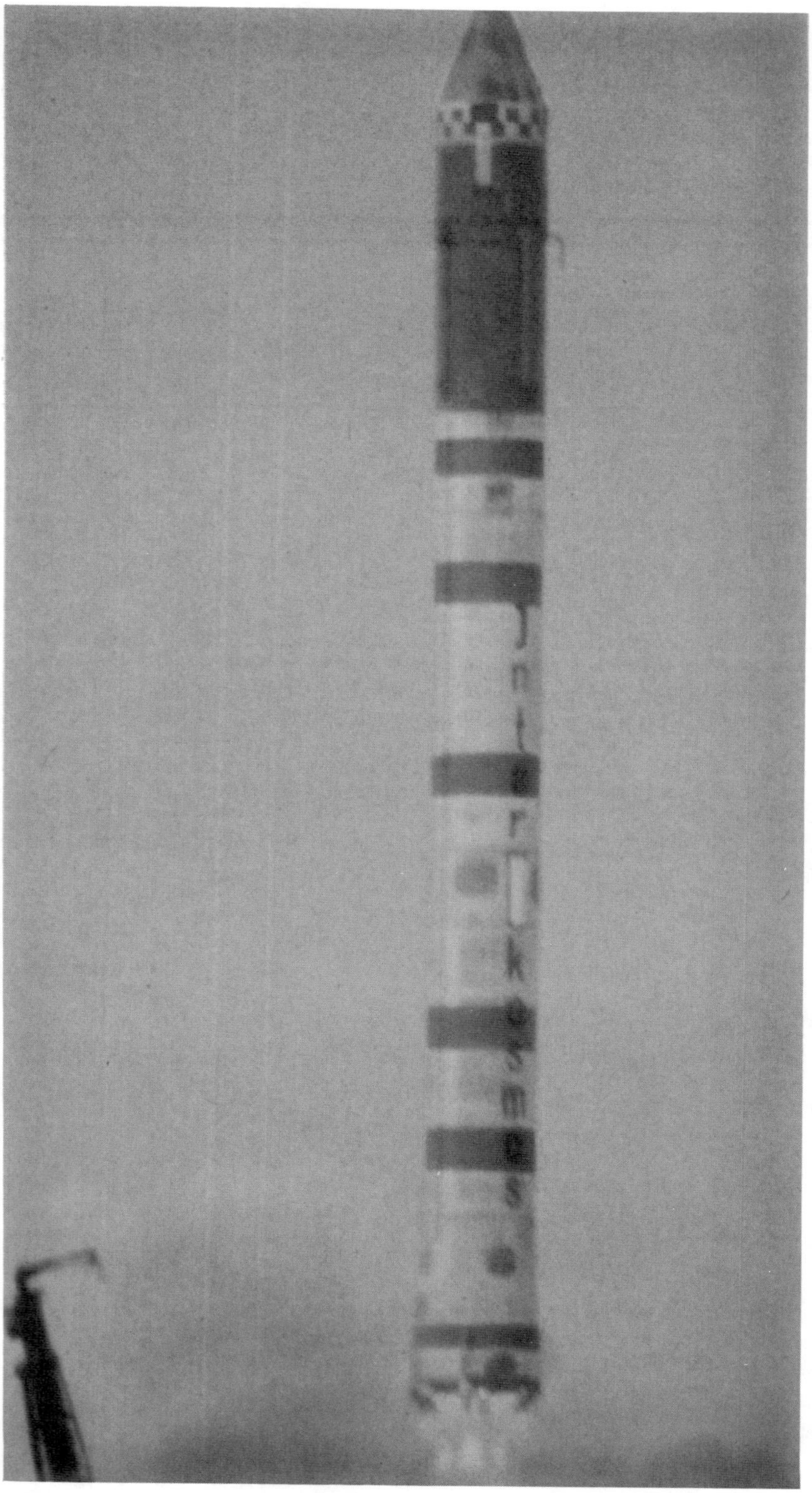

Photo Courtesy of Space Commerce Corp.

The Kosmos launch vehicle, the only vehicle in the CIS fleet to have been launched from all three launch sites in the former Soviet Union, is shown lifting off from the Plesetsk Cosmodrome.

Kosmos History

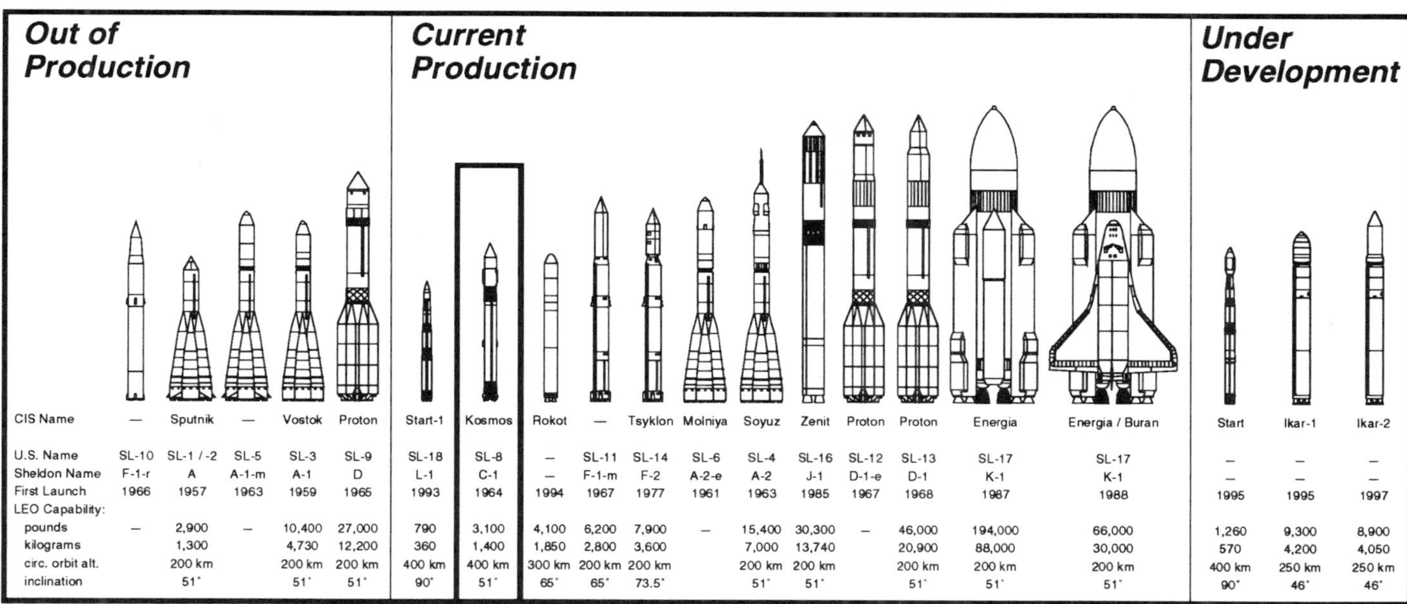

CIS Name	—	Sputnik	—	Vostok	Proton
U.S. Name	SL-10	SL-1 / -2	SL-5	SL-3	SL-9
Sheldon Name	F-1-r	A	A-1-m	A-1	D
First Launch	1966	1957	1963	1959	1965
LEO Capability:					
pounds	—	2,900	—	10,400	27,000
kilograms		1,300		4,730	12,200
circ. orbit alt.		200 km		200 km	200 km
inclination		51°		51°	51°

CIS Name	Start-1	Kosmos	Rokot	—	Tsyklon	Molniya	Soyuz	Zenit	Proton	Proton	Energia	Energia / Buran
U.S. Name	SL-18	SL-8	—	SL-11	SL-14	SL-6	SL-4	SL-16	SL-12	SL-13	SL-17	SL-17
Sheldon Name	L-1	C-1	—	F-1-m	F-2	A-2-e	A-2	J-1	D-1-e	D-1	K-1	K-1
First Launch	1993	1964	1994	1967	1977	1961	1963	1985	1967	1968	1987	1988
LEO Capability:												
pounds	790	3,100	4,100	6,200	7,900	—	15,400	30,300	—	46,000	194,000	66,000
kilograms	360	1,400	1,850	2,800	3,600		7,000	13,740		20,900	88,000	30,000
circ. orbit alt.	400 km	400 km	300 km	200 km	200 km		200 km	200 km		200 km	200 km	200 km
inclination	90°	51°	65°	65°	73.5°		51°	51°		51°	51°	51°

CIS Name	Start	Ikar-1	Ikar-2
U.S. Name	—	—	—
Sheldon Name	—	—	—
First Launch	1995	1995	1997
LEO Capability:			
pounds	1,260	9,300	8,900
kilograms	570	4,200	4,050
circ. orbit alt.	400 km	250 km	250 km
inclination	90°	46°	46°

Vehicle Description

CIS vehicles are identified either by their CIS, U.S., or Sheldon name. In the Soviet Union, it was standard practice to name a launch vehicle after its original payload (e.g., Kosmos, Proton). The U.S. names (developed by the U.S. Department of Defense) are alphanumeric designations based roughly on chronological appearance. The Sheldon names, a most commonly used system that was published by Dr. Charles Sheldon of the U.S. Library of Congress in 1968, emphasize the basic families of launch vehicles with special indicators for variants within a family.

Example: F-1-m

- e - earth escape or fourth stage
- m - maneuverable stage
- r - reentry stage
- Upper Stage (1, 2) - Note: "1" or "2" stage may vary among different families, e.g., A-1 upper stage different than B-1.
- Family (A, B, C, D, F, G, J, K)

Kosmos (C-1/SL-8) Based on the Skean SS-5, it has two stages with storable liquid propellant.

Historical Summary

The Kosmos (C-1, SL-8) was first launched in 1964. It is a two stage storable propellant vehicle with maximum capability to orbit of about 3,100 lb (1,400 kg). This bridged a payload gap between the B-1 rocket, 1,300 lb (600 kg) to orbit, and the A-class rockets, 10,400 lb (4,730 kg) to orbit. As with its predecessors, a military missile (in this case the SS-5 Skean IRBM) was utilized as the first stage. Both the SS-5 and Kosmos were designed by Mikhail Yangel at what is now the Ukrainian design bureau NPO Yuzhnoye, with production taking place at the PO Polyot facility in Omsk, Russia. Although they are now apparently separate entities, PO Polyot was once a branch of Yangel's design bureau. PO Polyot is currently responsible for production and marketing of the Kosmos.

Kosmos was first used in 1964 when it launched the triple payload of Kosmos 38, 39, and 40 out of Tyuratam. In addition to its more than 400 orbital launches, Kosmos has carried out more than 300 suborbital flights, primarily for hypersonics and reentry research. The most well-known of these suborbital flights were of the BOR-4 lifting body, flown under the names Kosmos 1374, 1445, 1517, and 1614 in 1982–1984. These subscale spaceplane missions were conducted for the Buran space shuttle program to test heat resistant reentry materials and gather additional transonic aerodynamic data. The four space missions were successful, culminating in recoveries in the Indian Ocean and the Black Sea. Although Kosmos is the only launcher to have used all three launch sites in the former Soviet Union, today all Kosmos launches are conducted at Plesetsk. It is currently used to launch primarily small communications and navigation satellites.

Reports in the Russian media have stated that production of the Kosmos has been discontinued; however, as of January 1995, Kosmos launches were still taking place. It should be pointed out that these two situations are not necessarily mutually exclusive due to the Russian tendency to stockpile completed rockets. Other reports indicate variously that the German firm OHB-System and the British company Plowshare Technology Ltd. have agreements to market excess launch capability.

The January 23, 1995, Kosmos launch was unique in that, in addition to its primary payload (a Russian-built Tsikada navigation satellite), it carried two smaller payloads: the U.S.-built Faisat-1, a store-and-dump communications satellite with the distinction of being the first U.S. satellite to be launched on a Russian booster; and the Swedish Astrid scientific microsat, the launch arrangements for which were made by Plowshare Technology Ltd.

Kosmos History

(continued)

Launch Record

The information on this page is based on the TRW Space Log and data provided by PO Polyot. In several instances, these two sources do not agree; in particular, the PO Polyot information indicates that a number of launches previously attributed in the west to the F-class vehicles (e.g., Tsyklon) were actually Kosmos launches; several instances also arose where launches attributed to Kosmos in the TRW Space Log were not reported as such by PO Polyot. PO Polyot also listed a number of Kosmos failures that were previously not included in western sources. In cases of disagreement, PO Polyot data was given preference.

The information presented here includes only orbital launches. In addition to the flights listed here, Kosmos has been used for over 300 suborbital launches, the most well-known of which are the four subscale BOR spaceplane flights of the early 1980s. Although they carried Kosmos designations, the BOR flights are not considered orbital launches by PO Polyot, and thus are not included here.

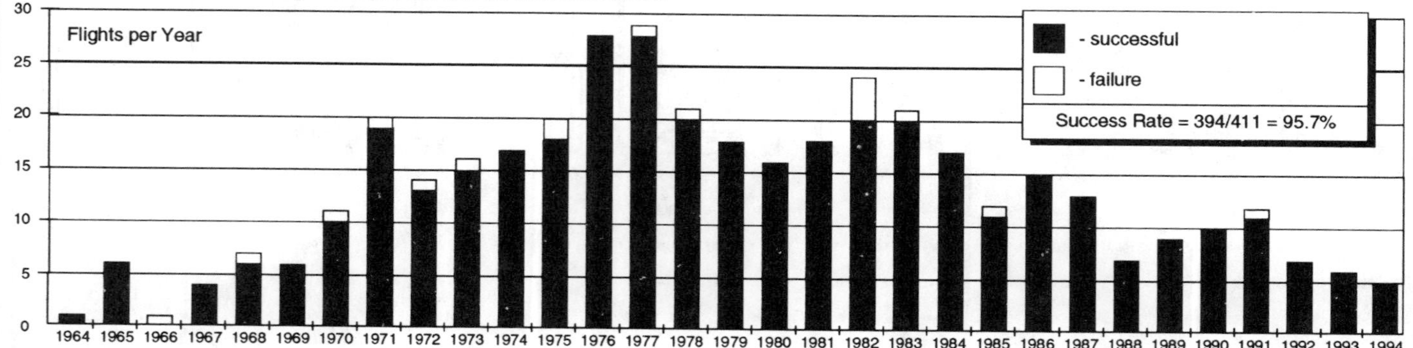

Year	Successful Flights				Failures
	Total	PL	KY	TT	
1964	1	–	–	1	–
1965	6	–	–	6	–
1966	–	–	–	–	1
1967	4	3	–	1	–
1968	6	5	–	1	1
1969	6	6	–	–	–
1970	10	10	–	–	1
1971	19	19	–	–	1
1972	13	13	–	–	1
1973	15	14	1	–	1
1974	17	16	1	–	–
1975	18	17	1	–	2
1976	28	27	1	–	–
1977	28	26	2	–	1
1978	20	19	1	–	1
1979	18	16	2	–	–
1980	16	15	1	–	–
1981	18	17	1	–	–
1982	20	17	3	–	4
1983	20	18	2	–	1
1984	17	15	1	1	–
1985	11	10	1	–	1
1986	15	15	–	–	–
1987	13	12	1	–	–
1988	7	7	–	–	–
1989	9	9	–	–	–
1990	10	9	–	1	–
1991	11	11	–	–	1
1992	7	7	–	–	–
1993	6	6	–	–	–
1994	5	5	–	–	–
TOTAL	394	364	19	11	17

SEE ADJACENT TABLE FOR LAUNCH RECORD PRIOR TO 1983

	YEAR	DATE	VEHICLE	SITE	PAYLOAD
278	1983	Jan-12	C-1	PL	Kosmos-1428
279		Jan-19	C-1	PL	Kosmos-1429 to 1436
280		Jan-15	C-1	??	unknown
		Failure			
281		Mar-24	C-1	PL	Kosmos-1447
282		Mar-30	C-1	PL	Kosmos-1448
283		Apr-6	C-1	PL	Kosmos-1450
284		Apr-12	C-1	PL	Kosmos-1452
285		Apr-19	C-1	PL	Kosmos-1453
286		May-6	C-1	PL	Kosmos-1459
287		May-19	C-1	PL	Kosmos-1463
288		May-24	C-1	PL	Kosmos-1464
289		May-26	C-1	KY	Kosmos-1465
290		Jul-6	C-1	PL	Kosmos-1473 to 1480
291		Aug-3	C-1	PL	Kosmos-1486
292		Aug-31	C-1	KY	Kosmos-1494
293		Sep-30	C-1	PL	Kosmos-1501
294		Oct-5	C-1	PL	Kosmos-1502
295		Oct-12	C-1	PL	Kosmos-1503
296		Oct-26	C-1	PL	Kosmos-1506
297		Nov-11	C-1	PL	Kosmos-1508
298		Dec-8	C-1	PL	Kosmos-1513
299	1984	Jan-5	C-1	PL	Kosmos-1522 to 1529
300		Jan-11	C-1	PL	Kosmos-1531
301		Jan-26	C-1	PL	Kosmos-1534
302		Feb-2	C-1	PL	Kosmos-1535
303		Feb-21	C-1	PL	Kosmos-1538
304		May-11	C-1	PL	Kosmos-1550
305		May-17	C-1	PL	Kosmos-1553
306		May-29	C-1	PL	Kosmos-1559 to 1566
307		Jun-8	C-1	PL	Kosmos-1570
308		Jun-21	C-1	PL	Kosmos-1574
309		Jun-27	C-1	PL	Kosmos-1577
310		Jun-28	C-1	KY	Kosmos-1578
311		Sep-13	C-1	PL	Kosmos-1598
312		Sep-27	C-1	TT	Kosmos-1601
313		Oct-11	C-1	PL	Kosmos-1605
314		Nov-15	C-1	PL	Kosmos-1610
315		Dec-20	C-1	PL	Kosmos-1615
316	1985	Jan-17	C-1	PL	Kosmos-1624
317		Feb-1	C-1	PL	Kosmos-1627
318		Feb-27	C-1	PL	Kosmos-1631
319		Mar-14	C-1	PL	Kosmos-1634
320		Mar-21	C-1	PL	Kosmos-1635 to 1642
321		May-30	C-1	PL	Kosmos-1655
322		Jun-19	C-1	PL	Kosmos-1662
323		Sep-4	C-1	PL	Kosmos-1680
324		Oct-2	C-1	KY	Kosmos-1688
325		Oct-23	C-1	??	unknown
		Failure			
326		Nov-28	C-1	PL	Kosmos-1704
327		Dec-19	C-1	PL	Kosmos-1709
328	1986	Jan-9	C-1	PL	Kosmos-1716 to 1723
329		Jan-16	C-1	PL	Kosmos-1725
330		Jan-23	C-1	PL	Kosmos-1727
331		Apr-18	C-1	PL	Kosmos-1741
332		May-23	C-1	PL	Kosmos-1745
333		Jun-6	C-1	PL	Kosmos-1748 to 1755
334		Jun-18	C-1	PL	Kosmos-1759
335		Jul-16	C-1	PL	Kosmos-1763
336		Sep-3	C-1	PL	Kosmos-1776
337		Sep-10	C-1	PL	Kosmos-1777
338		Oct-27	C-1	PL	Kosmos-1788
339		Nov-13	C-1	PL	Kosmos-1791
340		Nov-21	C-1	PL	Kosmos-1794 to 1801
341		Nov-25	C-1	PL	Kosmos-1802
342		Dec-17	C-1	PL	Kosmos-1808

	YEAR	DATE	VEHICLE	SITE	PAYLOAD
343	1987	Jan-21	C-1	PL	Kosmos-1814
344		Jan-22	C-1	KY	Kosmos-1815
345		Jan-29	C-1	PL	Kosmos-1816
346		Feb-18	C-1	PL	Kosmos-1821
347		Jun-9	C-1	PL	Kosmos-1850
348		Jun-16	C-1	PL	Kosmos-1852 to 1859
349		Jun-23	C-1	PL	Kosmos-1861
350		Jul-7	C-1	PL	Kosmos-1864
351		Jul-14	C-1	PL	Kosmos-1868
352		Oct-14	C-1	PL	Kosmos-1891
353		Dec-1	C-1	PL	Kosmos-1898
354		Dec-15	C-1	PL	Kosmos-1902
355		Dec-23	C-1	PL	Kosmos-1904
356	1988	Mar-11	C-1	PL	Kosmos-1924 to 1931
357		Mar-22	C-1	PL	Kosmos-1934
358		Apr-5	C-1	PL	Kosmos-1937
359		Jun-21	C-1	PL	Kosmos-1954
360		Jul-14	C-1	PL	Kosmos-1958
361		Jul-19	C-1	PL	Kosmos-1959
362		Jul-28	C-1	PL	Kosmos-1960
363	1989	Jan-26	C-1	PL	Kosmos-1992
364		Feb-14	C-1	PL	Kosmos-2002
365		Feb-22	C-1	PL	Kosmos-2004
366		Mar-24	C-1	PL	Kosmos-2008 to 2015
367		Apr-4	C-1	PL	Kosmos-2016
368		Jun-7	C-1	PL	Kosmos-2026
369		Jun-14	C-1	PL	Kosmos-2027
370		Jul-4	C-1	PL	Nadezhda 1
371		Jul-25	C-1	PL	Kosmos-2034
372	1990	Jan-18	C-1	PL	Kosmos-2056
373		Feb-6	C-1	TT	Kosmos-2059
374		Feb-27	C-1	PL	Nadezhda 2
375		Mar-20	C-1	PL	Kosmos-2061
376		Apr-6	C-1	PL	Kosmos-2064 to 2071
377		Apr-20	C-1	PL	Kosmos-2074
378		Apr-25	C-1	PL	Kosmos-2075
379		Aug-28	C-1	PL	Kosmos-2098
380		Sep-14	C-1	PL	Kosmos-2100
381		Dec-10	C-1	PL	Kosmos-2112
382	1991	Jan-29	C-1	PL	Informatr 1
383		Feb-5	C-1	PL	Kosmos 2123
384		Feb-12	C-1	PL	Kosmos 2125 to 2132
385		Feb-26	C-1	PL	Kosmos 2135
386		Mar-12	C-1	PL	Nadezhda 3
387		Mar-19	C-1	PL	Kosmos 2137
388		Apr-16	C-1	PL	Kosmos 2142
389		Jun-11	C-1	PL	Kosmos 2150
390		Jun-25	C-1	PL	unknown
		Failure			
391		Aug-22	C-1	PL	Kosmos 2154
392		Oct-10	C-1	PL	Kosmos 2164
393		Nov-27	C-1	PL	Kosmos 2173
394	1992	Feb-18	C-1	PL	Kosmos 2180
395		Mar-10	C-1	PL	Kosmos 2181
396		Apr-15	C-1	PL	Kosmos 2184
397		Jun-3	C-1	PL	Kosmos 2187 to 2194
398		Jul-2	C-1	PL	Kosmos 2195
399		Aug-12	C-1	PL	Kosmos 2208
400		Oct-29	C-1	PL	Kosmos 2218
401	1993	Jan-12	C-1	PL	Kosmos 2230
402		Feb-9	C-1	PL	Kosmos 2233
403		Apr-1	C-1	PL	Kosmos 2239
404		Jun-16	C-1	PL	Kosmos 2251
405		Oct-26	C-1	PL	Kosmos 2265
406		Nov-2	C-1	PL	Kosmos 2266
407	1994	Apr-26	C-1	PL	Kosmos 2279
408		Jul-14	C-1	PL	Nadezhda 4
409		Aug-2	C-1	PL	Kosmos 2285
410		Sep-27	C-1	PL	Kosmos 2292
411		Dec-20	C-1	PL	Kosmos 2298

Data Sources: TRW Space Log, PO Polyot

TT is Tyuratam (Baikonur)
PL is Plesetsk Cosmodrome
KY is Kapustin Yar

Kosmos General Description

C-1

Summary

Kosmos is based on Skean (SS-5) with a second stage added. Kosmos is the only Soviet launch vehicle to have flown from all three launch facilities; currently, it is launched from Plesetsk only. Nearly 400 Kosmos launchers have successfully achieved orbit. Use of the Kosmos is decreasing as a new generation of heavier satellites are being transferred to the Tsyklon. Primary payloads are now low altitude navigation and store/dump communications satellites.

Status

Operational since 1964

Key Organizations

Users - Russian Ministry of Defense, Russian Space Agency, commercial

Key Organizations -
- PO Polyot (manufacturer)
- NPO Energomash (stage 1 engine)
- KB Khimavtomatiki (stage 2 engine)

Vehicle

System Height	Up to 106 ft (32.4 m)
Payload Fairing Size	7.9 ft (2.4 m) diameter by 18.8 ft (5.72 m) height
Gross Mass	240,000 lb (109,000 kg)
Planned Enhancements	Kosmos-U (see *Notes* section)

Operations

Primary Missions	LEO
Compatible Upper Stages	–
First Launch	1964
Success / Flight Total	394 / 411 through 1994
Launch Sites	Plesetsk (62.8°N, 40.1°E), three pads Kapustin Yar (48.4°N, 45.8°E), two pads - has not been used since 1987 Tyuratam (45.6°N, 63.4°E) - no longer used
Launch Azimuth	Plesetsk ??°–??° (usual inclinations are 66°, 74°, 83°) Kapustin Yar ??° (usual inclination is 51°)
Nominal Flight Rate	7–10 / yr; each pad has a theoretical capability of 30 / yr
Planned Enhancements	None

Performance

216 nm (400 km) circ	2,400–3,100 lb (1,100–1,400 kg) depending on inclination and launch site
864 nm (1600 km) circ	1,300–1,800 lb (600– 800 kg) depending on inclination and launch site
Geotransfer Orbit, 51.6°	—
Geosynchronous Orbit	—

Financial Status

Estimated Launch Price: $10M (Plowshare Technology Ltd.)

Manifest

Not available

Remarks

Report in Russian media indicates that production of the Kosmos launcher may have been discontinued.

Kosmos Vehicle

Overall

Length	106 ft (32.4 m)
Gross Mass	240,000 lb (109,000 kg)
Thrust at Liftoff	334,000 lb (1,490,000 N)

Payload Fairing
Second Stage
First Stage
106 ft (32.4 m)
7.9 ft (2.4 m)

Aerofins
Thrust Chamber
Jet Vanes

Base view of RD-216 engine on first stage

Kosmos Stages

	Stage 1	Stage 2
Dimension:		
Length	73 ft (22.4 m)	22 ft (6.6 m)
Diameter	7.9 ft (2.4 m)	7.9 ft (2.4 m)
Mass:		
Propellant Mass	180.3K lb (81.8K kg)	41.9K lb (19.0K kg)
Gross Mass	192.0K lb (87.1K kg)	45.0K lb (20.4K kg)
Structure:		
Type	Skin-Stringer ?	Skin-Stringer ?
Material	Aluminum	Aluminum
Propulsion:		
Propellant	Nitric acid+27% N2O4 / UDMH	N2O4 / UDMH
Average Thrust (each)	334K lb (1490 kN) SL 391K lb (1740 kN) vac	35.2K lb (157 kN) vac total (main + verniers)
Engine Designation	RD-216/11D614	RD-??/11D49
Number of Engines	1 engine with 2 pumps + 4 chambers	1 + verniers
Isp	248 sec SL 291.3 sec vac	303 sec vac
Feed System	Gas generator	??
Chamber Pressure	1088 psia (75 bar)	?? psia (?? bar)
Mixture Ratio (O/F)	??	??
Throttling Capability	100% only	100% only
Expansion Ratio	??	??
Restart Capability	No	2 starts
Tank Pressurization	compressed gas	compressed gas
Control-Pitch,Yaw,Roll	Aero fins & jet vanes	verniers, ACS
Events:		
Nominal Burn Time	130 sec	375 sec main engine, 500 sec small verniers
Stage Shutdown	Burn to depletion	command shutdown
Stage Separation	??	4 spring pushers

Remarks:

The first stage, which uses storable liquid propellants, has a diameter of about 7.9 ft (2.4 m) with a base section that flares to a diameter of about 9.35 ft (2.85 m). Stage 1 employs four stabilization fins; the span from fin-tip to fin-tip is 14.4 ft (4.4 m). The first stage propulsion system consists of 2 turbopumps and four chambers, with the entire unit carrying the designation RD-216 (this propulsion system is also sometimes designated 11D614). The second stage engine apparently consists of one main chamber with a number of verniers for control during powered flight. This engine can perform on on-orbit restart, thus enabling Kosmos to place payloads into high circular orbits. The RD number of stage 2's engine is not known; the designation 11D49 has been reported. Stage 2 also uses a low thrust ACS to provide 3-axis control during the coast period between its first and second burns. The propellant supply for the ACS is kept in the long, narrow tanks external to stage 2.

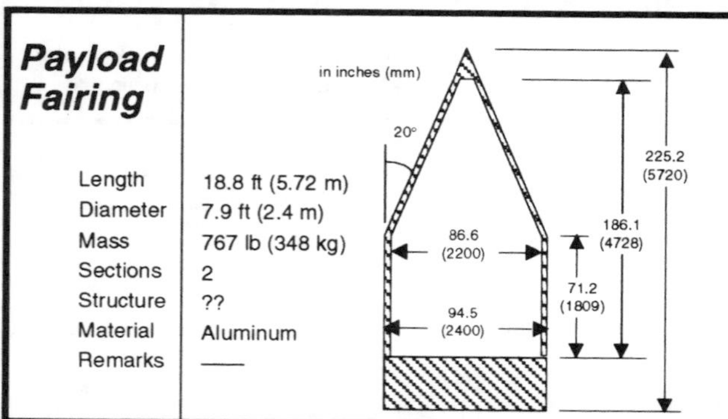

Payload Fairing

Length	18.8 ft (5.72 m)
Diameter	7.9 ft (2.4 m)
Mass	767 lb (348 kg)
Sections	2
Structure	??
Material	Aluminum
Remarks	——

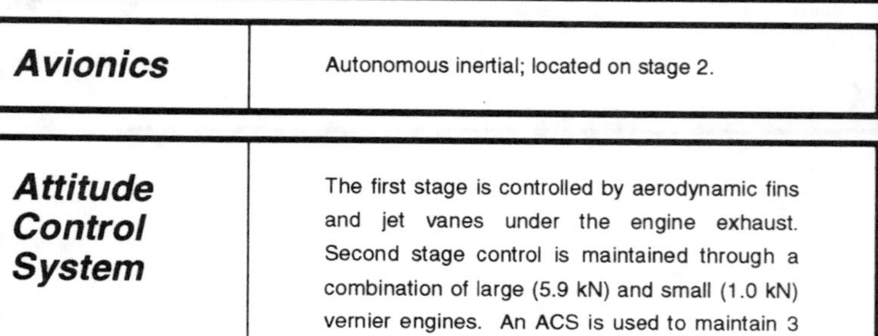

Avionics

Autonomous inertial; located on stage 2.

Attitude Control System

The first stage is controlled by aerodynamic fins and jet vanes under the engine exhaust. Second stage control is maintained through a combination of large (5.9 kN) and small (1.0 kN) vernier engines. An ACS is used to maintain 3 axis control during stage 2 coast periods.

Kosmos Performance

Performance Characteristics of Kosmos:

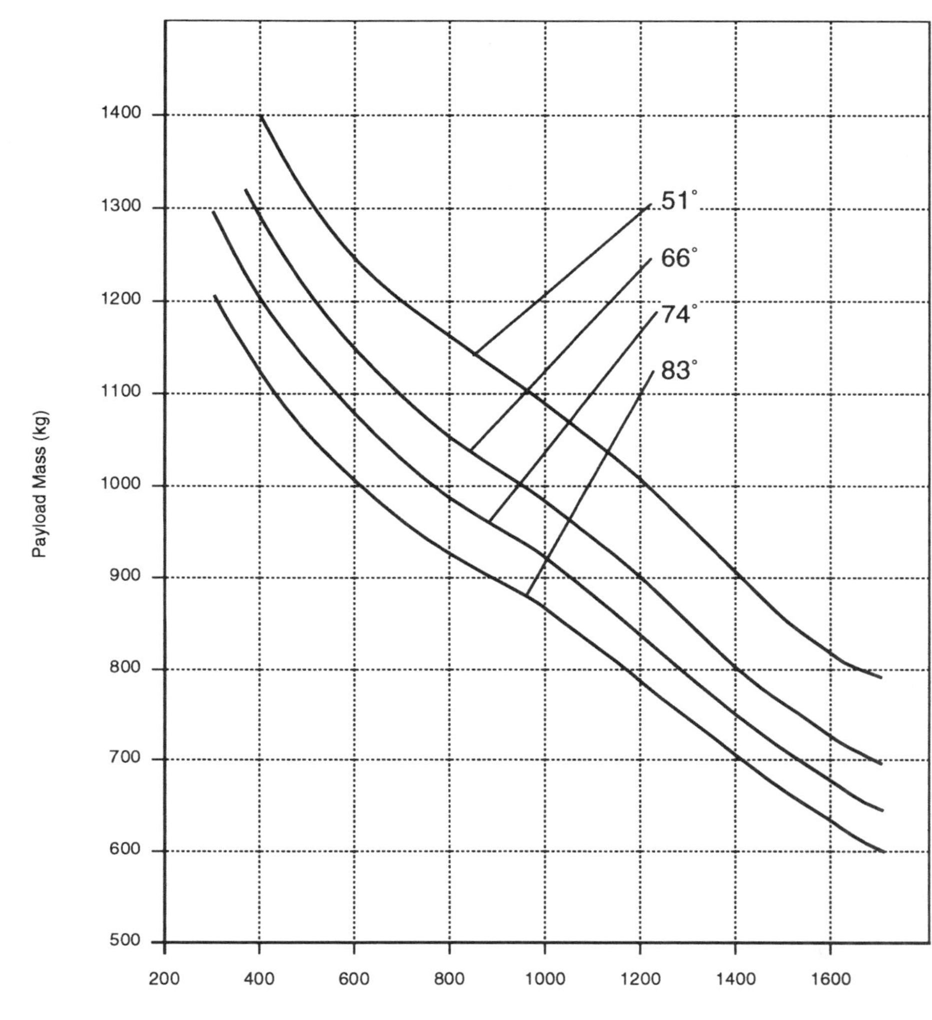

Launches into inclination of 51° take place from Kapustin Yar; other inclinations are reached from Plesetsk. No launches have taken place at Kapustin Yar since 1987; however, PO Polyot reports that two pads still exist there.

Kosmos Operations

Launch Site

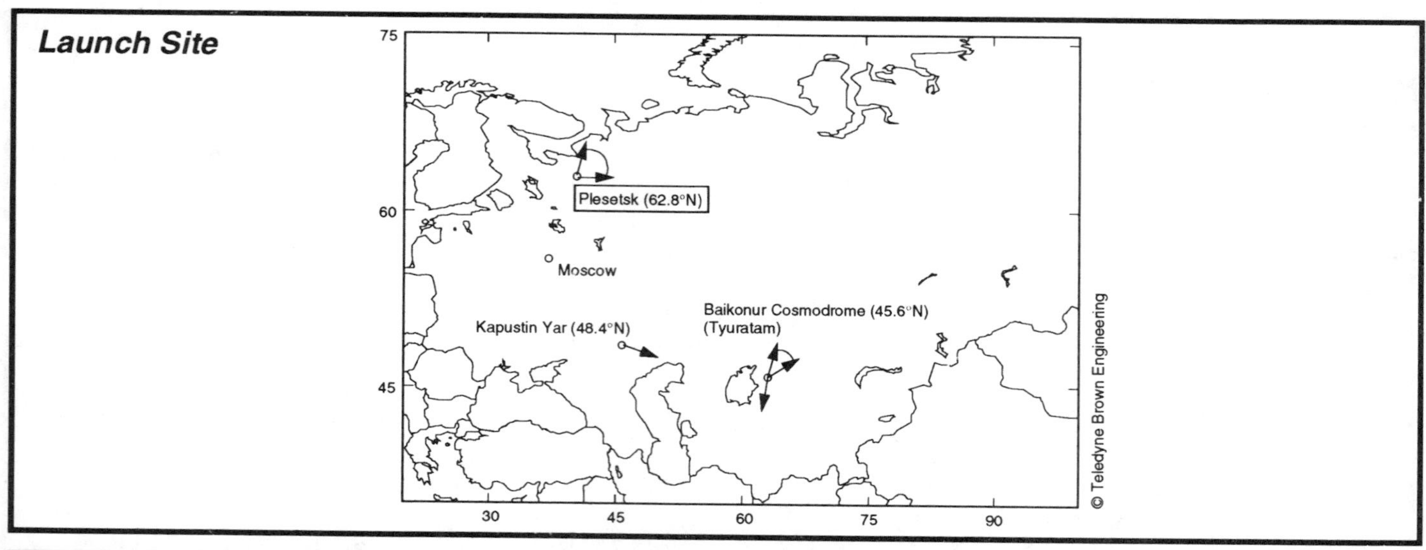

Launch Facilities

Kosmos Vehicle Tilted into the Launch Pad Service Tower

Launch Processing

Kosmos launches have been conducted from Tyuratam, Plesetsk, and Kapustin Yar. Currently, the Kosmos is launched only from Plesetsk. Historically, Plesetsk has been the world's busiest spaceport; it is also the former Soviet Union's primary military launch site, equivalent to Vandenberg AFB in the U.S. It is located 105 mi (170 km) south of Archangel in a heavily wooded area. Two pads are being maintained at Kapustin Yar; Kosmos is no longer launched from Tyuratam. The most recent Kosmos launches from Kapustin Yar and Tyuratam occurred in 1987 and 1990, respectively.

As with most CIS launch vehicles, the Kosmos can be launched under severe weather conditions, in temperatures ranging from -40° to 50° C and with surface winds of up to 20 m/sec. Vehicle preparation prior to transport to the pad involves 105 people working for 34–36 hours. Vehicle assembly, payload integration, and end-to-end systems testing are carried out inside the vehicle assembly building; temperatures inside this building are maintained between 15° C and 25° C. Both the assembly building and the pad possess 380 volt, 50 Hz power supplies. As is the norm for CIS vehicles, the Kosmos is horizontally transported to the pad and erected for launch. During transport and pad operations, the temperature inside the fairing is held to between 5° C and 35° C. On-pad processing requires 120–135 people and 8–10 hours. Vehicle fueling begins 4–5 hours prior to launch; at this point, all personnel are required to vacate the pad area. Air conditioning to the payload is stopped 1 hour before launch. The launch itself requires a crew of 20–25.

Flight Sequence

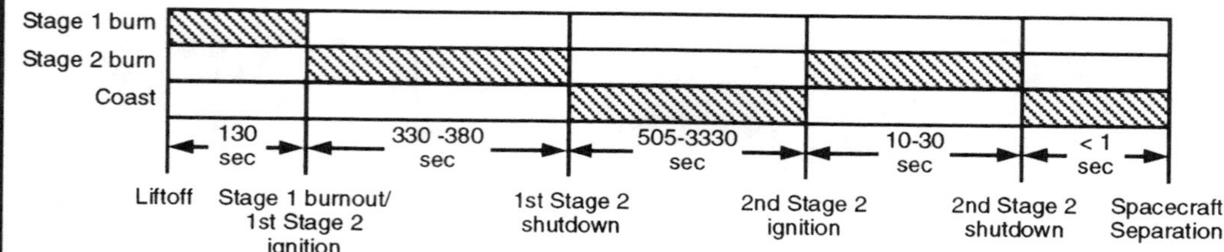

In order to reach high circular orbits, Kosmos employs two stage 2 burns. The first burn injects stage 2 and the payload into an elliptical orbit with a perigee altitude of ~108 nm (200 km) and an apogee altitude equal to the desired circular altitude. After coasting out to apogee, stage 2 burns for a second time to circularize. Exact sequence of events varies from mission to mission.

Kosmos Payload Accommodations

Payload Compartment	
Maximum Payload Diameter	86.6 in (2200 mm)
Maximum Cylinder Length	71.2 in (1809 mm)
Maximum Cone Length	114.9 in (2919 mm)
Payload Adapter	
Interface Diameter	41.7 in (1060 mm)
Payload Integration	
Nominal Mission Schedule Begins	T-12 months
Launch Window	
Latest Countdown Hold Not Requiring Recycling	T-0 min
On-Pad Storage Capability	?? hrs for a fueled vehicle
Latest Access to Payload	T-5 hrs through access doors
Environment	
Maximum Load Factors	+10 g axial, ±1.0 g lateral (flight) ±1.2 g axial, 1±0.3 g lateral (during ground transport)
Minimum Lateral / Longitudinal Payload Frequency	10 Hz / 10 Hz
Maximum Overall Acoustic Level	140 dB rms max
Maximum Flight Shock	?? g from ??-?? hz
Dynamic Pressure at Fairing Jettison	2.9 lb/ft^2 (140 N/m^2) at ~75 km altitude
Maximum Pressure Change in Fairing	?? psi/s (?? KPa/s)
Cleanliness Level in Fairing (Prior to Launch)	Class ??
Payload Delivery	
Orbit Injection Accuracy	(see table below)
Attitude Accuracy (3 sigma)	All axes: ±?? deg, ±3.5 deg/sec pitch/yaw, 5.5 deg/sec roll for 1,000 kg (2,200 lb) spacecraft
Nominal Payload Separation Rate	Velocity separation = $\sqrt{0.6 + 1{,}000/\text{mass payload}}$ (in m/sec)
Deployment Rotation Rate Available	0 rpm
Loiter Duration in Orbit	?? hrs
Maneuvers (Thermal / Collision / Telemetry)	solid rocket motor fires for collision avoidance

Orbit Injection Accuracy:

Nominal Orbit:	1,000 km circ.	1,600 km circ.	200 x 2,000 km
Perigee:	±40 km	±80 km	±45 km
Apogee:	±40 km	±80 km	±70 km
Period:	±30 sec	±70 sec	±60 sec
Eccentricity:	<0.005	<0.006	—

Kosmos Notes

Publications

Technical Publications

"Soviet Commercial Launch Vehicles", A. Dula, Space Commerce Corp., B. Gubanov, NPO Energia, Y. Smetanin, Glavkosmos, AIAA-89-2743, AIAA 25th Joint Propulsion Conference, Monterey, CA, July 10-12, 1989.

TRW Space Log, 1957-1991, 1992, 1993 TRW, El Segundo, CA.

The Soviet Year in Space 1986, 1988, 1989, Nicholas Johnson, Teledyne Brown Engineering, Colorado Springs, CO.

Soviet Space Programs: 1981-1987, United States Senate, Committee on Commerce, Science, and Transportation - Part 1 - May 1988.

Jane's Spaceflight Directory 1987, Jane's Publishing Inc., New York, NY.

Jane's Space Directory 1994-95, Jane's Information Group Inc., Alexandria, VA.

"Soviet Launch Vehicle Designations", Ralph Gibbons, Spaceflight, pg. 54-60,80, February 19, 1977.

Space Directory of Russia, February 1993 Edition, Euroconsult/Sevig Press, Paris, France.

The Russian Space Directory 1994, European Space Report, Munich, Germany.

"Euromir Research Starts; Russians Prepare GOMS", Aviation Week and Space Technology, October 10, 1994, pg. 23.

"Russia Launches Its First U.S. Satellite", Aviation Week and Space Technology, January 30, 1995, pg. 68.

"Giving Up the State Space Monopoly Leads the Field to a Commercial Orbit", The Financial News No. 36, August 18-24, 1994, Moscow (translated from Russian).

The Cosmos Rocket System - A User's Guide, Design Bureau Polyot, Omsk, Russia (translated into English by Plowshare Technology Ltd, Horsham, UK), August 1994.

Acronyms

ACS - Attitude Control System
CIS - Commonwealth of Independent States
GEO - geosynchronous orbit
GTO - geosynchronous transfer orbit
KB - Design Bureau
LEO - low Earth orbit
N2O4 - nitrogen tetroxide
NPO - Scientific Production Organization
PO - Production Organization
UDMH - unsymmetrical dimethylhydrazine
USSR - Union of Soviet Socialist Republics

Other Notes—Suborbital Capability

In addition to its more than 400 orbital launches, Kosmos has flown suborbitally more than 300 times. Its suborbital flight applications have included high speed (up to Mach 28) aerodynamics testing, reentry materials evaluation, and hypersonic gas dynamics experimentation. The most famous of these flights are the four BOR subscale spaceplane tests performed in the early 1980s to evaluate Buran shuttle technology and design elements. By using stage 2's second burn to accelerate the payload during descent into the atmosphere, a wide range of flight conditions can be achieved. Flight path angle and reentry velocity can be tailored to meet experiment requirements.

Suborbital Flight Characteristics

Payload mass: 500–1500 kg
Payload separation altitude: 105–1100 km
Payload apogee: 200–4450 km
Mission elapsed time at atmospheric reentry: 430–3350 sec
Coast from stage 2 second shutdown to atmospheric reentry: 5–2850 sec

Kosmos Upgrade Possibilities

PO Polyot is currently looking for investors to help finance a series of upgrades to the venerable Kosmos vehicle; the proposed upgraded vehicle is designated Kosmos-U. PO Polyot estimates the cost of upgrading to Kosmos-U at 12–15 million U.S. dollars. The current status of the Kosmos-U program is unknown.

The Kosmos-U upgrades can be classified in five groups: 1) replacement of out-of-date control and instrumentation systems on Kosmos; 2) reduction in the amount of residual propellants left in stage 1 at separation; 3) 3 sec increase in stage 1 and 2 specific engine impulse; 4) small increase (~0.7m) in stage 1 length; 5) upgrading of ground support and launch facilities.

The projected benefits associated with these improvements include: a 330–400 lb (150–180 kg) increase in payload capability; stage 2 out-of-plane maneuvering capability, which would permit access to a greater range of inclinations; less severe environmental effects due to reduced stage 1 residuals and smaller impact area; improved orbital injection accuracy; improved stage 2 collision avoidance capability following spacecraft separation, possibly including stage 2 deorbit.

Proton

Worldwide Marketing Representative:
Lockheed-Khrunichev-Energia
International (LKE)
San Jose, CA 95110
Phone: (408) 436-9771
Fax: (408) 436-7968

Photo Courtesy of Space Commerce Corp.

The Proton launch vehicle is used to launch communication satellites, space station module, and interplanetary missions.

Proton History

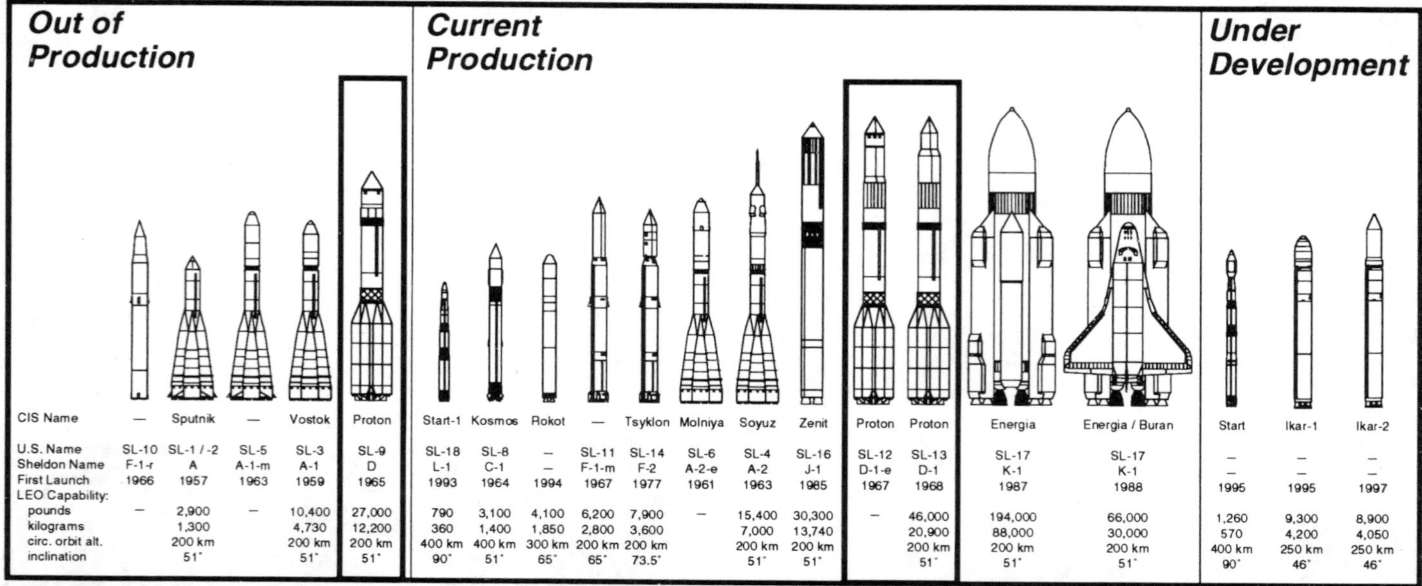

Out of Production

CIS Name	—	Sputnik	—	Vostok	Proton
U.S. Name	SL-10	SL-1 / -2	SL-5	SL-3	SL-9
Sheldon Name	F-1-r	A	A-1-m	A-1	D
First Launch	1966	1957	1963	1959	1965
LEO Capability:					
pounds	—	2,900	—	10,400	27,000
kilograms		1,300		4,730	12,200
circ. orbit alt.		200 km		200 km	200 km
inclination		51°		51°	51°

Current Production

CIS Name	Start-1	Kosmos	Rokot	—	Tsyklon	Molniya	Soyuz	Zenit	Proton	Proton	Energia	Energia / Buran
U.S. Name	SL-18	SL-8	–	SL-11	SL-14	SL-6	SL-4	SL-16	SL-12	SL-13	SL-17	SL-17
Sheldon Name	L-1	C-1	–	F-1-m	F-2	A-2-e	A-2	J-1	D-1-e	D-1	K-1	K-1
First Launch	1993	1964	1994	1967	1977	1961	1963	1985	1967	1968	1987	1988
LEO Capability:												
pounds	790	3,100	4,100	6,200	7,900	–	15,400	30,300	–	46,000	194,000	66,000
kilograms	360	1,400	1,850	2,800	3,600		7,000	13,740		20,900	88,000	30,000
circ. orbit alt.	400 km	400 km	300 km	200 km	200 km		200 km	200 km		200 km	200 km	200 km
inclination	90°	51°	65°	65°	73.5°		51°	51°		51°	51°	51°

Under Development

CIS Name	Start	Ikar-1	Ikar-2
U.S. Name	–	–	–
Sheldon Name	–	–	–
First Launch	1995	1995	1997
LEO Capability:			
pounds	1,260	9,300	8,900
kilograms	570	4,200	4,050
circ. orbit alt.	400 km	250 km	250 km
inclination	90°	46°	46°

Vehicle Description

CIS vehicles are identified either by their CIS, U.S., or Sheldon name. In the Soviet Union, it was standard practice to name a launch vehicle after its original payload (e.g., Kosmos, Proton). The U.S. names (developed by the U.S. Department of Defense) are alphanumeric designators based roughly on chronological appearance. The Sheldon names, a most commonly used system that was published by Dr. Charles Sheldon of the U.S. Library of Congress in 1968, emphasize the basic families of launch vehicles with special indicators for variants within a family.

Example: D-1-e

- e - earth escape or fourth stage
- m - maneuverable stage
- r - reentry stage
- Upper Stage (1, 2) - Note: "1" or "2" stage may vary among different families, e.g., A-1 upper stage different than B-1.
- Family (A, B, C, D, F, G, J, K)

D (SL-9) — Two stage vehicle using N2O4 and UDMH liquid propellant. The first stage has six liquid strap-ons that provide all the thrust. Second stage has four liquid engines.

D-1 (SL-13) — Same as D except the addition of a N2O4 and UDMH third stage for increase performance.

D-1-e (SL-12) — Same as the D-1 except the addition of a LOX and kerosene fourth stage for GEO and interplanetary missions.

Historical Summary

Details of the Soviet/CIS space program until recently have not been publicly available. Not until the launch of the two Vega probes to Halley's comet in December 1984 did the world receive its first complete view of the Proton launch vehicle. Additional data has since become available as a result of glasnost and the Russian effort to compete on the world commercial market. With the Energia launch vehicle program currently suspended, the Proton is the CIS's largest fully operational booster.

There are three variants to the Proton—a two, three, and four stage launch vehicle. The first three stages of the Proton were designed in the early 1960s by the design bureau headed by V. N. Chelomei. This organization, which eventually became KB Salyut, was combined with the Khrunichev factory (which actually manufactures the Proton hardware), at the behest of Russian president Boris Yeltsin, to form Khrunichev State Research and Production Space Center in 1993. The fourth stage, named Block-DM, was designed and is built by the Korolev bureau (now Rocket Space Corporation Energia). All stages of the Proton are built at various facilities in the Moscow area and transported by rail to the Baikonur Cosmodrome in Kazakhstan.

The original version of the Proton is known in the West as the D or SL-9 vehicle, and consisted of two stages: the first with a cluster of six engines and the second with a cluster of four. The first stage had a separate fuel tank for each engine, with a larger diameter oxidizer tank in the core crossfed to each engine, a design similar to that of the U.S. Saturn 1B. The SL-9 vehicle, which is no longer in use, was used on only four missions to launch satellites in the Proton series, thus giving the booster its name. It first flew in 1965 and was retired the next year. The low Earth orbit payload capability was 27,000 lb (12,200 kg).

The current three- and four-stage versions of the Proton (designated SL-13/D-1 and SL-12/D-1-e, respectively) possess virtually identical stages 1–3. Stage 1 is the same as that of the SL-9; the current stage 2 is essentially a stretched version of the SL-9's stage 2. The single engine stage 3 uses the same UDMH/N2O4 propellants as the first two stages.

The three-stage vehicle is used for launching ~20 metric ton payloads to low Earth orbit, and has launched all of the Soviet/CIS Salyut space stations, as well as all elements of the current Mir space station. It is also used to launch the large Almaz series of satellites. The four-stage adds the LOX/kerosene Block DM for high energy orbits such as geosynchronous, planetary, and high circular orbits. It has launched payloads such as Ekran, Raduga, Gorizont, the Zond and Luna lunar probes, Venera to Venus, Vega to Halley's comet and Mars, and Phobos to Mars. It is the only CIS launch vehicle currently

Proton History

(continued)

Historical Summary

(continued)

capable of placing satellites into geosynchronous orbit. Nearly 90% of Proton launches have used the D-1-e version. All Proton launches are conducted at the Baikonur Cosmodrome near Tyuratam, located just east of the Aral Sea in Kazakhstan (see *Operations* section).

It is interesting to note that the Soviets scheduled a Zond mission atop a Proton to carry Aleksei Leonov and Oleg Makarov around the Moon and back to Earth in December 1968 before the U.S. Apollo 8 mission. But that mission was scrubbed when an unmanned Zond 6 suffered a cabin depressurization the month before. Currently, there are no known plans to man-rate the Proton for cosmonauts.

Flight rate of the Proton has steadily grown from an initial 6 launches in 1970 to a peak of 13 in 1985. The Proton booster was considered operational in 1970 despite failing on seven of nine missions in 1969. From 1983 to 1986, the Proton had its longest string of 43 consecutive successes. In recent years, about 10% of all Soviet launches are Proton launches.

In 1983, the Soviet Union unsuccessfully offered Proton commercial launch services to the international Inmarsat organization. One result of this action was that the U.S. restricted American built payloads from being flown from the Soviet Union for reasons of technology transfer, despite assurances from the Soviets that they would not access the payload. More recently, a set of quotas have been negotiated between the U.S. and Russian governments for the launch on Proton of geosynchronous spacecraft containing U.S. components. The launch of nongeosynchronous payloads will be reviewed by the U.S. governement on a case-by-case basis.

In December of 1992 a joint venture company was formed between Lockheed Commercial Space Company, a subsidiary of Lockheed Missiles and Space Company of the U.S., Khrunichev State Research and Production Space Center of Russia, and Rocket Space Corporation Energia of Russia. This joint venture was approved by both the U.S. and Russian governments, and its incorporation completed in early 1993. Lockheed-Khrunichev-Energia International (LKE) currently serves as the focal point for the worldwide sale of the Proton launch vehicle and for customer interface and management of contracted launch services and integration programs.

The Proton launch vehicle permits delivery of 46,000 lb (20,900 kg) to LEO, 12,500 lb (5,700 kg) to lunar transfer trajectories, 11,700 lb (5,300 kg) to Venus transfer trajectories, 10,100 lb (4,600 kg) to Mars transfer trajectories, 12,100 lb (5,500 kg) geostationary transfer orbit (GTO), 4,850 lb (2,200 kg) to geostationary orbit (GEO), and 6,200 lb (2,800 kg) to sun synchronous polar orbit.

Proton Launch Vehicle Family

Proton D

Proton D-1

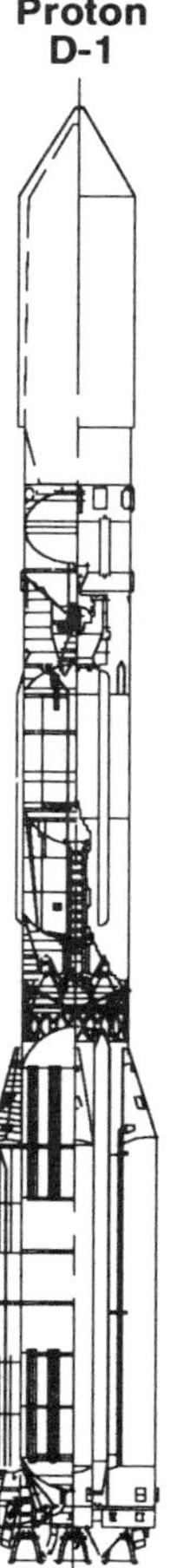

Proton D-1-e

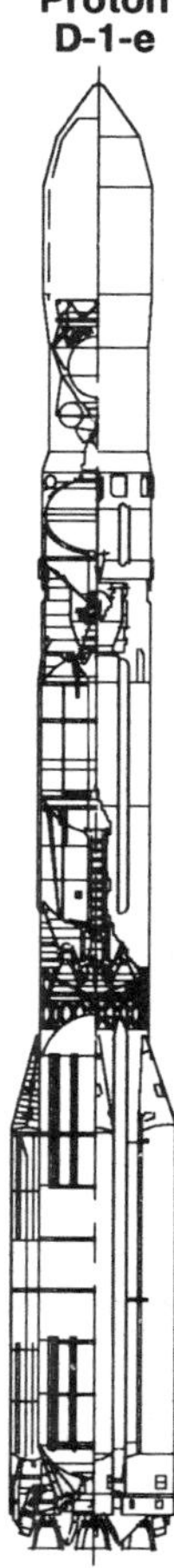

Proton History

(continued)

Launch Record

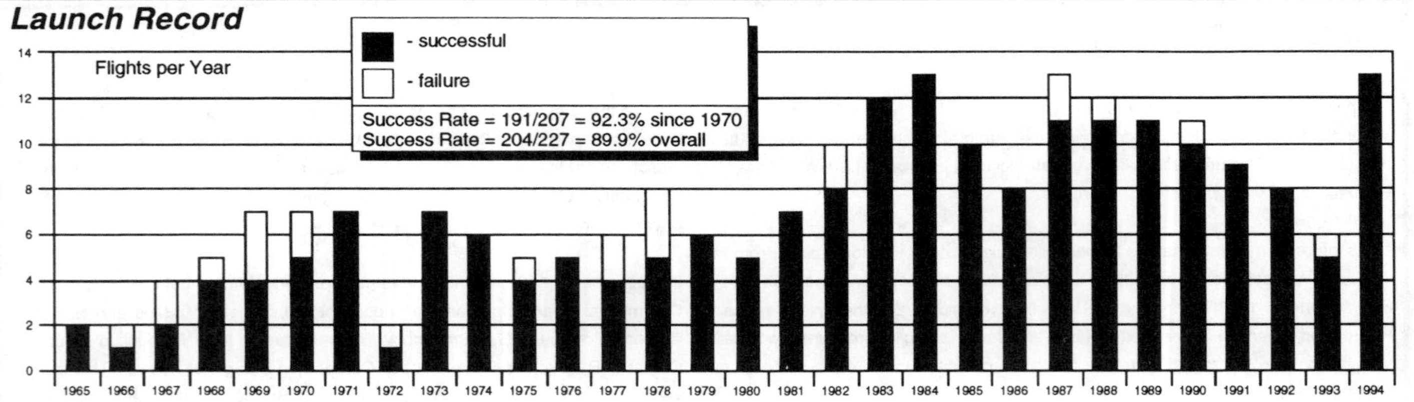

#	YEAR	DATE	VEHICLE	SITE	PAYLOAD
		ALL FLIGHTS PRIOR TO 1970 WERE CONSIDERED TEST FLIGHTS 1965-69: 13 SUCCESSFUL FLIGHTS, 7 FAILURES			
1	1970	Feb-6	D-1-e	TT	ASR Luna-16C;"Kosmos" SLM
		Failure - Stage 1 command destruct			
2		Aug-8	D-1	TT	Kosmos
		Failure			
3		Sep-12	D-1-e	TT	Luna-16-D
4		Oct-20	D-1-e	TT	Zond-8
5		Nov-10	D-1-e	TT	Luna-17, Lunakhod-1
6		Nov-24	D-1	TT	Kosmos-379
7		Dec-2	D-1	TT	Kosmos-382, Soviet Lunar Module, SLM flight test
8	1971	Feb-26	D-1	TT	Kosmos-398
9		Apr-19	D-1	TT	Salyut-1
10		May-10	D-1-e	TT	Kosmos-419 Mars 2A
11		May-19	D-1-e	TT	Mars-2
12		May-28	D-1-e	TT	Mars-3
13		Sep-2	D-1-e	TT	Luna-18
14		Sep-28	D-1-e	TT	Luna-19
15	1972	Feb-14	D-1-e	TT	Luna-20
16		Jul-29	D-1	TT	Salyut-2A
		Failure - Stage 2			
17	1973	Jan-8	D-1-e	TT	Luna 21, Lunakhod-2
18		Apr-3	D-1	TT	Salyut-2C
19		May-11	D-1	TT	Kosmos-557, Salyut-3A
20		Jul-21	D-1-e	TT	Mars-4
21		Jul-25	D-1-e	TT	Mars-5
22		Aug-5	D-1-e	TT	Mars-6
23		Aug-9	D-1-e	TT	Mars-7
24	1974	Mar-26	D-1-e	TT	Kosmos-637
25		May-29	D-1-e	TT	Luna-22
26		Jun-24	D-1	TT	Salyut-3B
27		Jul-29	D-1-e	TT	Molniya 1-S
28		Oct-28	D-1-e	TT	Luna-23
29		Dec-26	D-1	TT	Salyut-4
30	1975	Jun-6	D-1-e	TT	Venera-9
31		Jun-14	D-1-e	TT	Venera-10
32		Oct-8	D-1-e	TT	Kosmos-775
33		Oct-16	D-1-e	TT	ASR Luna 24A
		Failure - Stage 4			
34		Dec-22	D-1-e	TT	Raduga-1
35	1976	Jun-22	D-1	TT	Salyut-5
36		Aug-9	D-1-e	TT	Luna-24-B
37		Sep-11	D-1-e	TT	Raduga-2
38		Oct-26	D-1-e	TT	Ekran-1
39		Dec-15	D-1	TT	Double Kosmos-881-2 Space Planes
40	1977	Jul-17	D-1	TT	Kosmos-929 Star-Mod.
41		Jul-23	D-1-e	TT	Raduga-3
42		Aug-4	D-1	TT	Double Kosmos-X-X Space Planes
		Failure - Stage 2			
43		Sep-20	D-1-e	TT	Ekran-2
44		Sep-29	D-1	TT	Salyut-6
45		Oct-14	D-1-e	TT	unknown
		Failure			
46	1978	Mar-30	D-1	TT	Double Kosmos-997-8 Space Planes
47		May-27	D-1-e	TT	Ekran-3A
		Failure - Stage 1			
48		Jul-18	D-1-e	TT	Raduga-4
49		Aug-17	D-1-e	TT	Ekran-3B
		Failure - Stage 2			
50		Sep-9	D-1-e	TT	Venera-11
51		Sep-14	D-1-e	TT	Venera-12
52		Oct-17	D-1-e	TT	Ekran-3C
		Failure - Stage 2			
53		Dec-19	D-1-e	TT	Gorizont-1
54	1979	Feb-21	D-1-e	TT	Ekran-3-D
55		Apr-25	D-1-e	TT	Raduga-5
56		May-22	D-1	TT	Double Kosmos-1100-01 Space Planes
57		Jul-5	D-1-e	TT	Gorizont-2
58		Oct-3	D-1-e	TT	Ekran-4
59		Dec-28	D-1-e	TT	Gorizont-3
60	1980	Feb-2	D-1-e	TT	Raduga-6
61		Jun-14	D-1-e	TT	Gorizont-4
62		Jul-15	D-1-e	TT	Ekran-5
63		Oct-5	D-1-e	TT	Raduga-7
64		Dec-26	D-1-e	TT	Ekran-6
65	1981	Mar-18	D-1-e	TT	Raduga-8
66		Apr-25	D-1	TT	Kosmos-1267 Star-M
67		Jun-26	D-1-e	TT	Ekran-7
68		Jul-30	D-1-e	TT	Raduga-9
69		Oct-9	D-1-e	TT	Raduga-10
70		Oct-30	D-1-e	TT	Venera-13
71		Nov-4	D-1-e	TT	Venera-14
72	1982	Feb-5	D-1-e	TT	Ekran-8
73		Mar-15	D-1-e	TT	Gorizont-5
74		Apr-19	D-1	TT	Salyut-7
75		May-17	D-1-e	TT	Kosmos-1366
76		Jul-23	D-1-e	TT	Ekran-9A
		Failure - Stage 1			
77		Sep-16	D-1-e	TT	Ekran-9-B
78		Oct-12	D-1-e	TT	Kosmos-1413-15 (Glonass)
79		Oct-20	D-1-e	TT	Gorizont-6
80		Nov-26	D-1-e	TT	Raduga-11
81		Dec-24	D-1-e	TT	Raduga-12A
		Failure - Stage 2			
82	1983	Mar-2	D-1	TT	Kosmos-1443 Star-M
83		Mar-12	D-1-e	TT	Ekran-10
84		Mar-23	D-1-e	TT	Astron-1
85		Apr-8	D-1-e	TT	Raduga-12-B
86		Jun-2	D-1-e	TT	Venera-15
87		Jun-6	D-1-e	TT	Venera-16
88		Jul-1	D-1-e	TT	Gorizont-7
89		Aug-10	D-1-e	TT	Kosmos-1490-92 (Glonass)
90		Aug-25	D-1-e	TT	Raduga-13
91		Sep-29	D-1-e	TT	Ekran-11
92		Nov-30	D-1-e	TT	Goriziont-8
93		Dec-29	D-1-e	TT	Kosmos-1519-21 (Glonass)
94	1984	Feb-15	D-1-e	TT	Raduga-14
95		Mar-2	D-1-e	TT	Kosmos-1540 GEO
96		Mar-16	D-1-e	TT	Ekran-12
97		Mar-29	D-1-e	TT	Kosmos-1546 GEO
98		Apr-22	D-1-e	TT	Gorizont-9
99		May-19	D-1-e	TT	Kosmos-1554-56 (Glonass)
100		Jun-22	D-1-e	TT	Raduga-15
101		Aug-1	D-1-e	TT	Gorizont-10
102		Aug-24	D-1-e	TT	Ekran-13
103		Sep-4	D-1-e	TT	Kosmos-1593-95 (Glonass)
104		Sep-28	D-1-e	TT	Kosmos-1601 Elint
105		Dec-15	D-1-e	TT	Vega-1
106		Dec-21	D-1-e	TT	Vega-2
107	1985	Jan-18	D-1-e	TT	Gorizont-11
108		Feb-21	D-1-e	TT	Kosmos-1629 GEO
109		Mar-22	D-1-e	TT	Ekran-14
110		May-18	D-1-e	TT	Kosmos-1650-52 (Glonass)
111		May-30	D-1-e	TT	Kosmos-1656 Elint
112		Aug-9	D-1-e	TT	Raduga-16
113		Sep-27	D-1	TT	Kosmos-1686 Star-M
114		Oct-25	D-1-e	TT	Kosmos-1700 GEO
115		Nov-15	D-1-e	TT	Raduga-17
116		Dec-25	D-1-e	TT	Kosmos-1710-12 (Glonass)
117	1986	Jan-17	D-1-e	TT	Raduga-18
118		Feb-19	D-1	TT	Mir-1/Salyut-8
119		Apr-4	D-1-e	TT	Kosmos-1738 GEO
120		May-24	D-1-e	TT	Ekran-15
121		Jun-10	D-1-e	TT	Gorizont-12
122		Sep-16	D-1-e	TT	Kosmos-1778-80 (Glonass)
123		Oct-25	D-1-e	TT	Raduga-19
124		Nov-18	D-1-e	TT	Gorizont-13
125	1987	Jan-30	D-1-e	TT	Kosmos-1817, Ekran-16A
		Failure - Stage 4 failed to start			
126		Mar-19	D-1-e	TT	Raduga-20
127		Mar-31	D-1	TT	Kvant-1 (Quantum-1) for MIR
128		Apr-24	D-1-e	TT	Kosmos-1838-40 (Glonass)
		Failure - Premature stage 4 shutdown			
129		May-11	D-1-e	TT	Gorizont-14
130		Jul-25	D-1	TT	Kosmos-1870
131		Sep-3	D-1-e	TT	Ekran-16B
132		Sep-16	D-1-e	TT	Kosmos-1883-85 (Glonass)
133		Oct-1	D-1-e	TT	Kosmos-1888
134		Oct-28	D-1-e	TT	Kosmos-1894
135		Nov-26	D-1-e	TT	Kosmos-1897
136		Dec-10	D-1-e	TT	Raduga-21
137		Dec-27	D-1-e	TT	Ekran-17
138	1988	Feb-17	D-1-e	TT	Kosmos-1917-19 (Glonass)
		Failure - Stage 4 failed to separate			
139		Mar-31	D-1-e	TT	Gorizont-15
140		Apr-26	D-1-e	TT	Kosmos-1940
141		May-6	D-1-e	TT	Ekran-18
142		May-21	D-1-e	TT	Kosmos-1946-48 (Glonass)
143		Jul-7	D-1-e	TT	Phobos-1
144		Jul-12	D-1-e	TT	Phobos-2
145		Aug-1	D-1-e	TT	Kosmos-1961
146		Aug-18	D-1-e	TT	Gorizont-16
147		Sep-16	D-1-e	TT	Kosmos-1970-72 (Glonass)
148		Oct-20	D-1-e	TT	Raduga-22
149		Dec-10	D-1-e	TT	Ekran-19
150	1989	Jan-10	D-1-e	TT	Kosmos-1987-89 (2 Glonass, 1 geodetic)
151		Jan-26	D-1-e	TT	Gorizont-17
152		Apr-14	D-1-e	TT	Raduga-23
153		May-31	D-1-e	TT	Kosmos-2022-24 (2 Glonass, 1 geodetic)
154		Jun-21	D-1-e	TT	Raduga-24
155		Jul-5	D-1-e	TT	Gorizont-18
156		Sep-28	D-1-e	TT	Gorizont-19
157		Nov-26	D-1	TT	Kvant-2 (Quantum-2) for MIR
158		Dec-1	D-1-e	TT	Granat
159		Dec-15	D-1-e	TT	Raduga-25
160		Dec-27	D-1-e	TT	Kosmos-2054
161	1990	Feb-15	D-1-e	TT	Raduga-26
162		May-19	D-1-e	TT	Kosmos-2079-81 (Glonass)
163		May-31	D-1	TT	Kristall
164		Jun-20	D-1-e	TT	Gorizont-20
165		Jul-18	D-1-e	TT	Kosmos-2085
166		Aug-9	D-1-e	TT	unknown
		Failure - Stage 3			
167		Nov-3	D-1-e	TT	Gorizont-21
168		Nov-23	D-1-e	TT	Gorizont-22
169		Dec-8	D-1-e	TT	Kosmos-2109-11 (Glonass)
170		Dec-20	D-1-e	TT	Raduga-26
171		Dec-27	D-1-e	TT	Raduga 1-2
172	1991	Feb-14	D-1-e	TT	Kosmos-2133
173		Feb-28	D-1-e	TT	Raduga-27
174		Mar-31	D-1	TT	Almaz-1
175		Apr-4	D-1-e	TT	Kosmos-2139-41 (Glonass)
176		Jul-1	D-1-e	TT	Gorizont-23
177		Sep-13	D-1-e	TT	Kosmos-2155
178		Oct-23	D-1-e	TT	Gorizont-24
179		Nov-22	D-1-e	TT	Kosmos-2172
180		Dec-19	D-1-e	TT	Raduga-28
181	1992	Jan-29	D-1-e	TT	Kosmos-2177-79 (Glonass)
182		Apr-2	D-1-e	TT	Gorizont-25
183		Jul-14	D-1-e	TT	Gorizont-26
184		Jul-30	D-1-e	TT	Kosmos-2204-06 (Glonass)
185		Sep-10	D-1-e	TT	Kosmos-2209
186		Oct-30	D-1-e	TT	Ekran-20
187		Nov-27	D-1-e	TT	Gorizont-27
188		Dec-17	D-1-e	TT	Kosmos-2224
189	1993	Feb-17	D-1-e	TT	Kosmos-2234-36 (Glonass)
190		Mar-17	D-1-e	TT	Raduga-29
191		May-27	D-1-e	TT	Gorizont
		Failure - Stage 2 propulsion system			
192		Sep-30	D-1-e	TT	Raduga-30
193		Oct-28	D-1-e	TT	Gorizont-28
194		Nov-18	D-1-e	TT	Gorizont-29
195	1994	Jan-20	D-1-e	TT	Gals-1
196		Feb-5	D-1-e	TT	Raduga-1 (3)
197		Feb-18	D-1-e	TT	Raduga-31
198		Apr-11	D-1-e	TT	Kosmos 2275-77 (Glonass)
199		May-20	D-1-e	TT	Gorizont
200		Jul-7	D-1-e	TT	Kosmos 2282
201		Aug-11	D-1-e	TT	Kosmos 2287-89 (Glonass)
202		Sep-21	D-1-e	TT	Kosmos 2291
203		Oct-13	D-1-e	TT	Express
204		Oct-31	D-1-e	TT	Electro
205		Nov-20	D-1-e	TT	Kosmos 2294-96 (Gloanass)
206		Dec-16	D-1-e	TT	Luch
207		Dec-28	D-1-e	TT	Raduga-32

Data Sources: TRW Space Log, LKE International data

TT is the Baikonur Cosmodrome near Tyuratam

Proton General Description

D-1	D-1-e

Summary

The Proton was developed between 1961 and 1965. There are two basic versions of the launch vehicle; the three stage (D-1, SL-13) and four stage (D-1-e, SL-12). The first three stages carry payloads to low Earth orbit; the optional fourth stage is used to insert the payload into geosynchronous orbit as well as interplanetary trajectories. The first three stages use UDMH / N2O4 and the fourth stage LOX / kerosene. The Proton was offered commercially for the first time to Inmarsat in 1983. Since late 1992 Proton launch services have been offered commercially through the U.S. / Russian joint venture LKE International.

Status

D-1 & D-1-e - Operational since 1970

Key Organizations

Users - Russian Ministry of Defense, Russian Academy of Sciences, commercial

Launch Service Agency - For commercial users, LKE International

- Khrunichev State Research and Production Space Center (manufacturer of stages 1–3)
- RSC Energia (manufacturer of stage 4)
- NPO Energomash (stage 1 engines)
- KB Khimavtomatiki (stage 2 and 3 engines)
- NIIAP (guidance and control systems)

Vehicle

System Height	Up to 200 ft (61 m)
Payload Fairing Size	D-1: Two payload fairings (Models 1 and 2) See *Proton Vehicle* section for dimensions D-1-e: Three existing fairings (Models A, B, and C); standard commercial fairing under development See *Proton Vehicle* section for dimensions
Gross Mass	1,519,000 lb (689,000 kg)
Planned Enhancements	Proton-KM (see *Other Notes* section)

Operations

Primary Missions	LEO, GTO, GEO, interplanetary
Compatible Upper Stages	D-1-e fourth stage (Block DM)
First Launch	1967 for D-1 and D-1-e
Success / Flight Total	24 / 27 for D-1, 154 / 167 for D-1-e since 1970
Launch Site	Baikonur Cosmodrome (45.6°N, 63.4°E) near Tyuratam. Three of four pads are operational.
Launch Azimuth	62.6° for 51.6° inclination 37.5° for 64.8° inclination 25.2° for 72.7° inclination
Nominal Flight Rate	8–18 / yr
Planned Enhancements	—

Performance

108 nm (200 km) circ, 51.6°	46,000 lb (20,900 kg) for D-1
100 nm (185 km) circ, 90°	—
Geotransfer Orbit, 28.5°	12,100 lb (5,500 kg) for D-1-e (Note: 28.5° incl)
Geosynchronous Orbit	4,850 lb (2,200 kg) for D-1-e

Financial Status

Estimated Launch Price: $50–70M (U.S. DoT estimate)

Manifest: Not available

Remarks: —

Proton Vehicle

Overall

Length	200 ft (61 m)
Gross Mass	1,519,000 lb (689,000 kg)
Thrust at Liftoff	2,000,000 lb (9,000,000 N)

3-Stage Proton (D-1, SL-13) Exploded View

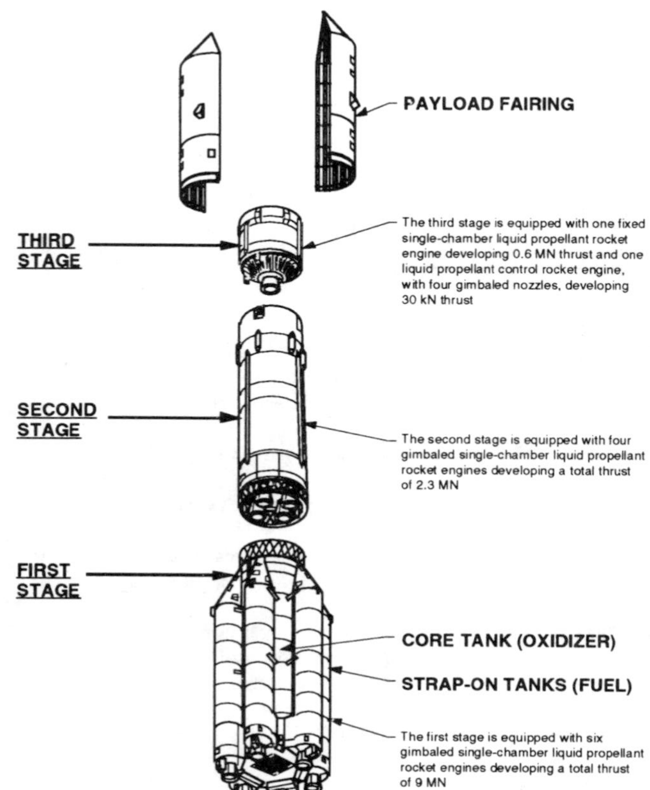

4-Stage Proton (D-1-e, SL-12) Exploded View

PAYLOAD

PAYLOAD FAIRING

PAYLOAD ADAPTER

FOURTH STAGE

The fourth stage is equipped with one liquid propellant rocket engine developing 86 kN thrust, and two "micro" engine clusters for attitude control and ullage maneuvers

FORWARD FOURTH STAGE SHROUD

AFT FOURTH STAGE SHROUD

THIRD STAGE

The third stage is equipped with one fixed single-chamber liquid propellant rocket engine developing 0.6 MN thrust and one liquid propellant control rocket engine, with four gimbaled nozzles,developing 30 kN thrust

SECOND STAGE

The second stage is equipped with four gimbaled single-chamber liquid propellant rocket engines developing a total thrust of 2.3 MN

FIRST STAGE

CORE TANK (OXIDIZER)

STRAP-ON TANKS (FUEL)

The first stage is equipped with six gimbaled single-chamber liquid propellant rocket engines developing a total thrust of 9 MN

Stage 4 (Block DM) Detailed View

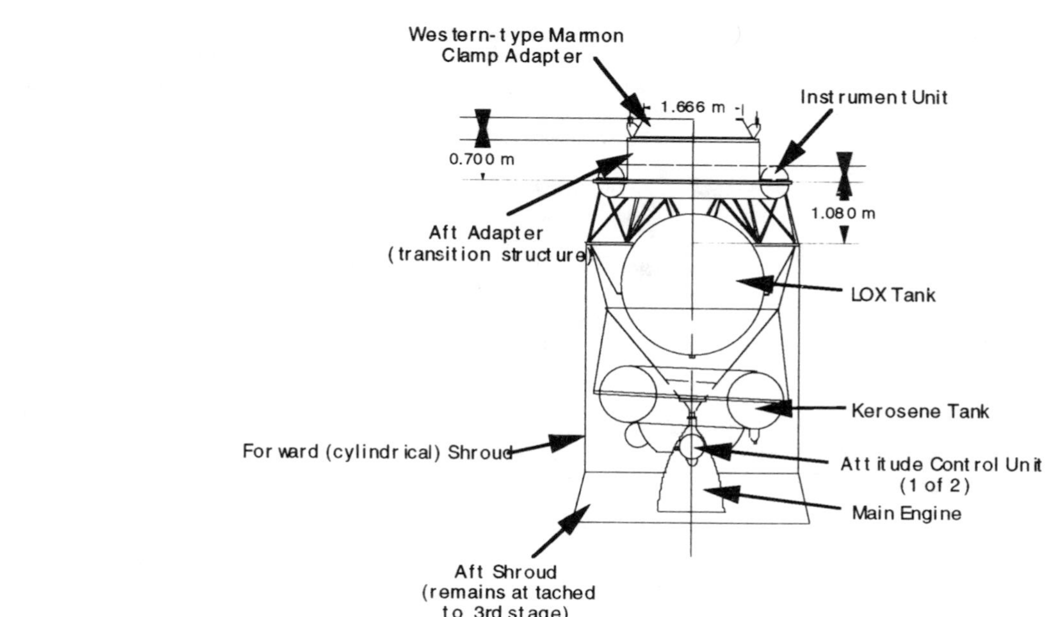

Proton Vehicle
(continued)

Proton Stages

	Stage 1	Stage 2	Stage 3	Stage 4 (D-1-e only) (Block D)
Dimension:				
Length	69.5 ft (21.2 m)	55.9 ft (17.1 m)	13.5 ft (4.1 m)	23.2 ft (7.1 m)
Diameter	24 ft (7.4 m)	13.4 ft (4.1 m)	13.4 ft (4.1 m)	12.1 ft (3.7 m)
Mass: (each)				
Propellant Mass	924.4K lb (419.4K kg)	344.1K lb (156.1K kg)	102.6K lb (46.6K kg)	33.2K lb (15.1K kg)
Gross Mass	992.7K lb (450.4K kg)	369.9K lb (167.8K kg)	116.0K lb (52.6K kg)	40.6K lb (18.4K kg)
Structure:				
Type	Skin-Stringer	Skin-Stringer	Skin-Stringer	Skin-Stringer
Material	Aluminum	Aluminum	Aluminum	Aluminum
Propulsion:				
Propellant	N2O4 / UDMH	N2O4 / UDMH	N2O4 / UDMH	LOX / RP-1
Average Thrust (each)	331K lb (1474 kN) SL 392K lb (1745 kN) vac	131K lb (582 kN) vac	131K lb (582 kN) vac 1.7K lb (7.7 kN) vac each vernier	18.8K lb (83.5 kN) vac
Engine Designation	RD-253	RD-0210	RD-0210	11D58M
Number of Engines	6	4	1 core engine, 4 verniers	1
Isp	267 sec SL 316 sec vac	326.5 sec vac	326.5 sec vac	361 sec vac
Feed System	Staged combustion	Staged combustion	Staged combustion	Staged combustion
Chamber Pressure	2,130 psia (147 bar)	2,175 psia (148 bar)	2,175 psia (148 bar)	1,123 psia (76.4 bar)
Mixture Ratio (O/F)	2.69	??	??	2.6
Throttling Capability	100% only	100% only	100% only	100% only
Expansion Ratio	26:1	??	??	189:1
Restart Capability	No	No	No	1–7 restarts
Tank Pressurization	Gas Generator	Gas Generator	Gas Generator	Stored pressurized gas
Control-Pitch, Yaw, Roll	Gimbal 6 nozzles	Gimbal 4 engine nozzles	4 verniers	Gimballing & ACS
Events:				
Nominal Burn Time	150 sec	200 sec	250 sec	680 sec
Stage Shutdown	Burn to depletion	Burn to depletion	Predetermined velocity	Predetermined velocity
Stage Separation	Stage 2 ignition	6 solid retros	Stage 3 retro	ACS/Spring ejection

Remarks:

The Proton launch vehicle is made up of three or four tandem stages. Proton is constructed of an all-aluminum alloy and is covered by an external thermal finishing paint. Proton's propellant tanks are internally coated with an anticorrosive coating. The oxidizer also has a passive abator mixed into it to reduce its corrosiveness. The first stage has a separate fuel tank for each engine and a larger diameter oxidizer tank at the core for attachment. Both the second and third stages have a common bulkhead dome tank. The fourth stage uses a spherical LOX tank and toroidal kerosene tank connected by trusses.

Six RD-253 engines, developed between 1961 and 1965, are installed in the Proton's first stage. After the fuel goes through the pumps, the major portion of the fuel, approximately 75%, enters the gas generator welded to the turbine housing. The remaining fuel is delivered to the regenerative cooling channel of the chamber. The gas after the turbine is delivered through a gas line to the combustion chamber, where it is burned with the liquid component that has passed through the chamber cooling channels. For reliable cooling of the chamber, its fire wall is protected by a refractory ceramic coating and gas-liquid film formed by delivery of the fuel component from the cooling channel to the wall through openings. The total power of the turbopump assemblies of all the RD-253 engines contained in the propulsion plant exceeds 150,000 hp. The maximum fuel pressure in the main lines of the engine reaches 400 atmospheres. The engine is fired by gravity flow of the fuel without the use of special starting devices, since the use of self-igniting hypergolic propellant eliminates the need for an ignition system. The engine is fired and cut off by nine pyrotechnic valves. A regulator and choke operating from electric drives are installed in the main lines to control the thrust and ratio of fuel component flow rates during flight of the engine. Supercharging assemblies produce gases for pressurizing the rocket fuel tanks. The extensive application of welding ensures the structural integrity of the RD-253 rocket engine. Heat shielding around the engines protect the engine assemblies against the reaction gas jet influence. The gimbal assemblies, which attach the engine to the rocket, provide for the rotation of the main thrust chambers in a plane parallel to the rocket's longitudinal axis so that flight direction and attitude of the booster can be controlled. Four single high pressure combustion chamber RD-0210 engines are mounted on the second stage. A single RD-0210 engine and four conventional vernier liquid engines are installed on the third stage. The third stage of Proton cannot be restarted or throttled, but like all stages on Proton, the burn time can be varied according to the mission.

The fourth stage, or Block DM, is used to transfer payloads from low Earth orbit into geostationary orbit or interplanetary trajectories. It is equipped with attitude and stabilization systems and control equipment for telecommunication, telemetry, and electric power generation to support the payload during boost and until separation on station. Two variants of the Block DM exist—the standard fully equipped version and a version without the pressurized Instrument Unit (which contains avionics and guidance systems) located at the top of the stage. This second version is sometimes designated "Block D". Typically, the Block DM is used for geosynchronous and other high-energy Earth orbital missions, while the Block D is employed for planetary missions. Use of the Block D, while increasing performance, requires that the payload provide all control functions normally handled by the Instrument Unit. Up to five engine restarts have been demonstrated, and seven are possible within 48 hours of launch. It has a single 11D58M engine that burns kerosene and LOX. The LOX tank of the fourth stage is thermally insulated. Proton has an 82-pin electrical cable and five pipelines for connecting the fourth stage and its payload to equipment on the ground.

Proton Vehicle
(continued)

Payload Fairing

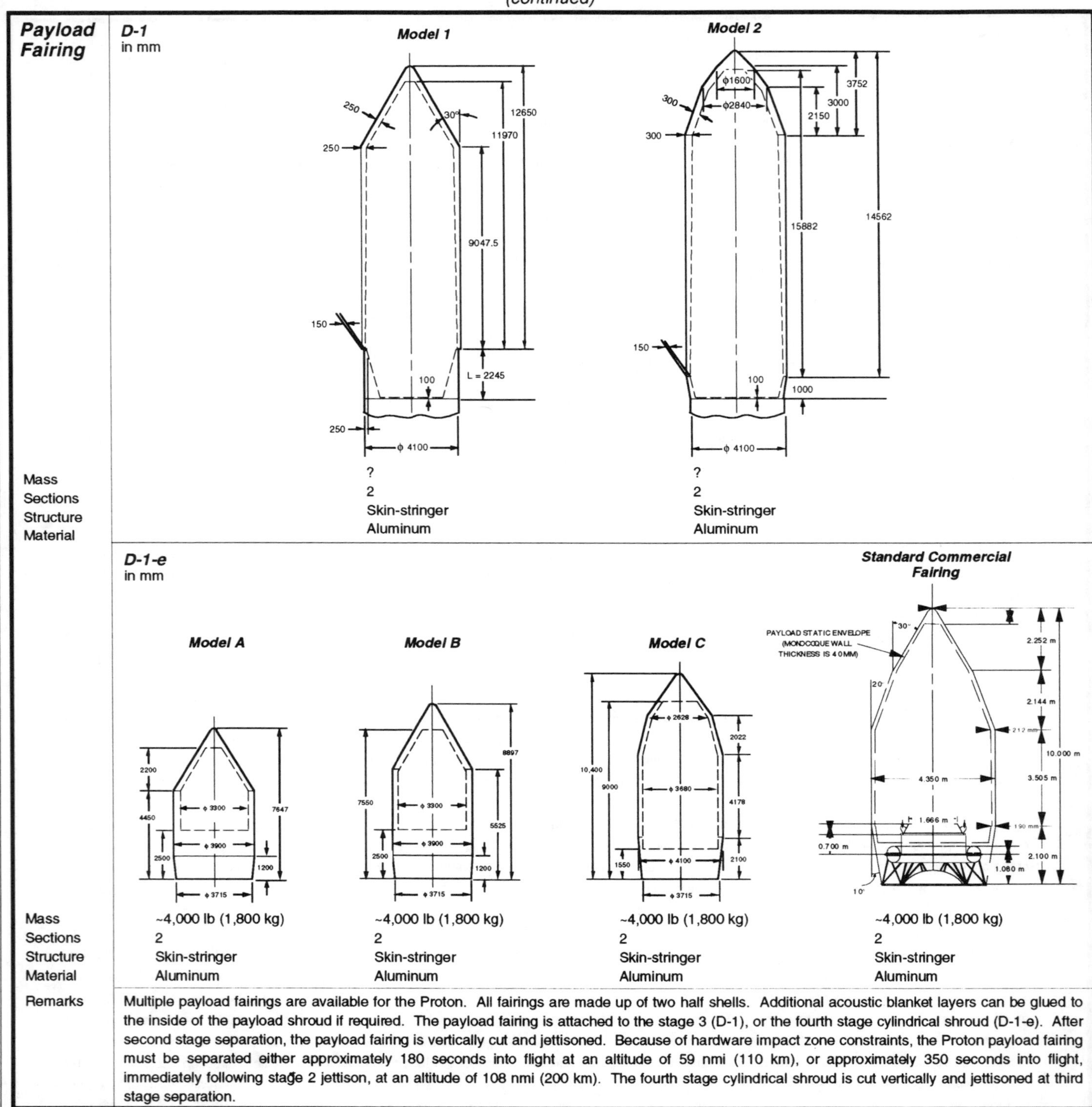

D-1 (in mm)

	Model 1	Model 2
Mass	?	?
Sections	2	2
Structure	Skin-stringer	Skin-stringer
Material	Aluminum	Aluminum

D-1-e (in mm)

	Model A	Model B	Model C	Standard Commercial Fairing
Mass	~4,000 lb (1,800 kg)	~4,000 lb (1,800 kg)	~4,000 lb (1,800 kg)	~4,000 lb (1,800 kg)
Sections	2	2	2	2
Structure	Skin-stringer	Skin-stringer	Skin-stringer	Skin-stringer
Material	Aluminum	Aluminum	Aluminum	Aluminum

Remarks

Multiple payload fairings are available for the Proton. All fairings are made up of two half shells. Additional acoustic blanket layers can be glued to the inside of the payload shroud if required. The payload fairing is attached to the stage 3 (D-1), or the fourth stage cylindrical shroud (D-1-e). After second stage separation, the payload fairing is vertically cut and jettisoned. Because of hardware impact zone constraints, the Proton payload fairing must be separated either approximately 180 seconds into flight at an altitude of 59 nmi (110 km), or approximately 350 seconds into flight, immediately following stage 2 jettison, at an altitude of 108 nmi (200 km). The fourth stage cylindrical shroud is cut vertically and jettisoned at third stage separation.

Avionics

The third stage is equipped with a triple redundant closed loop guidance system for the first three stages of the launch vehicle control.

Within its Instrument Unit, the fourth stage carries an inertial measurement unit and a triple redundant closed loop avionics package.

Attitude Control System

On stage 1, each of the six RD-253 engines can gimbal in a single plane tangential to the stage's circumference. Thus, all six engines together provide full yaw, pitch, and roll control. Second stage control is achieved by gimbaling the four RD-0210 engines. Stage 3 is controlled by four vernier rocket engines spaced around the single fixed RD-0210.

While the 11D58M main engine is burning, stage 4 uses a combination of main engine gimbaling and turbopump bleed gas for three axis control. During coast periods, the self-contained ACS maintains three axis control using a collection of small thrusters burning hypergolic N2O4 and UDMH propellants. Small thrusters are also used for on-orbit propellant settling prior to main engine ignition.

Proton D-1 Performance to 108 nmi (200 km) Circular Orbit

Note: D-1 can launch directly into these inclinations only. Interpolation between these points is not valid.

Inclination	Payload
51.6˚	46,000 lb (20,900 kg)
64.8˚	43,400 lb (19,700 kg)
72.7˚	18,800 lb (41,400 kg)

Proton D-1-e Performance to Selected Orbits

Note: Performance values correspond to Model C payload fairing and 350 sec fairing jettison time. Performance may change based on specific fairing and jettison time used.

Orbit	Payload
Geostationary	4,850 lb (2,200 kg)
Sunsynchronous	5,100 lb (2,300 kg)
Lunar transfer	9,000 lb (4,100 kg)
Mars transfer	8,800 lb (4,000 kg)
Glonass (64.8˚, 19,000 km circular)	6,800 lb (3,100 kg)

Proton D-1-e Performance to Circular Orbits as a Function of Inclination

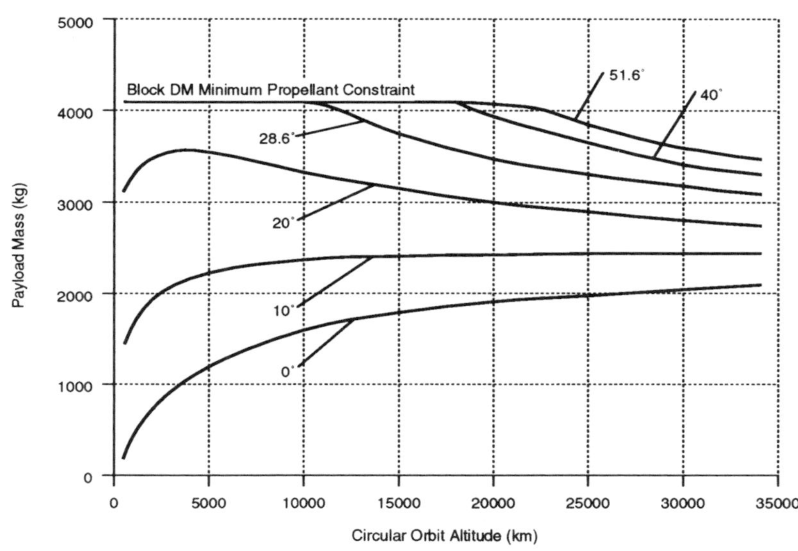

Vehicle:	Proton D-1-e
Payload Fairing:	Model C
Payload Fairing Jettison Time:	350 sec
Park Orbit Inclination:	51.6˚
Transfer Orbit Perigee Altitude:	200 km
Transfer Orbit Inclination for Final Inclinations < 48˚	48˚
Final Orbit Inclination	As Shown
3-Sigma Margin Reserved	

Proton D-1-e Performance to Earth Escape Trajectories

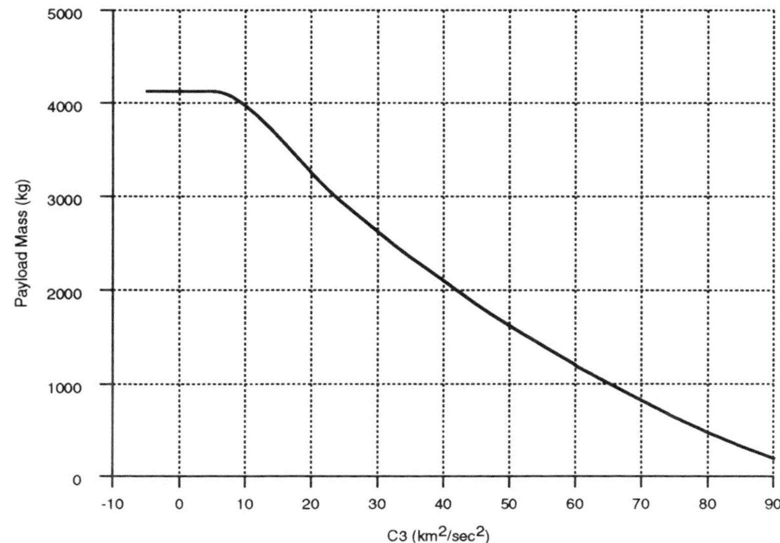

Vehicle:	Proton D-1-e
Payload Fairing:	Model C
Payload Fairing Jettison Time:	350 sec
Final Orbit Perigee Altitude:	200 km
Final Orbit Inclination:	51.6˚
3-Sigma Margin Reserved	

All performance values courtesy Lockheed-Khrunichev-Energia International

Proton Operations

Launch Site

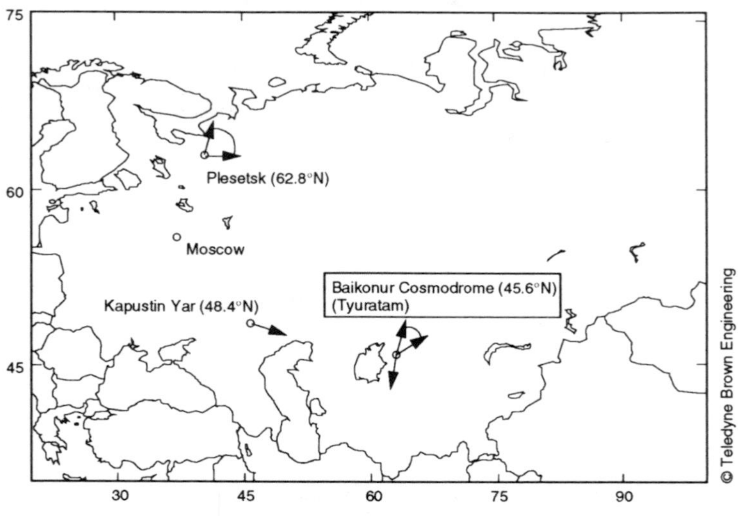

Launch Facilities

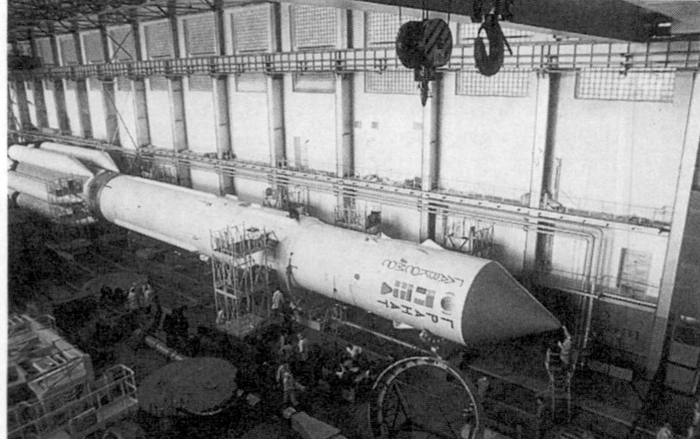

Proton Preparation at the Horizontal Assembly Facility

Transportation by Rail to the Launch Pad

View of Proton Launch Site Layout

Photos Courtesy of Space Commerce Corp.

Proton Operations

(continued)

Launch Processing

Tyuratam. Proton is launched from the Baikonur Cosmodrome located just east of the Aral Sea in Kazakhstan, north of Afghanistan. This site is usually referred to in the West as Tyuratam, a reference to a nearby railway station. Tyuratam, the CIS equivalent to the U.S. Cape Canaveral launch site, is where all manned, geosynchronous, lunar, interplanetary, antisatellite, and tactical ocean surveillance missions are launched. It is the only facility capable of launching the Proton, Zenit, and Energia. All other CIS vehicles can be launched from Tyuratam except the small Kosmos launcher. In total, 30 to 40 launches per year take place at Baikonur, approximately 25% of which are Proton launches. It is not unusual for two missions to be launched at Tyuratam less than 24 hours apart. Two Protons have been launched as little as four days apart.

Since the breakup of the Soviet Union, Russia has been in the difficult situation of having its largest space launch facility located in a foreign country (i.e., Kazakhstan). Despite political turmoil, the Baikonur Cosmodrome has remained in operation virtually without interruption. In the summer of 1994, Russia and Kazakhstan signed a long-term agreement whereby Russia would lease Baikonur from Kazakhstan, with day-to-day operations of the Cosmodrome remaining under the control of the Russian Space Forces.

The location of Baikonur (45.6° N) precludes launch directly into orbital inclinations of less than 45.6°; however, due to range safety restrictions (in particular, avoiding overflight of China), Proton is not launched into inclinations less than approximately 51°. The fact that Baikonur (unlike Cape Canaveral) is located at a landlocked site drives additional range safety constraints. Thus, to insure that jettisoned hardware lands in designated locations inside Russia and Kazakhstan, Proton is launched directly into only three discrete inclinations: 51.6°, 64.8°, and 72.7°. Payloads that require placement into inclinations significantly different than these must perform on-orbit plane change maneuvers. The D-1-e's Block DM is often used for this purpose.

The climate at Baikonur is hot in the summer, with temperatures rising as high as 120°F (50°C), and prone to violent snowstorms and -40°F (-40°C) lows in winter. A unique feature of the CIS space program is the ability for its vehicles to launch in the harsh climate conditions. The large Y-shaped complex covers about 55 mi (88 km) north to south and 100 mi (160 km) east to west. Each type of vehicle has its own processing and launch facilities, with the Proton area located in the northwest corner of the Baikonur complex. The vehicle processing and launch areas are connected to each other and the city of Leninsk by nearly 300 mi (500 km) of railroad lines.

Leninsk and Tyuratam are the residential areas for the employees of the Cosmodrome Space Facility and their families. Leninsk is located 1,300 mi (2,100 km) southeast of Moscow. Lush vegetation was painstakingly developed in spite of the harsh winters and desert-like summers. The driving time from Leninsk to the spacecraft processing area is about 30 to 45 minutes and to the launch site is approximately 1.5 hrs.

The main building of the Proton technical complex is the integration and testing facility. Assembly and integration of the Proton launch vehicle stages are carried out with the vehicle in a horizontal position. The first stage's core oxidizer tank and six fuel tank/engine assemblies are transported to Baikonur separately. Assembly of stage 1 is accomplished by placing the core tank in a fixture that rotates the tank about its longitudinal axis. After the first fuel tank/engine assembly is attached from underneath the horizontal oxidizer tank, the whole assembly is rotated one-sixth of a revolution to allow for the integration of the next fuel tank/engine module. The process is repeated until all six fuel tank/engine assemblies are attached.

Following completion of its assembly, stage 1 is moved by crane to the assembly integration erector trolley, where it is mated to stages 2 and 3. The assembled Proton then undergoes integration tests. Up to four Proton launch vehicles can be processed simultaneously, and the assembly building can accommodate as many as six boosters at one time. Each vehicle requires two weeks for assembly, followed by one week of end-to-end tests.

Once the spacecraft and fourth stage (for D-1-e missions) are assembled and encapsulated in the satellite preparation building, they are transported by rail 56 mi (90 km) to the Proton launch vehicle horizontal assembly building. Loads experienced during transport do not exceed launch loads. After the encapsulated spacecraft and upper stage are integrated to the first three stages, the launch vehicle is transported horizontally by rail to the launch pad and then raised into a vertical position. Thermal control inside the payload fairing is provided by an air conditioning system.

There are two Proton launch complexes each consisting of two launch pads 2,000 ft (600 m) apart. Three operational launch pads are currently available; the fourth pad is currently undergoing refurbishment. Two of the launch pads were built in the mid 1960s, and the other two during the mid 1970s.

It takes four hours to erect and install Proton on the launch pad; typically, this takes place three to five days prior to launch. Preliftoff maintenance and service operations are carried out with the aid of a mobile service tower mounted on railway trucks. No cable or umbilical tower is incorporated into the launch complex. Many of the services that would be provided by such an umbilical tower are provided by the mobile service tower. Other functions, including fuel loading, are performed by a special servicing mechanism that connects to the Proton at the base of stage 1. All fuel and compressed gas loading operations, including the attaching and removing of connections, are fully automated. The propellant for the vehicle is stored near the launch pad in underground bunkers. If a launch scrub occurs, it requires six to eight hours to defuel the booster and an additional four hours to lower it to a horizontal position for roll back to the horizontal assembly building.

Approximately eight hours before liftoff, the payload's onboard systems are checked out, and the satellite's readiness for launch is confirmed by checking its telemetry data. Individual safety devices are removed no less than 30 minutes before launch. Recharging of the onboard chemical batteries is stopped 1.5 hrs before launch, although trickle charging can continue until five minutes before liftoff. Launch pad personnel are evacuated from the pad area 1.5 hrs before launch. Shortly thereafter, the mobile service tower is rolled back to a safe distance from the launch pad. Launch go-ahead must be provided by the payload customer no less than two minutes before launch.

Launch is controlled from a command center located about 1 mi (1.5 km) from the launch complex. At the moment of liftoff, the launch pad servicing mechanism rises with the launch vehicle, tracking the vehicle's movement for the first few fractions of a second. After approximately 2 cm of motion, this mechanism is separated and withdrawn by a pneumatic actuator and secured behind an armored steel fire wall cover. This steel cover then helps form part of the launch pad flame deflector.

The Proton launch complex has undergone evolutionary modifications brought about by changes in the booster's applications. These upgrades include a cryogenic liquid oxygen handling system to support stage 4, a new compressed gas delivery system, and improved on-pad payload thermal control systems. In addition, the Proton launch complex underwent a general modernization between 1975 and 1980.

Proton Operations

(continued)

Typical Launch Processing Timeline

T-60 days	Delivery of satellite, mounting equipment, and other support equipment to the cosmodrome.
T-59 days	Unloading of aircraft.
T-58 days	Transportation of satellite and equipment to the technical zone of the horizontal assembly building.
T-57 days	Unloading of the spacecraft from transport containers.
T-56 days	Installation of the spacecraft in the "clean room" air chamber. Placement of the spacecraft on a work stand in the "clean room".
T-55 days	Opening of containers with the spacecraft platform and ground-based equipment. Placement of the spacecraft platform and ground-based equipment in the workplace.
T-52 days	Mating of the spacecraft and the spacecraft platform.
T-51 days	Installation, checkout of EGSE, preparation for working with the satellite.
	Connecting ESGE to the spacecraft.
	Electrical and electronic testing of the communications systems of the spacecraft.
T-39 days	Pneumatic and pressurization tests of the spacecraft.
T-36 days	Testing of solar panels, batteries, and systems deployment mechanisms.
T-18 days	Preparation of the spacecraft for fueling.
	Transportation of spacecraft to fueling area.
T-17 days	Fueling of spacecraft with propellants.
	Transportation of spacecraft from fueling area.
T-14 days	Checking of pyrotechnical devices by energizing setups using low voltage.
T-12 days	Final operations on the spacecraft.
T-11 days	Mating of spacecraft with launch vehicle adapter.
T-10 days	Mating of spacecraft with final stage adapter/shroud of the launch vehicle.
	Testing of the transit cable after assembly of the payload section.
	Loading of the payload section on the transporter.
	Transportation of the payload section to the launch vehicle last stage refueling area.
T-9 days	Fueling of the launch vehicle last stage.
	Transportation of the payload section to the launch vehicle assembly area.
T-6 days	Preparation of the payload section for mating with the launch vehicle; mating of the payload section with the launch vehicle; checking of the electrical connection of the payload section with the launch vehicle (testing of the transit cables).
T-6 days	Transfer of the launch vehicle with the payload section onto the erector transporter.
	Roll out from Horizontal Assembly Building and installation on the launch pad.
	General checkout of spacecraft systems on launch pad after service tower is positioned.
T-5 days	General testing and preparation of spacecraft and launch vehicle.
T-3 days	Composite testing of the launch vehicle and spacecraft.
	Simulation (dry run) of the launch vehicle fueling.
	Cycling of the onboard chemical batteries of the spacecraft.
T-2 days	Possible flight readiness review.
	Reserve time cycling of the onboard chemical batteries of the spacecraft.

HOURS:MIN	
T-24:00	Satellite final integration and closeout of preflight testing.
	General operations for preparation of the launch vehicle for launch, (6–8 hours required), and general systems checkout.
	Cycling of the onboard chemical batteries (static charge).
T-10:00 - 8:00	Final checkout test of the spacecraft onboard systems (begins 8 hours prior to launch).
	Final operations on the payload section including the spacecraft.
T-8:00	Preparation and final closeout of the spacecraft (8 hours prior to launch).
	Confirmation of the satellite's readiness for launch through telemetry data.
	Beginning of final operations tasks on the launch vehicle.
	Power-up satellite 6 to 8 hours prior to launch.
	Launch vehicle prepared for fueling.
T-7:00	Launch vehicle propellant loading.
	Cycling of onboard batteries.
T-2:30 - 2:10	Turn off the air conditioning system prior to removal of the protective safety devices.
T-1:50	Removal of the spacecraft's pyrotechnic safety devices.
T-1:40 - 1:30	End prelaunch spacecraft checkout on the launch pad and switch on the spacecraft's electrical power.
	Removal of the spacecraft's optical devices' protective covers.
T-1:40 - 1:20	Evacuate personnel from the servicing tower and pad area.
T-1:10 - 0:30	Service tower gantry is pulled back from the launch vehicle and secured after the fourth stage's cryogenic propellants are topped-off.
T-1:05	Payload is ready for launch.
T-0:60	Evacuation of personnel from the launch area.
T-0:45	Launch vehicle final preparation.
	Final countdown begins.
T-0:45 - 0:30	Completion of first stage propellant topoff.
T-2 sec	Startup, primary ignition.
T-0 sec	Liftoff.

Typical D-1-e Flight Sequence

Proton D-1-e GEO Mission

Flight Time (hr:min:sec)	Event
0:00:00	Liftoff.
	Advancement of first stage engines to nominal full thrust range.
0:00:01	Separation of liftoff service mechanism which rises with the vehicle for a fraction of a second.
0:00:10	A.P.U. startup.
0:00:21	Roll program.
0:01:00	Maximum dynamic pressure.
0:02:02	Second stage engine start to medium thrust range. [Preparatory command]
0:02:07	Fire in the hole ignition Second stage.
0:03:03	First stage engine shutdown, staging stage separation command, and advancement to nominal thrust range.
0:05:34	Payload fairing jettison command (option 1).
0:05:37	Four steering engines of third stage startup [Preparatory command]
0:05:38	Second stage engine shutdown.
0:05:41	Second and third stage separation. Second stage retrofire six solid motors.
0:06:10	Third stage main engine start.
0:09:38	Payload fairing jettison command (option 2).
0:09:52	Third stage main engine shutdown.
0:09:57	Third stage steering engine shutdown.
0:10:00	Third stage retrofire and separation from fourth stage.
0:10:05	Fourth stage cylindrical shroud separation and orbital insertion.
0:25:00	Execution of programmed orientation turns to align stage 4 for the first main engine ignition.
0:40:00	Three-axes stabilized passive flight.
	180° roll maneuver.
0:42:30	Guidance system compensation.
1:20:00	Three-axes stabilized passive flight.
1:27:30	First stage 4 main engine ignition for transfer into an elliptical transfer orbit.
	First shutdown of fourth stage engine and transfer orbit injection.
1:42:30	Execution of programmed orientation turns to align stage 4 for the second main engine ignition.
3:59:10	Completion of the programmed turns.
4:24:00	Compensating 180º roll maneuver of 2.5 min. duration, marking the midpoint of the transfer orbit coast.
6:58:00	Completion of the compensation turn.
7:09:20	Second stage 4 main engine ignition.
7:09:22	Second shutdown of fourth stage engine.
7:09:36	Pulse the satellite transmitters 14 sec. prior to separation.
	Spacecraft separation from fourth stage.

Proton Payload Accommodations

Payload Compartment	
Maximum Payload Diameter	D-1: Model 1 - 161 in (4.10 m) Model 2 - 161 in (4.10 m) D-1-e: Model A - 130 in (3.30 m) Model B - 130 in (3.30 m) Model C - 145 in (3.68 m) Standard Commercial Fairing - 156 in (3.97 m)
Maximum Cylinder and Cone Length	D-1: Model 1 - 586 in (14.9 m) total (cylinder and cone) Model 2 - 613 in (15.6 m) total (cylinder and cone) D-1-e: Model A - 163 in (4,150 mm) total (cylinder and cone) Model B - 199 in (5,050 mm) total (cylinder and cone) Model C - 293 in (7,450 mm) total (cylinder and cone) Standard Commercial Fairing - 287 in (7,300 mm) total (cylinder and cone)
Payload Adapter	
Interface Diameter	Both separation nut (truss type) and Marmon clamp (shell type) adapters available in standard diameters, including 37 in (937 mm), 47 in (1,194 mm) and 66 in (1,666 mm)
Payload Integration	
Nominal Mission Schedule Begins	T-24 months
Launch Window	
Latest Countdown Hold Not Requiring Recycling	T-70 min (In case of launch failure, next launch possible in 50–60 days)
On-Pad Storage Capability	48 hrs for a fueled vehicle
Latest Access to Payload	T-1.5 hrs through access doors
Environment	
Maximum Load Factors	+3.65 g axial, ±1.5 g lateral
Minimum Lateral/Longitudinal Payload Frequency	15 Hz / 30 Hz
Maximum Overall Acoustic Level	134 dB (full octave)
Maximum Flight Shock	2,500 g from 1,500–5,000 hz, typical value during payload separation
Maximum Dynamic Pressure on Fairing	778 lb/ft^2 (3,800 kgf/m^2), typical value
Maximum Pressure Change in Fairing	0.2 psi/s (1.5 kPa/s)
Cleanliness Level in Fairing (Prior to Launch)	Class 100,000 (No air conditioning during 4 hr erection onto pad and 70 min before launch).
Payload Delivery	
Standard Orbit and Accuracy (3 sigma)	LEO at 108 nm (200 km) circular at 51.6 deg is ±8.1 nmi (15 km), ±1.5 min GEO is ±192 - 390 nmi (355–723 km), ±15 min
Attitude Accuracy (3 sigma)	All axes: ±2 deg, ±0.2 deg/sec
Nominal Payload Separation Rate	0.15–0.5 ft/s (0.5–1.5 m/s)
Deployment Rotation Rate Available	1.5 rpm
Loiter Duration in Orbit	24 hrs for Block DM
Maneuvers (Thermal / Collision / Telemetry)	Yes

Proton Notes

Publications

User's Guides

Mission Planner's Manual - Proton Commercial Launch Vehicle, Space Commerce Corporation, Issue No. 1, June 1989.

Proton User's Guide, Lockheed-Khrunichev-Energia International, Inc., Rev. 1 Issue 1, December 1993.

Technical Publications

TRW Space Log, 1957-1991, 1992, 1993 TRW, El Segundo, CA.

The Soviet Year in Space - 1986, 1988, 1989 Nicholas Johnson, Teledyne Brown Engineering, Colorado Springs, CO.

Launch Vehicle Catalogue, European Space Agency, December 1989.

"Perestroika and Glasnost in the Soviet Space Program", Space Policy, November 1989.

Soviet Space Programs: 1981-1987, United States Senate, Committee on Commerce, Science, and Transportation - Part 1 - May 1988.

Jane's Spaceflight Directory 1987, Jane's Publishing Inc., New York, NY.

Jane's Space Directory 1994-95, Janes's Information Group, Inc., Alexandria, VA.

Space Directory of Russia, February 1993 Edition, Euroconsult/Sevig Press, Paris, France.

The Russian Space Space Directory 1994, European Space Report, Munich, Germany.

Acronyms

ACS - attitude control system
CIS - Commonwealth of Independent States
GEO - geosynchronous orbit
GTO - geosynchronous transfer orbit
KB - Design Bureau
LEO - low Earth orbit
LH2 - liquid hydrogen
LKE - Lockheed-Khrunichev-Energia
LOX - liquid oxygen
NIIAP - Scientific Research Institute for Automation and Instrument Engineering
N2O4 - nitrogen tetroxide
NPO - Scientific Production Organization
RP-1 - kerosene
RSC - Rocket Space Corporation
TT - Tyuratam
UDMH - unsymmetrical dimethylhydrazine
USSR - Union of Soviet Socialist Republics

Other Notes—Proton Upgrade Possibilities

Proton is currently undergoing a modernization process. This process appears to consist of two main components: upgrading stages 1–3, and replacing the Block DM with a more capable liquid oxygen / liquid hydrogen upper stage. This upgraded vehicle is referred to as as the Proton-M (also Proton-KM).

The upgrades to stages 1–3 involve evolutionary improvements to a number of the vehicle's major systems. This effort is already underway. The primary goals of this modernization are to increase reliability and reduce Proton's impact on the environment. Some performance increase will also be realized. The avionics system will be most affected; it is likely that the aging Proton avionics suite will be replaced by a derivative of the Zenit's system. It is also possible that the engines on one or more of the stages will be uprated to some extent, although the basic structure and propulsion systems will remain essentially unchanged.

The desire to reduce Proton's environmental impact is driven by the fact that Proton launches from a landlocked site. Uninhabited areas inside Kazakhstan and Russia are set aside as designated hardware impact areas. Currently, these areas are rather large to allow for dispersions in the uncontrolled descent of the spent hardware. By improving Proton's guidance system, it will become possible to reduce the size of these areas, thus allowing some of the land to be put to more productive use. In addition, these upgrades will allow for the reduction of the amount of unused, poisonous N2O4 and UDMH propellants left in the stage 1 and 2 tanks after jettison, further reducing potential environmental hazards.

A hydrogen/oxygen stage 4 is currently under development by Khrunichev Enterprise, with completion expected in 1996. Very few details are available concerning the mass, propellant load, and propulsive characteristics of this stage. It is anticipated that the Proton-M with LOX/LH2 upper stage will possess geosynchronous performance nearly double that of the current Proton D-1-e.

Proton-KM

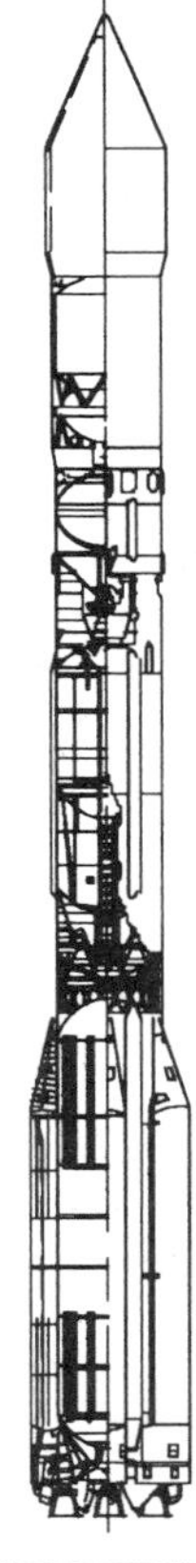

Launch Services Point of Contact:
Khrunichev State Research and Production Space Center
18 Novozovodskaya St.
Moscow
Russia, 121309
Phone: (7095) 145-8343
Fax: (7095) 142-5900

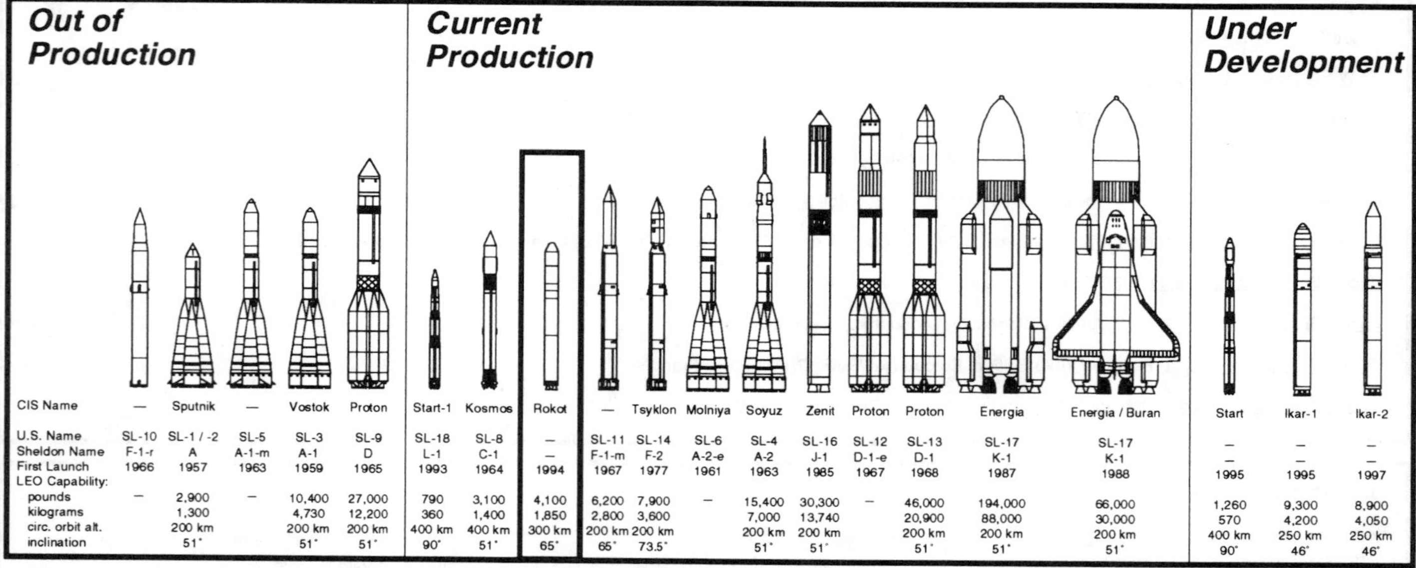

Out of Production

CIS Name	—	Sputnik	—	Vostok	Proton
U.S. Name	SL-10	SL-1 / -2	SL-5	SL-3	SL-9
Sheldon Name	F-1-r	A	A-1-m	A-1	D
First Launch	1966	1957	1963	1959	1965
LEO Capability:					
pounds	—	2,900	—	10,400	27,000
kilograms		1,300		4,730	12,200
circ. orbit alt.		200 km		200 km	200 km
inclination		51˚		51˚	51˚

Current Production

CIS Name	Start-1	Kosmos	Rokot	—	Tsyklon	Molniya
U.S. Name	SL-18	SL-8	—	SL-11	SL-14	SL-6
Sheldon Name	L-1	C-1	—	F-1-m	F-2	A-2-e
First Launch	1993	1964	1994	1967	1977	1961
LEO Capability:						
pounds	790	3,100	4,100	6,200	7,900	—
kilograms	360	1,400	1,850	2,800	3,600	
circ. orbit alt.	400 km	400 km	300 km	200 km	200 km	
inclination	90˚	51˚	65˚	65˚	73.5˚	

CIS Name	Soyuz	Zenit	Proton	Proton	Energia	Energia / Buran
U.S. Name	SL-4	SL-16	SL-12	SL-13	SL-17	SL-17
Sheldon Name	A-2	J-1	D-1-e	D-1	K-1	K-1
First Launch	1963	1985	1967	1968	1987	1988
LEO Capability:						
pounds	15,400	30,300	—	46,000	194,000	66,000
kilograms	7,000	13,740		20,900	88,000	30,000
circ. orbit alt.	200 km	200 km		200 km	200 km	200 km
inclination	51˚	51˚		51˚	51˚	51˚

Under Development

CIS Name	Start	Ikar-1	Ikar-2
U.S. Name	—	—	—
Sheldon Name	—	—	—
First Launch	1995	1995	1997
LEO Capability:			
pounds	1,260	9,300	8,900
kilograms	570	4,200	4,050
circ. orbit alt.	400 km	250 km	250 km
inclination	90˚	46˚	46˚

Historical Summary

The Rokot ("Rumble") launch vehicle is based on the SS-19 Stiletto (Russian designation RS-18) ICBM. The SS-19 is a silo-based, liquid-propellant two-stage missile. Development of the SS-19 began in the mid 1960s, with the first test launch taking place in 1973. Operational deployment at sites in Russia and Ukraine began in 1975; this missile is still actively deployed. The SS-19, along with the SS-17 and -18 missiles, represents the former Soviet Union's first true multiple independently targeted warhead capability. The SS-19 is approximately 89 ft (27 m) long, 8.2 ft (2.5 m) in diameter, and has a liftoff mass in excess of 220,000 lb (100 metric tons). This missile is hot-launched, meaning that its first stage main engines are ignited in the silo. The SS-19 was developed by the Chelomei design bureau, which has since evolved into Khrunichev State Research and Production Space Center.

Rokot, with a payload capacity of up to 4,400 lb (2,000 kg) to LEO, fills the gap between the well-proven Kosmos (up to 1,500 kg to LEO) and Tsyklon (3,600 kg to LEO) vehicles. Technical information on Rokot is scarce, probably owing, at least in part, to the SS-19's active deployment status. Rokot consists of the SS-19's first and second stages, plus an additional third stage designated Briz ("Breeze"). Briz is apparently a new stage; other applications of this stage have been proposed, including use as a fifth stage for Proton. It appears that Rokot is launched from existing SS-19 silos and utilizes existing SS-19 ground facilities to the greatest extent possible. Although the only Rokot launch as of the end of 1994 occurred at the Baikonur Cosmodrome, there are indications that Plesetsk launch capability may also exist. It is unclear whether Rokot vehicles are to be produced from decommissioned SS-19 missiles, manufactured as new vehicles, or if some combination of these approaches will be utilized. It appears that Khrunichev has entered into a partnership with Daimler-Benz Aerospace of Germany to market Rokot worldwide.

The first orbital flight of the Rokot took place at the Baikonur Cosmodrome on December 26, 1994, with an amateur radio satellite payload. Although Rokot successfully delivered its payload, stage 3 exploded some time after payload separation.

Launch Site

Note: The only Rokot launch as of the end of 1994 took place at the Baikonur Cosmodrome; Plesetsk launch capability is probable.

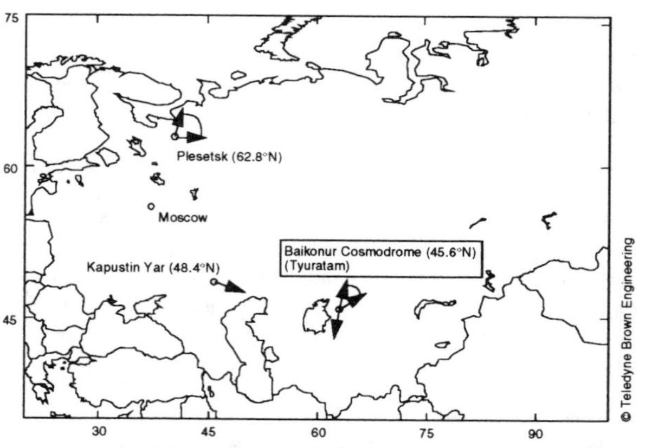

Launch Record

Success Rate = 1 / 1 = 100%

	YEAR	DATE	VEHICLE	SITE	PAYLOAD
1	1994	Dec-26	Rokot	Baikonur	Radio-ROSTO

Rokot

General Description

Summary — Rokot is a three-stage liquid-propellant small-class launch vehicle based on the SS-19 ICBM.

Status — Operational

Key Organizations — Users - Russian Space Agency, Russian Ministry of Defense, commercial

Responsible Organizations -
• Khrunichev SRPSC (Manufacturer)

Vehicle

System Height	88–92 ft (27–28 m)
Payload Fairing Size	~ 8.2 ft (2.5 m) outer diameter
Gross Mass	~ 235K lb (107K kg)
Planned Enhancements	??

Operations

Primary Missions	Low Earth Orbit (LEO)
Compatible Upper Stages	none
Launch Site	Baikonur, possibly Plesetsk
Inclination Constraints	??
Nominal Flight Rate	??
Planned Enhancements	none

Performance

162 nm (300 km) circ, 62°	4,100 lb (1,850 kg)
Geotransfer Orbit, 28.5°	—
Geosynchronous Orbit	—

Financial Status

Estimated Launch Price	$5–7M (U.S. DoT estimate)

Manifest — ??

Remarks — —

Rokot Vehicle

Very little technical information is available on Rokot. Stages 1 and 2 burn storable unsymmetrical dimethylhydrazine (UDMH) and nitrogen tetroxide (N2O4) propellants. It is not known what types of propellants Briz employs, although it is believed that it burns storable propellants as well. The stage 1 propulsion system consists of four combustion chambers; it is not known if these chambers are built as a single four-chamber engine (as with many other Russian multichamber propulsion system designs). Stage 2 and Briz each have single combustion chambers. It appears that Rokot is able to deliver payloads to circular orbits as high as 1,100 nmi (2,000 km); thus, it is likely that Briz is capable of multiple on-orbit starts. No information is available concerning the type of avionics package used or its location on the vehicle. The means by which the various stages steer (engine gimbaling, verniers, etc.) is also not known.

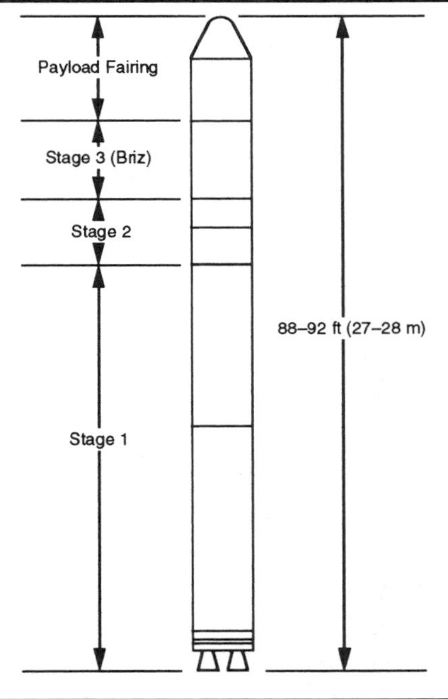

Publications

Technical Publications

Jane's Space Directory 1994-95, Jane's Information Group Inc., Alexandria, VA.

Jane's Strategic Weapons Systems, Jane's Information Group Inc., Alexandria, VA, 1994.

Space Directory of Russia, February 1993 Edition, Euroconsult/Sevig Press, Paris, France.

The Russian Space Directory 1994, European Space Report, Munich, Germany.

Launch Services Point of Contact:
Central Specialized Design Bureau
18 Pskovskaya Street
Samara
The Russian Federation, 443009
Phone: (78462) 22-28-14
Fax: (78462) 27-20-70
Telex: 214153 NIKA SU

Soyuz
Molniya

Photo Courtesy of Space Commerce Corp

The A-class launch system, of which the Soyuz and Molniya vehicles are the only currently active variants, is still the workhorse of the CIS space launch program with more launches than any other vehicle in the world. Much of the vehicle design has remained unchanged since the launch of Sputnik in 1957.

Soyuz / Molniya History

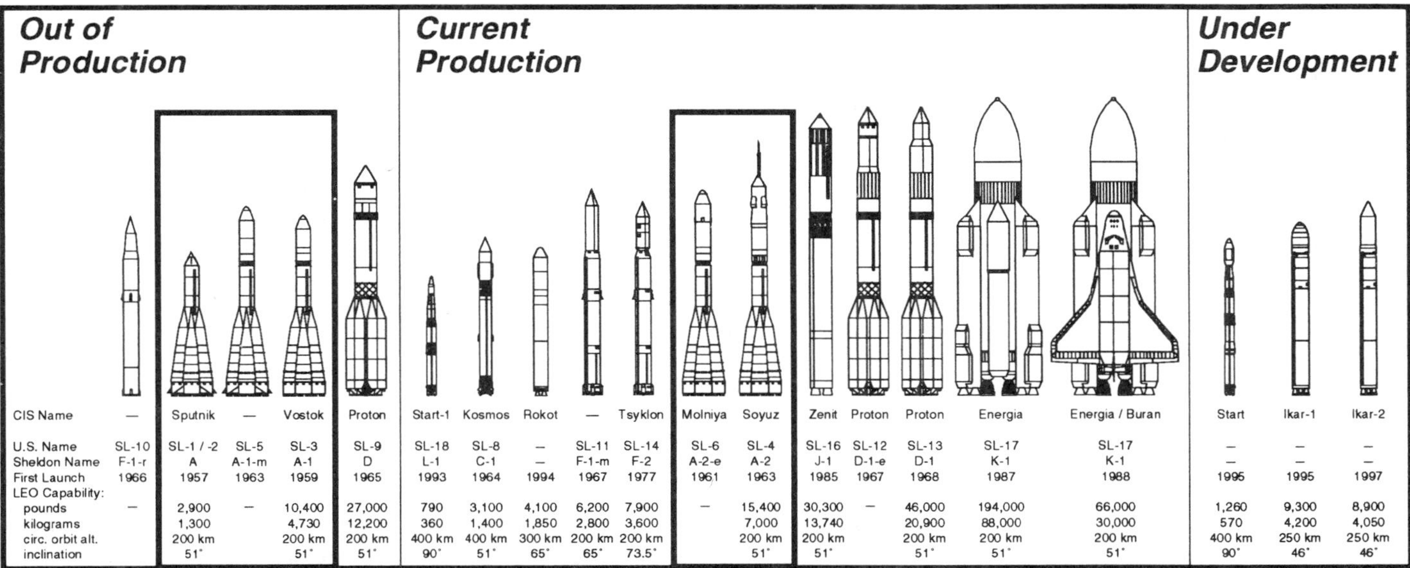

CIS Name	—	Sputnik	—	Vostok	Proton
U.S. Name	SL-10	SL-1 / -2	SL-5	SL-3	SL-9
Sheldon Name	F-1-r	A	A-1-m	A-1	D
First Launch	1966	1957	1963	1959	1965
LEO Capability:					
pounds	—	2,900	—	10,400	27,000
kilograms		1,300		4,730	12,200
circ. orbit alt.		200 km		200 km	200 km
inclination		51˚		51˚	51˚

CIS Name	Start-1	Kosmos	Rokot	—	Tsyklon	Molniya	Soyuz
U.S. Name	SL-18	SL-8	–	SL-11	SL-14	SL-6	SL-4
Sheldon Name	L-1	C-1	–	F-1-m	F-2	A-2-e	A-2
First Launch	1993	1964	1994	1967	1977	1961	1963
LEO Capability:							
pounds	790	3,100	4,100	6,200	7,900	—	15,400
kilograms	360	1,400	1,850	2,800	3,600		7,000
circ. orbit alt.	400 km	400 km	300 km	200 km	200 km		200 km
inclination	90˚	51˚	65˚	65˚	73.5˚		51˚

CIS Name	Zenit	Proton	Proton	Energia	Energia / Buran	Start	Ikar-1	Ikar-2
U.S. Name	SL-16	SL-12	SL-13	SL-17	SL-17	–	–	–
Sheldon Name	J-1	D-1-e	D-1	K-1	K-1	–	–	–
First Launch	1985	1967	1968	1987	1988	1995	1995	1997
LEO Capability:								
pounds	30,300	—	46,000	194,000	66,000	1,260	9,300	8,900
kilograms	13,740		20,900	88,000	30,000	570	4,200	4,050
circ. orbit alt.	200 km		200 km	200 km	200 km	400 km	250 km	250 km
inclination	51˚		51˚	51˚	51˚	90˚	46˚	46˚

Vehicle Description

CIS vehicles are identified either by their CIS, U.S., or Sheldon name. In the Soviet Union, it was standard practice to name a launch vehicle after its original payload (e.g., Kosmos, Proton). The U.S. names (developed by the U.S. Department of Defense) are alphanumeric designators based roughly on chronological appearance. The Sheldon names, a commonly used system that was published by Dr. Charles Sheldon of the U.S. Library of Congress in 1968, emphasize the basic families of launch vehicles with special indicators for variants within a family.

Example: A-1-e

Upper Stage (1, 2) - Note: "1" or "2" stage may vary among different families, e.g., A-1 upper stage different than B-1.

e - earth escape or fourth stage
m - maneuverable stage
r - reentry stage

Family (A, B, C, D, F, G, J, K)

A (SL-1/2)	Based on the Sapwood SS-6 ICBM, it has four symmetrically arranged strap-ons around a core stage, all burning LOX / kerosene propellants.
Vostok (A-1/SL-3)	Same as 'A' except addition of a LOX / kerosene core second stage.
A-1-m (SL-5)	Same as Vostok except addition of a maneuverable stage.
Soyuz (A-2/SL-4)	Same as Vostok except replacement of the core second stage with a more powerful second stage. The second stage is also LOX / kerosene.
Molniya (A-2-e/SL-6)	Same as Soyuz except addition of a LOX / kerosene third stage.

Historical Summary

The A-class of CIS launch vehicles is based on the original Soviet intercontinental ballistic missile (ICBM), the SS-6 Sapwood (Russian designation R-7), which was first flown in August 1957. Its first use as a launch vehicle took place on October 4 of that same year, indicating that this rocket was designed from its inception to serve both as an ICBM and a space launch vehicle. Although the SS-6 was unsuccessful as an ICBM (only four were operationally deployed), its direct decendants, in the form of the Soyuz and Molniya, are still in use as space launch vehicles today. With over 1,500 launches since 1957, the A-class launch vehicle family is by far the most frequently flown line of rockets in the world.

The original 'A' vehicle (or SL-1/-2) is classified as either a one and one-half or two stage vehicle. The vehicle consisted of a central core with four strap-on boosters. The Russians class this as a two stage assembly, but since the engines of the strap-ons and the central sustainer ignite simultaneously at liftoff, this parallel staging arrangement is generally described in the West as being a 1 1/2-stage configuration. Sergei Korolev's OKB-1 design bureau (which has since evolved into RSC Energia) was responsible for the development of the original 'A' vehicle. This 'A' vehicle (sometimes referred to as "Sputnik", after its first payload) launched the first three Sputnik satellites in 1957 and 1958. The third and largest Sputnik had a mass of 2,925 lb (1,327 kg) and was delivered to a 122 x 1,016 nm (226 x 1,881 km) orbit. All 'A' vehicle launches were from Tyuratam (Baikonur).

In 1958, Dmitri Kozlov, a close associate of Korolev, was placed in charge of establishing the A-class production facility in the city of Kuibyshev (now Samara), which lies about 550 mi (880 km) southeast of Moscow on the Volga River. On July 23, 1959, the A-class rocket Serial Production Design Department was officially established at the MZ "Progress" factory in Kuibyshev, and one year later it was incorporated into OKB-1 as its Kuibyshev branch office. By 1961, all design, engineering, flight testing, and operations responsibility for the the A-class boosters had been transferred from OKB-1 in Moscow to Kuibyshev. In 1974, this branch was split out into a separate organization and named Central Specialized Design Bureau (CSDB), with Dmitri Kozlov as chief designer. Today, CSDB, with Kozlov still

Soyuz / Molniya History

(continued)

Historical Summary

(continued)

serving as chief designer, reports directly to the Russian Space Agency. Production of the A-class boosters still takes place at the MZ Progress factory in Samara. Currently, MZ Progress production facilities are shared by RSC Energia and CSDB.

Over the years, the desire for increased launch vehicle capability led to the development of a series of more capable rockets based on the original 'A' vehicle design. The first step was to add a second core stage to the basic 'A' vehicle 1 1/2-stage configuration. Three failed launch attempts were made in 1958, with the first successful launch of a 2 1/2-stage vehicle taking place in 1959. This vehicle was designated A-1/SL-3 in the West; the Soviets assigned it the name Vostok (East). It consisted of an SS-6 core stage and strap-ons, with a second core stage attached by a truss structure. The initial application of the SL-3 vehicle was to launch the Luna moon payloads; it was later used to launch the manned Vostok capsule and first generation recoverable Kosmos satellites. Vostok maintained a steady launch rate of ~6/year, primarily high-inclination weather and ELINT payloads launched from Plesetsk, into the early 1980s. Use of the Vostok rapidly declined in the mid 1980s, and production of the vehicle was halted in 1985. On March 17, 1988, the first Soviet commercial launch occurred aboard a Vostok with an Indian remote sensing spacecraft, IRS-1A. The final Vostok flight carried the Indian IRS-1B into sun synchronous orbit from Baikonur on August 29, 1991. During its 30+ years of active use, Vostok underwent several systems upgrades, including a major redesign in 1964. Throughout its history, this vehicle was launched successfully 150 times out of 165 attempts.

The "Molniya" ("Lightning") launch vehicle (A-2-e, SL-6) was introduced in 1960; its first successful launch came in 1961. It dramatically increased the capability of the Soviet space program to launch high-energy planetary and lunar missions and place payloads into highly elliptical Earth orbits. Molniya was used to launch Luna 4 through 14, Venera 1 through 8, and the Soviet Union's first three Mars probes. Following the development of the much larger Proton (D-1-e, SL-12) in the mid 1960s, Molniya's planetary launch applications dwindled. The Molniya incorporates two core stages in addition to the 'A' vehicle core and strap-ons, resulting in a 3 1/2-stage vehicle. Molniya's second core stage is three times the size of Vostok's, and its main engine generates five times as much thrust. The small third core stage (designated Block L) is restartable on orbit, allowing for additional mission flexibility. Molniya was launched exclusively from Baikonur until 1970, when launch capability at Plesetsk was realized. The Molniya launch rate at Baikonur tapered off during the late 1980s, with the last Baikonur launch occurring in 1989. Today, this launch vehicle is operated exclusively from Plesetsk, supporting only the Molniya communication satellite and the Kosmos early warning satellite programs, which require highly elliptical, semisynchronous orbits. Through the end of 1994, 261 Molniya vehicles out of 293 launches successfully attained orbit.

The 2 1/2-stage A-2/SL-4, designated "Soyuz" ("Union") by the Soviets, premiered in 1963. This vehicle employs the same basic core first stage and strap-ons as Vostok and Molniya, and a core second stage very similar to that of Molniya. The larger core second stage provides a considerable increase over Vostok's payload capability to low Earth orbit. All Soviet and CIS manned missions have relied on the SL-4 since 1964, but the largest program it supports involves military and civilian recoverable photographic reconnaissance flights. Significant upgrades to the vehicle took place in 1973 and 1982. It is used today for launching Soyuz, Progress, and Biosats, as well as Kosmos observation satellites. It also launched the Voskhod manned vehicles. Over 1,000 SL-4 vehicles have been flown successfully from Tyuratam and Plesetsk. In December 1990, the first commercial launch of a private citizen, a Japanese journalist, took place using a Soyuz booster. Currently, 15–25 Soyuz vehicles are launched per year, down from a peak rate of 40–45 times per year in the 1970s and 1980s. Despite this drop in launch rate, the Soyuz is still the most frequently flown launch vehicle in the world. Its overall launch success rate, as of the end of 1994, is 96.8% (1,011 successful launches out of 1,044 attempts).

Although the A-class boosters have remained essentially unchanged in external appearance since their inception, their systems have been subjected to a constant stream of modifications over the years. In addition, subtle differences exist between superficially identical components. For example, the basic core first stage and strap-ons may vary slightly from vehicle to vehicle in such details as propellant loading and main engine thrust, depending on the vehicle version and mission application. Also, the second core stages used by Soyuz and Molniya, although very similar to one another in propellant capacity and engine performance, differ in that the control system for the upper stages resides on the second core stage for the Soyuz, and on the third core stage for Molniya. Soyuz can also employ a number of different payload fairings, depending on the mission at hand. One aspect of the A-class boosters that has remained unchanged is the exclusive use of LOX and kerosene propellants in all stages; however, this may change in the future with the advent of near-term vehicle upgrades.

A small number of A-class booster launches (15) using configurations classified as neither Sputnik, Vostok, Molniya, nor Soyuz were conducted in the 1960s and 1970s. Among these variants was the A-1-m/SL-5 vehicle, which was essentially a Vostok with a small additional stage.

The launch record of the A-class vehicles is unmatched, with a historical launch sucess rate of nearly 95% in more than 1,500 launches. Although more than half of the A-class failures occurred prior to 1968, information recently made available indicates that the A-class vehicles experienced an average of 1.5 launch failures per year in the 1980s. However, the high launch rates and relatively short times between failures and the next launch attempt suggest both a robust vehicle design and an effective system for failure investigation and resolution.

The Russian Space Agency (RSA) is reportedly funding the development of an upgraded A-class vehicle to replace the current Soyuz and Molniya boosters. The Russian name for this upgraded vehicle is Soyuz-2; the commonly reported name "Rus" actually refers to the overall development program, not the vehicle itself. The vehicle will presumably have a 2 1/2-stage variant for low Earth orbit applications and a 3 1/2-stage version for Molniya-type missions. Current plans include strap-on and core first stage propulsion upgrades, general modernization of the core second stage, a new flight control system, and larger payload fairings. It is also possible that the current Molniya third core stage will be replaced by the Fregat, a 6.5-metric ton upper stage derived from the Phobos spacecraft propulsion system and developed by the NPO Lavochkin organization. Soyuz-2 may fly as soon as 1996–1997. See *Notes* section for more information.

Soyuz / Molniya History

(continued)

Launch Record

Summary table covering 1957–1994 and bar graph are based on launch history data provided by the Central Specialized Design Bureau. Detailed table covering launches during 1991–1994 is based on TRW Space Log. Many discrepancies exist between Western sources and CSDB data for the years prior to 1991. In general, Western sources appear to underreport A-class vehicle failures. For example, TRW Space Log reports that the last A-class failure occurred in 1983; however, CSDB reports that nine failures took place from 1984–1994.

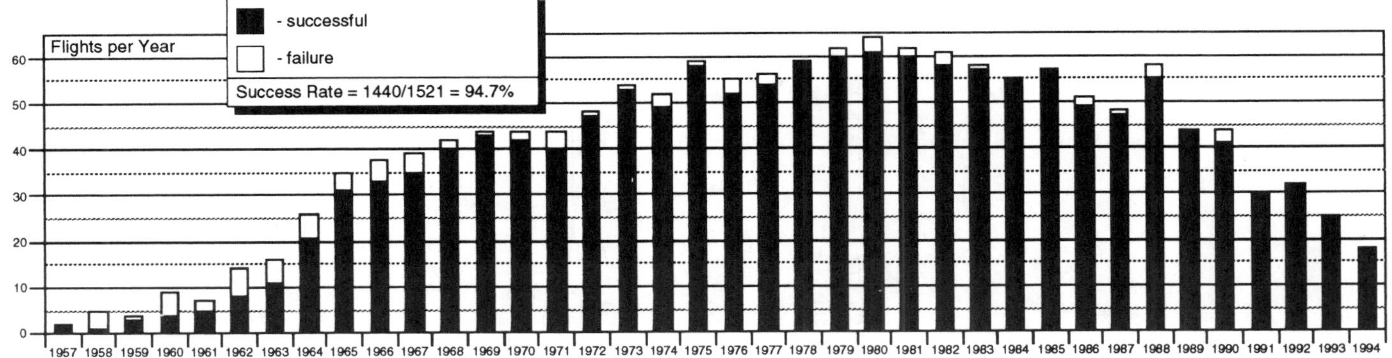

SEE TABLE BELOW FOR LAUNCH RECORD PRIOR TO 1991

	YEAR	DATE	VEHICLE	SITE	PAYLOAD
1417	1991	Jan-14	A-2	TT	Progress M-6
1418		Jan-17	A-2	PL	Kosmos 2121
1419		Feb-7	A-2	PL	Kosmos 2124
1420		Feb-15	A-2	TT	Kosmos 2134
1421		Feb-15	A-2-e	PL	Molniya 1-80
1422		Mar-6	A-2	TT	Kosmos 2136
1423		Mar-19	A-2	TT	Progress M-7
1424		Mar-22	A-2-e	PL	Molniya 3-40
1425		Mar-26	A-2	TT	Kosmos 2138
1426		May-18	A-2	TT	Soyuz TM-12
1427		May-21	A-2	PL	Resurs-F 10
1428		May-24	A-2	PL	Kosmos 2149
1429		May-30	A-2	TT	Progress M-8
1430		Jun-18	A-2-e	PL	Molniya 1-81
1431		Jun-28	A-2	PL	Kosmos 2018
1432		Jul-9	A-2	PL	Foton 2
1433		Jul-10	A-2	PL	Kosmos 2019
1434		Jul-23	A-2	TT	Kosmos 2020
1435		Aug-1	A-2	TT	Kosmos 2021
1436		Aug-20	A-2	PL	Resurs F-1, Pion-1,2
1437		Aug-21	A-2	PL	Kosmos 2025
1438		Aug-29	A-2-e	PL	Molniya 3-35
1439		Sep-17	A-2	TT	Kosmos 2028
1440		Sep-19	A-2	PL	Resurs F-2
1441		Oct-2	A-2	PL	Kosmos 2029
1442		Oct-4	A-2	PL	Kosmos 2030
1443		Oct-9	A-2	PL	Resurs F-3, Pion-3,4
1444		Oct-17	A-2	TT	Kosmos 2031
1445		Nov-20	A-2	PL	Kosmos 2032
1446		Dec-17	A-2	PL	Kosmos 2035
1447	1992	Jan-21	A-2	PL	Resurs F-3
1448		Jan-24	A-2	PL	Kosmos 2036
1449		Jan-25	A-2	TT	Progress M-1
1450		Mar-4	A-2	TT	Soyuz TM-8
1451		Mar-17	A-2	PL	Resurs F-4
1452		Apr-1	A-2	PL	Kosmos 2044
1453		Apr-8	A-2	TT	Kosmos 2045
1454		Apr-19	A-2-e	PL	Molniya 1-76
1455		Apr-29	A-2	PL	Kosmos 2047
1456		Apr-29	A-2	PL	Kosmos 2048
1457		May-28	A-2	PL	Kosmos 2049
1458		Jun-23	A-2-e	PL	Kosmos 2050
1459		Jun-30	A-2-e	PL	Molniya 3-36
1460		Jul-8	A-2	PL	Kosmos 2052
1461		Jul-24	A-2	TT	Progress M-2
1462		Jul-27	A-2	TT	Soyuz TM-15
1463		Jul-30	A-2	PL	Kosmos 2207
1464		Aug-6	A-2-e	PL	Molniya 1-84
1465		Aug-15	A-2	TT	Progress M-14
1466		Aug-19	A-2	PL	Resurs-F 16
1467		Sep-22	A-2	PL	Kosmos 2210
1468		Oct-8	A-2	PL	Foton 5
1469		Oct-14	A-2-e	PL	Molniya 3-42
1470		Oct-21	A-2-e	PL	Kosmos 2217
1471		Oct-27	A-2	TT	Progress M-15
1472		Nov-15	A-2	PL	Resurs 500
1473		Nov-20	A-2	PL	Kosmos 2220
1474		Nov-25	A-2-e	PL	Kosmos 2222
1475		Dec-2	A-2-e	PL	Molniya 3-43
1476		Dec-9	A-2	TT	Kosmos 2223
1477		Dec-22	A-2	TT	Kosmos 2225
1478		Dec-29	A-2	PL	Kosmos 2229
1479	1993	Jan-13	A-2-e	PL	Molniya 1-85
1480		Jan-19	A-2	PL	Kosmos 2231
1481		Jan-24	A-2	TT	Soyuz TM-16
1482		Jan-26	A-2-e	PL	Kosmos 2232
1483		Feb-21	A-2	TT	Progress M-16
1484		Mar-31	A-2	TT	Progress M-17
1485		Apr-2	A-2	PL	Kosmos 2240
1486		Apr-6	A-2-e	PL	Kosmos 2241
1487		Apr-21	A-2-e	PL	Molniya 3-44
1488		Apr-27	A-2	TT	Kosmos 2243
1489		May-21	A-2	PL	Resurs-F 17
1490		May-22	A-2	TT	Progress M-18
1491		May-26	A-2-e	PL	Molniya 1-86
1492		Jun-25	A-2	PL	Resurs-F 18
1493		Jul-1	A-2	TT	Soyuz TM-17
1494		Jul-14	A-2	PL	Kosmos 2259
1495		Jul-22	A-2	PL	Kosmos 2260
1496		Aug-4	A-2-e	PL	Molniya 3-45
1497		Aug-10	A-2-e	PL	Kosmos 2261
1498		Aug-10	A-2	TT	Progress M-19
1499		Aug-24	A-2	PL	Resurs-F 19
1500		Sep-7	A-2	TT	Kosmos 2262
1501		Oct-11	A-2	TT	Progress M-20
1502		Nov-5	A-2	TT	Kosmos 2267
1503		Dec-22	A-2-e	PL	Molniya 1-87
1504	1994	Jan-8	A-2	TT	Soyuz TM-18
1505		Jan-28	A-2	TT	Progress M-21
1506		Mar-17	A-2	PL	Kosmos 2274
1507		Mar-22	A-2	TT	Progress M-22
1508		Apr-28	A-2	TT	Kosmos 2280
1509		May-22	A-2	TT	Progress M-23
1510		Jun-7	A-2	PL	Kosmos 2281
1511		Jun-14	A-2	PL	Foton-9
1512		Jul-1	A-2	TT	Soyuz TM-19
1513		Jul-20	A-2	PL	Kosmos 2283
1514		Jul-29	A-2	TT	Kosmos 2284
1515		Aug-5	A-2-e	PL	Kosmos 2286
1516		Aug-23	A-2-e	PL	Molniya 3-46
1517		Aug-25	A-2	TT	Progress M-24
1518		Oct-3	A-2	TT	Soyuz TM-20
1519		Nov-11	A-2	TT	Progress M-25
1520		Dec-14	A-2-e	PL	Molniya 1-88
1521		Dec-29	A-2	TT	Kosmos 2305

PL is Plesetsk Cosmodrome
TT is Tyuratam (Baikonur)

YEAR	FLIGHTS BY VEHICLE (Successes / Failures)										TOTAL	
	Sputnik / A		Vostok / A-1		Soyuz / A-2		Molniya / A-2-e		Other 'A' Variants		Successes	Failures
	TT	PL	TT	PL	TT	PL	TT	PL	TT	PL		
1957	2 / 0	–	–	–	–	–	–	–	–	–	2	0
1958	1 / 1	–	0 / 3	–	–	–	–	–	–	–	1	4
1959	–	–	3 / 1	–	–	–	–	–	–	–	3	1
1960	–	–	4 / 3	–	–	–	0 / 2	–	–	–	4	5
1961	–	–	4 / 1	–	–	–	1 / 1	–	–	–	5	2
1962	–	–	7 / 1	–	–	–	1 / 5	–	–	–	8	6
1963	–	–	8 / 2	–	1 / 0	–	1 / 3	–	1 / 0	–	11	5
1964	–	–	12 / 0	–	5 / 0	–	3 / 5	–	1 / 0	–	21	5
1965	–	–	9 / 1	–	12 / 0	–	9 / 3	–	1 / 0	–	31	4
1966	–	–	8 / 1	3 / 0	11 / 1	3 / 1	7 / 2	–	1 / 0	–	33	5
1967	–	–	2 / 0	6 / 0	10 / 1	11 / 2	6 / 1	–	–	–	35	4
1968	–	–	–	2 / 0	19 / 0	14 / 1	5 / 1	–	–	–	40	2
1969	–	–	–	2 / 1	18 / 0	19 / 0	4 / 0	–	–	–	43	1
1970	–	–	–	5 / 0	14 / 0	16 / 1	2 / 1	4 / 0	1 / 0	–	42	2
1971	–	–	–	5 / 0	15 / 1	14 / 3	3 / 0	–	3 / 0	–	40	4
1972	–	–	–	5 / 0	13 / 0	16 / 1	5 / 0	6 / 0	2 / 0	–	47	1
1973	–	–	–	3 / 0	12 / 0	26 / 1	3 / 0	7 / 0	2 / 0	–	53	1
1974	–	–	–	6 / 0	16 / 1	18 / 2	–	7 / 0	2 / 0	–	49	3
1975	–	–	–	6 / 0	17 / 1	23 / 0	1 / 0	11 / 0	–	–	58	1
1976	–	–	–	5 / 0	17 / 0	20 / 1	4 / 0	5 / 2	1 / 0	–	52	3
1977	–	–	2 / 0	5 / 0	13 / 2	24 / 0	2 / 0	8 / 0	–	–	54	2
1978	–	–	–	5 / 0	18 / 0	27 / 0	1 / 0	8 / 0	–	–	59	0
1979	–	–	1 / 0	7 / 0	10 / 0	35 / 2	–	7 / 0	–	–	60	2
1980	–	–	0 / 1	6 / 0	13 / 0	32 / 0	2 / 0	8 / 2	–	–	61	3
1981	–	–	1 / 0	5 / 0	21 / 1	20 / 0	1 / 0	12 / 1	–	–	60	2
1982	–	–	–	5 / 0	23 / 1	20 / 1	1 / 1	9 / 0	–	–	58	3
1983	–	–	–	4 / 0	14 / 1	28 / 0	3 / 0	8 / 0	–	–	57	1
1984	–	–	–	–	16 / 0	28 / 0	–	11 / 0	–	–	55	0
1985	–	–	1 / 0	–	14 / 0	26 / 0	2 / 0	14 / 0	–	–	57	0
1986	–	–	–	–	20 / 1	16 / 0	2 / 0	11 / 1	–	–	49	2
1987	–	–	–	–	23 / 0	20 / 1	–	4 / 0	–	–	47	1
1988	–	–	2 / 0	–	22 / 2	20 / 1	1 / 0	10 / 0	–	–	55	3
1989	–	–	–	–	14 / 0	24 / 0	1 / 0	5 / 0	–	–	44	0
1990	–	–	–	–	12 / 0	18 / 2	–	11 / 1	–	–	41	3
1991	–	–	1 / 0	–	11 / 0	13 / 0	–	5 / 0	–	–	30	0
1992	–	–	–	–	11 / 0	13 / 0	–	8 / 0	–	–	32	0
1993	–	–	–	–	10 / 0	7 / 0	–	8 / 0	–	–	25	0
1994	–	–	–	–	11 / 0	4 / 0	–	3 / 0	–	–	18	0
TOTAL	3 / 1	–	65 / 14	85 / 1	456 / 13	555 / 20	71 / 25	190 / 7	15 / 0	–	1440	81

Data Sources: TRW Space Log, Central Specialized Design Bureau

Soyuz / Molniya General Description

Soyuz (SL-4) **Molniya (SL-6)**

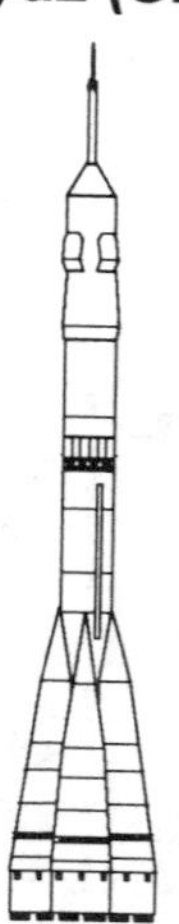

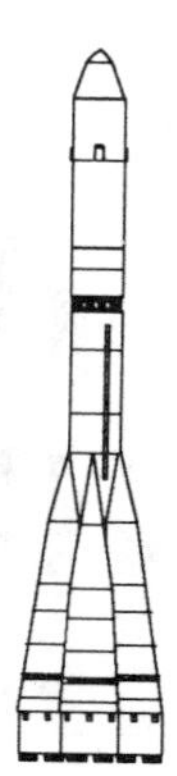

Summary

In the early days of the Soviet space program, numerous precursors to the current Soyuz and Molniya vehicles were identified by various names by the Soviets. The first major upgrade to the original A-class rocket was the 2 1/2-stage SL-3. This vehicle first flew successfully in 1959; it was assigned the name Vostok (East) in 1962. Vostok last flew in 1991. In 1960, the 3 1/2-stage SL-6 vehicle was introduced and almost immediately christened Molniya (Lightning). The 2 1/2-stage SL-4 was first launched in 1963. This vehicle was essentially a Molniya without the third core stage; its larger core second stage allowed it to handle heavier payloads than Vostok. Beginning in 1966, vehicles of this type were designated Soyuz (Union).

Status

Operational

Key Organizations

Users - Russian Space Agency, Russian Ministry of Defense, Russian Academy of Sciences, commercial

Responsible Organizations -
- Central Specialized Design Bureau (Manufacturer)
- NPO Energomash (Core Stage 1 and Strap-on Engines)
- KB Khimavtomatiki (Stage 3 Engine)

Vehicle

System Height	166 ft (50.7 m) for Soyuz with escape tower 142 ft (43.4 m) for Molniya
Payload Fairing Size (usable volume)	Soyuz - 7.7 ft (2.35 m) diameter by 29.5 ft (9.0 m height Molniya - 7.5 ft (2.3 m) diameter by 12.1 ft (3.7 m) height
Gross Mass	683,000 lb (310,000 kg) for Soyuz 672,000 lb (305,000 kg) for Molniya
Planned Enhancements	Soyuz-2 (see *Notes* section)

Operations

Primary Missions	Soyuz - LEO Molniya - highly elliptical, sun synchronous
Compatible Upper Stages	A-2-e Molniya third stage
First Launch	1961 for SL-6 (Molniya type), 1963 for SL-4 (Soyuz type)
Success / Flight Total	Through 1994, manufacturer states 725 / 744 for Soyuz, 241 / 253 for Molniya
Launch Site	Baikonur Cosmodrome (45.6°N, 63.4°E) with at least two A-class pads available. Plesetsk Cosmodrome (62.8°N, 40.1°E)
Launch Azimuth	Baikonur: 62°, 37°, 29° (corresponding to inclinations of 52°, 65°, 70°) Plesetsk: 90°, 59°, 40°, 18° (corresponding to inclinations of 63°, 67°, 73°, 82°)
Nominal Flight Rate	15–25/yr for Soyuz, and 5–10/yr for Molniya
Planned Enhancements	None

Performance

108 nm (200 km) circ, 51.6°	15,400 lb (7,000 kg) for Soyuz
443 nm (820 km) circ, 99°	3,970 lb (1,800 kg) for Molniya
'Molniya' orbit with apogee of 21600 nm (36000 km), 63°	4,400 lb (2,000 kg) for Molniya
Geotransfer Orbit, 51.6°	—
Geosynchronous Orbit	—

Financial Status

Estimated Launch Price	Soyuz - Molniya $12–25M (U.S. DoT estimate)

Manifest Not available

Remarks —

Soyuz / Molniya Vehicle

Overall

Soyuz

Length	166.2 ft (50.67 m)
Gross Mass	683,000 lb (310,000 kg)
Thrust at Liftoff	906,450 lb (4,031,900 N)

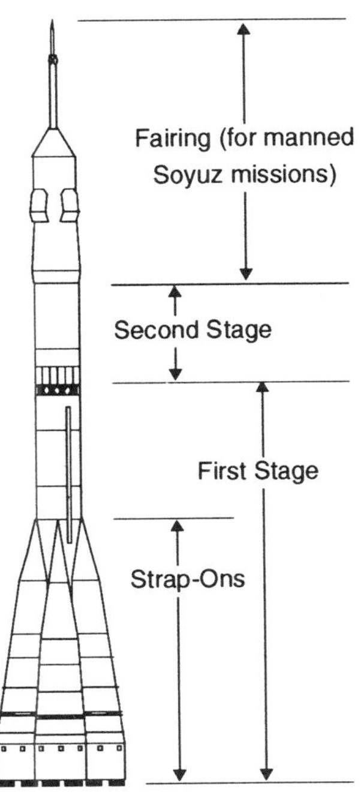

View of 20 Main and 12 Vernier Chambers on Base of Vehicle

Molniya

Length	142.5 ft (43.44 m)
Gross Mass	672,000 lb (305,000 kg)
Thrust at Liftoff	906,450 lb (4,031,900 N)

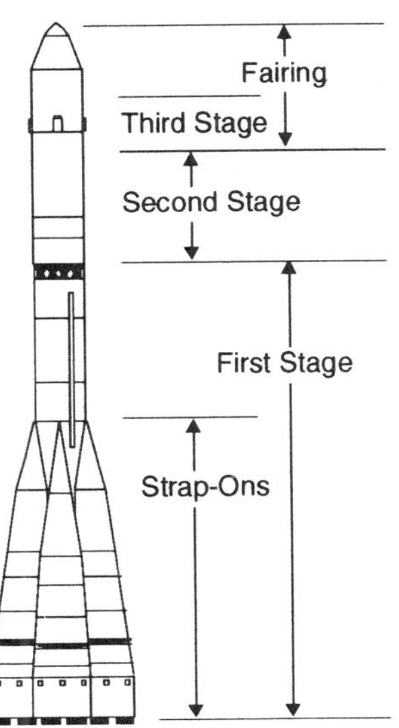

Launch of A-class Rocket

Photos Courtesy of Space Commerce Corp.

Soyuz / Molniya Vehicle

(continued)

Stages

	Strap-Ons	Core Stage 1	Core Stage 2	Stage 3 for Molniya
Dimension:				
Length	64.3 ft (19.6 m)	91.2 ft (27.8 m)	22.1 ft (6.74 m)	8.7 ft (2.64 m)
Diameter	8.8 ft (2.68 m) at the base	7.1-9.7 ft (2.15-2.95 m)	8.7 ft (2.66 m)	7.9 ft (2.41 m)
Mass: (each)				
Propellant Mass	86.4K lb (39.2K kg), variable	208K lb (94.5K kg), variable	50.7K lb (23.0K kg), variable	7.6K lb (3.45K kg)
Gross Mass	94.2K lb (42.75K kg), variable	223K lb (101.0K kg), variable	56.0K lb (25.4K kg), variable	9.9K lb (4.5K kg)
Structure:				
Type	Skin-Stringer	Skin-Stringer	Skin-Stringer	??
Material	Aluminum	Aluminum	Aluminum	Aluminum
Propulsion:				
Propellant	LOX / kerosene	LOX / kerosene	LOX / kerosene	LOX / kerosene
Average Thrust (each)	182.8K lb (813.2 kN) SL* 223.0K lb (991.7 kN) vac*	175.1K lb (778.9 kN) SL* 224.1K lb (997.1 kN) vac*	67.0K lb (298.2 kN) vac*	15.0K lb (66.7 kN) vac
Engine Designation	RD-107	RD-108	RD-461	RD-??
Number of Engines	1 pump + 4 chambers + 2 verniers per strap-on	1 pump + 4 chambers + 4 verniers	1 pump + 4 chambers + 4 verniers	1
Isp	257 sec SL 314 sec vac	248 sec SL 315 sec vac	330 sec vac	340 sec vac
Feed System	Gas generator	Gas generator	Gas generator	Gas generator
Chamber Pressure	848 psia (58.5 bar)	740 psia (51.0 bar)	?? psia (?? bar)	?? psia (?? bar)
Mixture Ratio (O/F)	2.47	2.39	??	??
Throttling Capability	100% only	100% only	100% only ?	??
Expansion Ratio	??	??	??	??
Restart Capability	No	No	??	??
Tank Pressurization	??	??	??	??
Control-Pitch, Yaw, Roll	aero fins & 8 verniers plus stage 1 verniers	4 verniers	4 verniers	3-axis ACS system
Events:				
Nominal Burn Time	118 sec	286 sec	230–250 sec	180 sec
Stage Shutdown	Burn to depletion	Burn to depletion	Command shutdown	Command shutdown
Stage Separation	??	??	Spring ejection	Spring ejection

* - thrust includes verniers

Remarks:

The rocket engines for the strap-ons and first core stage are the RD-107 and the RD-108, respectively. Both engines were developed between 1954 and 1957 by what is now NPO Energomash and use LOX and kerosene for propellants. At liftoff, the A-class vehicle has 20 main and 12 vernier chambers in action simultaneously. The propellant loading in the strap-ons and core can vary, depending on mission requirements. Values given above represent maximum capacity. The core second stage is attached to the top of the central sustainer core by an open truss structure. The Soyuz and Molniya version of this stage are virtually identical except for the avionics system (see below). Both use a single engine with four thrust chambers and four swiveling verniers. The third stage atop the Molniya has one engine and is enclosed in the payload fairing. The Molniya third stage, designated "Block L", burns two solid propellant settling motors prior to main engine ignition on orbit. The masses in the table above do not include these settling motors. Soyuz and Molniya also carry 16,000 lb (7,300 kg) of hydrogen peroxide (to power their engines' gas generators) and 3,500 lb (1,600 kg) of nitrogen. These masses are not included in the table above.

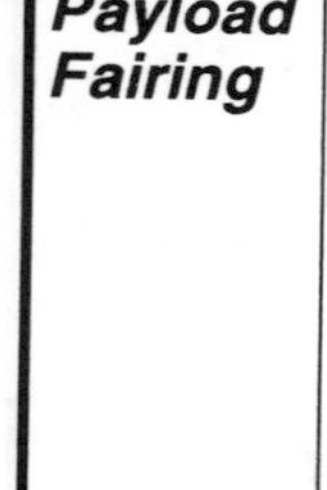

Payload Fairing

in inches (mm)

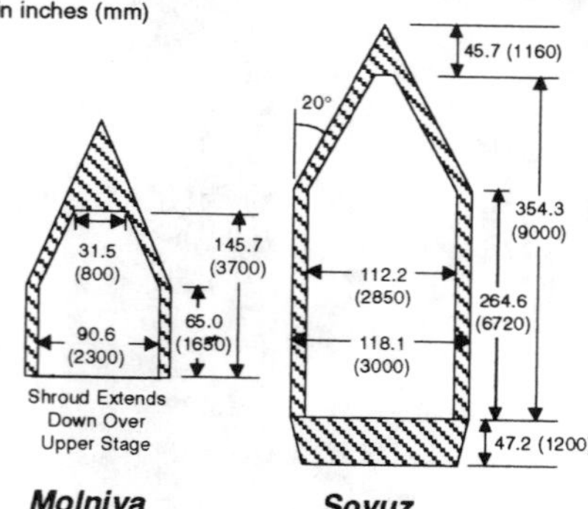

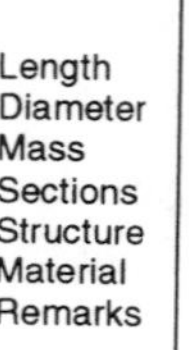

	Molniya	Soyuz
Length	25.6 ft (7.8 m)	37.3 ft (11.36 m)
Diameter	8.9 ft (2.7 m)	9.8 ft (3.00 m)
Mass	?? lb (?? kg)	9,900 lb (4,500 kg)
Sections	2	2
Structure	Skin-stringer ??	Skin-stringer ??
Material	Aluminum	Aluminum
Remarks	——	Used for Progress spacecraft

Note: At least three other fairings are used on Soyuz:

- 3.0 m external diameter, 16.8 m long (including escape system); for manned Soyuz spacecraft missions
- 2.7 m external diameter, 9.0 m long; for Bion, Foton, and Resurs-F spacecraft
- 3.3 m external diameter, 9.5 m long; for Kosmos-type spacecraft

Avionics

Two avionics suites are used. One system is located on core stage 1 and controls that stage and the strap-ons. A second avionics package, which controls the upper stage(s), resides on either core stage 2 (Soyuz) or the core stage 3 (Molniya).

Attitude Control System

The four strap-ons and core first stage burn in parallel and the vehicle is controlled by the two gimbaling verniers on each strap-on and four verniers on the core. The core second stage is controlled by four verniers. Molniya's Block L core stage 3 uses a 3-axis attitude control system.

Soyuz / Molniya Performance

Performance Characteristics of Soyuz and Molniya:

Generic performance curves are unavailable for these vehicles. General payload capability for Soyuz and Molniya is provided below; this information is based on historical data and/or manufacturer's information, as indicated. Performance listed in the *General Description* section was provided by CSDB and may not agree with historical data. All possible mission types may not be represented in the data given here. Historical data may not represent maximum performance. All inclinations are approximate and may vary by several tenths of a degree from flight to flight.

Soyuz

Orbits Accessible from Baikonur (from CSDB):

Inclinations:	52°, 65°, 70°
Perigee altitudes:	103 to 135 nm (190 to 250 km)
Apogee altitudes:	135 to 540 nm (250 to 1,000 km)

Orbits Accessible from Plesetsk (from CSDB):

Inclinations:	63°, 67°, 73°, 82°
Perigee altitudes:	97 to 135 nm (180 to 250 km)
Apogee altitudes:	135 to 540 nm (250 to 1,000 km)

Performance (from historical data):

Baikonur		
	108–162 nm (~200–300 km) altitude, 51.6°:	7,250 kg (16,000 lb)
	108–162 nm (~200–300 km) altitude, 65°:	7,000 kg (15,400 lb)
	108–162 nm (~200–300 km) altitude, 70°:	6,500 kg (14,300 lb)
Plesetsk		
	108–162 nm (~200–300 km) altitude, 63°:	6,500 kg (14,300 lb)
	108–162 nm (~200–300 km) altitude, 67°:	6,500 kg (14,300 lb)
	108–162 nm (~200–300 km) altitude, 73°:	6,300 kg (13,900 lb)
	108–162 nm (~200–300 km) altitude, 82°:	6,300 kg (13,900 lb)

Molniya

Performance (from historical data except where noted):

The Molniya launch vehicle can deliver up to 4,400 lb, (2,000 kg) to Molniya-type orbits, with perigees of 216–486 nm (400–4,000 km), apogees of 21,300–21,600 nm (39,000–40,000 km), and inclinations of 63°–68°, from either Baikonur or Plesetsk. However, since the late 1980s, all Molniya launches have originated from Plesetsk.

Based on the masses of the Venera, Luna, and Mars spacecraft, Molniya is capable of delivering 2,650 lb (1,200 kg) onto a Venus escape trajectory, 3,530 lb (1,600 kg) onto a lunar trajectory, and 1,980 lb (900 kg) to Mars.

Information from CSDB indicates that Molniya can place 4,000 lb (1,800 kg) into a 443 nm (820 km) circular, sun synchronous orbit at 99° inclination, launched from either Baikonur or Plesetsk. As of January 1995, no such mission had ever been flown.

Soyuz / Molniya Operations

Launch Site

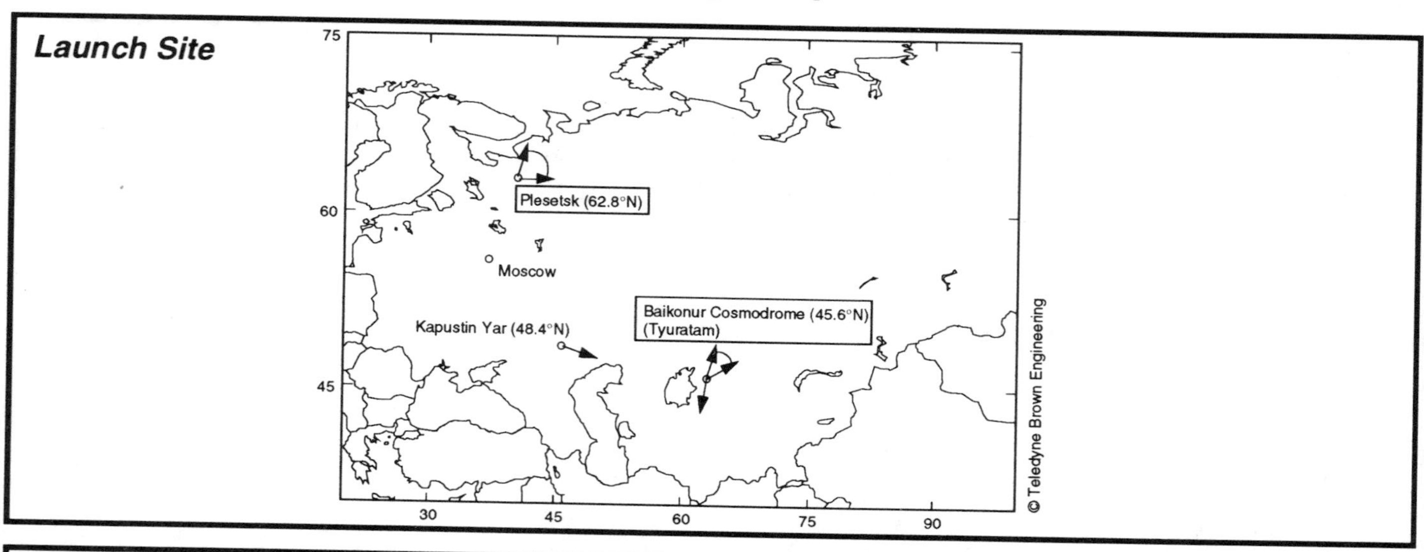

Launch Facilities

Delivery of Vehicle to Assembly Building in Stages

Horizontal Integration of Stages

View of Soyuz Vehicle on Launch Pad

Photos Courtesy of Space Commerce Corp.

Soyuz / Molniya Operations

(continued)

Launch Processing

Launch Sites. Two sites are used by the A-class vehicles: Baikonur Cosmodrome (i.e., Tyuratam) and Plesetsk Cosmodrome. Molniya has been launched exclusively from Plesetsk since 1989, while Soyuz launches are split fairly evenly between the two sites. Baikonur Cosmodrome, located in Kazakhstan, is where the first launches of Soviet artificial Earth satellites took place. This, the largest and most versatile of the Russian launch sites, is near the rail stop village of Tyuratam, the name commonly given to the launch site by Western analysts. Tyuratam may be considered to be analogous to Cape Canaveral and the Eastern Range in the United States. All launches associated with piloted spaceflight, lunar, planetary, and geosynchronous spacecraft originate from Tyuratam. At least two operational launch pads are available for Soyuz.

Since the breakup of the Soviet Union, Russia has been in the difficult situation of having its largest space launch facility located in a foreign country (i.e., Kazakhstan). Despite political turmoil, the Baikonur Cosmodrome has remained in operation virtually without interruption. In the summer of 1994, Russia and Kazakhstan signed a long-term agreement whereby Russia would lease Baikonur from Kazakhstan, with day-to-day operations of the cosmodrome remaining under the control of the Russian Space Forces.

The Plesetsk Cosmodrome is situated in a heavily wooded area close to the Arctic Circle near the town of Plesetsk, which is on the railway line from Moscow to Archangelsk. Historically, it is the world's most used launch site. Although used for some civilian communications, meteorological, and international scientific launches, most satellites launched from Plesetsk have military roles and it may be considered as the Russian equivalent of Vandenberg Air Force Base. Orbital inclinations attained from Plesetsk range from 62° to 83°; to date, direct launches into polar and sun synchronous launches have not been conducted from Plesetsk. Plesetsk was first officially acknowledged by the Soviet Union in 1983 after more than 900 launches. Construction on the first of several launch pads for the A-class launch vehicle began at Plesetsk in 1957 to support the Soviet ICBM program and was completed in December 1959. Nevertheless, the first space launch did not occur until May 1966. Five of the eleven CIS launch systems are operated from Plesetsk, with the A-class boosters being the largest.

Processing. Spacecraft and launch vehicle assembly and prelaunch processing for Soyuz and Molniya follow the same general procedure. The processing flow for the Baikonur Cosmodrome is outlined below; a similar procedure is probably employed at the Plesetsk site.

After arrival at the cosmodrome, the payload undergoes checkout and fueling. It is then transported to the Space Vehicle Assembly Building to be integrated with the launch vehicle, the components of which have already been transported to the cosmodrome by rail. The integration process can follow one of two approaches, depending on payload requirements. In one option, the payload, adapter, interstage, and, for Molniya, core stage 3, are integrated vertically. The two halves of the payload fairing are then attached to either the interstage (Soyuz) or core stage 3 (Molniya) and connected together. This assembly (referred to as the "head block" in Russian terminology) is then rotated to a horizontal position, moved to a transporter/erector railroad car, and integrated horizontally with the rest of the launch vehicle. In the other option, the payload, interstage, core stage 3 (Molniya), and core stage 2 are integrated vertically, after which the assembly is rotated to a horizontal position and the complete payload fairing is rolled into place over the payload. This stack is then transferred to a transporter/erector car and attached to the previously assembled core stage 1 and strap-ons. Throughout this process, temperatures within the MIK are maintained between 5° and 35° C. Once the payload is integrated with the launch vehicle, temperatures within the payload fairing can be maintained between 15° and 25° C.

Following integrated testing in the MIK, the launch vehicle is transported horizontally to the launch pad. The maximum speed during the 3.1 mi (5 km) rail trip is 10 km/hr. Upon arrival at the pad, the vehicle is raised into a vertical position and installed onto the pad. During the erection process, payload thermal control is interrupted for not more than 3.5 hours.

The launch pad consists of a fixed rectangular platform, in the center of which is a square pit. Once in place on the pad, the vehicle rests inside this pit on a rotating structure, with its aft end approximately 23 ft (7 m) below the level of the launch platform. The rotating structure allows the vehicle to be aligned to the proper flight azimuth prior to launch. Attached to the rotating structure are four supporting trusses, along with service towers and umbilical masts. The support trusses act to secure the vehicle in position prior to launch; they are held in place by the weight of the vehicle itself. The entire launch pad structure is suspended over a large flame trench.

Direct access to the launch vehicle (above the level of the launch platform) and payload is gained through service towers positioned around the vehicle; these are withdrawn 50 minutes before launch. Retractable platforms permit access to the launch vehicle below the level of the launch platform until one hour prior to launch. Payload fairing environmental control is halted 40 minutes before launch; by launch time, the temperature inside the fairing can vary from -35° to +50° C. An umbilical mast, supplying electrical, pneumatic, and hydraulic connections to core stages 2 and (for Molniya) 3, as well as the payload, is detached 10 minutes prior to ignition, at which time these vehicle elements are switched to internal power. A cable mast connected to the control system of core stage 1 and the strap-ons is disconnected at ignition. After 2–3 in (~5–7 cm) of vehicle vertical motion, the support trusses retract, under the influence of large counterweights, like petals to allow the vehicle to clear the pad.

The entire processing flow, from launch vehicle and payload arrival at the cosmodrome until launch, requires 121 hours, including 18 hours for on-pad activities. The A-class vehicles are able to perform under severe weather conditions, including dense fog, wind, rain, snow, and the wide temperature variations experienced at Baikonur (-40º–50ºC). For example, in March 1988, a Vostok launcher carrying an Indian satellite lifted off the pad at Tyuratam on schedule in the midst of a driving snowstorm.

Flight Sequence

A representative flight sequence for a Soyuz mission to low Earth orbit is provided at right. Event times may vary due to the specifics of the mission at hand. Molniya's flight profile is probably very similar to that of Soyuz through low Earth orbit insertion. Following core stage 2 jettison in park orbit, Molniya's Block L can coast under 3-axes stabilization for up to one hour before igniting its main engine to place the payload in its final orbit.

Typical Flight Sequence: Soyuz LEO Mission

MIN:SEC	
0:00	Liftoff
1:58	Strap-ons separate
2:30-2:50	Payload fairing jettison
4:46	Core stage 1 jettison; core stage 2 ignition
8:40-9:00	Orbital insertion

Soyuz / Molniya Payload Accommodations

Payload Envelope	
Maximum Diameter	Soyuz - 112.2 in (2850 mm) Molniya - 104.3 in (2650 mm)
Maximum Cylinder Length	Soyuz - 264.6 in (6720 mm) Molniya - 65.0 in (1650 mm)
Maximum Cone Length	Soyuz - 89.7 in (2280 mm) Molniya - 80.7 in (2050 mm)
	Note: Soyuz payload envelopes apply to Progress spacecraft fairing. At least 3 other fairings are used on Soyuz (see *Vehicle* section)
Payload Adapter	
Interface Diameter	?? in (?? mm)
Payload Integration	
Nominal Mission Schedule Begins	T-12 months
Launch Window	
Latest Countdown Hold Not Requiring Recycling	T-?? min
On-Pad Storage Capability	27 hrs for a fueled vehicle
Latest Access to Payload	T-50 min (service towers withdrawn) Ground equipment disconnected at T-10 min
Environment	
Maximum Load Factors	+4.8 g axial, ±1.2 g lateral
Minimum Lateral/Longitudinal Payload Frequency	15 Hz /25 Hz
Maximum Overall Acoustic Level	142 dB rms
Maximum Flight Shock	350 g at 1,000+ hz
Maximum Dynamic Pressure at Fairing Jettison	5.0 lb/ft^2 (240 N/m^2)
Maximum Pressure Change in Fairing	0.3 psi/s (2.0 KPa/s)
Cleanliness Level in Fairing (Prior to Launch)	Class ??
Payload Delivery	
Orbit Insertion Accuracy (3 sigma)	±22 s (period), ±0.1 deg (inclination)
Attitude Accuracy (3 sigma)	All axes: ±?? deg, ±3.5 deg/sec for Soyuz Longitudinal axis: ±?? deg, ± 1.1 deg/sec; other axes: ±?? deg, ± 9 deg/sec for Molniya
Nominal Payload Separation Rate	3.3 ft/s (1 m/s) for Soyuz, 5.6 ft/s (1.7 m/s) for Molniya
Deployment Rotation Rate Available	?? rpm
Loiter Duration in Orbit	1 hr for Molniya
Maneuvers (Thermal / Collision / Telemetry)	collision avoidance following separation

Publications

Technical Publications

"Soviet Commercial Launch Vehicles", A. Dula, Space Commerce Corp., B. Gubanov, NPO Energia, Y. Smetanin, Glavkosmos, AIAA-89-2743, AIAA 25th Joint Propulsion Conference, Monterey, CA, July 10-12, 1989.

TRW Space Log, 1957-1991, 1992, 1993 TRW, El Segundo, CA.

The Soviet Year in Space 1986, 1988, 1989, Nicholas Johnson, Teledyne Brown Engineering, Colorado Springs, CO.

Launch Vehicle Catalogue, European Space Agency, December 1989.

"Opportunities from Soviet Space Industry, A Commercial User's Guide", A. Dula, Space Commerce Corp., A.I. Dunayev, Glavkosmos, Paris Air Show 1989.

Soviet Space Programs: 1981-1987, United States Senate, Committee on Commerce, Science, and Transportation - Part 1 - May 1988.

Jane's Spaceflight Directory 1987, Jane's Publishing Inc., New York, NY.

Jane's Space Directory 1994-95, Jane's Information Group Inc., Alexandria, VA.

"Soviet Launch Vehicle Designations", Ralph Gibbons, Spaceflight, pg. 54-60,80, February 19, 1977.

Space Directory of Russia, February 1993 Edition, Euroconsult/Sevig Press, Paris, France.

The Russian Space Directory 1994, European Space Report, Munich, Germany.

Acronyms

ACS - Attitude Control System
CIS - Commonwealth of Independent States
CSDB - Central Specialized Design Bureau
GEO - geosynchronous orbit
GTO - geosynchronous transfer orbit
KB - Design Bureau
LEO - low Earth orbit
LOX - liquid oxygen
MIK - Space Vehicle Assembly Building
MZ - Machine-building Factory
NPO - Scientific Production Organization
RSC - Rocket Space Corporation
USSR - Union of Soviet Socialist Republics

Other Notes—A-Class Growth Possibilities

An upgraded A-class vehicle designated Soyuz-2 is reportedly under development, with first flight scheduled for the 1996–1997 time frame. It appears that two versions of the Soyuz-2 will be developed: a 2 1/2-stage SL-4 upgrade, and a 3 1/2-stage SL-6 upgrade. The term Rus, commonly reported as the name of the upgraded A-class launch vehicle, actually refers to the upgrade program, not the vehicle itself.

The Soyuz-2 will preserve the basic A-class configuration. The strap-ons and core first stage will remain externally unchanged, although the venerable RD-107 and RD-108 engines reportedly will be modified. The core second stage is to be substantially upgraded with the incorporation of a modern Zenit-based guidance system for improved accuracy. Molniya's Block L core stage 3 will be replaced by the slightly larger Fregat stage, developed by NPO Lavochkin as the propulsion module of the Phobos Mars spacecraft. If it is incorporated into the Soyuz-2 system, the UDMH/N2O4 powered Fregat will be the first A-class stage to use propellants other than LOX and kerosene. Larger payload fairings, based on NPO Lavochkin's Proton SL-12 fairing designs, will be available, with payload envelope diameters of 12.1 ft (3.7 m) and 13.5 ft (4.1 m).

Estimates of the performance of Soyuz-2 are given below:

Low altitude, 51.6° performance: 17,600 lb (8,000 kg)
900 km, 99° sun synchronous orbit performance: 11,000 lb (5,000 kg)
Geosynchronous transfer orbit (200 x 36,000 km, 51.6°) performance: 6,000 lb (2,700 kg)
Lunar transfer orbit performance: 3,700 lb (1,700 kg)
Mars / Venus transfer orbit performance: 2,600 lb (1,200 kg)

Start

Launch Services Point of Contact:
STC Complex
Berjozovaja Ave., 10/1
Moscow
Russia, 127276
Phone: (7095) 402-7321
Fax: (7095) 402-8229

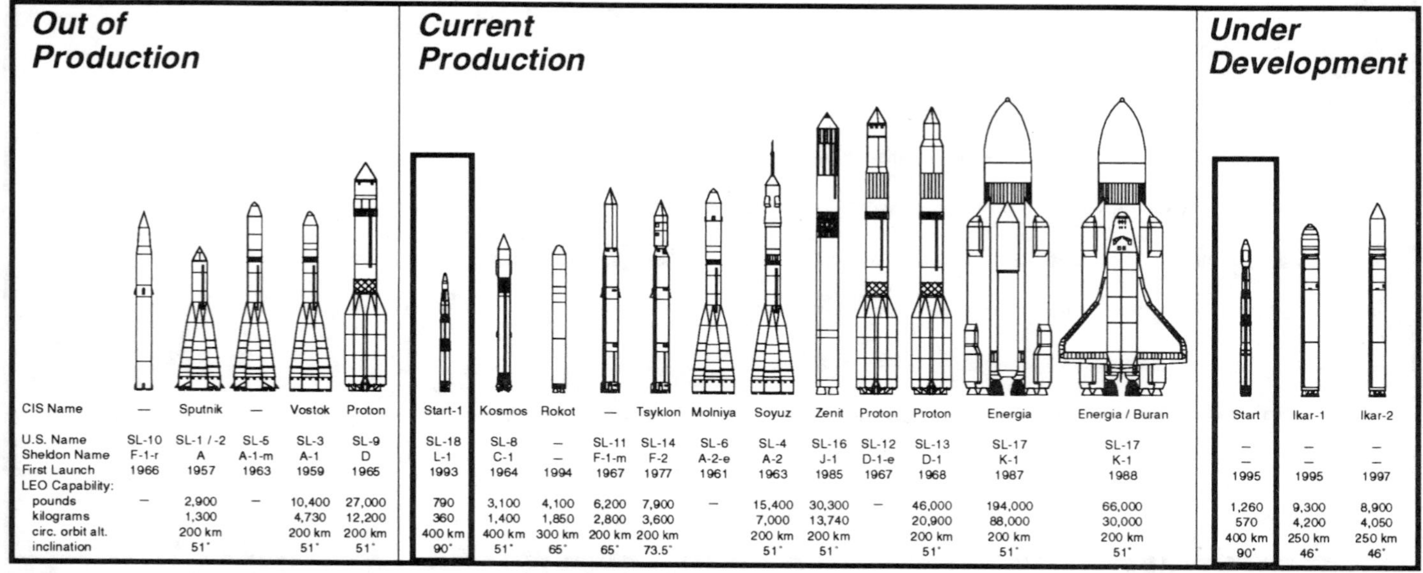

Out of Production

CIS Name	—	Sputnik	—	Vostok	Proton
U.S. Name	SL-10	SL-1 / -2	SL-5	SL-3	SL-9
Sheldon Name	F-1-r	A	A-1-m	A-1	D
First Launch	1966	1957	1963	1959	1965
LEO Capability:					
pounds	—	2,900	—	10,400	27,000
kilograms		1,300		4,730	12,200
circ. orbit alt.		200 km		200 km	200 km
inclination		51°		51°	51°

Current Production

CIS Name	Start-1	Kosmos	Rokot	—	Tsyklon	Molniya	Soyuz	Zenit	Proton	Proton	Energia	Energia / Buran
U.S. Name	SL-18	SL-8	—	SL-11	SL-14	SL-6	SL-4	SL-16	SL-12	SL-13	SL-17	SL-17
Sheldon Name	L-1	C-1	—	F-1-m	F-2	A-2-e	A-2	J-1	D-1-e	D-1	K-1	K-1
First Launch	1993	1964	1994	1967	1977	1961	1963	1985	1967	1968	1987	1988
LEO Capability:												
pounds	790	3,100	4,100	6,200	7,900	—	15,400	30,300	—	46,000	194,000	66,000
kilograms	360	1,400	1,850	2,800	3,600		7,000	13,740		20,900	88,000	30,000
circ. orbit alt.	400 km	400 km	300 km	200 km	200 km		200 km	200 km		200 km	200 km	200 km
inclination	90°	51°	65°	65°	73.5°		51°	51°		51°	51°	51°

Under Development

CIS Name	Start	Ikar-1	Ikar-2
U.S. Name	—	—	—
Sheldon Name	—	—	—
First Launch	1995	1995	1997
LEO Capability:			
pounds	1,260	9,300	8,900
kilograms	570	4,200	4,050
circ. orbit alt.	400 km	250 km	250 km
inclination	90°	46°	46°

Vehicle Description

CIS vehicles are identified either by their CIS, U.S., or Sheldon name. In the Soviet Union it was standard practice to name a launch vehicle after its original payload (e.g., Kosmos, Proton). The U.S. names (developed by the U.S. Department of Defense) are alphanumeric designators based roughly on chronological appearance. The Sheldon names, a commonly used system that was published by Dr. Charles Sheldon of the U.S. Library of Congress in 1968, emphasize the basic families of launch vehicles with special indicators for variants within a family.

Start-1 L-1/SL-18	Based on the SS-25 Sickle (Russian designation RS-12M Topol) road-mobile ICBM. Consists of SS-25's three solid propellant stages plus a solid fourth stage. A small postboost stage (probably burning storable liquid propellants) is used for final orbit injection.
Start	Similar to Start-1; will employ five solid stages instead of four, with Start-1's stage 2 repeated as stage 3. Larger payload fairing will also be used. First launch planned for March 1995.

Historical Summary

The Start-1 and Start launch vehicles are based on the three-stage SS-25 Sickle (Russian designation RS-12M Topol) solid-propellant, road-mobile ICBM. Having first entered service in 1985, the SS-25 is one of Russia's most modern strategic weapons systems, and is still actively deployed. The SS-25 was developed by the Moscow Institute of Heat Technology (MIHT), which was formed in the late 1940s as KB Nadiradze and later renamed. This organization was responsible for the design and development of almost all of the the former Soviet Union's solid-propellant ground-based strategic missiles. The Start vehicles are marketed by STC Complex, a joint stock company formed by MIHT in 1991. Yuri Solomonov is the director of STC Complex and deputy chief designer of MIHT. The primary purpose of STC Complex is to convert military technology into viable commercial products; the focal point of their operations is the Start launch vehicle family.

Start-1 is the CIS's smallest launch vehicle, and the first known to use all solid propellants. Its first flight (and, as of February 1995, its only flight) occurred on March 25, 1993, from the Plesetsk site in Russia and was successful. Start-1 consists of the SS-25's three solid stages, plus an additional fourth solid stage. It also employs a small liquid-propellant postboost stage to increase the accuracy of final orbit injection. The first launch of the larger Start vehicle is planned for March 1995. This vehicle will use the same basic stages as Start-1; however, by essentially inserting a duplicate stage 2 between Start-1's stages 2 and 3, the vehicle is transformed into a five stage configuration. The five-stage Start vehicle will also use a larger payload fairing.

Launch Record

The Start-1 vehicle has been launched orbitally once, on March 25, 1993. This demonstration launch was successful. The SS-20 and SS-25 missiles, upon which the Start launch vehicles are based, have undergone in excess of 400 successful suborbital test launches since the early 1970s.

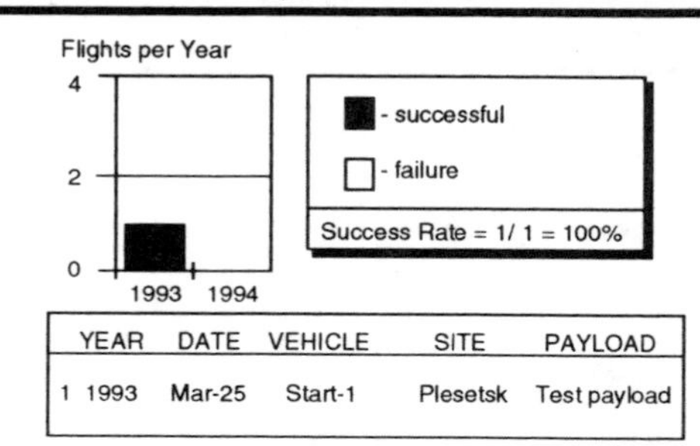

	YEAR	DATE	VEHICLE	SITE	PAYLOAD
1	1993	Mar-25	Start-1	Plesetsk	Test payload

Start General Description

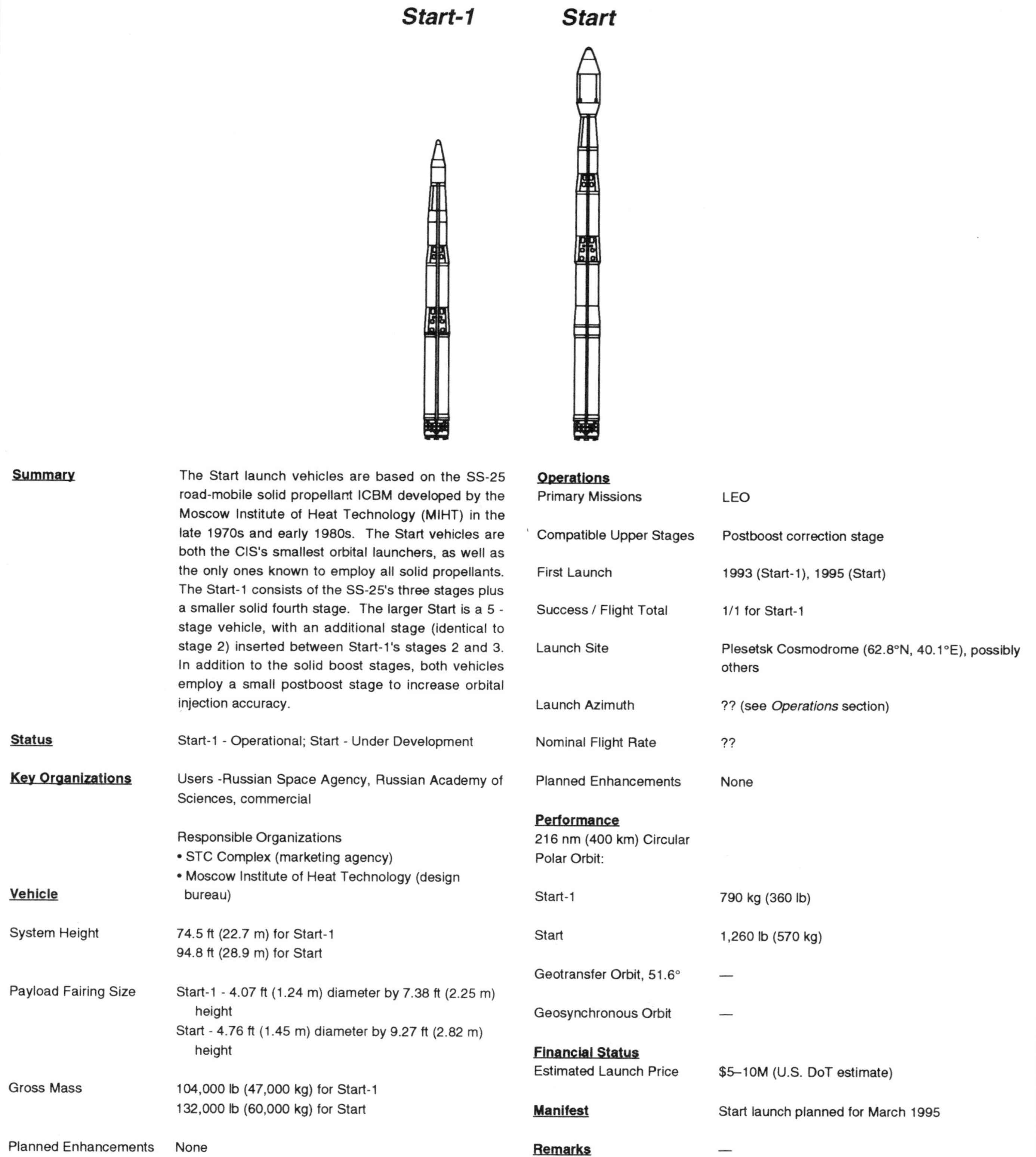

Summary

The Start launch vehicles are based on the SS-25 road-mobile solid propellant ICBM developed by the Moscow Institute of Heat Technology (MIHT) in the late 1970s and early 1980s. The Start vehicles are both the CIS's smallest orbital launchers, as well as the only ones known to employ all solid propellants. The Start-1 consists of the SS-25's three stages plus a smaller solid fourth stage. The larger Start is a 5 - stage vehicle, with an additional stage (identical to stage 2) inserted between Start-1's stages 2 and 3. In addition to the solid boost stages, both vehicles employ a small postboost stage to increase orbital injection accuracy.

Status

Start-1 - Operational; Start - Under Development

Key Organizations

Users -Russian Space Agency, Russian Academy of Sciences, commercial

Responsible Organizations
- STC Complex (marketing agency)
- Moscow Institute of Heat Technology (design bureau)

Vehicle

System Height	74.5 ft (22.7 m) for Start-1 94.8 ft (28.9 m) for Start
Payload Fairing Size	Start-1 - 4.07 ft (1.24 m) diameter by 7.38 ft (2.25 m) height Start - 4.76 ft (1.45 m) diameter by 9.27 ft (2.82 m) height
Gross Mass	104,000 lb (47,000 kg) for Start-1 132,000 lb (60,000 kg) for Start
Planned Enhancements	None

Operations

Primary Missions	LEO
Compatible Upper Stages	Postboost correction stage
First Launch	1993 (Start-1), 1995 (Start)
Success / Flight Total	1/1 for Start-1
Launch Site	Plesetsk Cosmodrome (62.8°N, 40.1°E), possibly others
Launch Azimuth	?? (see *Operations* section)
Nominal Flight Rate	??
Planned Enhancements	None

Performance

216 nm (400 km) Circular Polar Orbit:

Start-1	790 kg (360 lb)
Start	1,260 lb (570 kg)
Geotransfer Orbit, 51.6°	—
Geosynchronous Orbit	—

Financial Status

Estimated Launch Price — $5–10M (U.S. DoT estimate)

Manifest

Start launch planned for March 1995

Remarks

—

Start Vehicle

Overall

Start-1

Length	74.5 ft (22.7 m)
Gross Mass	104K lb (47K kg)
Thrust at Liftoff	? K lb (? kN)

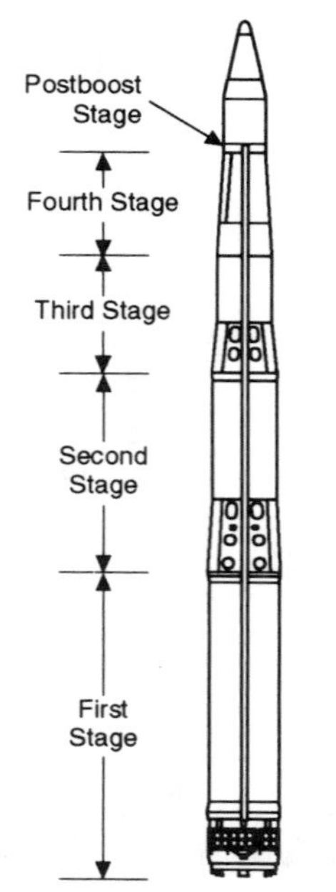

Start

Length	94.8 ft (28.9 m)
Gross Mass	132K lb (60K kg)
Thrust at Liftoff	? K lb (? kN)

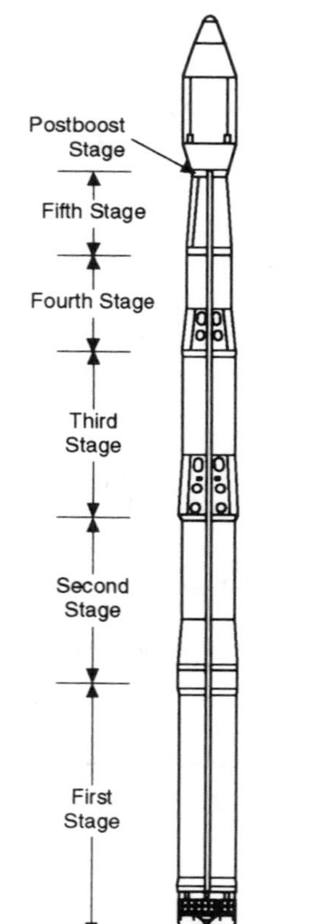

Stages

Very few details are available concerning the characteristics of the individual Start vehicle stages and systems. Because these launch vehicles are derived from a currently deployed ICBM, it is unlikely that details will be forthcoming. Available information indicates that each of the stages employs a single nozzle solid rocket motor for propulsion. It is known that stage 1 uses an array of moveable grid-type aerodynamic control surfaces around its base; however, it is not known whether nozzle gimbaling, ACS motors, spin stabilization, aerodynamic surfaces, or some combination is used for control beyond stage 1 flight. For the Start vehicle, it appears that stages 2 and 3 are identical, with this repeated stage being the main difference between it and the smaller Start-1. No information is available on the propulsive characteristics or mass properties.

Payload Fairings

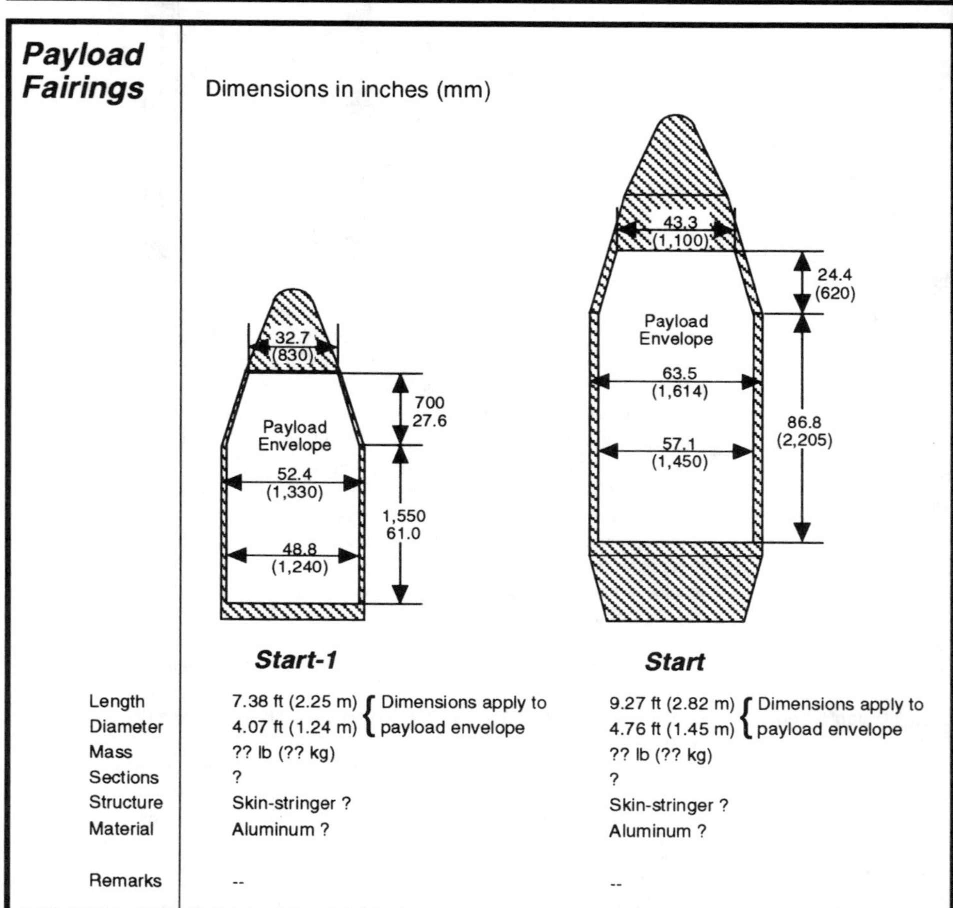

	Start-1	Start
Length	7.38 ft (2.25 m) Dimensions apply to	9.27 ft (2.82 m) Dimensions apply to
Diameter	4.07 ft (1.24 m) payload envelope	4.76 ft (1.45 m) payload envelope
Mass	?? lb (?? kg)	?? lb (?? kg)
Sections	?	?
Structure	Skin-stringer ?	Skin-stringer ?
Material	Aluminum ?	Aluminum ?
Remarks	--	--

Avionics

Unknown

Attitude Control System

Unknown

Performance Characteristics of Start-1 and Start Launch Vehicles (Performance courtesy STC Complex)

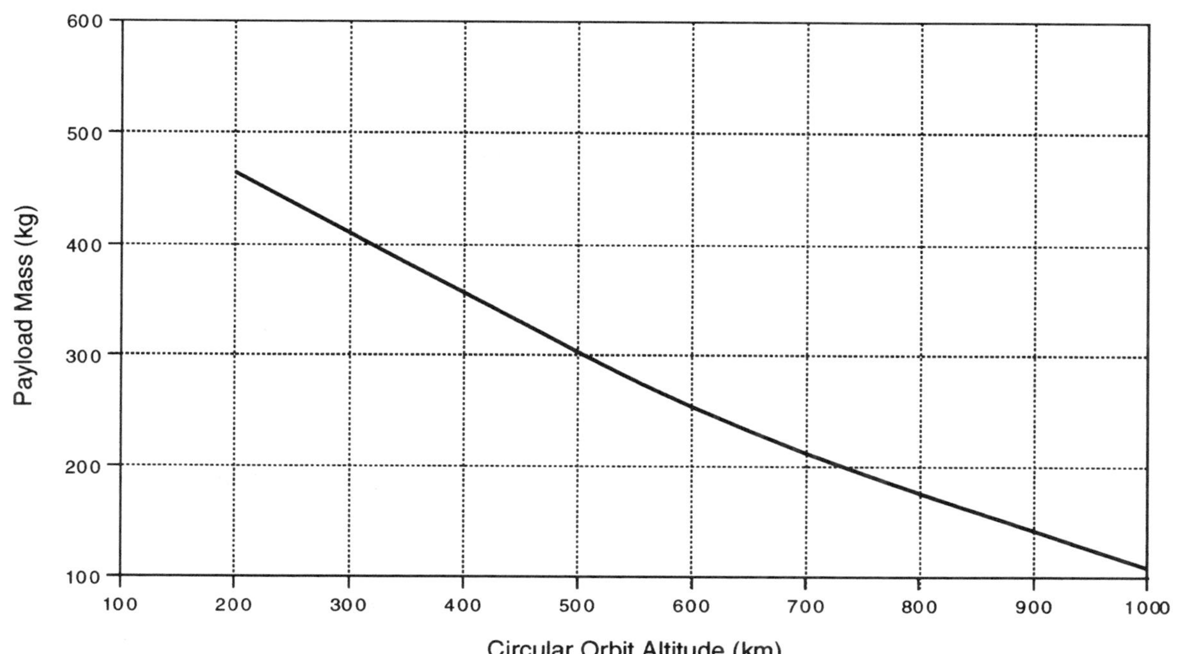

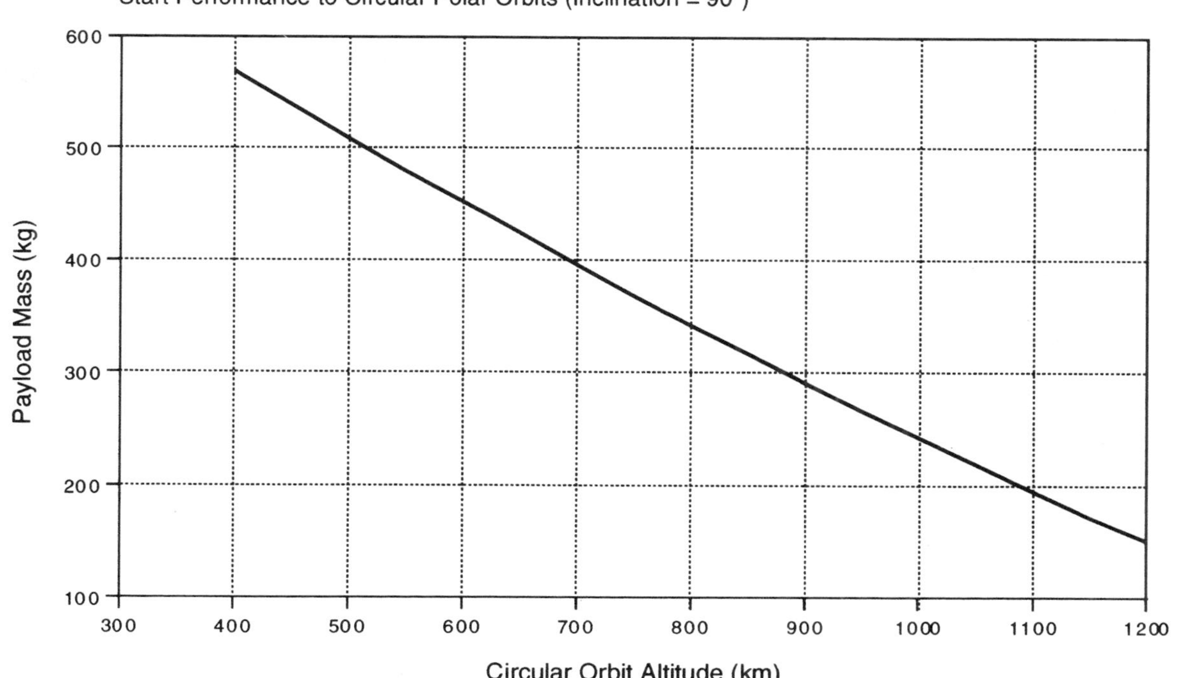

Start Operations

Launch Site

Note: STC Complex claims that Start vehicles can be launched from anywhere in world (within safety limits) due to the vehicles' road mobile missile heritage. To date, the only orbital launch of Start-1 took place at the Plesetsk Cosmodrome.

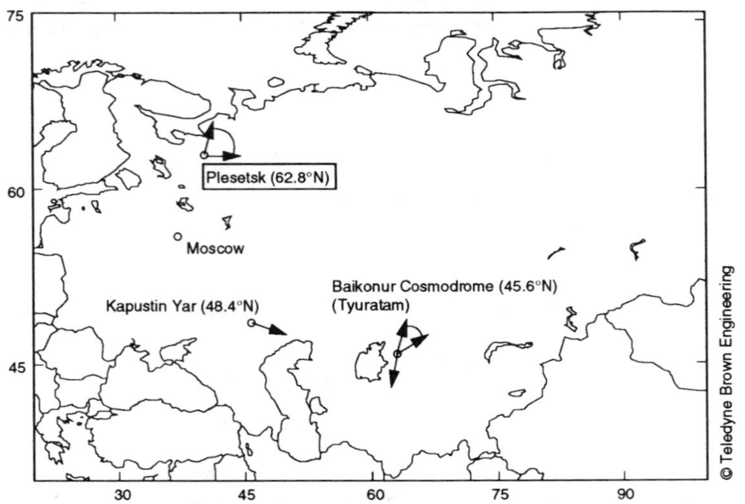

Launch Facilities

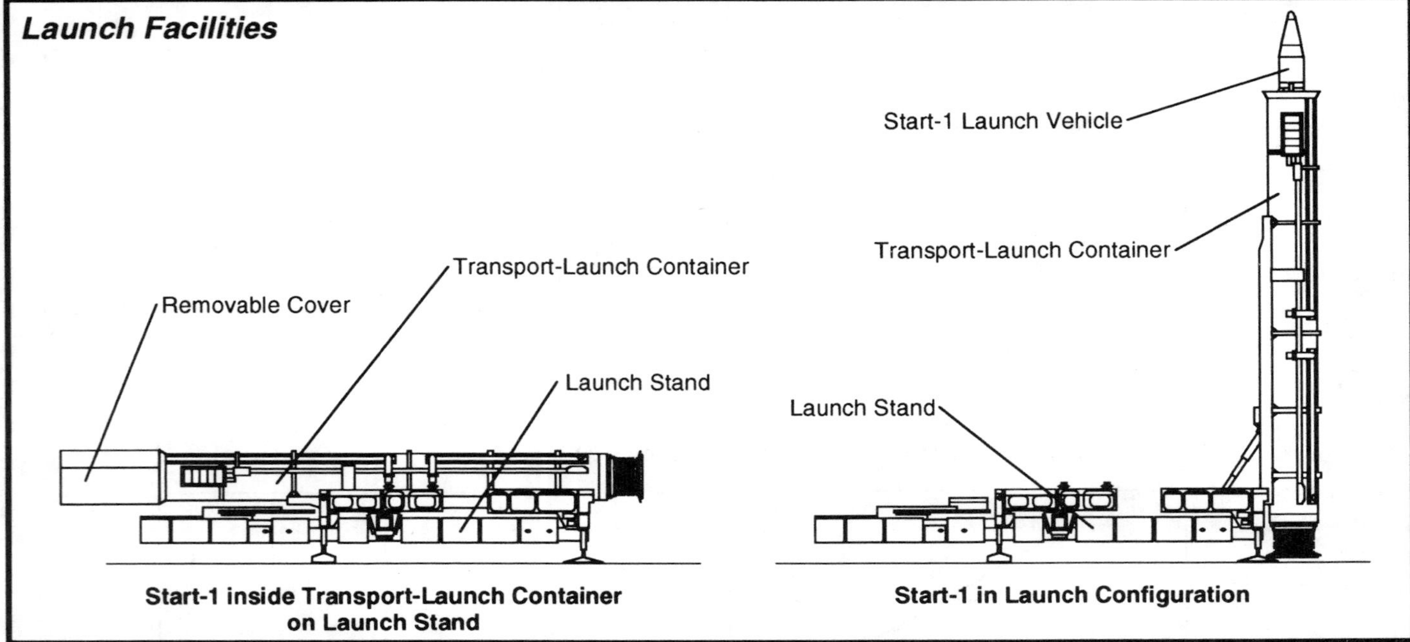

Start-1 inside Transport-Launch Container on Launch Stand

Start-1 in Launch Configuration

Launch Processing

The Start vehicles are based on the SS-25 road-mobile ICBM, and are designed to be launched from that missile's mobile launch platform. Thus, there are no fixed launch pads for Start at any of the CIS's cosmodromes. According to STC Complex, this arrangement allows the Start vehicle to be launched (in theory) from any suitable location in the world, provided that the country in which the launch takes place is a signatory of the Missile Technology Control Regime (MTCR), and both governments involved approve the launch. In reality, the only Start vehicle launch (a Start-1 launched in March 1993) to take place as of Febraury 1995 occurred at the Plesetsk Cosmodrome. An SS-25 launch platform was used for this launch.

A diagram of the Start mobile launch platform is shown above. Little details of the prelaunch processing flow are available; it is assumed that, like all other CIS launch vehicles, the Start vehicles are processed horizontally. It is also assumed that launches taking place at a CIS cosmodrome could make use of the prelaunch processing facilities located there. Remote launches would most likely require payload processing, encapsulation, and integration at an appropriate facility prior to transport to the launch site.

Following payload-launch vehicle integration, the complete vehicle is inserted into a Transport-Launch Container (TLC) and moved horizontally to the launch site, where the protective cover is removed. Shortly before launch, the TLC is rotated to a vertical position. Upon command, the vehicle is ejected from the TLC by means of a hot-gas generator, and after the vehicle clears the TLC, the stage 1 solid motor is ignited. In the West, this technique is referred to as a "cold launch".

Flight Sequence

Unknown

Start Notes

Payload Accommodations

Payload Acceleration Levels (Start-1 with 240 kg payload)

		Axial (g)	Lateral (g)
Phase of Flight:	Erection and Launch	2.8	1.25
	Stage 1 Flight	5.15	0.7
	Stage 2 Flight	6.5	0.6
	Stage 3 Flight	6.5	0.4
	Stage 4 Flight	9.0	0.5
	Post-Boost Stage Flight	0.1	0.03

Injection Accuracy (Start-1)

Perigee Altitude	±1 km
Apogee Altitude	±5 km
Inclination	±0.05°

Publications

Technical Publications

Jane's Space Directory 1994-95, Jane's Information Group Inc., Alexandria, VA.

Jane's Strategic Weapons Systems, Jane's Information Group Inc., Alexandria, VA, 1994.

Space Directory of Russia, February 1993 Edition, Euroconsult/Sevig Press, Paris, France.

The Russian Space Directory 1994, European Space Report, Munich, Germany.

Acronyms

CIS - Commonwealth of Independent States
GEO - geosynchronous orbit
GTO - geosynchronous transfer orbit
MIHT - Moscow Institute of Heat Technology
MTCR - Missile Technology Control Regime
ICBM - Intercontinental Ballistic Missile
KB - Design Bureau
LEO - low Earth orbit
TLC - Transport-Launch Container
STC - Scientific and Technical Corporation
USSR - Union of Soviet Socialist Republics

Launch Services Point of Contact:
NPO Yuzhnoye
Krivorozhskaya Street 3
Dniepropetrovsk, 8
Ukraine 320008
Phone: (70562) 42-00-22
Fax: (70562) 92-50-41
Telex: 143547 BRON

Photo Courtesy of Space Commerce Corp.

The Tsyklon operates from highly automated launch facilities which minimize manual tasks during erection, fueling, and final checkout.

Tsyklon History

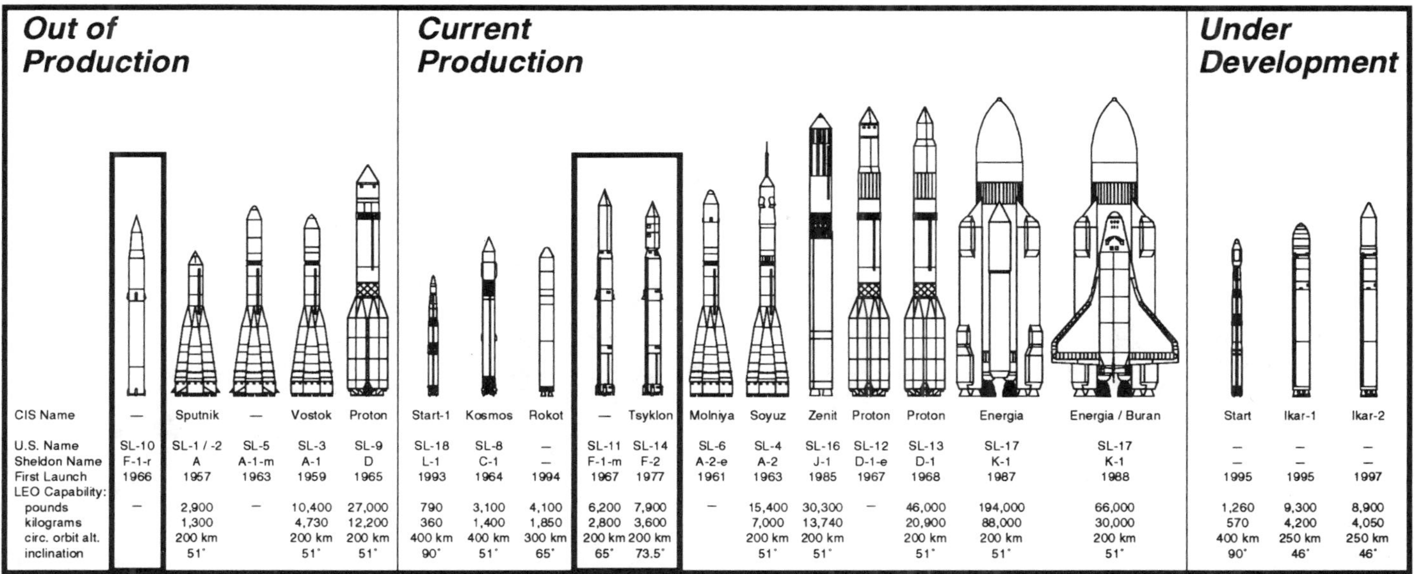

Out of Production

CIS Name	—	Sputnik	—	Vostok	Proton
U.S. Name	SL-10	SL-1 / -2	SL-5	SL-3	SL-9
Sheldon Name	F-1-r	A	A-1-m	A-1	D
First Launch	1966	1957	1963	1959	1965
LEO Capability:					
pounds	—	2,900	—	10,400	27,000
kilograms		1,300		4,730	12,200
circ. orbit alt.		200 km		200 km	200 km
inclination		51˚		51˚	51˚

Current Production

CIS Name	Start-1	Kosmos	Rokot	—	Tsyklon	Molniya	Soyuz	Zenit	Proton	Proton	Energia	Energia / Buran
U.S. Name	SL-18	SL-8	–	SL-11	SL-14	SL-6	SL-4	SL-16	SL-12	SL-13	SL-17	SL-17
Sheldon Name	L-1	C-1	–	F-1-m	F-2	A-2-e	A-2	J-1	D-1-e	D-1	K-1	K-1
First Launch	1993	1964	1994	1967	1977	1961	1963	1985	1967	1968	1987	1988
LEO Capability:												
pounds	790	3,100	4,100	6,200	7,900	—	15,400	30,300	—	46,000	194,000	66,000
kilograms	360	1,400	1,850	2,800	3,600		7,000	13,740		20,900	88,000	30,000
circ. orbit alt.	400 km	400 km	300 km	200 km	200 km		200 km	200 km		200 km	200 km	200 km
inclination	90˚	51˚	65˚	65˚	73.5˚		51˚	51˚		51˚	51˚	51˚

Under Development

CIS Name	Start	Ikar-1	Ikar-2
U.S. Name	–	–	–
Sheldon Name	–	–	–
First Launch	1995	1995	1997
LEO Capability:			
pounds	1,260	9,300	8,900
kilograms	570	4,200	4,050
circ. orbit alt.	400 km	250 km	250 km
inclination	90˚	46˚	46˚

Vehicle Description

CIS vehicles are identified either by their CIS, U.S., or Sheldon name. In the Soviet Union it was standard practice to name a launch vehicle after its original payload (e.g., Kosmos, Proton). The U.S. names (developed by the U.S. Department of Defense) are alphanumeric designators based roughly on chronological appearance. The Sheldon names, a most commonly used system that was published by Dr. Charles Sheldon of the U.S. Library of Congress in 1968, emphasize the basic families of launch vehicles with special indicators for variants within a family.

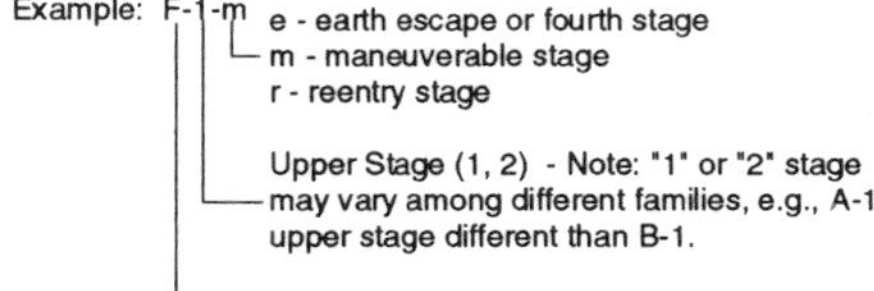

F-1-r (SL-10) — Based on the Scarp SS-9 ICBM, it has two stages with storable liquid propellant. Includes a reentry rocket which is actually part of the payload.

F-1-m (SL-11) — Same as F-1-r except includes a maneuverable stage which is actually part of the payload.

F-2 (SL-14) Tsyklon — Same as F-1-m except addition of small liquid third stage.

Historical Summary

The SS-9 ICBM, code-named Scarp by NATO, entered service in 1966 and was described as having intercontinental and orbital capability. The Russian designation for this missile is R-36. The SS-9, along with its subsequent derivative space launchers, was developed by what is now the Ukrainian Yuzhnoye organization; primary assembly takes place at the MZ Yuzhmash factory. Both facilities are located in Dniepropetrovsk, Ukraine.

The first space launch vehicle in the Tsyklon family was designated as the F-1 booster in the West. The F-1-r (SL-10) was introduced in 1966 and followed a year later by F-1-m (SL-11). The F-1 has been used primarily for military purposes. Its past applications include offensive weapons programs (FOBS), antisatellite interception (ASAT), and launching radar and electronic ocean reconnaissance satellites. The FOBS version is often referred to as F-1-r, with the "r" standing for a retro-rocket stage which is actually part of the payload. Similarly, the antisatellite interceptor and ocean reconnaissance launcher is referred to as F-1-m, the "m" standing for maneuverable stage, which is again actually part of the payload. F-1 variants are launched only from Tyuratam (Baikonur).

The F-2 (SL-14) which is merely the F-1 with a small third stage, was introduced in 1977. It has acquired a large number of missions previously flown on the Kosmos as well as the Vostok. These include communications, meteorology, remote sensing, science, geodesy, electronic intelligence, and minor military. The F-2 is launched from Plesetsk from a highly automated launch complex into orbits at inclinations around 73.5° and 82.5°.

The F-series vehicles account for about 10% of all the spacecraft launches in the CIS space program. Although the vehicles are manufactured in the now independent nation of Ukraine, they are still launched with some regularity from both Plesetsk (Russia) and Baikonur (Kazakhstan), apparently under the auspices of the Russian Space Forces (a branch of the Russian Ministry of Defense), which operates both launch sites. In addition, a significant fraction of the vehicle's systems and subsystems are of Russian manufacture. The exact nature of the complex relationship between the Russian and Ukrainian governments regarding the Tsyklon (and Ukrainian space-related products in general) is unclear.

Tsyklon History

(continued)

Launch Record

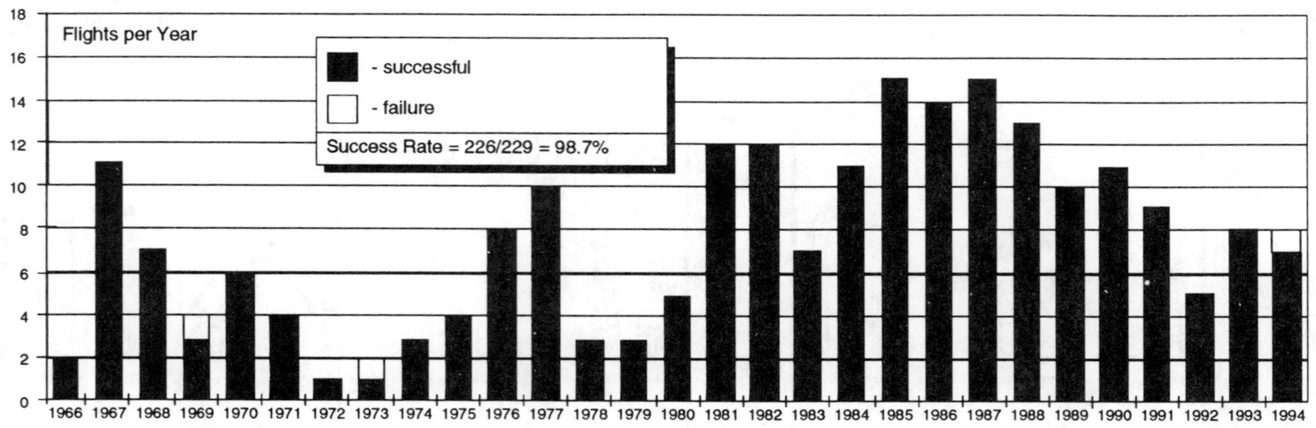

	YEAR	DATE	VEHICLE	SITE	PAYLOAD
1	1966	Sep-17	F-1-r	TT	None
2		Nov-2	F-1-r	TT	None
3	1967	Jan-25	F-1-r	TT	Kosmos 139
4		May-17	F-1-r	TT	Kosmos 160
5		Jul-17	F-1-r	TT	Kosmos 169
6		Jul-31	F-1-r	TT	Kosmos 170
7		Aug-8	F-1-r	TT	Kosmos 171
8		Sep-19	F-1-r	TT	Kosmos 178
9		Sep-22	F-1-r	TT	Kosmos 179
10		Oct-18	F-1-r	TT	Kosmos 183
11		Oct-27	F-1-m	TT	Kosmos 185
12		Oct-28	F-1-r	TT	Kosmos 187
13		Dec-27	F-1-m	TT	Kosmos 198
14	1968	Mar-22	F-1-m	TT	Kosmos 209
15		Apr-24	F-1-m	TT	Kosmos 217
16		Apr-25	F-1-r	TT	Kosmos 218
17		Oct-2	F-1-r	TT	Kosmos 244
18		Oct-19	F-1-m	TT	Kosmos 248
19		Oct-20	F-1-m	TT	Kosmos 249
20		Nov-1	F-1-m	TT	Kosmos 252
21	1969	Jan-24	F-1	TT	?
		Failure			
22		Aug-6	F-1-m	TT	Kosmos 291
23		Sep-15	F-1-r	TT	Kosmos 298
24		Dec-23	F-1-m	TT	Kosmos 316
25	1970	Jul-28	F-1-r	TT	Kosmos 354
26		Sep-25	F-1-r	TT	Kosmos 365
27		Oct-3	F-1-m	TT	Kosmos 367
28		Oct-20	F-1-m	TT	Kosmos 373
29		Oct-23	F-1-m	TT	Kosmos 374
30		Oct-30	F-1-m	TT	Kosmos 375
31	1971	Feb-25	F-1-m	TT	Kosmos 397
32		Apr-4	F-1-r	TT	Kosmos 404
33		Aug-8	F-1-r	TT	Kosmos 433
34		Dec-3	F-1-m	TT	Kosmos 462
35	1972	Aug-21	F-1-m	TT	Kosmos 516
36	1973	Apr-25	F-1-m	TT	?
		Failure			
37		Dec-27	F-1-m	TT	Kosmos 626
38	1974	May-15	F-1-m	TT	Kosmos 651
39		May-17	F-1-m	TT	Kosmos 654
40		Dec-24	F-1-m	TT	Kosmos 699
41	1975	Apr-2	F-1-m	TT	Kosmos 723
42		Apr-7	F-1-m	TT	Kosmos 724
43		Oct-29	F-1-m	TT	Kosmos 777
44		Dec-12	F-1-m	TT	Kosmos 785
45	1976	Feb-16	F-1-m	TT	Kosmos 804
46		Apr-13	F-1-m	TT	Kosmos 814
47		Jul-2	F-1-m	TT	Kosmos 838
48		Jul-21	F-1-m	TT	Kosmos 843
49		Oct-17	F-1-m	TT	Kosmos 860
50		Oct-21	F-1-m	TT	Kosmos 861
51		Nov-26	F-1-m	TT	Kosmos 868
52		Dec-27	F-1-m	TT	Kosmos 886
53	1977	May-23	F-1-m	TT	Kosmos 910
54		Jun-17	F-1-m	TT	Kosmos 918
55		Jun-24	F-2	PL	Kosmos 921
56		Aug-24	F-1-m	TT	Kosmos 937
57		Sep-16	F-1-m	TT	Kosmos 952
58		Sep-18	F-1-m	TT	Kosmos 954
59		Sep-24	F-2	PL	Kosmos 956
60		Oct-26	F-1-m	TT	Kosmos 961
61		Dec-21	F-1-m	TT	Kosmos 970
62		Dec-27	F-2	PL	Kosmos 972
63	1978	May-19	F-1-m	TT	Kosmos 1009
64		Jun-28	F-2	PL	Kosmos 1025
65		Oct-26	F-2	PL	Kosmos 1045, Radio 1-2
66	1979	Feb-12	F-2	PL	Kosmos 1076
67		Apr-18	F-1-m	TT	Kosmos 1094
68		Apr-25	F-1-m	TT	Kosmos 1096
69	1980	Jan-23	F-2	PL	Kosmos 1151
70		Mar-14	F-1-m	TT	Kosmos 1167
71		Apr-18	F-1-m	TT	Kosmos 1174
72		Apr-29	F-1-m	TT	Kosmos 1176
73		Nov-4	F-1-m	TT	Kosmos 1220
74	1981	Feb-2	F-1-m	TT	Kosmos 1243
75		Mar-5	F-1-m	TT	Kosmos 1249
76		Mar-14	F-1-m	TT	Kosmos 1258
77		Mar-20	F-1-m	TT	Kosmos 1260
78		Apr-21	F-1-m	TT	Kosmos 1266
79		Aug-4	F-1-m	TT	Kosmos 1286
80		Aug-24	F-1-m	TT	Kosmos 1299
81		Aug-24	F-2	PL	Kosmos 1300
82		Sep-14	F-1-m	TT	Kosmos 1306
83		Sep-21	F-2	PL	Oreol 3
84		Sep-30	F-2	PL	Kosmos 1312
85		Dec-3	F-2	PL	Kosmos 1328
86	1982	Feb-11	F-1-m	TT	Kosmos 1337
87		Mar-25	F-2	PL	Meteor 2-8
88		Apr-29	F-1-m	TT	Kosmos 1355
89		May-14	F-1-m	TT	Kosmos 1365
90		Jun-1	F-1-m	TT	Kosmos 1372
91		Jun-10	F-2	PL	Kosmos 1378
92		Jun-18	F-1-m	TT	Kosmos 1379
93		Aug-30	F-1-m	TT	Kosmos 1402
94		Sep-4	F-1-m	TT	Kosmos 1405
95		Sep-16	F-2	PL	Kosmos 1408
96		Sep-24	F-2	PL	Kosmos 1410
97		Oct-2	F-1-m	TT	Kosmos 1412
98	1983	Apr-23	F-2	PL	Kosmos 1455
99		May-7	F-1-m	TT	Kosmos 1461
100		Jun-23	F-2	PL	Kosmos 1470
101		Sep-28	F-2	PL	Kosmos 1500
102		Oct-29	F-1-m	TT	Kosmos 1507
103		Nov-24	F-2	PL	Kosmos 1510
104		Dec-15	F-2	PL	Kosmos 1515
105	1984	Feb-8	F-2	PL	Kosmos 1536
106		Mar-15	F-2	PL	Kosmos 1544
107		May-30	F-1-m	TT	Kosmos 1567
108		Jun-29	F-1-m	TT	Kosmos 1579
109		Jul-5	F-2	PL	Meteor 2-11
110		Aug-7	F-1-m	TT	Kosmos 1588
111		Aug-8	F-2	PL	Kosmos 1589
112		Sep-28	F-2	PL	Kosmos 1602
113		Oct-18	F-2	PL	Kosmos 1606
114		Oct-31	F-1	TT	Kosmos 1607
115		Nov-27	F-2	PL	Kosmos 1612
116	1985	Jan-15	F-2	PL	Kosmos 1617 to 1622
117		Jan-23	F-1	TT	Kosmos 1625
118		Jan-24	F-2	PL	Kosmos 1626
119		Feb-6	F-2	PL	Meteor 2-12
120		Mar-5	F-2	PL	Kosmos 1633
121		Apr-18	F-1-m	TT	Kosmos 1646
122		Jun-14	F-2	PL	Kosmos 1660
123		Jul-8	F-2	PL	Kosmos 1666
124		Aug-1	F-1-m	TT	Kosmos 1670
125		Aug-8	F-2	PL	Kosmos 1674
126		Aug-23	F-1-m	TT	Kosmos 1677
127		Sep-19	F-1-m	TT	Kosmos 1682
128		Oct-9	F-2	PL	Kosmos 1690 to 1695
129		Oct-24	F-2	PL	Meteor 3-1
130		Nov-22	F-2	PL	Kosmos 1703
131		Dec-12	F-2	PL	Kosmos 1707
132		Dec-26	F-2	PL	Meteor 2-13
133	1986	Jan-17	F-2	PL	Kosmos 1726
134		Feb-11	F-2	PL	Kosmso 1732
135		Feb-19	F-2	PL	Kosmos 1733
136		Feb-27	F-1-m	TT	Kosmos 1735
137		Mar-21	F-1-m	TT	Kosmos 1736
138		Mar-25	F-1-m	TT	Kosmos 1737
139		May-15	F-2	PL	Kosmos 1743
140		May-27	F-2	PL	Meteor 2-14
141		Jun-12	F-2	PL	Kosmos 1758
142		Jul-28	F-2	PL	Kosmos 1766
143		Aug-4	F-1-m	TT	Kosmos 1769
144		Aug-20	F-1-m	TT	Kosmos 1771
145		Sep-30	F-2	PL	Kosmos 1782
146		Dec-2	F-2	PL	Kosmos 1803
147		Dec-10	F-2	PL	Kosmos 1805
148		Dec-18	F-2	PL	Kosmos 1809
149	1987	Jan-5	F-2	PL	Meteor 2-15
150		Jan-14	F-2	PL	Kosmos 1812
151		Feb-1	F-1-m	PL	Kosmos 1818
152		Feb-20	F-2	PL	Kosmos 1823
153		Mar-3	F-2	PL	Kosmos 1825
154		Mar-13	F-2	PL	Kosmos 1827 to 1832
155		Apr-8	F-1	TT	Kosmos 1834
156		Apr-27	F-2	PL	Kosmos 1842
157		Jun-18	F-1	TT	Kosmos 1860
158		Jul-1	F-2	PL	Kosmos 1862
159		Jul-10	F-1	TT	Kosmos 1867
160		Jul-16	F-2	PL	Kosmos 1869
161		Aug-18	F-2	PL	Meteor 2-16
162		Sep-7	F-2	PL	Kosmos 1875 to 1880
163		Oct-10	F-1-m	TT	Kosmos 1890
164		Oct-20	F-2	PL	Kosmos 1892
165		Dec-12	F-1-m	TT	Kosmos 1900
166	1988	Jan-6	F-2	PL	Kosmos 1908
167		Jan-15	F-2	PL	Kosmos 1909 to 1914
168		Jan-30	F-2	PL	Meteor 2-17
169		Mar-14	F-1	TT	Kosmos 1932
170		Mar-15	F-2	PL	Kosmos 1933
171		May-28	F-1	TT	Kosmos 1949
172		May-30	F-2	PL	Kosmos 1950
173		Jun-14	F-2	PL	Kosmos 1953
174		Jul-5	F-2	PL	Okean 1
175		Jul-26	F-2	PL	Meteor 3-2
176		Oct-11	F-2	PL	Kosmos 1975
177		Nov-18	F-1	TT	Kosmos 1979
178		Dec-23	F-2	PL	Kosmos 1985
179	1989	Feb-10	F-2	PL	Kosmos 1994 to 1999
180		Feb-28	F-2	PL	Meteor 2-18
181		Jul-24	F-1	TT	Kosmos 2033
182		Aug-28	F-2	PL	Kosmos 2037
183		Sep-14	F-2	PL	Kosmos 2038 to 2043
184		Sep-27	F-1	TT	Kosmos 2046
185		Sep-28	F-2	PL	Interkosmos 24 (Activny)
186		Oct-24	F-2	PL	Meteor 3-3
187		Nov-24	F-1	TT	Kosmos 2051
188		Dec-27	F-2	PL	Kosmos 2053
189	1990	Feb-28	F-2	PL	Okean 2
190		Mar-14	F-1	TT	Kosmos 2060
191		Jun-26	F-2	PL	Meteor 2-19
192		Jul-30	F-2	PL	Kosmos 2088
193		Aug-8	F-2	PL	Kosmos 2090 to 2095
194		Aug-23	F-1	TT	Kosmos 2096
195		Sep-28	F-2	PL	Meteor 2-20
196		Nov-14	F-1	TT	Kosmos 2103
197		Nov-28	F-2	PL	Kosmos 2106
198		Dec-4	F-1	TT	Kosmos 2107
199		Dec-22	F-2	PL	Kosmos 2114 to 2119
200	1991	Jan-18	F-1	TT	Kosmos 2122
201		Apr-24	F-2	PL	Meteor 3-4
202		May-17	F-2	PL	Kosmos 2143 to 2148
203		Jun-4	F-2	PL	Okean 3
204		Jun-13	F-2	PL	Kosmos 2151
205		Aug-15	F-2	PL	Meteor 3-5, TOMS
206		Sep-28	F-2	PL	Kosmos 2157 to 2162
207		Nov-12	F-2	PL	Kosmos 2165 to 2170
208		Dec-18	F-2	PL	Interkosmos 25, Magion 3
209	1992	Jul-13	F-2	PL	Kosmos 2197 to 2202
210		Oct-20	F-2	PL	Kosmos 2211 to 2216
211		Nov-24	F-2	PL	Kosmos 2221
212		Dec-22	F-2	PL	Kosmos 2226
213		Dec-25	F-2	PL	Kosmos 2228
214	1993	Mar-30	F-1	TT	Kosmos 2238
215		Apr-16	F-2	PL	Kosmos 2242
216		Apr-28	F-1	TT	Kosmos 2244
217		May-11	F-2	PL	Kosmos 2245 to 2250
218		Jun-24	F-2	PL	Kosmos 2252 to 2257
219		Jul-7	F-1	TT	Kosmos 2258
220		Aug-31	F-2	PL	Meteor 2-21, Temisat
221		Sep-17	F-1	TT	Kosmos 2264
222	1994	Jan-25	F-2	PL	Meteor 3-6
223		Feb-12	F-2	PL	Kosmos 2268 to 2273
224		Mar-2	F-2	PL	AUOS-SM-KI
225		May-25	F-2	PL	?
		Failure			
226		Oct-11	F-2	PL	Okean-01
227		Nov-2	F-1	TT	Kosmos 2293
228		Nov-29	F-2	PL	Kosmos 2299
229		Dec-27	F-2	PL	Kosmos 2299 to 2304

Data Sources: TRW Space Log, NPO Yuzhnoye

TT is Baikonur Cosmodrome near Tyuratam
PL is Plesetsk Cosmodrome

Tsyklon General Description

SL-11 **SL-14 (Tsyklon)**

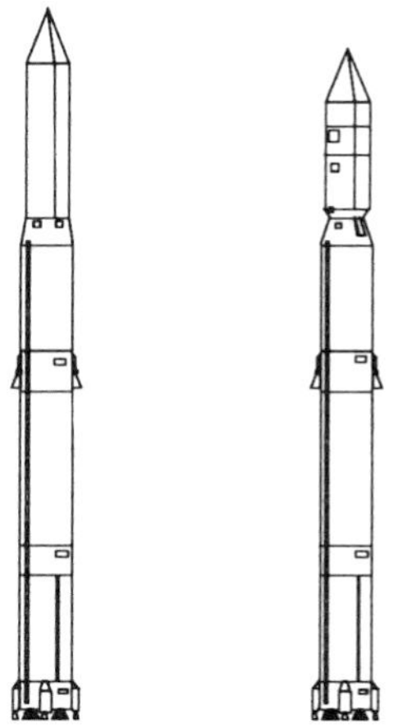

Summary

For low Earth orbit payloads which exceed the capabilities of the Kosmos but do not require the Soyuz, the two F-class variants are available. The F-1 vehicle was introduced in 1966 as the F-1-r and a year later the F-1-m was added. In 1977, a similar vehicle with a restartable third stage, the F-2 or Tsyklon, entered service. The Tsyklon is a three-stage launch vehicle that is used to launch the Kosmos and Meteor series of satellites. All three stages are liquid-fueled with N2O4 UDMH. The launch vehicle has tandem staging.

Status

Operational

Key Organizations

Users - Russian Ministry of Defense, Russian Academy of Sciences, Russian Space Agency, commercial

Key Organizations -
- NPO Yuzhnoye (design bureau)
- MZ Yuzhmash (manufacturer)
- NPO Energomash (stage 1 and 2 engines)

Vehicle

System Height	129.9 ft (39.58 m) for F-1-m 128.8 ft (39.27 m) for F-2
Payload Fairing Size	F-1-m - 7 ft (2.13 m) diameter by 46.4 ft (14.14 m) height F-2 - 8.9 ft (2.7 m) diameter by 31.3 ft (9.54 m) height
Gross Mass	397,000 lb (180,000 kg) for F-1-m 418,000 lb (190,000 kg) for F-2
Planned Enhancements	None

Operations

Primary Missions	LEO
Compatible Upper Stages	F-2 third stage
First Launch	1966 for F-1, 1977 for F-2
Success / Flight Total	120/122 for F-1 and 108/109 for F-2 through November 29, 1994.
Launch Site	F-1-m at Baikonur Cosmodrome (46.0°N, 62.9°E) F-2 at Plesetsk Cosmodrome (62.9°N, 40.8°E)
Launch Azimuth	F-1-m - 35°–39.7°(Max. inclination 65°) F-2 - 14°–36° (Has only flown inclinations between 73–83°)
Nominal Flight Rate	3–5 / year for F-1, and 7–11 / year for F-2
Planned Enhancements	None

Performance

108 nm (200 km) circ	6,200 lb (2,800 kg) at 65° inclination for F-1 7,900 lb (3,600 kg) at 73.5° inclination for F-2
108 nm (200 km) circ, 98°	—
Geotransfer Orbit, 51.6°	—
Geosynchronous Orbit	—

Financial Status

Estimated Launch Price: $10–15 million (NPO Yuzhnoye)

Manifest: Not available

Remarks: —

Tsyklon Vehicle

Overall

F-1-m

Length	129.9 ft (39.58 m)
Gross Mass	397K lb (180K kg)
Thrust at Liftoff	617K lb (2,745K N)

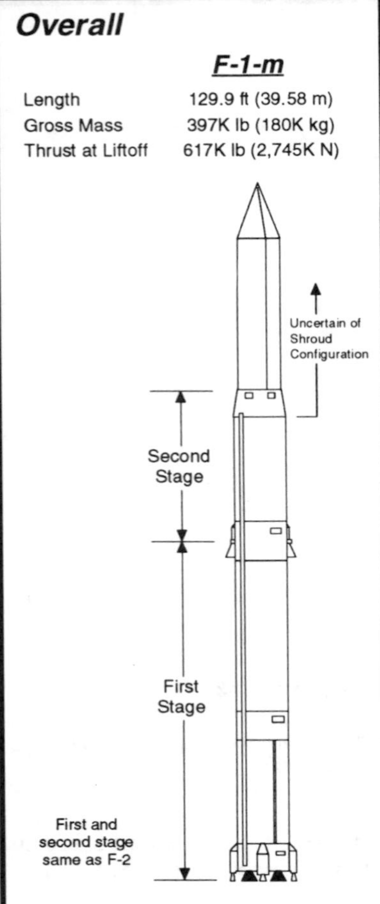

F-2

Length	128.8 ft (43.3 m)
Gross Mass	418K lb (190K kg)
Thrust at Liftoff	617K lb (2,745K N)

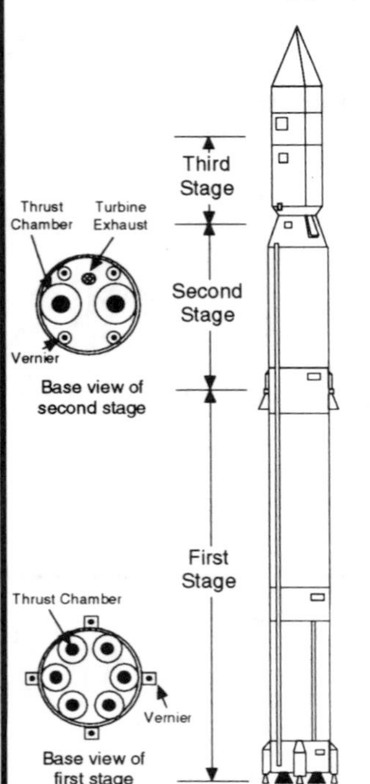

F-1-m / F-2 Stages

	Stage 1	Stage 2	Stage 3 (F-2 only)
Dimension:			
Length	61.5 ft (18.75 m)	33.1 ft (10.08 m)	9.02 ft (2.75 m)
Diameter	9.8 ft (3.0 m)	9.8 ft (3.0 m)	7.2 ft (2.2 m)
Mass: (each)			
Propellant Mass	261.7K lb (118.7K kg)	106.9K lb (48.5K kg)	6.6K lb (3.0K kg)
Gross Mass	280.0K lb (127.0K kg)	117.5K lb (53.3K kg)	10.1K lb (4.6K kg)
Structure:			
Type	Skin-Stringer	Skin-stringer	Monocoque
Material	Aluminum	Aluminum	Aluminum
Propulsion:			
Propellant	N2O4 / UDMH	N2O4 / UDMH	N2O4 / UDMH
Average Thrust (total)	681K lb (3,032 kN) vac *	227K lb (1,010 kN) vac *	17.5K lb (79.6 kN) vac *
Engine Designation	RD-251	RD-252	11D25
Number of Engines	3 pumps + 6 chambers + 4 verniers	1 pump + 2 chambers + turbine exhaust + 4 verniers	1 pump + 1 chamber + turbine exhaust
Isp	269.6 sec SL	317.6 sec vac	317 sec vac *
Feed System	turbopump	turbopump	turbopump
Chamber Pressure	1,209 psia (83.4 bar)	1,294 psia (89.3 bar)	1,287 psia (88.8 bar)
Mixture Ratio (O/F)	2.6	2.6	2.1
Throttling Capability	+5%/-10%	+5%/-10%	+5%/-10%
Expansion Ratio	14.7	46.1	112.4
Restart Capability	No	No	2 restarts
Tank Pressurization	gas generator	gas generator	helium
Control-Pitch, Yaw, Roll	4 verniers	4 verniers	8 steering nozzles, 8 ACS thrusters
Events:			
Nominal Burn Time	120 sec	160 sec	125 sec
Stage Shutdown	Command shutdown	Command shutdown	Command shutdown
Stage Separation	??	retrothrusters	Spring ejection

* thrust includes verniers

Remarks:

The Tsyklon comprises three tandem-arranged stages and a shroud. The third stage and the payload are enclosed within the shroud. Each Tsyklon stage is powered by liquid-propellant rocket engines using UDMH (fuel) and N2O4 (oxidizer). The first stage has six thrust chambers and four verniers for thrust and the second stage has two thrust chambers and four verniers for thrust. The F-2 can boost several spacecraft into orbit simultaneously and inject them subsequently into their individual orbits using up to three low-thrust firings of the third stage. The third-stage sustainer engine can be restarted twice under weightless conditions, which permits spacecraft to be placed into various orbits required for particular space missions. Typically, single starting of the third-stage sustainer is used for injecting spacecraft into orbits ranging from 108–135 nm (200–250 km) in altitude; double-starting of the third-stage sustainer is used for injecting spacecraft into orbits higher than 135 nm (250 km).

Payload Fairing

	F-1-m	F-2
Length	46.4 ft (14.14 m)	31.3 ft (9.54 m)
Diameter	7 ft (2.13 m)	8.9 ft (2.7 m)
Mass	?? lb (?? kg)	9,900 lb (4,500 kg)
Sections	2	2
Structure	Skin-stringer ?	Skin-stringer ?
Material	Aluminum	Aluminum
Remarks	Part of F-2 fairing similar to Vostok fairing.	

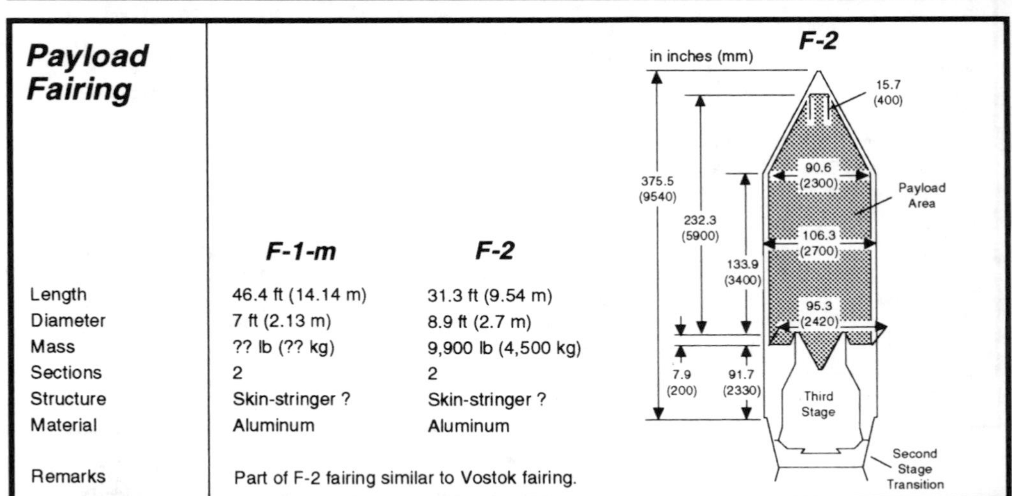

Avionics

Strap-down inertial unit.

Attitude Control System

First and second stage each use four verniers. Third stage uses eight fixed turbine exhaust nozzles during powered flight and an eight-thruster ACS package during coast phases.

Tsyklon Performance

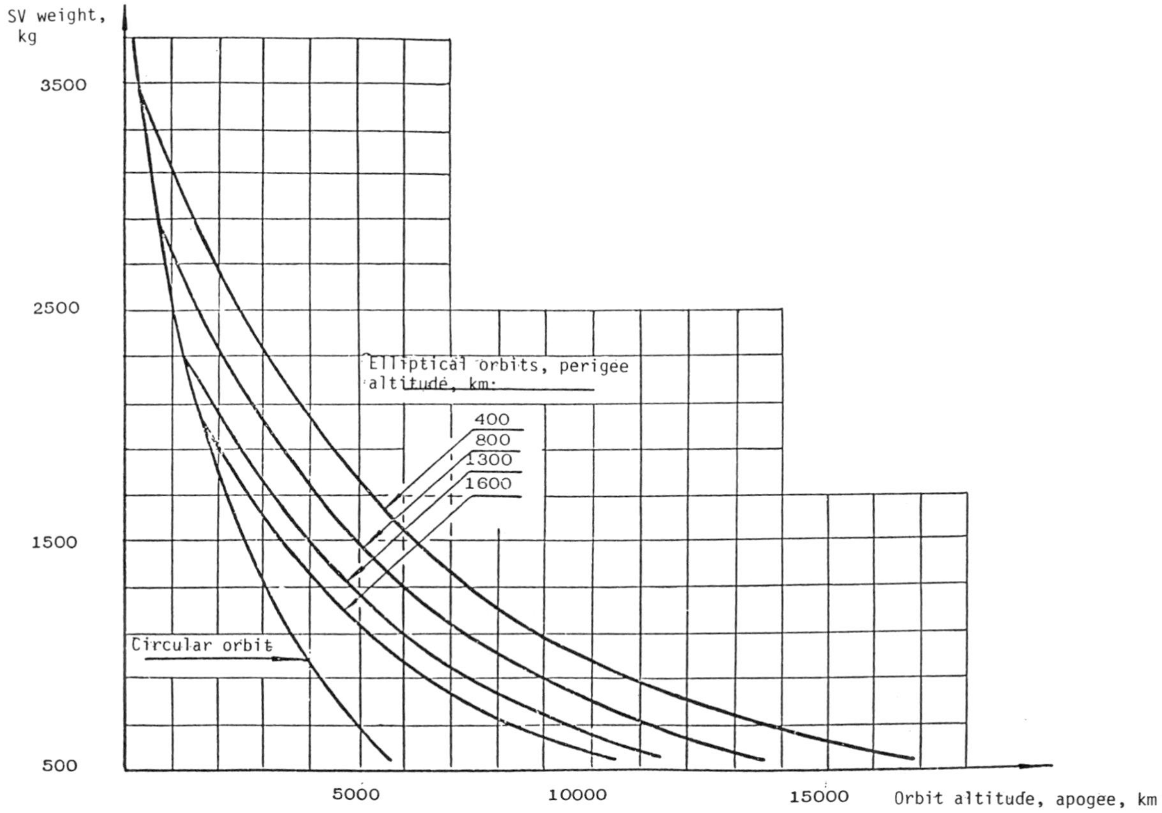

SV weight, kg
3500
2500
1500
500
Elliptical orbits, perigee altitude, km:
400
800
1300
1600
Circular orbit
5000
10000
15000
Orbit altitude, apogee, km

Elliptical Orbits of 82.5° Inclination

SV weight, kg
3500
2500
1500
Elliptical orbits, perigee altitude, km:
400
800
1200
1600
Circular orbit
5000
10000
15000
Orbit altitude, apogee, km

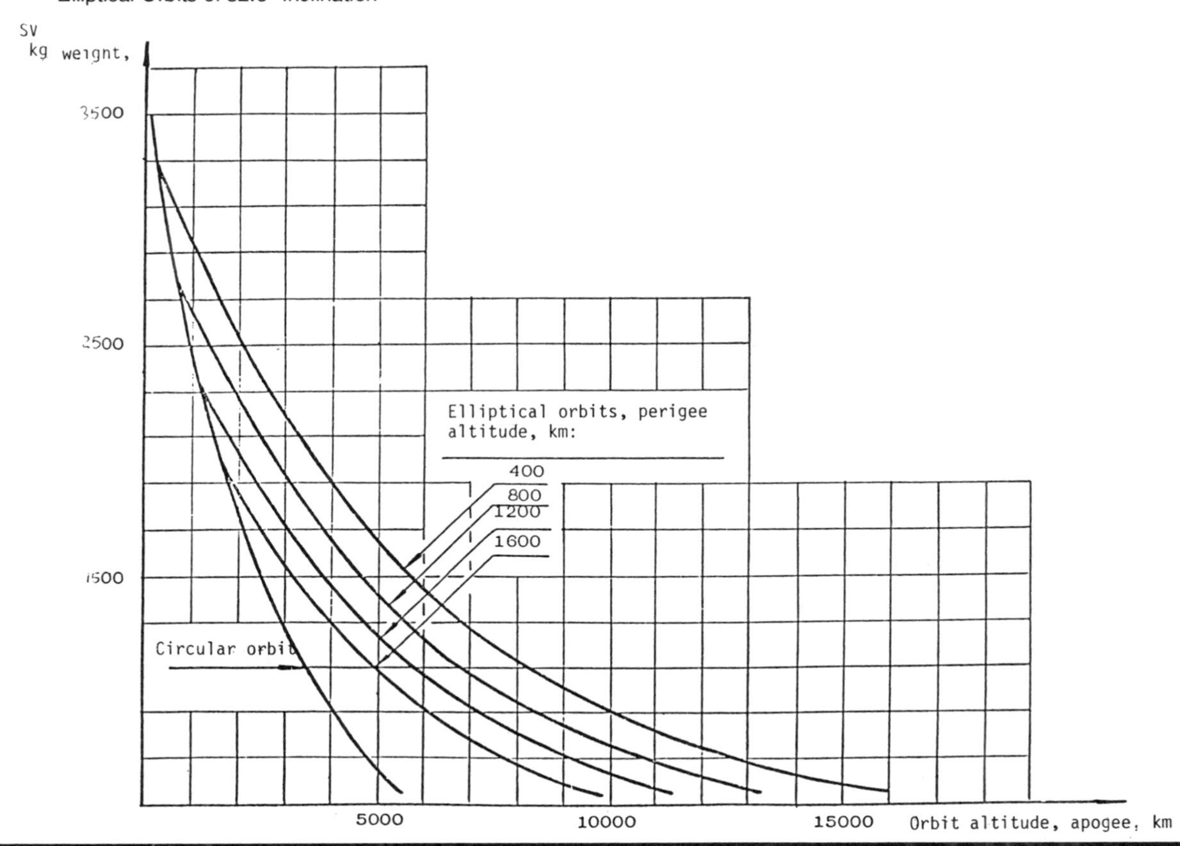

Tsyklon Operations

Launch Site

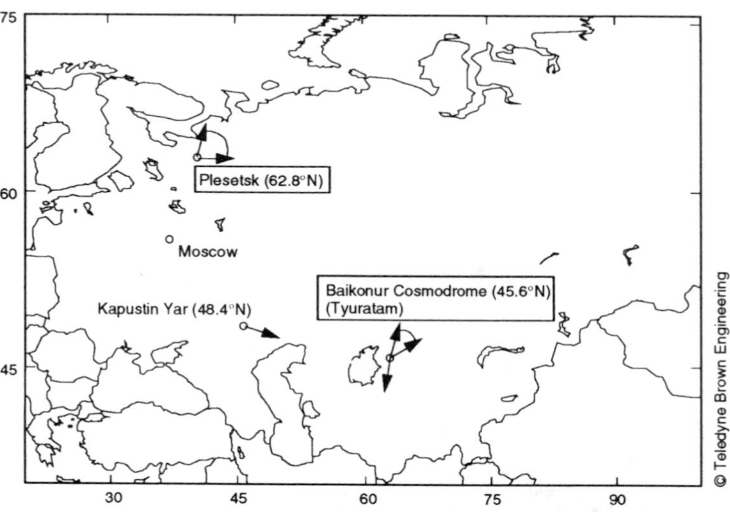

Launch Processing

Although the F-series boosters are manufactured in Ukraine, they continue to be launched from the same sites in Russia and Kazakhstan as they were before the breakup of the Soviet Union (albeit at a somewhat lower rate). An attitude of cooperation exists between the Ukrainian Space Agency, the Russian Space Agency, and the Russian Space Forces (the branch of the Russian Ministry of Defense that actually operates the launch facilities), due in large part to the interdependency left over from the days of Soviet rule. Any plans by the Ukrainian government to develop an independent space launch capability (i.e., a Ukrainian-controlled launch site) are undefined at this time.

The F-1-m is launched from the Baikonur Cosmodrome (also called Tyuratam, the name of a nearby railway stop) and F-2 from Plesetsk. Tyuratam, located in Kazakhstan, is where the first launches of Soviet artificial Earth satellites took place. This is the largest and most versatile of the former Soviet launch sites. Tyuratam may be considered to be analogous to Cape Canaveral and the Eastern Range in the United States. All launches associated with piloted spaceflight, lunar, planetary, and geosynchronous spacecraft originate from Tyuratam. The F-series launch pad is located in the northwest corner of Baikonur, near the Proton facilities. Although only the F-1 is currently launched from Baikonur, it has been speculated that, because of the similarities between the two vehicles, the F-2 could be launched from Baikonur with little or no facilities modifications.

The Plesetsk Cosmodrome is situated in a heavily wooded area close to the Arctic Circle near the town of Plesetsk, which is on the railway line from Moscow to Archangelsk. Historically, it has been the world's most used launch site with more launches than the rest of the world combined. Although used for some civilian communications, meteorological, and international scientific launches, most satellites launched from Plesetsk have military roles and it may be considered as the CIS equivalent of the Vandenberg Air Force Base. Both the F-1-m and F-2 can be accommodated at Plesetsk, although currently only the F-2 is launched from this site. Two F-series pads exist at Plesetsk, but it is not known if both are operational.

The Tsyklon is assembled, mated with the spacecraft, and transported to the pad in a horizontal position. When the rocket is placed vertically on the launch pad, manual access to the rocket and spacecraft is unavailable. The ground servicing system of the Tsyklon provides for a high degree of automation of prelaunch and launch operations.

Launch Facilities

No illustrations available.

Flight Sequence

F-2 Single-Burn Third Stage Flight Sequence

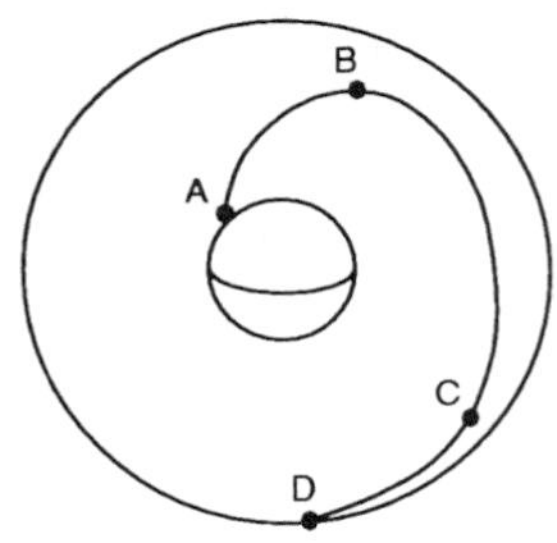

AB - Active segment of the flight trajectory of the first and second stages with a duration of approximately 280 sec.

BC - First passive segment of the third stage flight trajectory with a duration of not less than 40 sec.

CD - Active segment of the third stage flight trajectory with a duration of 5–118 sec.

F-2 Double-Burn Third Stage Flight Sequence

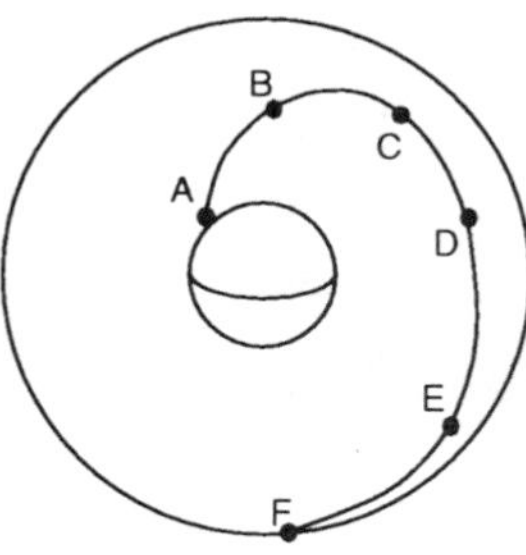

AB - Active segment of the flight trajectory of the first and second stages with a duration of approximately 280 sec.

BC - First passive segment of the third stage flight trajectory with a duration of not less than 40 sec.

CD - First active segment of the third stage flight trajectory with a duration of 5–113 sec.

DE - Second passive segment of the third stage flight trajectory with a duration of not less than 150 sec.

EF - Second active segment of the third stage flight trajectory with a duration of 5–113 sec.

Tsyklon Payload Accommodations

Payload Compartment	
Maximum Payload Diameter	F-1-m - 63.8 in (1,620 mm) F-2 - 90.6–95.3 in (2,300–2,420 mm)
Maximum Cylinder and Cone Length	F-1-m - 244.1 in (6,200 mm) F-2 - 133.9 in (3,400 mm) for cylinder, 98.4 in (2,500 mm) for cone
Payload Adapter	
Interface Diameter	42.5 in (1,080 mm)
Payload Integration	
Nominal Mission Schedule Begins	T-12 months
Launch Window	
Latest Countdown Hold Not Requiring Recycling	T-?? min
On Pad Storage Capability	2 hours for a fueled vehicle
Latest Access to Payload	T-30 hours
Environment	
Maximum Load Factors	+13.5 g axial (F-1-m), +10 g axial (F-2), ±0.2 g lateral (both)
Minimum Lateral / Longitudinal Payload Frequency	10-20 Hz all axes
Maximum Overall Acoustic Level	150 dB (full octave)
Maximum Flight Shock	1,000 g from 1,000–2,000 hz
Dynamic Pressure at Fairing Jettison	~0 lb/ft^2 (0 N/m^2)
Maximum Pressure Change in Fairing	0.3 psi/s (2.0 kPa/s)
Cleanliness Level in Fairing (Prior to Launch)	Class 100,000
Payload Delivery	
Standard Orbit and Accuracy (3 sigma)	Orbit: 351 nm (650 km) / 513 nm (950 km) / 810 nm (1,500 km) Radius: ±8.1 nm (15 km) / ±10.3 nm (19 km) / ±13.5 nm (25 km) Period: ±5 sec / ±8 sec / ±12 sec Inclination: ±3 min / ±3 min / ±3 min
Attitude Accuracy (3 sigma)	All axes: ±2 deg, ±0.3 deg/sec After separation: ±3 deg, ±3 deg/sec
Nominal Payload Separation Rate	4.1 ±0.8 ft/s (1.25 ±0.25 m/s)
Deployment Rotation Rate Available	?? rpm
Loiter Duration in Orbit	2.25 hours
Maneuvers (Thermal / Collision / Telemetry)	No

Tsyklon Notes

Publications

User's Guide

"Tsyklon" Carrier Rocket Preliminary Technical Requirements for Utilization, Glavkosmos/Space Commerce Corp., June 1989.

"Tsyklon User's Guide", NPO Yuzhnoye, Dniepropetrovsk, Ukraine.

Technical Publications

"Soviet Commercial Launch Vehicles", A. Dula, Space Commerce Corp., B. Gubanov, NPO Energia, Y. Smetanin, Glavkosmos, AIAA-89-2743, AIAA 25th Joint Propulsion Conference, Monterey, CA, July 10-12, 1989.

TRW Space Log, 1957-1991, 1992, 1993 TRW, El Segundo, CA.

The Soviet Year in Space 1986, 1988, 1989, Nicholas Johnson, Teledyne Brown Engineering, Colorado Springs, CO.

"Opportunities from Soviet Space Industry, A Commercial User's Guide", A. Dula, Space Commerce Corp., A.I. Dunayev, Glavkosmos, Paris Air Show 1989.

Soviet Space Programs: 1981-1987, United States Senate, Committee on Commerce, Science, and Transportation - Part 1 - May 1988.

Jane's Spaceflight Directory 1987, Jane's Publishing Inc., New York, NY.

Janes's Space Directory 1994-95, Jane's Information Group Inc., Alexandria, VA.

"Soviet Launch Vehicle Designations", Ralph Gibbons, Spaceflight, pg. 54-60,80, February 19, 1977.

Space Directory of Russia, February 1993 Edition, Euroconsult/Sevig Press, Paris, France.

The Russian Space Directory 1994, European Space Report, Munich, Germany.

Acronyms

CIS - Commonwealth of Independent States
FOBS - Fractional Orbit Bombardment System
GEO - geosynchronous orbit
GTO - geosynchronous transfer orbit
LEO - low Earth orbit
MZ - Machine Building Factory
N2O4 - nitrogen tetroxide
NPO - Scientific Production Organization
UDMH - unsymmetrical dimethylhydrazine
USSR - Union of Soviet Socialist Republics

Zenit

Launch Services Point of Contact:
NPO Yuzhnoye
Krivorozhskaya Street 3
Dniepropetrovsk, 8
Ukraine 320008
Phone: (70562) 42-00-22
Fax: (70562) 92-50-41
Telex: 143547 BRON

Photo Courtesy of Space Commerce Corp.

Zenit, one of the newest launch vehicles in the CIS rocket fleet, is being transported by a rail transport erector to its highly automated launch pad at the Baikonur Cosmodrome.

Zenit History

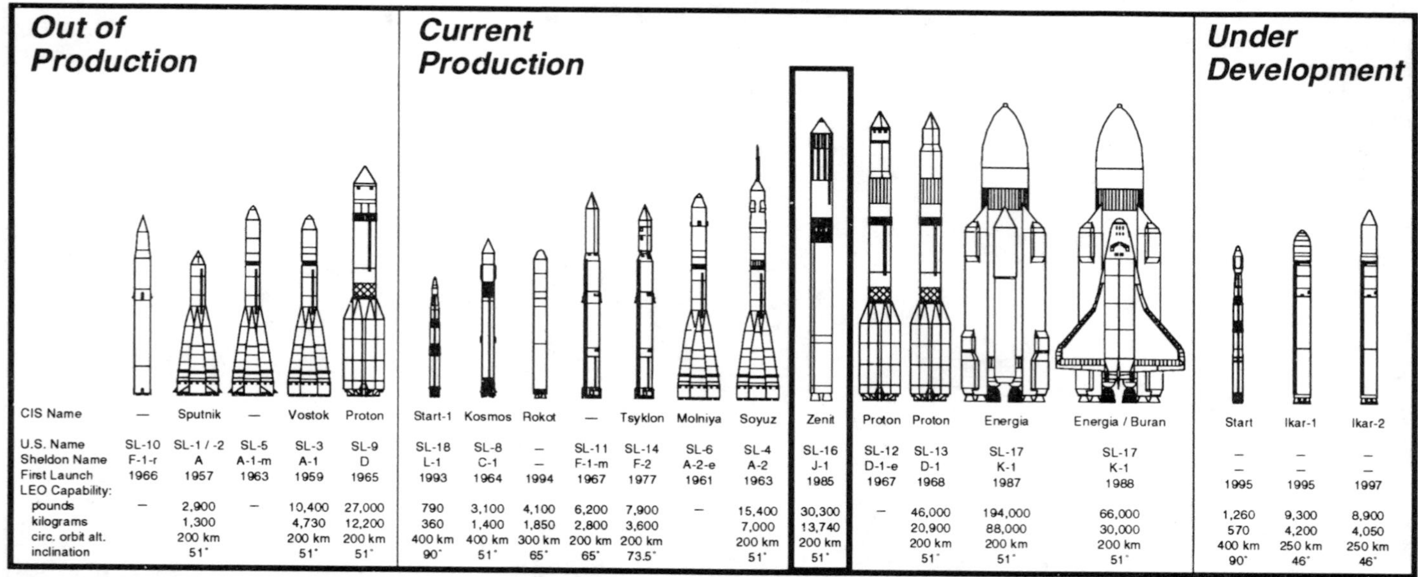

Out of Production

CIS Name	—	Sputnik	—	Vostok	Proton
U.S. Name	SL-10	SL-1 / -2	SL-5	SL-3	SL-9
Sheldon Name	F-1-r	A	A-1-m	A-1	D
First Launch	1966	1957	1963	1959	1965
LEO Capability:					
pounds	—	2,900	—	10,400	27,000
kilograms		1,300		4,730	12,200
circ. orbit alt.		200 km		200 km	200 km
inclination		51°		51°	51°

Current Production

CIS Name	Start-1	Kosmos	Rokot	—	Tsyklon	Molniya	Soyuz	Zenit	Proton	Proton	Energia	Energia / Buran
U.S. Name	SL-18	SL-8	—	SL-11	SL-14	SL-6	SL-4	SL-16	SL-12	SL-13	SL-17	SL-17
Sheldon Name	L-1	C-1	—	F-1-m	F-2	A-2-e	A-2	J-1	D-1-e	D-1	K-1	K-1
First Launch	1993	1964	1994	1967	1977	1961	1963	1985	1967	1968	1987	1988
LEO Capability:												
pounds	790	3,100	4,100	6,200	7,900	—	15,400	30,300	—	46,000	194,000	66,000
kilograms	360	1,400	1,850	2,800	3,600		7,000	13,740		20,900	88,000	30,000
circ. orbit alt.	400 km	400 km	300 km	200 km	200 km		200 km	200 km		200 km	200 km	200 km
inclination	90°	51°	65°	65°	73.5°		51°	51°		51°	51°	51°

Under Development

CIS Name	Start	Ikar-1	Ikar-2
U.S. Name	—	—	—
Sheldon Name	—	—	—
First Launch	1995	1995	1997
LEO Capability:			
pounds	1,260	9,300	8,900
kilograms	570	4,200	4,050
circ. orbit alt.	400 km	250 km	250 km
inclination	90°	46°	46°

Vehicle Description

CIS vehicles are identified either by their CIS, U.S., or Sheldon name. In the Soviet Union it was standard practice to name a launch vehicle after its original payload (e.g., Kosmos, Proton). The U.S. names (developed by the U.S. Department of Defense) are alphanumeric designators based roughly on chronological appearance. The Sheldon names, a most commonly used system that was published by Dr. Charles Sheldon of the U.S. Library of Congress in 1968, emphasize the basic families of launch vehicles with special indicators for variants within a family.

Example: D-1-e

- e - earth escape or fourth stage
- m - maneuverable stage
- r - reentry stage
- Upper Stage (1, 2) - Note: "1" or "2" stage may vary among different families, e.g., A-1 upper stage different than B-1.
- Family (A, B, C, D, F, G, J, K)

Zenit-2 (J-1/SL-16)	Two stage vehicle using LOX and kerosene liquid propellant. The first stage is also used on Energia as strap-ons. (The "-2" in Zenit-2 refers to the number of stages).
Zenit-3 (J-1/SL-16)	Same as Zenit-2 except for addition of a third stage using LOX and kerosene propellants. The third stage is based on the Proton's Block DM fourth stage.

Historical Summary

Zenit (J-1, SL-16), which first appeared in 1985 with two suborbital and two orbital tests, was the first totally new Soviet launch vehicle in 20 years. It was officially announced to the world as "Zenit" by NPO Yuzhnoye in May 1989. The basic Zenit vehicle consists of two LOX / kerosene stages, with the first stage being very similar to the Energia launch vehicle's strap-on boosters. A three stage variant, employing a third stage based on Proton's stage 4, is under development; the first Zenit-3 is expected to be launched in 1997. All launches are currently from the Baikonur Cosmodrome near Tyuratam; two pads are under construction at Plesetsk (the first may be available as early as 1995). In 1990–1992, Zenit experienced three consecutive failures, including one which resulted in the destruction of a launch pad; since then, Zenit has had eight successful launches with no failures.

The relationship between the Ukrainian and Russian governments regarding Ukrainian-built space hardware is unclear. Although Zenit is manufactured by the Ukrainian NPO Yuzhnoye organization, many of its systems, including main propulsion, are designed and built in Russia. The primary user of the Zenit appears to be the Russian military, although it is slated to carry manned and unmanned spacecraft associated with the International Space Station in coming years.

A number of international ventures designed to market Zenit commercially have appeared since the late 1980s, but none have come to fruition. In 1989, the privately owned Cape York Space Agency selected Zenit-3 for commercial launch from a proposed new launch site located on the eastern coast of Australia's Cape York. Under this arrangement, the former Soviet space marketing agency Glavkosmos was to have supplied the Zenit vehicles and provided training for the launch site crew, with marketing, launch site management, and operations to have been handled by CYSA and United Technologies of the United States. However, due to lack of investor support, this plan eventually faded away. More recently, a proposal has been put forth by RSC Energia, NPO Yuzhnoye, and Boeing to launch Zenit from a floating platform at sea. This plan is currently being evaluated by the involved parties for technical feasibility, as well as political and economic viability.

Zenit History

(continued)

Launch Record

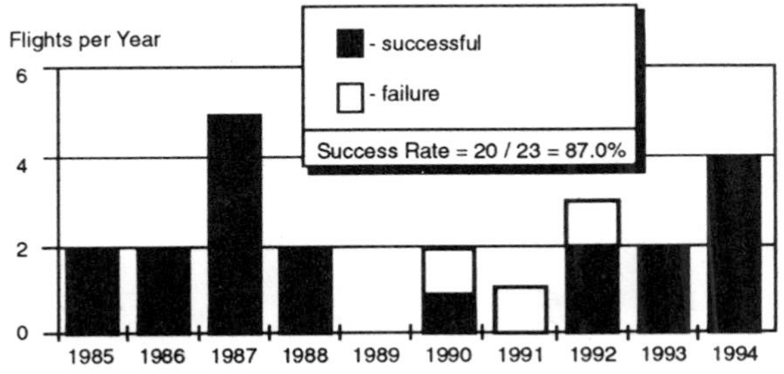

	YEAR	DATE	VEHICLE	SITE	PAYLOAD
*	1985	Apr-13	Zenit-2	Tyuratam	Suborbital test launch
*		Jun-21	Zenit-2	Tyuratam	Suborbital test launch
1		Oct-22	Zenit-2	Tyuratam	Kosmos-1967 (71°)
2		Dec-28	Zenit-2	Tyuratam	Kosmos-1714 (71°)
3	1986	Jul-30	Zenit-2	Tyuratam	Kosmos 1767 (65°)
4		Oct-22	Zenit-2	Tyuratam	Kosmos 1786 (65°)
5	1987	Feb-14	Zenit-2	Tyuratam	Kosmos 1820 (65°)
6		Mar-18	Zenit-2	Tyuratam	Kosmos 1833 (71°)
7		May-13	Zenit-2	Tyuratam	Kosmos 1844 (71°)
8		Aug-1	Zenit-2	Tyuratam	Kosmos 1871 (97°)
9		Aug-28	Zenit-2	Tyuratam	Kosmos 1873 (65°)
10	1988	May-15	Zenit-2	Tyuratam	Kosmos 1943 (71°)
11		Nov-23	Zenit-2	Tyuratam	Kosmos 1980 (71°)
12	1990	May-23	Zenit-2	Tyuratam	Kosmos 2082 (71°)
13		Oct-4	Zenit-2	Tyuratam	??
		Failure: Explosion over pad, first stage engine failure			
14	1991	Aug-30	Zenit-2	Tyuratam	??
		Failure: second stage engine failure			
15	1992	Feb-5	Zenit-2	Tyuratam	??
		Failure: second stage engine failure			
16		Nov-17	Zenit-2	Tyuratam	Kosmos 2219 (71°)
17		Dec-25	Zenit-2	Tyuratam	Kosmos 2227 (71°)
18	1993	Mar-26	Zenit-2	Tyuratam	Kosmos 2237 (71°)
19		Sep-16	Zenit-2	Tyuratam	Kosmos 2263 (71°)
20	1994	Apr-23	Zenit-2	Tyuratam	Kosmos 2278 (71°)
21		Aug-26	Zenit-2	Tyuratam	Kosmos 2290 (64.8°)
22		Nov-4	Zenit-2	Tyuratam	Resurs-O1 (98°)
23		Nov-24	Zenit-2	Tyuratam	Kosmos 2297 (71°)

* Suborbital tests not counted as flights

Data Sources: TRW Space Log, NPO Yuzhnoye

Zenit General Description

Zenit-2 Zenit-3

Summary

Zenit is a a two-stage (Zenit-2) or three-stage (Zenit-3) vehicle that utilizes LOX and kerosene. The first stage is essentially the strap-on boosters used for Energia. The second stage has a sustainer engine with four vernier engines for control. The optional third stage is essentially the Proton Block-DM fourth stage. Zenit-2 is used for circular orbits up to 810 nm (1,500 km) high; Zenit-3 is planned for high circular (including geostationary) and elliptical orbits, as well as interplanetary missions. Zenit-2 was first launched in 1985 and Zenit-3 is expected to be launched in 1997. All launches occur from the Baikonur Cosmodrome near Tyuratam. Currently, Zenit supports primarily a small military (ELINT) program, but within several years, it is slated to carry manned and unmanned spacecraft associated with the International Space Station, remote sensing and scientific satellites, and possible photographic reconnaissance platforms. Numerous international commercial marketing plans utilizing alternate launch sites have been proposed.

Status

Operational

Key Organizations

Users - Russian Ministry of Defense, Russian Space Agency, National Space Agency of Ukraine, commercial

Responsible Organizations -
- NPO Yuzhnoye (design/development)
- MZ Yuzhmash (manufacturer)
- NPO Energomash (stage 1 and 2 main engines)
- RSC Energia (stage 3)

Vehicle

System Height	187 ft (57.0 m) for Zenit-2 196 ft (59.6 m) for Zenit-3
Payload Fairing Size	Zenit-2: Short Fairing - 12.8 ft (3.9 m) diameter by 36.6 ft (11.2 m) height Long Fairing - 12.8 ft (3.9 m) diameter by 44.8 ft (13.6 m) height Zenit-3: 13.5 ft (4.1 m) diameter by 37.2 ft (11.4 m) height
Gross Mass	1,012K lb (459K kg) for Zenit-2 1,034K lb (469K kg) for Zenit-3
Planned Enhancements	Multiple payloads

Operations

Primary Missions	LEO, SSO, polar for Zenit-2 HEO, GTO, GEO, interplanetary for Zenit-3
Compatible Upper Stages	Zenit-3 third stage based on Proton's Block DM
First Launch	1985 for Zenit-2, projected 1997 for Zenit-3
Success / Flight Total	20 / 23 for Zenit-2, 0 / 0 for Zenit-3
Launch Site	Baikonur Cosmodrome (45.6°N, 63.4°E) 2 pads: one operational, one destroyed.
Launch Azimuth	Nominally 35°–194.4° (inclinations of 51°–99°)
Nominal Flight Rate	1–5 / yr
Planned Enhancements	Two pads under construction at Plesetsk for possible 1995 availability; various commercial plans under consideration

Performance

108 nm (200 km) circ, 51°	30,300 lb (13,740 kg) for Zenit-2 from Tyuratam
108 nm (200 km) circ, 99°	25,090 lb (11,380 kg) for Zenit-2 from Tyuratam
Geotransfer Orbit 189 nm (350 km) x 19,300 nm (35,800 km), 43.3°	11,420 lb (5,180 kg) for Zenit-3 from Tyuratam
Geosynchronous Orbit 19,300 nm (35,800 km) circ, 0°	3,384 lb (1,535 kg) for Zenit-3 from Tyuratam
High Circular Orbit 10,800 nm (20,000 km) circ, 64.8°	6,020 lb (2,730 kg) for Zenit-3 from Tyuratam
High Elliptical Orbit 810 nm (1,500 km) x 21,600 nm (40,000 km) circ, 62.8°	8,710 lb (3,950 kg) for Zenit-3 from Tyuratam

Financial Status

Estimated Launch Price	Zenit-2: $35–45 million (NPO Yuzhnoye) Zenit-3: $50–70 million (NPO Yuzhnoye)

Manifest

Not available

Remarks

—

Zenit Vehicle

Overall

	Zenit-2	Zenit-3
Length	187 ft (57.0 m)	196 ft (59.6 m)
Gross Mass	1,012,000 lb (459,000 kg)	1,034,000 lb (469,000 kg)
Thrust at Liftoff	1,632,000 lb (7,259,000 N)	1,611,000 lb (7,169,100 N)

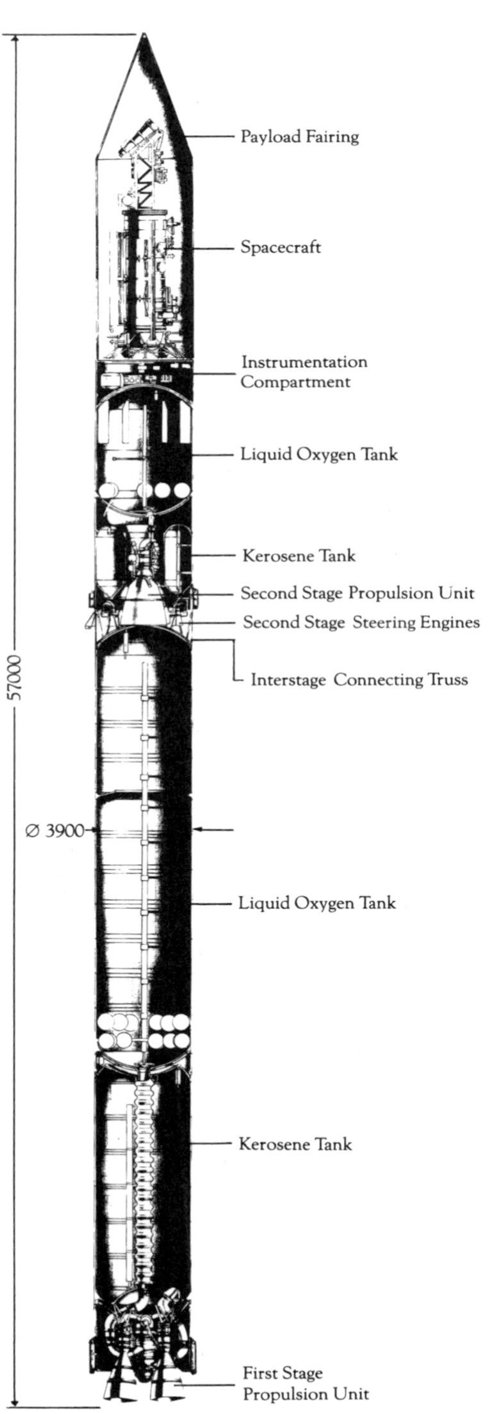

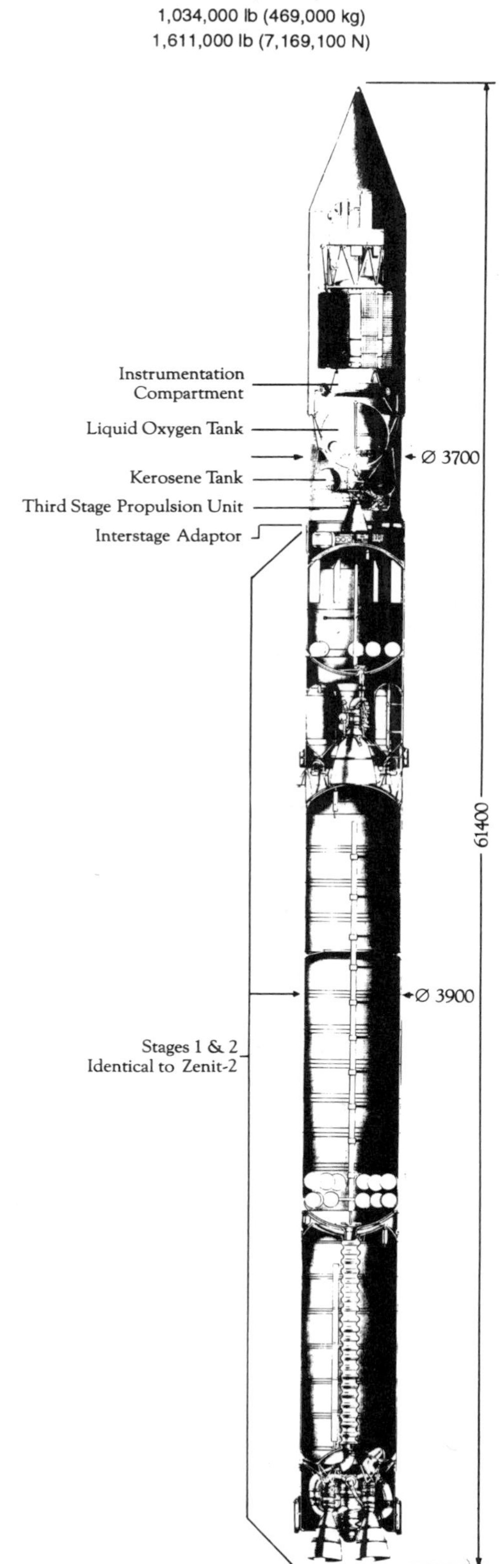

Zenit Vehicle

(continued)

Zenit Stages	Stage 1 (Zenit-2)	Stage 1 (Zenit-3)	Stage 2	Stage 3 for Zenit-3 only
Dimension:				
Length	108 ft (32.9 m)	108 ft (32.9 m)	37.7 ft (11.5 m)	18 ft (5.5 m)
Diameter	12.8 ft (3.9 m)	12.8 ft (3.9 m)	12.8 ft (3.9 m)	12 ft (3.7 m)
Mass: (each)				
Propellant Mass	719.4K lb (326.3K kg)	703K lb (326.3K kg)	180.8K lb (82.0K kg)	31.3 lb (14.2K kg)
Gross Mass	780.9K lb (354.2K kg)	778K lb (354.2K kg)	199.3K lb (90.4K kg)	38.1K lb (17.3K kg)
Structure:				
Type	Wafer	Wafer	Wafer	Skin-Stringer
Material	Aluminum	Aluminum	Aluminum	Aluminum
Propulsion:				
Propellant	LOX / kerosene	LOX / kerosene	LOX / kerosene	LOX / kerosene
Average Thrust (each)	1,632K lb (7,259 kN) SL 1,779K lb (7,911 kN) vac	1,723K lb (7,669 kN) SL 1,838K lb (8,181 kN) vac	187K lb (834 kN) vac sustainer 18K lb (78 kN) vac vernier	16K lb (71 kN) vac
Engine Designation	RD-170	RD-173	RD-120 sustainer+RD-8 vernier	11D58M (?)
Number of Engines	1 turbopump + 4 chambers	1 turbopump + 4 chambers	1 sustainer + 1 vernier with 1 turbopump + 4 chambers	1
Isp	309 sec SL 337 sec vac	311 sec SL 337 sec vac	350 sec vac sustainer 342 sec vac vernier	361.0 sec vac
Feed System	Staged combustion	Staged combustion	Staged combustion sustainer	Staged combustion
Chamber Pressure	3,556 psia (245 bar)	3,700 psia (255 bar)	2,364 psia (163 bar) sustainer 1,117 psia (77 bar) vernier	1,124 psia (77.5 bar)
Mixture Ratio (O/F)	2.63	2.63 (?)	2.58	2.53
Throttling Capability	49–102%	49–102%	85–100% sustainer 100% vernier	100% only
Expansion Ratio	37:1	37:1 (?)	106:1 sustainer 104:1 vernier	189:1
Restart Capability	No	No	No (vernier no)	7 restarts
Tank Pressurization	Stored pressurized gas	Stored pressurized gas	Stored pressurized gas	Stored pressurized gas
Control-Pitch,Yaw,Roll	Gimbal 4 nozzles (±5°)	Gimbal 4 nozzles (±5°)	4 chamber vernier	Powered - gimbal engine Coast - ACS
Events:				
Nominal Burn Time	140–150 sec	140–150 sec	200–315 sec sustainer 300–1100 sec vernier	0.8–660 sec
Stage Shutdown	Command shutdown	Command shutdown	Command shutdown	Command shutdown
Stage Separation	Retro-rockets	Retro-rockets	Retro-rockets	Spring ejection

Remarks:

The Zenit launch vehicle is made up of either two or three tandem stages. The first stage has a nested dome tank with the LOX tank above the kerosene tank. The RD-170 first stage engine, with its four thrust chambers, is the most powerful liquid propellant rocket engine ever built. This engine has been qualified for up to 10 reuses as part of the Energia launch vehicle program. The Zenit-3 will employ an uprated version of the RD-170, designated RD-173. The second stage has a cylindrical LOX tank and a toroidal tank for kerosene. The instrumentation compartment is located atop the final stage (second stage for Zenit-2 and third stage for Zenit-3).

The third stage used on Zenit-3 is located inside an interstage and has a spherical and a toroidal tank connected by trusses. The third stage is used to transfer payloads into geostationary orbit, highly elliptical orbits, and interplanetary trajectories. This stage is based on Proton's Block DM stage, with major differences being a reduced nominal propellant load and downrated engine thrust. It is equipped with attitude and stabilization systems and control equipment for telecommunication, telemetry, and electric power generation to support the payload during boost and until separation on station. Up to five engine restarts have been demonstrated on Proton, and seven are possible within 48 hours of launch. It has a single engine that burns kerosene and LOX.

Zenit Vehicle
(continued)

Payload Fairing

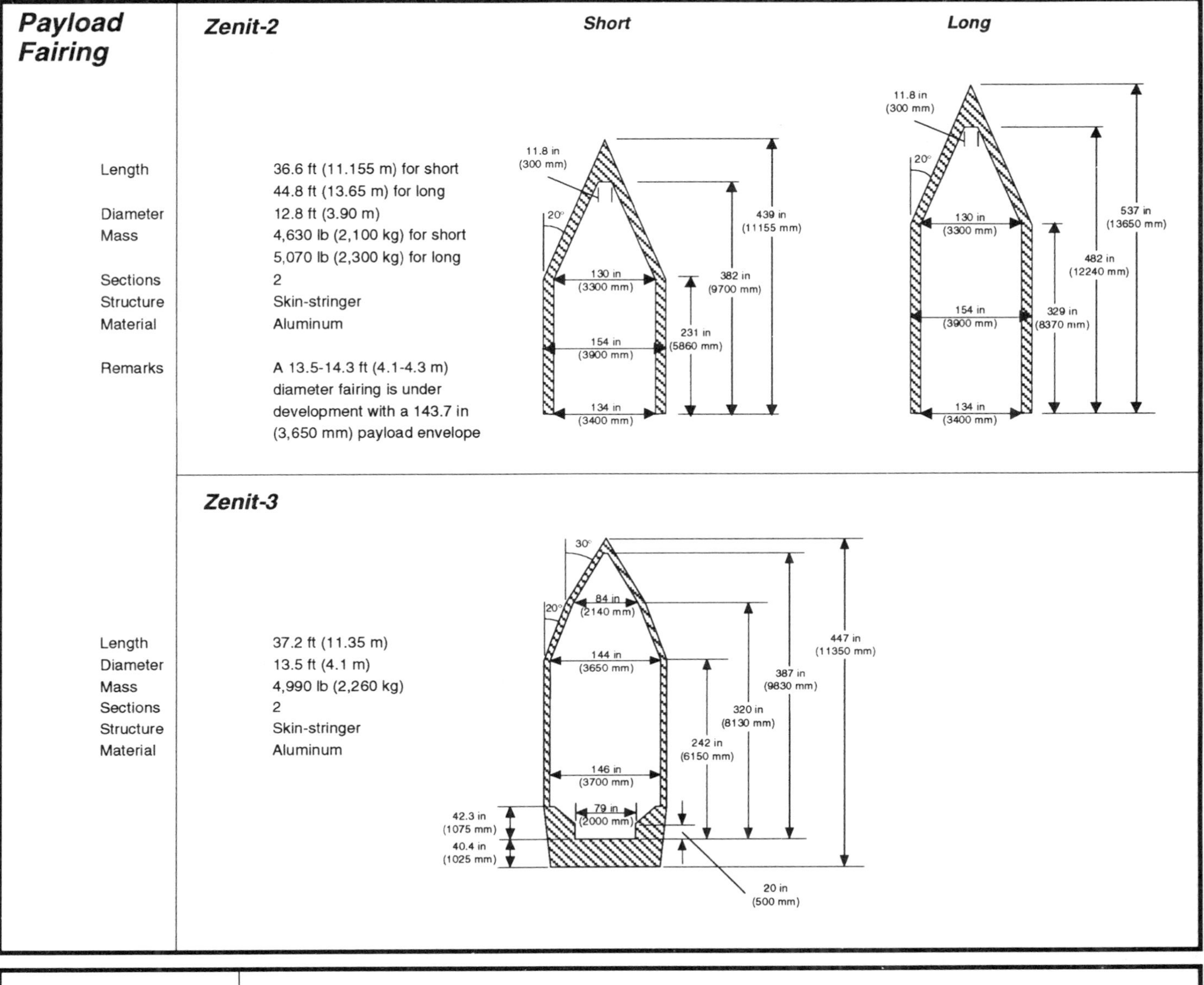

Zenit-2

Length	36.6 ft (11.155 m) for short 44.8 ft (13.65 m) for long
Diameter	12.8 ft (3.90 m)
Mass	4,630 lb (2,100 kg) for short 5,070 lb (2,300 kg) for long
Sections	2
Structure	Skin-stringer
Material	Aluminum
Remarks	A 13.5-14.3 ft (4.1-4.3 m) diameter fairing is under development with a 143.7 in (3,650 mm) payload envelope

Zenit-3

Length	37.2 ft (11.35 m)
Diameter	13.5 ft (4.1 m)
Mass	4,990 lb (2,260 kg)
Sections	2
Structure	Skin-stringer
Material	Aluminum

Avionics

The final stage (either second stage for Zenit-2 or third stage for Zenit-3) is equipped with a digital flight control system to control the vehicle.

Attitude Control System

The four thrust chambers of the first stage can be individually gimballed for control. The second stage has a fixed sustainer engine and a four-chamber vernier. The third stage uses main engine gimbaling for control during powered flight, and a self-contained ACS for three-axes attitude control during coast periods. The ACS employs a collection of small thrusters which burn hypergolic N2O4 and UDMH propellants. It is also possible to spin up the third stage to 0.7 rpm with this system. A higher spin rate would require incorporation of a new or upgraded spin jet system.

Zenit Performance

Performance Characteristics of Zenit-2 Launched from Baikonur Cosmodrome (courtesy NPO Yuzhnoye)
Note: Although technically feasible, launch into 46.2° inclination is not within the scope of normal operations.

Zenit-2 Performance as a Function of Circular Orbit Altitude for Various Inclinations

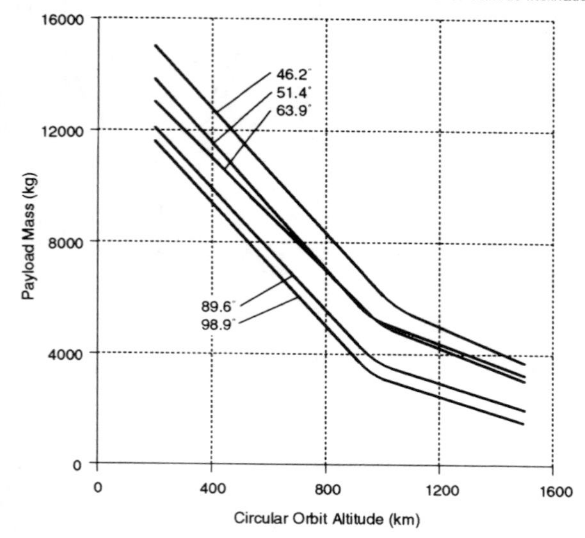

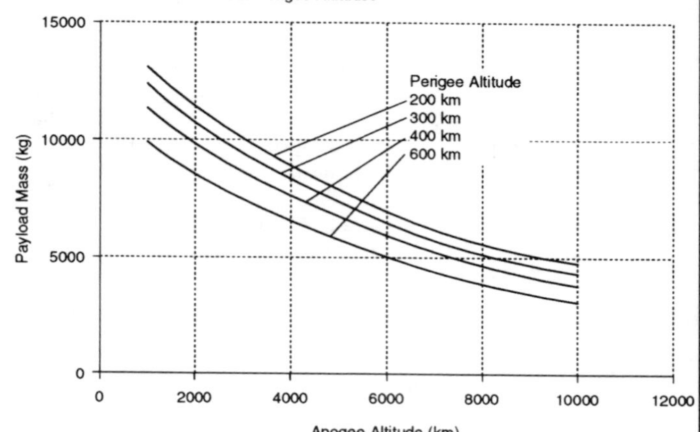

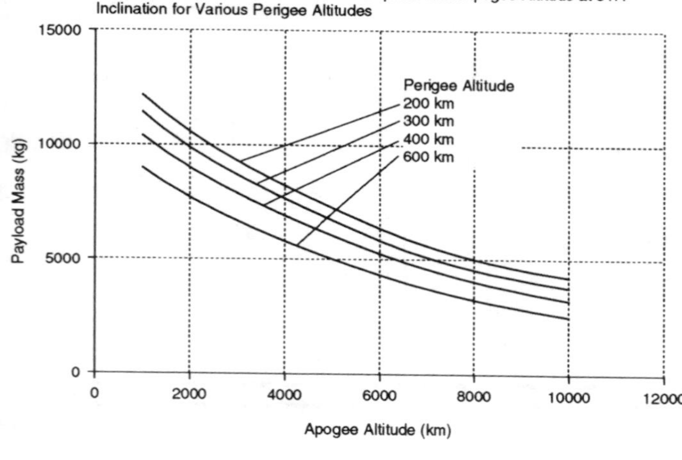

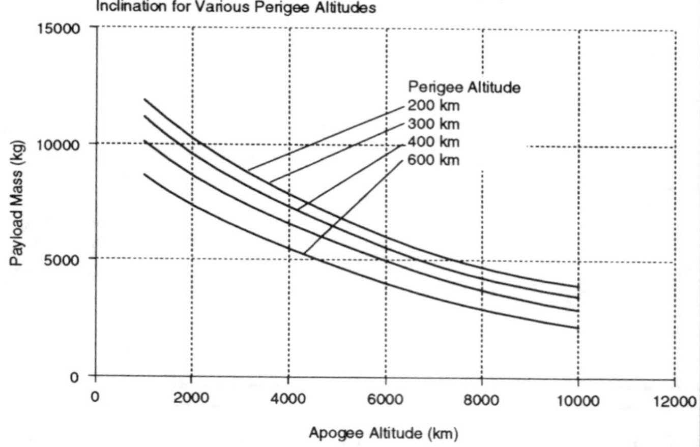

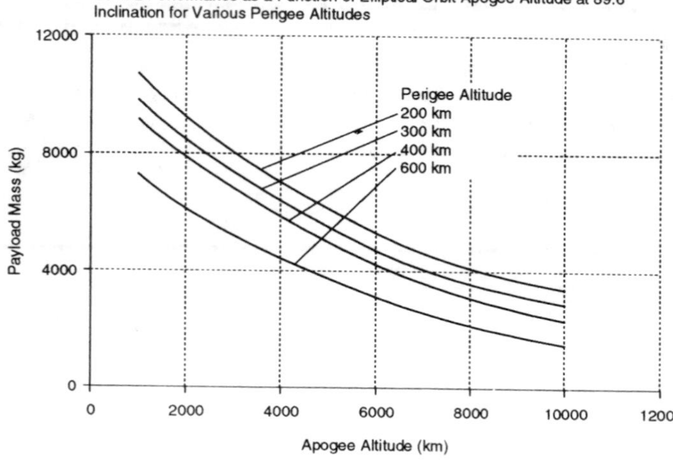

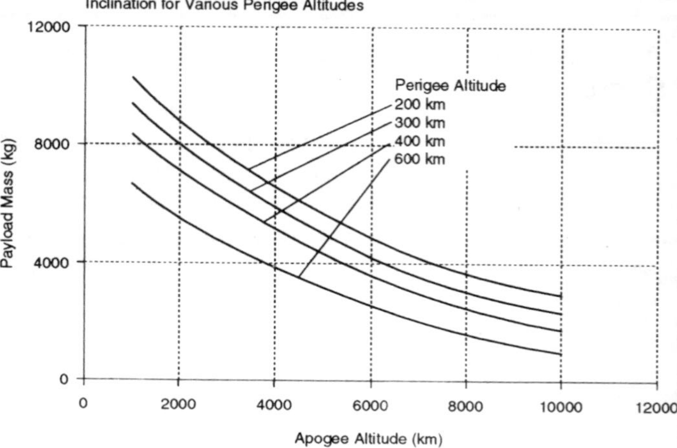

Zenit Operations

Launch Site

Note: Plesetsk launch pads under construction, with the first scheduled to become operational in 1995.

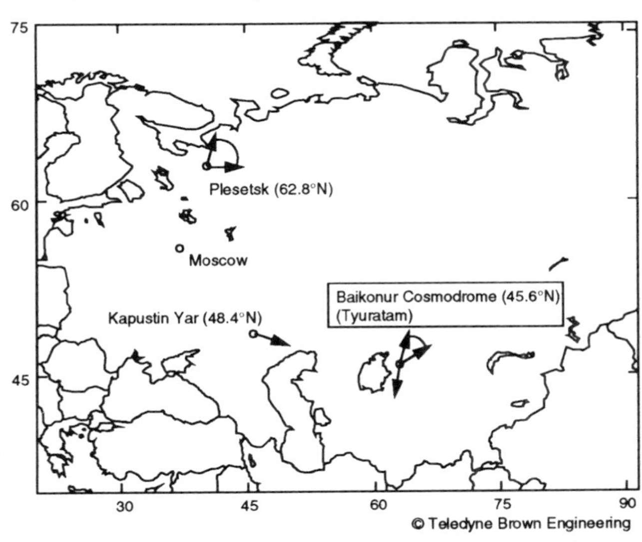

Launch Facilities

Tyuratam

Zenit Launch Pad Layout at Tyuratam

NPO Energia Concept for Dual Floating Platform Zenit Sea-Launch System

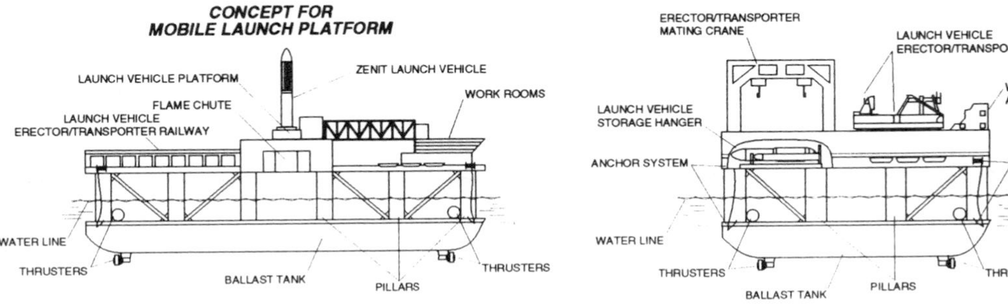

Zenit Operations

(continued)

Launch Processing

Although the Zenit vehicle is produced in Ukraine, it continues to be launched from the same site, and apparently by the same organizations, as it was before the breakup of the Soviet Union. A general attitude of cooperation seems to exist between the Ukrainian Space Agency, the Russian Space Agency, the government of Kazakhstan, and the Russian Space Forces (the branch of the Russian Ministry of Defense that actually operates the launch facilities). Ukraine has consistently expressed a desire for independence in space operations, but its lack of a launch site within its own borders mandates cooperation with other states of the former Soviet Union.

Tyuratam. Zenit is launched from the Baikonur Cosmodrome located just east of the Aral Sea in Kazakhstan, north of Afghanistan. This site is usually referred to in the West as Tyuratam, a reference to a nearby railway station. Tyuratam, the CIS equivalent to the U.S. Cape Canaveral launch site, is where all manned, geosynchronous, lunar, interplanetary, antisatellite, and tactical ocean surveillance missions are launched. It is the only facility capable of launching the Proton, Zenit, and Energia, although one Zenit pad is scheduled to enter service at Plesetsk in 1995. All other CIS vehicles can be launched from Tyuratam except the Tsyklon and Kosmos launchers. Typically, 30 to 40 launches per year are flown there. It is not unusual for two missions to be launched at Tyuratam less than 24 hours apart. Tyuratam is the only site which has conducted launches in retrograde orbits. Zenit has never been launched from the Baikonur Cosmodrome into inclinations less than 51° due to range safety restrictions, particularly those concerning China. A due east launch into an inclination of 46.2° is possible, with all planned hardware drops occurring inside Kazakhstan; however, such a launch is not within the scope of normal operations and would require extreme circumstances.

The climate at Baikonur is hot in the summer, with temperatures rising as high as 120°F (50°C), and prone to violent snowstorms and -40°F (-40°C) lows in winter. A unique feature of the CIS space program is the ability for its vehicles to launch in these harsh climate conditions. The large Y-shaped complex covers about 55 mi (88 km) north to south and 100 mi (160 km) east to west. Each type of vehicle has its own processing and launch facilities, with the Zenit area located on the eastern end of the Baikonur complex. The vehicle processing and launch areas are connected to each other and the city of Leninsk by nearly 300 mi (500 km) of railroad lines.

For Zenit, all of the launcher assembly, payload integration and launch preparation phases are described as "highly automated". Assembly can be completed in 80 hours (or 10 days at one eight hour shift per day). Once the vehicle is assembled, it can be stored for up to a year. Payload installation adds 4.5 days to the processing for a total of 14.5 days. The vehicle is then moved to the launch pad and positioned upright.

Like all other CIS launchers, Zenit is assembled horizontally. Interfaces between onboard electrical, pneumatic, and hydraulic systems and their corresponding ground-support systems are fully automatic, a term also applied to the continuous system integrity control. A stable thermal environment for the payload is maintained throughout the preparation phase. The checkout and testing of Zenit is performed in the systems integration and checkout building with the launch vehicle in horizontal position. Testing of the vehicle's onboard systems is conducted in both ground and simulated-flight modes.

Prior to launch, the vehicle is transported horizontally to the dual-pad launch site. Currently, only one of the two Zenit pads is operational, the other having been destroyed when a Zenit exploded at liftoff on October 4, 1990. Zenit launch pad operations, including erection, propellant loading, and checkout, require minimal human intervention. Zenit can be launched with 21 hours of arrival at the launch pad.

Plesetsk. Multiple Zenit pads are under construction at the Russian Plesetsk Cosmodrome; the first of these pads may enter service as early as 1995. When Zenit launch capability from Plesetsk is realized, the Zenit will become by far the largest rocket launched from this site.

Plesetsk is located in a heavily wooded area near the Arctic Circle at 62.8° N latitude, on the rail line between Moscow and Archangelsk. Historically, it has been the most prolific launch site in the world, although the total number of launches from Plesetsk has been declining in recent years. The majority of the launches from Plesetsk are high-inclination military payloads.

Cape York. In 1989, the privately owned Cape York Space Agency developed a proposal to operate a commercial launch site, using Zenit-3 vehicles, on the east coast of Cape York peninsula in north Queensland, Autstralia. The near-equatorial location of the chosen site (12° south latitude) is especially advantageous for geosynchronous launches. Zenit-3 vehicles, along with necessary ground crew training, were to have been provided by the Soviet Glavkosmos agency, with operations to have been supported by United Technologies, Inc. of the United States. However, a lack of investment brought the operation to a halt before construction began. Low-level investigations of this concept are ongoing.

Sea Launch. A proposed partnership between Boeing, NPO Yuzhnoye, and RSC Energia to launch Zenit from a floating sea platform was announced in June 1994. As with the Cape York proposal, the commercial goal of this venture would be to use the Zenit-3 to enter into the potentially lucrative, albeit highly competitive, geosynchronous orbital launch service market. In this partnership, the major components of the Zenit-3 would be manufactured by the appropriate organizations in Russia and Ukraine, with final vehicle assembly and payload integration to take place in the United States. The Kvaerner Group, a Norwegian firm experienced in the design of ocean oil drilling platforms, would provide the floating launch pad. The floating launch platform would be towed into international waters (presumably near the equator to maximize performance) for launches. Although Boeing had already begun discussion of this proposal with U.S. government officials, at the end of 1994 the concept was still characterized as in the study phase by all involved parties.

Flight Sequence

No data available.

Zenit Payload Accommodations

Payload Compartment	
Maximum Payload Diameter	Zenit-2 Short - 130-134 in (3,300–3,400 mm) Long - 130-134 in (3,300–3,400 mm) Zenit-3 - 144-146 in (3,650–3,700 mm)
Maximum Cylinder Length	Zenit-2 Short - 231 in (5,860 mm) Long - 329 in (8,370 mm) Zenit-3 - 200 in (5,075 mm)
Maximum Cone Length	Zenit-2 Short - 151 in (3,840 mm) Long - 152 in (3,870 mm) Zenit-3 - 145 in (3,680 mm)
Payload Adapter	
Interface Diameter	142.5 in (3,620 mm) for Zenit-2 131.9 in (3,350 mm) for Zenit-3 (spacecraft mass greater than 4.5 T) 78.7 in (2,000 mm) for Zenit-3 (spacecraft mass less than 4.5 T)
Payload Integration	
Nominal Mission Schedule Begins	dependent on payload
Launch Window	
Latest Countdown Hold Not Requiring Recycling	T-15 min (Next available launch opportunity is 2 hours)
On-Pad Storage Capability	4 hours for a fueled vehicle
Latest Access to Payload	T-30 hours or T-3 hours through access doors
Environment	
Maximum Load Factors	4-6 g axial, 0.6 g lateral
Minimum Lateral / Longitudinal Payload Frequency	15 Hz / 25 Hz
Maximum Overall Acoustic Level	140 dB from 20–20,000 hz
Maximum Flight Shock	500 g from 1,000–2,000 hz
Maximum Dynamic Pressure at Fairing Jettison	0 lb/ft^2 (0 N/m^2)
Maximum Pressure Change in Fairing	0.7 psi/s (5 KPa/s)
Cleanliness Level in Fairing (Prior to Launch)	Class 100,000
Payload Delivery	
Standard Orbit and Accuracy (3 sigma)	Zenit-2 into 108 nm (200 km) circular orbit Altitude ±1.9 nm (3.5 km) Inclination ±2 min Period ±2.5 sec Zenit-3 into geostationary orbit Altitude ±162 nm (300 km) Inclination ±42 min Period ±900 sec
Attitude Accuracy (3 sigma)	All axes: ±3 deg, ±3 deg/sec for Zenit-2
Nominal Payload Separation Rate	11.5 ft/s (3.5 m/s) for Zenit-2
Deployment Rotation Rate Available	0.7 rpm for Zenit-2 and -3
Loiter Duration in Orbit	0 hours for Zenit-2, 15 min to 23.9 hours for Zenit-3
Maneuvers (Thermal / Collision / Telemetry)	Yes

Zenit Notes

Publications

User's Guides

1993 Zenit User's Guide (in Russian), NPO Yuzhnoye, Dniepropetrovsk, Ukraine.

Technical Publications

"Soviets Hunting for Cause of Zenit Explosion Above Baikonur", Space News, pg. 1, 28, October 15-21, 1990.

"Zenit to Fly from Cape York", Spaceflight, pg. 148-149, May 1990.

"Zenit May Replace Soviet Proton", Space Business News, pg. 6, April 16, 1990.

Zenit (brochure), Cape York Space Agency, 1990.

"Cape York International Spaceport Information Sheet", Cape York Space Agency, 1990.

Space Transportation Propulsion USSR Launcher Technology, Rockwell International, BC-90-71, June 1990.

"Zenit Booster Revealed", Phillip S. Clark, Space Markets, pg. 176-179, 3rd quarter. 1989.

"Soviet Commercial Launch Vehicles", A. Dula, Space Commerce Corp., B. Gubanov, NPO Energia, Y. Smetanin, Glavkosmos, AIAA-89-2743, AIAA 25th Joint Propulsion Conference, Monterey, CA, July 10-12, 1989.

TRW Space Log, 1957-1991, 1992, 1993 TRW, El Segundo, CA.

The Soviet Year in Space 1986, 1988, 1989, Nicholas Johnson, Teledyne Brown Engineering, Colorado Springs, CO.

"USSR RD-170 Liquid Rocket Engine", NASA Johnson Space Center, August 25, 1989.

"Opportunities from Soviet Space Industry, A Commercial User's Guide", A. Dula, Space Commerce Corp., A.I. Dunayev, Glavkosmos, Paris Air Show 1989.

"Dzis I Jutro Systems Energia", Astronautika NR 6, (616), 1989.

Soviet Space Programs: 1981-1987, United States Senate, Committee on Commerce, Science, and Transportation - Part 1 - May 1988.

Jane's Space Directory 1994-95, Jane's Information Group, Inc., Alexandria, VA.

Space Directory of Russia, February 1993 Edition, Euroconsult/Sevig Press, Paris, France.

The Russian Space Directory 1994, European Space Report, Munich, Germany.

"Russian, Ukrainian, U.S. Giants Plan Zenit Venture", *Space News*, pg. 1, June 20-26, 1994.

Acronyms

ACS - attitude control system
CIS - Commonwealth of Independent States
CYSA - Cape York Space Agency
GEO - geosynchronous orbit
GTO - geosynchronous transfer orbit
LEO - low Earth orbit
LH2 - liquid hydrogen
LOX - liquid oxygen
NPO - Scientific Production Organization
RSC - Rocket Space Corporation
USSR - Union of Soviet Socialist Republics

United States

Table of Contents

Atlas

Government Point of Contact:
Atlas System Program Office (CLM)
U.S. Air Force Space Systems Division
P.O. Box 92960
Los Angeles AFB, CA 90009-2960, USA
Phone: (619) 363-3952

Industry Point of Contact:
Lockheed Martin
International Launch Services
101 West Broadway Suite 2000
San Diego, CA 92101
Phone: (619) 974-3800

On July 25, 1990, an Atlas I launched the CRRES payload into orbit for NASA and the U.S. Air Force.

Atlas History

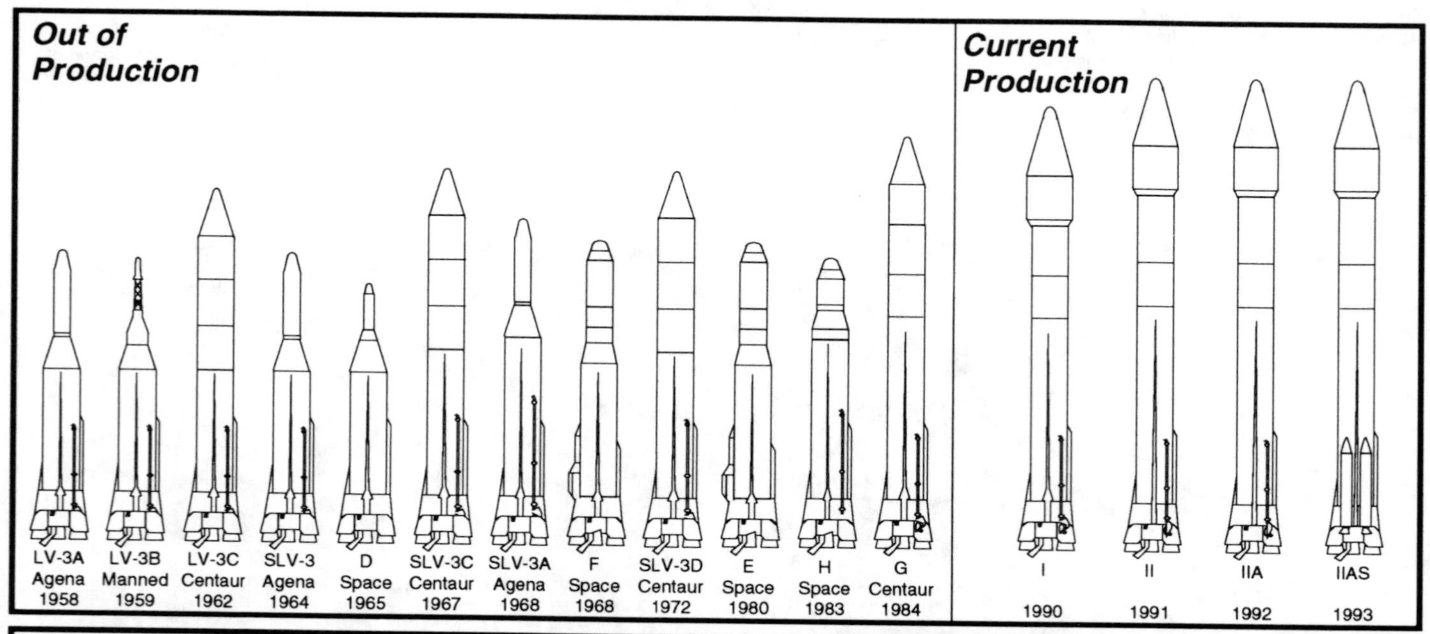

Vehicle Description

A	ICBM single stage test vehicle
B, C	ICBM 1-1/2 stage test vehicle
D	ICBM and later space launch vehicle
E, F	First an ICBM (1960), then a reentry test vehicle (1964), then a space launch vehicle (1968)
LV-3A	Same as D except Agena upper stage
LV-3B	Same as D except man-rated for project Mercury
SLV-3	Same as LV-3A except reliability improvements
SLV-3A	Same as SLV-3 except stretched 117 in (2.97 m)
LV-3C	Launched with Centaur D upper stage
SLV-3C	Same as LV-3C except stretched 51 in (1.30 m)
SLV-3D	Same as SLV-3C except Centaur uprated to D-1A and Atlas electronics integrated with Centaur
G	Same as SLV-3D but Atlas longer 81 in (2.06 m)
H	Same as SLV-3D except with E / F avionics and no Centaur upper stage.
I	Same as G except strengthened for 14 ft (4.27 m) payload fairing, and ring laser gyro added
II	Same as I except Atlas lengthened 108 in (2.74 m), engines uprated, add hydrazine roll control, fixed foam insulation, delete verniers, and Centaur stretched 36 in. (0.9 m)
IIA	Same as II except Centaur RL-10s engines uprated to 20K lbs (88K N) thrust and 6.5 sec Isp increase from extendable RL-10 nozzles
IIAS	Same as IIA except 4 Castor IVA strap-ons added

Atlas Genealogy

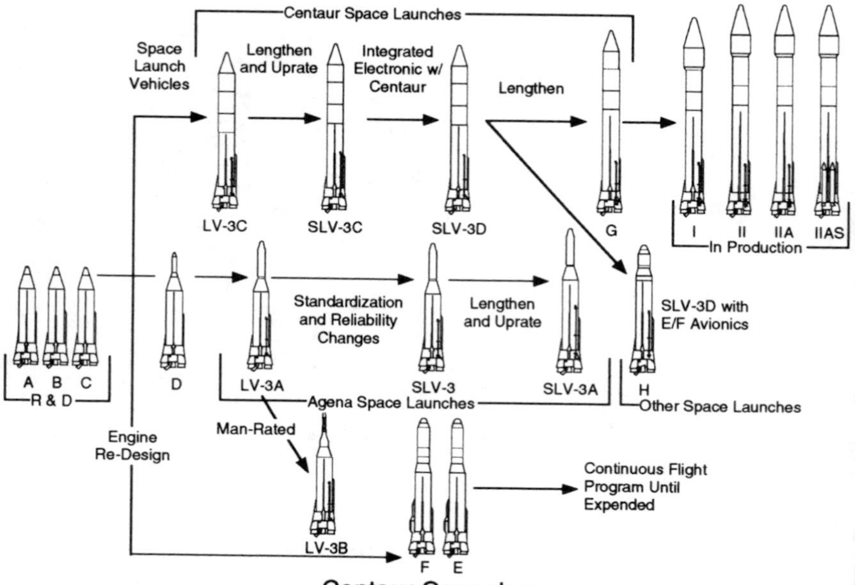

Centaur Genealogy

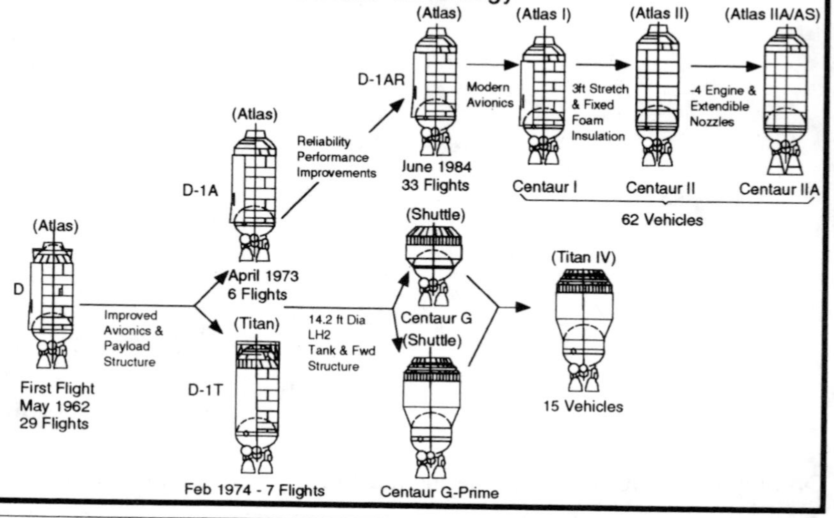

Atlas History

(continued)

Historical Summary

Atlas is produced by Lockheed Martin, which purchased the General Dynamics Space Systems Division and its Atlas and Centaur programs in May 1994.

The Atlas space launch vehicles evolved from the successful Atlas Intercontinental Ballistic Missile (ICBM). The basic 1 1/2-stage vehicle has changed little over the years and has been used for a variety of space missions.

Atlas has been involved in many prominent NASA projects but was considered an Air Force launch vehicle. More than 75% of its flights have been performed for the Air Force. In the early to mid 1960s, the Atlas D, E, and F vehicles were operational as ICBMs, and as many as 159 vehicles were deployed at U.S. missile sites. As Minuteman missiles replaced Atlas ICBMs in the late 1960s, Atlas vehicles were withdrawn and converted for space launch.

The first of several notable space launch missions performed by an Atlas vehicle was Project SCORE, which was the world's first communications satellite. Launched in 1958, this satellite transmitted President Eisenhower's Christmas message. The same year, Project Mercury, the first U.S. manned space program, was initiated, and the Atlas launch vehicle was chosen for this prestigious program. On February 20, 1962, after a successful launch on the man-rated Atlas D, John Glenn became the first U.S. astronaut to orbit Earth.

The Atlas space launch vehicle also was involved in the early lunar exploration missions. All three of the unmanned lunar exploration programs (Ranger, Lunar Orbiter, and Surveyor) used Atlas vehicles. Finally, Mariner probes to Mars, Venus, and Mercury and the Pioneer probes to Jupiter, Saturn, and Venus were launched by Atlas Centaur vehicles.

The Atlas ICBM project was initiated by General Dynamics as project MX-774 for the Air Force in 1945. After being cancelled in 1947 for lack of funds, it was reinstated four years later. The ICBM underwent a major scaling down in 1955 due to breakthroughs in thermonuclear weapons and made its first test flight two years later. From 1957 to 1959, research and development on Atlas produced the A, B, and C versions. A modified Atlas B was used for Project SCORE on one of its 10 successful tests during 1958 and 1959. Development continued with an improved guidance system on the Atlas C. The Atlas A, B, and C versions had a total of 23 research and development flights and led to the first operational Atlas flight using the Atlas D in 1959.

The Atlas D could be called the "granddaddy" of the current operational system. It was launched more times (123) than any other version of Atlas and was man-rated for use on the Mercury program. The Atlas D used a cluster of three engines (two boosters, one sustainer) to comprise its one and one-half stages; this staging scheme was used on all the following Atlas vehicles.

The Atlas D was the basis for two distinct branches of Atlas vehicles. One branch contained the Atlas E and F ICBMs which were used with the Atlas D in the U. S. missile silos. The other branch grew from the use of the Atlas D as a space launch vehicle. The Atlas D was modified, man-rated, and called the LV-3B for the Mercury missions. The 94.5 ft (28.8 m) tall Atlas used the same basic Atlas D system with the addition of a 3000 lb (1400 kg) manned Mercury capsule on top. Unlike Project SCORE, payload and Atlas sustainer remained attached, the Mercury payloads were separated to fly independently after orbit was achieved. Seven of the ten Mercury flights were successful, including all four manned missions.

As needed, additional Atlas Ds were converted to Atlas LV-3A or LV-3C by modifying vehicle structure and subsystems for each mission to be flown. This led to a successful set of launch vehicles and upper stages (LV-3A used Agena and LV-3C used Centaur upper stages).

The LV-3A was first launched in 1958 for Project SCORE and was last used in July 1965. In 1962, management for both the Atlas Agena and Atlas Centaur programs was transfered to the NASA Lewis Research Center. The LV-3A was involved heavily in the Ranger and Mariner programs; it made a total of 43 successful launches in 53 attempts. The LV-3Cs 11 successes in 12 attempts consisted of research and development flights for the Centaur and Surveyor lunar landers.

Unfortunately, the mission tailoring that was required to convert Atlas missiles to space launch vehicles caused long lead times that detracted from their low cost. As a result, a contract was awarded to General Dynamics in 1962 to develop a standardized launch vehicle (SLV).

The SLV line began with the SLV-3. This vehicle, like its predecessor the LV-3A, used primarily Agena upper stages. From its first launch in August 1964 to its final launch in August 1968, the SLV-3 was successful on 49 of 51 orbital launch attempts, including all 5 Lunar Orbiter flights.

In 1965, Convair Division of General Dynamics won a contract to improve the performance of its vehicles, and the SLV-3A and -3C were introduced. Both vehicles had been lengthened to add propellant, had increased engine thrust, and had reduced vehicle weight. They used the Rocketdyne MA-5 engine system with a total thrust of 431,300 lb (196,000 kg). As with the previous Atlas vehicles, these vehicles were considered one and one-half stages with their configuration of two booster engines and one sustainer engine. All three engines ignited at liftoff, but the booster section was jettisoned midway through flight, and the sustainer section continued thrusting to fuel depletion (SLV-3C) or until guidance-commanded cutoff (SLV-3A). The radio-guided SLV-3A stood 78.7 ft (24.0 m) tall by itself (9.75 ft (3.0 m) taller than the SLV-3) and 118 ft (36.0 m) tall with an Agena upper stage and payload fairing; it could deliver 8,500 lb (3,900 kg) of payload into a 100 nm (180 km) circular orbit with the Agena. The SLV-3A was used primarily for classified missions and was successful on 11 of 12 flights through its final launch in 1978. The SLV-3C was similar to the SLV-3A but was designed for use with the Centaur D upper stage. It successfully completed all of its 17 missions.

While the Centaur was emerging as an exceptional upper stage, the SLV-3C evolved into the SLV-3D. Unlike its predecessor, the SLV-3D was integrated electronically with the new Centaur D-1A upper stage. Thus, the Centaur's autopilot and guidance systems were used to control the launch vehicle, as opposed to earlier vehicles that were guided by an independent Atlas autopilot. Most other systems remained the same as those in preceding SLVs, including the same engine thrust as the SLV-3C. The SLV-3D stood 69.5 ft (21.2 m) by itself and 131 ft (59.5 m) with the Centaur and payload fairing added. All 32 SLV-3D launches used Centaur D-1As; its 30th success took place on the last launch in May 1983.

As the Atlas was progressing as a space launch vehicle, it also was developing as an ICBM. Atlas Es and Fs were being developed along with Atlas Ds as U.S. ICBMs in the late 1950s. The Atlas E and F were virtually identical one and one-half stage vehicles that used inertial guidance. A major difference between the E/F vehicle and the SLVs was that the E/F used a Rocketdyne MA-3 engine system instead of the MA-5. The Atlas E and F were deployed in U.S. missile silos for much of the 1960s until they were replaced by the Minuteman in 1965. The Atlas F then was used primarily for the Advanced Ballistic Reentry System (ABRES) program from 1965 to 1974. By the late 1970s, the remaining Atlas Es and Fs were converted for space launch. In Jaunuary 1967, launch responsibility for Atlas E and F was returned to General Dynamics by the Air Force. Since then, 83 of 90 flights have been successful, including all of the last 19.

Atlas History

(continued)

Historical Summary

(continued)

The Atlas E or F could place 1,750 lb (795 kg) into polar LEO with no upper stage. In 1981, the Canoga Overhaul Program (COP) was started because of Atlas E propulsion system failures. The program involved a complete overhaul and re-hot-firing of Atlas MA-3 engines. All COP engines either have flown satisfactorily or have been installed on the two remaining Atlas Es.

Two vehicles emerged from the SLV-3D: the Atlas H and the Atlas G. The Atlas H used most of the basic SLV systems but employed the radio guidance and avionics of the Atlas E. The MA-5 engines used LOX/RP-1 propellants and provided over 439,300 lb (1,954,000N) of thrust, 377,500 lb (1,679,000N) with two booster engines, 60,500 lb (269,000N) with a sustainer, and 1,338 lb (5,950N) with two verniers. In place of the Centaur, the Atlas H used a solid propellant kick motor as a second stage to propel up to 4,400 lb (2,000 kg) into polar LEO. The Atlas H was successful on all five of its launches; the last launch occurred in 1987, and there are no plans for future Atlas H launches.

The Atlas G was a stretched version (72.7 ft (22.2 m) tall) of its predecessor, the SLV-3D. Its MA-5 engine provided 7,500 lb (33,400 kg) more thrust than the SLV-3D's MA-5. The Atlas G was designed for use with the Centaur D-1A upper stage, including use of the Centaur guidance system. The Atlas G/ Centaur combination stood 137 ft (41.8 m) and was capable of delivering 5,200 lb (2,400 kg) to GTO.

As a result of an unprecedented string of launch failures and a related decision to remove commercial payloads from the Shuttle manifest, in 1987 General Dynamics decided to develop and build 18 Atlas/Centaurs (designated Atlas I) for commercial sale without having firm contracts for their purchase. Two new metal payload fairings were introduced with 11 ft (3.3 m) and 14 ft (4.2 m) diameters. This inhouse production plan would enable General Dynamics to have a commercial vehicle on the market by early 1990.

In May 1988, the Air Force chose General Dynamics to develop the Atlas II vehicle, primarily to launch Defense Satellite Communications System (DSCS) payloads. Subsequently, General Dynamics decided to scale back the Atlas I program to 11 vehicles and use the excess assets for their other Atlas programs. Four commercial vehicles are now offered: the Atlas I, the Atlas II and two enhanced versions, the Atlas IIA and Atlas IIAS.

The Atlas II is a further growth of the proven Atlas/Centaur family. This configuration was developed in response to a DoD requirement for a medium-lift launch vehicle (i.e., MLV-2) to boost 10 DSCS satellites and one STP payload into orbit. The Atlas booster has been stretched 9 ft (2.7 m) to increase the amount of propellant (LOX/RP-1) the booster can carry. It employs an improved Rocketdyne engine set, the MA-5A. The booster engine set incorporates flight-proven components (thrust chamber and turbomachinery) to increase sea level thrust to 414,000 lb (1,842,000N). The sustainer engine was not modified; thus, it provides the nominal 60,500 lb of thrust (269,000N). The vernier engines were replaced by hydrazine roll control modules mounted on the interstage adapter. In addition, the Centaur upper stage was stretched by 3 ft (0.9 m) but still uses two unmodified Pratt & Whitney RL10 engines. Burning LOX/LH2, these engines provide a total thrust of 33,300 lb (148,000N). Finally, the fiberglass honeycomb insulation panels for the Centaur were replaced with polyvinyl chloride fixed foam insulation panels. The baseline Atlas II uses the 11 ft (3.3 m) diameter PLF. This configuration is capable of placing 6,100 lb (2,770 kg) into GTO. The 14 ft (4.2 m) diameter PLF can be used, if desired, with a reduced performance of 5,900 lb (2,680 kg) to GTO.

The two final configurations of the Atlas/Centaur family—Atlas IIA and Atlas IIAS—are further modifications to the Atlas II. The major difference is that an upgraded RL10 engine, providing 20,250 lb (90,000N) of thrust, and extendable nozzles to provide 6.5 sec added specific impulse is employed in both versions, while the Atlas IIAS adds four Thiokol Castor IVA solid rocket motors (SRMs) to the booster stage. Each SRM provides and average thrust of approximately 97,500 lb (434,000N).

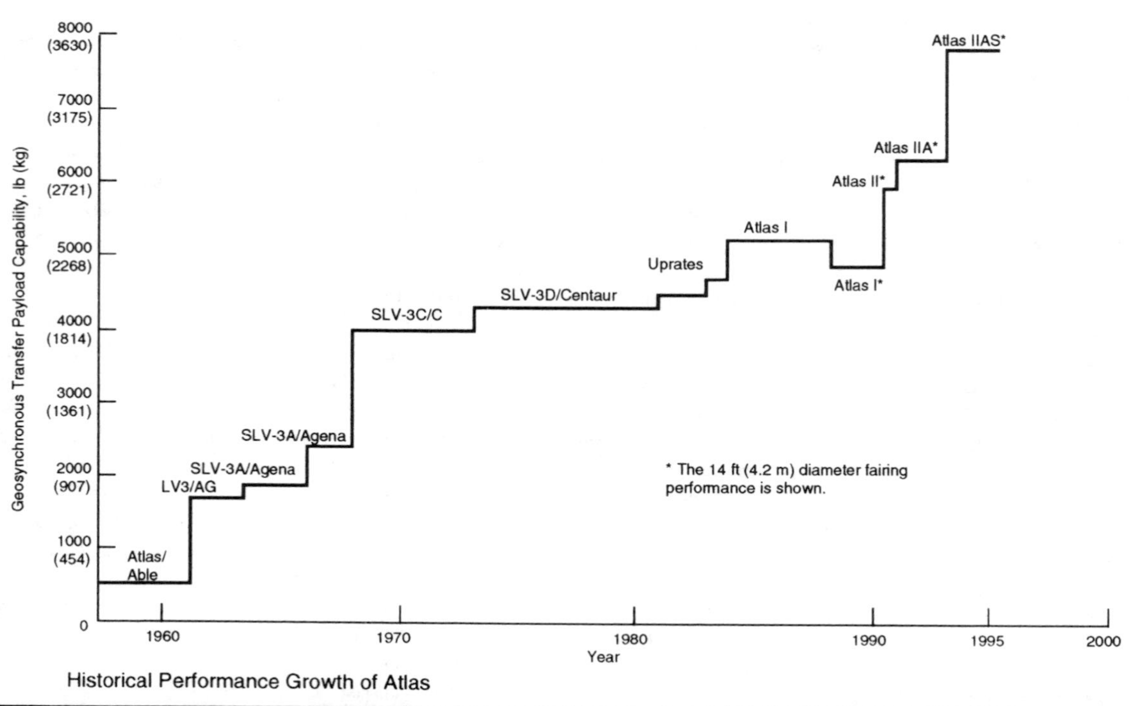

Historical Performance Growth of Atlas

Atlas History

(continued)

Launch Record

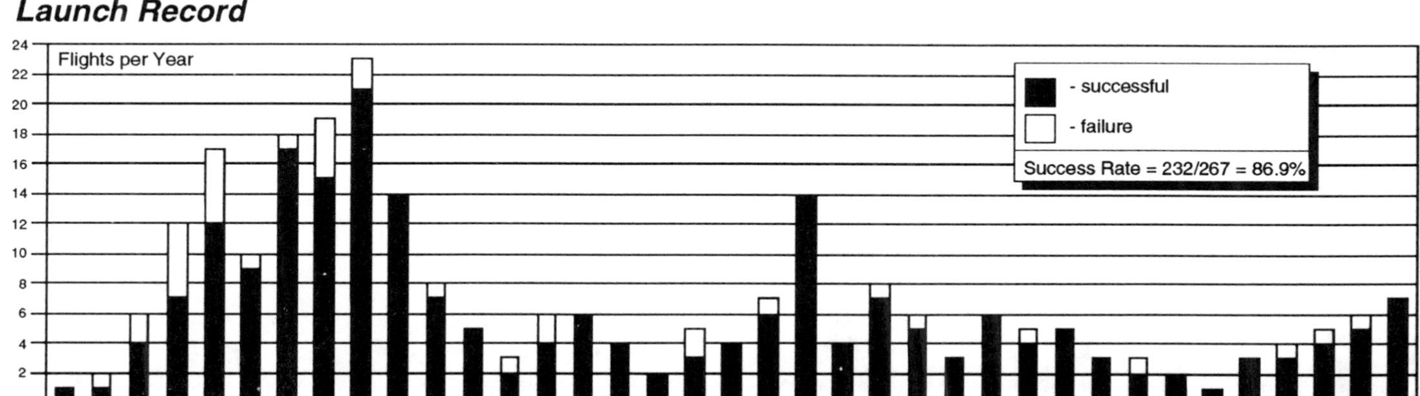

#	YEAR	DATE	VEHICLE	SITE	PAYLOAD *
1	1958	Dec-18	LV-3A	LC-11	Score
2	1959	Sept-9	LV-3B	LC-14	Mercury 1
		Failure-Electrical			
3		Nov-26	LC-3A	LC-14	Able
		[x] Failure-Agena shroud broke away T+45 sec			
4	1960	Feb-26	LC-3A	LC-14	Midas 1
		[x] Failure-Agena failed to separate			
5		May-24	LC-3A	LC-14	Midas 2
6		Jul-29	LV-3B	LC-14	Mercury 2
		Failure-Airframe			
7		Sep-25	LV-3A	LC-12	Able 5
8		Oct-11	LV-3A	PALC 1-1	Samos 1
		[x] Failure-Agena guidance			
9		Dec-15	LV-3A	LC-12	Able 5B
		Failure-Airframe; exploded T+70 sec			
10	1961	Jan-31	LV-3A	PALC 1-1	Samos 2
11		Feb-21	LV-3B	LC-14	Transit 3B
12		Apr-25	LV-3B	LC-14	Mercury 3
		Failure-Flight control			
13		Jul-12	LV-3A	PALC 1-2	Midas 3
14		Aug-23	LV-3A	LC-12	Ranger 1
15		Sep-9	LV-3A	PALC 1-1	Samos 3
		Failure-Electrical			
16		Sep-13	LV-3B	LC-14	Mercury 4
17		Oct-21	LV-3A	PALC 1-2	Midas 4
		Failure-Flight control			
18		Nov-18	LV-3A	LC-12	Ranger 2
19		Nov-22	LV-3A	PALC 1-1	
		Failure-Flight control			
20		Nov-29	LV-3B	LC-14	Mercury 5
21		Dec-22	LV-3A	PALC 1-2	
		Failure-Flight control			
22	1962	Jan-26	LV-3A	LC-12	Ranger 3
		Failure-Guidance			
23		Feb-20	LV-3B	LC-14	Mercury 6 [M]
24		Mar-7	LV-3A	PALC 1-2	
25		Apr-9	LV-3A	PALC 1-2	
		Failure-Propulsion			
26		Apr-23	LV-3A	LC-12	Ranger 4
27		Apr-26	LV-3A	PALC 1-1	
28		May-8	LV-3C	LC-36A	Suborbital [T]
		Failure-Centaur ignition failure			
29		May-24	LV-3B	LC-14	Mercury 7 [M]
30		Jun-17	LV-3A	PALC 1-1	
31		Jul-18	LV-3A	PALC 1-1	
32		Jul-22	LV-3A	LC-12	Mariner 1
		Failure-Guidance			
33		Aug-5	LV-3A	PALC 1-1	
34		Aug-27	LV-3A	LC-12	Mariner 2
35		Oct-3	LV-3B	LC-14	Mercury 8 [M]
36		Oct-18	LV-3A	LC-12	Ranger 5
37		Nov-11	LV-3A	PALC 1-1	
38		Dec-17	LV-3A	PALC 1-2	ERS 3&4
		Failure-Hydraulics			
39	1963	May-9	LV-3A	PALC 1-2	ERS 5&6
40		May-15	LV-3B	LC-14	Mercury 9 [M]
41		Jun-12	LV-3A	PALC 1-2	ERS 7&8
		Failure-Hydraulics			
42		Jul-12	LV-3A	PALC 2-3	
43		Jul-18	LV-3A	PALC 1-2	ERS 9&10
44		Sep-6	LV-3A	PALC 2-3	
45		Oct-16	LV-3A	LC-13	Vela 1&2
46		Oct-25	LV-3A	PALC 2-3	
47		Nov-27	LV-3C	LC-36A	None [T]
48		Dec-16	LV-3A	PALC 2-3	
49	1964	Jan-30	LV-3A	LC-12	Ranger 6
50		Feb-25	LV-3A	PALC 2-3	
51		Mar-11	LV-3A	PALC 2-3	
52		Apr-23	LV-3A	PALC 2-3	
53		May-19	LV-3A	PALC 2-3	
54		Jun-30	LV-3C	LC-36A	None [T]
		Failure-Centaur hydraulic			
55		Jul-6	LV-3A	PALC 2-3	
56		Jul-17	LV-3A	LC-13	Vela 3
57		Jul-28	LV-3A	LC-12	Ranger 7
58		Aug-14	SLV-3	PALC 2-4	
59		Sep-4	LV-3A	LC-12	OGO-1
60		Sep-23	SLV-3	PALC 2-4	
61		Oct-8	SLV-3	PALC 2-4	
62		Oct-23	LV-3A	PALC 2-3	
63		Nov-5	LV-3A	LC-13	Mariner 3
		[x] Failure-Agena shroud			
64		Nov-28	LV-3A	LC-12	Mariner 4
65		Dec-4	SLV-3	PALC 2-4	
66		Dec-11	LV-3C	LC-36A	None [T]
67	1965	Jan-21	D	576-B3	OV1-1
		Failure-Propulsion			
68		Jan-23	SLV-3	PALC 2-3	
69		Feb-17	LV-3A	LC-12	Ranger 8
70		Mar-2	LV-3C	LC-36A	None [T]
		Failure-Fuel pre-valve inadvertently closed			
71		Mar-12	SLV-3	PALC 2-3	
72		Mar-21	LV-3A	LC-12	Ranger 9
73		Apr-3	SLV-3	PALC 2-4	Snapshot
74		Apr-28	SLV-3	PALC 2-4	
75		May-27	SLV-3	PALC 2-4	
76		May-27	D	ABRES B3	OV1-3
		Failure-Propulsion			
77		Jun-25	SLV-3	PALC 2-4	
78		Jul-12	SLV-3	PALC2-4	
		Failure-Flight control			
79		Jul-20	LV-3A	LC-13	Vela 5 & 6
80		Aug-3	SLV-3	PALC 2-4	
81		Aug-11	LV-3C	LC-36B	None [T]
82		Sep-30	SLV-3	PALC 2-4	
83		Oct-5	D	ABRES B3	OV1-2
84		Oct-25	SLV-3	LC-14	Gemini 6/Target
85		Nov-8	SLV-3	PALC 2-4	
86	1966	Jan-19	SLV-3	PALC 2-4	
87		Feb-15	SLV-3	PALC 2-4	
88		Mar-16	SLV-3	LC-14	Gemini 8/Target
89		Mar-18	SLV-3	PALC 2-4	
90		Mar-30	D	ABRES B3	OV1 4 &5
91		Apr-7	LV-3C	LC-36B	None [T]
		Failure-Centaur propellant leak			
92		Apr-8	SLV-3	LC-12	OAO 1
93		Apr-19	SLV-3	PALC 2-4	
94		May-14	SLV-3	PALC 2-4	
95		May-17	SLV-3	LC-14	Gemini9/Target A
		Failure-Flight control			
96		May-30	LV-3C	LC-36A	Surveyor 1
97		Jun-1	SLV-3	LC-14	Gemini9/Target B
98		Jun-3	SLV-3	PALC 2-4	
99		Jun-6	SLV-3	LC-12	OGO-3
100		Jun-9	SLV-3	PALC 1-2	ERS-16
101		Jul-12	SLV-3	SLC-4E	
102		Jul-13	D	ABRES B3	OV1 7&8
103		Jul-18	SLV-3	LC-14	Gemini 10
104		Aug-10	SLV-3	LC-13	Lunar Orbiter 1
105		Aug-16	SLV-3	SLC-4E	
106		Aug-19	SLV-3	SLC-3E	
107		Sep-12	SLV-3	LC-14	Gemini 11/Target
108		Sep-16	SLV-3	SLC-4E	
109		Sep-20	LV-3C	LC-36A	Surveyor 2
110		Oct-5	SLV-3	SLC-3E	EGRS-8
111		Oct-12	SLV-3	SLC-4E	
112		Oct-26	LV-3C	LC-36B	None [T]
113		Nov-2	SLV-3	SLC-4E	
114		Nov-6	SLV-3	LC-13	Lunar Orbiter 2
115		Nov-11	SLV-3	LC-14	Gemini 12/Target
116		Dec-5	SLV-3	SLC-4E	
117		Dec-6	SLV-3	LC-12	ATS-1
118		Dec-11	D	ABRES B3	OV1 9&10
119	1967	Feb-2	SLV-3	SLC-4E	
120		Feb-4	SLV-3	LC-13	Lunar Orbiter 3
121		Apr-5	SLV-3	LC-12	ATS 2
122		Apr-17	LV-3C	LC-36B	Surveyor 3
123		May-4	SLV-3	LC-13	Lunar Orbiter 4
124		May-22	SLV-3	SLC-4E	
125		Jun-4	SLV-3	SLC-4E	
126		Jun-14	SLV-3	LC-12	Mariner 5
127		Jul-14	LV-3C	LC-36A	Surveyor 4
128		Jul-27	D	ABRES B3	OV1 11&12
129		Aug-1	SLV-3	LC-13	Lunar Orbiter 5
130		Sep-8	SLV-3C	LC-36B	Surveyor 5
131		Nov-5	SLV-3	LC-12	ATS 3
132		Nov-7	SLV-3C	LC-36B	Surveyor 6
133	1968	Jan-7	SLV-3C	LC-36A	Surveyor 7
134		Mar-4	SLV-3A	LC-13	OGO-5
135		Apr-6	F	ABRES A2	OV1 13&14
136		Jul-11	F	ABRES A2	OV1 15&16
137		Aug-6	SLV-3A	ESMC	
138		Aug-10	SLV-3C	LC-36A	ATS 4
		Failure-Centaur oxidizer leak; no restart			
139		Aug-16	SLV-3	SLC-3E	10 payloads
		[x] Failure-Burner II failure			
140		Dec-7	SLV-3C	LC-36B	OAO A2
141	1969	Feb-24	SLV-3C	LC-36B	Mariner 6
142		Mar-17	F	ABRES A2	OV1 17,18,19
143		Mar-27	SLV-3C	LC-36A	Mariner 7
144		Apr-12	SLV-3A	ESMC	
145		Aug-12	SLV-3C	LC36A	ATS 5
146	1970	Jun-19	SLV-3A	ESMC	
147		Aug-31	SLV-3A	ESMC	
148		Nov-30	SLV-3C	LC-36B	OAO-B
		Failure-Shroud failed to separate			
149	1971	Jan-25	SLV-3C	LC-36A	Intelsat IV F-2
150		May-8	SLV-3C	LC-36A	Mariner H
		Failure-Centaur electronic signal ended			
151		May-30	SLV-3C	LC-36B	Mariner 9
152		Aug-6	F	BMRS A2	OV1 20,21
153		Dec-4	SLV-3A	ESMC	
		Failure-Sustainer engine turbine			
154		Dec-19	SLV-3C	LC-36A	Intelsat IV F-3
155	1972	Jan-22	SLV-3C	LC-36B	Intelsat IV F-4
156		Mar-2	SLV-3C	LC-36A	Pioneer 10
157		Jun-13	SLV-3C	LC-36B	Intelsat IV F-5
158		Aug-21	SLV-3C	LC-36B	OAO-C
159		Oct-2	F	BMRS A1	SESP
160		Dec-20	SLV-3A	ESMC	
161	1973	Mar-6	SLV-3A	ESMC	
162		Apr-5	SLV-3D	LC-36B	Pioneer 11
163		Aug-23	SLV-3D	LC-36A	Intelsat IV F-7
164		Nov-3	SLV-3D	LC-36B	Mariner 10
165	1974	Jul-13	F	SLC-3W	STP P73-3
166		Nov-21	SLV-3D	LC-36B	Intelsat IV F-8
167	1975	Feb-20	SLV-3D	LC-36A	Intelsat IV F-6
		Failure-Staging electrical disconnect			
168		Apr-12	F	SLC-3W	STP P72-2
		Failure-External explosion in flame bucket			
169		May-22	SLV-3D	LC-36A	Intelsat IV F-1
170		Jun-18	SLV-3A	ESMC	
171		Sep-25	SLV-3D	LC-36B	Intelsat IVA F-1
172	1976	Jan-29	SLV-3D	LC-36B	Intelsat IVA F-2
173		Apr-30	F	WSMC	NOSS-1, SSU
174		May-13	SLV-3D	LC-36A	Comstar 1
175		Jul-22	SLV-3D	LC-36B	Comstar 2
176	1977	May-23	SLV-3A	ESMC	
177		May-26	SLV-3D	LC-36A	Intelsat IVA F-4
178		Jun-23	F	SLC-3W	Navstar GPS
179		Aug-12	SLV-3D	LC-36B	HEAO A
180		Sep-29	SLV-3D	LC-36A	Intelsat IVA F-5
		Failure-Booster gas generator hot gas leak			
181		Dec-8	F	WSMC	NOSS-3, SSU
182		Dec-11	SLV-3A	ESMC	
183	1978	Jan-6	SLV-3D	LC-36B	Intelsat IVA F-3
184		Feb-9	SLV-3D	LC-36A	Fltsatcom 1
185		Feb-22	F	SLC-3E	Navstar GPS
186		Mar-31	SLV-3D	LC-36B	Intelsat IVA F-6
187		Apr-7	SLV-3A	ESMC	
188		May-13	F	SLC-3E	Navstar GPS
189		May-20	SLV-3D	LC-36A	Pioneer Venus
190		Jun-26	F	SLC-3W	Seasat-A
191		Jun-29	SLV-3D	LC-36B	Comstar 3
192		Aug-8	SLV-3D	LC-36A	Pioneer Venus
193		Oct-6	F	SLC-3E	Navstar GPS
194		Oct-13	F	SLC-3W	TIROS-NOAA
195		Nov-13	SLV-3D	LC-36B	HEAO B
196		Dec-10	F	SLC-3E	Navstar GPS
197	1979	Feb-19	F	SLC-3W	STP 78-1
198		May-4	SLV-3D	LC-36A	Fltsatcom 3
199		Jun-27	F	SLC-3W	NOAA-A
200		Sep-20	SLV-3D	LC-36B	HEAO C
201	1980	Jan-17	SLV-3D	LC-36A	Fltsatcom
202		Feb-9	F	SLC-3E	Navstar GPS
203		Mar-3	F	WSMC	NOSS 3, SSU
204		Apr-26	F	SLC-3E	Navstar GPS
205		May-29	F	SLC-3W	NOAA-B
206		Oct-30	SI.V-3D	LC-36A	Fltsatcom 4
207		Dec-6	SLV-3D	LC-36B	Intelsat V F2
208		Dec-8	E	WSMC	
		Failure-Booster lube oil flow loss			
209	1981	Feb-21	SLV-3D	LC-36A	Comstar 4
210		May-23	SLV-3D	LC-36B	Intelsat V F-1
211		Jun-23	F	SLC-3W	NOAA-C
212		Aug-6	SLV-3D	LC-36A	Fltsatcom 5
213		Dec-15	SLV-3D	LC-36B	Intelsat V F-3
214		Dec-18	E	SLC-3E	Navstar GPS
		Failure-Booster gas gen. cooling plugged			
215	1982	Mar-4	SLV-3D	LC-36A	Intelsat V F-4
216		Sep-28	SLV-3D	LC-36B	Intelsat V F-5
217		Dec-20	E	SLC-3W	DMSP F-6
218	1983	Feb-9	H	WSMC	
219		Mar-28	E	SLC-3W	NOAA E (ATN)
220		May-19	SLV-3D	LC-36A	Intelsat V F-6
221		Jun-13	H	WSMC	
222		Jul-14	E	SLC-3W	Navstar GPS
223		Nov-17	E	SLC-3W	DMSP F-7
224	1984	Feb-5	H	WSMC	
225		Jun-9	G	LC-36B	Intelsat V F-9
		Failure-Centaur structural failure			
226		Jun-13	E	WSMC	Navstar GPS
227		Sep-8	E	WSMC	Navstar GPS
228		Dec-12	E	WSMC	NOAA-F
229	1985	Mar-12	E	SLC-3W	Geosat
230		Mar-22	G	LC-36B	Intelsat VA F-10
231		Jun-29	G	LC-36B	Intelsat VA F-11
232		Sep-28	G	LC-36B	Intelsat VA F-12
233		Oct-8	E	SLC-3W	Navstar GPS
234	1986	Feb-9	H	WSMC	
235		Sep-17	E	SLC-3W	NOAA-G
236		Dec-4	G	LC-36B	Fltsatcom 7
237	1987	Mar-26	G	LC-36B	Fltsatcom 6
		Failure-Lightning strike of guidance			
238		Jun-19	E	SLC-3W	DMSP-9
239		May-15	H	SLC	
240	1988	Feb-3	E	SLC-3W	DMSP-8
241		Sep-24	E	SLC-3W	NOAA-D
242	1989	Sep-25	G	LC-36B	Fltsatcom
243	1990	Apr-11	E	SLC-3W	Stacksat
244		Jul-25	I	LC-36B	CRRES
245		Dec-1	E	SLC-3W	DMSP-10
246	1991	Apr-18	I	36B	BS-3H
		Failure-Centaur engine			
247		May-14	E	3W	NOAA-D
248		Nov-28	E	3W	DMSP II
249		Dec-7	II	36B	Eutelsat
250	1992	Feb-10	II	36A	DSCS IIIB
251		Mar-13	I	36B	Galaxy V
252		Jun-9	IIA	36B	Intelsat-K
253		Jul-2	II	36A	DSCS IIIB
254		Aug 22	I	36B	Galaxy 1R
		Failure-Centaur engine			
255	1993	Mar-25	I	36B	UHF-F1
		Failure-Atlas engine			
256		Jul-19	II	36A	DSCS IIIB
257		Aug-9	E	3W	NOAA-I
258		Sep-3	I	36B	UHF-F2
259		Nov-28	II	36A	DSCS IIIB
260		Dec-15	IIAS	36B	Telstar 401
261	1994	Apr-13	I	36B	GOES-I
262		Jun-24	I	36B	UHF-F3
263		Aug-3	IIA	36A	DBS-2
264		Aug-29	E	3W	DMSP
265		Oct-6	IIAS	36B	Intelsat-703
266		Nov-29	IIA	36A	Orion-1
267		Dec-30	E	3W	NOAA-J

* Some Atlas launches are classified and full launch details are not disclosed
[x] - not counted as failure against vehicle

[M] - Manned Payload
[T] - Test Launch

LC-12, 14, 36, ESMC are at Cape Canaveral, Florida
SLC-3,4, PALC, ABRES, BMRS, WSMC are at Vandenberg, California

Atlas General Description

	Atlas E	Atlas I	Atlas II	Atlas IIA	Atlas IIAS
Summary	The Atlas E is a DoD launch vehicle, presently used to launch smaller payloads to polar orbit from VAFB. Atlas Es are decommissioned ICBM modified for space launch. Only two remain as of September 1994.	As a result of an unprecedented string of launch vehicle failures and a related decision to remove commercial payloads from the Shuttle manifest, in 1987 General Dynamics decided to develop and build 18 Atlas/Centaurs (designated Atlas I) for commercial sale without having firm contracts for their purchase. This inhouse production plan would enable General Dynamics to have a commercial vehicle on the market by early 1990. In May 1988, the Air Force chose General Dynamics to develop the Atlas II vehicle, primarily to launch Defense Satellite Communications System (DSCS) payloads. Subsequently, General Dynamics decided to scale back the Atlas I program to 11 vehicles and use the excess assets for their other Atlas programs. Lockheed Martin now offers four commercial vehicles: Atlas I, Atlas II, and two enhanced versions, Atlas IIA and Atlas IIAS.			
Status	Operational	Operational	Operational	Operational	Operational
Key Organizations	User - DoD, NOAA	User - commercial	User - DoD, commercial	User - commercial	User - commercial
	Launch Service Agency - Air Force Space Systems Division	Launch Service Agency - Lockheed Martin Commercial Launch Services	Launch Service Agency - Lockheed Martin Commercial Launch Services and U.S. Air Force	Launch Service Agency - Lockheed Martin Commercial Launch Services and U.S. Air Force	Launch Service Agency - Lockheed Martin Commercial Launch Services
	Prime Contractor - Lockheed Martin (Airframe, Assembly and Test, and Launch Operations)	Prime Contractor - Lockheed Martin (Airframe, Assembly and Test, and Launch Operations)	Prime Contractor - Lockheed Martin (Airframe, Assembly and Test, and Launch Operations)	Prime Contractor - Lockheed Martin (Airframe, Assembly and Test, and Launch Operations)	Prime Contractor - Lockheed Martin (Airframe, Assembly and Test, and Launch Operations)
	Principal Subcontractors Rocketdyne (Atlas engine - MA-3) General Electric (radio guidance)	Principal Subcontractors Rocketdyne (Atlas engine - MA-5) Pratt & Whitney (Centaur engine- RL-10) Honeywell (Avionics)	Principal Subcontractors Rocketdyne (Atlas engine - MA-5) Pratt & Whitney (Centaur engine- RL-10) Honeywell (Avionics)	Principal Subcontractors Rocketdyne (Atlas engine - MA-5) Pratt & Whitney (Centaur engine- RL-10) Honeywell (Avionics)	Principal Subcontractors Rocketdyne (Atlas engine - MA-5) Pratt & Whitney (Centaur engine- RL-10) Honeywell (Avionics) Thiokol (Castor IVA solid strap-ons)
Vehicle					
System Height	Up to 92.2 ft (28.1 m)	Up to 144 ft (43.9 m)	Up to 156 ft (47.5 m)	Up to 156 ft (47.5 m)	Up to 156 ft (47.5 m)
Payload Fairing Size	7 ft (2.1 m) diameter by 24 ft (7.4 m) height.	Medium fairing: 10.8 ft (3.3 m) diameter by 34.1 ft (10.4 m) height. Large fairing: 13.75 ft (4.2 m) diameter by 39.4 ft (12.0 m) height.	Medium fairing: 10.8 ft (3.3 m) diameter by 34.1 ft (10.4 m) height. Large fairing: 13.75 ft (4.2 m) diameter by 39.4 ft (12.0 m) height.	Medium fairing: 10.8 ft (3.3 m) diameter by 34.1 ft (10.4 m) height. Large fairing: 13.75 ft (4.2 m) diameter by 39.4 ft (12.0 m) height.	Medium fairing: 10.8 ft (3.3 m) diameter by 34.1 ft (10.4 m) height. Large fairing: 13.75 ft (4.2 m) diameter by 39.4 ft (12.0 m) height.
Gross Mass	266,700 lb (121,000 kg)	362,200 lb (164,300 kg)	413,500 lb (187,600 kg)	413,800 lb (187,700 kg)	515,900 lb (234,000 kg)
Planned Enhancements	None	None	None	None	None

Atlas General Description

(continued)

	Atlas E	Atlas I	Atlas II	Atlas IIA	Atlas IIAS
Operations					
Primary Missions	low polar orbit	LEO, GTO	LEO, GTO	LEO, GTO	LEO, GTO
Compatible Upper Stages	Burner II, OV 1, TE-M-364-4 SGS I and II	Centaur I	Centaur II	Centaur IIA	Centaur IIA
First Launch	1980	1990	1991	1992	1993
Success / Flight Total	19 / 21	5 / 8	5 / 5	3 / 3	2 / 2
Launch Site	VAFB - SLC-3W (34.7°N, 120.6°W)	CCAFS - LC-36B (28.5°N, 81.0°W)	CCAFS - LC-36A (28.5°N, 81.0°W)	CCAFS - LC-36B (28.5°N, 81.0°W)	CCAFS - LC-36B (28.5°N, 81.0°W)
Launch Azimuth	SLC-3W - 158°–301°	LC-36B - 90°–108°	LC-36A - 90°–108°	LC-36B - 90°–108°	LC-36B - 90°–108°
Nominal Flight Rate	2 / yr	5 / yr including IIA, IIAS	4 / yr	5 / yr including I, IIAS	5 / yr including I, IIA
Planned Enhancements	None	None	None	None	None
Performance		For vehicle I, II, IIA, and IIAS descriptions above, the 14 ft (4.2 m) payload fairing is assumed. The 11 ft (3.3 m) payload fairing will increase performance by about 450 lb (205 kg) for LEO missions, 200 lb (90 kg) for GTO missions. Note: the current maximum payload limit is 8,000 lb (3,630 kg) based on structural limits, but an upgrade is in development.			
100 nm (185 km) circ, 28°	—	—	14,500 lb (6,580 kg)	16,050 lb (7,280 kg)	19,050 lb (8,640 kg)
100 nm (185 km) circ, 90°	1,800 lb (820 kg)	*not available	12,150 lb (5,510 kg) *not currently available	13,600 lb (6,170 kg) *not currently available	16,100 lb (7,300 kg) *not currently available
Geotransfer Orbit, 28°	—	4,970 lb (2,255 kg)	6,200 lb (2,810 kg)	6,700 lb (3,039 kg)	7,950 lb (3,606 kg)
Financial Status					
Estimated Launch Price	$50M (Lockheed Martin)	$65–75M (Lockheed Martin)	$75–85M (Lockheed Martin)	$80–90M (Lockheed Martin)	$95–105M (Lockheed Martin)
Orders:	1	3	11	6	7
Payloads:	STP /USAF	GOES-J, K SAX	UHF (7) MLV (4)	Orion MSAT Galaxy-3R Inmarsat (2) Eutelsat	Intelsat (3) JCSat SOHO Telstar Loral DBS

Manifest

Site	Vehicle	1995	1996	1997
SLC-3W	Atlas E	1	—	—
LC-36B	Atlas I	1	1	—
LC-36A	Atlas II	4	1	4
LC-36B	Atlas IIA	2	3	—
LC-36B	Atlas IIAS	3	2	—

Remarks

Atlas E: Only one vehicle remains with no production capability for new procurement.

Lockheed Martin has commited to build 62 vehicles including 11 Atlas Is. Eventually Atlas Is will cease production in favor of the Atlas II family.

Atlas Vehicle

Overall

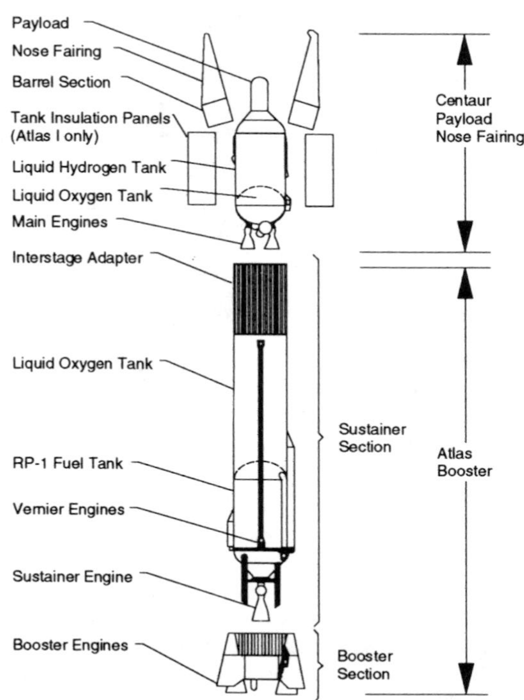

	Atlas E	**Atlas I**	**Atlas II**	**Atlas IIA**	**Atlas IIAS**
Length	Up to 92.2 ft (28.1 m)	Up to 144 ft (43.9 m)	Up to 156 ft (47.5 m)	Up to 156 ft (47.5 m)	Up to 156 ft (47.5 m)
Gross Mass	266.7K lb (121.0K kg)	362.2K lb (164.3K kg)	413.5K lb (187.6K kg)	413.8K lb (187.7K kg)	515.9K lb (234.0K kg)
Thrust at Liftoff	392K lb (1.74M N)	439.3K lb (1.95M N)	474.5K lb (2.11M N)	474.5K lb (2.11M N)	669.5K lb (2.98M N)

SRM

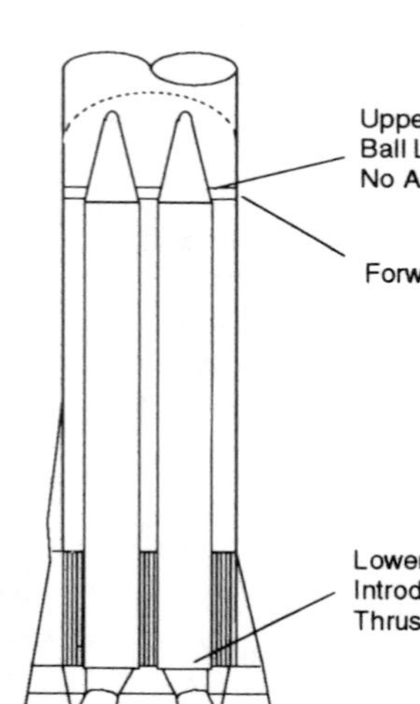

Jettison Thrusters

	Atlas IIAS (Castor IVA)
Dimension:	
Length	30.0 ft (9.12 m)
Diameter	3.3 ft (1.0 m)
Mass: (each)	
Propellant Mass	22.3K lb (10.1K kg)
Gross Mass	25.5K lb (11.6K kg)
Structure:	
Type	Monocoque
Case Material	Steel
Propulsion:	
Propellant	HTPB
Average Thrust (each)	108.7K lb (478.3K N) vac 97.5K lb (432.9K N) SL
Number of Motors	4
Number of Segments	1
Isp	229 sec SL
Chamber Pressure	691 psia (4.77M Pa)
Expansion Ratio	8.29:1
Control	Fixed 11° and 7° cant (booster for control)
Events:	
Nominal Burn Time	56.2 sec
Stage Shutdown	Burn to depletion
Stage Separation	Pyro thrusters

Remarks:

The Castor IVAs used for the Atlas IIAS are virtually identical to the off-the-shelf specification, with the exception of the thrust ball point. The interface is a ball-and-cup joint. The vertical location of the SRMs was dictated by plume impingement on the Atlas tank and the plume mixing point. An ablative shield has been added to the Atlas tank to prevent excessive heating from SRM plume impingement. Several structural modifications are made to the Atlas to accommodate the additional loads and attach points from the SRMs. The primary axial load-carrying member is the thrust section, where most of the Atlas modifications have been made. For the sustainer stage, a tank ring was added where the SRM forward attach point is placed. Operationally, two SRMs will be ignited on the pad along with the Atlas engines. In the start up sequence the Atlas engines attain full thrust, then two SRMs are ignited. After burnout, the two SRMs are jettisoned and four seconds later the other two SRMs ignite and jettison after burnout.

Atlas Vehicle
(continued)

Booster	Atlas E	Atlas I	Atlas II	Atlas IIA	Atlas IIAS
Dimension:					
Length	69.8 ft (21.3 m)	72.7 ft (22.2 m)	81.7 ft (24.9 m)	81.7 ft (24.9 m)	81.7 ft (24.9 m)
Diameter	10 ft (3.05 m)	10 ft (3.05 m)	10 ft (3.05 m)	10 ft (3.05 m)	10 ft (3.05 m)
Mass:					
Propellant Mass	248.8K lb (112.9K kg)	305.5K lb (138.3K kg)	344.5K lb (155.9K kg)	344.5K lb (155.9K kg)	344.5K lb (155.9K kg)
Gross Mass	266.7K lb (121.0K kg)	321.1K lb (145.7K kg)	365.3K lb (165.7K kg)	366.5K lb (166.2K kg)	368.3K lb (167.1K kg)
Structure:					
Type	Monocoque pressure-stabilized	Monocoque pressure-stabilized	Monocoque pressure-stabilized	Monocoque pressure-stabilized	Monocoque pressure-stabilized
Material	Stainless steel	Stainless steel	Stainless steel	Stainless steel	Stainless steel
Propulsion:					
Propellant	LOX-RP1	LOX-RP1	LOX-RP1	LOX-RP1	LOX-RP1
Average Thrust	330K lb (1.47M N) SL for booster 60K lb (267K N) SL for sustainer 670 lb (3.0K N) SL for each of two verniers	377.5K lb (1.68M N) SL for booster 60.5K lb (269K N) SL for sustainer 670 lb (3.0K N) SL for each of two verniers	414.0K lb (1.84M N) SL for booster 60.5K lb (269K N) SL for sustainer	414.0K lb (1.84M N) SL for booster 60.5K lb (269K N) SL for sustainer	414.0K lb (1.84M N) SL for booster 60.5K lb (269K N) SL for sustainer
Engine Designation	MA-3	MA-5	MA-5A	MA-5A	MA-5A
Number of Engines	3 thrust chamber plus 2 turbine driven pumps plus 2 verniers	3 thrust chamber plus 2 turbine driven pumps plus 2 verniers	3 thrust chamber plus 2 turbine driven pumps	3 thrust chamber plus 2 turbine driven pumps	3 thrust chamber plus 2 turbine driven pumps
Isp	252 sec SL booster 214 sec SL sustainer 172 sec SL verniers	259.1 sec SL booster 220.4 sec SL sustainer 186.7 sec SL verniers	261.1 sec SL booster 220.4 sec SL sustainer	263.0 sec SL booster 220.4 sec SL sustainer	263.0 sec SL booster 220.4 sec SL sustainer
Feed System	Gas Generator	Gas Generator	Gas Generator	Gas Generator	Gas Generator
Chamber Pressure	580 psia (40.0 bar) for booster 710 psia (49.0 bar) for sustainer 258 psia (17.8 bar) for verniers	639 psia (44.1 bar) for booster 735 psia (50.7 bar) for sustainer 258 psia (17.8 bar) for verniers	639 psia (44.1 bar) for booster 735 psia (50.7 bar) for sustainer	639 psia (44.1 bar) for booster 735 psia (50.7 bar) for sustainer	639 psia (44.1 bar) for booster 735 psia (50.7 bar) for sustainer
Mixture Ratio (O / F)	2.24 booster 2.27 sustainer 1.8 verniers	2.25 booster 2.27 sustainer 1.8 verniers	2.25 booster 2.27 sustainer	2.25 booster 2.27 sustainer	2.25 booster 2.27 sustainer
Throttling Capability	100% only	100% only	100% only	100% only	100% only
Expansion Ratio	8:1 booster 25:1 sustainer 6:1 verniers	8:1 booster 25:1 sustainer 6:1 verniers	8:1 booster 25:1 sustainer	8:1 booster 25:1 sustainer	8:1 booster 25:1 sustainer
Restart Capability	No	No	No	No	No
Tank Pressurization	High-pressure helium gas	High-pressure helium gas	High-pressure helium gas	High-pressure helium gas	High-pressure helium gas
Control - Pitch, Yaw	Hydraulic gimbaling	Hydraulic gimbaling	Hydraulic gimbaling	Hydraulic gimbaling	Hydraulic gimbaling
Roll	2 verniers	2 verniers	Hydrazine ACS	Hydrazine ACS	Hydrazine ACS
Events:					
Nominal Burn Time	120 sec booster 309 sec sustainer	174 sec booster 266 sec sustainer	172 sec booster 283 sec sustainer	172 sec booster 283 sec sustainer	172 sec booster 283 sec sustainer
Stage Shutdown	Guidance command	Burn to depletion	Burn to depletion	Burn to depletion	Burn to depletion
Stage Separation	10 pneumatic latches for booster, 8 retro-rockets for sustainer	10 pneumatic latches for booster, 8 retro-rockets for sustainer	10 pneumatic latches for booster, 8 retro-rockets for sustainer	10 pneumatic latches for booster, 8 retro-rockets for sustainer	16 pyro actuated separation latches

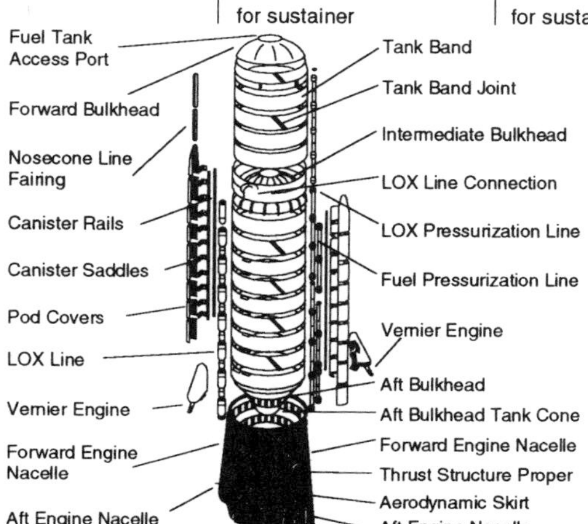

Remarks:

The Atlas booster propellant tanks are of thin-wall, fully monocoque, corrosion-resistant stainless steel construction. The fuel tank, which contains RP-1, and the oxidizer tank, which contains LOX, are separated by an ellipsoidal intermediate bulkhead. Structural integrity of the tanks is maintained in flight by the pressurization system and on the ground by either internal tank pressure or by application of mechanical stretch. Atlas booster propulsion is provided by the Rocketdyne engine system, which includes the sustainer, two vernier, and two booster engines. For Atlas II, IIA, and IIAS, the vernier engines are replaced with a hydrazine roll control system. All engines are ignited prior to liftoff. The section containing the booster engines is jettisoned and flight continues powered by the sustainer engine until propellant depletion. The Atlas is integrated with the Centaur vehicle by the interstage adapter that is *not* included in the mass and size data above. The interstage weighs 1,067 lb (482 kg) and is 10 ft (3.05 m) diameter by 13 ft (4 m) length. This aluminum skin/stringer frame structure provides the structural link between the Atlas and Centaur vehicles. The Atlas vehicle is separated from the Centaur vehicle by a pyrotechnic flexible linear shaped charge system attached to the forward ring of the interstage adapter.

Atlas Vehicle

(continued)

Centaur

Forward Door Opening
Forward Bulkhead
Tank Ring
Cylindrical Skins
LH_2
Double-Walled Intermediate Bulkhead
Tank Ring
LO_2
Thrust Barrel
Aft BulkHead

	Atlas I	Atlas II	Atlas IIA & IIAS
Dimension:			
Length	30 ft (9.15 m)	33 ft (10.1 m)	33 ft (10.1 m)
Diameter	10 ft (3.05 m)	10 ft (3.05 m)	10 ft (3.05 m)
Mass:			
Propellant Mass	30K lb (13.9K kg)	37K lb (16.7K kg)	37K lb (16.7K kg)
Gross Mass	34.3K lb (15.6K kg)	41.5K lb (18.8K kg)	41.8K lb (19.0K kg)
Structure:			
Type	Monocoque	Monocoque	Monocoque
Material	pressure-stabilized Stainless steel	pressure-stabilized Stainless steel	pressure-stabilized Stainless steel
Propulsion:			
Propellant	LOX-LH2	LOX-LH2	LOX-LH2
Average Thrust	33K lb (147K N)	33K lb (147K N)	41.6K lb (185K N)
Engine Designation	RL10A-3-3A	RL10A-3-3A	RL10A-4N
Number of Engines	2	2	2
Isp	444.4 sec vac	442.4 sec vac	448.9 sec vac
Feed System	Split expander	Split expander	Split expander
Chamber Pressure	465 psia (32.1 bar)	465 psia (32.1 bar)	465 psia (32.1 bar)
Mixture Ratio (O/F)	5.0	5.5	5.5
Throttling Capability	100% only	100% only	100% only
Expansion Ratio	61:1	61:1	85:1
Restart Capability	Multiple	Multiple	Multiple
Tank Pressurization	Helium gas and autogenous hydrogen	Helium gas and autogenous hydrogen	Helium gas and autogenous hydrogen
Control-Pitch,Yaw, Roll	Hydraulic gimbaling (2 nozzles)	Hydraulic gimbaling (2 nozzles)	Hydraulic gimbaling (2 nozzles)
Events:			
Nominal Burn Time	402 sec	488 sec	392 sec
Stage Shutdown	Burn to depletion or command shutdown	Burn to depletion or command shutdown	Burn to depletion or command shutdown
Stage Separation	Flexible linear shaped	Flexible linear shaped	Flexible linear shaped

Remarks:

Centaur was developed as the world's first high-energy propellant stage. Since its first launch, it has gone through several performance and reliability upgrades, particularly in the areas of electronics and software. The Centaur uses LH2 and LOX propellants separated by a double-wall, vacuum-insulated intermediate bulkhead. The propellant tanks, like those of the Atlas lower stage, are constructed of thin-wall fully monocoque, corrosion-resistant stainless steel. Tank stabilization is maintained at all times by either internal pressurization or application of mechanical stretch. The stub adapter and equipment module are attached to the forward end of the Centaur and provide for payload compartment attachment. The aluminum skin stringer stub adapter is bolted to the forward ring of the Centaur tank and supports the equipment module, payload fairing, and spacecraft adapter. The equipment module provides mounting arrangements for the Centaur avionics packages and the spacecraft adapter. For insulation, Atlas I uses fiberglass honeycomb insulation panels and the Atlas II family uses polyvinyl chloride fixed foam.

Atlas Vehicle

(continued)

Payload Fairing

	7 ft Diameter (Atlas E Only)	Medium Payload Fairing	Large Payload Fairing
	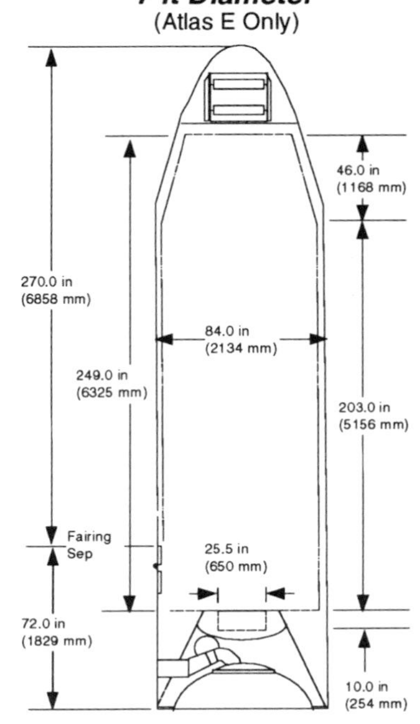	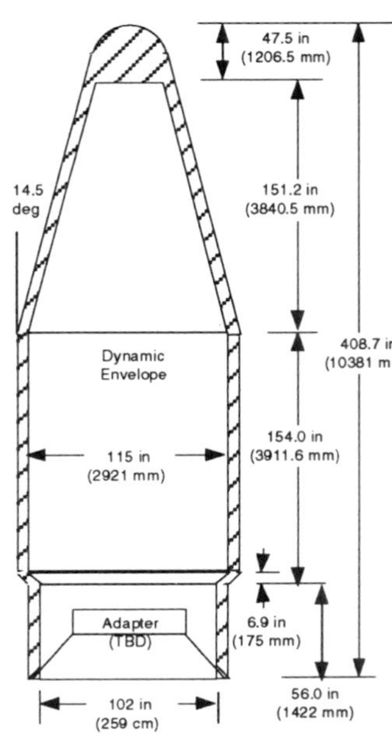	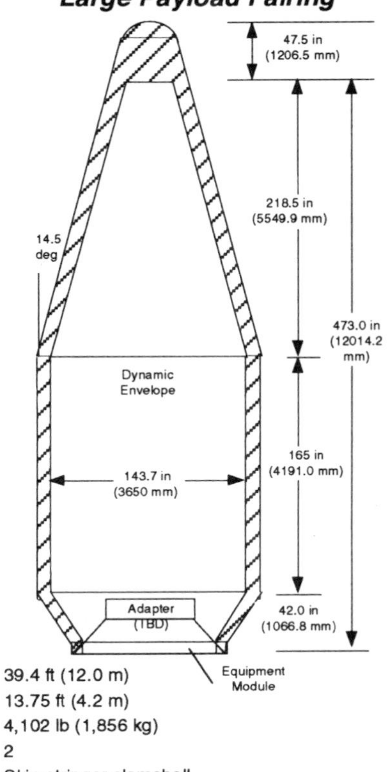
Length	22.5 ft (6.9 m)	34.1 ft (10.4 m)	39.4 ft (12.0 m)
Diameter	7 ft (2.1 m)	10.8 ft (3.3 m)	13.75 ft (4.2 m)
Mass	1,619 lb (735 kg)	3,027 lb (1,373 kg)	4,102 lb (1,856 kg)
Sections	2	2	2
Structure	Skin-stringer clamshell	Skin-stringer clamshell	Skin-stringer clamshell
Material	Aluminum	Aluminum	Aluminum
Remarks	7 ft fairing is only available for Atlas E and is no longer in production.	The payload fairing protects the spacecraft from time of encapsulation through atmospheric ascent. The Atlas user has a choice between the large or medium payload fairing configuration. For either fairing, a thermal shield or acoustic blanket can be added should the spacecraft require more benign environments. In addition, the use of a spacer on the payload fairing to increase the payload compartment length is being developed as a mission-peculiar option.	

Avionics

The Atlas is integrated with the Centaur avionics system for guidance, flight control, and sequencing functions. An external equipment pod houses Atlas systems such as range safety, propellant utilization pneumatics, and instrumentation. Centaur avionics packages mounted on the equipment module provide control and monitoring of all vehicle functions. The Inertial Navigation Unit (INU) performs the inertial guidance and attitude control computations for both Atlas and Centaur phases of flight and also provides control for Centaur tank pressures and propellant use. The baseline Atlas II avionics are improved by replacing the Sequence Control Unit and the Servo Inverter Unit with a new Remote Control Unit (RCU) and Power Distribution Unit (PDU). The RCU provides 128 channels of solid state vehicle/spacecraft sequencing (versus 96 relay switches for the Sequence Control Unit) under INU software control vis a MIL-STD-1553B data bus. Arm-safing of critical sequence functions is also included as well as provisions for 400 Hz power for vehicle engine control servo loop use. The PDU provides power changeover capabilities from ground power to internal main-vehicle battery power and busing connectors to meet power distribution requirements. Both packages represent significant weight, cost, and reliability improvements over the units they replace.

Attitude Control System

The Atlas attitude control is provided by hydraulically gimbaling the three nozzles for pitch / yaw and by the two verniers for roll on Atlas E and I and hydrazine ACS on Atlas II, IIA, and IIAS.

The Centaur attitude or reaction control system consists of four settling (S) thrusters and four pitch (P) and four yaw (Y) thrusters. All of the reaction control thrusters use hydrazine (N2H4) as the propellant which is pressure-fed from a common storage bottle(s). During parking orbit coast, propellant settling is accomplished using a 4S On mode and propellant retention using a 2S On mode. Prior to and during pre-Centaur second main engine start pressurization of the propellant tank, the 4S On mode is again used. During chilldown, the propellants issuing from the main engines provide the main percentage of the total longitudinal thrust; and gimbaling the Centaur main engines provides the primary pitch and yaw attitude control, with the P and Y thrusters providing roll control and backup pitch and yaw control. During the first main engine burn, the P and Y thrusters are turned on for three seconds to warm them prior to parking orbit coast.

Current missions may operate the four S thrusters during the second Centaur burn to burn excess N2H4 (if any) and, thus, increase performance. Various mission-peculiar combinations of reaction control thruster operation are used for spacecraft separation attitude and for the Centaur Collision/Contamination Avoidance Maneuver (CCAM).

Atlas Performance

CCAFS / Atlas Direct Ascent to Circular Orbits

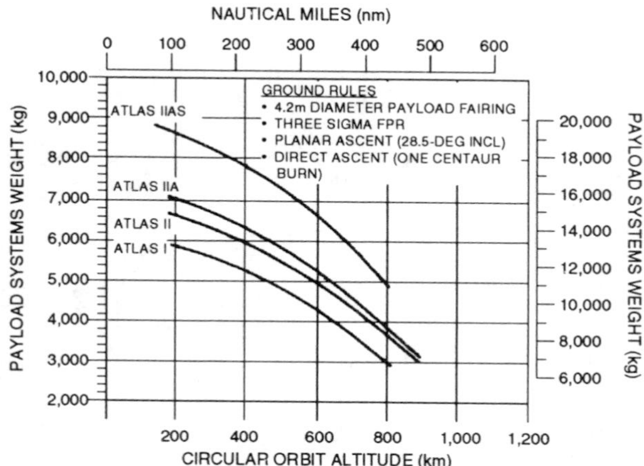

CCAFS / Atlas Parking Orbit Ascent to Circular Orbits

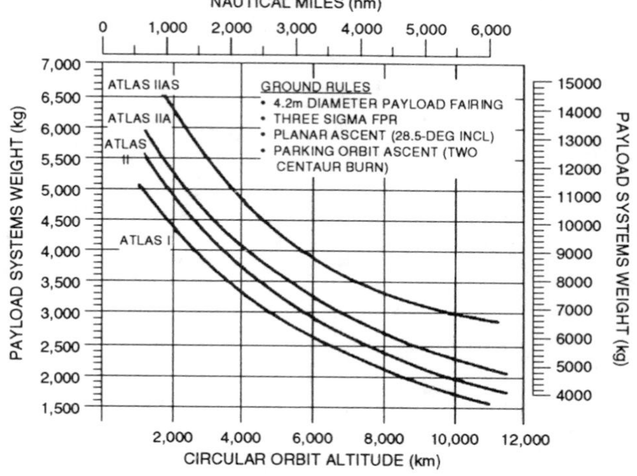

CCAFS / Atlas Elliptical Transfer Orbit

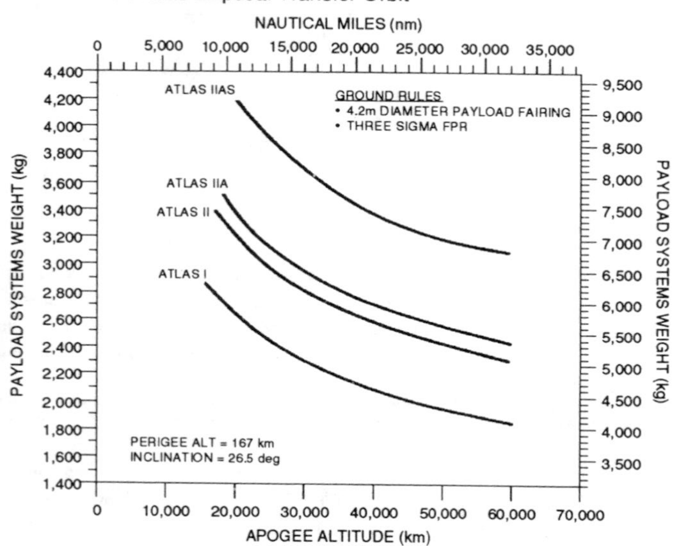

CCAFS / Atlas I at Various Geotransfer Orbit Inclinations

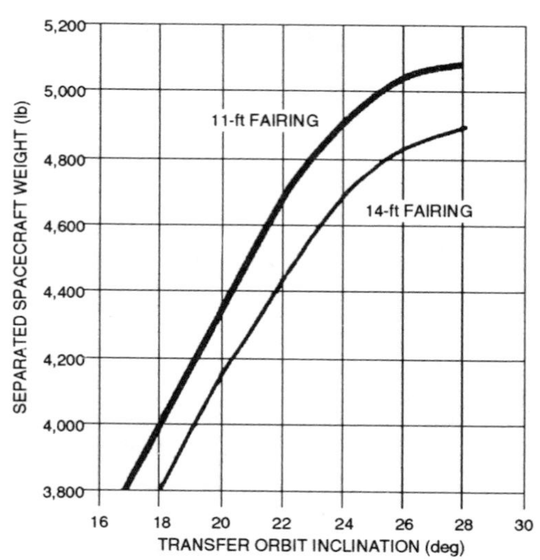

VAFB / Atlas E to Low Polar Orbit

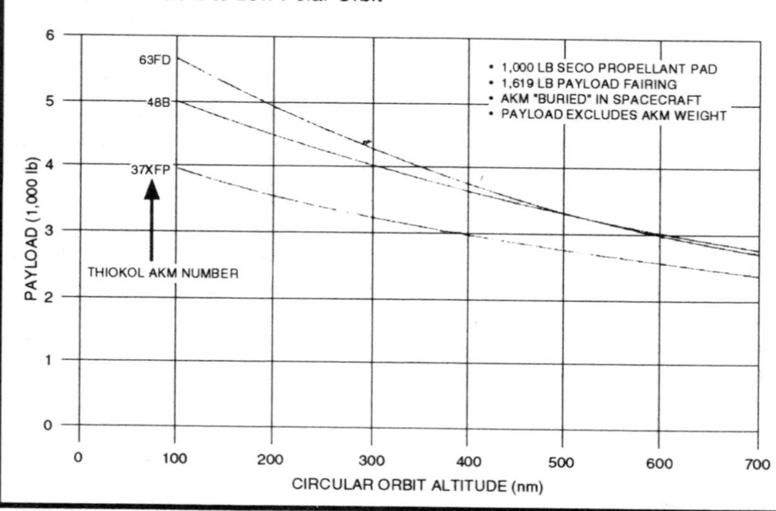

CCAFS / Atlas Earth-escape

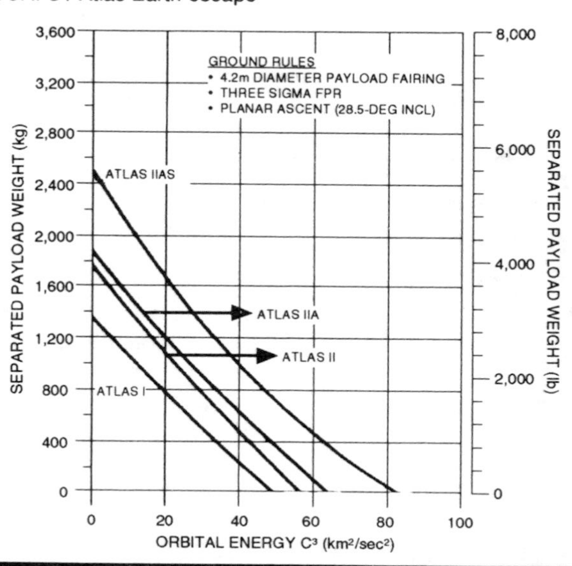

Atlas Operations

Launch Site

Vandenberg Air Force Base (34.7°)

Cape Canaveral Air Force Station (28.6°N)

SLC-2W (Delta II)
SLC-3W (Atlas E)
SLC-4E / W
(Titan II, Titan IV)
SLC-5 (Scout)
SLC-6
(LLV)
KSC
LC-39B (STS)
LC-39A (STS)
CCAFS
LC-41 (Titan IV)
LC-40
(Titan III, Titan IV)
LC-36A / B
(Atlas I, II)
LC-17A / B
(Delta, Delta II)

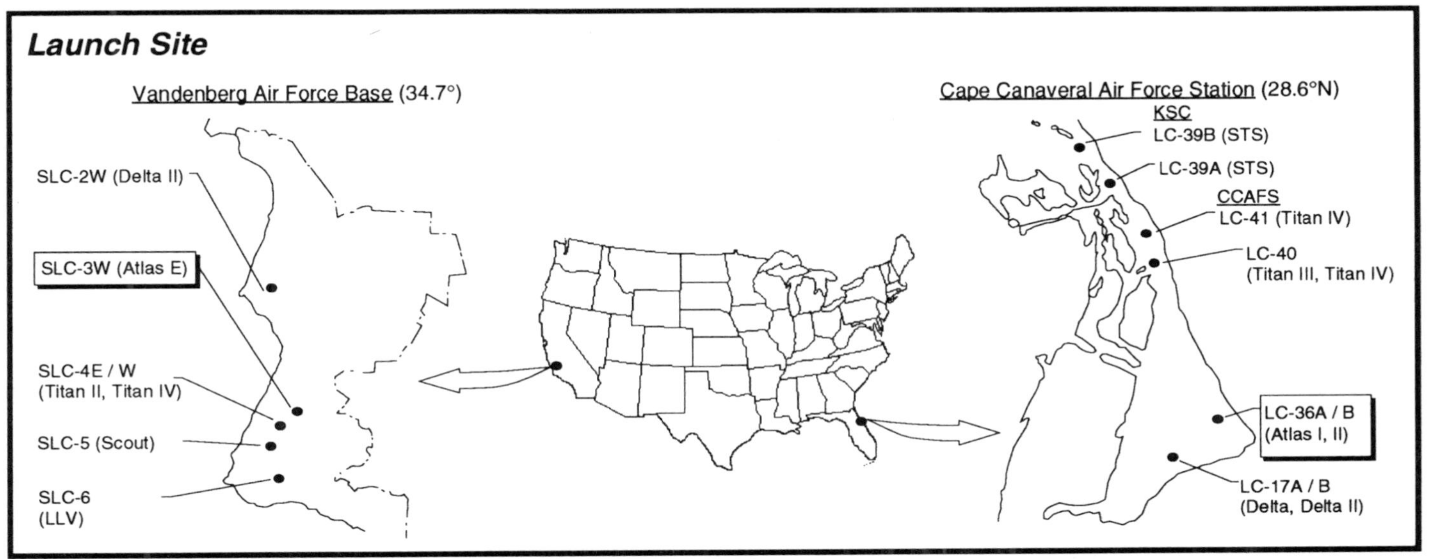

Launch Facilities

Cape Canaveral Air Force Station (CCAFS)

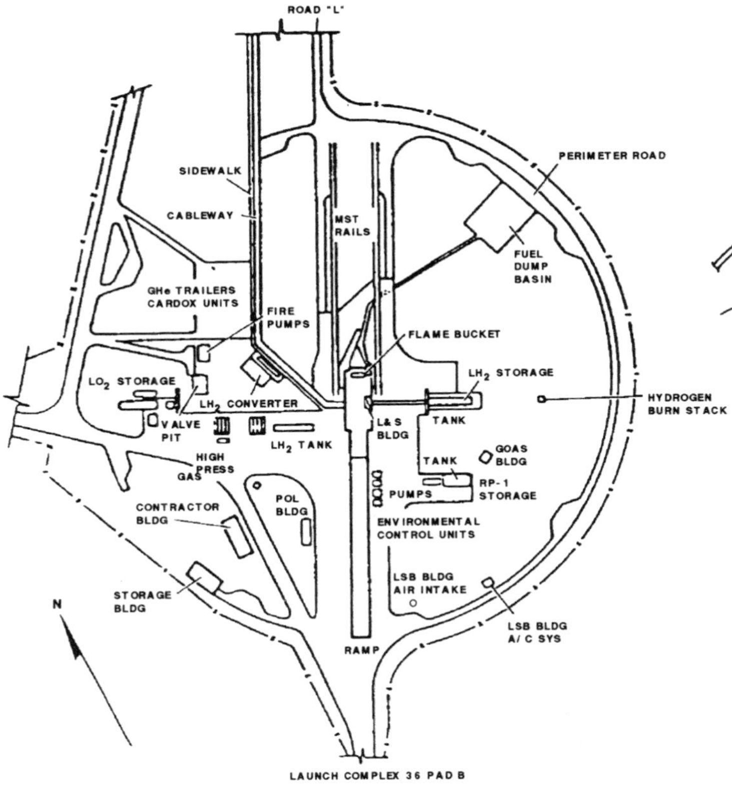

ROAD "L"
SIDEWALK
CABLEWAY
MST
RAILS
PERIMETER ROAD
FUEL
DUMP
BASIN
GHe TRAILERS
CARDOX UNITS
FIRE
PUMPS
FLAME BUCKET
LH2 STORAGE
LO2 STORAGE
LH2 CONVERTER
HYDROGEN
BURN STACK
VALVE
PIT
L&S
BLDG
TANK
LH2 TANK
HIGH
PRESS
GAS
GOAS
BLDG
TANK
PUMPS
RP-1
STORAGE
CONTRACTOR
BLDG
POL
BLDG
ENVIRONMENTAL
CONTROL UNITS
LSB BLDG
AIR INTAKE
STORAGE
BLDG
N
LSB BLDG
A/C SYS
RAMP
LAUNCH COMPLEX 36 PAD B

LC-36 Layout

Vandenberg Air Force Base (VAFB)

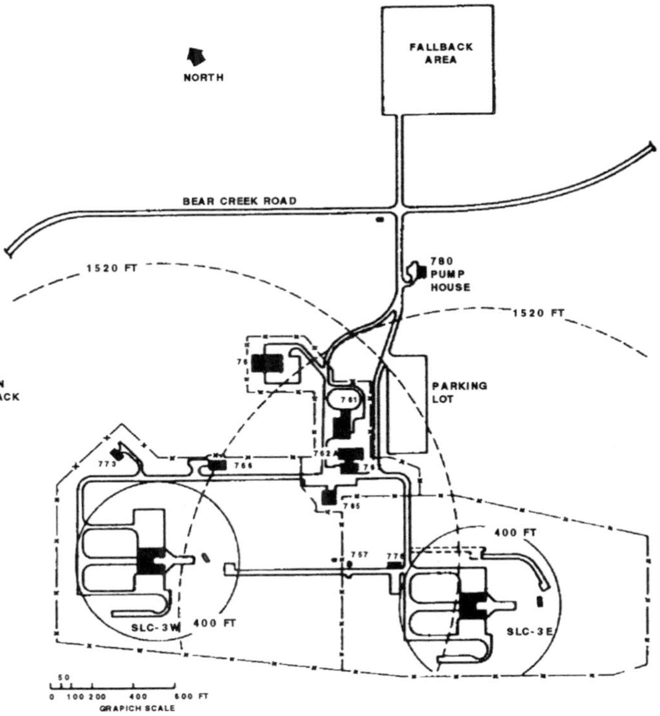

FALLBACK
AREA
NORTH
BEAR CREEK ROAD
1520 FT
780
PUMP
HOUSE
1520 FT
PARKING
LOT
773
400 FT
SLC-3W
400 FT
SLC-3E
0 100 200 400 600 FT
GRAPICH SCALE

SLC-3 Layout

Atlas Operations

(continued)

Launch Processing

Cape Canaveral Air Force Station (CCAFS). Production - Atlas launch vehicles are primarily manufactured at four U.S. facilities, supported by a team of international subsystem and component suppliers. Tanks for both the Centaur and sustainer stages are manufactured at Plant 19 in San Diego, CA. Beginning life as 32 in. (0.81 m) rolls of thin, corrosion-resistant steel, the Atlas and Centaur stages are fabricated by first welding the stainless steel into 10 ft (3.05 m) diameter rings. These rings are then stacked, stove-pipe fashion, and welded edge-to-edge to the required tank lengths. The thickness of the stainless steel varies from 0.046 in. (1.2 mm) near the bottom of the Atlas stage to 0.015 in. (0.4 mm) at the top of the Centaur. Following tank fabrication, the tanks are trucked to the Space Systems Division Kearny Mesa facility in San Diego for engine integration and final assembly. Harlingen, TX, is the site for manufacturing the payload fairing, spacecraft adapters, interstage adapter, and thrust section of the booster half stage. Upon completion of final assembly, Atlas and Centaur are flown to Cape Canaveral Air Force Station (CCAFS) for on-pad mating.

During 1995, Atlas and Centaur final assembly will be transitioned from the San Diego facility to a new facility within the Lockheed Martin complex in Denver. Tank fabrication in San Diego and current Harlingen activities will not be changed.

Factory Tests—Flight vehicle acceptance (or factory) tests of the Atlas vehicles are performed after launch vehicle final assembly operations are complete. Testing typically consists of system-level leak and functional tests of systems, low-pressure and leak checks of the main propellant tanks and intermediate bulkhead cavity, checkout of propellant-level sensing probes mounted inside the main tanks, verification of all electrical harnesses, and high-pressure pneumatic checks.

Launch Site Facilities—The Atlas launch facility is Launch Complex 36, located at CCAFS. The major facilities include two mobile service towers (MST), two umbilical towers (UT), and the blockhouse.

The MST is an open steel structure with an interior enclosure that contains retractable vehicle servicing and checkout levels/platforms. The tower contains an electric, trolley-mounted 10-ton (9,072 kg) overhead bridge crane used to hoist spacecraft fairings, and the upper stage vehicle into position. Two elevators serve all MST levels. The entire MST assembly is on a rail system, which allows it to be moved for launch. RF cabling and reradiating antennas are mounted on the service tower for spacecraft use.

The umbilical tower (UT) is a fixed structural steel tower extending above the launch pad. Retractable service booms are attached to the UT. The booms provide electrical power, instrumentation, propellants, pneumatics, and conditioned air or gaseous nitrogen (GN2) to the vehicle and spacecraft. These systems also provide quick-disconnect mechanisms at the respective vehicle interface and permit boom retraction at vehicle launch. The UT also provides a wind damper system that reduces wind-induced lateral oscillations of the vehicle during periods of service tower removal. The ground wind damper mechanism is released from the vehicle by the launch sequence.

The launch complex is serviced by GN2, gaseous helium (GHe), and propellant storage facilities within the complex area. Environmental Control Systems (ECS) exist for both the launch vehicle and the spacecraft.

The blockhouse serves as the operations and communications center for the launch complex. It contains all necessary control and monitoring equipment. The launch control, electrical, landline instrumentation, and ground computer systems are the major systems in this facility.

The launch control provides consoles and cabling for control of the launch complex systems. The landline instrumentation system (coupled with the closed-circuit TV system) monitors and records safety and performance data during test and launch operations. The ground computer system consists of redundant computer-controlled launch sets (CCLS) and a telemetry ground station. The CCLS provides control and monitoring of the vehicle guidance, navigation, and control systems and monitors vehicle instrumentation for potential anomalies during test and launch operations.

Launch Site Prelaunch Operations—Upon arrival at the launch site, all launch vehicle items are given a receiving inspection prior to erection on the launch pad in the missile service tower. Following erection of the Atlas and connection of ground umbilical lines, subsystem and system-level tests are performed to verify compatibility between airborne systems and associated ground support equipment in preparation for subsequent integrated system tests.

Payload fairing sections can be readied for launch in one of several locations depending upon the customer's choice for spacecraft final preparations and encapsulation. Remote encapsulation in one of the Hazardous Processing Facilities is the baseline. The payload fairing is mated to the Atlas for a payload fairing rotation test. This test is performed to verify that the total nose fairing assembly will open and rotate freely on its hinges for inflight jettison. Following this test, the payload fairing is removed and returned to the clean room for final cleaning and preparation for spacecraft encapsulation.

Launch Site Integrated Test Operations—To achieve flight readiness of the Atlas launch vehicle to enter countdown, two major integrated tests are performed at Complex 36 prior to launch day:

Wet Dress Rehearsal (WDR)—Simulates a major part of the launch countdown and demonstrates the capability of the ground and airborne systems to: load and maintain Atlas propellant levels; charge the airborne helium bottles to flight pressures; perform chilldown of main engines; verify proper operations of all air conditioning systems; and demonstrate proper operations of all systems under cryogenic conditions. The test is performed with the Atlas in a fully assembled configuration and with the service tower withdrawn. Installation of the payload fairing is optional. The spacecraft is typically not installed for this test.

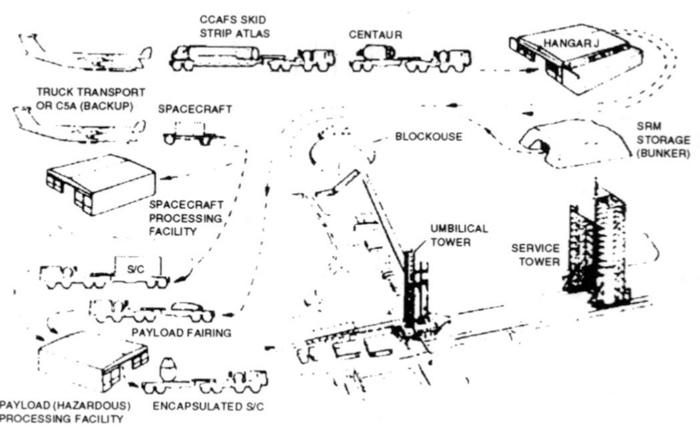

Launch Processing Flow at CCAFS

Atlas Operations

(continued)

Launch Processing

(continued)

Launch Vehicle Simulated Flight (SMFLT)—Verifies, on an integrated basis, that all Atlas ground and airborne electrical systems are capable of properly combined operation throughout a simulated launch countdown and flight sequence. Demonstrates spacecraft mate to launch vehicle; proper operation of the launch ladder release sequence; capability to operate on internal power with ground umbilicals pulled; and proper operation of all pyrotechnics by use of squib simulators. The test is performed with the Atlas in a fully assembled configuration except for the nose fairing and spacecraft. The service tower is in place around the vehicle throughout the test. The umbilicals are pulled at T-0 time.

T-71 days:	Erect Atlas
T-70 days:	Install Interstage Adapter
T-69 days:	Erect Centaur
T-45 days:	Mate Insulation Panels
T-40 days:	Payload Fairing Rotation Test
T-29 day:	Terminal Countdown Demonstration
T-15 day:	Flight Events Demonstration
T-13/8 day:	Erect, Mate, and Checkout Space Vehicle. Install Fairing
T-4/1 day:	Launch Readiness
T-0 day:	Launch Countdown Operations

Atlas Vehicle Integration Operations at CCAFS

Launch Countdown Operations—The Atlas launch countdown consists of an approximate 9 to 10-hour count which includes two built-in holds—one at T-90 minutes (for 30 minutes) and the second at T-5 minutes (for 10 minutes)—to enhance the launch-on-time capability.

In addition to the scheduled 30-minute and 10-minute countdown holds, additional hold time can be scheduled for up to two additional hours or until the end of the scheduled launch window, whichever comes first.

Prior to T-4 seconds (when the upper stage aft panel is ejected), the launch vehicle has a 24-hour turnaround capability following a launch abort due to a nonlaunch vehicle/GSE problem. If the abort occurs after securing the interstage adapter area, access to it will be required to service the N2H4 system.

A launch abort after T-4 seconds and prior to T-0.7 seconds requires a 48-hour recycle. The principal reason for a 48-hour recycle versus a 24-hour recycle is the added time requirement for replacing the upper stage aft panel (ejected at T-4 seconds) and the removal and replacement of the propellant pressurization lines' pyro valves (fired at T-2 seconds).

Vandenberg Air Force Base (VAFB). Atlas E Launch Vehicle—The Atlas E SLV is created by refurbishing an Atlas ICBM. The refurbishment process conducted by Lockheed Martin Space Systems Division as prime contractor begins when the missiles are removed from storage at Norton AFB, CA where they have been maintained under a corrosion contract program managed by the Air Force Logistics Center (AFLC). The Atlas is sent to the Lockheed Martin Space Systems Division facility where a ringweld modification is performed in order to allow spacecraft and upper stage adapters to be attached where the reentry vehicle was originally mounted. Next they are shipped to VAFB for processing under the Vandenberg Atlas Modification Program (VAMP). Refurbishment is accomplished by installing overhauled and hot-fired engines (Rocketdyne) along with new guidance, destruct, telemetry, electrical, and hydraulic systems. All wiring is replaced and any necessary structural repairs are accomplished.

Once the VAMP process is completed and the booster is ready for active launch processing, the mission support team (180 people) and the launch crews (118 people) generate the typical processing "waterfall" schedule.

One Atlas launch pad (SLC-3W) is currently operational; a second pad (SLC-3E) is presently being modified to provide capability to launch the full Atlas II, IIA, and IIAS family, and will be operational in 1996. Several support facilities located at SLC-3 include technical support buildings, launch operations building, vehicle support building, and launch service building. The propellants used in the Atlas vehicle are stored onsite, and can be directly loaded onto the vehicle. Atlas vehicles are delivered to the launch areas and are erected on the launch pad, serviced, and launched.

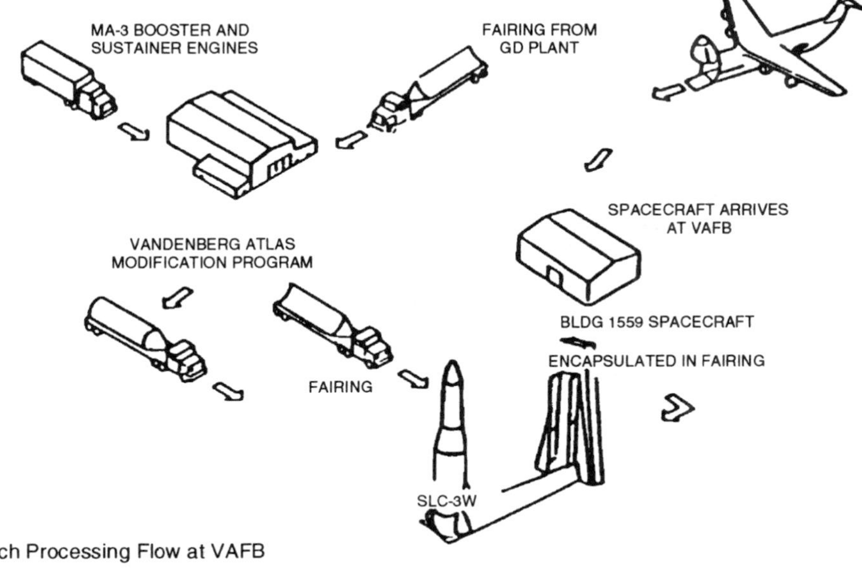

Atlas E Launch Processing Flow at VAFB

Atlas Operations

(continued)

Flight Sequence

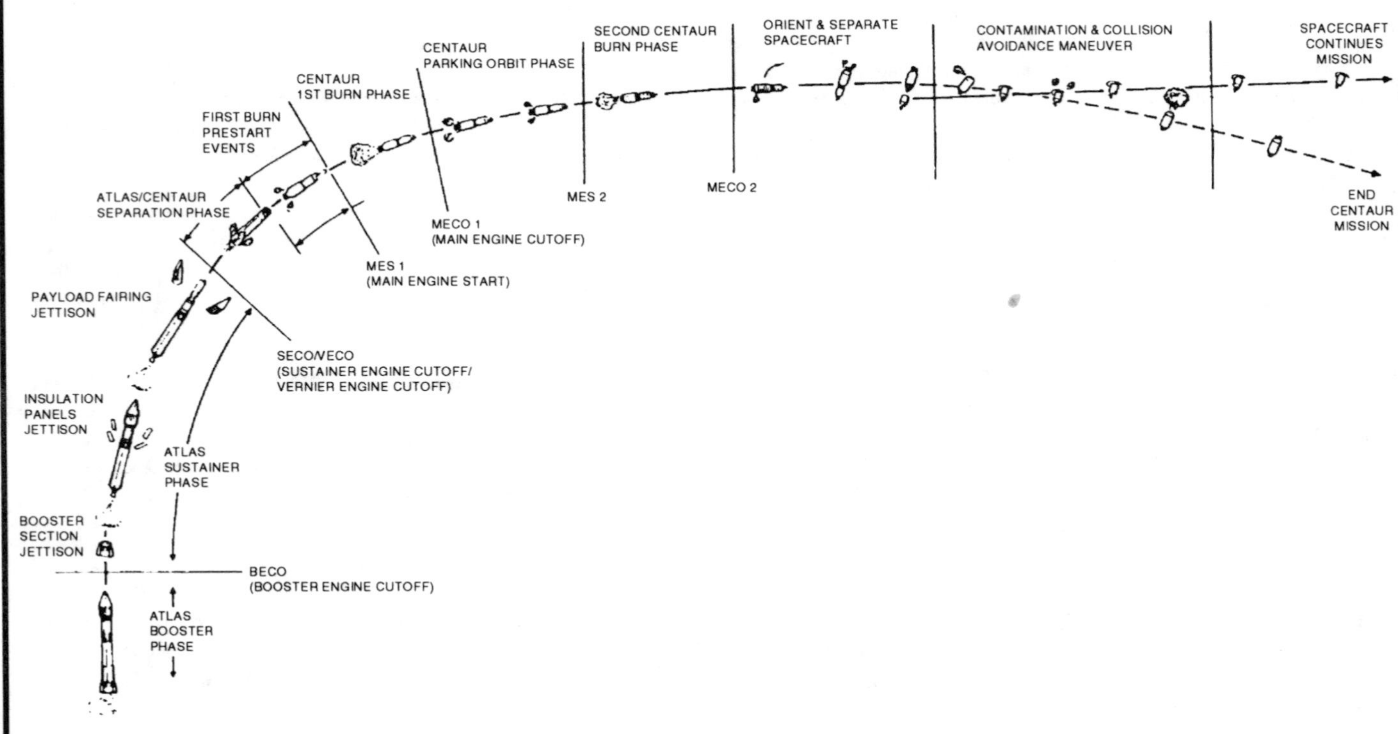

Event	Flight Time (min:sec) Atlas I	Atlas II	Atlas IIA	Atlas IIAS
Liftoff/SRM ignition (1st pair)	00:00	00:00	00:00	00:00
SRM burnout (1st pair)	—	—	—	00:56.0
SRM ignition (2nd pair)	—	—	—	01:06.2
SRM jettison (1st pair)	—	—	—	01:31.7
SRM burnout (2nd pair)	—	—	—	02:02.5
SRM jettison (2nd pair)	—	—	—	02:06.2
Atlas booster engine cutoff (BECO)	02:35.5	02:52.4	02:48.7	02:47.8
Atlas booster package jettison	02:38.5	02:55.5	02:51.8	02:50.9
Insulation panel jettison	03:00.5	—	—	—
Payload fairing jettison	03:40.8	03:46.0	03:52.4	03:37.5
Atlas sustainer engine cutoff (SECO)	04:27.0	04:38.2	04:38.9	04:43.4
Atlas sustainer jettison	04:29.0	04:40.2	04:40.9	04:45.4
Centaur first main engine start (MES1)	04:39.5	04:50.7	04:51.4	04:55.9
Centaur first main eingne cutoff (MECO1)	09:53.2	11:13.3	09:55.1	09:48.1
Centaur second main engine start (MES2)	24:08.7	24:32.5	24:09.4	23:40.4
Centaur second main engine cutoff (MECO2)	25:42.5	26:17.3	25:39.5	25:21.1
Alignment to separation attitude and spin-up	25:44.5	26:19.3	25:41.5	25:23.1
Separate spacecraft	27:57.5	28:32.3	27:54.5	27:36.1
Collision and contamination avoidance maneuver	37:57.5	38:32.3	37:54.5	37:36.1
Satellite first apogee arrival	5.7 hrs	5.7 hrs	5.7 hrs	5.7 hrs

Atlas Payload Accommodations

	Atlas I (except where noted)
Payload Compartment	
Maximum Payload Diameter	115 in (2,921 mm) for Medium PLF 143.7 in (3,650 mm) for Large PLF
Maximum Cylinder Length	210.0 in (5,334 mm) for Medium PLF 207.0 in (5,258 mm) for Large PLF
Maximum Cone Length	151.2 in (3,840.5 mm) for Medium PLF 218.5 in (5,550 mm) for Large PLF
Payload Adapter	
Interface Diameter	36.9 in (937 mm) for Type A 45.2 in (1,147 mm) for Type B 65.6 in (1,666 mm) for Type D
Payload Integration	
Nominal Mission Schedule Begins	T-24 months
Launch Window	
Latest Countdown Hold Not Requiring Recycling	T-5 min
On-Pad Storage Capability	Months without propellant
Latest Access to Payload	T-3 hours
Environment	
Maximum Load Factors	+6.0 g axial, ±2.0 g lateral (not at same time)
Minimum Lateral / Longitudinal Payload Frequency	10 Hz / 15 Hz
Maximum Overall Acoustic Level	137 dB (1 / 3 octave) for Medium PLF 138 dB (1 / 3 octave) for Large PLF [0.4 dB higher for Atlas II, IIA, and IIAS]
Maximum Flight Shock	2,000 g at 1,500 hz
Maximum Dynamic Pressure on Fairing	700 lb/ft^2 (33,520 N/m^2)
Maximum Pressure Change in Fairing	0.8 psi/s (5.4 KPa/s)
Cleanliness Level in Fairing (Prior to Launch)	Class 100,000
Payload Delivery	
Standard Orbit and Accuracy (3 sigma)	LEO: 220 nm (407km) circular ±3.5 nm (6.5 km), 28.5±0.011° inclination GTO: 90 nm (167 km) ±1.01 nm (1.87 km) perigee 19,391 nm (35,912 km) ±28 nm (52 km) apogee 26.4 ±0.020° inclination
Attitude Accuracy (3 sigma)	Pitch: ±0.5 deg; ±0.2 deg/sec Yaw: ±0.5 deg; ±0.2 deg/sec Roll: ±0.5 deg; ±0.5 deg/sec
Nominal Payload Separation Rate	1.0–3.0 ft/s (0.3 - 0.9 m/s)
Deployment Rotation Rate Available	0–4.7 rpm
Loiter Duration in Orbit	Variable
Maneuvers (Thermal / Collision / Telemetry)	Yes

Atlas Notes

Publications

User's Guide

Mission Planner's Guide for the Atlas Launch Vehicle Family, General Dynamics Commercial Launch Services, Inc., Rev. 4, July 1993.

Technical Publications

"Status and Review of the Commercial Atlas Program" by M. D. Patzer, R. C. White and T. Bohlen, General Dynamics Commercial Launch services, AIAA-94-0902, AIAA International Communications Satellite Conference, San Diego, March 1994.

"The Commercial Atlas Today" by M. D. Patzer and R. C. White, General Dynamics Commercial Launch Services, AIAA-90-2715, AIAA 26th Joint Propulsion Conference, Orlando, FL, July 16-18, 1990.

"Atlas Family Update" by R. C. White and M. D. Patzer, General Dynamics Commercial Launch Services, AIAA-90-0827, AIAA Aerospace Engineering Conference, Los Angeles, CA, February, 1990.

"A Historical Look at United States Launch Vehicles: 1967-Present", ANSER, STDN 90-4, second edition, February 1990.

"Continued Enhancements to the Commercial Atlas Launch System", by Barry Matsumori and Greg Wong, General Dynamics Space Systems Division, IAF-89-197, 40th International Astronautical Congress, Malaga, Spain, Oct 7-12, 1989.

"The Atlas and Centaur 'Steel Balloon' Tanks A Legacy of Karel Bossart", by Richard E. Martin, General Dynamics Space Systems Division, IAF-89-738, 40th International Astronautical Congress, Malaga, Spain, Oct 7-12, 1989.

"Commercial Atlas/Centaur Update", by L. R. Scherer and R. C. White, General Dynamics Space Systems Division, AIAA-88-0854, 1988.

"Atlas and Centaur Capabilities, Status, and Plans for Commercial Launch Services" by L. R. Scherer and R. C. White, General Dynamics Space Systems Division, AIAA-87-1798, AIAA 23th Joint Propulsion Conference, San Diego, CA, June 29-July 2, 1987.

Acronyms

ACS - attitude control system
ABRES - Advanced Ballistic Reentry System
AKM - apogee kick motor
BMRS - Ballistic Missile Reentry System
CCAFS - Cape Canaveral Air Force Station
CCLS - Computer Controlled Launch Sets
COP - Canoga Overhaul Program
DMSP - Defense Meteorological Satellite Program
DSCS - Defense Satellite Communications Systems
ECS - environmental control systems
GEO - geosynchronous orbit
GERTS - General Electric Radio Tracking System
GHE - gaseous helium
GN2 - gaseous nitrogen
GSE - ground support equipment
GTO - geosynchronous transfer orbit
ICBM - Intercontinental Ballistic Missile
INU - Inertial Navigation Unit
LC - launch complex
LEO - low Earth orbit
LH2 - liquid hydrogen
LOX - liquid oxygen
MLV - Medium Launch Vehicle
MST - Mobile Service Tower
NASA - National Aeronautics and Space Administration
PALC - Point Arguello Launch Complex
PDU - Power Distribution Unit
PLF - payload fairing
RF - radio frequency
RCU - Remote Control Unit
RP1 - kerosene
SL - sea level
SLC - Space Launch Complex
SLV - Standardized Launch Vehicle
SMFLT - Launch Vehicle Simulated Flight
SRM - solid rocket motor
STP - Space Test Program
UT - umbilical tower
VAC - vacuum
VAFB - Vandenberg Air Force Base
VAMP - Vandenberg Atlas Modification Program
WDR - Wet Dress Rehearsal

Other Notes - Atlas Growth Possibilities

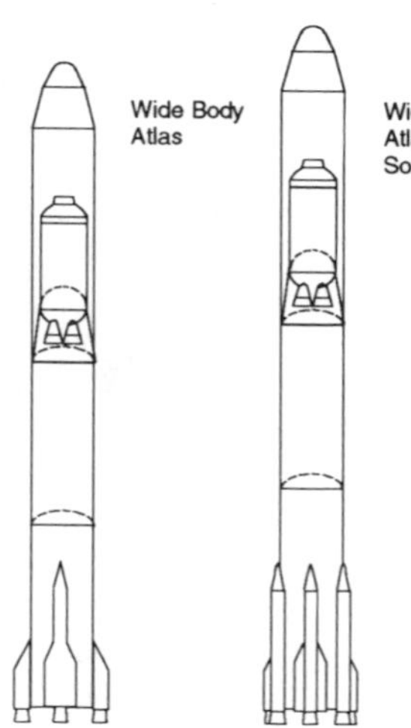

Possible next steps in the Atlas growth plan include improved booster engines, stretched graphite epoxy solid rocket motors, extended fuel tanks, single engine Centaur, or a wide body Atlas. A wide body Atlas is powered by five Rocketdyne RS-27 engines that produce more than 1,000,000 lbs (4.4 MN) of thrust at liftoff. The wide body Atlas has more than twice the propellant capacity of Atlas I contained in a 14 ft (4.3 m) diameter tank and would be capable of about 11,000 lb (5,500kg) performance to GTO and 30% more with solid strap-ons.

A no Centaur version of Atlas II, designated JII, could be available for west coast launches. Such a vehicle could be flown with a variety of upper stages such as Payload Assist Module-Delta (PAM-D).

Conestoga

Industry Point of Contact:
EER Systems Corporation
1593 Spring Hill Road
Vienna, Virginia 22182
Phone: (703) 847-5750

Artist Conception of Conestoga

Conestoga History

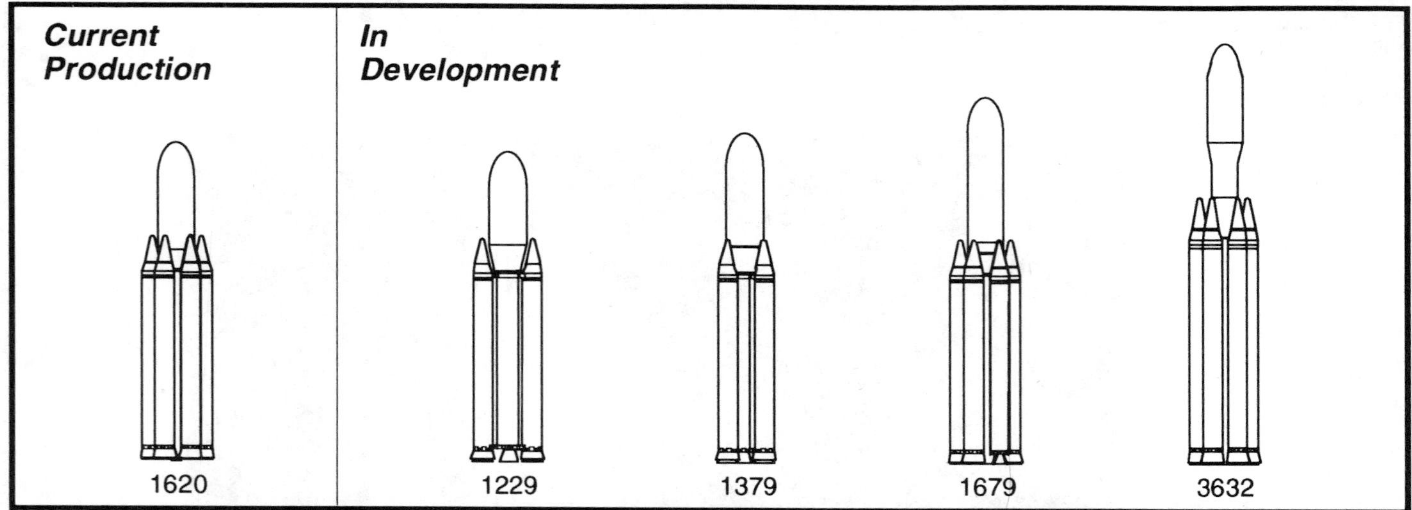

Vehicle Description

The Conestoga launch vehicle is modular in design and supports a wide range of payload and orbit requirements by using a simple building block approach of strap-on motors and upper stages. The booster stage rockets consist of one core CASTOR solid rocket motor (SRM) surrounded by two to six strap-on CASTOR IVA and/or IVB SRMs. An upper stage combination of one to two motors from the STAR 37, 48, or 63 series can be added directly above the core booster SRM.

The four digit designator used to identify Conestoga configurations is explained at right. The 1620 vehicle is used as an example.

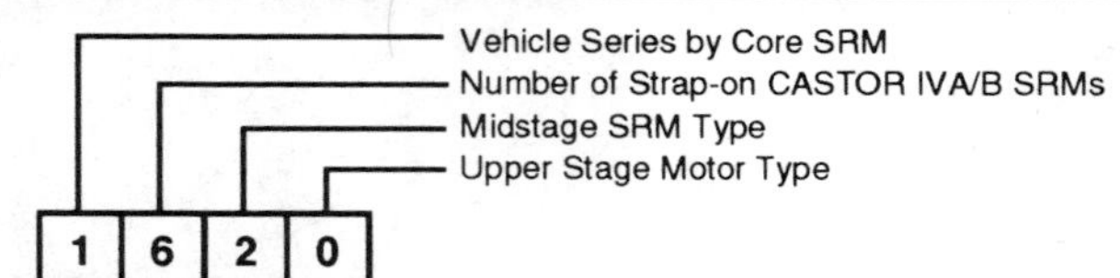

The first digit defines the ELV series by core motor type
- 1 - CASTOR IVB
- 2 - CASTOR IVA
- 3 - CASTOR IV AXL (8 foot 8 inch extension to CASTOR IVA)

The second digit indicates the number of strap-on CASTOR IVA or B SRMs. Being either 2, 3, 4, or 6.

The third and fourth digits indicate the type of motor being used for the mid stage and upper stage, respectively:
- 1 - STAR 37FM
- 2 - STAR 48V
- 3 - Orion 50
- 5 - STAR 48A
- 6 - STAR 63D
- 7 - STAR 63F
- 9 - Liquid Transfer Stage
- 0 - Upper Stage Only (no mid stage)

Historical Summary

In 1985, EER Systems assessed that the small satellite market (500–2,000 lb) would increase singificantly in the decade of the 1990s. Furthermore, no operational launch vehicle at that time targeted this range of payloads. Such a vehicle was being developed, however, by Space Services Inc. (SSI), of America. In November 1990, EER Systems purchased the assets of SSI. These assets included the Starfire suborbital launch vehicle, the Conestoga orbital launch vehicle, and the Consort contract for suborbital launches. SSI was integrated into EER Systems and currently operates as the Space Systems Group.

SSI privately funded the first commercial launch of a rocket in 1982 and was the first company to obtain a commercial launch license. In 1985, they received the first Department of Transportation mission approval. In 1986, they signed the industry's first agreement to use a U.S. government range as a commercial launch site (Wallops Flight Facility).

In 1988, SSI bid and won the Consort Program for four suborbital launches of the Starfire I vehicle. In 1989, they launched the first commercial suborbital launch in the United States—Consort 1. In December 1990, EER Systems was selected to be the launch vehicle provider for the COMET Program for five orbital launches. The negotiated contract was signed in March 1991. At the present time, the COMET launch vehicle (Conestoga Model 1620) is in production and will be launched in mid 1995.

The Conestoga vehicle is modular and supports a wide range of payload and orbit requirements using a simple building block approach of strap-on motors to augment thrust capabilities. Conestoga's upper stage combination of one to two motors provides the opportunity to select from a wide range of capabilities. The Conestoga makes full use of flight proven "off-the-shelf" propulsion units for all stages.

Launch Record

The first Conestoga launch will be mid-1995.

Conestoga General Description

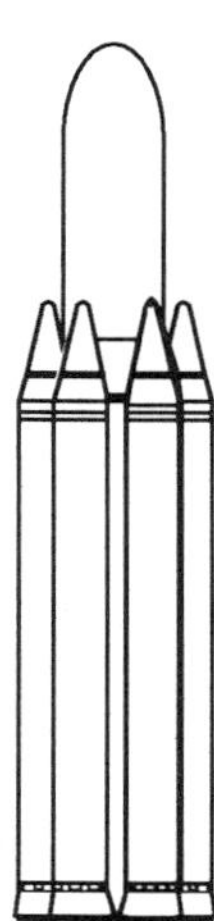

Summary — Conestoga is a four-stage, solid-propellant, inertially guided launch vehicle developed by EER Systems.

Status

Conestoga 1620 - First flight scheduled for mid 1995
Other Conestoga Family Members - In Development

Key Organizations

User - NASA, commercial, DoD

Launch Service Agency - EER Systems

Prime Contractor - EER Systems

Principle Subcontractors:
Thiokol (Solid Rocket Motors)
Tracor (Fairing)

Vehicle

System Height	50 ft (15.24 m)
Payload Fairing Size	Production - 6 ft (1.83 m) diameter by 16 ft (4.88 m) height Development - 6 ft (1.83 m) diameter by 24 ft (7.32 m) height
Gross Mass	192,700 lb (87,407 kg) for Conestoga 1620
Planned Enhancements	A wide variety of growth possibilities are available

Operations

Primary Missions	LEO (18° i to 65° i) orbit from Wallops Flight Facility
Compatible Upper Stages	Solid Rocket Motors (STAR 37, 48, or 63) Bipropellant Engines
First Launch	Mid 1995
Success/Flight Total	0 / 0
Launch Site	Wallops Flight Facility
Launch Azimuth	??
Nominal Flight Rate	Up to 6 / yr
Planned Enhancements	Other possible launch sites include, CCAFS, Vandenberg AFB, CA, Poker Flats, AL, and Churchill, Canada

Performance

250 nm (463 km) circ., 40°	1,960 lb (889 kg)
250 nm (463 km) circ., 90°	1,450 lb (658 kg)
Geotransfer Orbit	—
Geosynchronous Orbit	—

Financial Status

Estimated Launch Price	$18–20M (EER Systems)
Orders Payload / Agency	COMET / NASA

Planned Enhancement — EER Systems has under development a family of Conestoga Launch Vehicles that have payload capabilities in the 500 lb to 5,000 lb to LEO range.

Manifest — 1995 COMET

Remarks — None

Conestoga Vehicle

Overall

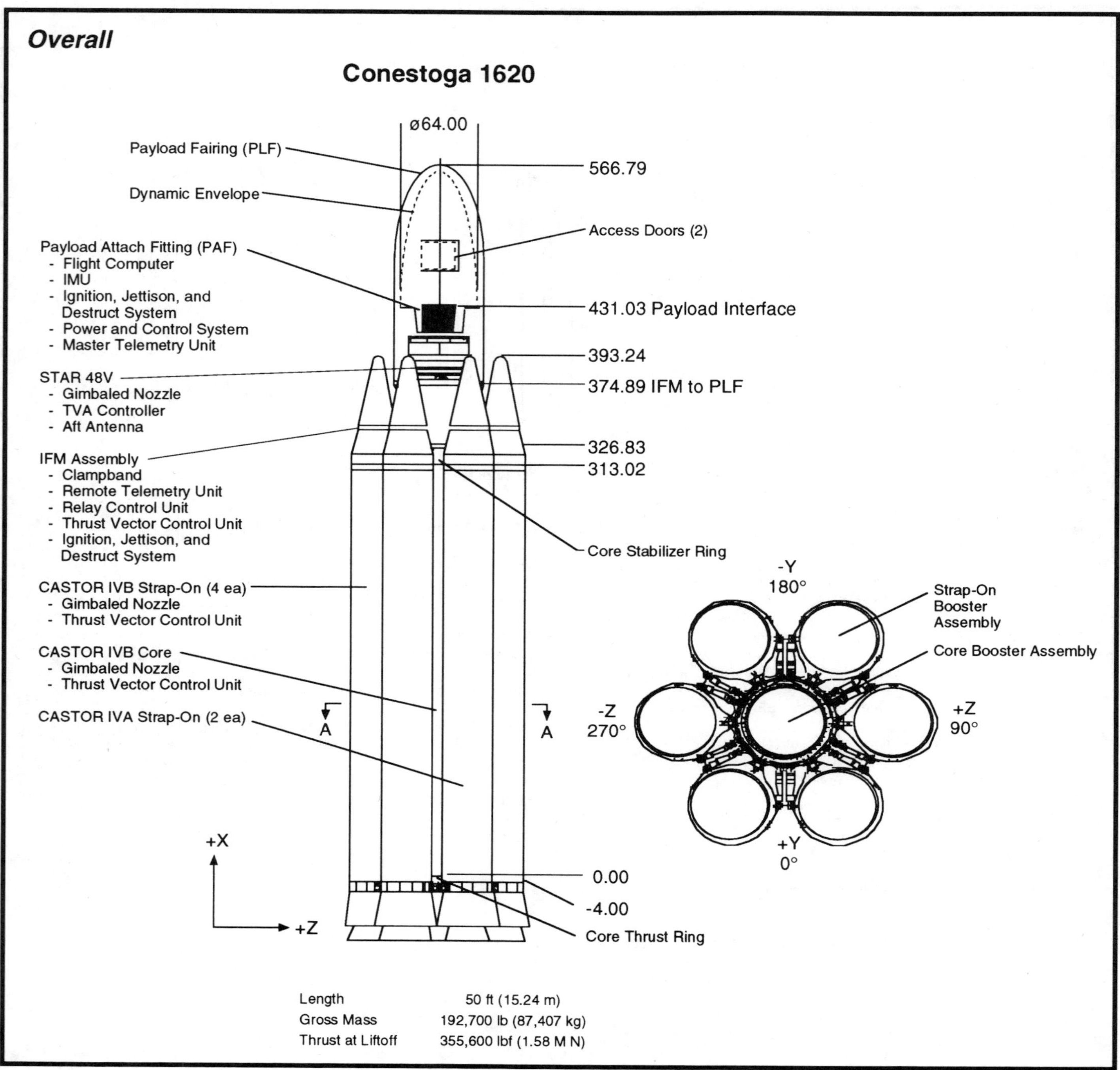

Length	50 ft (15.24 m)
Gross Mass	192,700 lb (87,407 kg)
Thrust at Liftoff	355,600 lbf (1.58 M N)

Conestoga Vehicle
(continued)

Booster SRMs

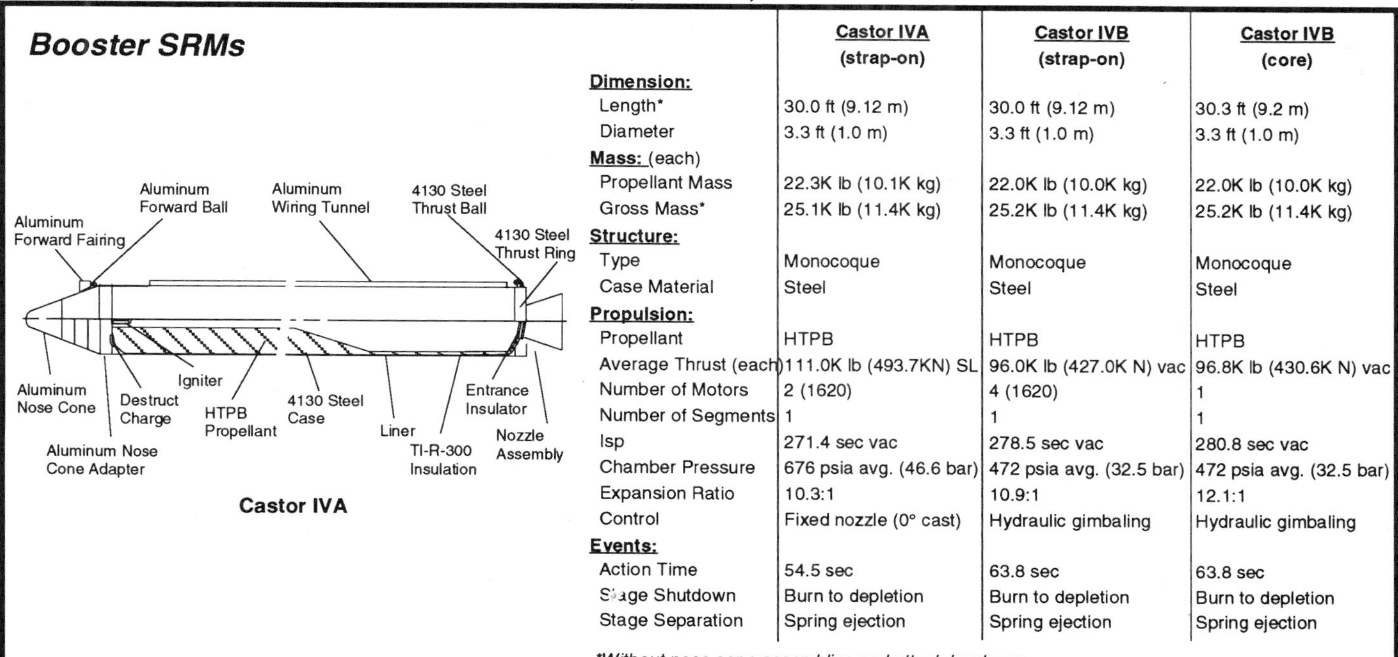

Castor IVA

	Castor IVA (strap-on)	Castor IVB (strap-on)	Castor IVB (core)
Dimension:			
Length*	30.0 ft (9.12 m)	30.0 ft (9.12 m)	30.3 ft (9.2 m)
Diameter	3.3 ft (1.0 m)	3.3 ft (1.0 m)	3.3 ft (1.0 m)
Mass: (each)			
Propellant Mass	22.3K lb (10.1K kg)	22.0K lb (10.0K kg)	22.0K lb (10.0K kg)
Gross Mass*	25.1K lb (11.4K kg)	25.2K lb (11.4K kg)	25.2K lb (11.4K kg)
Structure:			
Type	Monocoque	Monocoque	Monocoque
Case Material	Steel	Steel	Steel
Propulsion:			
Propellant	HTPB	HTPB	HTPB
Average Thrust (each)	111.0K lb (493.7KN) SL	96.0K lb (427.0K N) vac	96.8K lb (430.6K N) vac
Number of Motors	2 (1620)	4 (1620)	1
Number of Segments	1	1	1
Isp	271.4 sec vac	278.5 sec vac	280.8 sec vac
Chamber Pressure	676 psia avg. (46.6 bar)	472 psia avg. (32.5 bar)	472 psia avg. (32.5 bar)
Expansion Ratio	10.3:1	10.9:1	12.1:1
Control	Fixed nozzle (0° cast)	Hydraulic gimbaling	Hydraulic gimbaling
Events:			
Action Time	54.5 sec	63.8 sec	63.8 sec
Stage Shutdown	Burn to depletion	Burn to depletion	Burn to depletion
Stage Separation	Spring ejection	Spring ejection	Spring ejection

**Without nose cone assemblies and attach hardware*

Remarks:

The heart of the Conestoga family of solid rocket motor launch vehicles are the CASTOR IVA and CASTOR IVB manufactured by Thiokol, Huntsville. The use of these flight proven, reliable motors in a modular, building block approach provides low cost and high reliability.

The Conestoga 1620 launch vehicle uses two CASTOR IVAs and five CASTOR IVBs . Two CASTOR IVAs and two CASTOR IVBs are used for the first stage while two more CASTOR IVBs are used for the second stage of the 1620. The third stage, or core, is a single CASTOR IVB.

The internal design of the CASTOR IVA is identical to the 153 motors that have flown on the Delta II vehicle. The internal ballistics of the CASTOR IVB were modified from those of the CASTOR IVA to create a regressive pressure/time and thrust/time trace. This minimizes peak aerodynamic loads on the launch vehicle as well as reduces axial acceleration. The CASTOR IVB uses a TVC system based on Peacekeeper, SICBM, Trident, and Shuttle SRM. The two actuaors which vector the nozzle are hydraulically powered. Hydraulic pressure is supplied by a precharged pneumatic bottle that is actuated by an electroexplosive pyrotechnical valve. Proportional control valves, in concert with displacement transducers, control the nozzle angular displacement. The system has the capability to hold for over four minutes after system initiation. Maximum deflection of the nozzle is six degrees. The remainder of the CASTOR IVB design is common to the CASTOR family. As a performance enhancement, the expansion ratios of the CASTOR IVBs have been increased by extending the length of the nozzles.

STAR 48V Upper Stage

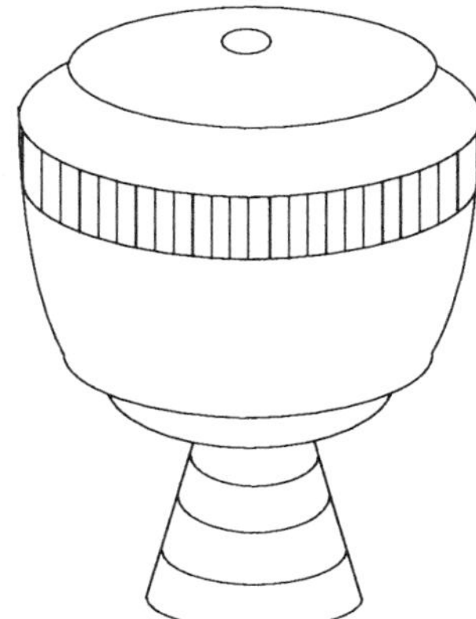

	STAR 48V
Dimension:	
Length	6.8 ft (2.07 m)
Diameter	4.1 ft (1.25 m)
Mass: (each)	
Propellant Mass	4,430 lb (2,010 kg)
Gross Mass	4,765 lb (2,161 kg)
Structure:	
Type	Monocoque
Case Material	Titanium
Propulsion:	
Propellant	HTPB
Average Thrust (each)	15,355 lb (68,300 N) vac
Number of Motors	1
Number of Segments	1
Isp	292.0 sec vac
Chamber Pressure	575 psia (39.7 bar)
Expansion Ratio	48.8:1
Control	3-axis stabilized
Events:	
Action Time	86.7 sec
Stage Shutdown	Burn to depletion
Stage Separation	Spring ejection

Remarks:

The upper stage of the Conestoga 1620 is a Thiokol STAR 48V. A straight forward development of the STAR 48 motor used on the Delta and Shuttle program uses Thrust Vector Control (TVC) to eliminate the necessity to spin the stage. The TVC system is based on the proven Flexseal approach Thiokol has developed and used on Peacekeeper and CASTOR IVB. The STAR 48V nozzle is essentially identical to the STAR 48B nozzle that has been flown or tested more than 30 times.

Conestoga Vehicle
(continued)

Payload Fairing

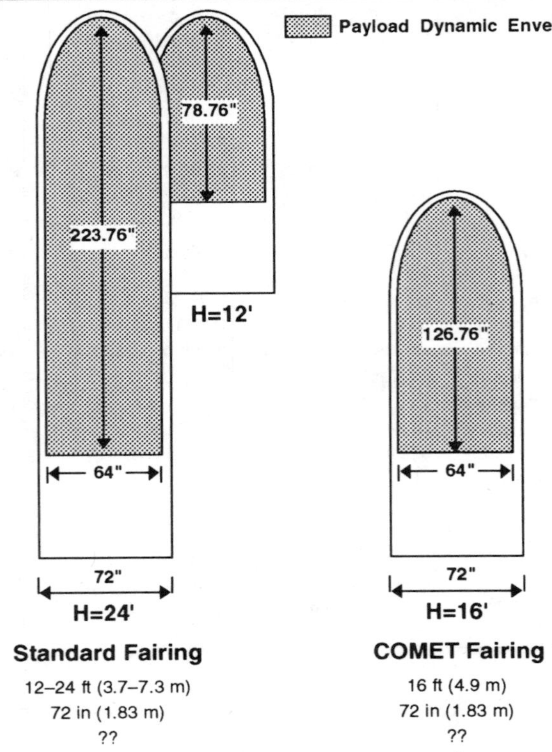

	Standard Fairing	COMET Fairing
Length	12–24 ft (3.7–7.3 m)	16 ft (4.9 m)
Diameter	72 in (1.83 m)	72 in (1.83 m)
Mass	??	??
Sections	2	2
Structure	??	??
Material	Composite honeycomb	Composite honeycomb

Remarks:
The Conestoga fleet of launch vehicles provides complete accommodations to meet user requirements with varying fairing sizes, standard or unique mechanical and electrical interfaces, and specified launch and ascent environments. While on the launch pad, the PLF provides payload specified environmental conditions such as conditioned air flow and required venting. Antennae mounted along the inner surface of the PLF provide for both S-band transmission and Flight Termination System (FTS) reception.

The standard PLF is 72 inches in external diameter with a 64-inch dynamic envelope and ranges in length from 12 to 24 feet, as shown. The standard PLF is also equipped with two 16 x 16-inch doors. One door accommodates the air conditioning system and the other can be used for payload access. Doors in various sizes and locations can also be incorporated into the PLF design, within structural capabilities, to meet specific mission unique access requirements.

Fairing separation is performed by a thrusting joint longitudinal separation system and a tension band horizontal separation system. The separation system has been tested and shown to be contamination free. The PLF is fabricated in near identical half-shells with the cylinder, forebody, and nose sections forming a continuous structure.

To provide thermal protection on the PLF baseline design, a spray-on silicon-based thermal protection is applied to the PLF outer surface.

Avionics System

The Conestoga avionics system consists of the following subsystems: power and control; telemetry; guidance, navigation, and control (GN&C); flight termination (FTS); and radio frequency (RF).

The power and control subsystem is located in the third and fourth stages. The fourth stage houses a Relay Control Unit (RCU) and a flight battery. The third stage houses an RCU and an ordnance battery.

The telemetry subsystem spans the third and fourth stages with sensors located throughout the entire vehicle. The fourth stage houses a master telemetry unit and various sensors while the third stage houses a Remote Telemetry Unit and additional sensors. Other sensors are located within the payload fairing. A telemetry interface can embed payload telemetry into the launch vehicle telemetry downlink. The telemetry system provides a continuous data link which is also used by range safety to track and monitor vehicle performance during flight. It also provides bandwidth to acquire all of the necessary guidance, navigation, and control; motor performance; acoustic noise; vibration; low-frequency trajectory maneuvers; modal frequencies; pressure; temperature; strain; and payload data during ground testing and flight.

The GN&C spans the entire vehicle and consists of the Flight Computer (FC) containing the flight software, the Inertial Measurement Unit (IMU), a TVA controller for the upper stage, and TVC controllers for the five CASTOR IVB SRMs.

The FC distributes digital computer commands for the command control of the TVC controllers, TVA actuators and for relays within the RCUs. It also controls the passing of telemetry data to the master telemetry unit for processing. The FC performs the guidance, navigation, and control functions based on open and closed-loop algorithms. It integrates the equations of motion based on IMU inputs of acceleration and rate and distributes digital commands to control the CASTOR IVB thrust vector control system.

The TVA controller on the fourth-stage motor nozzle receives digital pitch and yaw nozzle-position commands from the FC and provides electromechanical pitch and yaw commands to the nozzle. The TVA controller receives power from a thermal battery supplied with the motor which is located on its aft section.

The FTS is located on the fourth stage and consists of two 28V batteries, two UHF receivers, a relay control network, an antenna system, and a series of explosive charges located on each propulsion unit. The FTS is isolated within the launch vehicle avionics system and is used by the range safety officer to terminate the flight in the event the launch vehicle goes out of control during the launch and ascent phases.

The RF subsystem is located in the Payload Attach Fitting (PAF) and consists of an S-band transmitter and two omni antennas, one located in the upper fairing and one around the upper stage SRM. A C-band transponder is also carried for range safety tracking and the C-band antenna strip is collocated with the S-band antenna on the STAR 48V motor.

Attitude Control System

Unknown

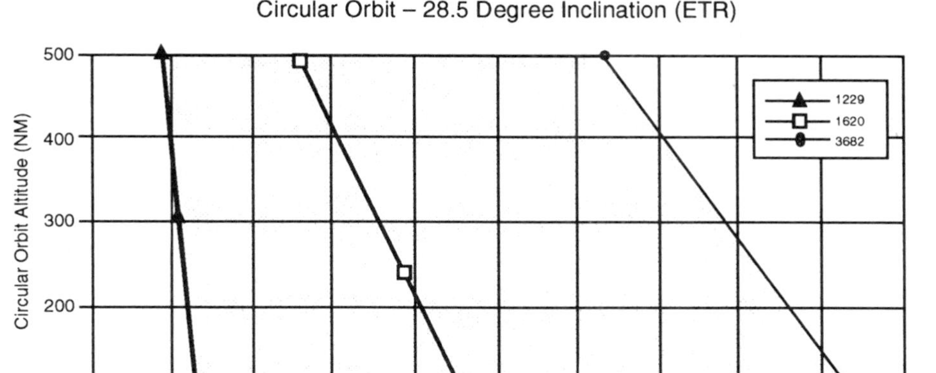
Circular Orbit – 28.5 Degree Inclination (ETR)
Circular Orbit Altitude (NM)
500
400
300
200
100
0
500
1,000
1,500
2,000
2,500
3,000
3,500
4,000
4,500
5,000
Payload Weight (lbs)
1229
1620
3682

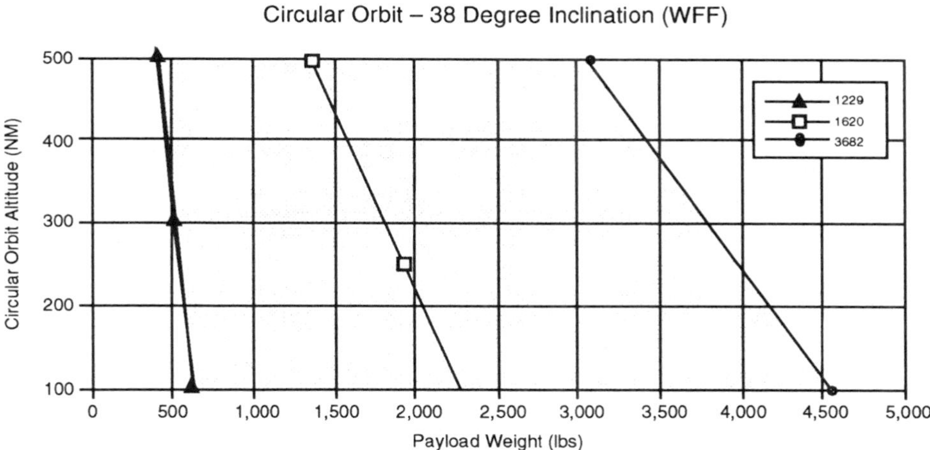
Circular Orbit – 38 Degree Inclination (WFF)
Circular Orbit Altitude (NM)
500
400
300
200
100
0
500
1,000
1,500
2,000
2,500
3,000
3,500
4,000
4,500
5,000
Payload Weight (lbs)
1229
1620
3682

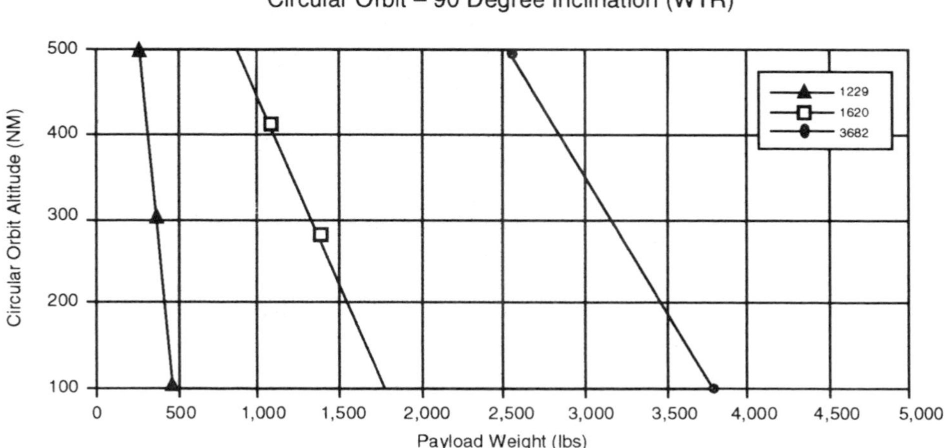
Circular Orbit – 90 Degree Inclination (WTR)
Circular Orbit Altitude (NM)
500
400
300
200
100
0
500
1,000
1,500
2,000
2,500
3,000
3,500
4,000
4,500
5,000
Payload Weight (lbs)
1229
1620
3682

Conestoga Operations
(continued)

Launch Site

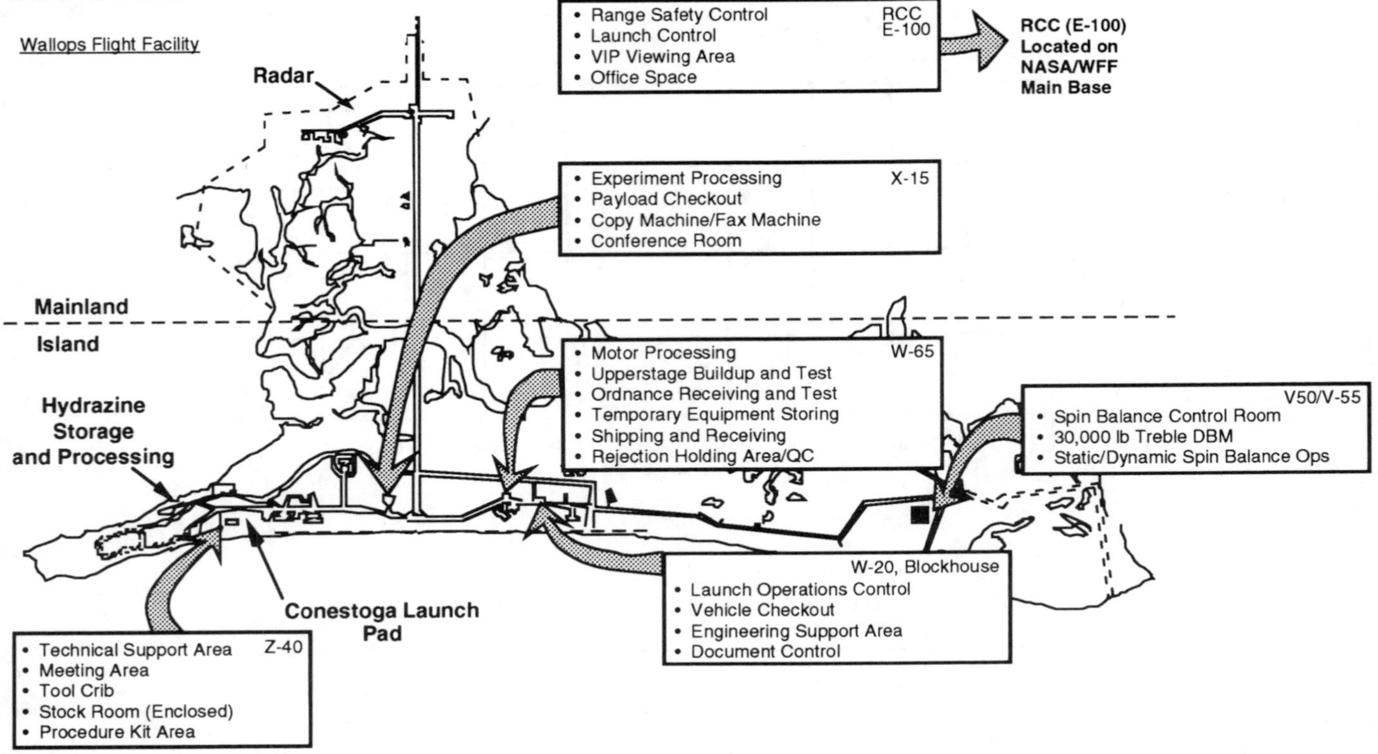

Wallops Flight Facility

Radar

- Range Safety Control
- Launch Control
- VIP Viewing Area
- Office Space

RCC
E-100

RCC (E-100)
Located on
NASA/WFF
Main Base

- Experiment Processing
- Payload Checkout
- Copy Machine/Fax Machine
- Conference Room

X-15

Mainland

Island

- Motor Processing
- Upperstage Buildup and Test
- Ordnance Receiving and Test
- Temporary Equipment Storing
- Shipping and Receiving
- Rejection Holding Area/QC

W-65

Hydrazine
Storage
and Processing

V50/V-55
- Spin Balance Control Room
- 30,000 lb Treble DBM
- Static/Dynamic Spin Balance Ops

W-20, Blockhouse
- Launch Operations Control
- Vehicle Checkout
- Engineering Support Area
- Document Control

Conestoga Launch
Pad

- Technical Support Area
- Meeting Area
- Tool Crib
- Stock Room (Enclosed)
- Procedure Kit Area

Z-40

Launch Facilities

- Clamshell Design
- Multilevel Access to Vehicle
- Clean Room
- Late Access to Payload (4 hours)
- Environmental Shelter
- Modular Construction
- Relocatable
- WFF/VAFB/CCAFS Compatible

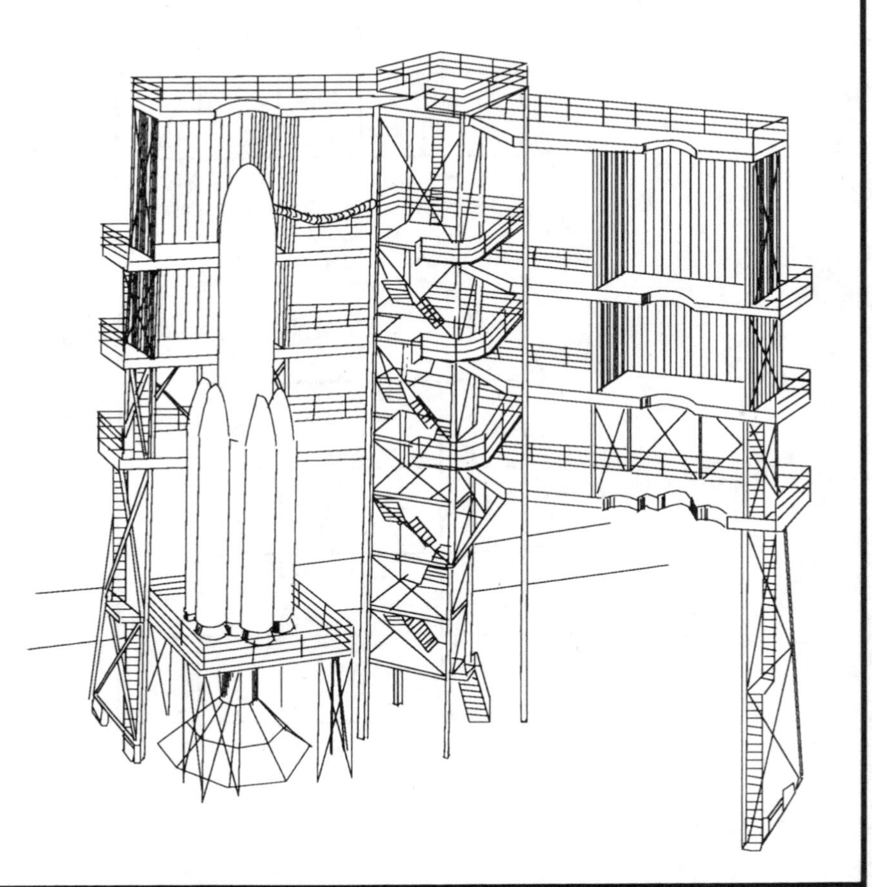

Conestoga Operations

Launch Processing

The initial launch complex for the Conestoga fleet, to support missions with orbital inclination requirements of 38–66 degrees, is in place at the NASA Wallops Flight Facility (WFF) , Virginia, which is situated on the Atlantic Coast at 37.9 degrees north latitude. Planning activities are also underway to establish additional launch capabilities at sites which can accommodate polar and sun synchronous orbital inclination requirements as well as mission requirements for inclinations less than 38 degrees.

The launch complex has a modular Portable Service Tower (PST) which can, if required, be dismantled into transportable modules and relocated to an alternate launch site. The PST is a clamshell design which provides multilevel vehicle access, clean room capabilities, environmental shelter and conditioning and late access to the payload. The initial mission of the Conestoga 1620, COMET-1, requires late access to the payload and thus was a key design element of the PST. The PST was designed and built by EER Systems and provides for vertical integration, checkout, and launch of the Conestoga fleet and its payloads.

Conestoga vehicle and payload preparations are conducted in a number of NASA-owned facilities which are leased to EER for performing commercial launch services. The Wallops Range Control Center (RCC) and the Range Tracking Radars are located on the mainland, with the remaining facilities located on the island, including the Payload Processing Facility, Spin Balance Facilities, Blockhouse, Vehicle Preparation Facility, Hydrazine Processing Facility, Launch Pad, Portable Service Tower, and the Pad Terminal Building.

The Wallops Range Control Center (RCC) provides range safety control and a VIP area to monitor the launch. This facility has a fully automated power backup system and uninterrupted power supply to all systems located at the facility. The Launch Pad supports the vehicle launch mount and the PST. Entrance roads to the pad and sufficient maneuvering area for cranes, trucks, and other vehicles are provided.

Flight Sequence

The nominal trajectory of the Conestoga 1620 is shown below:

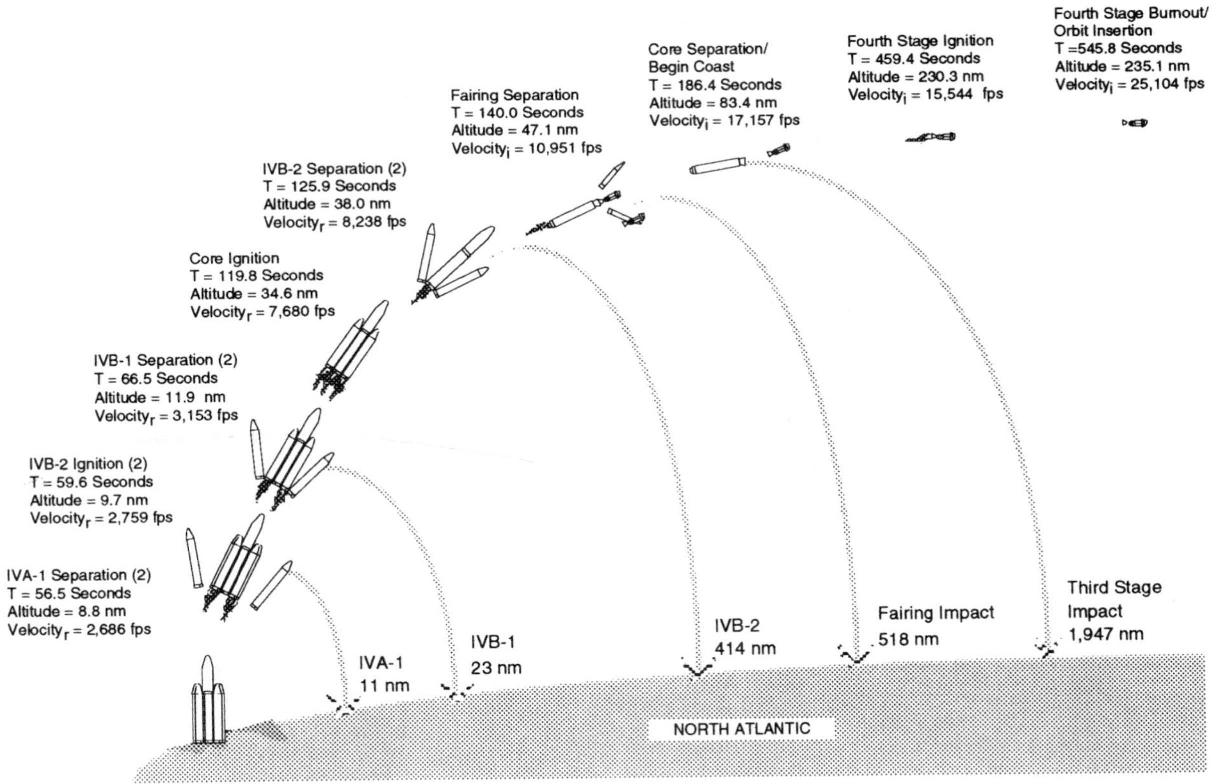

Conestoga Notes

Payload Accommodations

Payload Compartment	
Maximum Payload Diameter	64 in (1,626 mm)
Maximum Cylinder Length	?? in (?? mm)
Maximum Cone Length	?? in (?? mm)
Payload Adapter	
Interface Diameter	37 in (940 mm) or 46.1 in (1,171 mm)
Payload Integration	
Nominal Mission Schedule Begins	T-12 months
Launch Window	
Latest Countdown Hold Not Requiring Recycling	T-?? min
On-Pad Storage Capability	?? hrs for a fueled vehicle
Latest Access to Payload	T-4 hrs
Environment	
Maximum Load Factors	+11.0 g axial, ±2.7 g lateral
Minimum Lateral / Longitudinal Payload Frequency	?? Hz / ?? Hz
Maximum Overall Acoustic Level	128.5 dB (full octave)
Maximum Flight Shock	2,000 g from 1,000–4,000 hz
Maximum Dynamic Pressure on Fairing	?? lb/ft2 (?? N/m2)
Maximum Pressure Change in Fairing	?? psi/s (?? KPa/s)
Cleanliness Level in Fairing (Prior to Launch)	Class ??
Payload Delivery	
Standard Orbit and Accuracy (3 sigma)	Altitude ±?? nm (?? km) Inclination ±?? deg
Attitude Accuracy (3 sigma)	All axes: ±?? deg, ±??deg /sec
Nominal Payload Separation Rate	?? ft/s (?? m/s)
Deployment Rotation Rate Available	?? rpm
Loiter Duration in Orbit	?? hrs
Maneuvers (Thermal / Collision / Telemetry)	??

Publications

User's Guide

Conestoga Payload User's Guide, EER Systems Corporation, Release 1.0, August 1992.

Acronyms

CCAFS - Cape Canaveral Air Force Station
ELV - expendable launch vehicle
ETR - Eastern Test Range
FC - flight computer
FTS - flight termination system
GN&C guidance navigation and control
IMU - inertial measurement unit
LEO - low earth orbit
PAF - payload attach fitting
PLF - payload fairing
PST - portable service tower
RCC - range control center
RCU - relay control unit
RF - radio frequency
SRM - solid rocket motor
SSI - Space Services Incorporated
TVC - thrust vector control
VAFB - Vandenberg Air Force Base
WFF - Wallops Flight Facility
WTR - Western Test Range

Delta

Government Point of Contact:
Delta System Program Office (CLZ)
U.S. Air Force Space Systems Division
P.O. Box 92960
Los Angeles AFB, CA 90009-2960, USA
Phone: (310) 363-2723

Industry Point of Contact:
McDonnell Douglas
Space Systems Company
5301 Bolsa Avenue
Huntington Beach, CA 92647, USA
Phone: (714) 896-4131

The first Delta II launch occurred on February 14, 1989, successfully orbiting the first of a new generation of U.S. Air Force Global Positioning (GPS) satellites.

Delta History

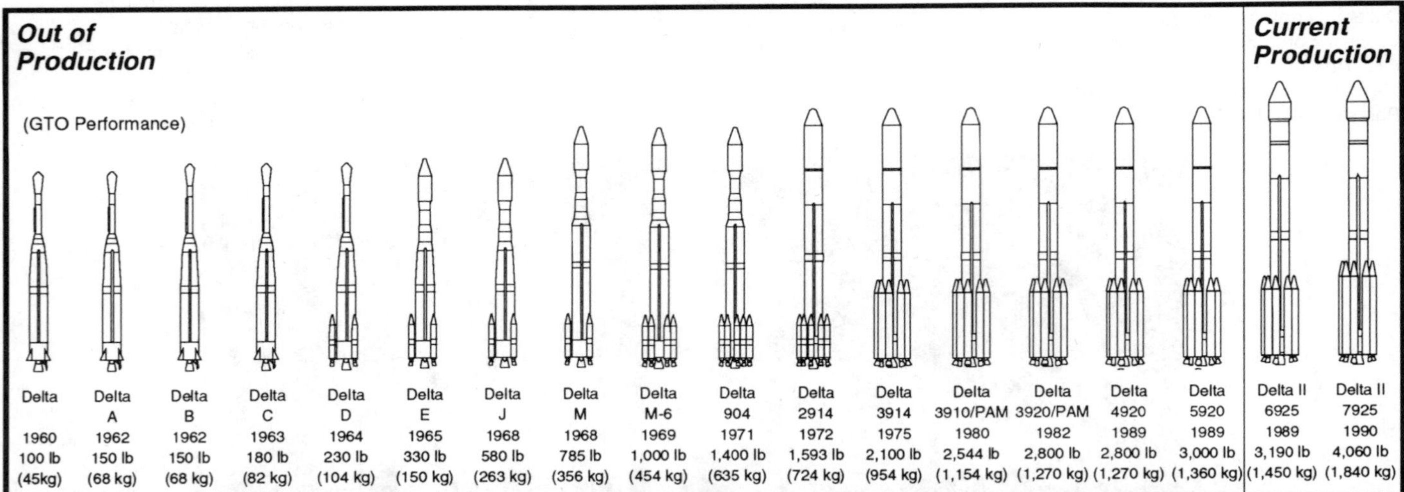

Vehicle Description

Vehicle	Modified Stage	Modification
Delta	1 2 3	Modified Thor. MB-3 Blk I engine Vanguard AJ10-118 propulsion system Vanguard X-248 motor
Delta A	1	Engine replaced with MB-3 Blk II
Delta B	2	Tanks lengthened higher energy oxidizer used
Delta C	3 PLF	Replaced with Scout X-258 motor Bulbous replaced low drag
Delta D	0	Added 3 Thor-developed solid rocket motors (Castor I)
Delta E	0 1 2 3 PLF	Castor II replaced Castor I MB-3 Blk III replaced Blk II Propellant tanks widened Replaced with USAF-developed FW-4 motor Fairing enlarged to 65 in (1.65 m) diameter (from Agena)
Delta J	3	TE-364-3 used
Delta L, M, N	1 3	Tanks lengthened RP-1 tank widened Varied - FW-4 (L). TE-364-3 (M). None (N)
Delta M-6, N-6	0	Six Castor II employed
Delta 900	0 2	Nine Castor II used Replaced with Transtage AJ10-118F engine
Delta 1604	0 3	Six Castor II used TE-364-4 employed
Delta 1910, 1913, 1914	0 3 PLF	Nine Castor II Varied - None (1910). TE-364-3 (1913). TE364-4 (1914) 96 in (2.44 m) diameter replaced 65 in (1.65 m)
Delta 2310, 2313	0 1 2 3	Three Castor II RS-27 replaced MB-3 TR-201 engine replaced AJ10-118F Varied - None (2310). TE-364-3(2313). TE 364-4 (2314).
Delta 2910, 2913, 2914	0 3	Nine Castor II Varied - None (2910). TE-364-3 (2913). TE-364-4 (2314).
Delta 3910, 3913, 3914	0 3	Nine Castor IV replaced Castor II Varied - None or PAM (3910).TE-364-3 (3913).TE-364-4 (3914)
Delta 3920, 3924	2 3	AJ10-118K engine replaced TR-201 Varied - None or PAM (3920). TE-364-4 (3924)
Delta 4920	0 1	Castor IVA replaced Castor IV MB-3 replaced RS-27
Delta 5920	1	RS-27 replaced MB-3 engine
Delta 6925	1 3 PLF	Tanks lengthened 12 ft (3.66 m) STAR 48B motor used Bulbous. 114 in (2.90 m) diameter used
Delta 7925	0 1	GEMs replaced Castor IVA RS-27A replaced RS-27 (12:1 expansion ratio)

Four Digit Designator

- First Digit - First StageType of Augmentation
- Second Digit - Number of Augmentation Motors
- Third Digit - Type of Second Stage
- Fourth Digit - Type of Third Stage

Example: Delta 6925

First Digit:
- 0 - Castor II, Long Tank, MB-3 Engine
- 1 - Castor II, Extended Long Tank, MB-3 Engine
- 2 - Castor II, Extended Long Tank, RS-27 Engine
- 3 - Castor IV, Extended Long Tank, RS-27 Engine
- 4 - Castor IVA, Extended Long Tank, MB-3 Engine
- 5 - Castor IVA, Extended Long Tank, RS-27 Engine
- 6 - Castor IVA, Extra Extended Long Tank, RS-27 Engine
- 7 - GEM, Extra Extended Long Tank, RS-27A Engine

Second Digit:
- 3 - Three Augmentation solid rocket motors
- 9 - Nine Augmentation solid rocket motors

Third Digit:
- 0 - AJ10-118 (Aerojet)
- 1 - TR-201 (TRW)
- 2 - AJ10-118K (Aerojet)

Fourth Digit:
- 0 - No Third Stage
- 3 - TE-364-3
- 4 - TE-364-4
- 5 - PAM-D Derivative (STAR 48B)

Historical Summary

The Delta Launch Vehicle family originated in 1959 when NASA Goddard Space Flight Center awarded a contract to Douglas Aircraft Company, now McDonnell Douglas Corporation, to produce and integrate 12 launch vehicles. The Delta, using components from the U.S. Air Force's Thor IRBM program and the U.S. Navy's Vanguard launch vehicle program, was available 18 months after go-ahead. On May 13, 1960, the first Delta was launched from Cape Canaveral Air Force Station with a 179 lb (81 kg) Echo I passive communications satellite. Although this first flight was a failure, the ensuing series of successful launches established Delta as one of the most reliable U. S. boosters.

In the years since the first vehicles were produced, the Delta has evolved to meet the ever increasing demands of its payloads—including weather, scientific, and communications satellites. Each Delta modification corresponded to an increase in payload capacity.

The 1960 Delta used a modified Thor booster with a Rocketdyne MB-3 engine as a first stage; the Vanguard second-stage propulsion, Aerojet's AJ10-118, as

Delta History

(continued)

Historical Summary

(continued)

a second stage; and the Vanguard X-248 solid rocket motor as the third stage. In 1962, the Delta A configuration evolved with an improved MB-3 engine of the Thor first stage, and then the Delta B added a lengthened second stage with a higher energy propellant. The Delta C configuration had a bulbous fairing and replaced the Vanguard-developed third stage motor with the Scout-developed X-258 motor. In 1964, three solid strap-on motors, Thiokol Castor Is, were used on the Delta D. The Delta E of 1965 showed three major changes: larger payload fairing from the Air Force Agena stage, larger diameter propellant tanks and restart capability on the second stage, and replacement of the X-258 third stage with the Air Force–developed FW-4. In addition, more powerful Castor II strap-on motors were being used.

In 1968, the third stage again was changed to the Thiokol TE-364-3 motor on the Delta J configuration. Evolutions in 1968 also included lengthening the first-stage propellant tanks and increasing the RP-1 tank diameter on the Delta L, M, and N vehicles. These configurations provided nearly identical lift capacity to orbit with varying third stages—the FW-4, TE-36, and none, respectively. The M-6 and N-6 Deltas raised the number of solid strap-ons to six in the early 1970s, increasing performance by approximately 27%. In 1972, however, the number of solid strap-ons went from six to nine, and the Titan Transtage engine (Aerojet AJ-10-118F) replaced the original Vanguard second-stage propulsion systems on the Delta 900.

In the early 1970s, eight variations of the 1000 series of Delta vehicles and the familiar Delta nomenclature emerged. All eight configurations incorporated some first-stage changes but varied in the number of Castor II strap-ons and the type of third stage used (if any). The later 1900 series vehicles possessed a larger payload fairing (from 65 inches to 96 inches). For example, the Delta 1914 had a larger payload fairing, the extended first-stage tank length, an isogrid structure in the first stage, and a longer third-stage motor.

McDonnell Douglas developed five types of Delta 2000 series vehicles. These vehicles, primarily flown between 1974 and 1978, again differed in the type of third stage and the number of solid Castor II SRMs. Two major evolutions occurred with the development of these vehicles. First, the 2000 series vehicles no longer employed the Aerojet AJ10-118F second-stage engine. Instead, McDonnell Douglas opted for the TRW TR-201 engine (a modification of the descent rocket of the lunar module). Second, the MB-3 main engine was replaced by the more powerful Rocketdyne RS-27 engine.

The Delta configurations of the late 1970s and early 1980s were designated the 3900 series principally the 3914 and 3910/PAM. The 3900 series is very similar to the 2900 series except the 2900 employed Castor II solid strap-ons, whereas the 3900 uses nine larger and more powerful Castor IV solid motors.

Each Castor IV produces 85,000 lb (38,600 kg) of thrust versus the 52,200 lb (23,700 kg) of the Castor II. The new component, PAM (developed by McDonnell Douglas), is used as the third stage. These vehicles are 116 ft (35 m) tall and 9 ft (2.7 m) wide and weigh approximately 420,000 lb (191,000 kg) at launch. The first stage is again powered by the LOX/RP-1 RS-27 engine that provides 207,000 lb (94,000 kg) of thrust at sea level. The second stages use the TRW TR-201 bipropellant N_2O_4/Aerozine-50 engine. This engine is pressurized by helium and provides approximately 9,800 lb (4,500 kg) of thrust. The 3914 and 3910 have different third stages—the Thiokol TE-364-4 and the Thiokol STAR-48 solid motors, respectively. Both the TE-364 and the PAM's STAR-48 provide approximately 15,000 lb (6,800 kg) of thrust. The PAM originally was developed for the Space Shuttle to transfer satellites from a low Earth parking orbit to their final orbits.

In 1979, McDonnell Douglas realized that the transition to the Shuttle would take longer than anticipated. Not only did this mean that more Delta launches would be required, but the Delta configuration had to evolve in order to accommodate payloads designed for the higher capability of the Shuttle; therefore, the Delta 3920 was developed. The TR-201 second-stage engine was replaced by a new Aerojet engine designated the AJ10-118K. This engine was designed by Aerojet for the second stage of the Japanese N-II launch vehicle, using the Titan Transtage engine with larger tankage. The Air Force also had funded a program to increase the performance of the Transtage injector known as the Improved Transtage Injector Program (ITIP). McDonnell Douglas decided to increase the 3920 performance with minimum risk by using the ITIP engine and the existing N-II stage tanks for the second stage. In addition, one operational procedure changed with the 3920 configuration. The solid motor ignition sequencing of five ground-lit/four air-lit used before was changed to a six/three sequence to enhance performance.

The unprecedented string of launch vehicle failures that occurred 1985 through 1987, seriously impeded U.S. space launch capability. In one of several steps to revitalize this, the Air Force held a competition for a medium launch vehicle that primarily would launch the Global Positioning System (GPS) satellites. The contract was awarded to McDonnell Douglas for their Delta II series—designated the 6925 and 7925. These configurations are growth versions of the 3920/PAM-D vehicles. This contract is for launches from the east coast only at LC-17A and B. In 1990, McDonnell Douglas received a contract from NASA that included west coast launches to help reestablish the west coast launch capability for Delta. In 1992, the 6920 series vehicle was phased out. Only the 7920 series is now available.

The launch of Delta Flight #219 (GPS-1), in March 1993, marked the first launch carrying a secondary payload aboard the vehicle's second stage. There have been two successful secondary payloads launched on subsequent Delta vehicles and plans to launch secondary and tertiary satellites aboard the P91-1 launch in early 1996.

The Delta Launch Vehicle is undergoing a multimillion dollar major avionics redesign spearheaded by new, state-of-the-art, Redundant Inertial Flight Control Assembly (RIFCA). This unit being designed and built by Allied Signal of New Jersey is scheduled for a first flight in mid 1995. This unit and its complementary hardware replace systems flown since the early days of Delta.

The Delta has evolved continuously over the years to meet the needs of its users. By providing both east (LC-17) and west (SLC-2) coast launches, McDonnell Douglas's reliable Delta has boosted a variety of payloads.

Delta History

(continued)

Launch Record

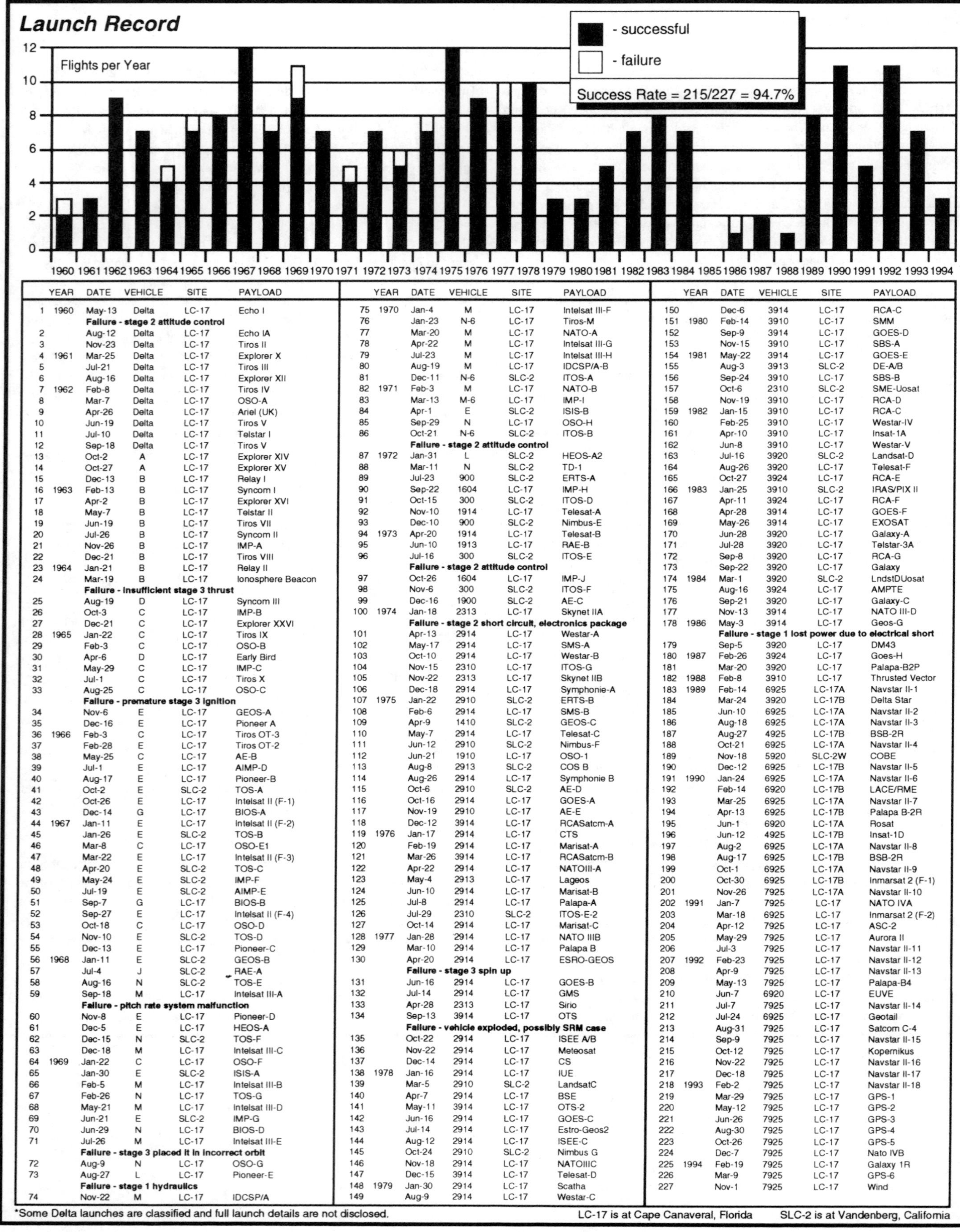

	YEAR	DATE	VEHICLE	SITE	PAYLOAD
1	1960	May-13	Delta	LC-17	Echo I
		Failure - stage 2 attitude control			
2		Aug-12	Delta	LC-17	Echo IA
3		Nov-23	Delta	LC-17	Tiros II
4	1961	Mar-25	Delta	LC-17	Explorer X
5		Jul-21	Delta	LC-17	Tiros III
6		Aug-16	Delta	LC-17	Explorer XII
7	1962	Feb-8	Delta	LC-17	Tiros IV
8		Mar-7	Delta	LC-17	OSO-A
9		Apr-26	Delta	LC-17	Ariel (UK)
10		Jun-19	Delta	LC-17	Tiros V
11		Jul-10	Delta	LC-17	Telstar I
12		Sep-18	Delta	LC-17	Tiros V
13		Oct-2	A	LC-17	Explorer XIV
14		Oct-27	A	LC-17	Explorer XV
15		Dec-13	B	LC-17	Relay I
16	1963	Feb-13	B	LC-17	Syncom I
17		Apr-2	B	LC-17	Explorer XVI
18		May-7	B	LC-17	Telstar II
19		Jun-19	B	LC-17	Tiros VII
20		Jul-26	B	LC-17	Syncom II
21		Nov-26	B	LC-17	IMP-A
22		Dec-21	B	LC-17	Tiros VIII
23	1964	Jan-21	B	LC-17	Relay II
24		Mar-19	B	LC-17	Ionosphere Beacon
		Failure - insufficient stage 3 thrust			
25		Aug-19	D	LC-17	Syncom III
26		Oct-3	C	LC-17	IMP-B
27		Dec-21	C	LC-17	Explorer XXVI
28	1965	Jan-22	C	LC-17	Tiros IX
29		Feb-3	C	LC-17	OSO-B
30		Apr-6	D	LC-17	Early Bird
31		May-29	C	LC-17	IMP-C
32		Jul-1	C	LC-17	Tiros X
33		Aug-25	C	LC-17	OSO-C
		Failure - premature stage 3 ignition			
34		Nov-6	E	LC-17	GEOS-A
35		Dec-16	E	LC-17	Pioneer A
36	1966	Feb-3	C	LC-17	Tiros OT-3
37		Feb-28	E	LC-17	Tiros OT-2
38		May-25	C	LC-17	AE-B
39		Jul-1	E	LC-17	AIMP-D
40		Aug-17	E	LC-17	Pioneer-B
41		Oct-2	E	SLC-2	TOS-A
42		Oct-26	E	LC-17	Intelsat II (F-1)
43		Dec-14	G	LC-17	BIOS-A
44	1967	Jan-11	E	LC-17	Intelsat II (F-2)
45		Jan-26	E	SLC-2	TOS-B
46		Mar-8	C	LC-17	OSO-E1
47		Mar-22	E	LC-17	Intelsat II (F-3)
48		Apr-20	E	SLC-2	TOS-C
49		May-24	E	SLC-2	IMP-F
50		Jul-19	E	SLC-2	AIMP-E
51		Sep-7	G	LC-17	BIOS-B
52		Sep-27	E	LC-17	Intelsat II (F-4)
53		Oct-18	C	LC-17	OSO-D
54		Nov-10	E	SLC-2	TOS-D
55		Dec-13	E	LC-17	Pioneer-C
56	1968	Jan-11	E	SLC-2	GEOS-B
57		Jul-4	J	SLC-2	RAE-A
58		Aug-16	N	SLC-2	TOS-E
59		Sep-18	M	LC-17	Intelsat III-A
		Failure - pitch rate system malfunction			
60		Nov-8	E	LC-17	Pioneer-D
61		Dec-5	E	LC-17	HEOS-A
62		Dec-15	N	SLC-2	TOS-F
63		Dec-18	M	LC-17	Intelsat III-C
64	1969	Jan-22	C	LC-17	OSO-F
65		Jan-30	E	SLC-2	ISIS-A
66		Feb-5	M	LC-17	Intelsat III-B
67		Feb-26	N	LC-17	TOS-G
68		May-21	M	LC-17	Intelsat III-D
69		Jun-21	E	SLC-2	IMP-G
70		Jun-29	N	LC-17	BIOS-D
71		Jul-26	M	LC-17	Intelsat III-E
		Failure - stage 3 placed it in incorrect orbit			
72		Aug-9	N	LC-17	OSO-G
73		Aug-27	L	LC-17	Pioneer-E
		Failure - stage 1 hydraulics			
74		Nov-22	M	LC-17	IDCSP/A
75	1970	Jan-4	M	LC-17	Intelsat III-F
76		Jan-23	N-6	LC-17	Tiros-M
77		Mar-20	M	LC-17	NATO-A
78		Apr-22	M	LC-17	Intelsat III-G
79		Jul-23	M	LC-17	Intelsat III-H
80		Aug-19	M	LC-17	IDCSP/A-B
81		Dec-11	N-6	SLC-2	ITOS-A
82	1971	Feb-3	M	LC-17	NATO-B
83		Mar-13	M-6	LC-17	IMP-I
84		Apr-1	E	SLC-2	ISIS-B
85		Sep-29	N	LC-17	OSO-H
86		Oct-21	N-6	SLC-2	ITOS-B
		Failure - stage 2 attitude control			
87	1972	Jan-31	L	SLC-2	HEOS-A2
88		Mar-11	N	SLC-2	TD-1
89		Jul-23	900	SLC-2	ERTS-A
90		Sep-22	1604	LC-17	IMP-H
91		Oct-15	300	SLC-2	ITOS-D
92		Nov-10	1914	LC-17	Telesat-A
93		Dec-10	900	SLC-2	Nimbus-E
94	1973	Apr-20	1914	LC-17	Telesat-B
95		Jun-10	1913	LC-17	RAE-B
96		Jul-16	300	SLC-2	ITOS-E
		Failure - stage 2 attitude control			
97		Oct-26	1604	LC-17	IMP-J
98		Nov-6	300	SLC-2	ITOS-F
99		Dec-16	1900	SLC-2	AE-C
100	1974	Jan-18	2313	LC-17	Skynet IIA
		Failure - stage 2 short circuit, electronics package			
101		Apr-13	2914	LC-17	Westar-A
102		May-17	2914	LC-17	SMS-A
103		Oct-10	2914	LC-17	Westar-B
104		Nov-15	2310	LC-17	ITOS-G
105		Nov-22	2313	LC-17	Skynet IIB
106		Dec-18	2914	LC-17	Symphonie-A
107	1975	Jan-22	2910	SLC-2	ERTS-B
108		Feb-6	2914	LC-17	SMS-B
109		Apr-9	1410	SLC-2	GEOS-C
110		May-7	2914	LC-17	Telesat-C
111		Jun-12	2910	SLC-2	Nimbus-F
112		Jun-21	1910	LC-17	OSO-1
113		Aug-8	2913	SLC-2	COS B
114		Aug-26	2914	LC-17	Symphonie B
115		Oct-6	2910	SLC-2	AE-D
116		Oct-16	2914	LC-17	GOES-A
117		Nov-19	2910	LC-17	AE-E
118		Dec-12	3914	LC-17	RCASatcm-A
119	1976	Jan-17	2914	LC-17	CTS
120		Feb-19	2914	LC-17	Marisat-A
121		Mar-26	3914	LC-17	RCASatcm-B
122		Apr-22	2914	LC-17	NATOIII-A
123		May-4	2913	LC-17	Lageos
124		Jun-10	2914	LC-17	Marisat-B
125		Jul-8	2914	LC-17	Palapa-A
126		Jul-29	2310	SLC-2	ITOS-E-2
127		Oct-14	2914	LC-17	Marisat-C
128	1977	Jan-28	2914	LC-17	NATO IIIB
129		Mar-10	2914	LC-17	Palapa B
130		Apr-20	2914	LC-17	ESRO-GEOS
		Failure - stage 3 spin up			
131		Jun-16	2914	LC-17	GOES-B
132		Jul-14	2914	LC-17	GMS
133		Apr-28	2313	LC-17	Sirio
134		Sep-13	3914	LC-17	OTS
		Failure - vehicle exploded, possibly SRM case			
135		Oct-22	2914	LC-17	ISEE A/B
136		Nov-22	2914	LC-17	Meteosat
137		Dec-14	2914	LC-17	CS
138	1978	Jan-16	2914	LC-17	IUE
139		Mar-5	2910	SLC-2	LandsatC
140		Apr-7	2914	LC-17	BSE
141		May-11	3914	LC-17	OTS-2
142		Jun-16	2914	LC-17	GOES-C
143		Jul-14	2914	LC-17	Estro-Geos2
144		Aug-12	2914	LC-17	ISEE-C
145		Oct-24	2910	SLC-2	Nimbus G
146		Nov-18	2914	LC-17	NATOIIIC
147		Dec-15	3914	LC-17	Telesat-D
148	1979	Jan-30	2914	LC-17	Scatha
149		Aug-9	2914	LC-17	Westar-C
150		Dec-6	3914	LC-17	RCA-C
151	1980	Feb-14	3910	LC-17	SMM
152		Sep-9	3914	LC-17	GOES-D
153		Nov-15	3910	LC-17	SBS-A
154	1981	May-22	3914	LC-17	GOES-E
155		Aug-3	3913	SLC-2	DE-A/B
156		Sep-24	3910	LC-17	SBS-B
157		Oct-6	2310	SLC-2	SME-Uosat
158		Nov-19	3910	LC-17	RCA-D
159	1982	Jan-15	3910	LC-17	RCA-C
160		Feb-25	3910	LC-17	Westar-IV
161		Apr-10	3910	LC-17	Insat-1A
162		Jun-8	3910	LC-17	Westar-V
163		Jul-16	3920	SLC-2	Landsat-D
164		Aug-26	3920	LC-17	Telesat-F
165		Oct-27	3924	LC-17	RCA-E
166	1983	Jan-25	3910	SLC-2	IRAS/PIX II
167		Apr-11	3924	LC-17	RCA-F
168		Apr-28	3914	LC-17	GOES-F
169		May-26	3914	LC-17	EXOSAT
170		Jun-28	3920	LC-17	Galaxy-A
171		Jul-28	3920	LC-17	Telstar-3A
172		Sep-8	3920	LC-17	RCA-G
173		Sep-22	3920	LC-17	Galaxy
174	1984	Mar-1	3920	SLC-2	LndstDUosat
175		Aug-16	3924	LC-17	AMPTE
176		Sep-21	3920	LC-17	Galaxy-C
177		Nov-13	3914	LC-17	NATO III-D
178	1986	May-3	3914	LC-17	Geos-G
		Failure - stage 1 lost power due to electrical short			
179		Sep-5	3920	LC-17	DM43
180	1987	Feb-26	3924	LC-17	Goes-H
181		Mar-20	3920	LC-17	Palapa-B2P
182	1988	Feb-8	3910	LC-17	Thrusted Vector
183	1989	Feb-14	6925	LC-17A	Navstar II-1
184		Mar-24	3920	LC-17B	Delta Star
185		Jun-10	6925	LC-17A	Navstar II-2
186		Aug-18	6925	LC-17A	Navstar II-3
187		Aug-27	4925	LC-17B	BSB-2R
188		Oct-21	6925	LC-17A	Navstar II-4
189		Nov-18	5920	SLC-2W	COBE
190		Dec-12	6925	LC-17B	Navstar II-5
191	1990	Jan-24	6925	LC-17A	Navstar II-6
192		Feb-14	6920	LC-17B	LACE/RME
193		Mar-25	6925	LC-17A	Navstar II-7
194		Apr-13	6925	LC-17B	Palapa B-2R
195		Jun-1	6920	LC-17A	Rosat
196		Jun-12	4925	LC-17B	Insat-1D
197		Aug-2	6925	LC-17A	Navstar II-8
198		Aug-17	6925	LC-17B	BSB-2R
199		Oct-1	6925	LC-17A	Navstar II-9
200		Oct-30	6925	LC-17B	Inmarsat 2 (F-1)
201		Nov-26	7925	LC-17A	Navstar II-10
202	1991	Jan-7	7925	LC-17	NATO IVA
203		Mar-18	6925	LC-17	Inmarsat 2 (F-2)
204		Apr-12	7925	LC-17	ASC-2
205		May-29	7925	LC-17	Aurora II
206		Jul-3	7925	LC-17	Navstar II-11
207	1992	Feb-23	7925	LC-17	Navstar II-12
208		Apr-9	7925	LC-17	Navstar II-13
209		May-13	7925	LC-17	Palapa-B4
210		Jun-7	6920	LC-17	EUVE
211		Jul-7	7925	LC-17	Navstar II-14
212		Jul-24	6925	LC-17	Geotail
213		Aug-31	7925	LC-17	Satcom C-4
214		Sep-9	7925	LC-17	Navstar II-15
215		Oct-12	7925	LC-17	Kopernikus
216		Nov-22	7925	LC-17	Navstar II-16
217		Dec-18	7925	LC-17	Navstar II-17
218	1993	Feb-2	7925	LC-17	Navstar II-18
219		Mar-29	7925	LC-17	GPS-1
220		May-12	7925	LC-17	GPS-2
221		Jun-26	7925	LC-17	GPS-3
222		Aug-30	7925	LC-17	GPS-4
223		Oct-26	7925	LC-17	GPS-5
224		Dec-7	7925	LC-17	Nato IVB
225	1994	Feb-19	7925	LC-17	Galaxy 1R
226		Mar-9	7925	LC-17	GPS-6
227		Nov-1	7925	LC-17	Wind

*Some Delta launches are classified and full launch details are not disclosed.

LC-17 is at Cape Canaveral, Florida SLC-2 is at Vandenberg, California

Delta General Description

6925 (reference only) **7925**

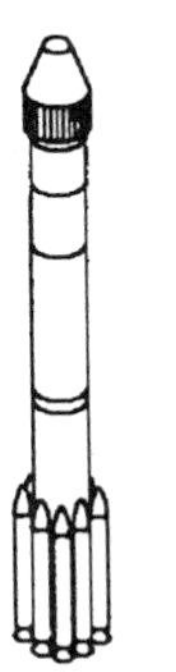

Summary

The Department of Defense's Delta II was begun following the Challenger accident as a means to offload the Global Positioning System (GPS) from the Shuttle manifest and accelerate the operational availability of GPS. The booster class was known as Medium Launch Vehicle (MLV) 1. Delta II was selected following an open competition as MLV-1. The initial version, 6925, initially flew February 14, 1989, and the upgraded version, 7925, initially flew on November 26, 1990. A two stage version exists for each Delta II, a 6920 and 7920, which are primarily flown for low Earth orbit missions.

Status

6925 - Operational; 7925 - Operational

Key Organizations

User - DoD, NASA, commercial

Launch Service Agency - McDonnell Douglas / U.S. Air Force

Prime Contractor - McDonnell Douglas Space Systems Company (Airframe, Assembly and Test, and Launch Operations)

Principal Subcontractors
- Rocketdyne (Stage 1 engine - RS-27)
- Aerojet (Stage 2 engine - AJ-10-118K)
- Thiokol (Strap-ons - Castor IVA) for 6925
- Hercules (Strap-ons - GEMs) for 7925
- Delco (Guidance computer)

Vehicle

System Height: Up to 125 ft (38.1 m) for both

Payload Fairing Size: 9.5 ft (2.9 m) diameter by 27.8 ft (8.47 m) height or 10.0 ft (3.05 m) diameter by 26.0 ft (7.92 m) height.

Gross Mass: 480,000 lb (218,000 kg) for 6925
506,000 lb (230,000 kg) for 7925

Planned Enhancements: —

Operations

Primary Missions: 6/7925 vehicle is primarily GTO or polar elliptical and 6/7920 is LEO or low polar

Compatible Upper Stages: PAM-D

Operations (continued)

First Launch: 1989 for 6925, 1990 for 7925

Success / Flight Total: 17 / 17 for 6925, 24 / 24 for 7925

Launch Site: CCAFS - LC-17A & B (28.5°N, 81.0°W)
VAFB - SLC-2W (34.7°N, 120.6°W)

Launch Azimuth: LC-17 - 57°-112° (max. inclination is 28.5°- 51°)
SLC-2W - 185°-270° (max. inclination is 63°-145°)

Nominal Flight Rate: 10 / yr

Planned Enhancements: Upgrade of launch site SLC-2W proposed for VAFB would be available for the polar launch in mid-1993. SLC-2W is counted against the 10 flight/yr since the launch crew is shared with the East Coast. SLC-2E is non-operational.

Performance

In all cases below, the 9.5 ft (2.9 m) payload fairing is assumed. The 10.0 ft (3.05 m) payload fairing will reduce performance by about 110 lb (50 kg) for the three stage vehicle and 270 lb (120 kg) for the two-stage vehicle.

100 nm (185 km) circ, 28°: 8,780 lb (3,990 kg) for 6920 vehicle
11,220 lb (5,089 kg) for 7920 vehicle

100 nm (185 km) circ, 90°: 6,490 lb (2,950 kg) for 6920 vehicle
8,575 lb (3,890 kg) for 7920 vehicle

Geotransfer Orbit, 28°: 3,190 lb (1,450 kg) for 6925 vehicle
2,100 lb (950 kg) for 6920 vehicle
4,060 lb (1,840 kg) for 7925 vehicle

Financial Status

Estimated Launch Price: $45–50M (McDonnell Douglas)

Manifest

Site	1995	1996	1997
LC-17	6	5	5
SLC-2W	2	3	6

Remarks

The 6920 series vehicle was phased out in 1992, only 7920 series vehicles now available.

Delta Vehicle

Overall

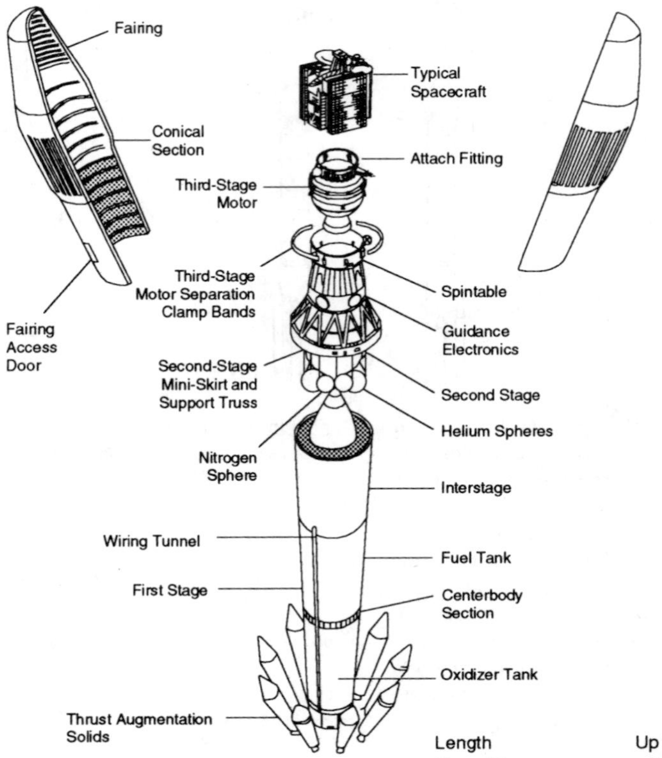

	6925	7925
Length	Up to 125 ft (38.1 m)	Up to 125 ft (38.1 m)
Gross Mass	480K lb (220K kg)	506K lb (230K kg)
Thrust at Liftoff	595K lb (2.62M N)	598K lb (2.63M N)

SRM

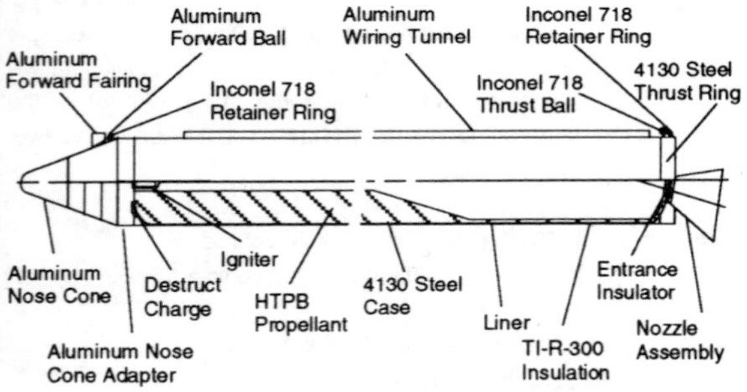

Castor IVA

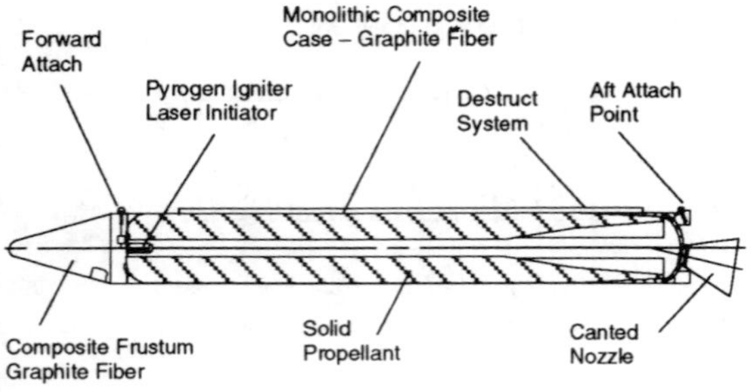

GEM

	6925 (Castor IVA)	7925 (GEM)
Dimension:		
Length	36.6 ft (11.2 m)	42.5 ft (13.0 m)
Diameter	3.3 ft (1.0 m)	3.3 ft (1.0 m)
Mass: (each)		
Propellant Mass	22.3K lb (10.1K kg)	25.8K lb (11.7K kg)
Gross Mass	25.8K lb (11.7K kg) Ground Lit 26.1K lb (11.9K kg) Air Lit	28.6K lb (13.0K kg) Ground Lit 28.8K lb (13.1K kg) Air Lit
Structure:		
Type	Monocoque	Monocoque
Case Material	Steel	Graphite Epoxy
Propulsion:		
Propellant	HTPB	HTPB
Average Thrust (each)	108.7K lb (478.3K N) vac 97.7K lb (427.1K N) SL	110.8K lb (487.6K N) vac 98.9K lb (435K N) SL
Number of Motors	9	9
Number of Segments	1	1
Isp	265.7 sec vac 237.3 sec SL	273.8 sec vac 245.7 sec SL
Chamber Pressure	691 psia (47.7 bar)	817 psia (56.3 bar)
Expansion Ratio	8.29:1	10.65:1
Control	Fixed 11 deg cant (Stage 1 provides control)	Fixed 10 deg cant (Stage 1 provides control)
Events:		
Nominal Burn Time	56.2 sec	63.0 sec
Stage Shutdown	Burn to depletion	Burn to depletion
Stage Separation	Spring ejection	Spring ejection

Remarks:

The Graphite Epoxy Motors (GEMs), developed by Hercules are six feet longer and provide more thrust than the Castor IVA (developed by Thiokol) and the graphite-epoxy case is lighter than the steel case used on the Castors.

Delta Vehicle
(continued)

Stage 1

	6925 *(reference only)*	7925
Dimension:		
Length	85.6 ft (26.1 m)	85.6 ft (26.1 m)
Diameter	8.0 ft (2.44 m)	8.0 ft (2.44 m)
Mass:		
Propellant Mass	211.3K lb (96.1K kg)	211.1K lb (96.0K kg)
Gross Mass	223.8K lb (101.7 kg)	224.2K lb (101.9 kg)
Structure:		
Type	Isogrid	Isogrid
Material	Aluminum	Aluminum
Propulsion:		
Propellant	LOX-RP1	LOX-RP1
Average Thrust	207K lb (911K N) SL 232K lb (1,020K N) vac	201K lb (884K N) SL 237K lb (1,043K N) vac
Engine Designation	RS-27	RS-27A
Number of Engines	1	1
Isp	263.2 sec SL 295.0 sec vac	255.6 sec SL 301.8 sec vac
Feed System	Gas Generator	Gas Generator
Chamber Pressure	702 psia (48.4 bar)	702 psia (48.4 bar)
Mixture Ratio (O/F)	2.24	2.24
Throttling Capability	100% only	100% only
Expansion Ratio	8:1	12:1
Restart Capability	No	No
Tank Pressurization	High-pressure nitrogen gas	High-pressure nitrogen gas
Control - Pitch, Yaw	Hydraulic gimbaling	Hydraulic gimbaling
Roll	2 vernier engines	2 vernier engines
Events:		
Nominal Burn Time	265 sec	265 sec
Stage Shutdown	Burn to depletion	Burn to depletion
Stage Separation	Spring ejection	Spring ejection

Remarks:

The first stage has an engine section that houses the Rocketdyne RS-27 main engine, two Rocketdyne LR101-NA-11 vernier engines, and provides the aft attachments for the strap-on solid propellant motors. The cylindrical isogrid RP-1 fuel and LOX tanks are extended 4.7 ft (1.43m) and 7.3 ft (2.23m), respectively, beyond the 3920 configuration. The two tanks are separated by a center body section that houses control electronics, ordnance sequencing equipment, and telemetry system. The RS-27 is a single start, liquid bipropellant rocket engine with a thrust rating of 207,000 lb (921K N) at sea level. The two vernier engines provide roll control during main-engine burn, and attitude control after cutoff and before second-stage separation. Thrust augmentation is provided by nine unsegmented solid propellant rocket motors, six ignited at liftoff and the remaining three ignited in flight. A rate gyro has been added to the first stage, forward of the center body section, to assure adequate stability margins with the extended tanks and larger fairing. The Delta interstage assembly extends from the top of the first stage to the second stage miniskirt. This 15.5 ft (4.72m) long isogrid structure carries loads from the second stage, third stage, spacecraft, and fairing to the first stage, and contains an exhaust vent and six spring-driven separation rods.

Delta Vehicle
(continued)

Stage 2

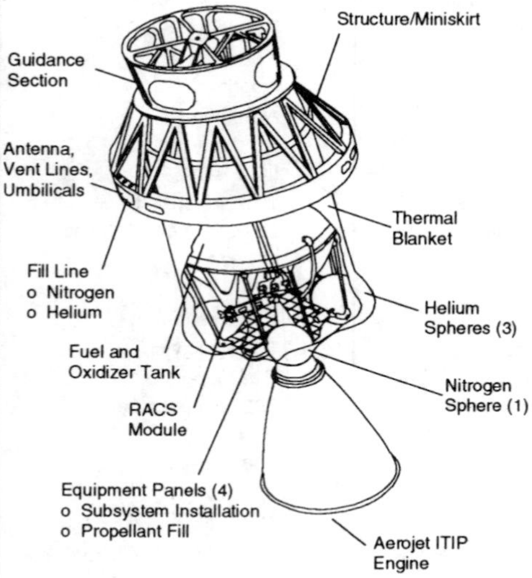

Structure/Miniskirt
Guidance Section
Antenna, Vent Lines, Umbilicals
Thermal Blanket
Fill Line
o Nitrogen
o Helium
Helium Spheres (3)
Fuel and Oxidizer Tank
Nitrogen Sphere (1)
RACS Module
Equipment Panels (4)
o Subsystem Installation
o Propellant Fill
Aerojet ITIP Engine

	6925 (reference only)	7925
Dimension:		
Length	19.6 ft (5.97 m)	19.6 ft (5.97 m)
Diameter	8.0 ft (2.44 m)	8.0 ft (2.44 m)
Mass:		
Propellant Mass	13,367 lb (6,076 kg)	13,367 lb (6,076 kg)
Gross Mass	15,394 lb (6,997 kg)	15,394 lb (6,997 kg)
Structure:		
Type	Monocoque (Common Dome)	Monocoque (Common Dome)
Material	Stainless Steel	Stainless Steel
Propulsion:		
Propellant	N2O4-A50	N2O4-A50
Average Thrust	9,645 lb (42,430 N) vac	9,645 lb (42,430 N) vac
Engine Designation	AJ-10-118K	AJ-10-118K
Number of Engines	1	1
Isp	319.4 sec vac	319.4 sec vac
Feed System	Pressure Fed	Pressure Fed
Chamber Pressure	827 psia (57.0 bar)	827 psia (57.0 bar)
Mixture Ratio (O/F)	1.8	1.8
Throttling Capability	100% only	100% only
Expansion Ratio	65:1	65:1
Restart Capability	Multiple	Multiple
Tank Pressurization	Gaseous helium	Gaseous helium
Control - Pitch, Yaw	Hydraulic gimbaling	Hydraulic gimbaling
Roll	Nitrogen cold gas	Nitrogen cold gas
Events:		
Nominal Burn Time	439.7 sec	439.7 sec
Stage Shutdown	Burn to depletion or command shutdown	Burn to depletion or command shutdown
Stage Separation	Cold-gas helium for retro	Cold-gas helium for retro

Remarks:

The second stage (also known as the Second Stage Propulsion System (SSPS)) uses the restartable Aerojet AJ10-118K engine developed for the USAF Improved Transtage Injector program, and uses N2O4 and Aerozine 50 (A50) storable propellants. Gaseous helium is used for pressurization, and a nitrogen cold-gas jet system provides attitude control during coast periods and roll control during powered flight. Hydraulically activated gimbals provide pitch and yaw control. An isogrid configuration equipment panel is attached to the aft section. The forward section of the second stage houses guidance and control equipment that provides guidance sequencing and stabilization signals for both first and second stages.

Stage 3

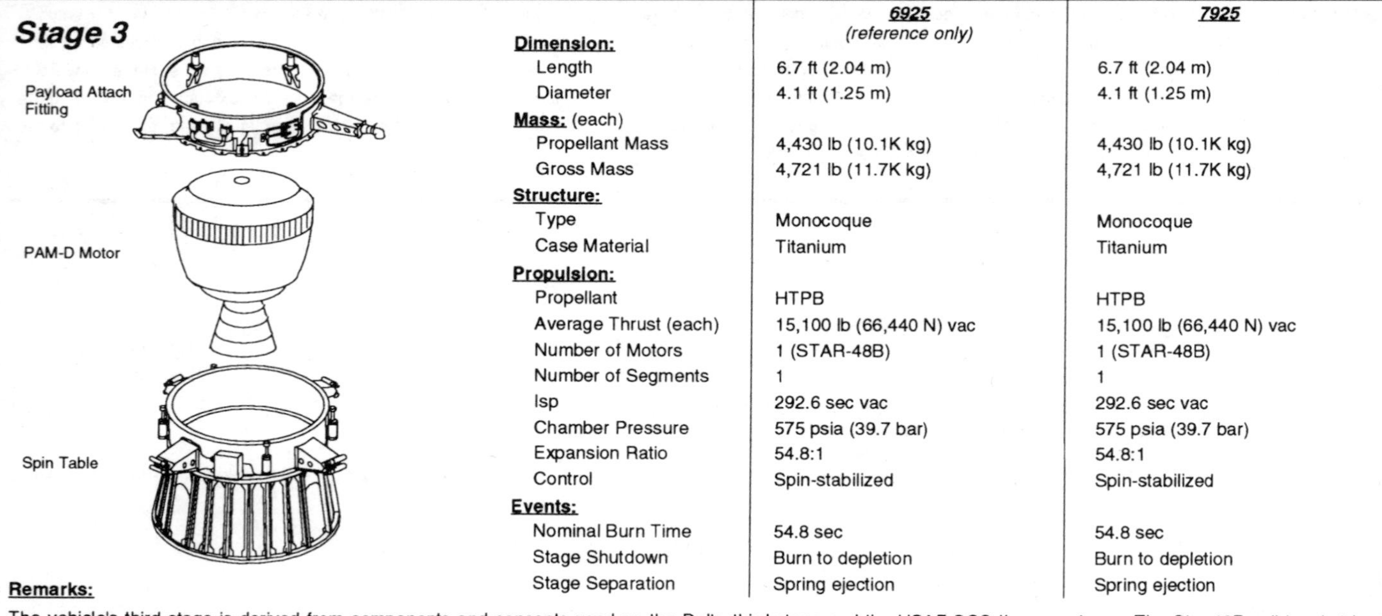

	6925 (reference only)	7925
Dimension:		
Length	6.7 ft (2.04 m)	6.7 ft (2.04 m)
Diameter	4.1 ft (1.25 m)	4.1 ft (1.25 m)
Mass: (each)		
Propellant Mass	4,430 lb (10.1K kg)	4,430 lb (10.1K kg)
Gross Mass	4,721 lb (11.7K kg)	4,721 lb (11.7K kg)
Structure:		
Type	Monocoque	Monocoque
Case Material	Titanium	Titanium
Propulsion:		
Propellant	HTPB	HTPB
Average Thrust (each)	15,100 lb (66,440 N) vac	15,100 lb (66,440 N) vac
Number of Motors	1 (STAR-48B)	1 (STAR-48B)
Number of Segments	1	1
Isp	292.6 sec vac	292.6 sec vac
Chamber Pressure	575 psia (39.7 bar)	575 psia (39.7 bar)
Expansion Ratio	54.8:1	54.8:1
Control	Spin-stabilized	Spin-stabilized
Events:		
Nominal Burn Time	54.8 sec	54.8 sec
Stage Shutdown	Burn to depletion	Burn to depletion
Stage Separation	Spring ejection	Spring ejection

Remarks:

The vehicle's third stage is derived from components and concepts used on the Delta third stage and the USAF SGS-II upper stage. The Star-48B solid-rocket is supported at the base of the motor on a spin table that mates to the top of the second stage guidance section. The payload attach fitting (PAF) is the structure that provides the transition from the top of the solid-rocket motor to the spacecraft interface. Before third stage deployment, the stage and spacecraft are spun-up using spin rockets rotating the assembly on a spin bearing. Variable spin rate is achieved by selecting rockets from an inventory of different size, qualified spin rockets. An ordnance sequencing system is used to release the third-stage/spacecraft after spin-up, to fire the Star-48B motor, and to separate the spacecraft following motor burn. The stage also contains a nutation control system (NCS) to suppress coning, and an S-band telemetry system. Mechanical and electrical interfaces with the spacecraft are identical to those established on the Delta 3920/PAM program.

Delta Vehicle
(continued)

Payload Fairing

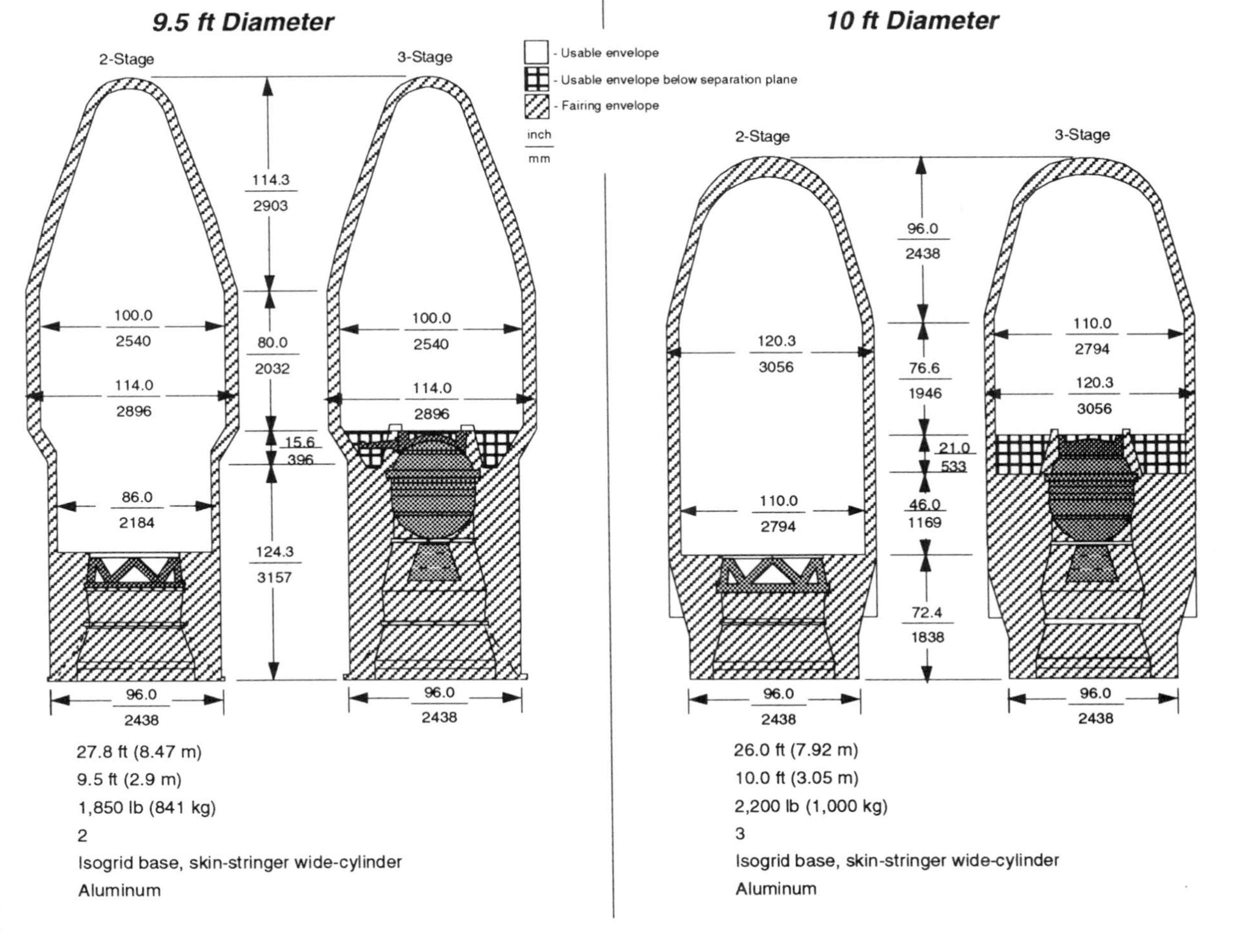

	9.5 ft Diameter	10 ft Diameter
Length	27.8 ft (8.47 m)	26.0 ft (7.92 m)
Diameter	9.5 ft (2.9 m)	10.0 ft (3.05 m)
Mass	1,850 lb (841 kg)	2,200 lb (1,000 kg)
Sections	2	3
Structure	Isogrid base, skin-stringer wide-cylinder	Isogrid base, skin-stringer wide-cylinder
Material	Aluminum	Aluminum

Remarks The payload fairing shields the payload from buffeting and aerodynamic heating while in the atmospheric phase of flight. The aluminum structure, which incorporates acoustic absorption blankets on its interior, accommodates the spacecraft envelope. Fairing halves are separated by a flight-proven contamination-free separation joint. The aft end is identical to the present Delta eight-foot isogrid fairing to maintain the same second stage interface. The center section, aluminum skin-stringer construction similar to fairings currently being constructed by McDonnell Douglas for Titan vehicles, increases the envelope to accommodate the GPS spacecraft, and also provides increased flexibility for the commercial user.

Avionics

The Delta inertial guidance system (DIGS) is a strap-down all inertial system consisting of a Delta redundant inertial measurement system (DRIMS) and a Delco guidance computer (GC). The DRIMS contains three gyros, four accelerometers, and conditioning electronics. DRIMS data is processed in the computer to obtain attitude reference and navigation information. The computer also issues preprogrammed sequence commands and provides control system stabilization logic for both powered and coast phases of flight.

Electronic packages in both first and second stages receive commands from the GC, and drive the servo simplifiers for engine gimbal and the switch amplifier for control jet (vernier or gas jet) operations. Both first and second stages have a battery-supplied DC power system. Separate batteries are used for the guidance and control, ordnance, and engine systems. The instrumentation and flight termination systems are powered by the same battery. The vehicle also contains a telemetry system and a range safety tracking system.

Attitude Control System

The attitude control system (ACS) is used to maintain the second stage in-flight attitude with roll, pitch, and yaw control plus propellant settling. Roll control is used in conjunction with the gimbal actuation system for vehicle course control during powered flight. During vehicle coast phase after first burn of the second stage engine, the ACS performs the necessary course correction maneuvers of pitch, yaw, and roll to keep the vehicle in its correct attitude for engine restart and subsequent third stage or payload separation. In near zero gravity, just to engine restart, the ACS propellant settling valves are energized to provide thrust, causing a forward acceleration to settle the propellants to the bottom of their respective tanks. This ensures propellant availability to the thrust chamber for engine restart. The ACS operates by controlled release of pressurized gaseous nitrogen through various nozzles which are aligned with specific axes of the vehicle. The gaseous nitrogen is stored in a high pressure spherical titanium alloy tank. For the three-stage configuration, the third stage is spin-stabilized, thus not requiring an ACS system.

Delta Performance

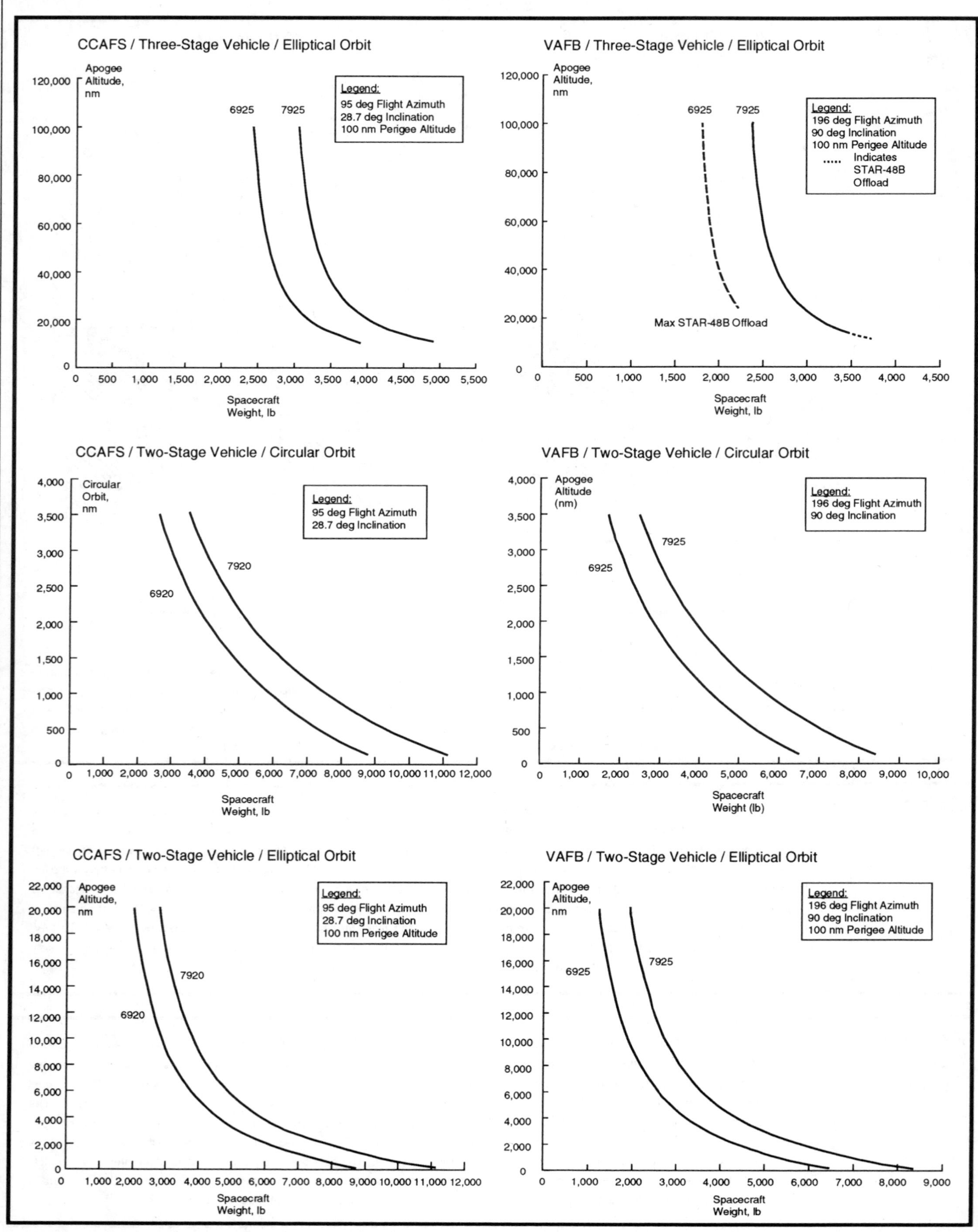

Delta Operations

Launch Site

Vandenberg Air Force Base (34.7°N)
SLC-2W (Delta II)
SLC-3W (Atlas E)
SLC-4E / W
(Titan II, Titan IV)
SLC-5 (Scout)
SLC-6 (LLV)
Cape Canaveral Air Force Station (28.5°N)
KSC
LC-39B (STS)
LC-39A (STS)
CCAFS
LC-41 (Titan IV)
LC-40
(Titan III, Titan IV)
LC-36A / B
(Atlas I, II)
LC-17A / B
(Delta, Delta II)

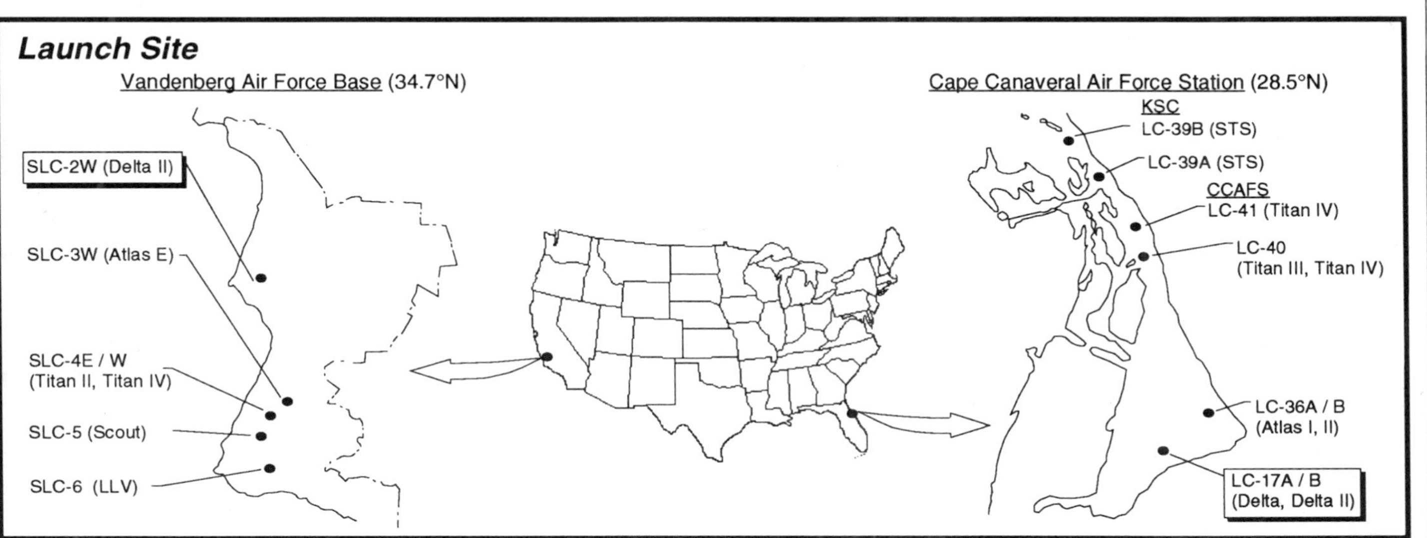

Launch Facilities

Cape Canaveral Air Force Station (CCAFS)

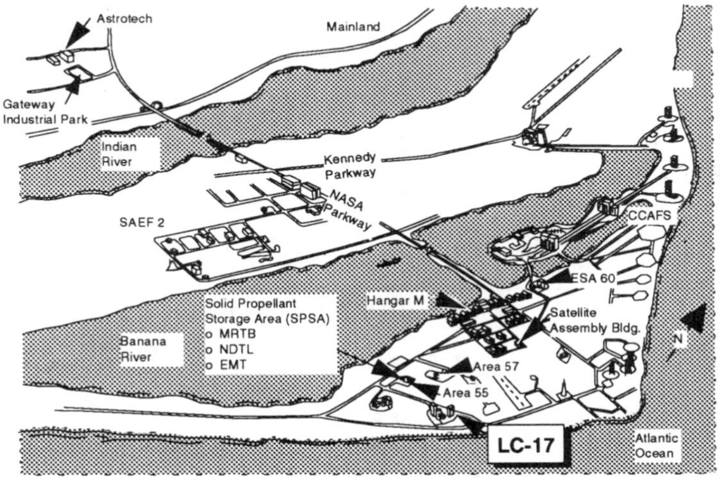

Astrotech
Mainland
Gateway
Industrial Park
Indian
River
Kennedy
Parkway
NASA
Parkway
SAEF 2
CCAFS
ESA 60
Hangar M
Satellite
Assembly Bldg.
Solid Propellant
Storage Area (SPSA)
o MRTB
o NDTL
o EMT
Banana
River
Area 57
Area 55
N
LC-17
Atlantic
Ocean

LC-17 Location

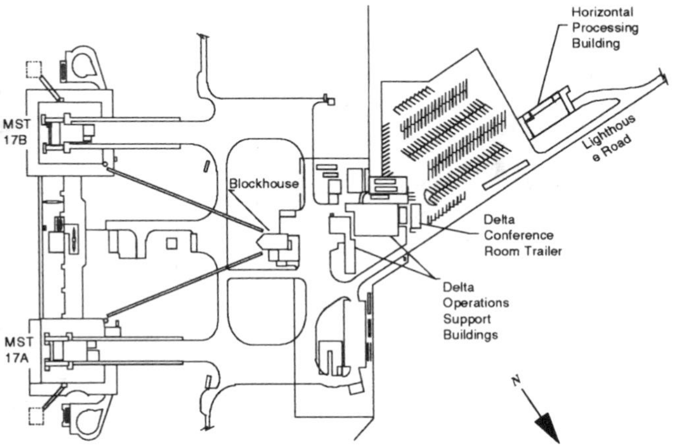

Horizontal
Processing
Building
MST
17B
Blockhouse
Lighthouse Road
Delta
Conference
Room Trailer
Delta
Operations
Support
Buildings
MST
17A
N

LC-17 Layout

Vandenberg Air Force Base (VAFB)

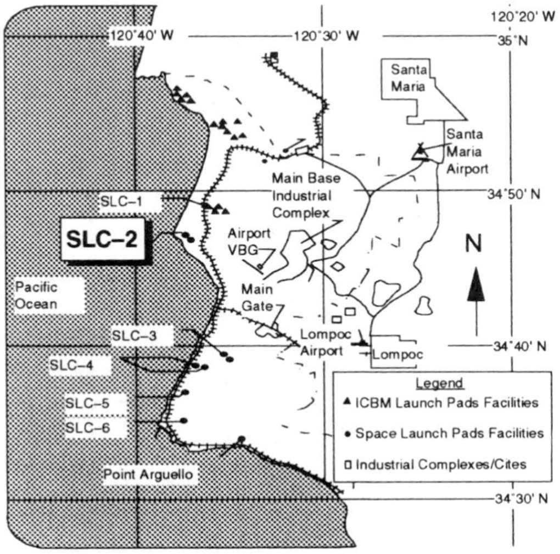

120°20' W
120°40' W
120°30' W
35°N
Santa
Maria
Santa
Maria
Airport
Main Base
Industrial
Complex
SLC–1
34°50' N
SLC–2
Airport
VBG
N
Pacific
Ocean
Main
Gate
SLC–3
Lompoc
Airport
Lompoc
34°40' N
SLC–4
Legend
▲ ICBM Launch Pads Facilities
• Space Launch Pads Facilities
□ Industrial Complexes/Cites
SLC–5
SLC–6
Point Arguello
34°30' N

SLC-2 Location

Tank
Farm
Solar Road
Lunar Road
Aero Road
Tangair Road
Launch Mount
Gate
House
Azimuth
Bldg
Pad 2W
Gantry
Terra Road
NASA Launch
Operations
Building
Launch Operations
Support Building
Blockhouse
0 30 60 90 120 150
Scale in m
0 500
Scale in ft
N
SLC-2E
(Nonoperational
Pad)

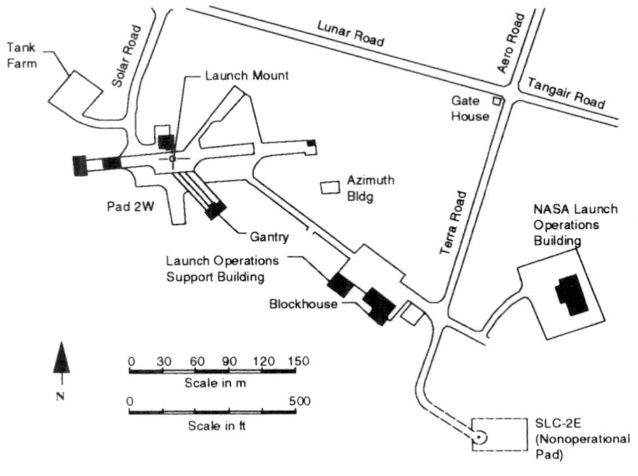

SLC-2 Layout

Delta Operations

(continued)

Launch Processing

Cape Canaveral Air Force Station (CCAFS). At CCAFS, Delta vehicles are launched from Launch Complex 17 (LC-17.) The launch complex contains two active pads, 17A and 17B. The two pads can be used for simultaneous buildup of two vehicles. The operations which take place on each of the pads include vehicle buildup and checkout, propellant servicing, spacecraft integration, and launch countdown.

Prior to vehicle assembly at the launch site, several facilities at CCAFS are used for the processing of vehicle, upper stage, and upper stage combinations. They include the Delta Mission Checkout (DMCO), Area 57, Missile Research Test Building (MRTB), Propellant Servicing Facility (PSF), Explosive Safe Area (ESA)-60A, and NAVSTAR Processing Facility (NPF). DMCO is located behind Hanger M in the CCAFS Industrial Area. DMCO is used for vehicle (first and second stage) electrical and pressure checkout. Area 57 is used for solid motor buildup and processing. The MRTB is used for ordnance receiving, inspection, and solid motor preflight preparations. ESA-60A is located approximately one mile northwest of the CCAFS Industrial Area. The ESA-60A facility is used for payload buildup, checkout, ordnance installation, and propellant loading. The NPF, like the PSF, is located in the Minuteman Area. This facility was modified to act in conjunction with the PSF to accommodate the processing of fueled spacecraft and upper stages (both hypergolic and solids), such as the NAVSTAR spacecraft.

Following is a detailed description of launch processing from LC-17. Delta II integration, checkout, and test activities are basically accomplished in two phases: (1) the offline operations which include DMCO, and SRM (thrust augmenting) processing in Area 57, and (2) online, or pad operations which consist of first and second stage checkout, GSE validation, upper stage/payload installation, battery trickle charging, fairing installation and launch. The basic philosophy of vehicle integration is to complete checkout of the vehicle using separate test equipment, offline from pad operations before committing the hardware to the online pad operations.

Delta II booster components are shipped to CCAFS separately for final integration and checkout before launch. Major components include the first stage, second stage, SRM, upper stage (PAM-D) components, fairing, and interstage.

The Delta first and second stages are received, unpacked and inspected at Hangar M prior to DMCO. Payload fairing and interstage are also received and stored until pad integration. DMCO operations on the first and second stages are a series of electrical checks on each booster subsystem prior to ordnance installation and integration at the pad. The following subsystems are checked: hydraulic, propulsion, telemetry, destruct, and control. After DMCO, the first and second stages are moved to the Horizontal Processing Facility (HPF) and Area

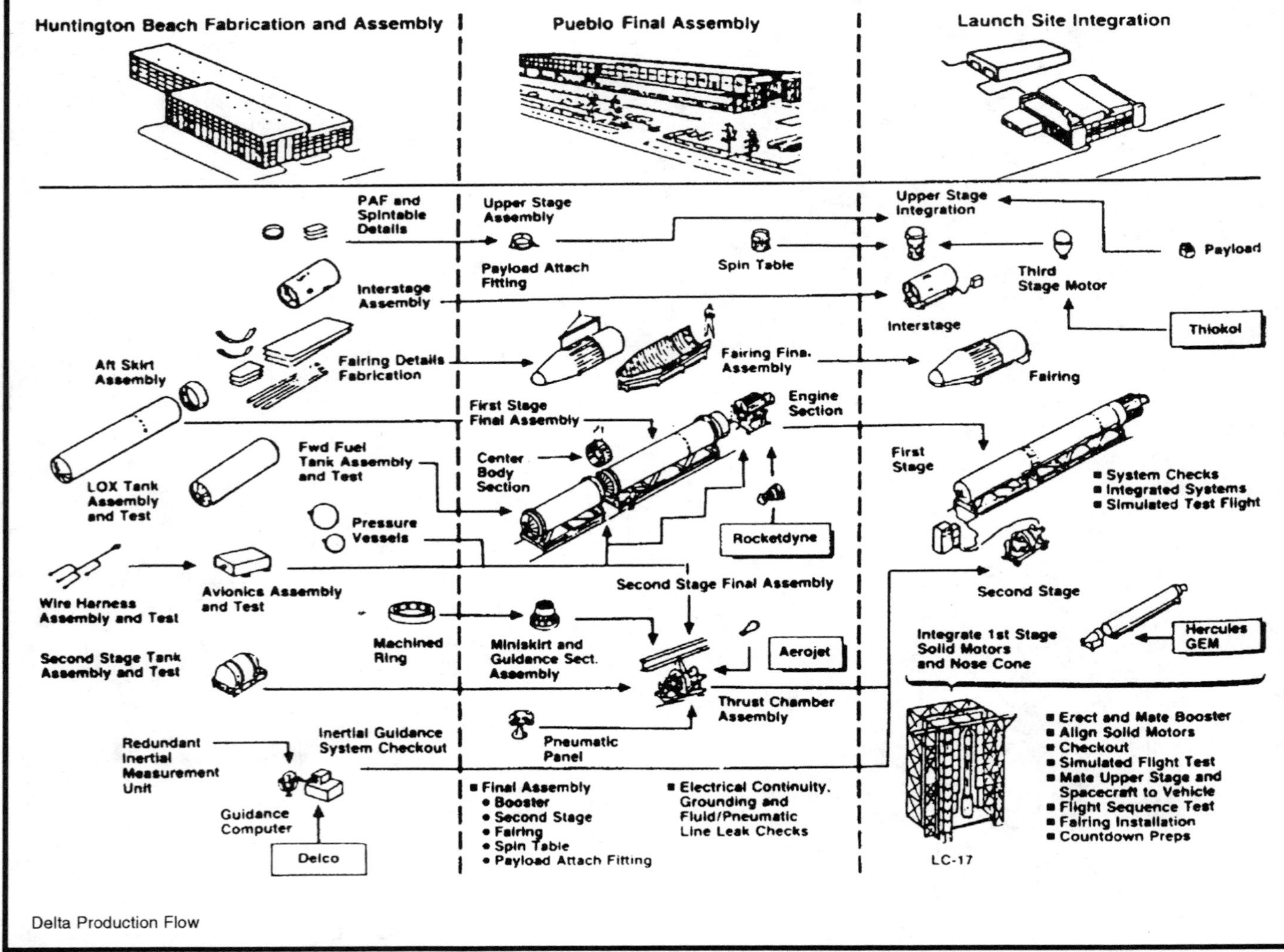

Delta Production Flow

Delta Operations
(continued)

Launch Processing
(continued)

55 for final preerection preparations and destruct charge installation. Besides destruct charge installation, ordnance to separate the Castor IVA SRMs is also installed on the first stage in the HPF. Second stage processing at Area 55 also includes propulsion leak checks, remote leak evaluation, harness installation, and nozzle extension installation.

The GEM SRMs are processed in Area 57. The SRMs are received, checked for leaks and stored until required. At that time, destruct igniters are installed prior to pad integration.

PAM-D components to build up the upper stage arrive at the NAVSTAR Processing Facility (NPF). The SRMs are shipped by the vendor and stored at the launch site until needed. The Payload Attach Fittings (PAF) and spacecraft-unique wire harnesses are built at the upper stage contractor's facility and shipped to the launch site. Upper stage ground operations begin with the arrival of the Star-48B solid motor via commercial ground transportation. Following X-ray inspection of the motor, the PAF is mated to the forward end of the motor. Items to be installed on the PAF include the safe and arm mechanism, ETAs, yo-weight, yo-weight cutters, test batteries, and wire harnesses. When the upper stage buildup is complete, the upper stage is moved to the Delta Spin Test Facility (DSTF) where it is aligned. After balancing, the upper stage is moved to NPF for test to verify the employment sequence circuitry, spin system, and flight sequence.

Pad activities include the erection of the Delta II booster and mating of the payload prior to launch. Final preparations on the Delta II first stage prior to erection are performed in the HPF after which the first stage is moved to SLC-17 for erection. After installing the interstage and nine SRMs, the second stage is then erected and mated. After preerection preparations on the fairing are completed at Hangar M, the fairing halves are erected and stowed on the MST.

Once the Delta II booster is stacked, system tests including propellant loading, flight pressurization, hydraulic, pneumatic and electrical control, and guidance checks, are performed in preparation for mating the PAM-D and payload. The fully integrated and checked out payload is received from the NPF and hoisted into the MST where it is mated to the second stage prior to removal of the handling canister. The MST sectionalized handling canister is then removed.

Integrated vehicle system testing starts with a flight sequence test to verify engine sequences, receipt of control, guidance, ordnance firing, and SV discrete commands. Finally, a flight program verification test is performed through the complete mission to verify flight program events on internal power. Preflight finally gets underway as the fairing is installed, the test equipment connections are removed and ordnance is installed and connected. The second stage propellants are loaded. The guidance computer, flight beacon, and range safety systems are checked, the spacecraft is finalized, fairing access doors are closed, and the MST is retracted. During the terminal countdown, the second stage systems are pressurized, first stage propellants are loaded, electrical hydraulic and control systems are verified prior to liftoff.

Vandenberg Air Force Base (VAFB). At VAFB, Delta vehicles are launched from Space Launch Complex 2 (SLC-2). The launch complex contains two pads: SLC-2E which is not active and was disassembled in 1972 and SLC-2W which is currently being refurbished for a Delta launch planned for this year. Vehicle and payload processing operations are performed at the launch complex and at Building 836 in South Vandenberg. The Delta rockets are delivered to the complex from Huntington Beach, CA, processed through DMCO, and are transported to Vandenberg AFB, where they are erected and serviced at the launch complex. SLC-2 activities are similar to the LC-17 description above.

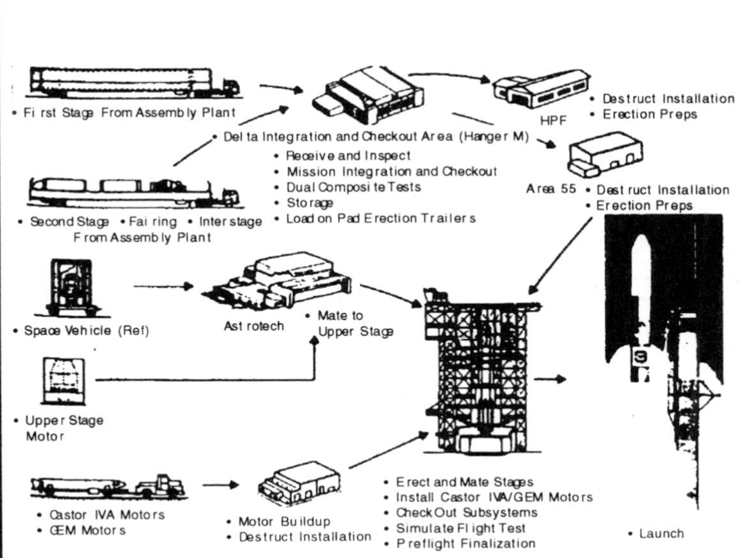

Delta Launch Site Operations Flow (CCAFS)

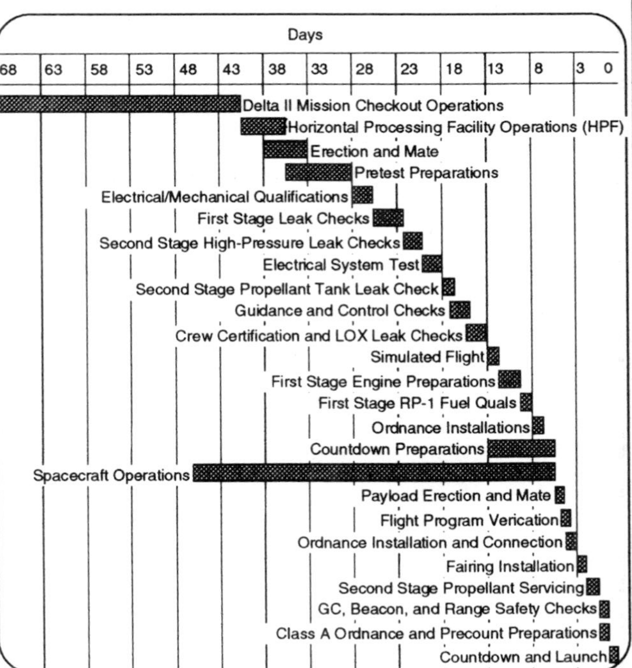

Delta Typical Launch Operations Timeline (CCAFS)

Delta Operations
(continued)

Flight Sequence

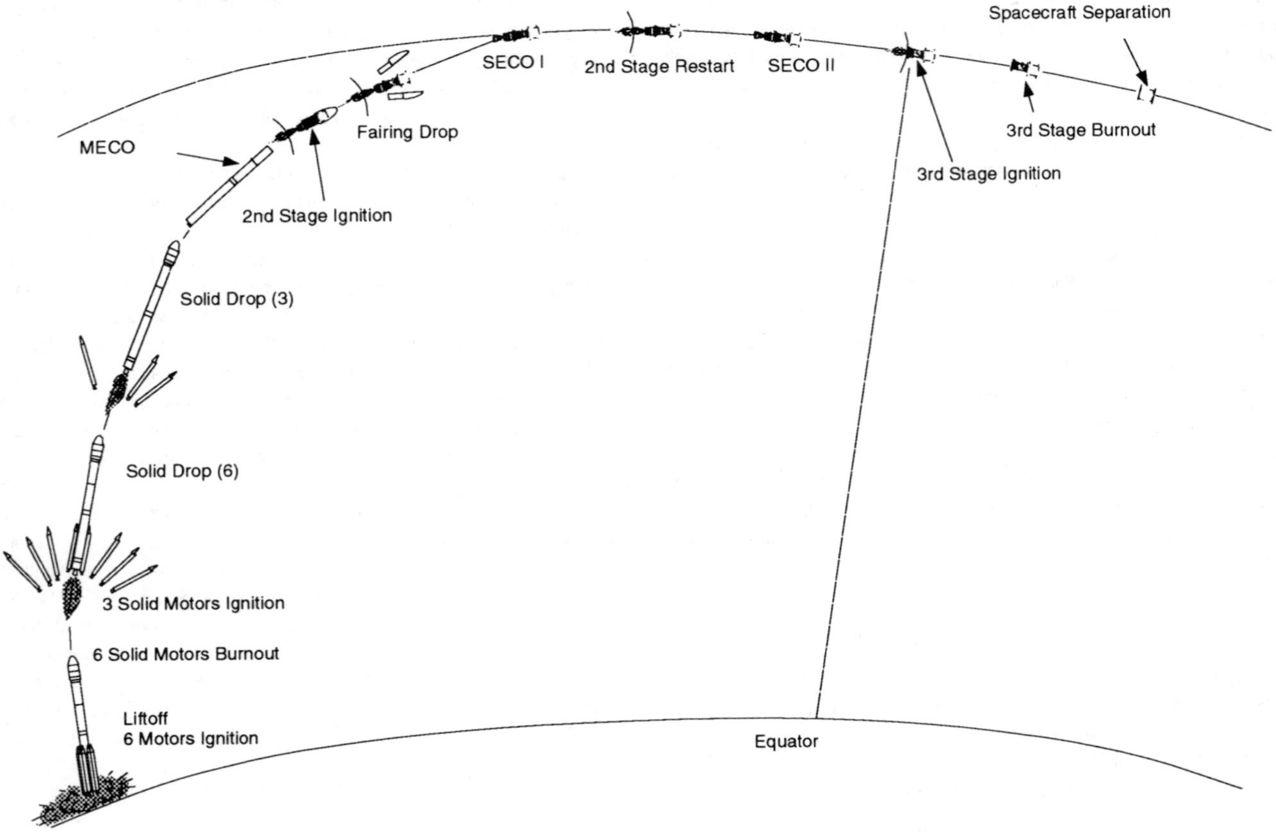

Event	6920	6925	7920	7925
First Stage				
Main engine ignition	T + 0 sec	T + 0	T + 0	T + 0
Solid motor ignition (6 solids)	T + 0	T + 0	T + 0	T + 0
Solid motor burnout (6 solids)	T+ 56	T + 56	T + 63	T + 63
Solid motor ignition (3 solids)	T + 61	T + 61	T + 66	T + 66
Solid motor separation (3/3 solids)	T + 62/63	T + 62/63	T + 67/68	T + 67/68
Solid motor burnout (3 solids)	T + 117	T + 117	T + 130	T + 130
Solid motor separation (3 solids)	T + 122	T + 122	T + 133	T + 133
MECO (M)	T + 265	T + 265	T + 265	T + 265
Second Stage				
Blow stage VII separation bolts	M + 8	M + 8	M + 8	M + 8
Stage II ignition	M + 13	M + 13	M + 13	M + 13
Fairing separation	M + 43	M + 43	M + 37	M + 37
SECO (S1)	M + 390	M + 415	M + 355	M + 360
Stage II engine restart	S1 + 2900	S1 + 610	S1-2900	S1+ 616
SECO (S2)	S1 + 2925	S1 + 631	S1-2956	S1+ 686
Third Stage				
Fire spin rockets, start stage III sequencer		S2 + 50		S2 + 50
Separate stage III		S2 + 53		S2 + 53
Stage III ignition		S2 + 90		S2 + 90
Stage III burnout		S2 + 177		S2 + 177
Spacecraft				
Spacecraft separation	S2 + 250	S2 + 290	S2 + 250	S2 + 290

Delta Payload Accommodations

	Delta II
Payload Compartment	
Maximum Payload Diameter	100.0 in (2,540 mm) for 9.5 ft PLF 110.0 in (2,794 mm) for 10 ft PLF
Maximum Cylinder Length	80.0 in (2,032 mm) plus additional 67.5 in (1,717 mm) length at 86.0 in (2,184 mm) diameter for 2 stage with 9.5 ft PLF 80.0 in (2,032 mm) for 3 stage with 9.5 ft PLF 143.62 in (3,648 mm) for 2 stage with 10 ft PLF 76.61 in (1,946 mm) for 3 stage with 10 ft PLF
Maximum Cone Length	103.8 in (2,637 mm) for 9.5 ft PLF 84.0 in (2,133 mm) for 10 ft PLF
Payload Adapter	
Interface Diameter	37.0 in (940 mm) for 3 stage vehicle 60.0 in (1,524 mm) for 2 stage vehicle
Payload Integration	
Nominal Mission Schedule Begins	T-30 months
Launch Window	
Latest Countdown Hold Not Requiring Recycling	T-60 min
On-Pad Storage Capability	??
Latest Access to Payload	T-17 hrs
Environment	
Maximum Load Factors	+6.0 g axial, ±2.0 g lateral
Minimum Lateral / Longitudinal Payload Frequency	15 Hz / 35 Hz
Maximum Overall Acoustic Level	144.5 dB (1/3 octave) for 792X with 10 ft PLF 139.6 dB (1/3 octave) for 792X with 9.5 ft PLF 144.0 dB (1/3 octave) for 692X with 10 ft PLF 139.1 dB (1/3 octave) for 692X with 9.5 ft PLF
Maximum Flight Shock	5,500 g at 4,000 hz for 2 stage vehicle 4,100 g at 1,500 hz for 3 stage vehicle
Maximum Dynamic Pressure on Fairing	1,230 lb/ft^2 (58,898 N/m^2)
Maximum Pressure Change in Fairing	0.5 psi/s (3.45 KPa/s)
Cleanliness Level in Fairing (Prior to Launch)	Class 10,000+
Payload Delivery	
Standard Orbit and Accuracy (3 sigma)	LEO: ±10 nm (18 km), ±0.5° inclination GTO: ±3 nm (6 km) perigee, ±400-600 nm (740–1,100 km) apogee, ±0.2-0.6° inclination
Attitude Accuracy (3 sigma)	±1–2.5°
Nominal Payload Separation Rate	1 ft/s (0.3 m/s) for 2 stage vehicle 2–8 ft/s (0.6–2.4 m/s) for 3 stage vehicle
Deployment Rotation Rate Available	0–4 rpm for 2 stage vehicle 30–100 rpm for 3 stage vehicle
Loiter Duration in Orbit	??
Maneuvers (Thermal / Collision / Telemetry)	Yes

Delta Notes

Publications

User's Guide

Payload Planner's Guide, McDonnell Douglas Commercial Delta, Inc., December 1989.

Technical Publications

"Delta Status Update" J. F. Meyers, McDonnell Douglas, AIAA-90-2714, AIAA 26th Joint Propulsion Conference, Orlando, FL, July 16-18, 1990.

"A Historical Look at United States Launch Vehicles: 1967-Present", ANSER, STDN 90-4, second edition, February 1990.

"Delta II - A New Era Under Way", J. F. Meyers, McDonnell Douglas, IAF-89-196, 40th International Astronautical Congress, Malaga, Spain, Oct 7-12, 1989.

"Delta II - A New Generation Begins", J. F. Meyers, McDonnell Douglas, AIAA-89-2740, AIAA 25th Joint Propulsion Conference, Monterey, CA, July 10-12, 1989.

"Graphite Epoxy Motors (GEM) for the Delta II Launch Vehicle" N. Vlahakis, Hercules and D. Van Dorn, McDonnell Douglas, AIAA-89-2313, AIAA 25th Joint Propulsion Conference, Monterey, CA, July 10-12, 1989.

"Caster Solid Rocket Motors for Launch Vehicle Propulsion", C. Bryant, Thiokol, AIAA-89-2422, AIAA 25th Joint Propulsion Conference, Monterey, CA, July 10-12,1989.

"Delta II - Commercial Space Transportation", J.F. Meyers, McDonnell Douglas, July 1988.

"Delta Launch Vehicle", J. P. Porter, et al, McDonnell Douglas, IAF-87-184, 38th International Astronautical Congress, Brighton UK, 10-17 October 1987.

Acronyms

A50 - Aerozine 50
ACS - attitude control system
AKM - apogee kick motor
CCAFS - Cape Canaveral Air Force Station
DMCO - Delta Mission Checkout
DSTF - Delta Spin Test Facility
ESA - Explosive Safe Area
FY - fiscal year
GEM - Graphite Epoxy Motors
GEO - geosynchronous orbit
GPS - Global Positioning System
GSE - Ground Support Equipment
GTO - geosynchronous transfer orbit
HPF - Horizontal Processing Facility
HTPB - Hydroxy terminated polybutadiene
IRBM - Intermediate Range Ballistic Missile
ITIP - Improved Transtage Injector Program
LC - Launch Complex
LEO - low Earth orbit
LOX - liquid oxygen
MLV - Medium Launch Vehicle
MRTB - Missile Research Test Building
MST - Mobile Service Tower
NASA - National Aeronautics & Space Administration
NPF - NAVSTAR Processing Facility
N204 - nitrogen tetroxide
PAF - Payload Attach Fittings
PAM - Payload Assist Module
PLF - payload fairing
PSF - Propellant Servicing Facility
RP1 - kerosene
SLC - Space Launch Complex
SRM - solid rocket motor
SSPS - second stage propulsion system
VAFB - Vandenberg Air Force Base

Other Notes—Delta Growth Possibilities

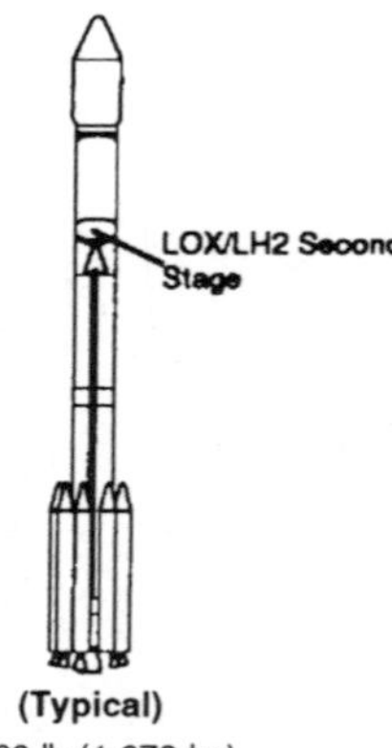

(Typical)

GPS	3,680 lb (1,670 kg)
GTO	5,730 lb (2,600 kg)
LEO	15,600 lb (7,080 kg)

Possible next steps in the Delta growth plan include improved booster engines, stretched graphite epoxy SRMs, extended fuel tanks, wider payload fairings, and cryogenic second stage. A third pad, LC-17C, could be available near LC-17A and B for the uprated Deltas.

McDonnell Douglas has also designed a possible heavy lift vehicle using Delta components. The vehicle uses seven Delta boosters in a cluster augmented by solid rocket boosters from the Space Shuttle. It could be topped by a 60-ft-high Titan IV payload fairing. The vehicle could loft 110,000 lb (50,000 kg) into low Earth orbit.

LLV

Industry Point of Contact:
Lockheed Martin
Missiles and Space
O/8M-01 B/154
1111 Lockheed Way
P.O. Box 3504
Sunnyvale, CA 94089-3504
Phone: (800) 552-8624

A mock-up of an LLV1 on SLC-6 at Vandenberg Air Force Base.

LLV History

In Development

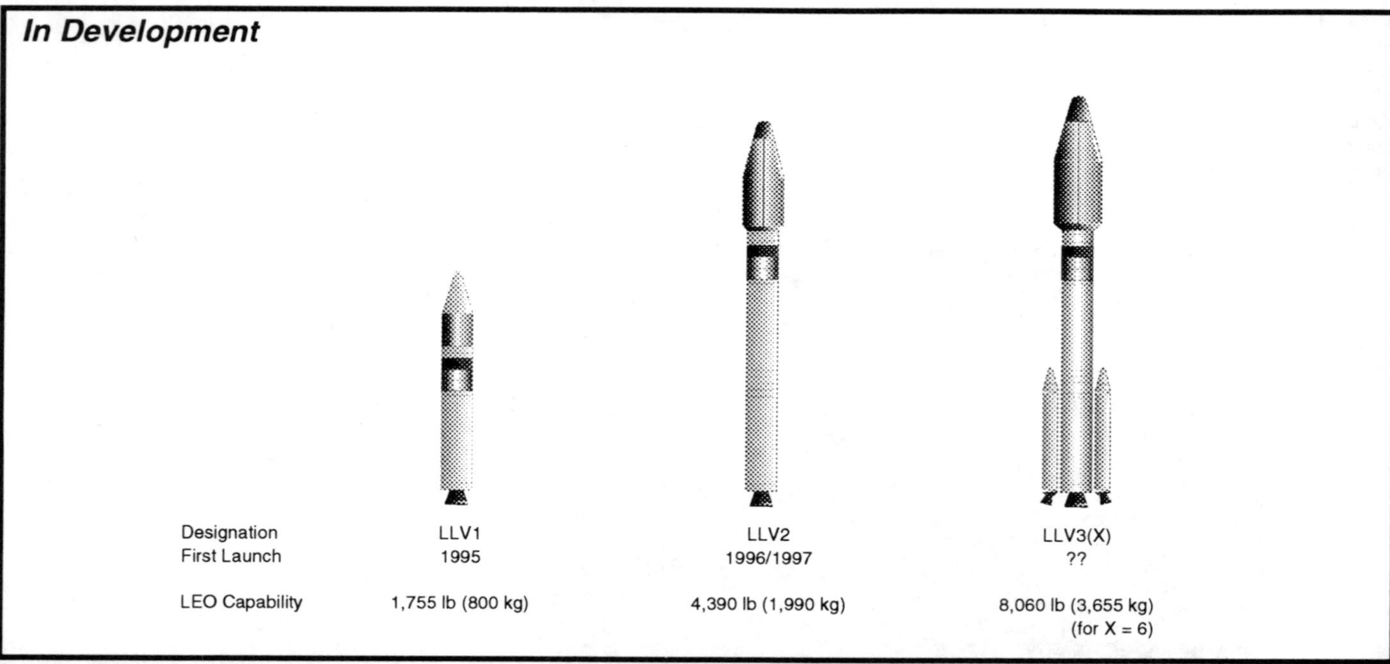

Vehicle Description

LLV1 — Two-stage launch vehicle with a single Castor 120 as the first stage and an Orbus 21D as the second stage with the ACS and avionics packaged in an Orbit Adjust Module (OAM).

LLV2 — Three-stage launch vehicle that adds an additional Castor 120 to the LLV1 configuration.

LLV3(X) — Three-stage launch vehicle that incrementally adds Castor IVA-XL strap-ons onto the LLV2 configuration. The number of strap-ons for each configuration is indicated in parenthesis. X = 2, 3, 4, or 6, i.e., LLV3(2) has two strap-ons.

Historical Summary

In May 1993, Lockheed announced the internally funded development of a low-cost series of small to medium launch vehicles. Through combinations of common elements, an incrementally increasing capability to LEO (185 km, 28.5°) from 800 kg to 3,655 kg is achieved. The concept is a new vehicle that makes use of existing rocket motor designs and other components to minimize development cost and risk. The principle building blocks are the Thiokol Castor 120, the United Technology Chemical Systems Division Orbus 21D, the Thiokol Castor IVA, and the Lockheed Orbit Adjust Module.

Launch Record

The first launch is scheduled for 1995. An LLV1 will carry a commercial payload.

LLV General Description

	LLV1	LLV2	LLV3(X)
Summary	Two-stage solid-propellant inertially guided launch vehicle. First launch is scheduled for early 1995.	Three-stage solid-propellant inertially guided launch vehicle.	Identical to the LLV2 with the addition of 2–6 strap-on solid motors.
Status	In Development	In Development	In Development
Key Organizations	User - DoD, NASA, commercial Launch Service Agency - Lockheed Martin and Space Company (LMMSC) Prime Contractor - LMMSC (Program Management, System Engineering/Integration, Final Assembly, Launch Operations) Subcontractors - Thiokol Corporation (Castor 120™ and IVA) United Technologies Corporation (Orbus® 21D) Olin Aerospace (Attitude Control System)		
Vehicle			
System Height (with 92 in fairing)	744.1 in (18.9 m)	1,110.2 in (28.2 m)	1,110.2 in (28.2 m)
Payload Fairing Size	92 in (2.34 m) diameter by 240 in (6.10 m) height	120 in (3.05 m) diameter by 363 in (9.22 m) height	141 in (3.58 m) diameter by 443 in (11.25 m) height
Gross Mass	145,200 lb (66,000 kg)	264,330 lb (120,150 kg)	318,560–425,370 lb (144,800–193,350 kg)
Operations			
Primary Missions	LEO	LEO	LEO
Compatible Upper Stages	Small integral solid motors	Small integral solid motors	Small integral solid motors
First Launch	1995	1996/1997	??
Success / Flight Total	0 / 0	0 / 0	0 / 0
Launch Site	VAFB SLC-6 (120.6°W, 34.7°N) CCAFS LC-46 (81°W, 28.5°N)	VAFB SLC-6 (120.6°W, 34.7°N) CCAFS LC-46 (81°W, 28.5°N)	VAFB SLC-6 (120.6°W, 34.7°N) CCAFS LC-46 (81°W, 28.5°N)
Launch Azimuth	VAFB 140° - 201° CCAFS 35° - 120°	VAFB 140° - 201° CCAFS 35° - 120°	VAFB 140° - 201° CCAFS 35° - 120°
Nominal Flight Rate	Current production capability can support 12 flights per year. Can be increased to 24 flights per year.		
Performance (with 92 in fairing)			
100 nm (185 km) circ, 28°	1,755 lb (800 kg)	4,390 lb (1,990 kg)	LLV3 (2) - 5,780 lb (2,620 kg) LLV3 (4) - 6,965 lb (3,160 kg) LLV3 (6) - 8,060 lb (3,655 kg)
100 nm (185 km) circ, 90°	1,140 lb (520 kg)	3,290 lb (1,490 kg)	LLV3 (2) - 4,455 lb (2,020 kg) LLV3 (4) - 5,375 lb (2,440 kg) LLV3 (6) - 6,295 lb (2,855 kg)
Geotransfer Orbit, 28°	—	—	—
Geosynchronous Orbit	—	—	—
Financial Status			
Estimated Launch Price (not including range and insurance costs)	$16M (LMSC).	$21M (LMSC).	$23–27M (LMSC).
Orders: Payload / Agency (firm / option)	Commercial / CTA (1/0) Navy GFO / Ball (1/0) SSTI Clark CTA (1/0) SSTI Lewis TWR (1/0)	—	—
Manifest	1995 - CTA; 1996 - BALL 1996 - SSTI / Clark; 1996 - SST I / Lewis	—	—

LLV Vehicle

Overall

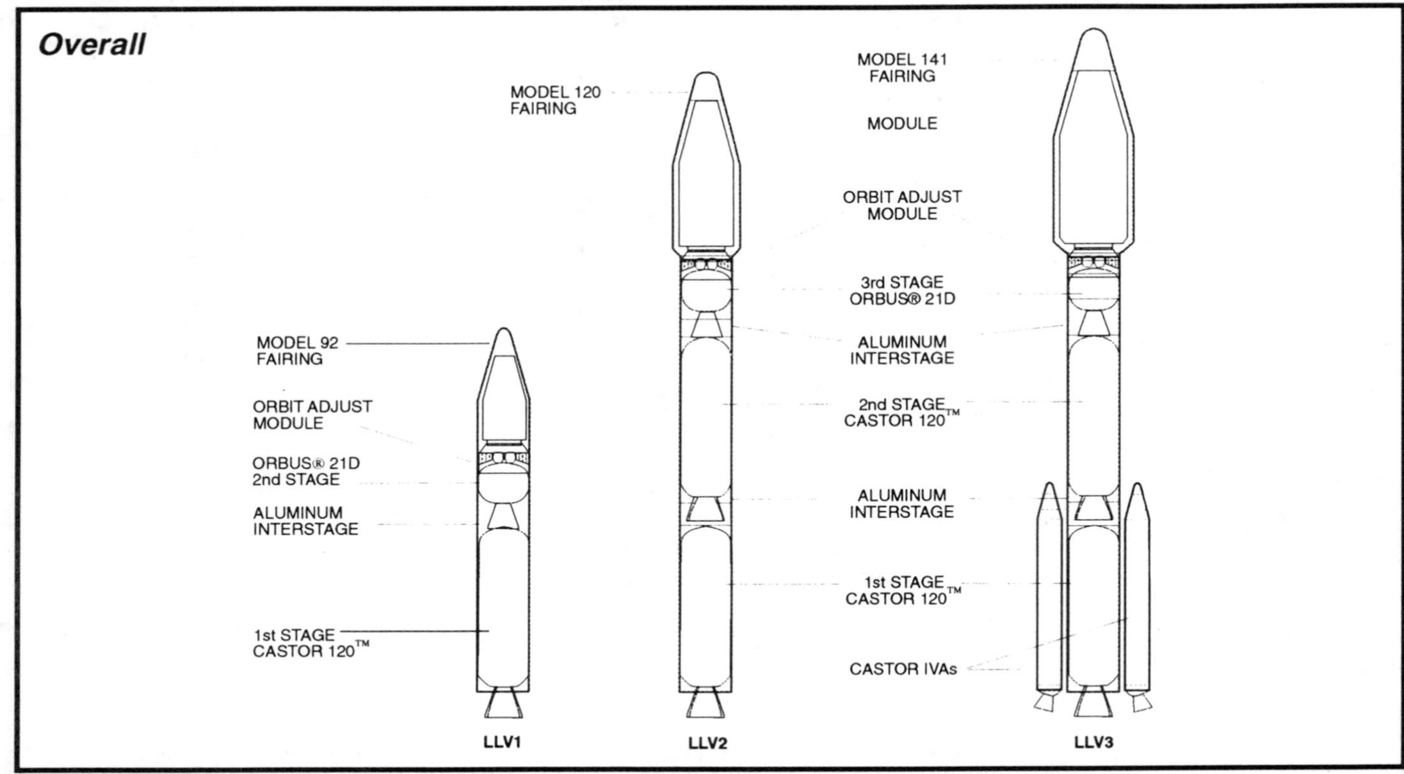

Stages

	Castor 120	Castor-120	Castor IVA	Orbus 21D
	LLV1 - Stage 1	—	—	LLV1 - Stage 2
	LLV2 - Stage 1	LLV2 - Stage 2	—	LLV2 - Stage 3
	LLV3(X) - Stage 1	LLV3(X) - Stage 2	LLV3(X) - Stage 1 Strap-ons	LLV3(X) - Stage 3
Dimension:				
Length	355.1 in (9.02 m)	355.1 in (9.02 m)	362.2 in (9.20 m)	125.2 in (3.18 m)
Diameter	92.9 in (2.36 m)	92.9 in (2.36 m)	40.1 in (101.85 cm)	92.0 in (233.7 cm)
Mass: (each)				
Propellant Mass	107,381 lb (48,809 kg)	107,381 lb (48,809 kg)	22,268 lb (10,131 kg)	21,560 lb (9,800 kg)
Gross Mass	116,275 lb (52,852 kg)	116,644 lb (53,020 kg)	25,562 lb (11,630 kg)	23,417 lb (10,644 kg)
Structure:				
Type	Filament Wound	Filament Wound	Monocoque	Filament Wound
Case Material	Graphite Epoxy	Graphite Epoxy	Steel	KEVLAR® Epoxy
Propulsion:				
Propellant	HTPB	HTPB	HTPB	HTPB
Average Thrust (each)	349,830 lb (1554.8 kN)	361,305 lb (1,605.8 kN)	11,830 lb (490.8 kN)	41,220 LB (183.2 kN)
Number of Motors	1	1	2, 3, 4, or 6	1
Number of Segments	1	1	1	1
Isp	280 sec	286 sec	266.5 sec	295.7 sec
Chamber Pressure	??	??	??	??
Expansion Ratio	17	24	8	63
Control-Pitch,Yaw	Cold-gas blowdown TVA	Cold-gas blowdown TVA	—	Electromechanical
Roll	Monopropellant ACS	Monopropellant ACS	Monopropellant ACS	Monopropellant ACS
Events:				
Nominal Burn Time	82.7 sec	82.7 sec	52.5 sec	154 sec
Stage Shutdown	Burn to depletion	Burn to depletion	Burn to depletion	Burn to depletion
Stage Separation	Linear shaped charge	Linear shaped charge	N/A	Zip

Orbus® 21D

Castor™ 120

LLV Vehicle

(continued)

Payload Fairing

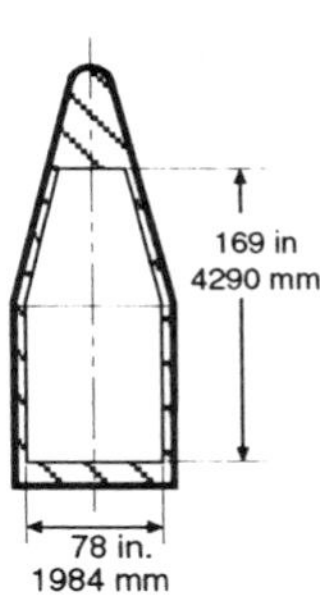

280 in
7101 mm
108 in.
2743 mm

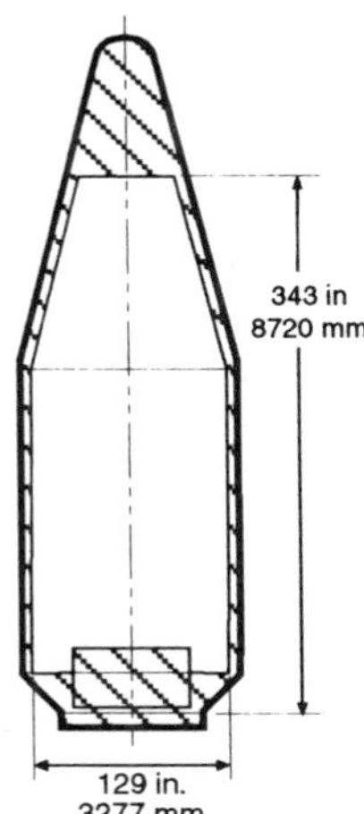

	MODEL 92 ALUMINUM 92 IN	MODEL 120 ALUMINUM 120 IN	MODEL 141 ALUMINUM 141 IN
Length	240 in (6.10 m)	363 in (9.22 m)	443 in (11.25 m)
Diameter	92 in (2.34 m)	120 in (3.05 m)	141 in (3.58 m)
Sections	2	2	2
Structures	Monocoque	Filament Wound	Filament Wound

Remarks

Three clamshell fairings ranging from 2.34 m diameter to 3.58 m diameter are available. The smallest is compatible with any of the vehicles, the larger two are restricted to the LLV2 and LLV3(X). Springs, located at the fairing base, and patented separation joints connecting the fairing halves along the longitudinal seam and the base to the OAM, provide contamination free separation just prior to ignition of the Orbus® 21D. Filtered air conditioning maintains the spacecraft environment at a negotiated temperature and humidity from encapsulation until launch. Seventy-four copper conductors of varying gauge and shielding and two fiber optic cables are available for spacecraft connection to support equipment through the T-0 umbilical. Ten analog and ten discrete monitors together with five signal conditioned continuity loops and five separation indicators are available as standard through the launch vehicle telemetry. The avionics computer can issue eight discrete commands for spacecraft control during ascent.

Payload volume: 2.34 m diameter fairing - 10.6m^3; 3.05 m diameter fairing - 29.5m^3; 3.58 diameter fairing - 56.5m^3

Avionics

Autopilot is hosted in a Lockheed designed RISC 3000 based avionics processor and uses a Litton LN-100L IMU with software modified to LLV requirements to provide incremental velocity and angle data to the Lockheed navigation software. The IMU is a strap-down system employing three nondithered Litton Zero-lock™ 18 cm path length ring laser gyros and three A-4 accelerometers. Avionics performance is monitored by the avionics processor and formatted for down link via the vehicle telemetry system.

Attitude Control System

Olin Aerospace Company will integrate the ACS into the Orbital Adjust Module (OAM) for each vehicle. Designed from flight proven components, the four-axial and six-pitch, yaw, and roll thrusters are manifolded to up to six composite overwrap pressure vessels in a six-pack design. Nitrogen is used to pressurize the blowdown system that feeds the hydrazine to the thrusters. An EED-activated pyro valve isolates the hydrazine until one second after first stage ignition.

Thrust: 4 times 222N
Burn Time: variable
Propellant Mass: 355 kg
Separation: N/A

Flight Sequence: ACS is activated one second after first stage ignition and is operated as required to maintain the mission profile.

LLV Performance

LLV1

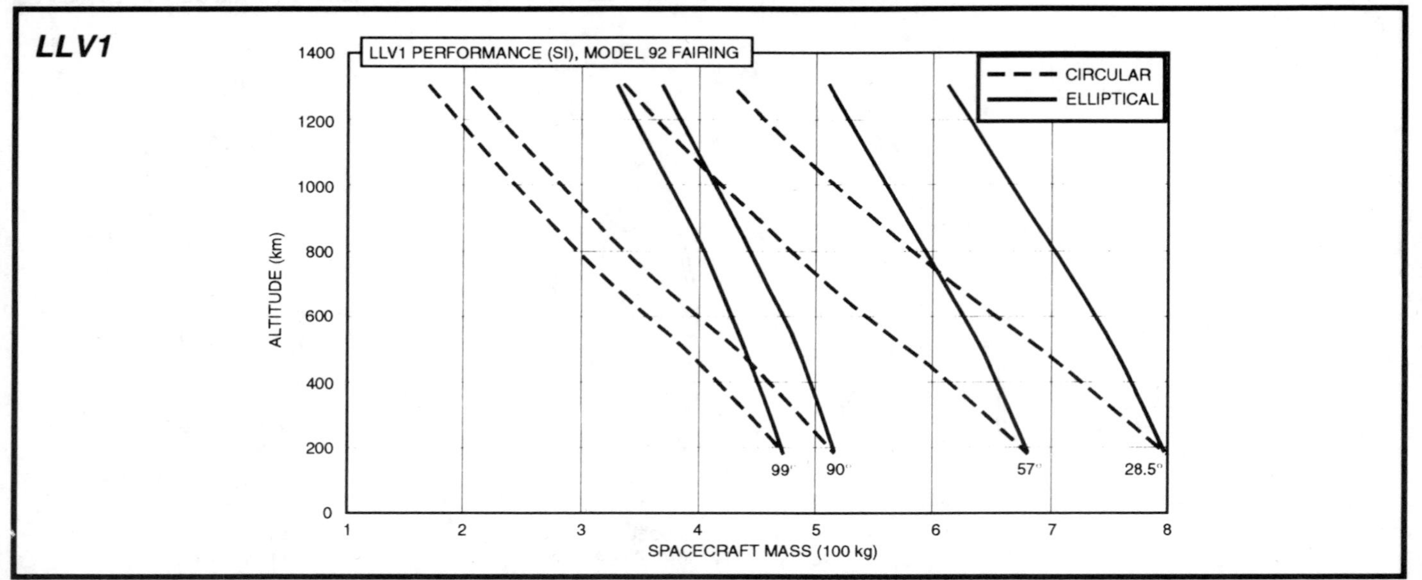

LLV2

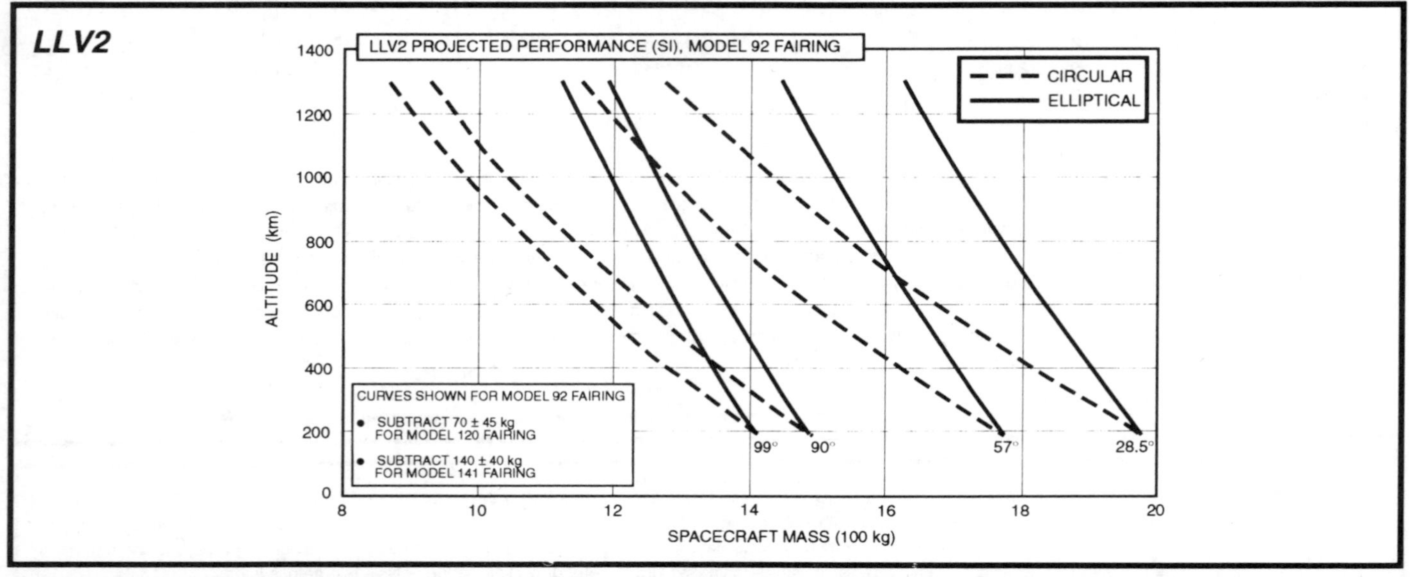

LLV3(6)

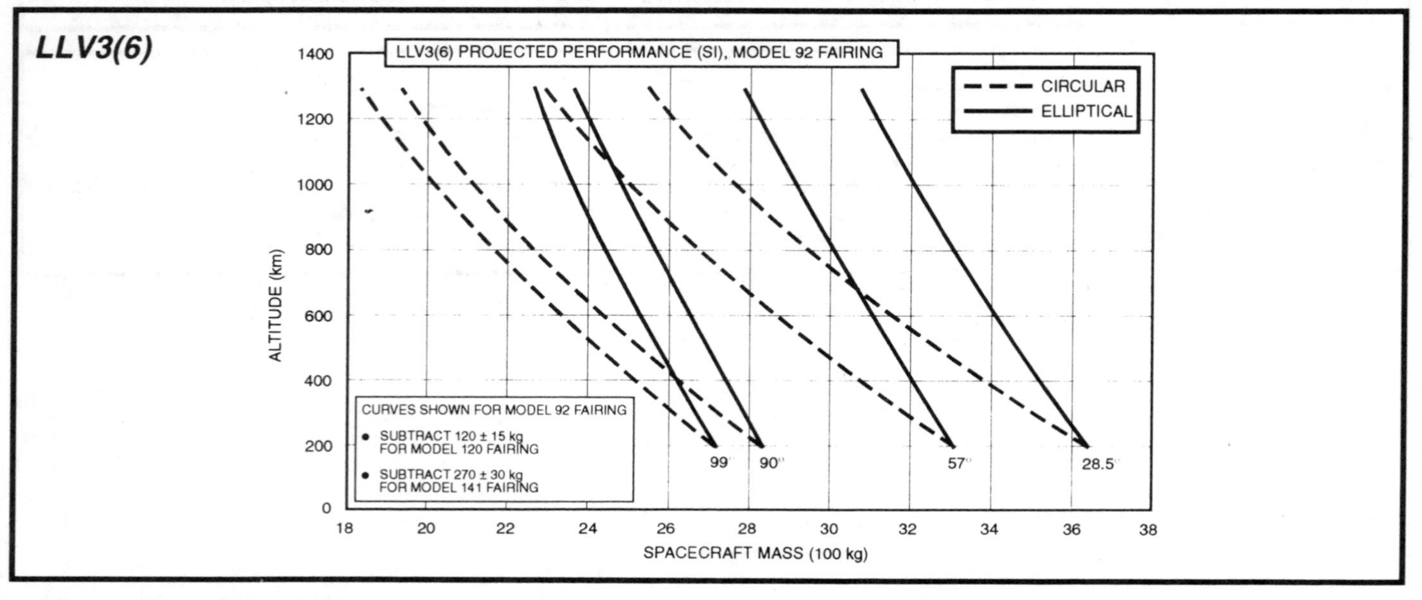

LLV Operations

Launch Site

Vandenberg
Air Force Base
(SLC-6)
Cape Canaveral
(LC-46)

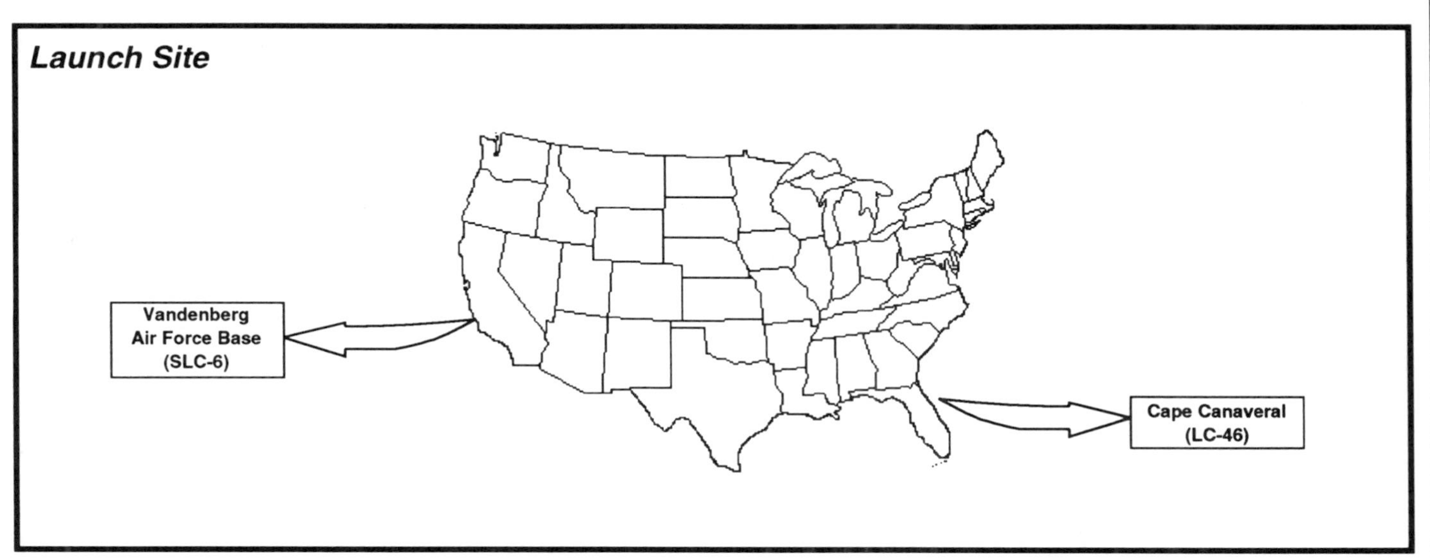

Launch Facilities

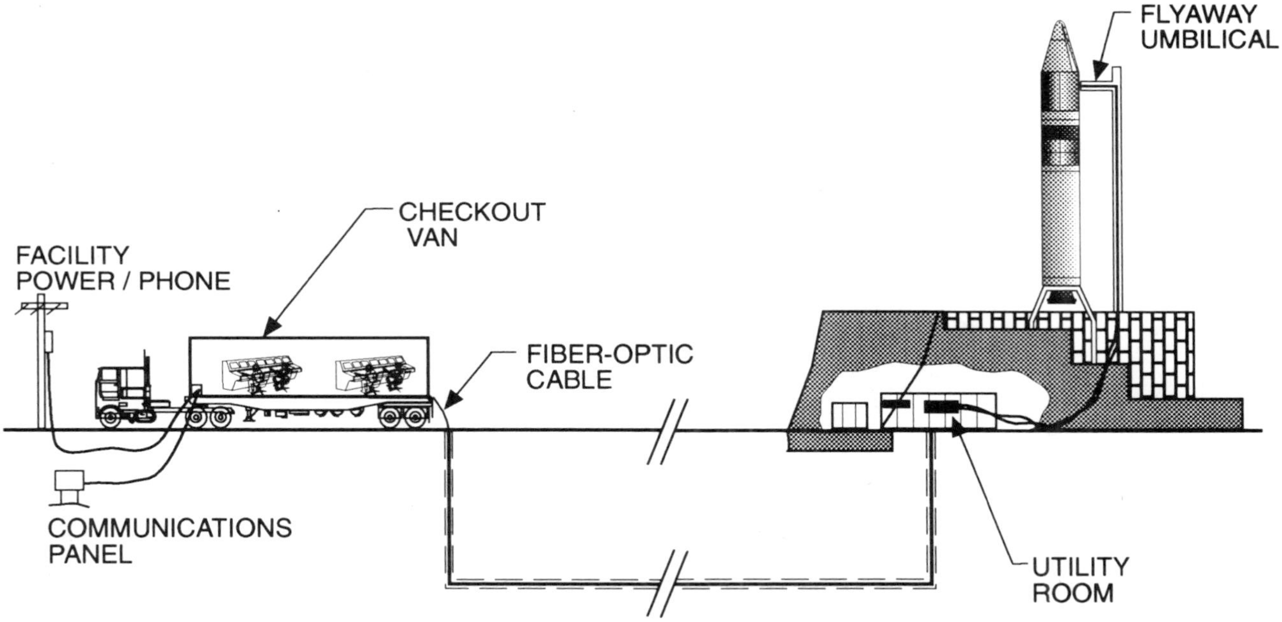

FLYAWAY
UMBILICAL
CHECKOUT
VAN
FACILITY
POWER / PHONE
FIBER-OPTIC
CABLE
COMMUNICATIONS
PANEL
UTILITY
ROOM

LLV Operations

(continued)

Launch Processing

The integration sequence provides maximum independence of payload processing from launch vehicle integration. Fully tested stages and modules are delivered to the launch site as late as 14 days before launch. For Western Range launches, VAFB SLC-6 will be used. For Eastern Range launches, LC-46 is planned to be used. The stages are integrated and interfaces verified on the pad using the same checkout van as used for factory test. The van, located outside the hazard zone, contains dedicated console positions for range safety, the payload launch conductor, and Lockheed launch personnel, and is connected to the umbilical tower via a fiber optic data system. Scaffolding provides vehicle access for assembly and checkout operations. The OAM is fueled offline at a hazardous processing facility, then brought to the pad for integration nine days before launch. The spacecraft processing is conducted offline in a payload processing facility. When the payload is ready, the adaptor is mated and spacecraft encapsulated in the fairing. Transportation to the pad can be as late as three days before launch. The scaffolding is removed on launch day following vehicle close-out operations. Countdown operations, including gyro compassing alignment of the IMU, are controlled from the checkout van.

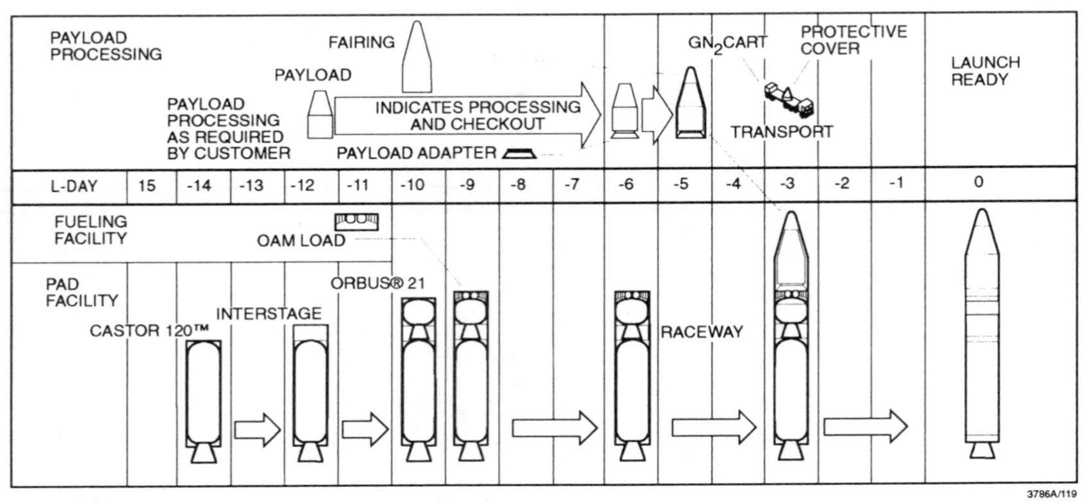

Flight Sequence

LLV1 Typical Flight Sequence

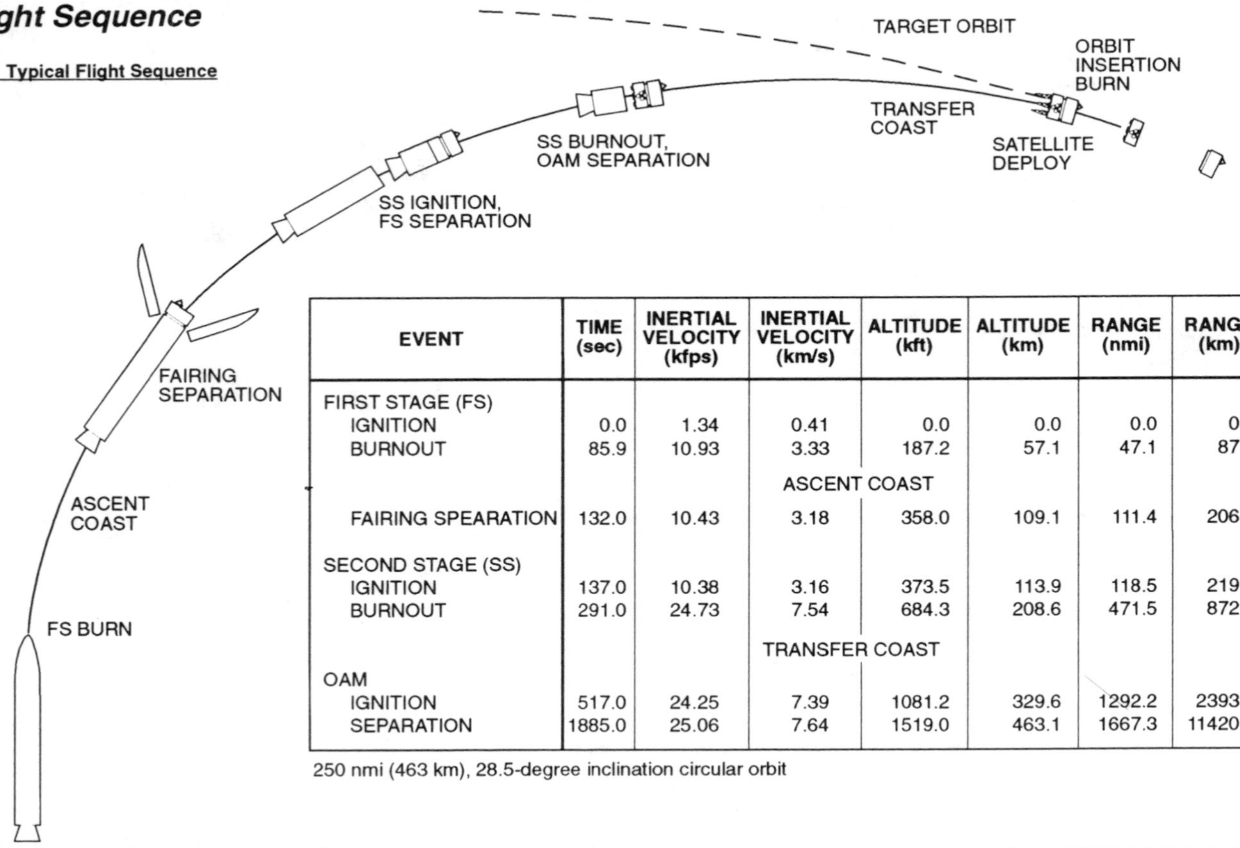

TARGET ORBIT

ORBIT INSERTION BURN

TRANSFER COAST

SATELLITE DEPLOY

SS BURNOUT, OAM SEPARATION

SS IGNITION, FS SEPARATION

FAIRING SEPARATION

ASCENT COAST

FS BURN

EVENT	TIME (sec)	INERTIAL VELOCITY (kfps)	INERTIAL VELOCITY (km/s)	ALTITUDE (kft)	ALTITUDE (km)	RANGE (nmi)	RANGE (km)
FIRST STAGE (FS)							
IGNITION	0.0	1.34	0.41	0.0	0.0	0.0	0.0
BURNOUT	85.9	10.93	3.33	187.2	57.1	47.1	87.2
ASCENT COAST							
FAIRING SPEARATION	132.0	10.43	3.18	358.0	109.1	111.4	206.3
SECOND STAGE (SS)							
IGNITION	137.0	10.38	3.16	373.5	113.9	118.5	219.4
BURNOUT	291.0	24.73	7.54	684.3	208.6	471.5	872.2
TRANSFER COAST							
OAM							
IGNITION	517.0	24.25	7.39	1081.2	329.6	1292.2	2393.0
SEPARATION	1885.0	25.06	7.64	1519.0	463.1	1667.3	11420.9

250 nmi (463 km), 28.5-degree inclination circular orbit

LLV Payload Accommodations

	LLV1 (92 in fairing)	LLV2 (120 in fairing)	LLV3(6) (141 in fairing)
Payload Compartment			
Maximum Payload Diameter	78.1 in (1,984 mm)	108.0 in (2,743 mm)	129.0 in (3,277 mm)
Maximum Cylinder Length	90.2 in (2,291 mm)	166.2 in (4,221 mm)	215.9 in (5,484 mm)
Maximum Cone Length	78.8 in (2,002 mm)	113.4 in (2,880 mm)	127.4 in (3,236 mm)
Payload Adapter			
Interface Diameter	66 in	66 in	66 in
Payload Integration			
Nominal Mission Schedule Begins	T-18 months	T-18 months	T-18 months
Launch Window			
Latest Countdown Hold Not Requiring Recycling	0 min	0 min	0 min
On-Pad Storage Capability	Indefinite	Indefinite	Indefinite
Latest Access to Payload	T-16 hrs	T-16 hrs	T-16 hrs
Environment			
Maximum Load Factors	Axial - +4 g / -8 g Lateral - ±2.5 g	Axial - +1 g / -8 g Lateral - ±2.5 g	Axial - +2 g / -7 g Laterial - ±2.5 g
Minimum Lateral / Longitudinal Payload Frequency	15 Hz / 30 Hz	12 Hz / 30 Hz	11 Hz / 30 Hz
Maximum Overall Acoustic Level	133.5 dB (transonic) / 129.3 dB (launch)	??	??
Maximum Flight Shock	1,800 g from 1,750–2,000 Hz	??	??
Maximum Dynamic Pressure on Fairing	3,500 lb / ft²	??	??
Maximum Pressure Change in Fairing	0.4 psi / sec	??	??
Cleanliness Level in Fairing (Prior to Launch)	Class 100,000	Class 100,000	Class 100,000
Payload Delivery			
Standard Orbit and Accuracy (3 sigma)	±4.5 nmi ±.06 inclination	±4.5 nmi ±.06 inclination	±4.5 nmi ±.06 inclination
Attitude Accuracy (3 sigma)	±1.0°	±1.0°	±1.0°
Nominal Payload Separation Rate	1–10 ft / sec	1–10 ft / sec	1–10 ft / sec
Deployment Rotation Rate Available	0–60 rpm	0–60 rpm	0–60 rpm
Loiter Duration in Orbit	100 min	??	??
Maneuvers (Thermal / Collision / Telemetry)	Yes	Yes	Yes

LLV Notes

Publications

Technical Publications

"Lockheed Launch Vehicle: Status, Capabilities, and Plans for Development", J. L. Cobb and D. E. Davis for AIAA Space Programs and Technologies Conference, September 27-29, 1994, Huntsville, AL.

"A New Family of Small and Medium Launch Vehicles Developed Through Integration of Existing Hardware", D. E. Davis, J. W. Angeli, A. J. MacLaren for 15th AIAA Internal Communications Satellite Systems Conference, February 28-March 3, 1994, San Diego, CA.

"A Status Report - Lockheed Launch Vehicle", D. E. Davis, J. W. Angeli, A. J. MacLaren for the 7th AIAA - Utah State University Conference on Small Satellites.

"Solid Rocket Motor Space Launch Vehicles", A. J. MacLaren and H. D. Trudeau, published in Acta Astronautics Vol. 30, pgs. 165-172, 1993.

Acronyms

ACS - attitude control system
CCAFS - Cape Canaveral Air Force Station
FS - first stage
HTPB - hydroxy terminated poloy butadinene
IMU - Inertial Measurement Unit
LC - launch complex
LEO - low Earth orbit
LLV - Lockheed Launch Vehicle
LMSC - Lockheed Missiles and Space Company
OAM - Orbit Assist Module
SLC - Space Launch Complex
SS - second stage
TVA - thrust vector alignment
VAFB - Vandenberg Air Foce Base

Pegasus / Taurus

Industry Point of Contact:
Orbital Sciences Corporation
12500 Fair Lakes Circle
Fairfax, VA 22033, USA
Phone: (703) 631-3600

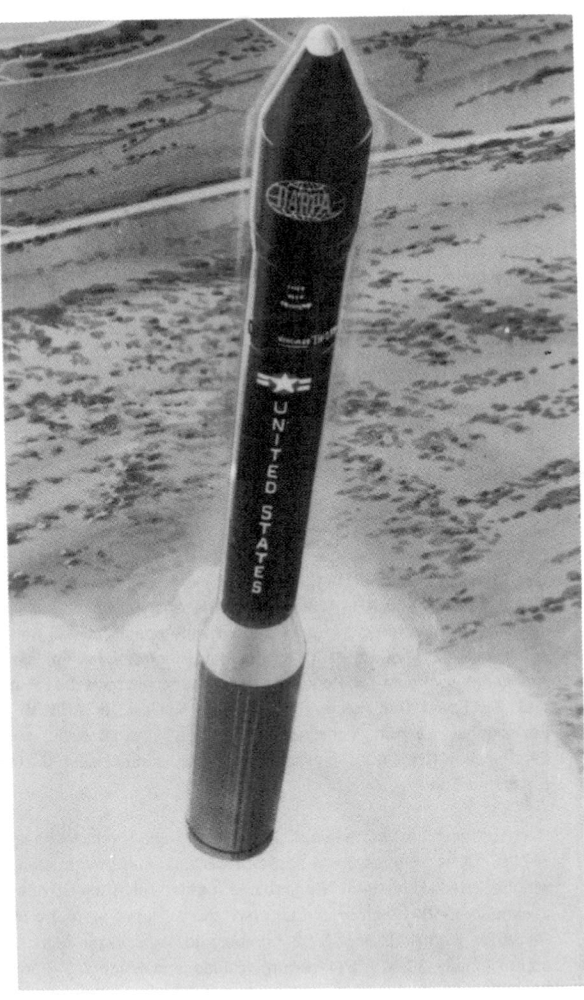

Artist drawing of the air-launched Pegasus (left) and ground-launched Taurus (right) small launch vehicles.

Pegasus / Taurus History

Current Production

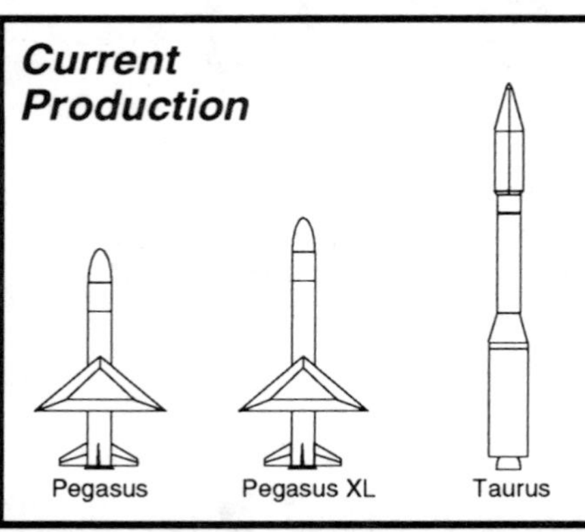

Vehicle Description

Pegasus	Three-stage, solid-propellant, inertially guided, all-composite winged-launch vehicle carried aloft by an aircraft.
Pegasus XL	Growth version of the Pegasus with lengthened Stage 1 and Stage 2, allowing for an increase in propellant of 24% and 30%, respectively.
Taurus	Four-stage, inertially guided three-axis stabilized solid-propellant launch vehicle that is fully road mobile. Stages two through four are derived from Pegasus.

Historical Summary

In 1987, Orbital Science Corporation (OSC) began development of a new, commercially funded space launch vehicle. Conceived by Dr. Antonio L. Elias, then chief engineer at OSC, the distinctive air-launched Pegasus space booster was intended to inexpensively launch small, low-cost payloads into space. OSC entered into a joint venture with Hercules Aerospace in 1988 for the development and production of the Pegasus vehicle; Hercules was responsible for the development of the three brand new solid rocket motors and the payload fairing, while OSC provided the remaining mechanical and avionics systems, ground and flight software, the carrier aircraft interface, mission integration, and overall system engineering. The total Pegasus development cost of over $50 million was split evenly between the joint venture partners.

In July 1988, OSC was awarded a $8.4 million firm fixed-price contract from the Defense Advanced Research Projects Agency (DARPA, now ARPA) providing for one Pegasus launch vehicle and fixed price options for five additional missions. All six missions have been exercised, with five having been flown as of December 1994. The contract provided for OSC and Hercules to retain proprietary rights to technology and data developed at private expense under the Pegasus program. Virtually all vehicle development was funded by OSC and Hercules, with ARPA serving as the anchor tenant for Pegasus launch services. The contract included performance specifications, not vehicle design specifications, allowing OSC to maintain control over vehicle development. Additional contracts for Pegasus launch services have been awarded by the U.S. Air Force (Air Force Small Launch Vehicle, or AFSLV), NASA (Small Expendable Launch Vehicles, or SELVs), the Ballistic Missile Defense Organization (OLS-600), and international customers such as the countries of Brazil and Spain.

Pegasus was the first all-new U.S. space launch vehicle designed since the 1970s. It is a three-stage, solid-propellant, inertially guided, all-composite winged space booster. Pegasus is carried aloft by a specially modified Lockheed L-1011 carrier aircraft (six initial flights were performed using a NASA/Air Force Boeing B-52 bomber) to level flight launch conditions of approximately 38,000 ft (11.8 km) altitude and Mach 0.5. Following release from the aircraft and ignition of its first stage motor, Pegasus follows a nearly vacuum optimized lifting ascent trajectory to orbit, carrying 600 lb (270 kg) payloads to 250 nm (480 km) polar orbits as well as proportional payloads to other altitudes and inclinations or suborbital trajectories.

Advanced propulsion, structural, and avionics technologies, coupled with an air-launched lifting trajectory give Pegasus approximately twice the performance of a similarly sized ground-launched vehicle. This increase is the result of a number of factors: potential and kinetic energy imparted by the carrier aircraft, reduced aerodynamic drag due to lower air density flight profile, improved propulsion efficiency due to higher motor expansion ratios, reduced gravity loss due to unique "S-shaped" trajectory and wing-generated lift, and reduced thrust direction losses due to reduced velocity vector turning.

Operational advantages over traditional ground-based vehicles operating from fixed pads include significantly reduced range safety concerns (drop from the carrier aircraft typically takes place 50 miles off shore), ability to achieve any launch azimuth without "dog legs" or out-of-plane maneuvering, and ability to fly over or around launch-constraining weather with the carrier aircraft.

The first Pegasus vehicle was rolled out at OSC's Vehicle Assembly Building (VAB) at the NASA Dryden Flight Research Facility at Edwards Air Force Base in August 1989. Fully flight-like with the exception of inert propellant in the three solid rocket motors, the vehicle was used for a series of integration tests and three captive carry flights with the NASA B-52 carrier aircraft. The first flight took place on April 5, 1990, and successfully placed two spacecraft into orbit, an ARPA/Navy experimental communications satellite and a NASA Goddard bus which remained attached to the Pegasus thrid stage and included two chemical release experiment canisters and a comprehensive payload environment instrumentation package.

The need for additional performance and precise injection accuracy led to the development of the Hydrazine Auxiliary Propulsion System (HAPS). HAPS is a restartable liquid monopropellant fourth stage for Pegasus which was first flown on the second mission on July 19, 1991. That mission, also flown for ARPA, placed seven Microsats in orbit. A staging anomaly during the separation of the spent first stage, unrelated to the HAPS, caused a velocity shortfall and resulted in a lower than desired orbit. The third mission, flown on February 9, 1993, represented the first commercial Pegasus mission. It also represented the first East Coast launch of Pegasus, as the vehicle and its Brazilian spacecraft were integrated at the VAB and ferried cross-country on the B-52 to Cape Canaveral Air Force Station (CCAFS) for launch operations. Mission control was performed from NASA Wallops Flight Facility. The seventh mission, the last to be flown using the B-52, was the first flight of OSC's PegaStar spacecraft bus, an integrated design which uses booster avionics and structural components to form the basis of a satellite. Launched August 3, 1994, that PegaStar vehicle carried photovoltaic flight experiments for the U.S. Air Force.

In order to satisfy the need for even greater payload performance, the Pegasus XL was conceived in 1991. The primary change from the original configuration was the lengthening of Stages 1 and 2, allowing for an increase in propellant of 24% and 30%, respectively. Structural and avionics upgrades were also incorporated, however the most visible external modification is the relocation of the two horizontal fins to a downward-canted location as required to clear the landing gear doors on OSC's L-1011 carrier aircraft. These modifications have also been retrofitted to the shorter standard vehicle for its use from the L-1011, as all subsequent flights will use this carrier aircraft. One final change in the Pegasus program is the relocation of the VAB from Dryden to Vandenberg Air Force Base (VAFB).

The first launch of the Pegasus XL took place on June 26, 1994. Representing the first use of the L-1011 as well as the first flight of the XL, all carrier aircraft and initial flight operations went smoothly. Unfortunately, vehicle control was lost at 35 seconds into flight, followed by loss of telemetry at 38 seconds and subsequent commanded flight termination. The accident investigation team determined that improperly modeled vehicle aerodynamics resulted in a failure

Pegasus / Taurus History

(continued)

Historical Summary

(continued)

of the autopilot to properly control the vehicle. Revised aerodynamic models, validated through wind tunnel testing, are in place for a second flight attempt in early 1995.

OSC was awarded a fixed price contract by ARPA in July 1989 for the demonstration launch of a new standard small launch vehicle (SSLV) and four optional launches. Called Taurus, the vehicle is a four-stage, inertially guided, all-solid-propellant ground launched space booster. The Taurus configuration mates derivatives of the Pegasus rocket motors and avionics with a Thiokol Castor 120 (the initial flight of Taurus used a Thiokol Peacekeeper first stage, as the Castor 120 was not yet available). A new graphite composite payload fairing and avionics section complete the vehicle. The Pegasus stages retain their Stage 1, 2, and 3 designations from that program, while the Castor 120 is designated as Stage 0. The fairing uses an encapsulated cargo element (ECE) design, where the payload is processed and mated to the Taurus payload adapter cone in an off-line facility. Following the completion of payload processing, the fairing is installed, encapsulating the payload in a sealed, environmentally controlled element. The ECE is transported to the launch site and mated to the launch vehicle as a unit, providing protection to the payload.

In response to the SSLV requirements, the Taurus launch system must be completely road transportable and allow setup at an austere site in five days. Following the setup, payload integration and launch must take place within 72 hours of call-up. The entire launch system is self-contained, including power, communications, vehicle command and monitoring, and integration and test equipment. The original SSLV performance requirements called for the ability to place a 1,000 lb (454 kg) payload in a 400 nm (740 km) circular polar orbit. Subsequent vehicle modifications increased that performance to 1,900 lbs (863 kg), while due east launches to 100 nm (185 km) can carry 3,100 lbs (1,400 kg).

The first Taurus vehicle was unveiled in August 1992 at OSC's Chandler, AZ, facility. As was the case with Pegasus, this vehicle was also fully flight-like with the exception of inert propellant in the four solid rocket motors. It was used in a series of four Pathfinder operations at Chandler and VAFB, where the Taurus rapid response operation procedures were demonstrated and validated, in addition to testing the function of the various vehicle and support equipment systems.

Processing of the first Taurus flight vehicle took place in the Missile Assembly Building at VAFB, with launch operations taking place from a bare concrete pad which had been constructed at launch site 576E, also at VAFB. Payloads for that mission were processed at the payload processing room at SLC-6, the former west coast space shuttle launch facility. This maiden flight successfully placed the Air Force's STEP M0 and ARPA's DARPASAT spacecraft into their intended orbit on March 13, 1994.

Launch Record

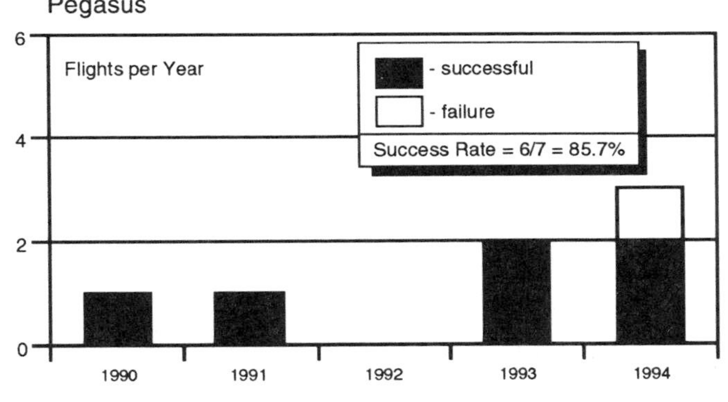

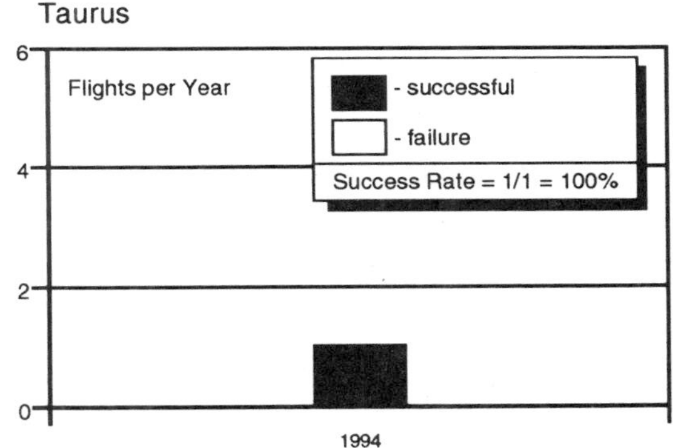

	YEAR	DATE	VEHICLE	SITE	PAYLOAD
X-1	1990	Apr-5	Pegasus	DFRF/B-52	PegSat/SECS
X-2*	1991	Jul-19	Pegasus/HAPS	DFRF/B-52	Microsats 1-7
F-3	1993	Feb-9	Pegasus	KSC/B-52	SCD-1/OPX-1
F-4		Apr-25	Pegasus	DFRF/B-52	Alexis/OPX-2
T-1	1994	Mar-13	Taurus	VAFB/576E	STEP M0/DARPASAT
F-5**		May-18	Pegasus/HAPS	DFRF/B-52	STEP M2
F-6		Jun-26	Pegasus XL	VAFB/L-1011	STEP M1
		Failure - autopilot error due to incorrect aerodynamics			
F-7		Aug-3	Pegasus	DFRF/B-52	APEX

* A staging anomaly during first stage separation caused a velocity shortfall and resulted in a lower than desired orbit.

** An off-nominal orbit was achieved. The first 3 stages performed nominally. Anomaly occured in HAPS stage.

Pegasus / Taurus General Description

	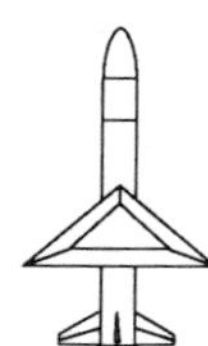	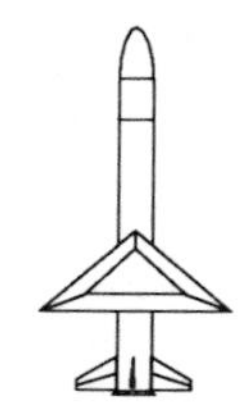	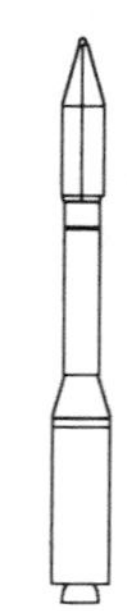
	Pegasus	***Pegasus XL***	***Taurus***
Summary	Pegasus is a three-stage, solid-propellant, inertially guided, graphite composite, winged air-launched space booster developed as a privately funded joint venture of OSC and Hercules Aerospace Company. Pegasus is carried aloft by a specially modified wide-body transport aircraft.	Reflecting the need for additional payload performance, Pegasus XL is a growth version of the basic Pegasus launch vehicle. The first and second stages of the basic Pegasus are stretched to hold 24% and 30% more propellant, respectively. Performance is increased approximately 50%. Structural and avionics improvements are also incorporated.	Taurus is a four stage, inertially guided, all solid propellant groundlaunched space booster, which mates derivatives of the Pegasus rocket motors and avionics with a Thiokol Castor 120. The Taurus launch system is designed for transportability and rapid setup and launch from an austere launch site.
Status	Operational	Operational	Operational
Key Organizations	User - DoD, commercial Launch Service Agency - OSC Launch Systems Group Prime Contractor OSC (System Integration, Avionics, Structures, Carrier Aircraft, Launch Operations) Hercules (Solid Rocket Motors, Payload Fairing) Principal Subcontractors Scaled Composites (Wing, Fins) Litton (INS) OR (Flight Computer) Parker (RCS, TVC, Fin Actuators)	User - DoD, commercial Launch Service Agency - OSC Launch Systems Group Prime Contractor OSC (System Integration, Avionics, Structures, Carrier Aircraft, Launch Operations) Hercules (Solid Rocket Motors, Payload Fairing) Principal Subcontractors Scaled Composites (Wing, Fins) Litton (INS) OR (Flight Computer) Parker (RCS, TVC, Fin Actuators)	User - DoD, commercial Launch Service Agency - OSC Launch Systems Group Prime Contractor OSC (System Integration, Avionics, Structures, Launch Operations) Principal Subcontractors Thiokol (Stage 0 Solid Rocket Motor) Hercules (Stage 1, 2, and 3 Solid Rocket Motors) Litton (INS) OR (Flight Computer) Parker (RCS, TVC) Allied Signal (TVC)
Vehicle			
System Height	50.9 ft (15.5 m)	57.5 ft (17.5 m)	90 ft (27.4 m)
Payload Fairing Size	4.17 ft (1.27 m) diameter by 14.53 ft (4.43 m) height	4.17 ft (1.27 m) diameter by 14.53 ft (4.43 m) height	5.25 ft (1.60 m) diameter by 11.67 ft (3.56 m) height
Gross Mass	42,000 lb (19,000 kg)	52,000 lb (24,000 kg)	160,000 lb (73,000 kg)
Planned Enhancements	None	Increased fairing diameter and length	7.5 ft (2.29 m) diameter fairing, use of Pegasus XL stages, addition of strap-ons to Stage 0
Primary Missions	LEO	LEO	LEO, GTO, Planetary
Compatible Upper Stage	Small integral solid motors, HAPS	Small integral solid motors, HAPS	STAR 37, HAPS
First Launch	1990	1994	1994
Success/Flight Total	6/7	0/1	1/1
Launch Site	VAFB/L-1011 (DFRF/B-52 for initial flights) CCAFS/L-1011	VAFB/L-1011 CCAFS/L-1011	VAFB - 576E, SLC-6 (34.7° N, 120.6° W) CCAFS - LC-46 (28.5° N - 81.0° W)

Pegasus / Taurus General Description

(continued)

	Pegasus	Pegasus XL	Taurus
Vehicle (continued)			
Launch Azimuth	0°–360°	0°– 360°	57°–112° (LC-46) 147°–270° (SLC-6) 185°–270° (576 E)
Nominal Flight Rate	Up to 12 / yr (combined Pegasus and Pegasus XL	Up to 12 / yr (combined Pegasus and Pegasus XL	Up to 6 / yr
Planned Enhancements	Launch operations from non-CONUS sites	Launch operations from non-CONUS sites	Launch operations from non-CONUS sites
Performance			
108 nm (200 km) circ, 28°	700 lb (375 kg)	1,015 lb (460 kg)	3,100 lb (1,400 kg)
108 nm (200 km) circ, 90°	—	760 lb (345 kg)	2,340 lb (1,060 kg)
Geotransfer Orbit, 28°	—	—	990 lb (450 kg) with spin-stabilized STAR 37
Geosynchronous Orbit	—	—	—
Financial Status			
Estimated Launch Price	$11M (OSC)	$12M (OSC)	$18–20M (OSC)
Orders: Payload/Agency (Firm/Option)	ARPA (6/0) SCD-1/Brazil (1/0) Orbcomm/OBC (1/0) MSTI 3/BMDO (1/4)	STP/USAF (3/37) SMEX/NASA (7/3) Orbcomm/OBC (3/1) Minisat/Spain (1/0	ARPA (1/4) Clementine 2/BMDO (1/4)
Manifest			
Remarks	To be phased out in favor of Pegasus XL	—	First flight used Peacekeeper first stage in place of Castor 120

Pegasus / Taurus Vehicle

Overall

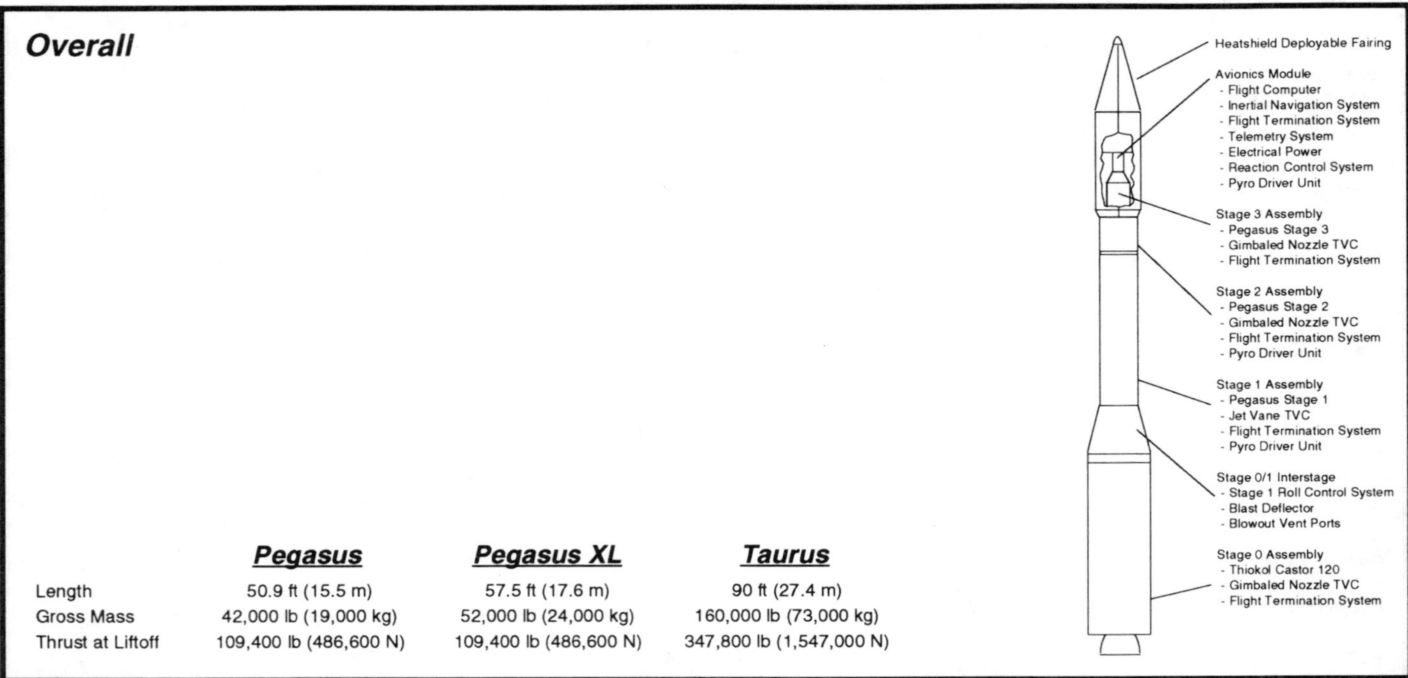

	Pegasus	**Pegasus XL**	**Taurus**
Length	50.9 ft (15.5 m)	57.5 ft (17.6 m)	90 ft (27.4 m)
Gross Mass	42,000 lb (19,000 kg)	52,000 lb (24,000 kg)	160,000 lb (73,000 kg)
Thrust at Liftoff	109,400 lb (486,600 N)	109,400 lb (486,600 N)	347,800 lb (1,547,000 N)

Taurus Stages

	Stage 0	**Stage 1**	**Stage 2**	**Stage 3**
Dimension:				
Length	29.59 ft (9.02 m)	29.1 ft (8.88 m)	8.92 ft (2.65 m)	4.39 ft (1.34 m)
Diameter	7.74 ft (2.36 m)	4.17 ft (1.27 m)	4.17 ft (1.27 m)	3.17 ft (0.97 m)
Mass: (each)				
Propellant Mass	108,000 lb (49,000 kg)	26,809 lb (12,160 kg)	6,670 lb (3,025 kg)	1,699 lb (771 kg)
Gross Mass	??	30,910 lb (14,020 kg)	7,430 lb (3,370 kg)	1,977 lb (897 kg)
Structure:				
Type	Filament Wound	Monocoque	Monocoque	Monocoque
Case Material	Graphite Epoxy	Graphite Epoxy	Graphite Epoxy	Graphite Epoxy
Propulsion:				
Propellant	HTPB	HTPB	HTPB	HTPB
Average Thrust (each)	379,400 lb (1,687,000 N) vac	109,020 lb (484,900 N) vac	26,580 lb (118,200 N) vac	7,175 lb (31.9 N) vac
Number of Motors	1	1	1	1
Number of Segments	1	1	1	1
Isp	280 sec SL	294.4 sec vac	292.3 sec vac	288.9 sec vac
Chamber Pressure	1,400 psia (96.5 bar)	847 psia (58.4 bar)	845 psia (58.3 bar)	549 psia (37.8 bar)
Expansion Ratio	17:1	40:1	65:1	60:1
Control - Pitch,Yaw	Hydraulic gimballing	Hydraulic Jet Vane	EMA (±3°)	EMA (±3°)
Roll	Nitrogen cold gas RCS	Hydraulic Jet Vane	Nitrogen cold gas RCS	Nitrogen cold gas RCS
Events:				
Nominal Burn Time	81 sec	72.4 sec	73.3 sec	68.4 sec
Stage Shutdown	Burn to depletion	Burn to depletion	Burn to depletion	Burn to depletion
Stage Separation	"Fire in hole" Stg 2 ignite	Linear shaped charge	Spring ejection	Spring ejection

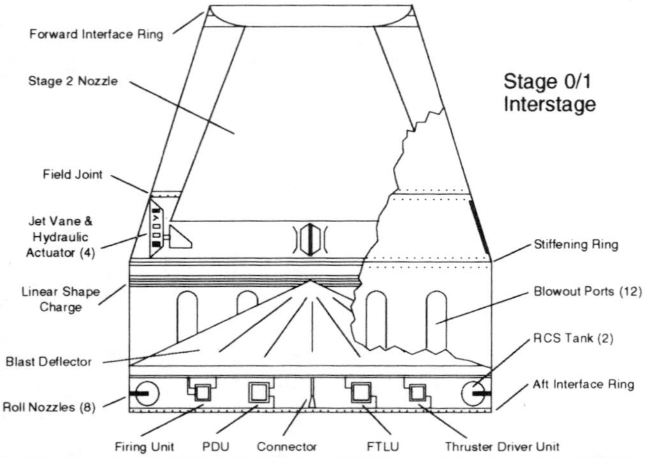

Stage 0/1 Interstage

Forward Interface Ring
Stage 2 Nozzle
Field Joint
Jet Vane & Hydraulic Actuator (4)
Linear Shape Charge
Blast Deflector
Roll Nozzles (8)
Stiffening Ring
Blowout Ports (12)
RCS Tank (2)
Aft Interface Ring
Firing Unit
PDU
Connector
FTLU
Thruster Driver Unit

Remarks:

Propulsion is based on proven motors currently in production for Pegasus and the new Castor 120. Stage 0 propulsion will be provided by the Castor 120. Stages 1, 2, and 3 consist of Pegasus Stage 1, 2, and 3 motors, respectively, that are structurally modified to carry higher loads associated with the larger payload flown on Taurus.

Pegasus / Taurus Vehicle
(continued)

Pegasus Stages

Stage 1

Stage 2

Stage 3

	Stage 1	Stage 2	Stage 3
Dimension:			
Length	29.1 ft (8.88 m)	8.92 ft (2.65 m)	4.39 ft (1.34 m)
Diameter	4.17 ft (1.27 m)	4.17 ft (1.27 m)	3.17 ft (0.97 m)
Wingspan	22.0 ft (6.71 m)	—	—
Mass: (each)			
Propellant Mass	26,809 lb (12,160 kg)	6,670 lb (3,025 kg)	1,699 lb (771 kg)
Gross Mass	29,586 lb (13,420 kg)	7,430 lb (3,370 kg)	1,977 lb (897 kg)
Structure:			
Type	Monocoque	Monocoque	Monocoque
Case Material	Graphite Epoxy	Graphite Epoxy	Graphite Epoxy
Propulsion:			
Propellant	HTPB	HTPB	HTPB
Average Thrust (each)	109,020 lb (484.9K N) vac	26,580 lb (118.2K N) vac	7,175 lb (31.9K N) vac
Number of Motors	1	1	1
Number of Segments	1	1	1
Isp	294.4 sec vac	292.3 sec vac	288.9 sec vac
Chamber Pressure	847 psia (58.4 bar)	845 psia (58.3 bar)	549 psia (37.8 bar)
Expansion Ratio	40:1	65:1	60:1
Control - Pitch,Yaw	3 hydraulic aero fins	EMA (±3°)	EMA (±3°)
Roll	3 hydraulic aero fins	Nitrogen cold gas RCS	Nitrogen cold gas RCS
Events:			
Nominal Burn Time	72.4 sec	73.3 sec	68.4 sec
Stage Shutdown	Burn to depletion	Burn to depletion	Burn to depletion
Stage Separation	Linear Shaped Charge	Spring ejection	Spring ejection

Remarks:

The Pegasus consists of three graphite-epoxy composite case solid-propellant rocket motors, a fixed delta platform composite wing, an aft skirt assembly including three composite control fins, an avionics section forward of the Stage 3 motor, and a two-piece composite payload fairing.

Pegasus main propulsion is provided by three new graphite-epoxy composite case solid propellant rocket motors developed by Hercules. The motors have been developed using a conservative design philosophy which includes the use of demonstrated component technology, maximum use of common components and tooling among stages, a 1.4 factor of safety for all structural components, and the use of class 1.3 propellant.

All three motor cases are wound using IM7 graphite fibers with aramid-filled, ethylene-propylene diene monomer (EPDM) material insulation similar to those of the Delta II strap ons. The nondetonating Class 1.3 propellant is an 88% solid hydroxy-terminated polybutadiene (HTPB)-based formulation identical to that used on the Pershing II missile. This propellant has mechanical electricity and slow burn rate properties appropriate for an air-launched rocket. All three stages have low erosion 3D carbon-carbon integral throat insets and carbon-phenolic nozzles with graphite epoxy overwraps. Thermal protection of the wing during hypervelocity flight is achieved by means of selectively applied additional layers of graphite composite, which are allowed to char and ablate, and approach made possible by the single-use nature of the airframe. The fin leading edges are protected with Korotherm, while artificial cork is used over less critical areas of the wing and fuselage.

The Stage 1 motor has a core-burning grain design with the igniter mounted on the forward dome. The design includes a fixed nozzle, an aluminum wing mounting saddle and an extended forward skirt. The forward skirt, which also serves as an interstage adapter, incorporates two linear shaped-charges for Stage 1/Stage 2 separation. The Stage 2 motor is also a core-burning design and uses a silicon elastomer flexseal nozzle and electromechanical actuators for thrust vector control (TVC). The Stage 3 motor incorporates a head-end grain design to maximize propellant density. Stage 3 also uses a flexseal nozzle and electromechanical actuators for TVC, and employs a toroidal igniter. A flight termination charge is mounted on the aft dome of each motor to satisfy both range safety and aircraft safety requirements. If initiated, it cuts through the graphite case, insulation, and propellant, compromising the structural integrity of the motor. If the case is pressurized (i.e, if the stage is ignited), then the case domes will rupture locally and the stage will immediately become nonpropulsive and begin to tumble.

In addition to the motors, the major structural elements of the flight vehicle are the wing, fins, aft structure, avionics structure, and payload fairing. The wing configuration is a truncated delta planform with a 45° sweptback leading edge and a 22 ft (6.7 m) span. The airfoil is double wedge with a 2 in. (5 cm) radius leading edge. The wing thickness is truncated to 8 in. (20 cm), with upper and lower parallel surfaces facilitating attachment to the ASE pylon adapter and motor case wing saddle respectively. The wind and fins are fabricated of lightweight graphite composite material. The aluminum aft structure support the three active fins, associated electromechanical fin actuators and high voltage battery system. It attached to the aft skirt extension of the Stage 1 motor case. The avionics structure mounted forward of Stage 3 motor is a graphite composite assembly that includes a conical section, a cylindrical section and a planar aluminum honeycomb avionics shelf.

Pegasus / Taurus Vehicle

(continued)

Pegasus XL Stages

	Stage 1	Stage 2	Stage 3
Dimension:			
Length	29.1 ft (8.88 m)	11.7 ft (3.58 m)	4.39 ft (1.34 m)
Diameter	4.17 ft (1.27 m)	4.17 ft (1.27 m)	3.17 ft (0.97 m)
Wingspan	22.0 ft (6.71 m)	—	—
Mass: (each)			
Propellant Mass	33,176 lb (15,048 kg)	8,633 lb (3,915 kg)	1,699 lb (771 kg)
Gross Mass	36,186 lb (16,412 kg)	9,551 lb (4,331 kg)	1,977 lb (897 kg)
Structure:			
Type	Monocoque	Monocoque	Monocoque
Case Material	Graphite Epoxy	Graphite Epoxy	Graphite Epoxy
Propulsion:			
Propellant	HTPB	HTPB	HTPB
Avg. Thrust (each)	132,410 lb (589.0K N) vac	34,515 lb (153.5K N) vac	70,435 lb (31.6K N) vac
Number of Motors	1	1	1
Number of Segments	1	1	1
Isp	293.2 sec vac	290.1 sec vac	290.2 sec vac
Chamber Pressure	1100 psia (7.58 bar)	1019 psia (7.03 bar)	656 psia (4.52 bar)
Expansion Ratio	39.5:1	58.6:1	67.5:1
Control - Pitch,Yaw	3 hydraulic aero fins	EMA (±3°)	EMA (±3°)
- Roll	3 hydraulic aero fins	Nitrogen cold gas RCS	Nitrogen cold gas RCS
Events:			
Nominal Burn Time	73.4 sec	72.5 sec	69.6 sec
Stage Shutdown	Burn to depletion	Burn to depletion	Burn to depletion
Stage Separation	Linear Shaped Charge	Spring ejection	Spring ejection

Pegasus / Taurus Vehicle

(continued)

Payload Fairing

Pegasus (in inches)

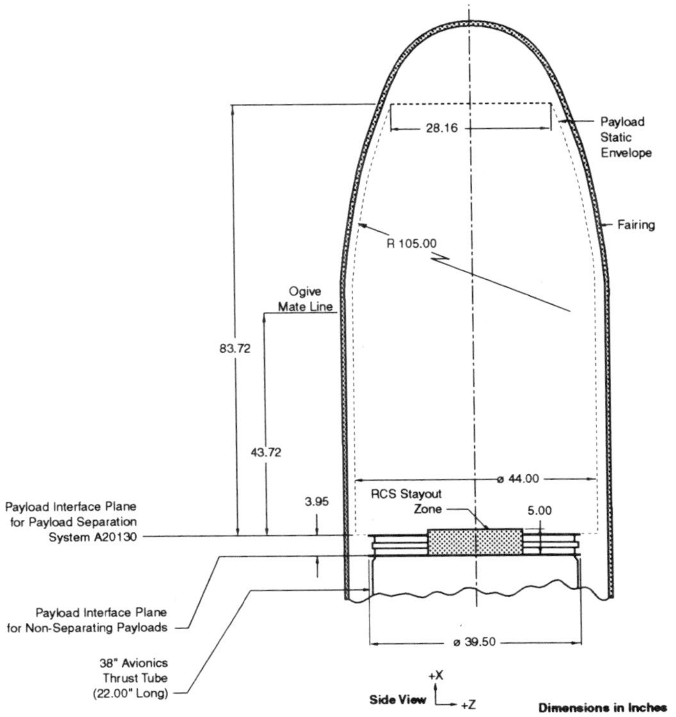

Payload Static Envelope
Fairing
28.16
R 105.00
Ogive Mate Line
83.72
43.72
3.95
ø 44.00
RCS Stayout Zone
5.00
Payload Interface Plane for Payload Separation System A20130
Payload Interface Plane for Non-Separating Payloads
38" Avionics Thrust Tube (22.00" Long)
ø 39.50
+X
+Z
Side View
Dimensions in Inches

Length	14.5 ft (4.42 m)
Diameter	4.17 ft (1.27 m)
Mass	244 lb (110 kg)
Sections	2
Structure	Composite sandwich
Material	Aluminum honeycomb / graphite epoxy

Remarks

The payload fairing is a two piece carbon composite structure consisting of 60 mil face sheets over 0.5 in. (13 mm) aluminum honeycomb. The fairing maintains the 50 in. (1,270 mm) outside diameter of the second stage motor and completely encloses the payload, avionics subsystem and smaller diameter third stage motor. Openings are provided for two sets of reaction control system (RCS) thruster pods, a payload access door, and pyrotechnic bolt cutters for separation of the forward fairing clamp ring. Honeycomb venting is provided through small holes in the inside face sheet. Bulk venting is provided by two cutouts near the base of the fairing. When on the ground the payload area is cooled and maintained under positive pressurization by the air conditioning cart. During captive flight dry nitrogen is purged through the fairing from tanks inside the carrier aircraft.

Taurus

50.8cm
(20")
279.4 cm
(110")
A
A
160 cm
(63")
137.2 cm
(54")
110.57 cm
(43.53")
Payload Dynamic Envelope

Payload Interface
60 0.25 Inch Fasteners
NAS 1351C4
Bolt Pattern
Per Tool
C80001-1
98.58 cm
(Ø38.81")

View A-A

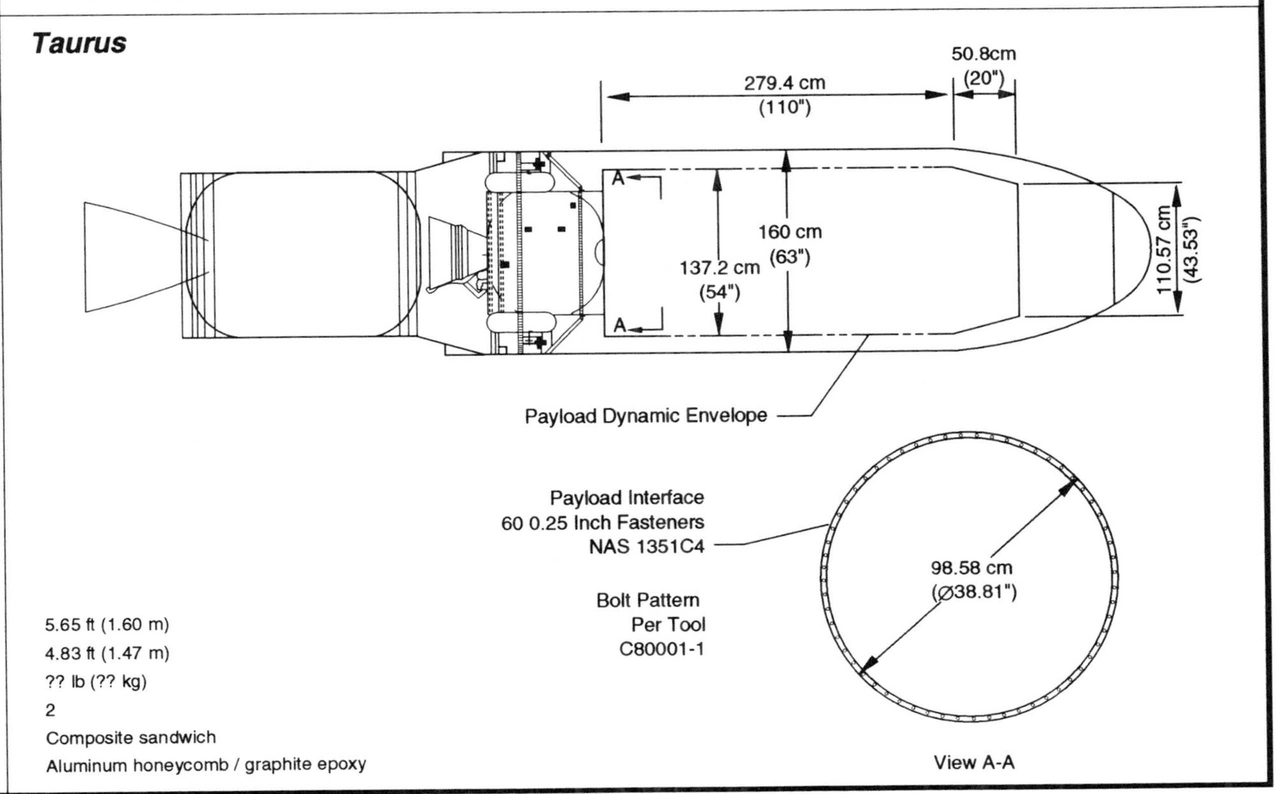

Length	5.65 ft (1.60 m)
Diameter	4.83 ft (1.47 m)
Mass	?? lb (?? kg)
Sections	2
Structure	Composite sandwich
Material	Aluminum honeycomb / graphite epoxy

Pegasus / Taurus Vehicle

(continued)

Avionics

Pegasus: The avionics subsystem is mounted to the third stage motor and serves as a mounting structure for most vehicle avionics. These include an inertial measurement unit (IMU), flight computer, telemetry transmitter, telemetry multiplexer, ordnance and thruster driver units, RCS thrusters, dual flight termination receivers, radar transponder, batteries, various other components and harness. The structure is comprised of a graphite conical and cylindrical section with an aluminum planar honeycomb deck. The avionics structure also provides a mechanical interface for the payload.

Pegasus is controlled by a multiprocessor 32-bit flight computer which communicates via serial interfaces with individual microprocessors in the vehicle's "smart" actuator and sensor assemblies. These distributed microprocessors manage the actuators, ordnance initiation devices, and telemetry data-gathering systems, including initialization, self-test and health reporting via telemetry.

The telemetry system doubles as a ground checkout unit. During flight all critical vehicle performance parameters are transmitted to the ground using a single 56 kbps S-band telemetry channel. A C-band radar transponder is provided to improve the ability of ground stations to track the vehicle during ascent. A fully redundant UHF flight termination system is provided to satisfy range safety requirements. Antenna systems for these three RF links are installed on both the second and third stage motors. The electrical power system on Pegasus consists of two individual battery and power distribution subsystems.

The electronic airborne support equipment (ASE) for the Pegasus is contained in a single pallet onboard the B-52 aircraft. The ASE controls the aircraft power supply interface to the launch vehicle, downloads mission data to the vehicle flight computer, enables the vehicle for drop, and provides reference data for vehicle IMU alignment. The Pegasus flight computer controls microprocessors to manage actuators, ordnance initiation devices, and telemetry data-gathering systems. Two IMUs are used in conducting a Pegasus flight. The vehicle contains an IMU internally that is initialized in flight by a precision IMU carried in the carrier aircraft's ASE.

Taurus: A slightly modified Pegasus avionics module mounts the avionics that provide electrical power, telemetry, sequencing, guidance, and control for the vehicle. It also hosts flight termination system electronics and radar tracking transponder, and provides payload mechanical and electrical interfaces. Most of the avionics components are identical to those used on Pegasus. New electrical interfaces to the Stage 0 and Stage 1 vector control actuators are derivatives of existing Pegasus hardware. In addition to hardware commonality, guidance and other software philosophy and implementation is based on proved Pegasus programs.

Attitude Control System

Pegasus: The wing and fins, mounted on the Pegasus first stage, aerodynamically provide pitch, yaw, and roll control during the first stage powered flight and the coast period after burnout. The second and third stages control pitch and yaw attitude by gimbaling their nozzles. Cold gas jets, located forward of the third stage, are used to provide roll control throughout second and third stage flight. The cold gas reaction system is also employed in every coast period to maintain three-axis control. The payload fairing provides openings for the two pods of reaction control system thrusters so that reaction control is available prior to payload fairing separation.

Following orbital insertion, the Pegasus third stage executes a series of prespecified commands contained in the mission data load to provide the desired initial payload attitude prior to payload separation. Either an inertially fixed or spin-stabilized attitude may be specified. For inertial attitudes the payload and third stage can be oriented to $\pm$ 2 degrees in angular position in each axis. For a spin-stabilized initial attitude, the maximum spin rate achievable depends on the payload and spent third stage combined spin-axis moment of inertia. RCS can provide up to 1,000 lb-in-sec total impulse for spin-up. Orientation of the payload/third-stage spin axis can be achieved to $\pm$ 2 degrees.

Taurus: Thrust vector control for pitch and yaw on Stage 1 is provided by a flexseal nozzle with a turbo hydraulic actuation system. Roll control during Stage 0 flight is provided by a cold gas system located in the Stage 0/Stage 1 interstage forward of the Stage 0 motor. The Stage 1 motor used government surplus, flight proven Pershing 1A jet vane assemblies in the top section of the Stage 0/Stage 1 interstage to provide thrust vector control in roll, pitch, and yaw. The remaining motors are used in the same configuration as on Pegasus.

Pegasus / Taurus Performance

Pegasus XL

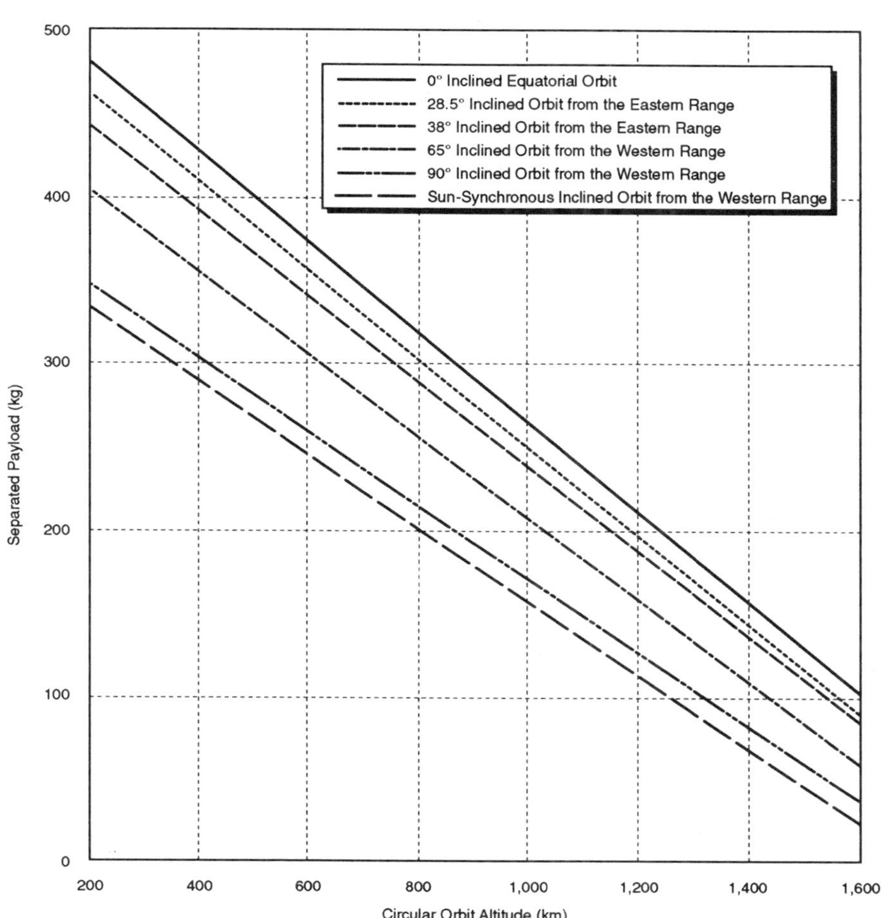

Pegasus XL Ground Rules

- Drop Conditions
  38,000 ft (11,590 Meters)
  770 ft/sec (235 m/sec)
- 220 ft/sec (67 m/sec) Guidance Reserve Maintained
- Nominal Fairing Separation at 0.01 psf
- Performance Numbers Reflect Separated Payload Mass Including 3.9 kg 38" Payload Attach Fitting Mass

Taurus

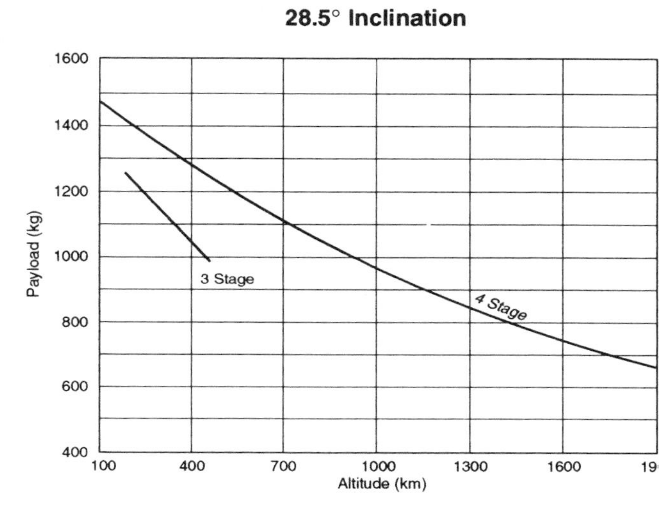

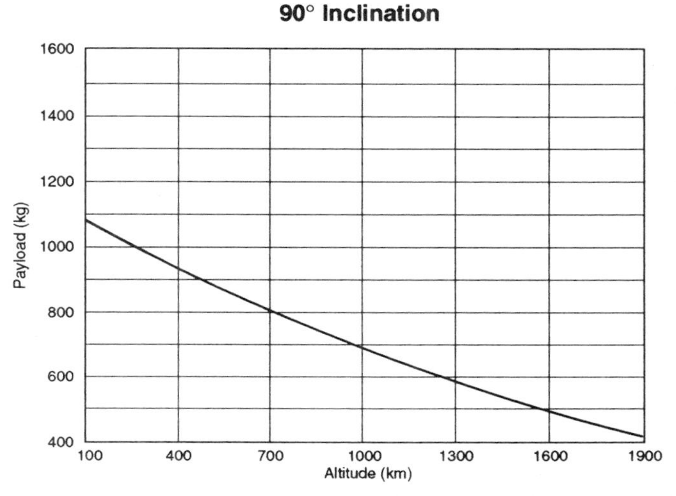

Pegasus / Taurus Operations

Launch Site

NASA Dryden Flight Research Facility
Pegasus - takeoff on a B-52

Vandenberg Air Force Base
Taurus - from a mobile launch pad
Pegasus - takeoff on L-1011

Cape Canaveral Air Force Station
Taurus - from a mobile launch pad
Pegasus - takeoff on L-1011

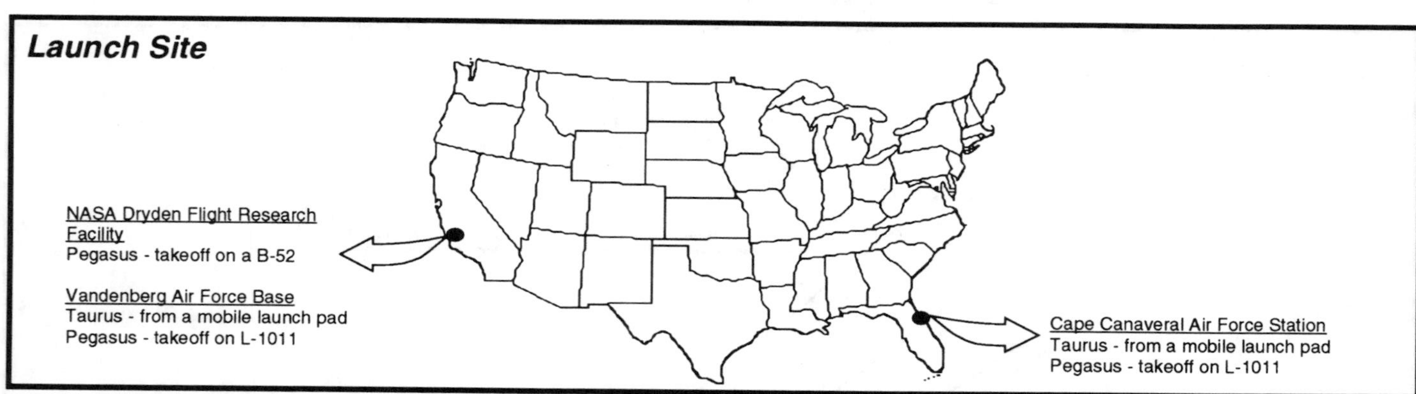

Launch Facilities

Pegasus

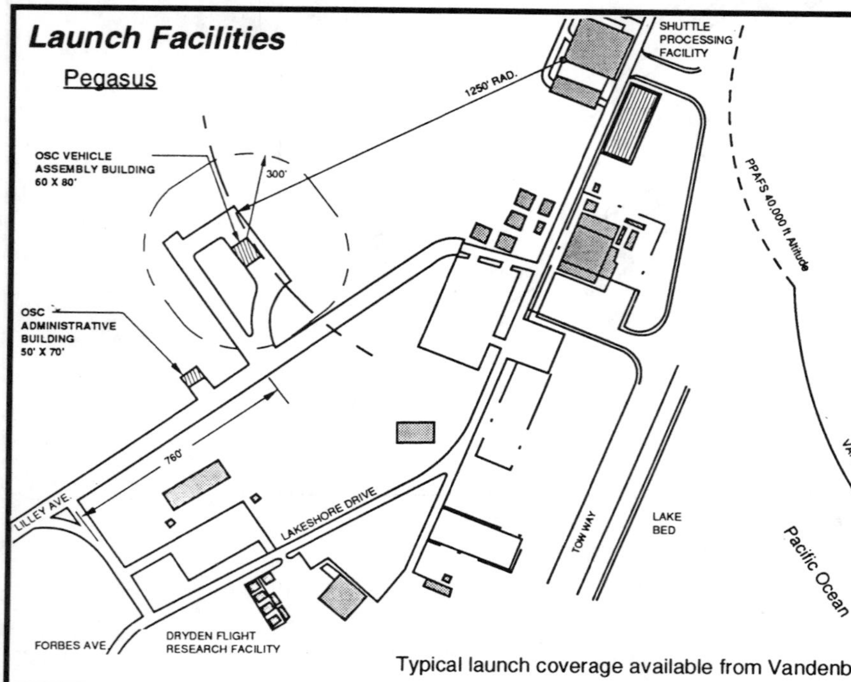

SHUTTLE PROCESSING FACILITY
1250' RAD.
OSC VEHICLE ASSEMBLY BUILDING 60 X 80'
300'
OSC ADMINISTRATIVE BUILDING 50' X 70'
760'
LILLEY AVE
LAKESHORE DRIVE
TOW WAY
LAKE BED
FORBES AVE
DRYDEN FLIGHT RESEARCH FACILITY
PPAFS 40,000 ft Altitude

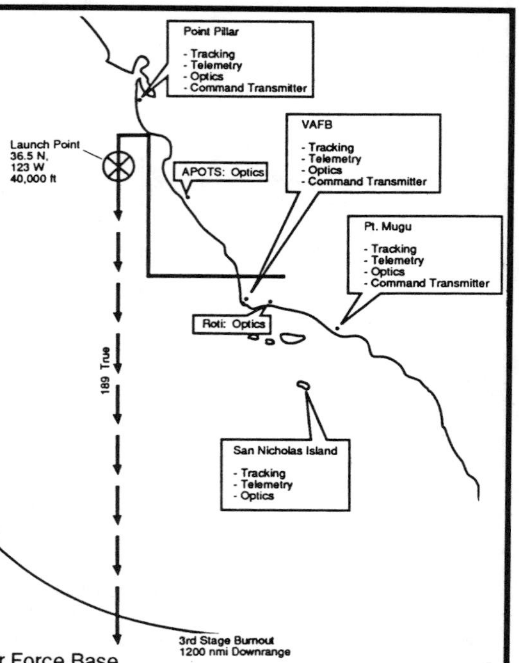

Point Pillar
- Tracking
- Telemetry
- Optics
- Command Transmitter
VAFB
- Tracking
- Telemetry
- Optics
- Command Transmitter
Launch Point
36.5 N,
123 W
40,000 ft
APOTS: Optics
Pt. Mugu
- Tracking
- Telemetry
- Optics
- Command Transmitter
Roti: Optics
189 True
San Nicholas Island
- Tracking
- Telemetry
- Optics
VAFB 40,000 ft Altitude
Pacific Ocean
3rd Stage Burnout
1200 nmi Downrange

Typical launch coverage available from Vandenberg Air Force Base

Taurus

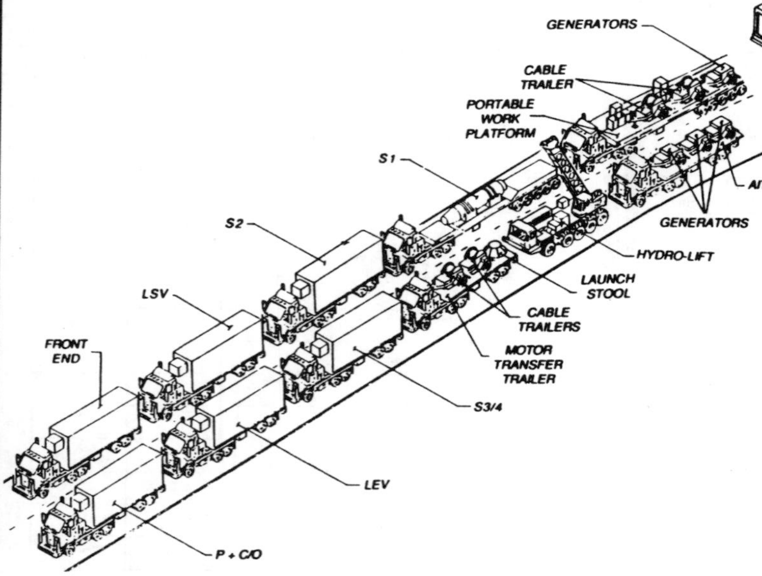

GENERATORS
CABLE TRAILER
PORTABLE WORK PLATFORM
S1
AIT
GENERATORS
S2
HYDRO-LIFT
LAUNCH STOOL
LSV
CABLE TRAILERS
MOTOR TRANSFER TRAILER
FRONT END
S3/4
LEV
P + C/O

Vehicle and launch support equipment

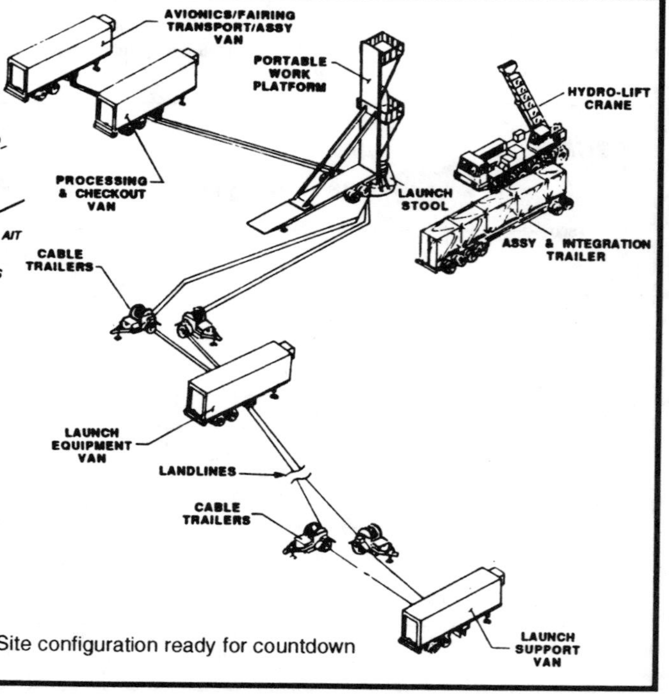

AVIONICS/FAIRING TRANSPORT/ASSY VAN
PORTABLE WORK PLATFORM
HYDRO-LIFT CRANE
PROCESSING & CHECKOUT VAN
LAUNCH STOOL
ASSY & INTEGRATION TRAILER
CABLE TRAILERS
LAUNCH EQUIPMENT VAN
LANDLINES
CABLE TRAILERS
LAUNCH SUPPORT VAN

Site configuration ready for countdown

Pegasus / Taurus Operations

(continued)

Launch Processing

Pegasus: Pegasus was designed to simplify and minimize field integration effort, facilities, and equipment. Build up of Pegasus begins with the delivery to the integration site of the solid rocket motor sections. Motors are shipped in standard ordnance transportation vans (TARVANs) on custom designed handling dollies. The motor sections remain on these handling dollies throughout the integration process which eliminates the need for lifting motors in the field. Upon arrival, stages are removed from the TARVANs and placed on a custom designed multifunction Assembly and Integration Trailer (AIT). The AIT has integral lifting jacks which allow it to be elevated to TARVAN bed height and the motor sections offloaded directly onto the AIT bed. The AIT is then lowered to floor level for vehicle integration. The avionics subsystem is delivered to the field completely integrated, acceptance tested, and ready for integration with the third stage motor. The wing, fins, and payload fairing are received with all thermal protection and instrumentation installed. Once the vehicle has been integrated and tested the payload is mated and the fairing installed. The AIT is then used to transport Pegasus to the carrier aircraft, elevate it, and align it for mating. The combined AIT and custom dolly system provides full six-degree-of-freedom movement capability for the finished vehicle. A portable air conditioning unit provides filtered conditioned air for the payload and avionics while the vehicle is being mated prior to takeoff.

No launch pad is required for Pegasus—initially only a 12,000 ft (3.7 km) by 200 ft (61 m) runway. (Commercial transport aircraft that could potentially carry the Pegasus can operate from shorter, narrower runways). Thus, no launch pad costs or launch pad bottlenecks are anticipated with the exception of the maintenance and availability of the carrier aircraft. In addition, because launch is initiated well off shore, range safety requirements and associated concerns may be reduced.

During the initial series of development flight, the NASA Dryden Flight Research Facility NB-52-008—one of two modified B-52 aircraft used during the X-15 rocket research plan program—will be the carrier aircraft, flown by a three-person crew. DARPA has negotiated the necessary agreements to lease this vehicle from NASA Dryden Flight Research Facility (DFRF) for the first six launches. This particular B-52 aircraft has performed nearly 450 drops, including test flights for the X-15 (similar in size and shape to the Pegasus vehicle). Generic compatibility (i.e, maximum gross weight) does exist with other models of B-52 aircraft, including the C, G, and H series, as well as with a variety of wide-body commercial aircraft.

Launch begins with the release of Pegasus from the carrier aircraft at 38,000 ft (11.6 km) altitude and 0.79 Mach. First stage ignition occurs five second after release from the carrier aircraft after Pegasus has fallen approximately 300 ft (91 m).

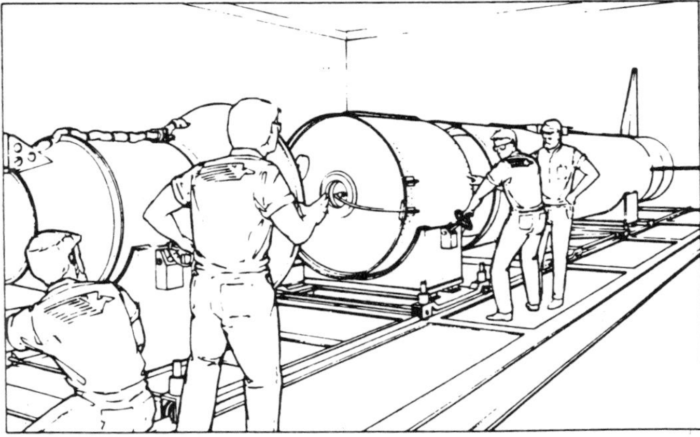

Pegasus Assembly and Test

Pegasus Payload Integration

T-14 days:	Motor Set, Wing, Fins, and Other Subassemblies Arrive "Just In Time"
T-9 days:	S1/Aft Skirt, S2/S3/AS Integrated and Tested. Harnesses Installed
T-6 days:	Vehicle Stacked. Wing Mated, N2 Tank Filled. Payload Arrives.
T-5 days:	Systems Checks Completed. Payload Checks Completed.
T-4 days:	Batteries Activated, Squibs Installed. Clean Tent Activated. Payload Installation Begins.
T-1 day:	Payload Installed. N2 Tank Topped Off. Fairing Closed.
T-12 hrs:	Trailer Airpack Activated.
T-8 hrs:	Carrier Aircraft Call-Up. Mating Operations Begin.
T-4 hrs:	Mating Operations Complete; Pegasus, Payload On B-52 Power. Reference IMU Initialized.
T-3 hrs:	B-52 Crew Entry, Engine Starts, Taxi.
T-2 hrs:	B-52 Takeoff.
T-45 min:	Operational Altitude, Drop Zone Reached.
T-20 min:	Mission Data Load Down-Loaded and Verified.
T-10 min:	IMU Alignment S-Turns Completed. Pegasus and Payload on Internal Power. Last Systems Checkout.
T + 0 sec:	Pegasus Release; Sequence Begins; FTS Armed.
T + 1 sec:	Autopilot Enabled. Safe and Arms Rotated to Arm.
T + 5 sec:	S1 Ignition. Pitch-Up Initiated.

Pegasus Preparation Timeline

Pegasus / Taurus Operations

(continued)

Launch Processing

(continued)

Taurus: A complement of road (also military aircraft) transportable Launch Support Equipment (LSE) is used to establish a launch site on a dry pad and platform Taurus launch operations. Major elements of LSE include motor and motor transfer trailers, assembly and integration trailer, launch equipment van, launch support van processing and checkout van, hydrolift crane and equipment, cable trailers, generators, launch stool and portable work platforms. Most of the LSE is derived from equipment used on the current Air Force Peacekeeper, Starbird, and Pegasus programs.

After arrival at the launch site, the first day is spent unloading LSE and positioning equipment on the launch site. During the second day, the launch stool is bolted on the dry pad, cables are connected, all consoles are checked out functionally and Stage 1 is moved to the transfer trailer. Day three begins with the attachment of the Stage 1/Stage 2 interstage lower section to the top of Stage 1. The combination is then erected onto the launch stool and covered, Finally, the Stage 1/Stage 2 interstage upper section is mounted to Stage 2 and the combination is rolled onto the covered air borne integration trailer.

Day four see Stage 2, 3, and 4 rolled onto the covered Assembly and Integration Trailor (AIT), aligned and mated. Functional checks are performed on the vehicle avionics. On day five, the lifting sling is attached to the Stage 2, 3, 4 stack and integrated system tests are completed. Finally, precountdown range, LSE, and vehicle checks are performed. The vehicle can now be held in this condition for a long period awaiting a launch command.

When the launch command is given, the 72 hr launch processing flow begins with the integration of the payload to the Stage 2, 3, 4 stack and performance of integrated tests. Next, umbilicals are connected, the payload fairing is attached and system level checks are performed. Following successful checks, the hydrolift crane lifts the Stage 2, 3, 4 and payload stack and it is mated to the already erected Stage 1. Finally, the work platforms are secured and the countdown begins for launch.

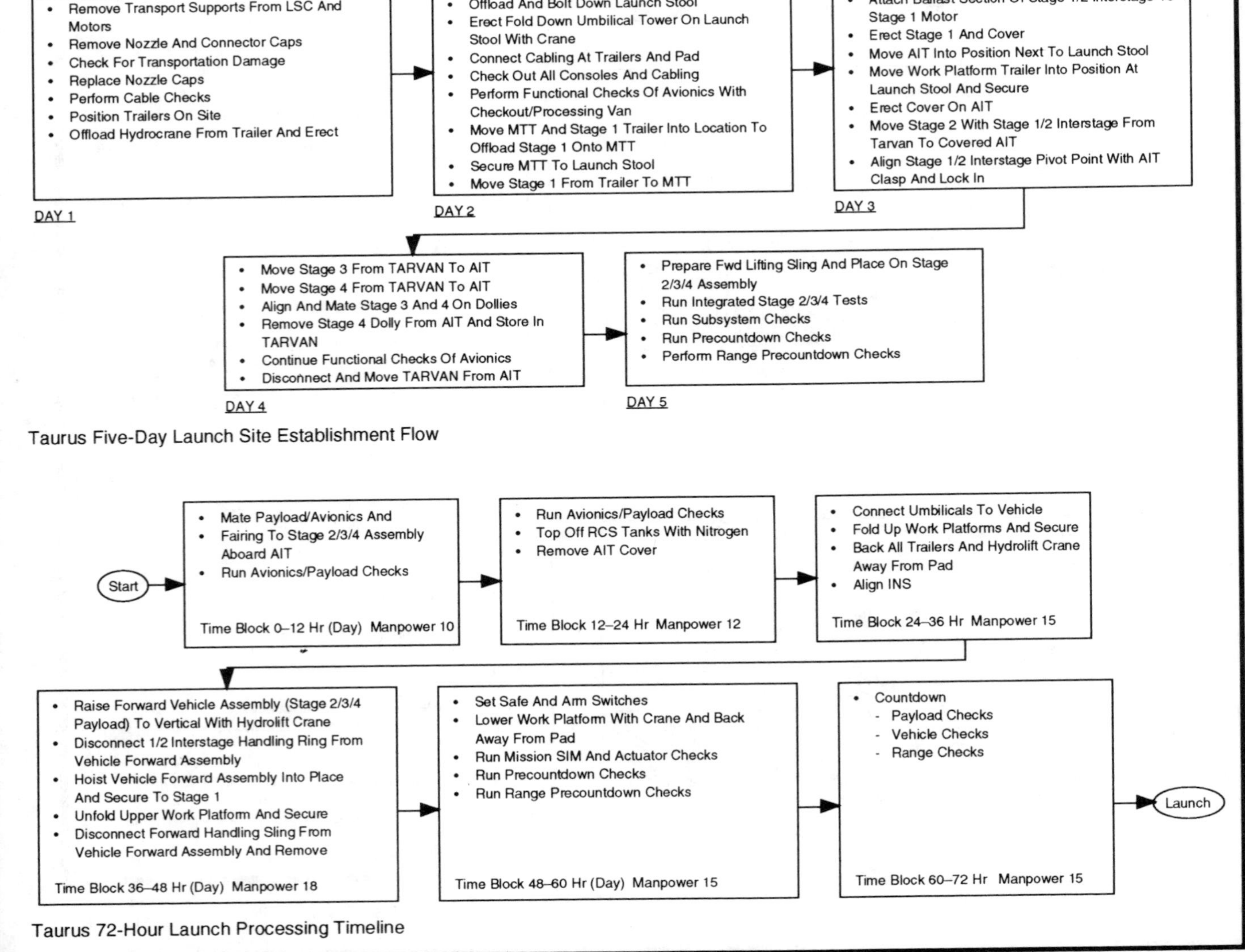

Taurus Five-Day Launch Site Establishment Flow

Taurus 72-Hour Launch Processing Timeline

Pegasus / Taurus Operations

(continued)

Flight Sequence

Pegasus Typical Flight Sequence

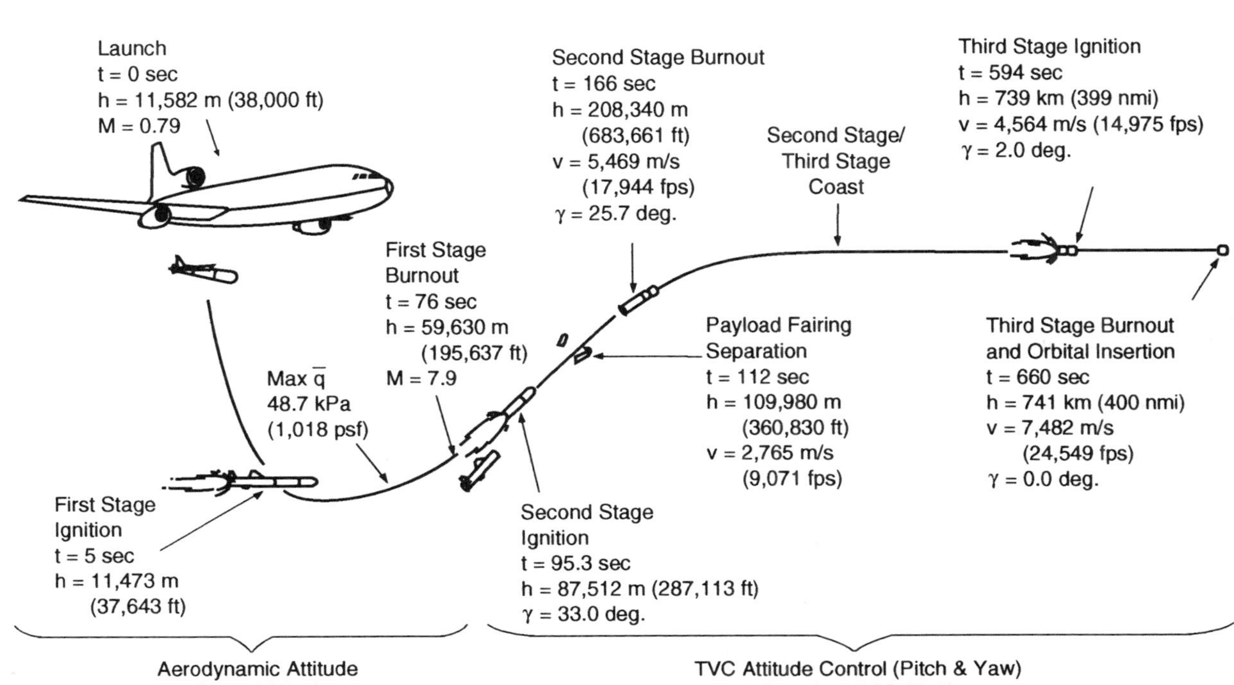

Taurus Typical Flight Sequence

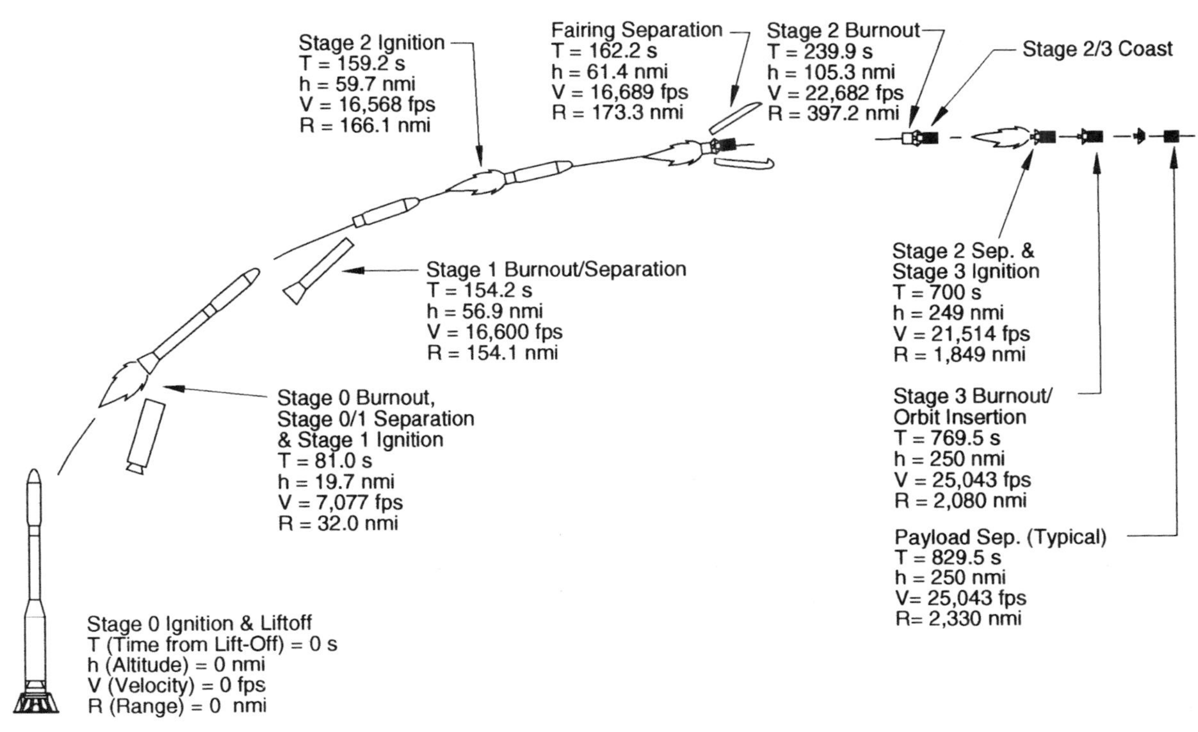

Pegasus / Taurus Payload Accommodations

	Pegasus	Taurus
Payload Compartment		
Maximum Payload Diameter	44.0 in (1118 mm)	54.0 in (1372 mm)
Maximum Cylinder Length	43.72 in (1110 mm)	110.0 in (2794 mm)
Maximum Cone Length	40.0 in (1016 mm)	20.0 in (508 mm)
Payload Adapter		
Interface Diameter	38.81 in (985.8 mm)	38.81 in (985.8 mm)
Payload Integration		
Nominal Mission Schedule Begins	T-104 weeks	T-80 weeks
Launch Window		
Latest Countdown Hold Not Requiring Recycling	T-?? min	T-?? min
On-Pad Storage Capability	Indefinite	Indefinite
Latest Access to Payload	T-24 hours	T-?? hours
Environment		
Maximum Load Factors	+13 g axial, ±6 g lateral	+11 g axial
Minimum Lateral / Longitudinal Payload Frequency	?? Hz / ?? Hz	?? Hz / ?? Hz
Maximum Overall Acoustic Level	133.5 dB from carrier aircraft 117 dB free flight	141 dB (one third octave)
Maximum Flight Shock	800 g from 1,000–10,000 hz	1,100 g from 1,500–10,000 hz
Maximum Dynamic Pressure on Fairing	?? lb/ft^2 (?? N/m^2)	?? lb/ft^2 (?? N/m^2)
Maximum Pressure Change in Fairing	?? psi/s (?? KPa/s)	?? psi/s (?? KPa/s)
Cleanliness Level in Fairing (Prior to Launch)	Class ??	Class ??
Payload Delivery		
Standard Orbit and Accuracy (3 sigma)	For 270 nm (500 km) circular polar orbit: ±60 nmi (111 km), ±0.2° inclination With hydrazine precision injection kit: ±15 nm (28 km), ±0.05° inclination	Taurus injection accuracies (95°, probability): ±30 nmi (55 km), ±0.2° inclination
Attitude Accuracy (3 sigma)	±4°	±??°
Nominal Payload Separation Rate	?? ft/s (?? m/s)	?? ft/s (?? m/s)
Deployment Rotation Rate Available	Up to 100 rpm	Up to 100 rpm
Loiter Duration in Orbit	40 min	40 min
Maneuvers (Thermal / Collision / Telemetry)	Yes	Yes

Pegasus / Taurus Notes

Publications

User's Guide

NASA SELVs Pegasus Launch System Payload Users Guide, Release 2.00, Orbital Sciences Corporation, June 1994.

Commercial Pegasus Launch System Payload Users Guide, Release 3.00, Orbital Sciences Corporation, October 1, 1993.

Commercial Taurus Launch System Payload Users Guide, Release 1.00, Orbital Sciences Corporation, August 10, 1992.

Technical Publications

"Pegasus, Taurus, and Glimpses of the Future", C. Schade, Orbital Sciences Corporation, AIAA 90-3573, AIAA Space Programs and Technologies Conference, Huntsville, AL, September 25-28, 1990.

"A Historical Look at United States Launch Vehicles: 1967-Present", ANSER, STDN 90-4, second edition, February 1990.

"Payload Interface Guide for The Pegasus Air-Launched Space Booster", by M. Mosier and G. Harris, Orbital Sciences Corporation, 1989.

"Design and Development of Pegasus Propulsion", by H. Carroll, J. Godfrey and J. Crum, Hercules Missiles and Space Group, AIAA 89-2314, AIAA 25th Joint Propulsion Conference, Monterey, CA, July 10-12, 1989.

Acronyms

AIT - Assembly and Integration Trailer
ASE - Airborne Support Equipment
CCAFS - Cape Canaveral Air Force Station
DARPA - Defense Advanced Research Projects Agency
DFRF - Dryden Flight Research Facility
EMA - electromechanical actuators
FTS - flight termination system
GTO - geosynchronous transfer orbit
HTPB - hydroxy terminated polybutadiene
IMU - inertial measurement unit
INS - inertial navigation system
LEO - low Earth orbit
LEV - Launch Equipment Van
LSE - Launch Support Equipment
LSV - Launch Support Van
MTT - Motor Transfer Trailer
NASA - National Aeronautics and Space Administration
N2 - nitrogen
OSC - Orbital Sciences Corporation
P+C/O - processing plus checkout
RCS - reaction control system
S1 - Stage 1
S2 - Stage 2
S3/4 - Stage 3 and 4
SSLV - Standard Small Launch Vehicle
TARVAN - transportation van
TVC - thrust vector control
VAFB - Vandenberg Air Force Base

Pegasus / Taurus Notes

(continued)

Other Notes - Taurus Growth Possibilities

Currently proposed is the Taurus II vehicle, a growth vehicle which makes use of vehicle elements and operational techniques developed on the Taurus I program to satisfy the launch needs of payloads between Taurus I and Delta II. The basic Taurus II is a three stage vehicle using the Thiokol Castor 120 as both the first and second stages. A new storable liquid bipropellant third stage, developed by a joint venture of Atlantic Research Corporation (ARC) and Deutsche Aerospace (DASA) couples with a 10 ft (3 m) composite fairing to complete the basic core vehicle configuration. Up to eight Thiokol Castor IVA strap-on solid rocket motors can be added to the first stage to tailor vehicle performance. With no strap-ons, lift capability is roughly equivalent to the Air Force's Titan II space launch vehicle; with all eight strap-ons, performance is comparable to the Delta II. Based on engines developed for the Ariane-5 launch vehicle, the liquid third stage is capable of several restarts, allowing Taurus II to deliver multiple payloads (such as LEO communications satellites) to different orbits on the same mission. A spin-stabilized solid-propellant upper stage, such as the STAR 37 or STAR 48, can also be added as a perigee kick motor for missions to GTO.

Summary	Designed to flexibly serve the payload range between Taurus I and Delta II, the Taurus II vehicle uses components and operational techniques developed on the Taurus I vehicle to provide a cost-effective commercial launch system. Taurus II is a three stage vehicle with solid-propellant first and second stages and a liquid third stage. Up to eight solid-propellant strap-on boosters can be added to the first stage
Status	In Development
Key Organizations	User - commercial Launch Service Agency - OSC Launch Systems Group Prime Contractor OSC (System Integration, Avionics, Structures, Launch Operations) Principal Subcontractors Thiokol (Stage 1 and 2 Solid Rocket Motors) ARC/DASA (Liquid Stage 3) Litton (INS) OR (Flight Computer) Allied Signal (TVC)
Vehicle	
System Height	100 ft (30.5 m)
Payload Fairing Size	10 ft (3.0m) diameter by 14.53 ft (4.43 m) height
Gross Mass	TII-0 - 275,000 lb (125,000 kg) TII-8 - 275,000 lb (125,000 kg)
Vehicle (continued)	
Planned Enhancements	None
Primary Missions	LEO, GTO, Planetary
Compatible Upper Stage	STAR 37, STAR 48
First Launch	??
Success/Flight Total	0/0
Launch Site	VAFB - SLC-6 (34.7° N, 120.6° W) CCAFS - LC-46 (28.5° N - 81.0° W)
Launch Azimuth	57°–112° (LC-46) 147°– 270° (SLC-6)
Nominal Flight Rate	Up to 6 / yr
Planned Enhancements	Launch operations from non-CONUS sites
Performance	
108 nm (200 km) circ, 28°	TII-0 - 5,000 lb (2,300 kg) TII-8 - 10,100 lb (4,600 kg)
108 nm (200 km) circ, 90°	TII-0 - 3,800 lb (1,700 kg) TII-8 - 8,700 lb (4,000 kg)
Geotransfer Orbit, 28°	TII-8 - 3,850 lb (1,750 kg) with PKM
Geosynchronous Orbit	—
Financial Status	
Estimated Launch Price	TII-0 - <$30M (OSC) TII-8 - <$40M (OSC)

Space Shuttle

Government Point of Contact:
Office of Space Flight
National Aeronautics and Space Administration
Washington, DC 20546, USA
Phone: (202) 453-1000

Industry Point of Contact:
Rockwell International
Space Transportation Systems Division
12214 Lakewood Blvd
Downey, CA 90241, USA
Phone: (213) 922-3344

About to clear the launch tower, NASA's Space Shuttle Columbia heads for orbit from Kennedy Space Center Launch Complex 39A. This is the first launch with the unpainted external tank.

Space Shuttle History

Orbiters Out of Service

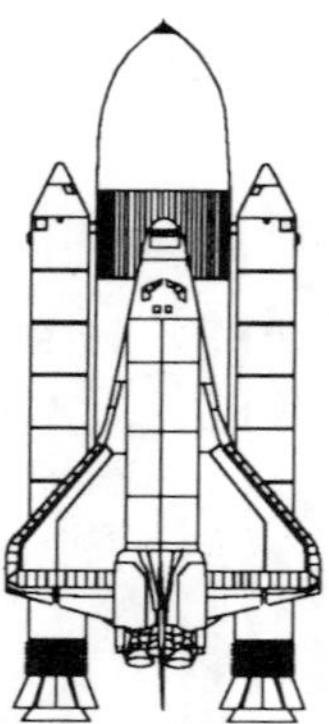

OV-101
Enterprise
(Nonspace Test Vehicle)

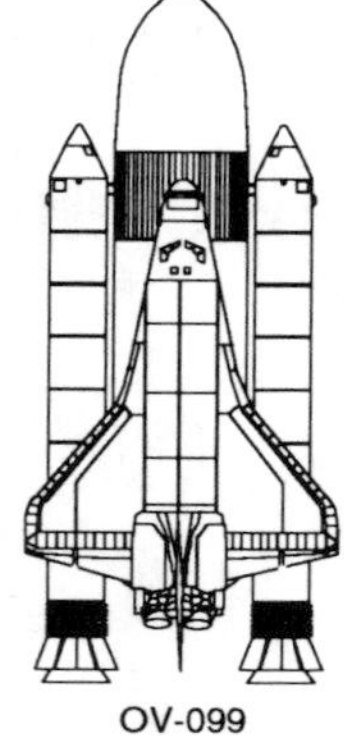

OV-099
Challenger
1983
(Destroyed in 1986)

Orbiters In Service

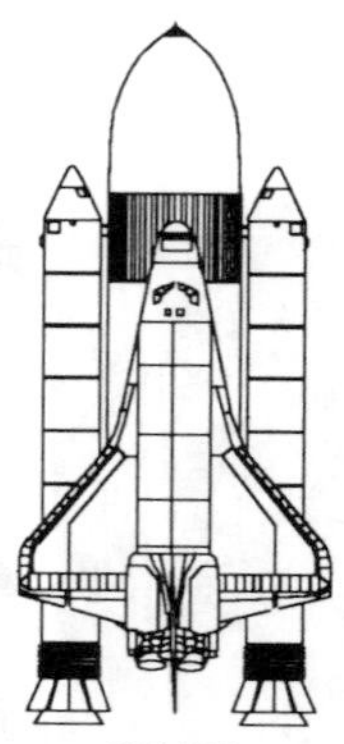

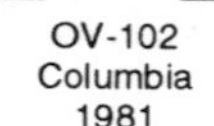

OV-102
Columbia
1981

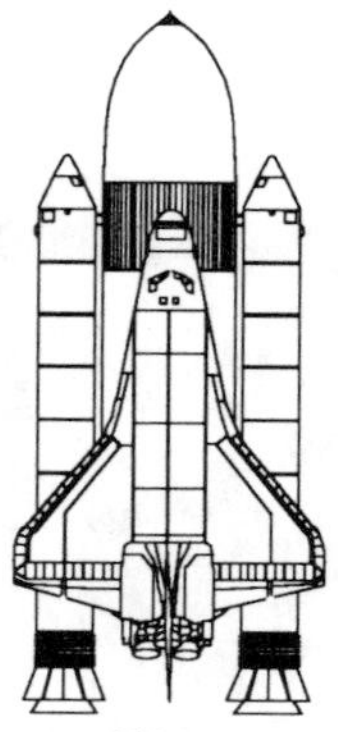

OV-103
Discovery
1984

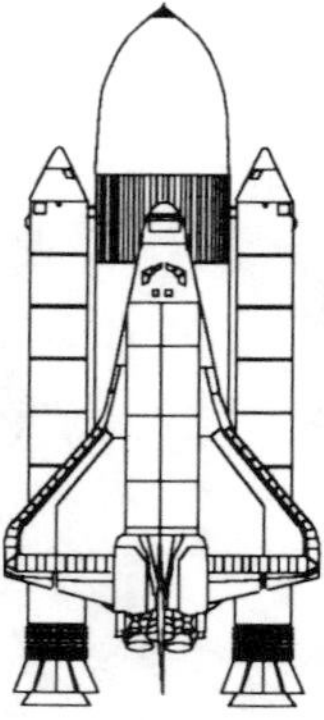

OV-104
Atlantis
1985

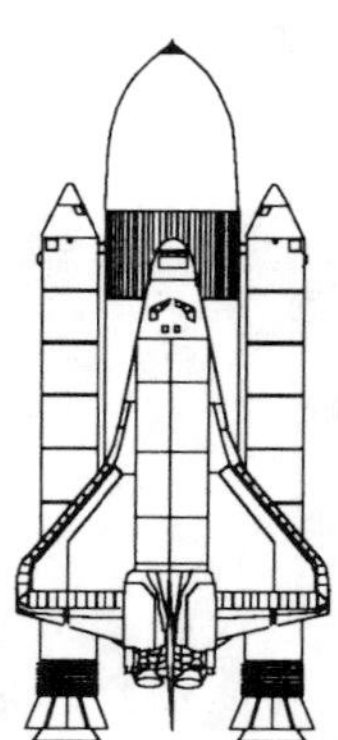

OV-105
Endeavour
1992

Vehicle Description

Enterprise	Test orbiter only. Flew atop a 747 to demonstrate atmospheric and landing tests. Also used for practice and verification fit check tool at launch sites.
Challenger	Originally a structural test article. Second operational orbiter. Destroyed in 1986 flight accident.
Columbia	First orbiter flown. 8,200 lbs (3,720 kg) heavier than other orbiters. Has extended duration orbiter (EDO) capability extending mission up to 16 days.
Discovery	Third operational orbiter.
Atlantis	Fourth operational orbiter.
Endeavour	Replacement for Challenger. Has extended duration orbiter (EDO) capability extending mission up to 16 days.

Historical Summary

In the early 1960s, virtually all U.S. aerospace companies conducted studies of recoverable space boosters. The experiments, studies and developmental efforts (such as the Dynasoar and the X-series of rocket-powered aircraft) conducted during this time explored the technological possibilities for an aerospace vehicle which would combine the performance of the expendable vehicles with reusability. These efforts supplied the background for studies in the late 1960s on reusable space transportation vehicle configurations.

In January 1969, the National Aeronautics and Space Administration (NASA) initiated four six-month Phase A feasibility study contracts on reusable vehicle concepts, called at that time ILRV (Integrated Launch and Reentry Vehicles). General Dynamics, Lockheed, McDonnell Douglas, and North American Rockwell (now Rockwell International) each received one of the $500,000 contracts. Though NASA had originally favored a stage-and-a-half design with external propellant tanks, the Manned Spacecraft Center (now the Johnson Space Center) began a study of two-stage, fully reusable concepts. By July the four contractor studies were oriented toward such a design, and when completed in November, the four studies emphasized the two-stage, fully reusable concept.

NASA established a Space Shuttle Task Group and this group reported to NASA in May 1969. Its report stated that the space shuttle system development should not be considered in terms of probability but rather in terms of essentiality for future space operations. The preferred concept was a fully or near-fully reusable system capable of controlled runway landings.

In March 1970, President Nixon redefined U.S. space goals. The president made it clear that, while the space program should not be allowed to stagnate, it had to be balanced with the critical problems facing the U.S. here on Earth. Proposed post-Apollo lunar missions and manned expeditions to Mars could not be mounted at that time. The administration recommended a NASA budget for fiscal year 1971 that was less than the budget for fiscal year 1970.

NASA revised its near-term goals. The post-Apollo civilian space program would now include only the space station and the Space Shuttle. NASA planned for tandem development of the Shuttle and the space station. The dual program was to cost about $5 billion each for the space station and Shuttle. When it became clear that concurrent development would not be funded, NASA was forced to choose between the two programs. Since it was not reasonable to build the space station without a low-cost supply system, NASA's only logical choice was to develop the Shuttle as the major program for the 1970s and to postpone the space station until after Shuttle expenditures peaked.

The Space Shuttle, as originally envisioned, was to be a fully reusable, two-stage vehicle. The booster stage was to be the size of a Boeing 747 and the orbital stage about the size of a Boeing 707. Both stages were to be rocket-powered, burning hydrogen and oxygen carried in internal fuel tanks. The two stages would be attached in parallel for a vertical takeoff. After launch, the booster would fly back to the launch site for a horizontal landing and would be refurbished for the next flight. The orbital stage would proceed to orbit and, upon completing its mission, return to Earth and land horizontally.

Historical Summary
(continued)

In July 1970, NASA awarded Phase B detailed design contracts for the Space Shuttle to North American Rockwell and McDonnell Douglas. The Shuttle was to be capable of placing a payload of 25,000 lb (11,340 kg) in a 240 nm (445 km), 55 degree orbit. Cross-range capability was to be 200–1500 nm (370–2,780 km). Both straight winged and delta winged designs were studied.

In a parallel effort, NASA conducted an inhouse review of the Shuttle program as well as contracting for three extended Phase A feasibility studies on alternate Shuttle concepts. These contracts went to Grumman/Boeing, Lockheed, and Chrysler. Though these studies pursued concepts which did not prove viable, they did impact the ultimate design of the orbiter by influencing design concepts and philosophies.

In January 1971, NASA changed the requirements for the Shuttle. The Shuttle now had a required capability to put 65,000 lb (29,500 kg) of payload into a 100 nm (185 km) due east orbit, 40,000 lb (18,150 kg) into a 55 degree orbit and 25,000 lb (11,340 kg) into a 277 nm (513 km) polar orbit. The projected development cost of this configuration was estimated to be approximately $9.9 billion.

Economic factors were important in reaching the decision on the final Shuttle configuration. As the Shuttle studies progressed, it became clear that the development would be more expensive than originally projected. The Office of Management and Budget (OMB) asked NASA to do an inhouse cost / benefit analysis of the Shuttle and also to contract out a more detailed, independent study.

NASA's internal cost/benefit study done in response to the OMB request showed a distinct advantage for the Shuttle over exisiting expendable rocket systems, possible future low-cost expendables, and a hybrid Shuttle system of a reusable orbiter, expendable booster. But the study also estimated development costs ranging from $6.4 billion to $9.6 billion (FY71 $).

Mathematica, Inc. of Princeton, NJ, estimated development costs of about $12.8 billion for the two-stage reusable Shuttle. In its final report, in which Mathematica considered a number of different possibilities for a space transportation system as well as different configurations of the Shuttle, Mathematica concluded, "...that the development of a thrust-assisted-orbiter Shuttle system (TAOS) was justified, within a level of space activities between 300 and 360 Shuttle flights in the 1979–1990 period...."

Political considerations also affected the development of the Shuttle. Development of the fully reusable Shuttle would have been technically difficult, which meant that (1) it would have been expensive to build and (2) it would have been opposed by many members of Congress and questioned by the administration due to the uncertainty of the required technology for the manned booster.

Opposition to the Shuttle in the Senate was reinforced by the opposition to the project by some of the scientific community. Some scientists believed that the anticipated mission model and payload savings projections would not materialize and that launch costs would be significantly higher than anticipated. Others believed that there would be extensive cost overruns and that a civilian space program dominated by manned missions would not be cost effective. Some critics attacked NASA's changing justification for the Shuttle (from a vehicle necessary to provide logistical support for large-scale manned space efforts to a vehicle which would revolutionize space activity by its availability as a utilitarian, cost effective transportation system for civilian space science and applications and for national defense missions). Still others fought the Shuttle on the basis that the Department of Defense (DoD), as the projected largest single user, should help finance the development.

The opposition to the Shuttle peaked in 1970 when an amendment to delete funds for the Shuttle project was defeated in the Senate by only four votes. Despite intense debate and continuing opposition in the Senate, the Congress did repeatedly vote for funding for the Shuttle program, though perhaps not at the level viewed as necessary by some proponents of the program.

The administration established a December 1971 deadline for final endorsement or cancellation of the Shuttle program. The administration, NASA, and congressional proponents of the Shuttle felt compelling pressure to make a decision by the publication of the fiscal 1973 budget in January 1972. They feared that opponents of the Shuttle would be able to defeat the Shuttle in Congress if it were to be deferred any longer. Also, further deferment of the Shuttle might have made it impossible to hold together the industrial teams which had been brought together.

Alternate booster concepts were studied to achieve a less expensive design for the Shuttle. NASA officially decided to scrap the fully reusable Shuttle design it had worked on for 18 months and find a new, more cost-effective Shuttle concept. NASA ultimately chose to pursue a thrust-assisted-orbiter Shuttle system (TAOS). The new Shuttle configuration involved less technological risk. The TAOS consisted of a manned orbiter vehicle, an expendable external propellant tank, and two recoverable solid fueled rockets. The only non-reusable part would be the external tank.

On January 5, 1972, President Nixon endorsed the Shuttle program based on the thrust-assisted orbiter concept and requested the development of the space transportation system to begin at once. The Shuttle, he noted, would enable the United States to achieve a working presence in space by making space transportation available routinely and by reducing costs and preparation time.

Having received the go-ahead, NASA was ready to begin acquisition and development of the Shuttle elements. Acquisition for the Space Shuttle was competitive. Separate contractors were selected for the design and manufacture of the orbiter, its main engines, the external tank and the solid rocket boosters. Rockwell International was selected as prime integrating contractor. A protest of the main engine contract award resulted in a one-year delay to the engine development effort.

By the end of 1972, estimated development costs for the Shuttle were $5.15 billion (FY71 $) and the estimated cost per flight was $10.5 million (FY71 $). Initial launch was anticipated for 1978 and the flight rate was projected to be up to 50 flights per year. By 1980, estimates of the development costs increased 20% to $6.2B (FY71 $), the cost per launch increased to $15.2 million (FY71 $), the initial launch was slipped to 1981, and the projected flight rate was reduced to 24 flights per year. These changes were due in part to fiscal constraints early in the development program and to schedule slips in the engine and thermal protection system development. Despite all these challenges, the first Shuttle launch occurred on April 12, 1981, and was spectacularly successful.

On January 28, 1986, the 25th Shuttle mission and the 10th for the orbiter Challenger ended in disaster 70 seconds into the flight when a burn-through of a Solid Rocket Booster (SRB) o-ring resulted in the rupturing of the external tank and the subsequent break up of the orbiter.

In addition to replacing the Challenger with a new orbiter (Endeavour), NASA implemented a comprehensive return-to-flight strategy. During the 32-month recovery period, 10 major reviews and analyses were undertaken for the Shuttle program, including Shuttle hardware, operations, and organization. Numerous improvements were made to the orbiter, including a new crew escape system, enhanced landing and deceleration systems, safety modifications to the gaseous oxygen flow control valve, and the redesign of the

Space Shuttle History

(continued)

Historical Summary

(continued)

orbiter's 17 in. (432 mm) quick disconnect valve. Also, the Space Shuttle Main Engine (SSME) underwent an extensive ground test program to recertify the engine and demonstrate its reliability through testing at and beyond its original design limits. Upgrades were also made to the flight computers and the orbiter braking systems. Finally, the SRBs were significantly redesigned and thoroughly tested.

As a part of the long-term systematic improvement approach, a number of key improvements were undertaken, including orbiter modifications to enable extended duration missions and refinements to extend the service life of the SSME. The nominal flight rate is 8 flights per year with an annual maximum of 12 flights, with a fleet of four orbiters. Currently only those payloads that require a manned presence or the unique capabilities of the Shuttle are allowed to be manifested on the Shuttle.

As a result of the design changes and safety modifications made during the return-to-flight effort, the payload capability of the Shuttle decreased from a maximum of 61,400 lbs (27,900 kg) due east to approximately 53,700 lbs (24,400 kg). The west coast launch and landing facility for Shuttle operations, developed by the DoD at Vandenberg AFB for launches into polar orbit, was placed in mothball status and Shuttle launches out of Vandenberg AFB were eliminated. A NASA program to develop and use a modified Centaur upper stage to lift 10,000 lbs (4,540 kg) to geosynchronous orbit was cancelled because of safety concerns. NASA had proposed an Advanced Solid Rocket Motor (ASRM) to increase payload capability approximately 8,000 lbs (3,630 kg). The program was funded for a time but was later cancelled.

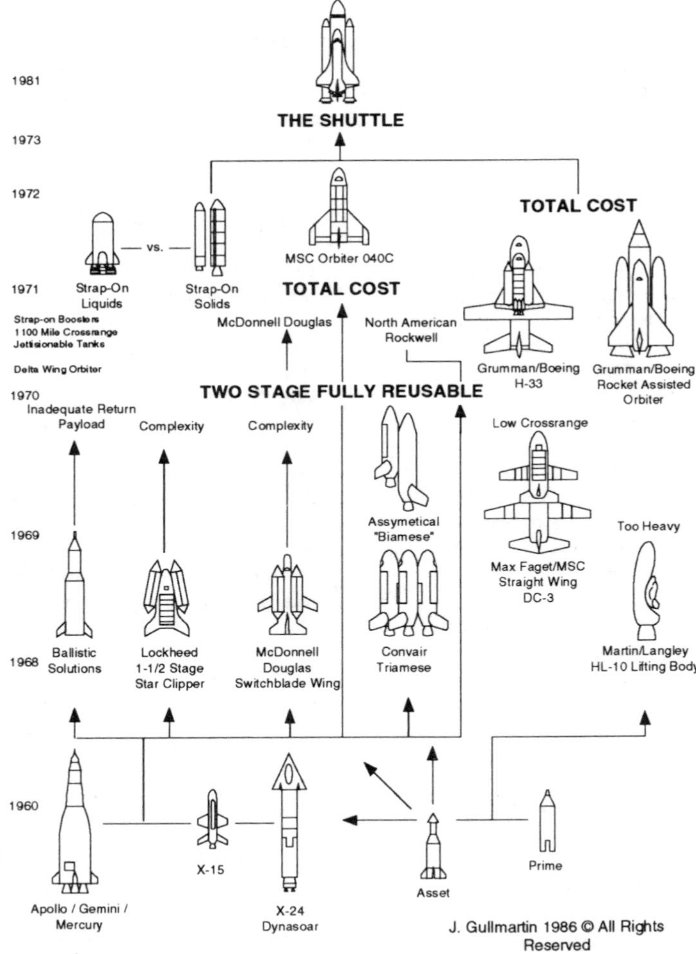

Historical design concepts for the Space Shuttle

Space Shuttle History
(continued)

Launch Record

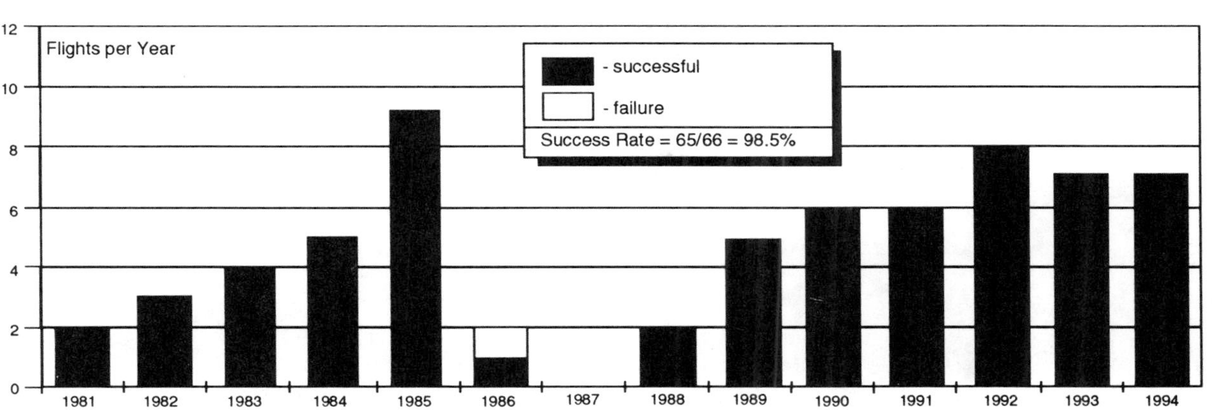

	YEAR	DATE	VEHICLE	ORBITER	SITE	PAYLOAD*
1	1981	Apr-12	STS-1 [T]	Columbia	KSC	Development Flight
						Instrumentation (DFI)
2		Nov-12	STS-2 [T]	Columbia	KSC	OSTA-1
						DFI
3	1982	Mar-22	STS-3 [T]	Columbia	KSC	OSS-1
						DFI
4		Jun-27	STS-4 [T]	Columbia	KSC	DOD
						DFI
5		Nov-11	STS-5	Columbia	KSC	SBS-C
						Telesat-E
6	1983	Apr-4	STS-6	Challenger	KSC	TDRS-A / IUS
7		Jun-18	STS-7	Challenger	KSC	Telesat-F / PAM-D
						Palapa B-1 / PAM-D
						SPAS-01
						OSTA-2
8		Aug-30	STS-8	Challenger	KSC	Insat 1-B / PAM-D
9		Nov-28	STS-9	Columbia	KSC	Spacelab 1
10	1984	Feb-3	41-B	Challenger	KSC	SPAS-01A
						Palapa-B-2 / PAM-D
						Westar-VI / PAM-D
11		Apr-6	41-C	Challenger	KSC	LDEF-1
						Solar Max Repair
						OAST-1
12		Aug-30	41-D	Discovery	KSC	SBS-D / PAM-D
						Telestar 3C / PAM-D
						Syncom IV-2 / Integral
13		Oct-5	41G	Challenger	KSC	ERBS
						OSTA-3
14		Nov-8	51-A	Discovery	KSC	Syncom IV-1 / Integral
						Telesat-H / PAM-D
						Retrieve Palapa / Westar
15	1985	Jan-24	51-C	Discovery	KSC	DOD / IUS
16		Apr-12	51D	Discovery	KSC	Syncom IV-3 / Integral
						Telesat-I / PAM-D
17		Apr-29	51-B	Challenger	KSC	Spacelab-3
						Nusat
18		Jun-17	51-G	Discovery	KSC	Arabsat-A / PAM-D
						Telstar-3D / PAM-D
						Morelos-A / PAM-D
19		Jul-29	51-F	Challenger	KSC	Spacelab-2
20		Aug-27	51-I	Discovery	KSC	Syncom IV-4 / Integral
						ASC-1 / PAM-D
						Aussat-1 / PAM-D
21		Oct-3	51-J	Atlantis	KSC	DOD
22		Oct-30	61-A	Challenger	KSC	Spacelab-D1
						GLOMR
23		Nov-26	61-B	Atlantis	KSC	Morelos-B / PAM-D
						Satcom Ku-2 / PAM-D2
						Aussat-1 / PAM-D
24	1986	Jan-12	61-C	Columbia	KSC	Satcom Ku-1 / PAM-D2
25		Jan-28	51-L	Challenger	KSC	TDRS-B / IUS
		Failure - SRB O-Ring Burnthrough				
26	1988	Sep-29	STS-26	Discovery	KSC	TDRS-C / IUS
27		Dec-2	STS-27	Atlantis	KSC	DOD

	YEAR	DATE	VEHICLE	ORBITER	SITE	PAYLOAD*
28	1989	Mar-13	STS-29	Discovery	KSC	TDRS-D / IUS
29		May-4	STS-30	Atlantis	KSC	Magellan / IUS
30		Aug-8	STS-28	Columbia	KSC	DOD
31		Oct-18	STS-34	Atlantis	KSC	Galileo / IUS
32		Nov-22	STS-33	Discovery	KSC	DOD
33	1990	Jan-9	STS-32	Columbia	KSC	Syncom IV-F5 / Integral
						LDEF Grapple
34		Feb-28	STS-36	Atlantis	KSC	DOD
35		Apr-24	STS-31	Discovery	KSC	Hubble Telescope
36		Oct-6	STS-41	Discovery	KSC	Ulysses
37		Nov-15	STS-38	Atlantis	KSC	DOD
38		Dec-2	STS-35	Columbia	KSC	Astro-01
						BBXRT-01
39	1991	Apr-5	STS-37	Atlantis	KSC	GRO
40		Apr-28	STS-39	Discovery	KSC	IBSS
						MPEC
41		Jun-5	STS-40	Columbia	KSC	SLS-1
42		Aug-2	STS-43	Atlantis	KSC	TDRS-05
43		Sep-12	STS-48	Discovery	KSC	UARS
44		Nov-24	STS-44	Atlantis	KSC	DSP
						IOCM
45	1992	Jan-22	STS-42	Discovery	KSC	IML-01
						IMAX-05
46		Mar-24	STS-45	Atlantis	KSC	ATLAS-01
47		May-7	STS-49	Endeavour	KSC	Intelsat-VI
48		Jun-25	STS-50	Columbia	KSC	USML-01
49		Jul-31	STS-46	Atlantis	KSC	TSS-01
50		Sep-12	STS-47	Endeavour	KSC	Spacelab-J
51		Oct-22	STS-52	Columbia	KSC	LAGEOS
						USMP-01
52		Dec-2	STS-53	Discovery	KSC	DOD-01
						ODERACS
53	1993	Jan-13	STS-54	Endeavour	KSC	TDRS-06
						DXS
54		Apr-8	STS-56	Discovery	KSC	SPARTAN-201
						ATLAS-02
55		Apr-26	STS-55	Columbia	KSC	Spacelab-D2
56	1993	Jun-21	STS-57	Endeavour	KSC	EURECA
						SpaceHab-01
57		Sept-12	STS-51	Discovery	KSC	ACTS
						ORFEUS
58		Oct-18	STS-58	Columbia	KSC	SLS-02
59		Dec-2	STS-61	Endeavour	KSC	Hubble Telescope Repair
60	1994	Feb-3	STS-60	Discovery	KSC	Wake Shield Facility
						SpaceHab-02
61		Mar-4	STS-62	Columbia	KSC	USMG-02
62		Apr-9	STS-59	Endeavour	KSC	SRL-02
63		Jul-8	STS-65	Columbia	KSC	IML-02
64		Sept-9	STS-64	Discovery	KSC	SAPRTAN-201
						SAFER
						SPIFEX
						ROMPS
65		Sept-30	STS-68	Endeavour	KSC	SRL-02
66		Nov-3	STS-66	Atlantis	KSC	ATLAS-3
						CRISTA-SPAS

* - Only list major or deployable payloads. Department of Defense (DOD) payloads are classified and full launch details are not disclosed. All Space Shuttle launches are manned.

[T] - Test Launch
KSC - Kennedy Space Center, Florida

Space Shuttle General Description

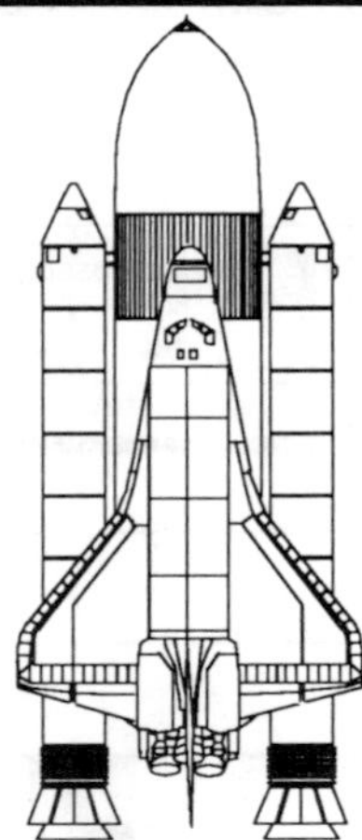

Summary

The Space Shuttle consists of a reusable delta-winged spaceplane called an orbiter; two solid propellant rocket boosters, which are recovered and reused; and an expendable external tank containing liquid propellants for the orbiter's three main engines.

Status — Operational

Key Organizations

User - DoD, NASA
Launch Service Agency - NASA
Prime Contractor - Rockwell International (Orbiter)
Associate Contractors
- Rocketdyne (SSME)
- Martin Marietta (External Tank)
- Thiokol (Solid Rocket Boosters)

Principal Orbiter Subcontractors
- Aerojet (OMS thrusters)
- Marquardt (RCS thrusters)
- Honeywell / IBM (Avionics)
- Fairchild Republic (Vertical tail)
- General Dynamics (Mid fuselage)
- Grumman (Wings)
- McDonnell Douglas (Aft propulsion system)
- United Technologies CSD (Fuel cell)
- Spar Aerospace (Remote manipulator system)

Vehicle

System Height	184.2 ft (56.14 m)
Payload Bay Size	15.5 ft (4.7 m) diameter by 60.9 ft (18.6 m) length
Gross Mass	4,500,000 lb (2,040,000 kg)
Planned Enhancements	Block II SSME, including an advanced turbopump, is planned for increased engine reliability. The Super Light Weight (SLWT) ET, using aluminum-lithium in place of aluminum, is under design and will provide ~8,000 lbs. of additional ascent performance. Various orbiter weight reductions will provide ~1,000 lbs. of increased ascent performance. The light-weight SRB will add ~1,200 lbs. of ascent performance.

Operations

Primary Missions	Easterly LEO with upper stages for other orbits.
Compatible Upper Stages	PAM-D, PAM-D2, IUS, TOS, integral motors
First Launch	1981
Success / Flight Total	65 / 66 (98.5%)
Launch Site	KSC - LC-39A & B (28.5°N, 81.0°W)
Launch Azimuth	LC-39 - 35°–120° (max. inclination is 28.5°– 57°)
Nominal Flight Rate	8 / year
Maximum Flight Rate	12 / year
Planned Enhancements	None

Performance

Subtract 8,000 lb (3,630 kg) for Columbia LEO performance.

110 nm (204 km) circ, 28°	53,800 lb (24,400 kg) (Subtract 100 lb (45 kg) for each 1 nm (1.8 km) of increased LEO altitude)
Geotransfer Orbit, 28°	13,000 lb (5,900 kg) with IUS or TOS 4,000 lb (1,800 kg) with PAM-D2 2,800 lb (1,300 kg) with PAM-D
Geosynchronous Orbit	5,200 lb (2,360 kg) with IUS

Financial Status

Estimated Launch Price (JSC)

Dedicated flights in FY88$ are: $245M for full cost (total budget divided by flight rate), $130M for commercial user, $109M for DoD user (excludes certain launch support costs), or $63M for marginal cost (i.e., next flight). The price charged to a customer for a shared ride is based on the larger percentage of the orbiter's weight or length (60 ft - 18.3 m) capability that the customer's payload uses. To determine these load factors, the total weight (including all support equipment) and the total length (including 6 in. (152 mm) for dynamic clearance) are divided by the orbiter's capabilities and then divided by 0.75 which is the average load factor. Hence,

Load factor = (payload length) / (bay length) / 0.75
or = (payload mass) / (Shuttle performance) / 0.75

Shared price = Dedicated price times larger load factor from above

Manifest

	1995	1996	1997	1998**
Columbia	1	3	1	1
Discovery	2	0	3	0
Atlantis	2	3	2	1
Endeavour	3	1	1	1
Total	8	7	7	3

** Change due to management reserve increase.*
*** 1998 manifest is incomplete*

Remarks

After the Challenger accident, it was decided that a DoD, NASA, and commercial payload would only be assigned to the Shuttle if it required manned presence; it required the unique capabilities of the Shuttle; or other compelling circumstances existed.

Space Shuttle Vehicle

Overall

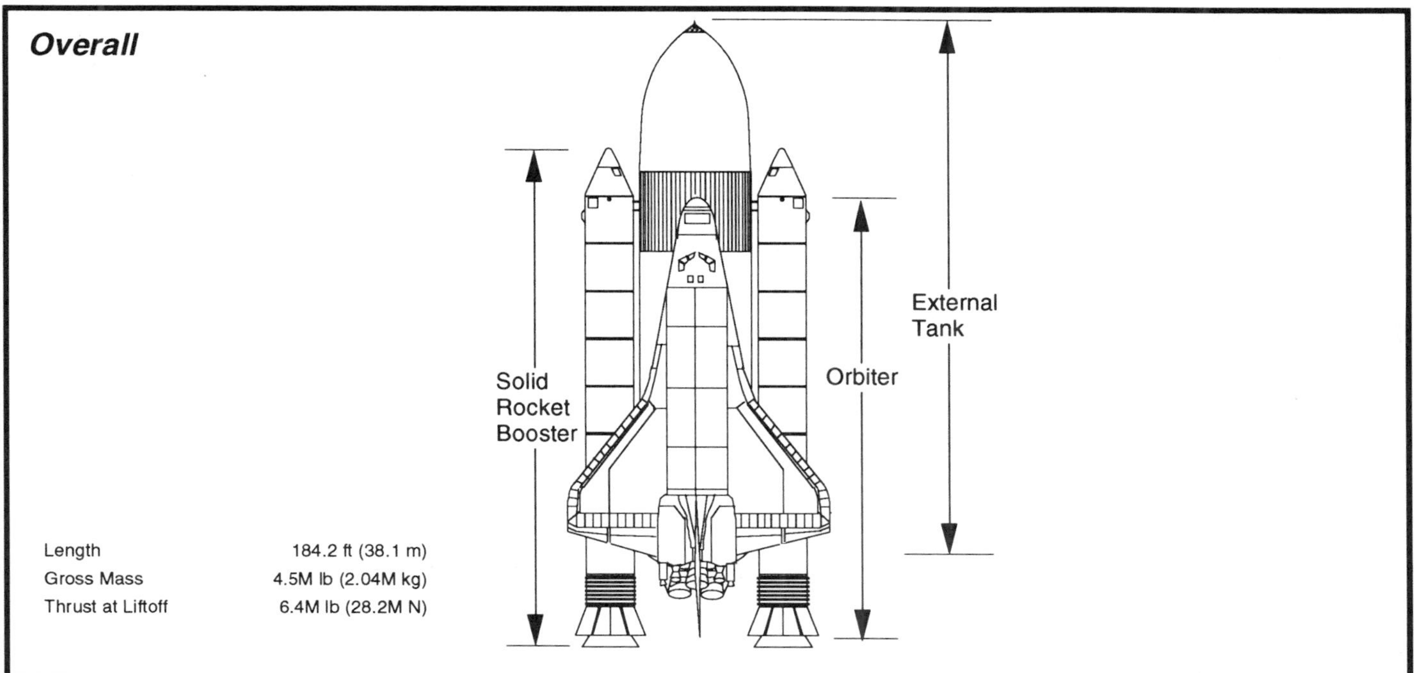

Length	184.2 ft (38.1 m)
Gross Mass	4.5M lb (2.04M kg)
Thrust at Liftoff	6.4M lb (28.2M N)

SRB

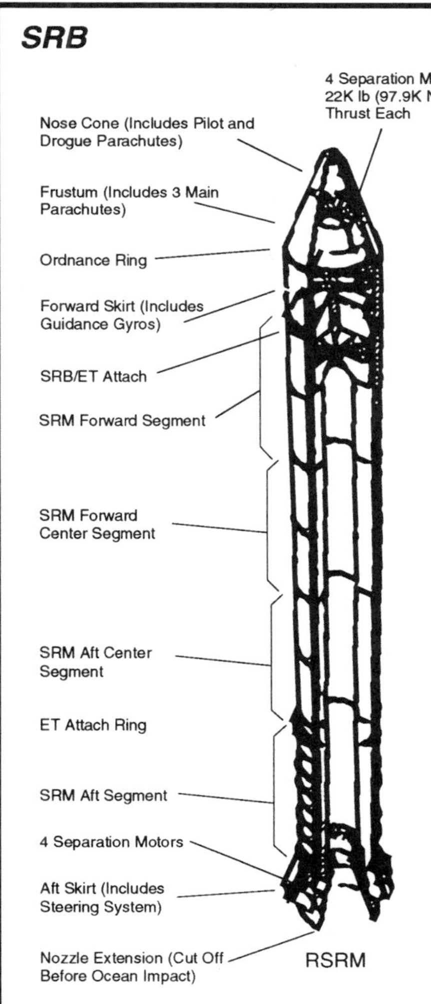

	RSRM
Dimension:	
Length	149.16 ft (45.46 m)
Diameter	12.38 ft (3.77 m)
Mass: (each)	
Propellant Mass	1.107M lb (502K kg)
Gross Mass	1.30M lb (590K kg)
Structure:	
Type	Monocoque
Case Material	Steel
Propulsion:	
Propellant	PBAN
Average Thrust (each)	2.65M lb (11.79M N) SL
Number of Motors	2
Number of Segments	4
Isp	267.3 sec vac ??? sec SL
Chamber Pressure	918 psia (63.3 bar)
Expansion Ratio	7.5:1
Control-Pitch, Yaw, Roll	Integral flexible bearing (±8°)
Events:	
Nominal Burn Time	123 sec
Stage Shutdown	Burn to depletion
Stage Separation	4 retro-rockets each fore and aft

Remarks:

Two Solid Rocket Boosters provide 80% of the Space Shuttle's total liftoff thrust. A number of reliability improvements have been made to the SRBs since the Challenger accident. As a result, the improved SRBs are also referred to as redesigned solid rocket motors (RSRM). The expended motors are parachuted back to Earth, retrieved in the Atlantic Ocean approximately 110 nm (200 km) from the launch site, towed back to Port Canaveral, disassembled, and then returned on railcars to Thiokol Space Operations in Utah. There the booster segments are refurbished, reloaded, and returned to the Kennedy Space Center (KSC) where they are reassembled for another launch. The steel case components of the RSRMs can be used 20 times.

Space Shuttle Vehicle

(continued)

External Tank

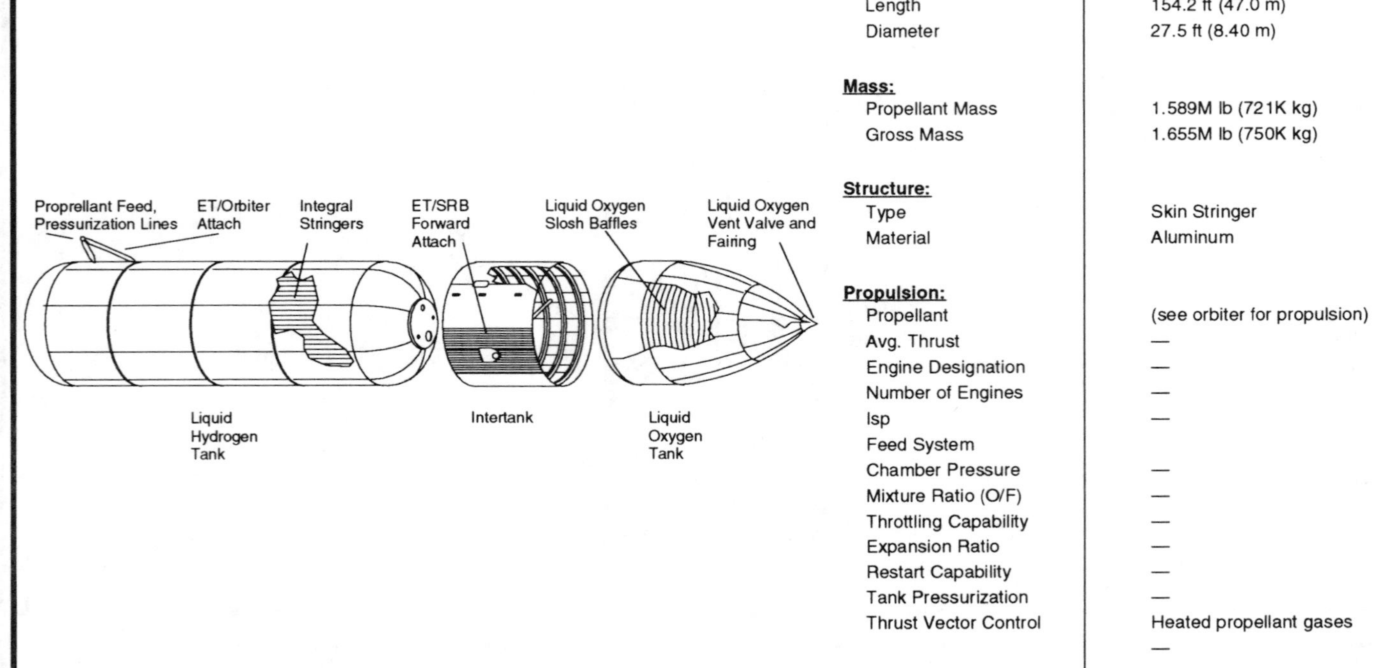

Dimension:	
Length	154.2 ft (47.0 m)
Diameter	27.5 ft (8.40 m)
Mass:	
Propellant Mass	1.589M lb (721K kg)
Gross Mass	1.655M lb (750K kg)
Structure:	
Type	Skin Stringer
Material	Aluminum
Propulsion:	
Propellant	(see orbiter for propulsion)
Avg. Thrust	—
Engine Designation	—
Number of Engines	—
Isp	—
Feed System	
Chamber Pressure	—
Mixture Ratio (O/F)	—
Throttling Capability	—
Expansion Ratio	—
Restart Capability	—
Tank Pressurization	—
Thrust Vector Control	Heated propellant gases
	—
Events:	
Nominal Burn Time	—
Stage Shutdown	—
Stage Separation	Pyrotechnic and tumbling system for reentry

Remarks:

The external tank (ET) contains the propellants for the three Space Shuttle Main Engines and forms the structural backbone of the Shuttle system in the launch configuration. At liftoff, the ET absorbs the total thrust loads of the three SSMEs and the two Solid Rocket Boosters. When the SRBs separate at an altitude of approximately 24 nm (44 km), the Orbiter, with the main engines still burning, carries the ET piggyback to near orbital velocity, approximtely 57 nm (106 km) above the Earth. There, 8.5 minutes into the mission, the now empty tank separates and falls in a preplanned trajectory into the Indian Ocean. The ET is the only major expendable element of the Space Shuttle.

The three main components of the ET are an oxygen tank, located in the forward position, an aft-positioned hydrogen tank, and a collar-like intertank, which connects the two propellant tanks, houses instrumentation and processing equipment, and provides the attachment structure for the forward end of the SRB. The hydrogen tank is 2.5 times larger than the oxygen tank. The skin of the ET is covered with a thermal protection system that is a nominal 1 in. (25 mm) thick coating of spray-on polyisocyanurate foam. The thermal protection system maintains the propellants at an acceptable temperature, protects the skin surface from aerodynamic heat, and minimizes ice formation on the external surface.

The ET includes a propellant feed system to duct the propellants to the orbiter engines, a pressurization and vent system to regulate the tank pressure, an environmental conditioning system to regulate the temperature and render the atmosphere in the intertank area inert, and an electrical system to distribute power and instrumentation signals and provide lightning protection. Most of the fluid control components (except for the vent valves) are located in the orbiter to minimize throwaway costs.

Space Shuttle Vehicle

(continued)

Orbiter

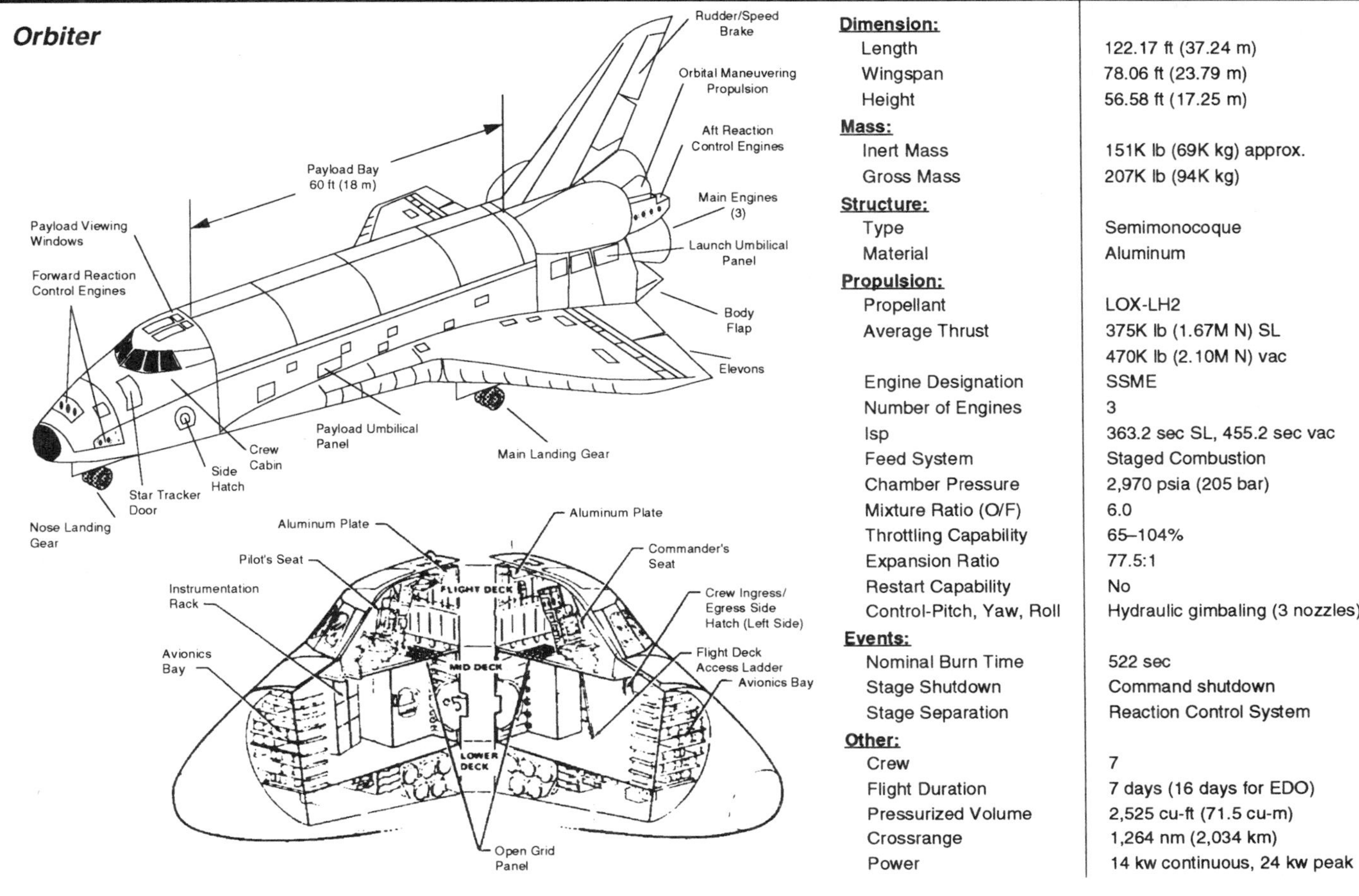

Dimension:	
Length	122.17 ft (37.24 m)
Wingspan	78.06 ft (23.79 m)
Height	56.58 ft (17.25 m)
Mass:	
Inert Mass	151K lb (69K kg) approx.
Gross Mass	207K lb (94K kg)
Structure:	
Type	Semimonocoque
Material	Aluminum
Propulsion:	
Propellant	LOX-LH2
Average Thrust	375K lb (1.67M N) SL 470K lb (2.10M N) vac
Engine Designation	SSME
Number of Engines	3
Isp	363.2 sec SL, 455.2 sec vac
Feed System	Staged Combustion
Chamber Pressure	2,970 psia (205 bar)
Mixture Ratio (O/F)	6.0
Throttling Capability	65–104%
Expansion Ratio	77.5:1
Restart Capability	No
Control-Pitch, Yaw, Roll	Hydraulic gimbaling (3 nozzles)
Events:	
Nominal Burn Time	522 sec
Stage Shutdown	Command shutdown
Stage Separation	Reaction Control System
Other:	
Crew	7
Flight Duration	7 days (16 days for EDO)
Pressurized Volume	2,525 cu-ft (71.5 cu-m)
Crossrange	1,264 nm (2,034 km)
Power	14 kw continuous, 24 kw peak

Remarks:

The Space Shuttle orbiter is designed as a space transport vehicle which can be reused for 100 missions. The crew compartment of the spacecraft accommodates up to seven crewmembers and can handle ten persons during emergency operations. The orbiter's 60 by 15 ft (18.3 by 4.5 m) cargo bay can ferry payloads to and from low-Earth orbit (100–330 nm or 185–600 km). It is similar in size and weight to modern transport aircraft. The three SSMEs located in the aft fuselage comprise the main propulsion system. Fuel for the orbiter's main engines is carried in the ET. Both the SRBs and the ET are jettisoned prior to orbital insertion. In orbit, the orbiter is maneuvered by the orbital maneuvering system (OMS) contained in two pods on the aft fuselage. The reaction control system, contained in the two OMS pods and in a module in the nose section of the forward fuselage, provides attitude control in space and during reentry and is used during rendezvous and docking maneuvers. After completing on-orbit operations, the orbiter reenters the Earth's atmosphere and glides to a runway landing. Nominal landing velocity is approximately 180 knots (330 km/hr). The orbiter is constructed primarily of aluminum and is protected from reentry heat by the thermal protection system. The principal substructures of the orbiter are the crew module and forward fuselage; mid fuselage and payload bay doors; aft fuselage and engine thrust structures; wing; and vertical tail.

The liquid hydrogen/liquid oxygen fueled SSME is a reusable high-performance rocket engine capable of operating at various thrust levels. Ignited on the ground prior to launch, the SSMEs operate in parallel with the SRBs during the initial ascent. After the boosters separate, the main engines become the sole propulsion element for the remainder of the ascent to orbit. The SSMEs develop thrust by using high-energy propellants in a staged combustion cycle. The propellants are partially combusted in dual preburners to produce high-pressure hot gas to drive the turbopumps. Combustion is completed in the main combustion chamber. The cycle ensures maximum performance by reducing parasitic losses. The SSME can be throttled over a thrust range of 65 to 104%, which provides for a high thrust level during liftoff and the initial ascent phase but allows thrust to be reduced to limit dynamic pressure in the Mach 1 regime and acceleration to 3gs during the final ascent phase. The engines are gimbaled to provide pitch, yaw, and roll control during the orbiter boost phase.

The orbiter cabin is designed as a combination working and living area. The pressurized crew compartment contains three levels. The flight deck contains the displays and controls used to pilot, monitor, and control the orbiter and the mission payloads. Seating for as many as four crewmembers can be provided on the flight deck. The mid deck contains passenger seating for three crew members, the living area, an airlock, the galley, sleeping compartments, the toilet, and avionics equipment compartments. An aft hatch in the airlock provides access to the payload bay. The lower deck contains the environmental control equipment and is readily accessible from above through removable floor panels. Located outside the crew module in the payload bay are provisions for a docking module and a transfer tunnel with an adapter to allow crew and equipment transfer for docking, Spacelab, and extravehicular operations. The environmental control and life support system (ECLSS) provides a comfortable shirt sleeve habitable environment 61–90°F (16–32°C) for the crew and a conditioned thermal environment (heat controlled) for the electronic components. The ECLSS bay, which includes air-handling equipment, lithium hydroxide canisters, water circulations pumps, and supply and waste water, is located in the mid deck of the orbiter and contains the pressurization system, the air revitalization system, the active thermal control system, and the water and waste management system.

Space Shuttle Vehicle

(continued)

Payload Bay

Length	60.9 ft (18.6 m)
Diameter	15.5 ft (4.7 m)
Structure	skin-stringer mid fuselage
Material	Aluminum
Remarks	The payload bay is enclosed by doors that open to expose the entire length and full width of the cargo bay. The orbiter cargo bay measures 15.0 by 60.0 ft (4.6 by 18.3 m). This volume is the maximum allowable payload dynamic envelope, including payload deflections. In addition, a nominal 3 in. (76 mm) clearance between the payload envelope and the orbiter structure is required to prevent orbiter deflection interference between the orbiter and the payload envelope.

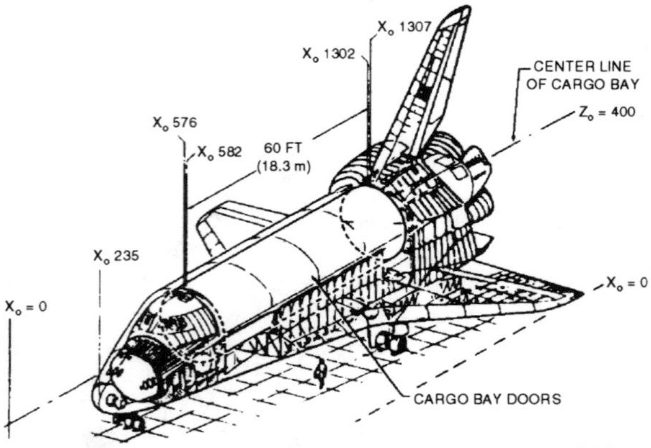

Avionics

The Shuttle avionics system control functions include guidance, navigation, control, and electrical power distribution for the orbiter, the ET, and the SRBs. In addition, the avionics control the communications equipment and can control payloads. Orbiter avionics automatically determine vehicle status and operational readiness and provide sequencing and control for the ET and the SRBs during launch and ascent. Automatic vehicle flight control can be used for every mission phase except docking. Manual control is also available at all times as a crew option.

The avionics are designed with redundant hardware and software to withstand multiple failures. The Space Shuttle avionics system consists of more than 200 electronic "black boxes" connected to a set of five computers through common party lines called data buses. The electronic black boxes offer dual or triple redundancy for every function.

The avionics system is closely interrelated with three other systems of the orbiter—the guidance, navigation, and control system; the controls and displays system; and the communications and data systems.

The orbiter has an electrical power and distribution system and a hydraulic power system. Electrical power is generated by three fuel cells that use cryogenically stored hydrogen and oxygen reactants. Hydraulic power is derived from three independent hydraulic pumps, each driven by its own hydrazine-fueled auxiliary power unit and cooled by its own ammonia spray boiler.

Attitude Control System

Launch. During launch, control is established through gimbaling of the SRB and SSME nozzles.

Orbital Maneuvering System (OMS). Two orbital maneuvering engines, located in external pods on each side of the aft fuselage, provide thrust for orbit insertion, orbit transfer, rendezvous, and deorbit. Up ot 24,000 lb (10,900 kg) of usable propellant can be loaded in the two OMS pods. Each pod contains a high-pressure helium storage bottle, the tank pressurization regulators and controls, a fuel tank, an oxidizer tank, and a pressure-fed regeneratively cooled rocket engine.

Each engine develops a vacuum thrust of 6,000 lb (26,700 N) using monomethylhydrazine (MMH) and nitrogen tetroxide (N2O4). They are burned at a nominal oxidizer to fuel ratio of 1.65 and a chamber pressure of 125 psia (860K Pa). The engine is designed for 100 missions with a service life of 10 years and is capable of sustaining 1,000 starts and 15 hours of cumulative firing time. Each engine is 77 in. (1,960 mm) long and weighs 260 lb (118 kg). The engine is gimbaled by pitch and yaw electromechanical actuators attached to the vehicle structure at the forward end of the combustion chamber. The controller for the actuators is mounted in the pod structure.

Reaction Control System. The orbiter reaction control system (RCS) provides the thrust for velocity changes along the axis of the orbiter and attitude control (pitch, yaw, and roll) during the orbit insertion, on-orbit, and reentry phases of flight. It has 38 bipropellant primary thrusters and 6 vernier thrusters. The primary thrusters are used for normal translation and attitude control. The vernier thrusters are used for fine attitude control and payload pointing where contamination or plume impingement are important considerations. Each primary thruster provides 870 lb (3,870 N) thrust. The vernier thrusters, which have no redundancy, are oriented to vector plumes away from the payload bay. Each vernier thruster provides 25 lb (110 N) thrust.

The reaction control system is grouped in three modules, one in the orbiter nose and one in each aft fuselage pod. Each module is independent and contains its own pressurization system and propellant tanks. The forward module contains 14 primary thrusters and 2 vernier thrusters. The multiple primary thrusters pointing in each direction provide redundancy for mission safety.

RCS propellants are MMH and N2O4. Total RCS propellant weight is 7,260 lb (3,300 kg). The design mixture ratio of 1.65 (oxidizer weight to fuel weight) permits the use of identical propellant tanks for both fuel and oxidizer. A system of heaters is used to maintain temperatures of the engines, propellant lines, and other components within operational limits.

An interconnect between the OMS and RCS in the aft pod permits the use of OMS propellants by the RCS for orbital maneuvers. In addition, the interconnect can be used for crossfeeding OMS and RCS propellants between the right and left pods.

Space Shuttle Performance

Kennedy Space Center (28.5° inclination)

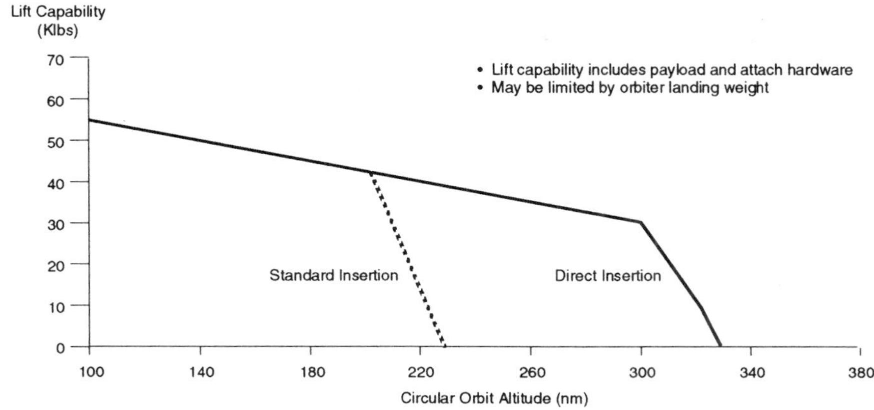

	Mission	Ascent Performance Capability (x1,000)	Landing Weight Capability (x1,000)
4,000 lbs (1,815 kg) Management Reserve	Max Performance - 28.5 deg	53.8 lb (24.4 kg)	56.6 lb (25.7 kg)
	Max Performance - 57.0 deg	39.8 lb (18.1 kg)	TBD
	Space Station - 51.6 deg	37.7 lb (17.1 kg)	TBD

Assumptions for each mission type:

Mission Configuration	Maximum Performance	Space Station
Altitude	110 nm (204 km)	220 nm (407 km)
Crew Size / Duration	5 man / 4 day	5 man / 7 day
Cryo (Hardware / Fluid Level)	3 tanks / 3 off-loaded	4 tanks / 3 full
Forward RCS	Off-loaded	Full
RMS	Off	On
Rendezvous	No	Yes

Note: Subtract 8,000 lb (3,630 kg) for orbiter Columbia. Also subtract approximately 100 lb (45 kg) for each 1 nm (1.8 km) of increased altitude.

Space Shuttle Operations

Launch Site

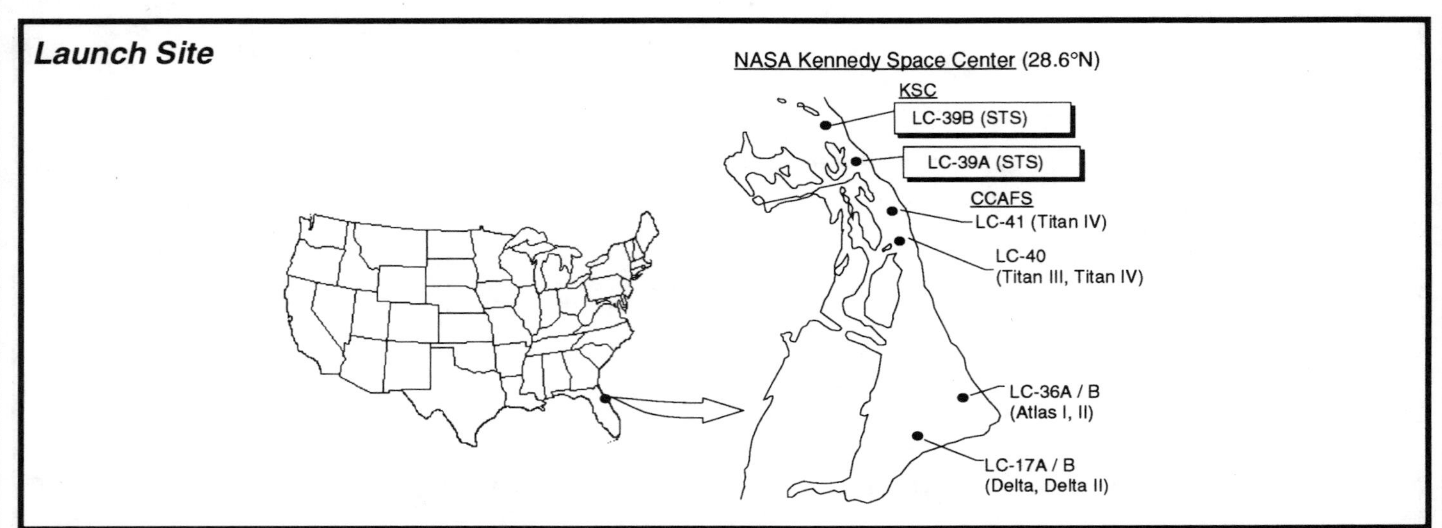

NASA Kennedy Space Center (28.6°N)

KSC
LC-39B (STS)
LC-39A (STS)
CCAFS
LC-41 (Titan IV)
LC-40
(Titan III, Titan IV)
LC-36A / B
(Atlas I, II)
LC-17A / B
(Delta, Delta II)

Launch Facilities

Kennedy Space Center

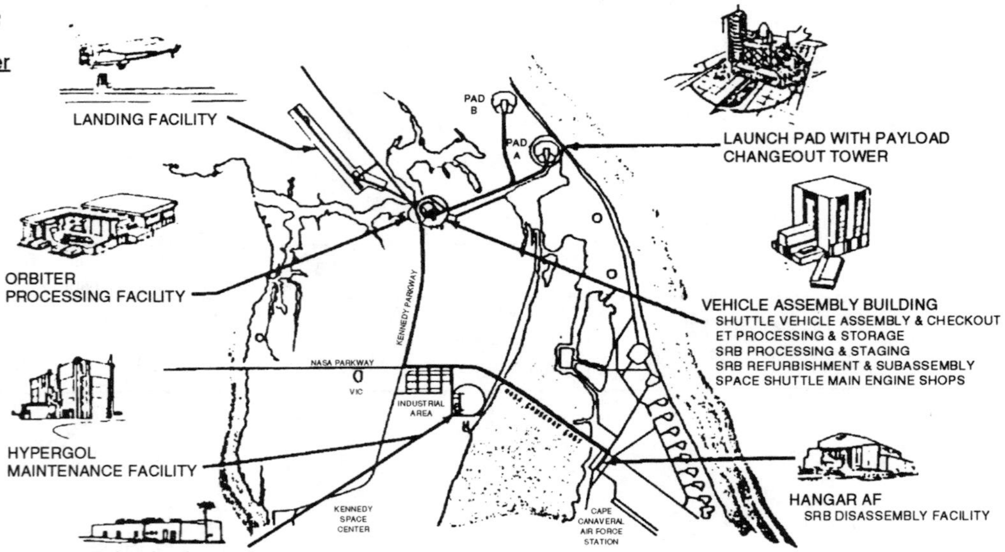

LANDING FACILITY
PAD B
PAD A
LAUNCH PAD WITH PAYLOAD
CHANGEOUT TOWER
ORBITER
PROCESSING FACILITY
KENNEDY PARKWAY
VEHICLE ASSEMBLY BUILDING
SHUTTLE VEHICLE ASSEMBLY & CHECKOUT
ET PROCESSING & STORAGE
SRB PROCESSING & STAGING
SRB REFURBISHMENT & SUBASSEMBLY
SPACE SHUTTLE MAIN ENGINE SHOPS
NASA PARKWAY
VIC
INDUSTRIAL AREA
HYPERGOL
MAINTENANCE FACILITY
HANGAR AF
SRB DISASSEMBLY FACILITY
KENNEDY SPACE CENTER
CAPE CANAVERAL AIR FORCE STATION
PARACHUTE FACILITY

Launch Facilities Location

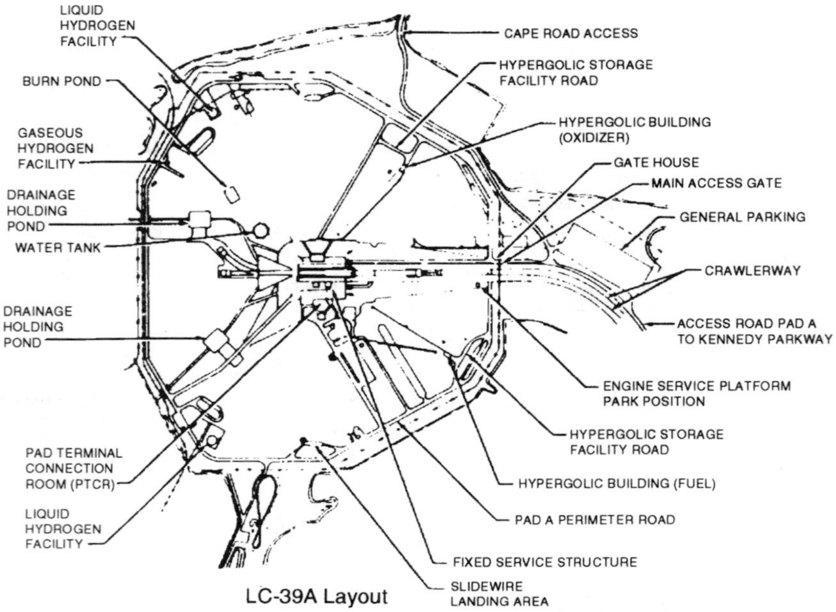

LIQUID
HYDROGEN
FACILITY
BURN POND
GASEOUS
HYDROGEN
FACILITY
DRAINAGE
HOLDING
POND
WATER TANK
DRAINAGE
HOLDING
POND
PAD TERMINAL
CONNECTION
ROOM (PTCR)
LIQUID
HYDROGEN
FACILITY
CAPE ROAD ACCESS
HYPERGOLIC STORAGE
FACILITY ROAD
HYPERGOLIC BUILDING
(OXIDIZER)
GATE HOUSE
MAIN ACCESS GATE
GENERAL PARKING
CRAWLERWAY
ACCESS ROAD PAD A
TO KENNEDY PARKWAY
ENGINE SERVICE PLATFORM
PARK POSITION
HYPERGOLIC STORAGE
FACILITY ROAD
HYPERGOLIC BUILDING (FUEL)
PAD A PERIMETER ROAD
FIXED SERVICE STRUCTURE
SLIDEWIRE
LANDING AREA

LC-39A Layout

Space Shuttle Operations

(continued)

Launch Processing

Space Shuttle launch operations requires the use of a wide variety of both advanced and routine technologies and interrelated subsystems. At KSC, Shuttle processing is carried out by contractors monitored by NASA employees. Lockheed Space Operations Company currently holds the prime contract for this work. Lockheed subcontracts with a variety of other companies, including Rockwell International, the prime contractor for the Shuttle orbiter. Approximately 6,550 contractors and 1,000 NASA employees directly support Shuttle launch operations at KSC. The Johnson Space Center (JSC) has responsibility for on-orbit mission operations and some launch operations, which involve about 5,675 contractors and 1,158 NASA employees. The Marshall Space Flight Center (MSFC) contributes engineering expertise for Shuttle modifications and supports the Shuttle with approximately 670 NASA employees and 11,000 prime contractor employees.

Payload Processing—Payloads for the Space Shuttle can be installed either horizontally or vertically. Horizontal payloads such as Spacelab are installed in the orbiter when it is still in the Orbiter Processing Facility, prior to being mated with the external tank and solid-fuel rocket boosters. Vertical payloads are installed in the Orbiter's payload after the fully assembled Shuttle arrives at the launch pad.

Payload owners have five options for processing payloads, ranging from minimum KSC involvement, essentially "ship and shoot," up to maximum KSC involvement. Payloads ready for launch upon arrival at KSC can be installed in the orbiter 2 to 50 days before launch with no servicing. Payloads requiring KSC participation for assembly must have all flight experiments and component hardware delivered to KSC up to a year in advance of the expected launch date to allow for assembly, integration, and testing.

Prelaunch processing—The basic processing flow is organized according to the integrate-transfer-launch (ITL) concept, which separates the major processing elements and allows certain functions to proceed in parallel until the vehicle is assembled in the Vehicle Assembly Building and transported to the launch pad.

Orbiter Processing—Orbiter processing constitutes the critical limit to the achievable flight rate. Refurbishing the orbiter Columbia for the second Shuttle flight consumed nearly 200 working days at three shifts per day. By the ill-fated flight of the Challenger in January 1986, this turnaround time had been reduced to 55 days. Numerous modifications to the orbiter refurbishment process, made as a result of the Challenger accident, have resulted in a nominal turn around time of 75 days.

In the Orbiter Processing Facility, NASA contractors check and refurbish every major system in the orbiter after each flight. They remove the Shuttle main engines and complete refurbishment of the SSME in the offline engine shop. Any of the 31,000 ceramic tiles of the thermal protection system that are missing or damaged during the flight are also replaced. Modifications to the orbiters are made during refurbishment. Finally, any horizontal payloads, such as Spacelab, are installed in the payload bay.

SRB Processing—The contractor (Thiokol) ships new and refurbished solid-fuel rocket motor segments and associated hardware, including the forward and aft closures, nozzle assemblies, and nozzle extensions by rail. When the segments arrive at KSC, they are moved into the Rotational Processing and Surge facility where they are inspected and stored until needed. The SRBs are stacked in the VAB on top of the mobile launch platform (MLP). Stacking operations take approximately 19 days.

ET Processing—The ET, which is manufactured by Martin Marietta at the Michoud Assembly Facility outside New Orleans, LA, is transported to KSC by sea barge. In the Vehicle Assembly Building (VAB), contractors inspect the external insulation and interfaces for ground support equipment connection. The electrical systems are checked and the fluid systems tested. A crane hoists the tank to a vertical position and transfers it to the mobile launcher platform, where it is mated with the twin SRBs.

Vehicle Assembly and Integration—After orbiter processing is complete, it is towed to the VAB High Bay, lifted to the vertical, and mated to the ET and SRBs. After mating all the sections of the Shuttle and connecting all umbilicals, engineers test each connection electrically and mechanically.

The computer-controlled launch processing system which is operating from the firing rooms of the launch control center, semi-automatically controls and checks out much of the Shuttle vehicle, both in the VAB and at Launch Complex 39. If any subsystem is found to be unsatisfactory, the computer will provide data that will help isolate the fault.

Transfer and Launch—When the Shuttle is fully assembled on the MLP, a crawler-transporter is positioned under the MLP and slowly (1 mi/hr or 1.6 km/hr) moves the Shuttle to Launch Complex 39A or B. Once at the pad, workers gain access to the Shuttle through the fixed service structure. The rotating service structure gives access to service fuel cells, to load and remove payloads, and to load hypergolic fuels for the orbital maneuvering system and the reaction control system. Those payloads to be installed vertically are transported to the rotating service structure in a protective payload canister.

After the Shuttle arrives at the pad, most checkout operations are controlled from the launch control center. After checkout operations are completed, power is applied to the orbiter and ground support equipment. Launch-readiness tests are performed and the tanks are prepared to receive their fuels. The Shuttle is now ready for the cryogenic propellants to be loaded and the flight crew to board.

During the final six or seven hours of the countdown and the liquid hydrogen and liquid oxygen are loaded into the ET. Finally, the flight crew and operations personnel complete all preparations and the Shuttle lifts off.

Mission Operations—Mission operations comprised all activities associated with planning and executing a mission. The primary focus is on gathering data, performing analyses, and developing the software required to meet the mission's objectives. Mission operations begin the day a payload is conceived, continue through the day of the launch, and end only after the mission is satisfactorily completed and the data analyzed. From beginning to end, mission operations for a Shuttle flight may take two years or more. JSC is the central control point for Shuttle missions.

Post-Launch Processing—Two minutes into the Shuttle's flight, the two solid-fuel rocket boosters are jettisoned and parachute into the Atlantic Ocean downrange from KSC. Two specially designed retrieval vessels recover the boosters and their components. The smaller components are hauled on board the ships and the boosters are towed back to the KSC Solid Rocket Booster Disassembly Facility. At this facility, the boosters and other components are washed, disassembled, cleaned, and stripped before they are shipped by rail to Thiokol for refurbishment.

Space Shuttle Operations

(continued)

Launch Processing

(continued)

Nominal completion of a Shuttle mission calls for a landing at the KSC Shuttle Landing Facility (SLF). When required, the Shuttle Orbiter can also land at Edwards Air Force Base in CA, or White Sands, NM. These landing facilities, along with others located in Zarogosa, Spain; Casablanca, Morocco; Rota, Spain; and Guam serve as emergency landing facilities for aborted launches. After landing, the orbiter must be drained of hazardous fuels and inspected for any exterior damage. Payload technicians remove any payloads brought back to Earth. If it lands anywhere but KSC, the orbiter must be lifted onto the back of a specially equipped Boeing 747 and ferried back to the SLF at KSC.

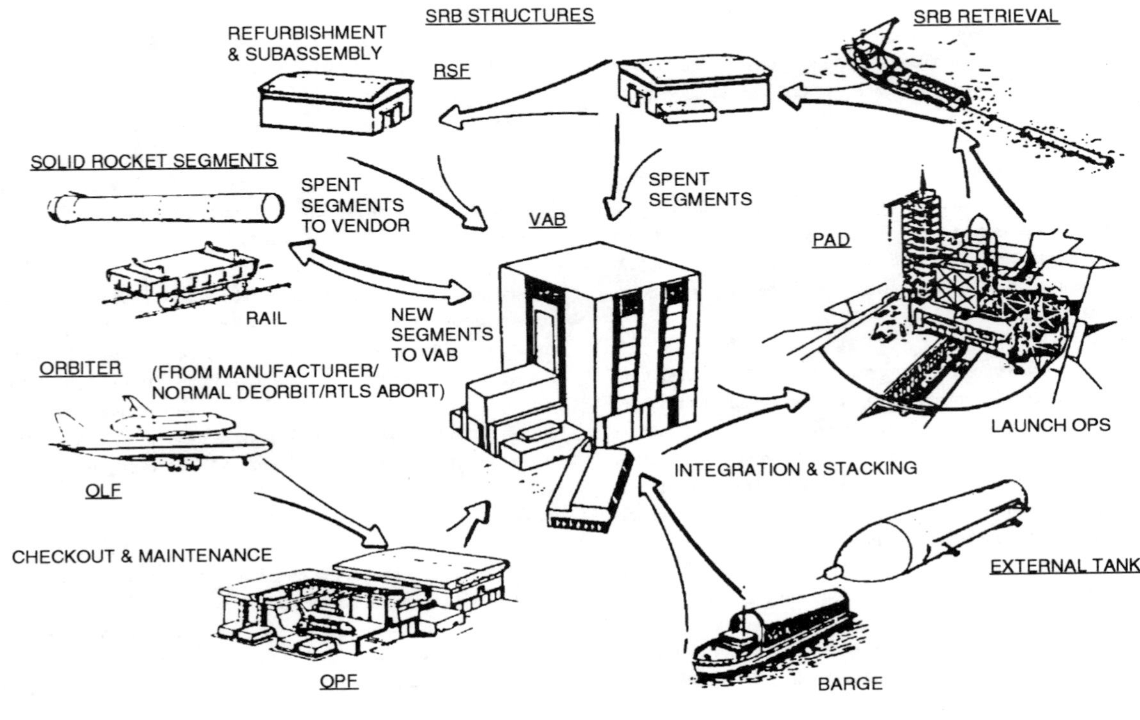

Space Shuttle Ground Turnaround Operations

Space Shuttle Operations
(continued)

Flight Sequence

Launch Countdown Sequence

Event	Function
T - 11 hr	Start retraction of rotating service structure (completed by T-7 hr 30 min)
T - 5 hr 30 min	Enter 6-hr built-in hold, followed by clearing of pad
T - 5 hr	Start countdown; begin chilldown of liquid oxygen/liquid hydrogen transfer system
T - 4 hr 30 min	Begin liquid oxygen fill of External Tank
T - 2 hr 50 min	Begin liquid hydrogen fill of External Tank
T - 2 hr 4 min	1-hr built-in hold, followed by crew entry operations
T - 1 hr 5 min	Crew entry complete; cabin hatch closed; start cabin leak check (completed by T-25 min)
T - 30 min	Secure white room; ground crew retires to fallback area by T-10 min; Range safety activation/Mission Control Center guidance update
T - 25 min	Mission Control Center/crew communications checks; crew given landing weather information for return-to-site abort or abort once around
T - 20 min	Load flight program
T - 9 min	10-min built-in hold (There is also a 5-min hold capability between T-9 and T-2 min and a 2-min hold capability between T-2 min and T-27 sec.)
T - 9 min	Go for launch/start launch processing system ground launch sequencer (automatic sequence)
T - 7 min	Start crew access arm retraction
T - 5 min	Activate orbiter hydraulic power units (APUs)
T - 4 min 30 sec	Orbiter goes to internal power
T - 3 min	Gimbal main engines to start position
T - 2 min 55 sec	External tank oxygen to flight pressure
T - 2 min 30 sec	Retract external tank gaseous oxygen vent arm
T - 1 min 57 sec	External Tank hydrogen to flight pressure
T - 27 sec a	Latest hold point if needed (Following any hold below the T-2 min mark, the countdown will be automatically recycled to T-9 min.)
T - 25 sec	Activate SRB hydraulic power units; initiative for management of countdown sequence assumed by onboard computers; ground launch sequencer remains online
T - 18 sec	Solid Rocket Booster nozzle profile conducted
T - 3.46 to 3.22 sec	Main engines start
T - 0	Main engines at 90% thrust
T + 2.64 sec	Solid Rocket Booster fire command/holddown bolts triggered
T + 3 sec	LIFTOFF

Space Shuttle Launch Events

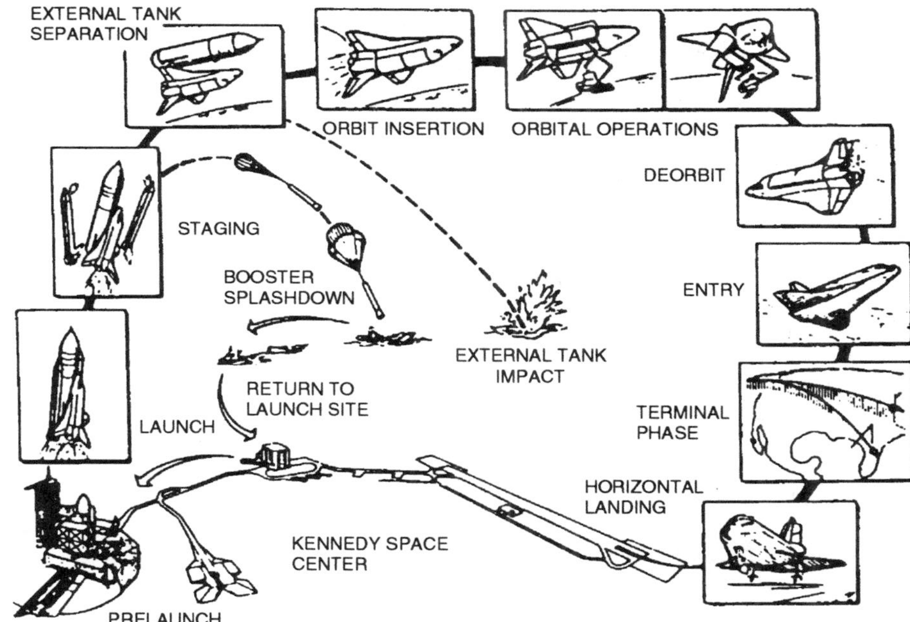

Event	Time min:sec	Geodetic Altitude mi (km) a		Inertial Velocity mph (km/hr)		Inertial Velocity mph (km/hr)	
SSME ignition	-00:03.46	184	(56) b	914	(1 471) c	0	(0)
SEB ignition	00:03	184	(56) b	914	(1 471) c	0	(0)
Begin pitchover	00:07	545	(166) b	917	(1 476)	0	(0)
Maximum dynamic pressure	01:09	8.3	(13.4)	1 654	(2 662)	4	(6.4)
SEB separation	02:04	29.4	(47.3)	3 438	(5 533)	23.7	(38.1)
Main engine cutoff	08:38	73	(117.5)	17 500	(28 163)	829.3	(1 335)
External Tank separation	08:50	73.5	(118.3)	17 498	(28 160)	886.6	(1 427)
OMS-1 ignition	10:39	78.3	(126)	17 479	(28 129)	1 380	(2 221)
OMS-1 cutoff	12:24	83.2	(133.9)	17 591	(28 309)	1 860	(2 993)
OMS-2 ignition	43.58	173.6	(279.4)	17 201	(27 682)	9 775	(15 731)
OMS-2 cutoff	45:34	174.2	(280.3)	17 321	(27 875)	10 269	(16 526)

a - Altitude reference to orbiter center-of-gravity above the geodetic representation of the Earth's surface.
b - in meters (feet)
c - Rotational velocity of Earth at KSC latitude of 28.5°N

Space Shuttle Payload Accommodations

Payload Compartment	
Maximum Payload Diameter	180.0 in (4,570 mm)
Maximum Cylinder Length	720.0 in (1,830 mm)
Maximum Cone Length	—
Payload Adapter	
Interface Diameter	Mission Unique
Payload Integration	
Nominal Mission Schedule Begins	T-36 to 48 months for primary payloads
Launch Window	
Latest Countdown Hold Not Requiring Recycling	T-31 sec
On-Pad Storage Capability	8 hrs fueled
Latest Access to Payload	T-17 to 48 hrs
Environment	
Maximum Load Factors	+3.2 g axial, ±2.5 g lateral, 4.2 g landing
Minimum Lateral/Longitudinal Payload Frequency	15 Hz / 35 Hz
Maximum Overall Acoustic Level	140 dB (one third octave)
Maximum Flight Shock	5,500 g at 4,000 hz
Maximum Dynamic Pressure on Vehicle	819 psf
Maximum Pressure Change in Fairing	0.5 psi/s (3.45 KPa/s)
Cleanliness Level in Fairing (Prior to Launch)	Class 10,000+
Payload Delivery	
Standard Orbit and Accuracy (3 sigma)	Circular orbit: ±10 nmi (18 km), ±0.5° inclination
Attitude Accuracy (3 sigma)	Primary RCS thrusters: ±0.1°, ±0.2 °/sec all axes Vernier RCS thrusters: ±0.1°, ±0.01 °/sec all axes
Nominal Payload Separation Rate	1 ft/s (0.3 m/s)
Deployment Rotation Rate Available	0 rpm (spin tables or special carrier required)
Loiter Duration in Orbit	7 days or 16 days with Extended Duration Orbiter
Maneuvers (Thermal / Collision / Telemetry)	Yes

Space Shuttle Notes

Publications

User's Guide

Space Transportation User Handbook, National Aeronautics & Space Administration, NASA-TM-84765, May 1982.

Space Shuttle News Reference, National Aeronautics & Space Administration.

STS Customer Accommodations, NASA, Johnson Space Center, JSC-21000-HBK, May 1986.

Space Shuttle - The History of Developing the National Space Transportation System, Dennis R. Jenkins, 1992.

Technical Publications

"The Advanced Solid Rocket Motor Project,", H. D. Trudeau, Lockheed Missiles & Space Co., 27th Space Congress, April 24-27, 1990.

"Evolutionary Transportation Concepts," Charles Teixeira, NASA Johnson Space Center, 27th Space Congress, April 24-27, 1990.

"A Historical Look at United States Launch Vehicles: 1967-Present", ANSER, STDN 90-4, second edition, February 1990.

"Space Shuttle Orbiter Update", S. Nagle, Rockwell International, AIAA-89-2406 25th Joint Propulsion Conference, Monterey, CA, July 10-12, 1989.

"Redesigned Solid Rocket Motor Enhancements", A. J. McDonald, Morton Thiokol, AIAA-89-2620, 25th Joint Propulsion Conference, Monterey, CA, July 10-12, 1989.

"Reducing Launch Operations Costs - New Technologies and Practices", U.S. Congress Office of Technology Assessment, OTA-TM-ISC-28, September 1988.

"Evolution of the Space Shuttle Design", J. P. Loftus, et al., NASA Johnson Space Center, 1986.

"Access to Space Study, Volume 1, NASA, 1993".

Acronyms

ASRM - advanced solid rocket motor
DOD - Department of Defense
EDO - extended duration orbiter
ET - external tank
FY - fiscal year
HTPB - hydroxy terminated polybutadiene
JSC - Johnson Space Center
KSC - Kennedy Space Center
LC - launch complex
LEO - low Earth orbit
LH2 - liquid hydrogen
LOX - liquid oxygen
LRB - liquid rocket booster
MLP - mobile launch platform
MMH - monomethyl hydrazine
MSFC - Marshall Space Flight Center
NASA - National Aeronautics and Space Administration
N204 - nitrogen tetroxide
OMS - orbital maneuvering system
OPF - Orbiter Processing Facility
PBAN - polybutadiene acrylonitrile acrylic acid
RCS - Reaction Control System
RSRM - redesigned solid rocket motor
SLC - space launch complex
SRB - solid rocket booster
SSME - space shuttle main engine
TAOS - thrust-assisted orbiter Shuttle system
VAB - Vehicle Assembly Building
VAFB - Vandenberg Air Force Base

Other Notes - Possible Shuttle Growth

Shuttle growth or evolution builds on the existing Space Shuttle system in an evolutionary, systematic program to provide improvements in safety, obsolescense mitigation, performance, cost reduction, and enhanced reliability. Changes could be incorporated in the existing fleet as modifications or retrofits.

Concepts for evolving the orbiter include many internal upgrades such as avionics upgrades and electromechanical actuators that would not impact the outer mold line of the orbiter. Major improvements are continuing with the SSME to provide better reliability, increased performance, and cost reductions. Improvements being considered for the ET include use of aluminum-lithium for the tanks and graphite epoxy for the intertank resulting in 8,000 lb (3,630 kg) of additional performance.

Additional items such as new liquid rocket boosters (LRBs) have also been considered. The LRBs would not only improve reliability by providing engine out capability, but would also allow improved emergency escape capability for the crew by permitting booster throttling and/or engine shutdown. Additionally, the LRBs would provide an increase of 20,000 lb (9,070 kg) lift capability and reduce ground processing time over the current SRBs. Several LRB concepts being traded include pressure-fed or pump-fed engines, LOX/RP-1 or LOX/LH2 propellant, and 15 to 18 ft (4.6 to 5.5 m) diameter boosters.

Titan

Government Point of Contact:
Titan System Program Office (ME)
U.S. Air Force Space & Missile Center
2400 El Segundo Blvd.
El Segundo, CA 90245
Phone: (310) 363-1110

Industry Point of Contact:
Lockheed Martin Astronautics
P.O. Box 179
Denver, CO 80201, USA
Phone: (303) 977-3000

The first Titan IV Centaur with the Milstar spacecraft taken on the morning of launch, February 7, 1994.

Titan History

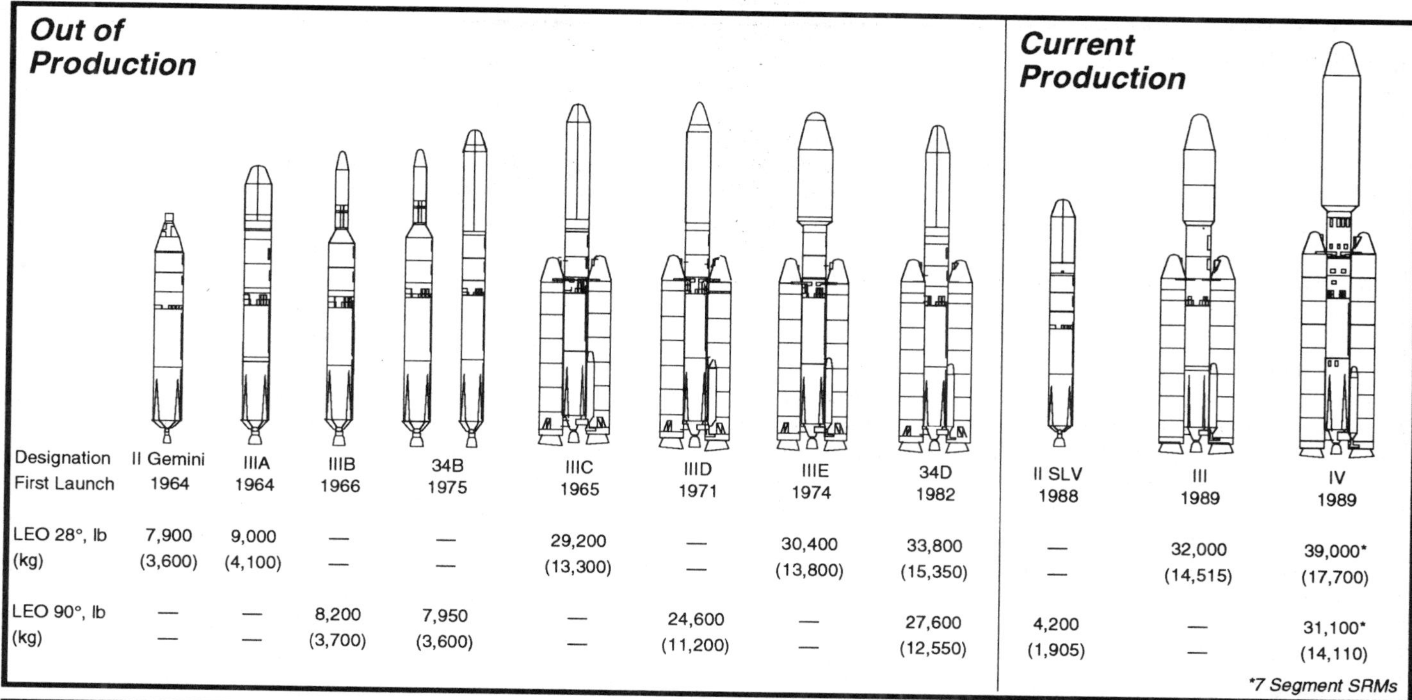

	Out of Production								Current Production		
Designation	II Gemini	IIIA	IIIB	34B	IIIC	IIID	IIIE	34D	II SLV	III	IV
First Launch	1964	1964	1966	1975	1965	1971	1974	1982	1988	1989	1989
LEO 28°, lb	7,900	9,000	—	—	29,200	—	30,400	33,800	—	32,000	39,000*
(kg)	(3,600)	(4,100)	—	—	(13,300)	—	(13,800)	(15,350)	—	(14,515)	(17,700)
LEO 90°, lb	—	—	8,200	7,950	—	24,600	—	27,600	4,200	—	31,100*
(kg)	—	—	(3,700)	(3,600)	—	(11,200)	—	(12,550)	(1,905)	—	(14,110)

*7 Segment SRMs

Vehicle Description

Titan II Gemini	Titan II ICBM converted to a man-rated space launch vehicle
Titan IIIA	Same as Titan II Gemini except stretched stage 1 and 2 and integral Transtage upper stage.
Titan IIIB	Same as Titan IIIA except Agena upper stage instead of Transtage.
Titan 34B	Same as Titan IIIA except stretched stage 1.
Titan IIIC	Same as Titan IIIA except five-segment solid rocket motors.
Titan IIID	Same as Titan IIIC except no upper stage.
Titan IIIE	Same as Titan IIID except Centaur upper stage and 14-ft (4.3 m) diameter payload fairing.
Titan 34D	Same as Titan 34B except a 5-1/2-segment solid rocket motor. Uses either Transtage or IUS upper stage.
Titan II Space Launch Vehicle	Refurbished Titan II ICBM with 10 ft (3.0 m) diameter payload fairing.
Titan III	Same as Titan 34D except stretched stage 2, single or dual carrier, enhanced liquid rocket engines and 13.1 ft (4.0 m) diameter payload fairing. Can use either a PAM-D2, Transtage, or TOS upper stage.
Titan IV	Same as Titan 34D except stretched stage 1 and stage 2, 7-segment solid rocket motor or three-segment solid rocket motor upgrade, and 16.7 ft (5.1 m) diameter payload fairing. Can use either a IUS or Centaur upper stage.

Historical Summary

The Titan family was established in October 1955, when the U.S. Air Force awarded the then Martin Company a contract to build a heavy duty space system. It became known as Titan I, the nation's first two-stage intercontinental ballistic missile (ICBM) and first underground silo-based ICBM. It proved many structural and propulsion techniques that were later incorporated into Titan II. The Titan II was a heavy-duty missile using storable propellants. In addition to providing technological legacy for Titan III, Titan II became a man-rated space booster for NASA's Gemini program. Twelve successful Gemini missions were flown on Titan II. Today, the Titan II is returning as a space launch vehicle with the old ICBMs converted to deliver payloads to orbit.

In 1961, representatives from both the Department of Defense (DoD) and NASA made an in-depth study of the nation's space-booster needs. Testimony presented to the group pointed to mission requirements which could not be met by any system then in existence or in the development stage. In addition, the study suggested that a high order of cost effectiveness and an increase in system reliability could be achieved by using a flexible standardized system instead of a variety of boosters, each modified specifically for the mission at hand. As a result, Titan III was born. The Titan III was to be the outgrowth of propulsion technology developed in both the Titan II and Minuteman ballistic missile programs, the first space launcher to combine the features of both liquid and solid propulsion systems. On December 1, 1962, the development program began.

Development of the Titan III was rapid. Following the go-ahead, development and fabrication of the booster began immediately at the Martin Marietta Denver installation. Dredging operations for the Titan III Integrate-Transfer-Launch (ITL) facilities began at Cape Canaveral, FL, on February 1, 1963. In the spring of 1963, AC Electronics Division of General Motors Corporation completed the basic design of the airborne and ground components for Titan III's inertial guidance system. The engines for Titan III's new upper stage, Transtage, were fired for the first time in July 1963, by Aerojet-General Corporation of Sacramento, CA. The same month, near Sunnyvale, CA,

Titan History

(continued)

Historical Summary

(continued)

Chemical Systems Division of United Technologies Corporation successfully fired the 1.2 million lb (53.4M N) thrust solid-propellant booster motors of the type used for Titan IIIC. In May 1964, the U.S. Air Force accepted the first completed Titan IIIA, a core only vehicle, and the following October the first Titan IIIC, a core plus five-segment solid strap-on vehicle. The maiden flights from Cape Canaveral were made on September 1, 1964, for Titan IIIA and June 18, 1965, for Titan IIIC.

Over the years, a number of other Titans were developed. The Titan IIIB was similar to Titan IIIA except the Titan IIIB used radio guidance and had an Agena upper stage instead of Titan IIIA's inertial guidance and Transtage upper stage. The Titan IIID was similar to Titan IIIC except without the Transtage. The Titan IIIE was a Titan IIID integrated with a Centaur D-1T upper stage and a 14-ft (4.3 m) diameter payload fairing. The Titan IIIE was used to launch the highly successful Viking spacecrafts to Mars and the Voyager spacecrafts to Jupiter, Saturn, Uranus, and Neptune. It is worth noting that in the mid 1960s, a Titan IIIM was being designed as a seven-segment solid motor, man-rated vehicle for launching the Manned Orbiting Laboratory. The Titan 34B is a stretched core version of the Titan IIIB. The Titan 34D is a stretched core and a five-and-a-half-segment solid motor version of the Titan IIID.

Today's Titan vehicles, Titan II, III, and IV, are derived from these earlier Titans. In 1984 the DoD called for a launch system that would complement the Space Shuttle and better ensure access to space for certain national security payloads. In 1985, the U.S. Air Force contracted with Martin Marietta Corporation for 10 launch vehicles, originally called the Complementary Expendable Launch Vehicle (CELV) or Titan 34D7 and later Titan IV. The Titan IV was derived from a Titan 34D, with a stretched core and seven segment solid motor (originally designed for Titan IIIM), and 16.7 ft (5.1 m) diameter payload fairing. The Titan IV program initially started as a short-term program to acquire and launch 10 Titan IV vehicles and Centaur upper stages off one East Coast pad at a cost of $2 billion. However, after the Challenger accident in 1986, the program has grown to 41 Titan IV vehicles with a mix of upper stages to eventually be launched off two East and one West Coast pads at an estimated cost of $10 billion. With the offloading of DoD payloads from Shuttle, Titan IV has become the DoD's main access to orbit for many of its heavy payloads. The first Titan IV (no upper stage) launch occurred successfully on June 14, 1989.

The U.S. Air Force initiated the Titan II Space Launch Vehicle (SLV) the same time as Titan IV. The Titan II SLV was developed from refurbished Titan II ICBMs with technology and hardware developed for the Titan III program incorporated. The goal of this U.S. Air Force sponsored program is to maximize the use of Titan II ICBM resources, which were deactivated from 1982 to 1987. Deactivated Titan II ICBMs are stored at Norton Air Force Base, CA, until selected for flight. Fifty-five missiles are available for refurbishment, the current U.S. Air Force contract is for 14 vehicles. The Titan II SLV is launched from Vandenberg Air Force Base (VAFB) and the first flight occurred on September 5, 1988.

Shortly after the Challenger accident in 1986 when the U.S. government decided to offload commercial payloads from the Space Shuttle, Martin Marietta announced plans to develop a Titan III commercial launch vehicle on its own funds. This Titan III is derived from the Titan 34D with a stretched second stage and a bulbous shroud for dual or dedicated payloads. On December 31, 1989, the first Titan III was launched with two communication satellites, one British and one Japanese.

Titan History
(continued)

Launch Record

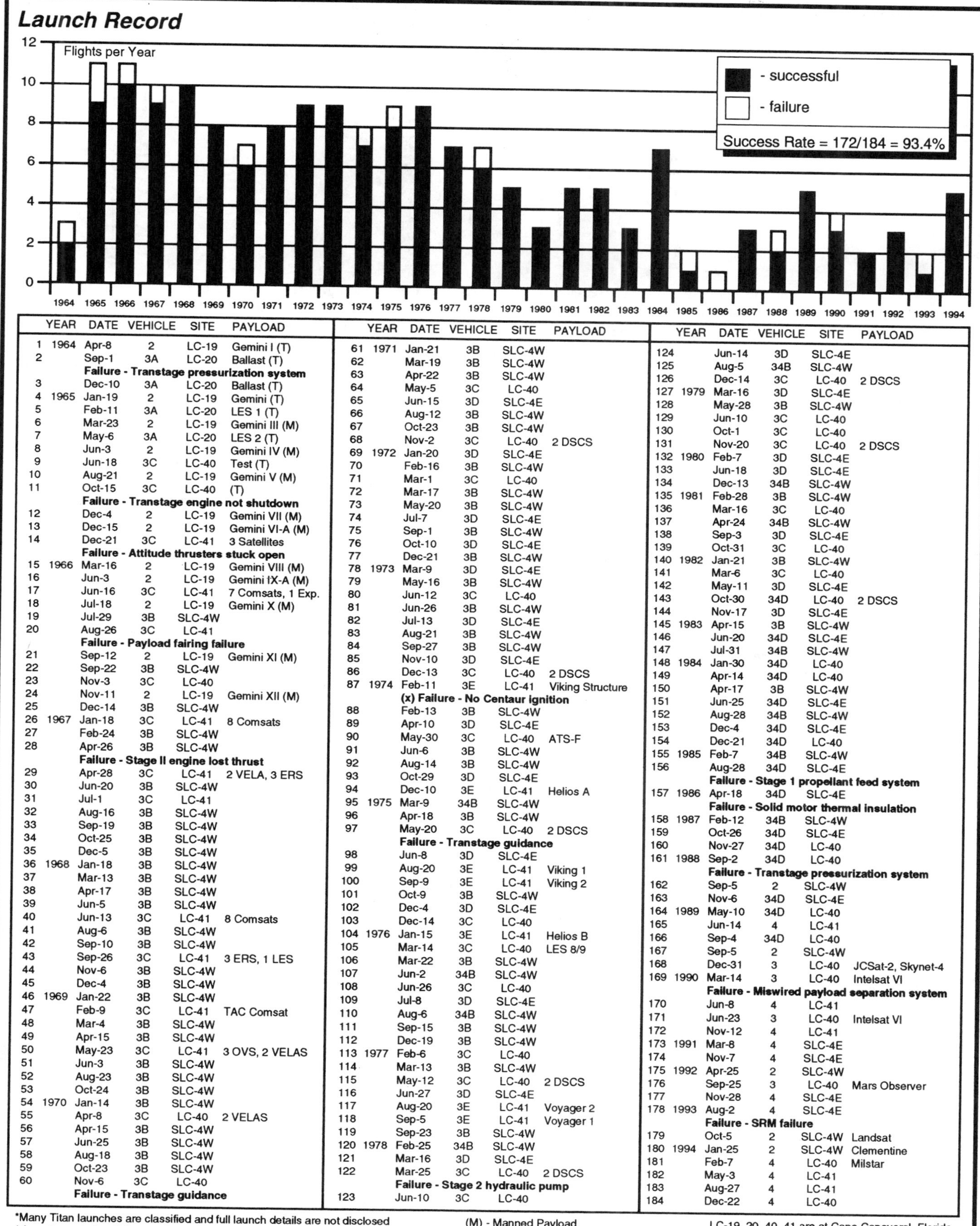

	YEAR	DATE	VEHICLE	SITE	PAYLOAD
1	1964	Apr-8	2	LC-19	Gemini I (T)
2		Sep-1	3A	LC-20	Ballast (T)
		Failure - Transtage pressurization system			
3		Dec-10	3A	LC-20	Ballast (T)
4	1965	Jan-19	2	LC-19	Gemini (T)
5		Feb-11	3A	LC-20	LES 1 (T)
6		Mar-23	2	LC-19	Gemini III (M)
7		May-6	3A	LC-20	LES 2 (T)
8		Jun-3	2	LC-19	Gemini IV (M)
9		Jun-18	3C	LC-40	Test (T)
10		Aug-21	2	LC-19	Gemini V (M)
11		Oct-15	3C	LC-40	(T)
		Failure - Transtage engine not shutdown			
12		Dec-4	2	LC-19	Gemini VII (M)
13		Dec-15	2	LC-19	Gemini VI-A (M)
14		Dec-21	3C	LC-41	3 Satellites
		Failure - Attitude thrusters stuck open			
15	1966	Mar-16	2	LC-19	Gemini VIII (M)
16		Jun-3	2	LC-19	Gemini IX-A (M)
17		Jun-16	3C	LC-41	7 Comsats, 1 Exp.
18		Jul-18	2	LC-19	Gemini X (M)
19		Jul-29	3B	SLC-4W	
20		Aug-26	3C	LC-41	
		Failure - Payload fairing failure			
21		Sep-12	2	LC-19	Gemini XI (M)
22		Sep-22	3B	SLC-4W	
23		Nov-3	3C	LC-40	
24		Nov-11	2	LC-19	Gemini XII (M)
25		Dec-14	3B	SLC-4W	
26	1967	Jan-18	3C	LC-41	8 Comsats
27		Feb-24	3B	SLC-4W	
28		Apr-26	3B	SLC-4W	
		Failure - Stage II engine lost thrust			
29		Apr-28	3C	LC-41	2 VELA, 3 ERS
30		Jun-20	3B	SLC-4W	
31		Jul-1	3C	LC-41	
32		Aug-16	3B	SLC-4W	
33		Sep-19	3B	SLC-4W	
34		Oct-25	3B	SLC-4W	
35		Dec-5	3B	SLC-4W	
36	1968	Jan-18	3B	SLC-4W	
37		Mar-13	3B	SLC-4W	
38		Apr-17	3B	SLC-4W	
39		Jun-5	3B	SLC-4W	
40		Jun-13	3C	LC-41	8 Comsats
41		Aug-6	3B	SLC-4W	
42		Sep-10	3B	SLC-4W	
43		Sep-26	3C	LC-41	3 ERS, 1 LES
44		Nov-6	3B	SLC-4W	
45		Dec-4	3B	SLC-4W	
46	1969	Jan-22	3B	SLC-4W	
47		Feb-9	3C	LC-41	TAC Comsat
48		Mar-4	3B	SLC-4W	
49		Apr-15	3B	SLC-4W	
50		May-23	3C	LC-41	3 OVS, 2 VELAS
51		Jun-3	3B	SLC-4W	
52		Aug-23	3B	SLC-4W	
53		Oct-24	3B	SLC-4W	
54	1970	Jan-14	3B	SLC-4W	
55		Apr-8	3C	LC-40	2 VELAS
56		Apr-15	3B	SLC-4W	
57		Jun-25	3B	SLC-4W	
58		Aug-18	3B	SLC-4W	
59		Oct-23	3B	SLC-4W	
60		Nov-6	3C	LC-40	
		Failure - Transtage guidance			
61	1971	Jan-21	3B	SLC-4W	
62		Mar-19	3B	SLC-4W	
63		Apr-22	3B	SLC-4W	
64		May-5	3C	LC-40	
65		Jun-15	3D	SLC-4E	
66		Aug-12	3B	SLC-4W	
67		Oct-23	3B	SLC-4W	
68		Nov-2	3C	LC-40	2 DSCS
69	1972	Jan-20	3D	SLC-4E	
70		Feb-16	3B	SLC-4W	
71		Mar-1	3C	LC-40	
72		Mar-17	3B	SLC-4W	
73		May-20	3B	SLC-4W	
74		Jul-7	3D	SLC-4E	
75		Sep-1	3B	SLC-4W	
76		Oct-10	3D	SLC-4E	
77		Dec-21	3B	SLC-4W	
78	1973	Mar-9	3D	SLC-4E	
79		May-16	3B	SLC-4W	
80		Jun-12	3C	LC-40	
81		Jun-26	3B	SLC-4W	
82		Jul-13	3D	SLC-4E	
83		Aug-21	3B	SLC-4W	
84		Sep-27	3B	SLC-4W	
85		Nov-10	3D	SLC-4E	
86		Dec-13	3C	LC-40	2 DSCS
87	1974	Feb-11	3E	LC-41	Viking Structure
		(x) Failure - No Centaur ignition			
88		Feb-13	3B	SLC-4W	
89		Apr-10	3D	SLC-4E	
90		May-30	3C	LC-40	ATS-F
91		Jun-6	3B	SLC-4W	
92		Aug-14	3B	SLC-4W	
93		Oct-29	3D	SLC-4E	
94		Dec-10	3E	LC-41	Helios A
95	1975	Mar-9	34B	SLC-4W	
96		Apr-18	3B	SLC-4W	
97		May-20	3C	LC-40	2 DSCS
		Failure - Transtage guidance			
98		Jun-8	3D	SLC-4E	
99		Aug-20	3E	LC-41	Viking 1
100		Sep-9	3E	LC-41	Viking 2
101		Oct-9	3B	SLC-4W	
102		Dec-4	3D	SLC-4E	
103		Dec-14	3C	LC-40	
104	1976	Jan-15	3E	LC-41	Helios B
105		Mar-14	3C	LC-40	LES 8/9
106		Mar-22	3B	SLC-4W	
107		Jun-2	34B	SLC-4W	
108		Jun-26	3C	LC-40	
109		Jul-8	3D	SLC-4E	
110		Aug-6	34B	SLC-4W	
111		Sep-15	3B	SLC-4W	
112		Dec-19	3B	SLC-4W	
113	1977	Feb-6	3C	LC-40	
114		Mar-13	3B	SLC-4W	
115		May-12	3C	LC-40	2 DSCS
116		Jun-27	3D	SLC-4E	
117		Aug-20	3E	LC-41	Voyager 2
118		Sep-5	3E	LC-41	Voyager 1
119		Sep-23	3B	SLC-4W	
120	1978	Feb-25	34B	SLC-4W	
121		Mar-16	3D	SLC-4E	
122		Mar-25	3C	LC-40	2 DSCS
		Failure - Stage 2 hydraulic pump			
123		Jun-10	3C	LC-40	
124		Jun-14	3D	SLC-4E	
125		Aug-5	34B	SLC-4W	
126		Dec-14	3C	LC-40	2 DSCS
127	1979	Mar-16	3D	SLC-4E	
128		May-28	3B	SLC-4W	
129		Jun-10	3C	LC-40	
130		Oct-1	3C	LC-40	
131		Nov-20	3C	LC-40	2 DSCS
132	1980	Feb-7	3D	SLC-4E	
133		Jun-18	3D	SLC-4E	
134		Dec-13	34B	SLC-4W	
135	1981	Feb-28	3B	SLC-4W	
136		Mar-16	3C	LC-40	
137		Apr-24	34B	SLC-4W	
138		Sep-3	3D	SLC-4E	
139		Oct-31	3C	LC-40	
140	1982	Jan-21	3B	SLC-4W	
141		Mar-6	3C	LC-40	
142		May-11	3D	SLC-4E	
143		Oct-30	34D	LC-40	2 DSCS
144		Nov-17	3D	SLC-4E	
145	1983	Apr-15	3B	SLC-4W	
146		Jun-20	34D	SLC-4E	
147		Jul-31	34B	SLC-4W	
148	1984	Jan-30	34D	LC-40	
149		Apr-14	34D	LC-40	
150		Apr-17	3B	SLC-4W	
151		Jun-25	34D	SLC-4E	
152		Aug-28	34B	SLC-4W	
153		Dec-4	34D	SLC-4E	
154		Dec-21	34D	LC-40	
155	1985	Feb-7	34B	SLC-4W	
156		Aug-28	34D	SLC-4E	
		Failure - Stage 1 propellant feed system			
157	1986	Apr-18	34D	SLC-4E	
		Failure - Solid motor thermal insulation			
158	1987	Feb-12	34B	SLC-4W	
159		Oct-26	34D	SLC-4E	
160		Nov-27	34D	LC-40	
161	1988	Sep-2	34D	LC-40	
		Failure - Transtage pressurization system			
162		Sep-5	2	SLC-4W	
163		Nov-6	34D	SLC-4E	
164	1989	May-10	34D	LC-40	
165		Jun-14	4	LC-41	
166		Sep-4	34D	LC-40	
167		Sep-5	2	SLC-4W	
168		Dec-31	3	LC-40	JCSat-2, Skynet-4
169	1990	Mar-14	3	LC-40	Intelsat VI
		Failure - Miswired payload separation system			
170		Jun-8	4	LC-41	
171		Jun-23	3	LC-40	Intelsat VI
172		Nov-12	4	LC-41	
173	1991	Mar-8	4	SLC-4E	
174		Nov-7	4	SLC-4E	
175	1992	Apr-25	2	SLC-4W	
176		Sep-25	3	LC-40	Mars Observer
177		Nov-28	4	SLC-4E	
178	1993	Aug-2	4	SLC-4E	
		Failure - SRM failure			
179		Oct-5	2	SLC-4W	Landsat
180	1994	Jan-25	2	SLC-4W	Clementine
181		Feb-7	4	LC-40	Milstar
182		May-3	4	LC-41	
183		Aug-27	4	LC-41	
184		Dec-22	4	LC-40	

*Many Titan launches are classified and full launch details are not disclosed
(x) - not counted as failure against vehicle

(M) - Manned Payload
(T) - Test Launch

LC-19, 20, 40, 41 are at Cape Canaveral, Florida
SLC-4E, 4W are at Vandenberg, California

Titan General Description

	Titan II	Titan III	Titan IV
Summary	The Titan II vehicles are decommissioned ICBMs that have been refurbished and equipped with hardware required for space launch. The Titan II was modified by DOD for launches of smaller payloads to polar orbit.	On August 19, 1986, Martin Marietta announced it would offer commercial launch services with a Titan III (also known as Commercial Titan), which is a modified Titan 34D. Payloads can be flown in either a single or dual carrier.	Titan IV program was initiated in 1985 by the Air Force as a means for launching Shuttle-class payloads. Titan IV will be the Air Force's workhorse throughout the 1990s. It can launch payloads with either a Centaur, IUS or no upper stage.
Status	Operational	Operational	Operational
Key Organizations	User - DoD, NASA, NOAA	User - Commercial, NASA	User - DoD, NASA
	Launch Service Agency - U.S. Air Force	Launch Service Agency - Lockheed Martin	Launch Service Agency - U.S. Air Force
	Prime Contractor - Lockheed Martin Astronautics (Airframe Refurbishment, Assembly and Test, and Launch Operations)	Prime Contractor - Lockheed Martin Astronautics (Airframe, Assembly and Test, and Launch Operations)	Prime Contractor - Lockheed Martin Astronautics Group (Airframe, Assembly and Test, and Launch Operations)
	Principal Subcontractors - McDonnell Douglas (Payload Fairings) Aerojet TechSystems (Liquid Rocket Engines) Delco Electronics (Guidance) SCI (Instrumentation) Cincinnati Electronics (Command Receivers)	Principal Subcontractors - Contraves AG (Payload Fairing) Dornier GmbH (Payload Carrier Assembly) United Technologies (Solid Rocket Motor) Aerojet TechSystems (Liquid Rocket Engines) Delco Electronics (Guidance) SCI (Instrumentation) Cincinnati Electronics (Command Receivers)	Principal Subcontractors - McDonnell Douglas (Payload Fairings) United Technologies (Solid Rocket Motor) Hercules (Solid Rocket Motor Upgrade) Aerojet TechSystems (Liquid Rocket Engines) Delco Electronics (Guidance) SCI (Instrumentation) Cincinnati Electronics (Command Receivers) Associate Contractor Boeing (Inertial Upper Stage)
Vehicle			
System Height	Up to 140.8 ft (42.9 m)	Up to 155 ft (47.3 m)	Up to 204 ft (62.2 m)
Payload Fairing Size	10 ft (3.0 m) diameter by 20 ft (6.1 m), 25 ft (7.6 m), or 30 ft (9.1 m) height	13.1 ft (4.0 m) diameter by 53.5 ft (16.3 m) height for dual carrier, 42.6 ft (13.0 m) for single carrier	16.7 ft (5.1 m) diameter by 50, 56, 66, 76 or 86 ft (15.2, 17.1, 20.1, 23.2 or 26.2 m) height
Gross Mass	340,000 lb (155,000 kg)	1,500,000 lb (680,000 kg)	1,900,000 lb (860,000 kg)
Planned Enhancements	Circulation system is available. Secondary Payloads may be flown. Upper stage configuration will be available.	None	Centaur-based Honeywell single-string avionics and the Hercules solid rocket motor upgrade (SRMU) by 1996.
Operations			
Primary Missions	High-inclination LEO missions.	Easterly LEO missions	Polar, LEO or GEO missions
Compatible Upper Stages	None	PAM-DII, Transtage, TOS	IUS, Centaur
First Launch	1988	1989	1989

Titan General Description

(continued)

	Titan II	Titan III	Titan IV
Operations (continued) Success / Flight Total	5 / 5	3 / 4	10 / 11
Launch Site	VAFB SLC-4W (34.7°N, 120.6°N)	CCAFS - LC-40 (28.5°N, 81.0°W)	CCAFS - LC-40 and 41 (28.5°N, 81.0°W) VAFB - SLC-4E (34.7°N, 120.6°W)
Launch Azimuth	148°–301° (SLC-4W)	93°–112° (LC-40)	93°–112° (LC-40/41) 147°–210° (SLC-4E)
Nominal Flight Rate	Up to 3 / yr	About 1 / yr	CCAF supports up to 6 / yr (incl. Titan III) and VAFB up to 2 / yr.
Planned Enhancements	Possible launches from CCAFS are being studied	None	Centaur Offline Processing Facility at CCAFS by 1995.
Performance 100 nm (185 km) circ, 28°	—	31,200 lb (14,150 kg) - dual carrier 32,000 lb (14,515 kg) - single carrier	39,000 lb (17,700 kg) - SRM 47,700 lb (21,640 kg) - SRMU
100 nm (185 km) circ, 90°	4,200 lb (1,905 kg)	—	31,100 lb (14,110 kg) - SRM 41,000 lb (18,600 kg) - SRMU
Geotransfer Orbit, 28°	—	4,080 lb (1,850 kg) with PAM-DII 9,500 lb (4310 kg) with Transtage - dual carrier 11,000 lb (5,000 kg) with TOS - single carrier	With appropriate spacecraft kick motor: 14,000 lb (6,350 kg) - SRM 19,000 lb (8,620 kg) - SRMU
Geosynchronous Orbit	—	With appropriate spacecraft kick motor: Up to 3,000 lb each (1,360 kg) - dual carrier Up to 5,500 lb (2,500 kg) - single carrier	5,250 lb (2,380 kg) with IUS - SRM With the Centaur upper stage: 10,000 lb (4,540 kg) - SRM 12,700 lb (5,760 kg) - SRMU (Centaur is structurally limited to 11,500 lb (5,220 kg)).
Financial Status Estimated Launch Price	$34M (Martin Marietta)	$130–150M dedicated (Martin Marietta) (Upper stage not included.)	$196M for Titan IV No Upper Stage (Martin Marietta) ?? for Titan IV IUS $248M for Titan IV Centaur (Martin Marietta)
Orders: Payload/Agency (firm / option)	DMSP / USAF (4 / 0) TIROS / NOAA (3 / 0) TBD 2 missions	None	Classified / USAF (38 / 8) Cassini / NASA (1 / 0) [Total vehicles 38/8]
Manifest	Site: SLC-4W — 1995: 1, 1996: 2, 1997: 1, 1998: 1, 1999: 1, 2000: 1	Site: SLC-40 — 1995: —, 1996: —, 1997: —, 1998: —, 1999: —, 2000: —	Up to 6 / yr at CCAFS. 2–4 / yr at VAFB. Actual manifest is classified.
Remarks	The Air Force has a continuing program for modification and launch of the Titan IIs. There were 53 in storage available for launch of which 14 are on order and 5 of these have already flown.	In June 1989, Martin Marietta announced a new approach to marketing Commercial Titan. Rather than selling half rides, only dedicated rides are offered. The customer provides one or two satellites at their discretion.	None

Titan Vehicle

Overall

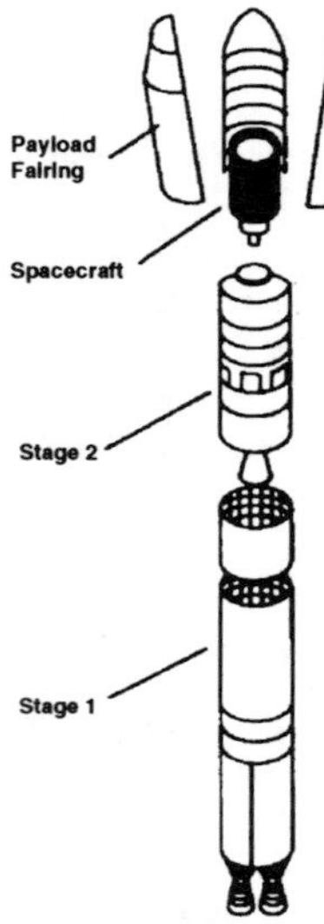

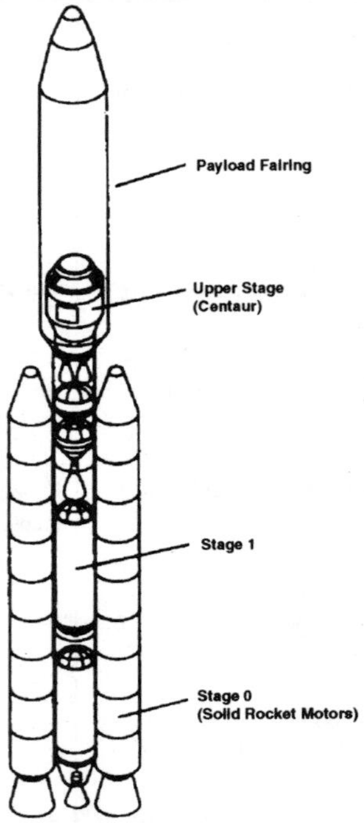

Payload Fairing
Spacecraft
Stage 2
Stage 1

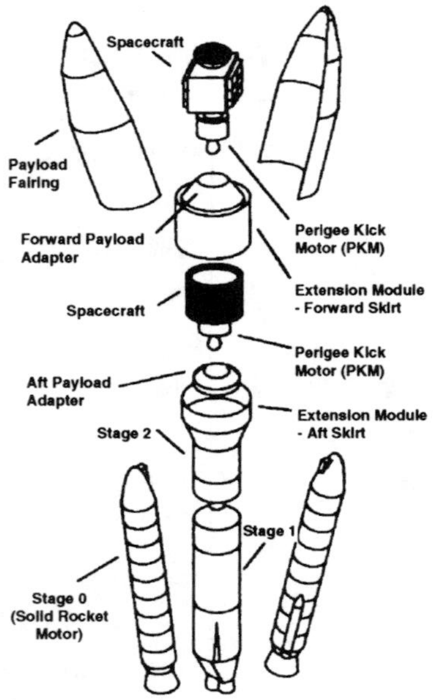

Spacecraft
Payload Fairing
Forward Payload Adapter
Perigee Kick Motor (PKM)
Extension Module - Forward Skirt
Spacecraft
Perigee Kick Motor (PKM)
Aft Payload Adapter
Extension Module - Aft Skirt
Stage 2
Stage 1
Stage 0 (Solid Rocket Motor)

Payload Fairing
Upper Stage (Centaur)
Stage 1
Stage 0 (Solid Rocket Motors)

Length:	Up to 140.8 ft (42.9 m)	Up to 155 ft (47.3 m)	Up to 204 ft (62.2 m)
Gross Mass:	340K lb (155K kg)	1.5M lb (680K kg)	1.9M lb (860K kg)
Thrust at Liftoff:	474K lb (2.09M N)	2.6M (11M N)	3.2M lb (14M N)

Stage 0

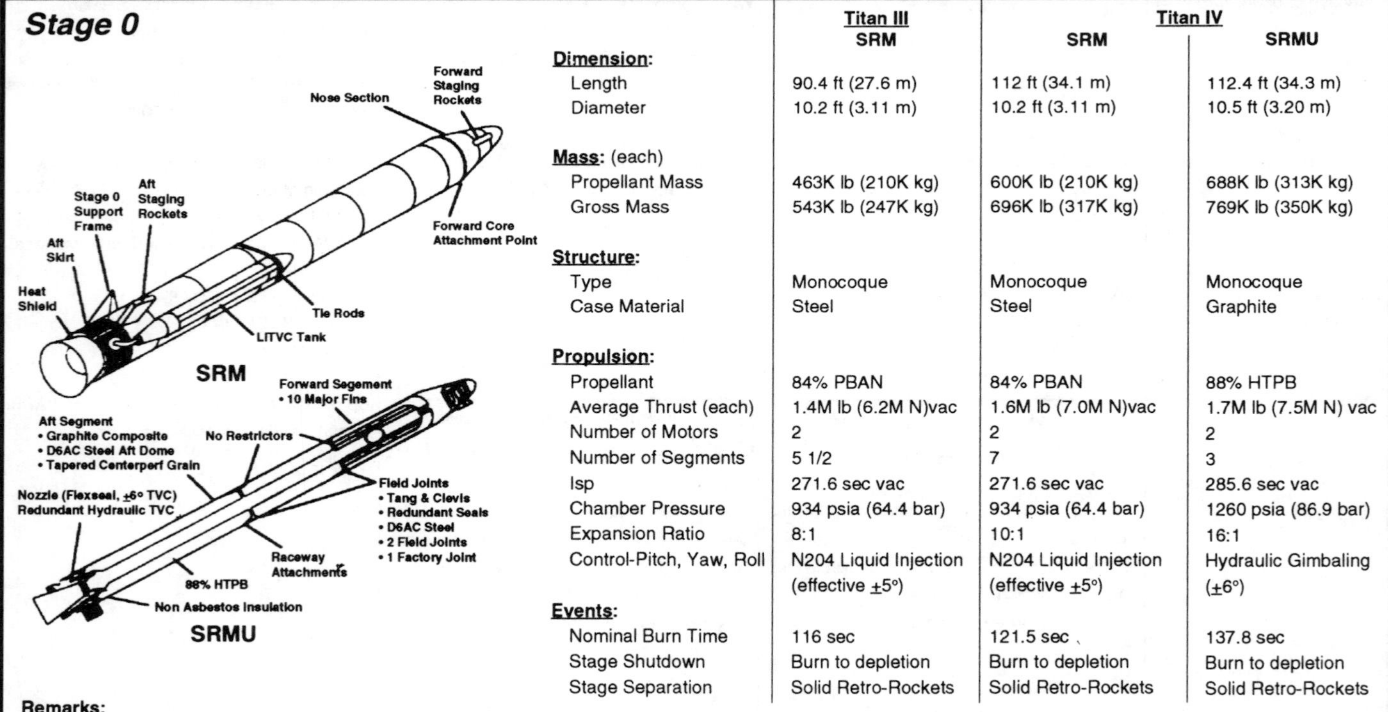

	Titan III	Titan IV	
	SRM	SRM	SRMU
Dimension:			
Length	90.4 ft (27.6 m)	112 ft (34.1 m)	112.4 ft (34.3 m)
Diameter	10.2 ft (3.11 m)	10.2 ft (3.11 m)	10.5 ft (3.20 m)
Mass: (each)			
Propellant Mass	463K lb (210K kg)	600K lb (210K kg)	688K lb (313K kg)
Gross Mass	543K lb (247K kg)	696K lb (317K kg)	769K lb (350K kg)
Structure:			
Type	Monocoque	Monocoque	Monocoque
Case Material	Steel	Steel	Graphite
Propulsion:			
Propellant	84% PBAN	84% PBAN	88% HTPB
Average Thrust (each)	1.4M lb (6.2M N)vac	1.6M lb (7.0M N)vac	1.7M lb (7.5M N) vac
Number of Motors	2	2	2
Number of Segments	5 1/2	7	3
Isp	271.6 sec vac	271.6 sec vac	285.6 sec vac
Chamber Pressure	934 psia (64.4 bar)	934 psia (64.4 bar)	1260 psia (86.9 bar)
Expansion Ratio	8:1	10:1	16:1
Control-Pitch, Yaw, Roll	N204 Liquid Injection (effective ±5°)	N204 Liquid Injection (effective ±5°)	Hydraulic Gimbaling (±6°)
Events:			
Nominal Burn Time	116 sec	121.5 sec	137.8 sec
Stage Shutdown	Burn to depletion	Burn to depletion	Burn to depletion
Stage Separation	Solid Retro-Rockets	Solid Retro-Rockets	Solid Retro-Rockets

Remarks:
The liftoff thrust of the Titan III and IV is provided solely by the two SRMs, which constitute Stage 0. The SRMs provide all the initial flight thrust and control. The nozzle is canted six degrees from the center line to insure aerodynamic stability of the vehicle during the initial ascent. Thrust vectoring to allow for steering of the vehicle is accomplished by injecting nitrogen tetroxide through 24 flow injector valves which surround the nozzle exit cone. The valves are actuated hydraulically, and they vary the flow rate by changing the orifice area. Injection of the nitrogen tetroxide creates an oblique shock wave that deflects the direction of the rocket's gases, providing steering for the booster stage. As SRMs burnout, the vehicle avionics sense the deceleration and command the core stage to ignite and SRM's are jettisoned. The goals of the SRMU, which is to be available in 1996, are to increase reliability, performance, and producibility. Key differences versus the SRM include the number of field joints (3 vs. 7), case material (composite vs. steel), propellant (HTPB vs. PBAN), size (12% heavier, 5% wider), thrust vector control (flexseal nozzle vs. secondary injection), and performance (25% increase).

Titan Vehicle

(continued)

Stage 1

1. Tank Entry Dome
2. Forward Dome
3. Oxidizer Tank Autogenous Pressurization Line
4. Forward Skirt Oxidizer Dome
5. Blast Ports
6. Oxidizer Tank Barrel
7. Aft Skirt Oxidizer Tank
8. Tension Splice
9. Fuel Tank Forward Tank Skirt
10. Fuel Tank Barrel
11. Tank Panel Longeron
12. Aft Cone
13. Fuel Suction Line
14. Air Scoop (TIIIB Only)
15. Oxidizer Suction Line
16. Oxidizer Suction Line
17. Internal Conduit
18. External Conduit
19. Fuel Tank Augotenous Pressurization Line
20. Forward Dome Fuel Tank
21. Aft Dome

	Titan II	Titan III	Titan IV
Dimension:			
Length	70.7 ft (21.5 m)	78.6 ft (24.0 m)	86.5 ft (26.4 m)
Diameter	10.0 ft (3.05 m)	10.0 ft (3.05 m)	10.0 ft (3.05 m)
Mass:			
Propellant Mass	260K lb (118K kg)	294K lb (134K kg)	340K lb (155K kg)
Gross Mass	269K lb (122K kg)	310 lb (141 kg)	359K lb (163K kg)
Structure:			
Type	Skin Stringer	Skin Stringer	Skin Stringer
Material	Aluminum	Aluminum	Aluminum
Propulsion:			
Propellant	N204 - Aerozine-50	N204 - Aerozine-50	N204 - Aerozine-50
Average Thrust	474K lb (2.09M N) vac	548K lb (2.41M N) vac	548K lb (2.41M N) vac
Engine Designation	LR87-AJ-5	LR87-AJ-11	LR87-AJ-11
Number of Engines	2	2	2
Isp	296 sec vacuum	302 sec vacuum	302 sec vacuum
Feed System	Gas Generator	Gas Generator	Gas Generator
Chamber Pressure	829 psia (57.2 bar)	829 psia (57.2 bar)	829 psia (57.2 bar)
Mixture Ratio (O/F)	1.9	1.9	1.9
Throttling Capability	100% only	100% only	100% only
Expansion Ratio	12:1	15:1	15:1
Restart Capability	No	No	No
Tank Pressurization	Autogenous	Autogenous	Autogenous
Control-Pitch, Yaw, Roll	Hydraulic gimbaling (2 nozzles)	Hydraulic gimbaling (2 nozzles)	Hydraulic gimbaling (2 nozzles)
Events:			
Nominal Burn Time	147 sec	164 sec	164 sec
Stage Shutdown	Burn to depletion	Burn to depletion	Burn to depletion
Stage Separation	"Fire in hole" stg 2 ignite	"Fire in hole" stg 2 ignite	"Fire in hole" stg 2 ignite

Remarks:

Stage 1 consists of an LR87 liquid-propellant rocket engine attached to an airframe which includes the fuel and oxidizer tanks, intertank structure, forward skirt and aft skirt. The fuel and oxidizer tanks are welded structures consisting of a forward dome, barrel section, and aft dome. The tanks are structurally independent, minimizing the hazard of propellant mixing should a leak develop in either tank. Also, this arrangement allows filling, draining, or pressurization of either tank without jeopardizing structural integrity. They are mounted in tandem with the oxidizer tank in front, and the fuel tank has an internal conduit to duct the oxidizer to the rocket engine. Both tanks have an access cover in the forward dome. The intertank structure and the skirts have welded frames to which the aerodynamic surface is riveted. Access doors are provided in the forward and aft sections and in the intertank structure, and four longerons on the aft skirt allow for Stage 0 attachment.

Semi-monocoque and monocoque construction is used for all sections (monocoque implies a structure in which the outer skin carries bending or shear stresses without frames or braces). Aluminum alloy is used throughout for skin areas, with gauges about twice as thick as that of Titan II. Milling of low-stress areas of tank interiors reduce vehicle weight. Integrally milled T-shape skin stringers extend longitudinally in each tank. Frames attached to the stringers, much like barrel hoops used internally, stabilize the structure when unpressurized. Thus, the vehicle can support its own weight in the unpressurized condition, either empty or loaded, when positioned vertically or horizontally.

A boattail heat shield encloses the Stage 1 engine compartment on Titan III/IV to protect Stage 1 engine components from radiant heat produced by the exhaust plumes of the SRMs. The boattail is aluminum and covers those portions of the engine above the thrust chamber assembly above the throat. Covers, made of refrasil sandwiched between two layers of Inconel, protect each thrust chamber from injection to aft end but are not part of the boattail. Exit closures are provided to shield the thrust chamber interior from heat. These closures are explosively separated when Stage 1 engine operation is initiated. Rubber covers over the turbine exhaust stacks are blown off by start cartridge gas pressure upon engine start. Thrust vector control is accomplished by gimbaling the engine thrust chamber to provide pitch, yaw, and roll corrections. Hydraulic actuators, driven from the engine turbopump and controlled by electrical signals from the guidance and flight control systems, provide the gimbal force. Stage 1 is shut down when either of the propellants is exhausted, by low level propellant sensors, or by command shutdown.

Both Stage 1 and 2 use storable hypergolic propellant of N2O4 oxidizer and Aerozine-50 fuel. N2O4 contains by weight approximately 30% nitrogen and 70% oxygen. Aerozine-50, which combines the stability of UDMH and higher performance of hydrazine, contains by weight approximately 50% UDMH and 50% hydrazine. Use of hypergolic propellants, which burn spontaneously when mixed, eliminates the need for an igniter thus improving staging ignition reliability. These propellants are storable and can remain in a launch ready state for extended periods. The use of propellants storable at ambient temperature and pressure eliminates holds and delays as when handling cryogenic liquids. This feature gives Titan the demonstrated capability of meeting critical launch windows. The propellant tanks are pressurized on the ground with nitrogen. An autogenous pressurization system using cooled fuel-rich turbine exhaust gas for the fuel tank and vaporized nitrogen tetroxide for the oxidizer tank maintains the in-flight pressure requirements.

Titan Vehicle

(continued)

Stage 2

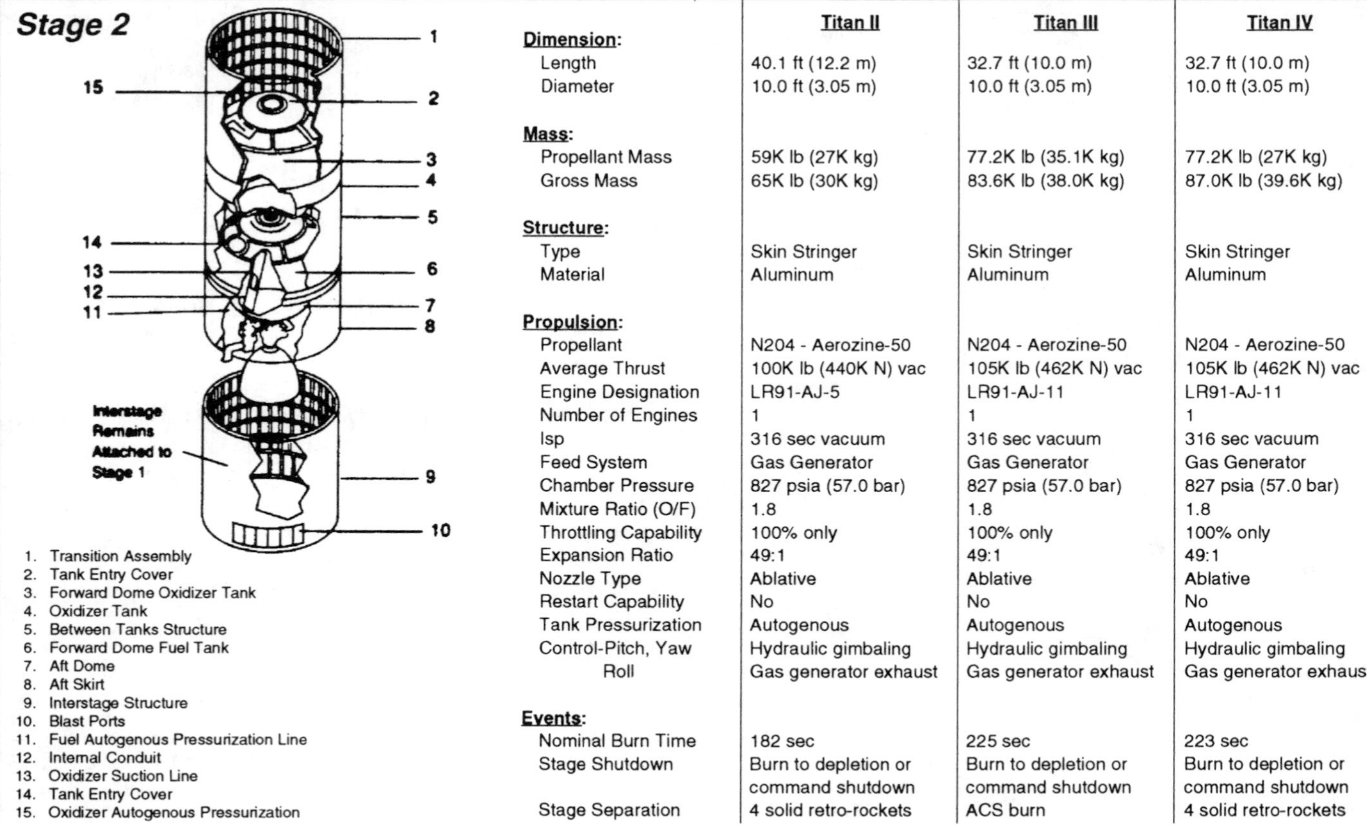

1. Transition Assembly
2. Tank Entry Cover
3. Forward Dome Oxidizer Tank
4. Oxidizer Tank
5. Between Tanks Structure
6. Forward Dome Fuel Tank
7. Aft Dome
8. Aft Skirt
9. Interstage Structure
10. Blast Ports
11. Fuel Autogenous Pressurization Line
12. Internal Conduit
13. Oxidizer Suction Line
14. Tank Entry Cover
15. Oxidizer Autogenous Pressurization

Interstage Remains Attached to Stage 1

	Titan II	Titan III	Titan IV
Dimension:			
Length	40.1 ft (12.2 m)	32.7 ft (10.0 m)	32.7 ft (10.0 m)
Diameter	10.0 ft (3.05 m)	10.0 ft (3.05 m)	10.0 ft (3.05 m)
Mass:			
Propellant Mass	59K lb (27K kg)	77.2K lb (35.1K kg)	77.2K lb (27K kg)
Gross Mass	65K lb (30K kg)	83.6K lb (38.0K kg)	87.0K lb (39.6K kg)
Structure:			
Type	Skin Stringer	Skin Stringer	Skin Stringer
Material	Aluminum	Aluminum	Aluminum
Propulsion:			
Propellant	N204 - Aerozine-50	N204 - Aerozine-50	N204 - Aerozine-50
Average Thrust	100K lb (440K N) vac	105K lb (462K N) vac	105K lb (462K N) vac
Engine Designation	LR91-AJ-5	LR91-AJ-11	LR91-AJ-11
Number of Engines	1	1	1
Isp	316 sec vacuum	316 sec vacuum	316 sec vacuum
Feed System	Gas Generator	Gas Generator	Gas Generator
Chamber Pressure	827 psia (57.0 bar)	827 psia (57.0 bar)	827 psia (57.0 bar)
Mixture Ratio (O/F)	1.8	1.8	1.8
Throttling Capability	100% only	100% only	100% only
Expansion Ratio	49:1	49:1	49:1
Nozzle Type	Ablative	Ablative	Ablative
Restart Capability	No	No	No
Tank Pressurization	Autogenous	Autogenous	Autogenous
Control-Pitch, Yaw	Hydraulic gimbaling	Hydraulic gimbaling	Hydraulic gimbaling
Roll	Gas generator exhaust	Gas generator exhaust	Gas generator exhaust
Events:			
Nominal Burn Time	182 sec	225 sec	223 sec
Stage Shutdown	Burn to depletion or command shutdown	Burn to depletion or command shutdown	Burn to depletion or command shutdown
Stage Separation	4 solid retro-rockets	ACS burn	4 solid retro-rockets

Remarks:

Stage 2 uses an LR91 liquid propellant rocket engine attached to an airframe similar in construction to that of Stage 1. A significant difference between the Titan III/IV Stage 2 and its counterpart in the Titan II vehicle is the removal of a major electronic equipment bay from the intertank section. The airframe consists of a transition assembly for attachment of an upper stage or payload, oxidizer tank, intertank structure, fuel tank, and aft skirt. An interstage structure connects Stages 1 and 2. Construction of the interstage structure is similar to that part of the nontank portions of the rest of the airframe, i.e., aluminum skin riveted to a welded frame. The Stage 2 propellant tanks are similar in structure to those of Stage 1. Thrust vector control is accomplished by gimbaling the chamber, but roll control, which is impossible using only one thrust unit, is provided by ducting pump turbine exhaust through a swiveled nozzle to produce thrust.

The Stage 1 and 2 separate from each other using the "fire-in-the-hole" technique. In short, when the first liquid stage depletes all the propellant and the onboard avionics sense the deceleration, the Stage 2 is commanded to ignite while still attached to first stage. The overpressure is vented through interstage blast ports until the pyros separate the two stages. This technique is the most efficient means for separating two stages in that no velocity is lost due to drift time. Depending upon the payload requirement, the Stage 2 is shut down by command upon achievement of payload target velocity. A signal from low level propellant sensors, or propellant exhaustion. For Titan III and some Titan II missions, the second stage remains with the payloads throughout deployment to act as a stable platform.

Upper Stages

PAM DII TOS IUS Centaur

	Titan III		Titan IV		
	PAM-DII	**TOS**	**IUS**		**Centaur**
Manufacturer	McDonnell Douglas	OSC/Martin Marietta	Boeing		Martin Marietta
Length	6.6 ft (2.0 m)	11 ft (3.3 m)	17 ft (5.2 m)		29.5 ft (9.0 m)
Diameter	5.25 ft (1.6 m)	11.2 ft (3.4 m)	9.5 ft (2.9 m)		14 ft (4.3 m)
			Stage 1	**Stage 2**	
Propellant Mass	7,140 lb (3,240 kg)	21,400 lb (9,710 kg)	21,400 lb (9,710 kg)	6,060 lb (2,750 kg)	44,880 lb (20,320 kg)
Gross Mass	7,695 lb (3,490 kg)	23,800 lb (10,800 kg)	23,960 lb (10,965 kg)	8,600 lb (3,900 kg)	52,600 lb (23,860 kg)
Propellant	Solid	Solid/HTPB	Solid/HTPB	Solid/HTPB	LOX/LH2
Engine Company	Thiokol	CSD	CSD	CSD	Pratt & Whitney
Engine Designation	STAR-63D	SRM-1	SRM-1	SRM-2	RL-10A-3-3A (2 unit)
Avg. Thrust	17.6K lb (78.3K N)	45K lb (200K N)	45K LB (200K N)	18.3K lb (81.2K N)	2 x 16.5K lb (73K N)
Isp	281.7 sec vac	294 sec vac	292.9 sec vac	300.9 sec vac	444 sec vac
Restart Capability	No	No	Multiple	No	Multiple
Control	Spin stabilized	1 nozzle and 3-axis ACS	1 nozzle and 3-axis ACS	1 nozzle and 3-axis ACS	2 nozzles and 3-axis ACS
Nominal Burn Time	121 sec	150 sec	153 sec	104 sec	600 sec

Remarks:

PAM-DII requires a spin table during deployment. IUS is two stages for its geosynchronous missions.

Titan Vehicle

(continued)

Payload Fairing

	Titan II	Titan III		Titan IV
	25 ft (7.6 m) Version	Dual Carrier Version		86 ft (26.2 m) Centaur Version
	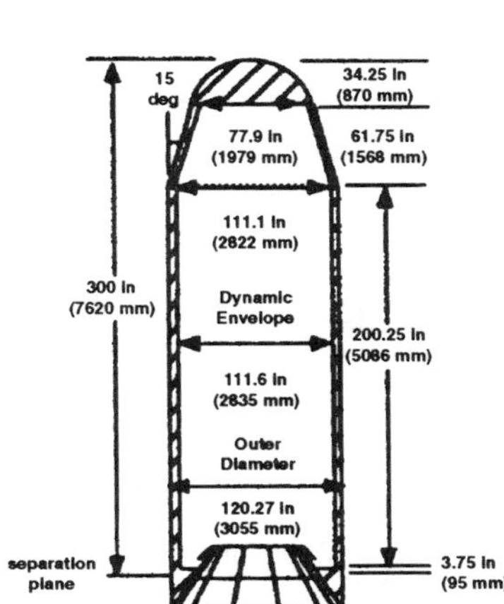	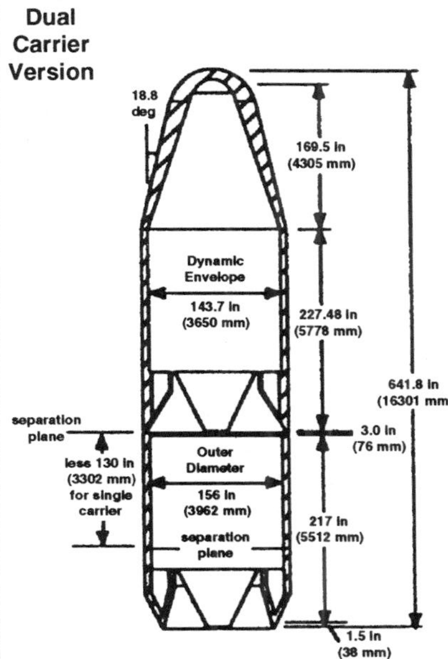		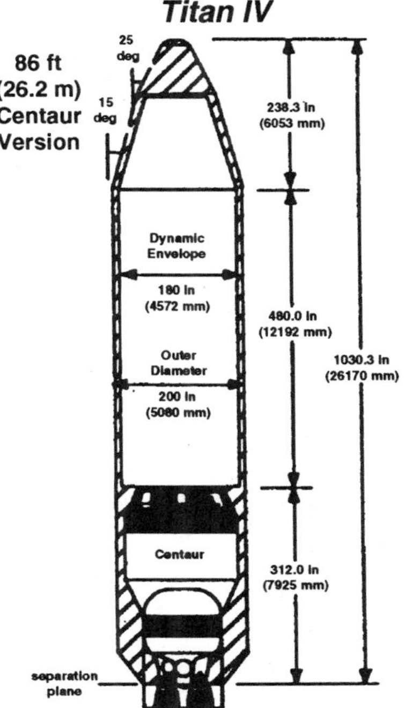
		Dual Payload Carrier	**Single Payload Carrier**	
Length	20, 25, or 30 ft (6.1, 7.6, or 9.1 m)	53.5 ft (16.3 m)	42.6 ft (13.0 m)	50, 56, 66, 76, or 86 ft (17.1, 20.1, 23.2, 26.2 m)
Diameter	10.0 ft (3.05 m)	13.1 ft (4.0 m)	13.1 ft (4.0 m)	16.7 ft (5.08 m)
Mass	1,435, 1,650, or 2,000 lb (652, 750, or 909 kg)	6,325 lb (2,875 kg)	4,990 lb (2,268 kg)	8, 11, 12, 13, 14K lb (3.6, 5.0, 5.5, 5.9, 6.3K kg)
Sections	3	2 for each compartment		3
Structure	Skin Stringer	Composite sandwich		Isogrid
Material	Aluminum	Aluminum honeycomb / graphite epoxy		Aluminum
Remarks	The fairing is an existing design used for Titan 34D called a universal payload fairing. The fairing has noncontaminating separation joints that contain any debris in longitudinal bellows along each trisector upon fairing jettison during flight.	Both payload carriers are based on designs used for Ariane IV. For dedicated flights, the upper portion of the extension module and its adapter are removed. For contamination protection to the payloads, a cover is secured between the fairing and extension module and another cover on bottom of extension module.		The fairing is based on a design used for Titan IIIE. The fairing is separated into three parts at a speed of 15 ft/s (4.57 m/s) by ignition of a mild detonating fuse in thrusting joints. Contamination is avoided by retaining smoke, gases, and metal particles in bellows which form an integral part of each thrust structure.

Avionics

The Titan avionics system is designed to provide the flexibility to adapt to a variety of upper stage vehicles. For Titan II, the guidance equipment is taken from the ICBM, and the electrical and ordnance taken from Titan 34D design. For Titan III and Titan IV, the avionics is based on earlier Transtage avionics using a four gimbaled carousel gyro. Starting with the 24th flight, Titan IV will use a Centaur-based single-string avionics system using a strap-down platform with ring laser gyros.

The basic avionics necessary to fly the lower stages are located in Stage 2 forward skirt for Titan III and IV and in Stage 2 intertank for Titan II. Flight computer, attitude sensors, load relief sensors, range safety provisions, redundant electrical power and associated harness are installed as standard equipment.

The digital flight computer is the heart of the avionics system and can be altered to accommodate a wide range of upper-body dynamic parameters. Its special purpose input/output directly interfaces the central processor with other avionics equipment to permit time-sharing for navigation, guidance, control, discrete sequencing, malfunction logic, prelaunch checkout, and guidance initialization. Inertial reference is provided and maintained by the inertial measurement unit.

The instrumentation and telemetry equipment includes a data bus pulse modulation encoder, and S-band transmitter. The tracking and flight safety capability has been designed for compatibility with range safety requirements. Electrical power is provided by silver-zinc batteries.

Attitude Control System

Stage 0 control is provided by the liquid injection thrust vector control, Stage 1 by gimbaling two nozzles, and Stage 2 by its single nozzle and turbine exhaust for roll control. The attitude control system (ACS) is contained in the Stage 2 forward skirt for Titan III and in Stage 2 intertank for Titan II. The ACS is a pressure-fed, positive expulsion, monopropellant system with multiple rocket engines. The system performs a wide range of propulsion functions including vehicle pitch, yaw, and roll control; orbit vernier control, and multipayload deployment. The anhydrous hydrazine (N_2H_4) propellant and pressurizing nitrogen gas are stored in a single tank separated by a rubber diaphragm. Each of the six rocket engine modules carries two thruster assemblies. Thrust varies between 16 lb (7.3 kg) and 27 lb (12 kg) during the mission. For Titan II, an enhanced ACS is available to provide circularization capability.

Titan Performance

Titan II

Circular Peformance: Enhanced Attitude Control System (EACS)

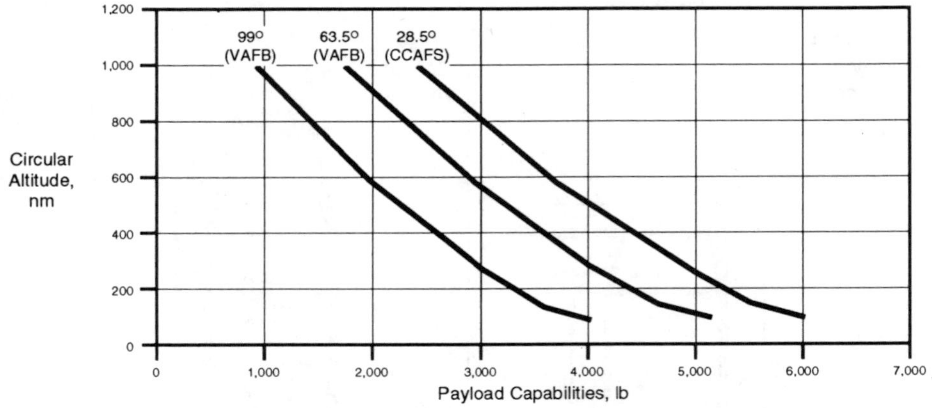

Elliptical Performance

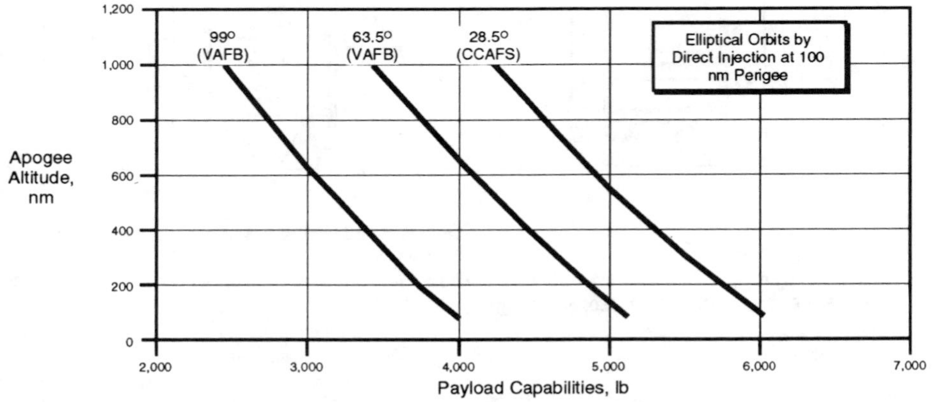

Titan III

Dual Payload Carrier from CCAFS

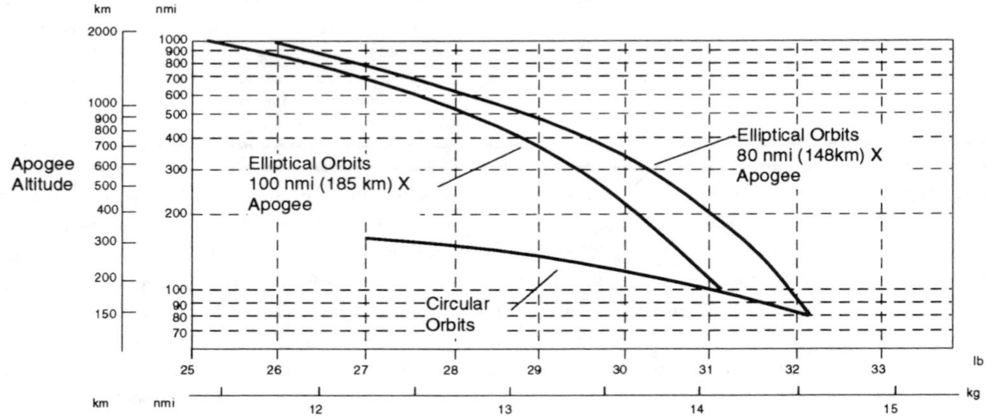

Single Payload Carrier from CCAFS

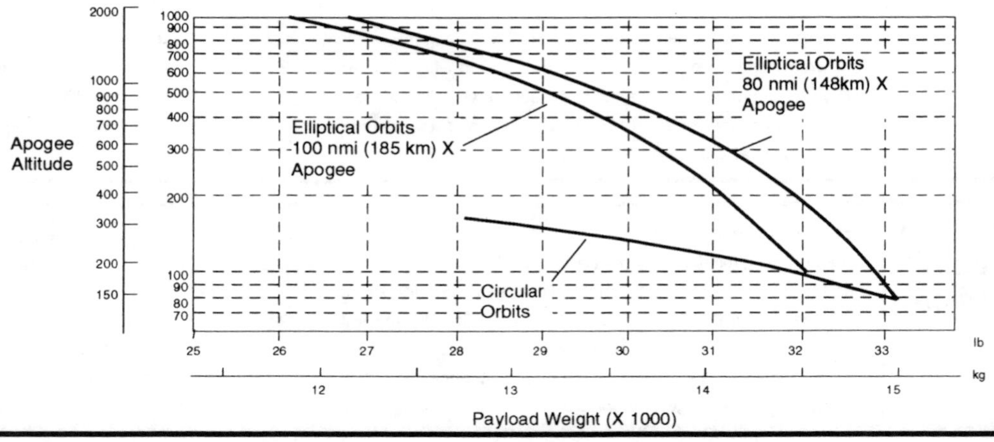

Titan Performance
(continued)

Titan IV

Titan IV / Centaur / SRM from CCAFS

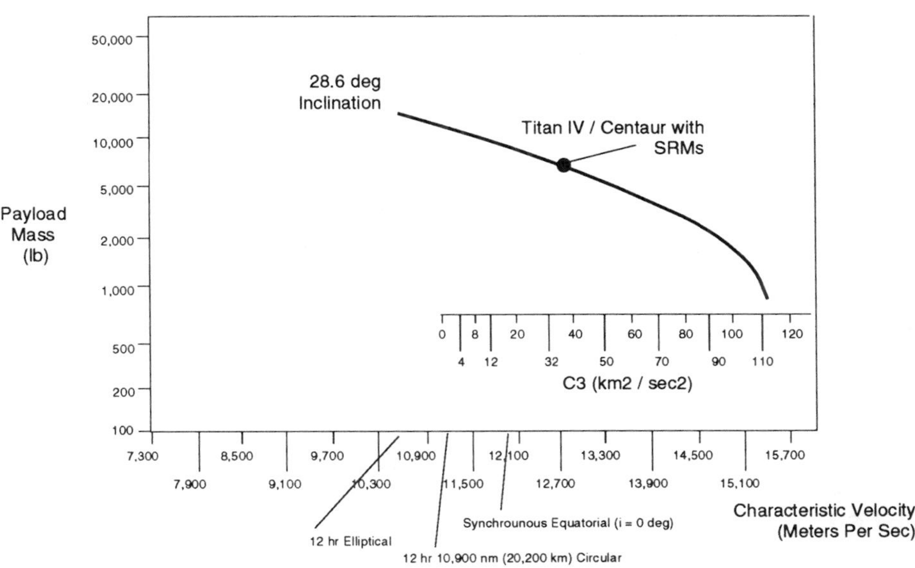

No Upper Stage / SRM from VAFB

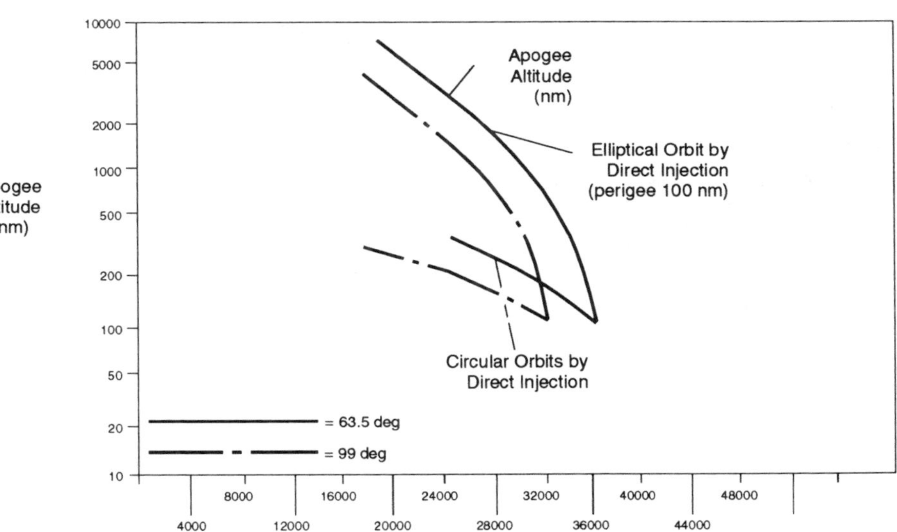

Titan Operations

Launch Site

Vandenberg Air Force Base (34.7°N) **Cape Canaveral Air Force Station (28.5°N)**

SLC-2W (Delta II)
SLC-3W (Atlas E)
SLC-4E / W
(Titan II, Titan IV)
SLC-5 (Scout)
SLC-6 (LLV)
KSC
LC-39B (STS)
LC-39A (STS)
CCAFS
LC-41 (Titan IV)
LC-40
(Titan III, Titan IV)
LC-36A / B
(Atlas I, II)
LC-17A / B
(Delta, Delta II)

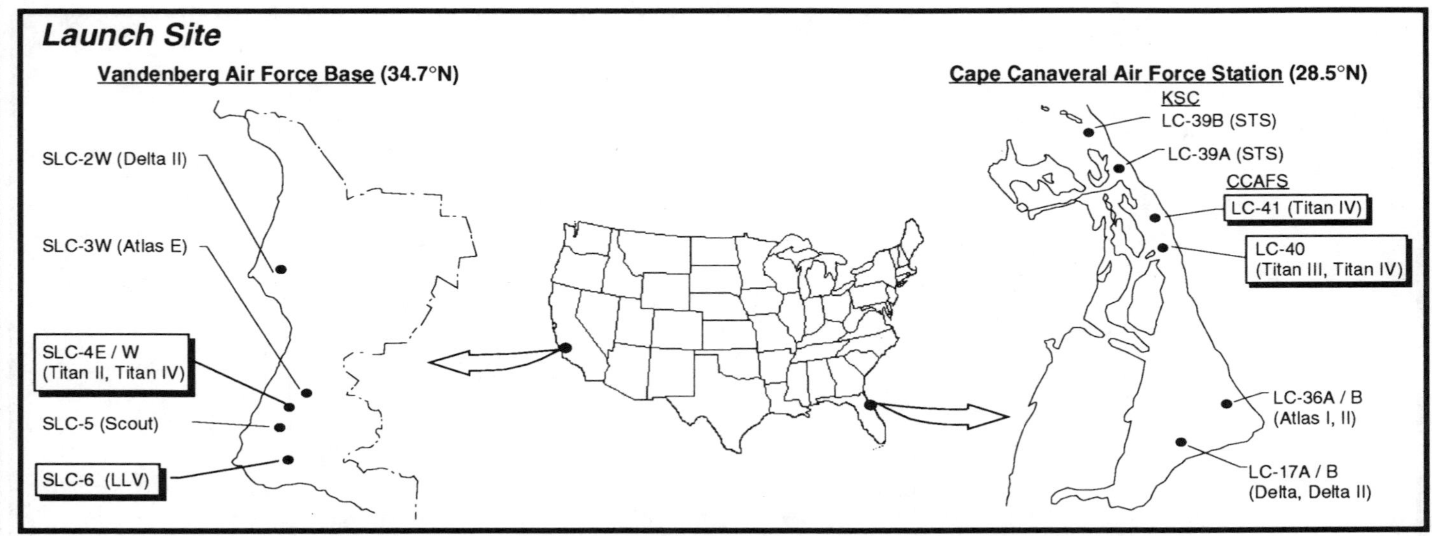

Launch Facilities

Cape Canaveral Air Force Station (CCAFS)

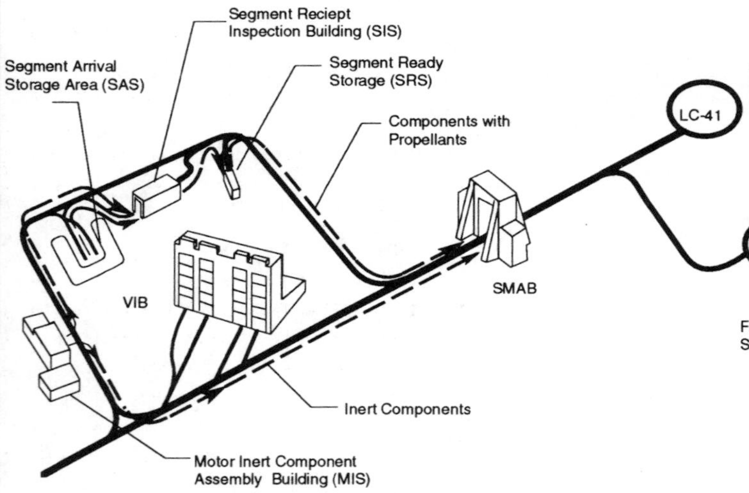

Segment Reciept
Inspection Building (SIS)
Segment Arrival
Storage Area (SAS)
Segment Ready
Storage (SRS)
Components with
Propellants
LC-41
LC-40
VIB
SMAB
Inert Components
Motor Inert Component
Assembly Building (MIS)

CCAFS LC-40/41 Launch Facilities Layout

Environmental
Control Shelter
Mobile
Service
Tower
Aerospace
Ground
Equipment
Building
Gas Storage
Area
Exhaust
Duct
Oxidizer
Vent Stack
Umbilical
Tower
Support
Building
(Underground)
Oxidizer
Holding
Area
Fuel Vent
Stack
Fuel Holding
Pond
Protective
Clothing
Building
Fuel Holding
Area
Launch
Platform

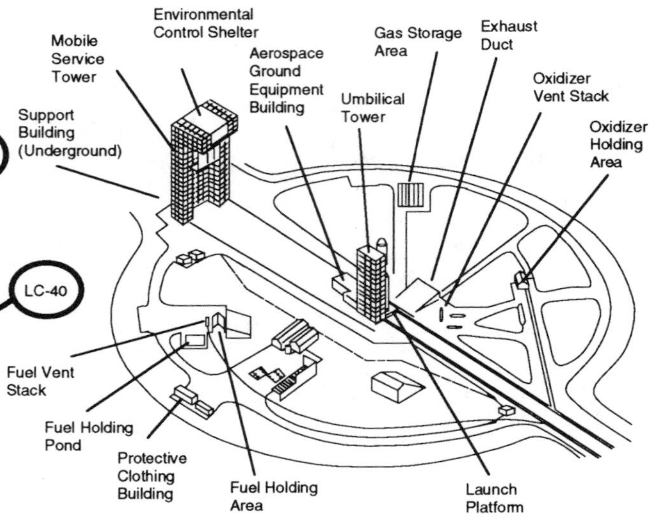

Typical Titan CCAFS Launch Pad Layout

Vandenberg Air Force Base (VAFB)

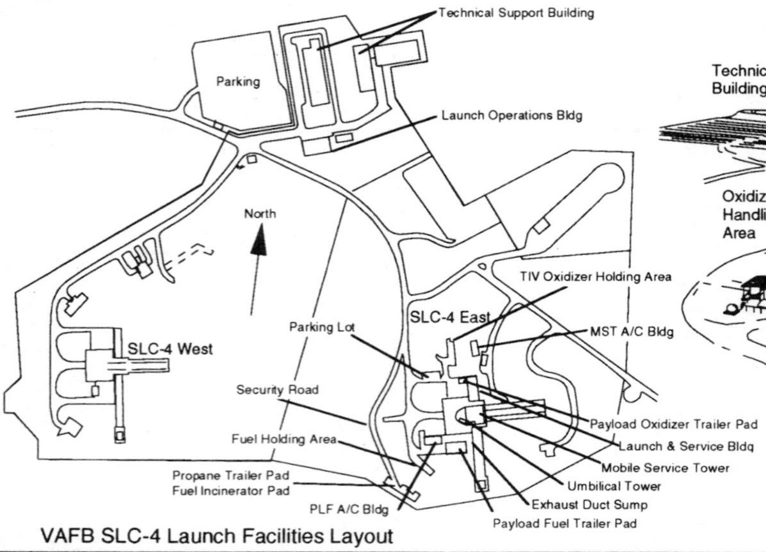

Technical Support Building
Parking
Launch Operations Bldg
North
TIV Oxidizer Holding Area
SLC-4 East
Parking Lot
MST A/C Bldg
SLC-4 West
Security Road
Payload Oxidizer Trailer Pad
Fuel Holding Area
Launch & Service Bldg
Mobile Service Tower
Propane Trailer Pad
Fuel Incinerator Pad
Umbilical Tower
Exhaust Duct Sump
PLF A/C Bldg
Payload Fuel Trailer Pad

VAFB SLC-4 Launch Facilities Layout

Meteorology
Survey Tower
Mobile
Service
Tower
Launch
Operations Bldg
To
SLC-4E
Technical Support
Buildings No. 1 & No. 2
Power
Substation
Oxidizer
Handling
Area
Vehicle
Support Bldg
Launch
Services
Bldg
Umbilical
Tower
Pad Support
Building
LN2
Storage
Ordnance
Magazine
Fuel
Handling
Area

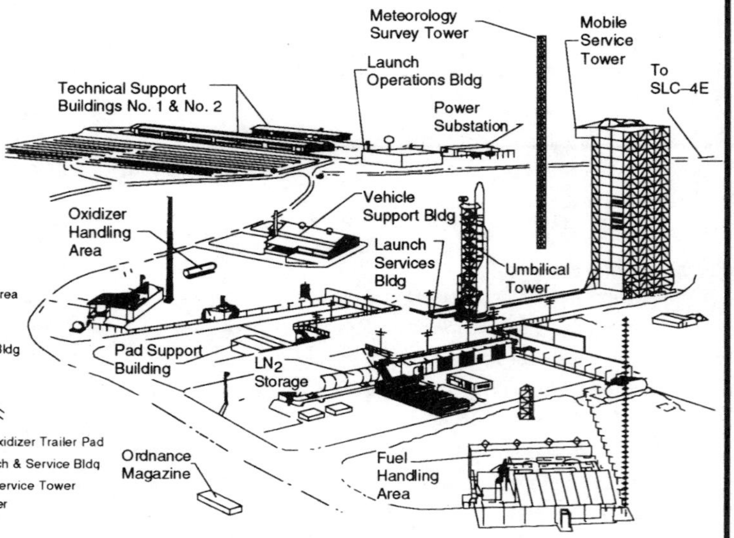

Typical Titan VAFB Launch Pad Layout

Titan Operations

(continued)

Launch Processing

Cape Canaveral Air Force Station (CCAFS). Titan III and Titan IV launched from CCAFS are processed using an Integrate-Transfer- Launch (ITL) concept. This permits maximum use of the launch complex since most prelaunch activities are conducted at a remote integration building. ITL consists of the SRM Processing Area, Vertical Integration Building (VIB), Solid Motor Assembly Building (SMAB), Launch Complex (LC) 40 and 41, a Titan transporter, and a railway system. The primary purpose of the SRM Processing Area is to receive, inspect, assemble, and test inert components, and store and prepare SRMs. The VIB is used to build-up and mate the core stages. In the SMAB, both SRMs are stacked and mated to the core vehicle. For upper stages, the Centaur is processed in one cell of the VIB and the IUS in the SMAB. The Titan transporter not only moves the launch vehicle to the pad (via railway) but also serves as the launch platform. Initial spacecraft operations are performed in a building assigned in the industrial area or an offsite location for commercial payloads.

The VIB is used to inspect, erect, and checkout the Titan core (Stage 1 and 2). The VIB has four cells for assembly and checkout, a center section for an inspection area, and a launch control center for monitoring vehicle operations. Cells 1, 2, and 4 are used for Titan and cell 3 for Centaur checkout. Stage 1 is removed from its transportation trailer and positioned on a rotation fixture in the VIB low bay area for receiving inspection. After engine installation, Stage 1 is hoisted to a vertical configuration and placed on a transporter in a cell. Stage 1 components are then installed. The Stage 2 horizontal operations in the VIB low bay are similar to Stage 1. Stage 2 is then hoisted and positioned on top of the Stage 1. Vertical alignment of the stages and stage electrical interface checks are then conducted.

A low bay area near each VIB cell provides space for checkout and instrumentation of Aerospace Ground Equipment (AGE) vans which are attached to the transporter. After checkout, the core vehicle is prepared for transport to the SMAB. The transporter moves at 3–4 mph (5–6 km/hr) and it takes 45–60 minutes to reach the SMAB.

The SRMs are processed in parallel with the core in the SRM processing area. The segments arrive via tractor trailer in the Segment Arrival Storage Area (SAS). Inert parts are inspected, assembled, and tested in the Motor Inert Storage (MIS) building. The segments are received and visually inspected for damage. In the Ready Inspection Building (RIS), they undergo laser and video imaging, dimensional measurements, ultrasonic inspection, infrared imaging, and finally x-ray inspection. And the Segment Ready Storage Building (SRS) provides a controlled environment for SRM components before they are individually transferred by rail to the SMAB. The segments are then individually moved by rail to the SMAB to mate with the core arriving from the VIB.

The SMAB has two high and two low bays. One high bay is used for SRM assembly, and another for core/SRM mate and checkout. One low bay is used for shuttle payload integration, and another for IUS assembly and checkout. The SMAB is capable of stacking all 5-1/2 segments for Titan III, but only five of the seven SRM segments for Titan IV due to transporter limitations. In 1993, the Solid Motor Assembly and Readiness Facility (SMARF) will be available for stacking all 5-1/2 segments of the SRM and three segments of the new SRMU. The remaining segments are mated at the launch pad with the Mobile Service Tower (MST). After stacking, the SRMs are mated to the Titan core vehicle. The vehicle is now transported by rail to the launch pad.

In parallel to booster processing, the Titan IV PLF is received and processed in the VIB Annex. In the annex, all prelaunch processing for the PLF is accomplished. This includes mechanical preparations, thermal coating application, electrical functional checks, final cleaning, and weighing. The forward and aft PLF sections are then transported to the pad and stored in the environmental shelter.

Titan III is launched from Launch Complex-40 (LC-40) and Titan IV from LC-41 (and LC-40 by 1993). The launch complex consists of the launch pad, Mobile Service Tower (MST), umbilical tower, aerospace ground equipment building, air conditioning shelter, gas storage area, propellant holding areas, and other service facilities. For Titan III, the payload fairing encapsulates the already mated spacecraft and upper stage offsite, and is then transported and hoisted atop the core. For Titan IV, the spacecraft, upper stage (except for IUS which is mated at the SMAB), and payload fairing are separately hoisted and mated at the launch pad. The MST facilitates mating of payload and payload fairing with the vehicle. An environmental shelter above the vehicle is provided to ensure a clean and secure area for spacecraft activities.

In preparation for actual launch, a combined systems test (CST) is performed which includes verification checks, a terminal countdown, and simulated launch and flight. After a successful launch CST data review, the readiness countdown is started. Some of the major events in the readiness countdown are: power-on payload test, MST and umbilical tower prelaunch preparations, install flight ordnance and checkout, connect core and SRM batteries, load core oxidizer then fuel, load SRM N2O4, pressurize propellant tanks, and Titan guidance system launch preparations.

The launch countdown begins with the launch status preparations and includes manual activities and an automatic sequence for launching the vehicle in accordance with the established launch time. Some key events that occur are: remove core/SRM destruct and ignition safety pins, flight control preparations, move MST to park position, range clearance for launch and upper stage conversion to internal power, Mission Director Go/No Go, Launch Director Go/No Go, start automatic sequence, ignite SRMs and liftoff.

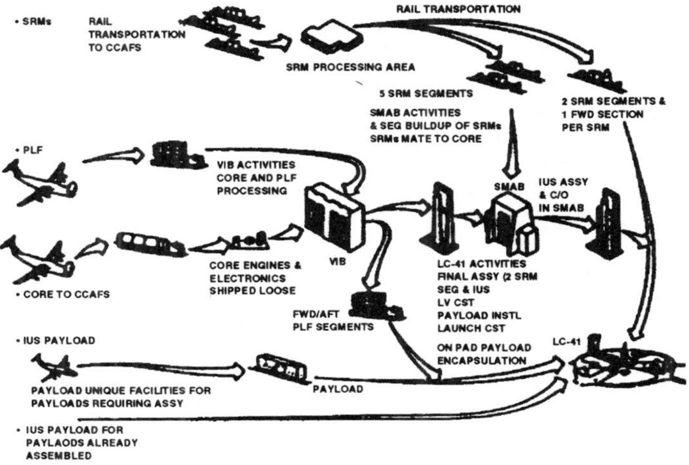

Illustration of CCAFS Titan IV/IUS Processing

Titan Operations

(continued)

Launch Processing

(continued)

Vandenberg Air Force Base (VAFB). Titan II and Titan IV launched from VAFB are processed using a build-on-pad concept. The Space Launch Complex-4 (SLC-4) is part of a two launch pad configuration that shares a common Launch Operations Building, which includes the launch control center. Each launch pad area contains a mobile service tower (MST), a propellant storage area, a high-pressure gas storage area, a launch service building on which a fixed umbilical tower is mounted, a launch mount, and a MST track system.

Titan IV is launched from SLC-4E. The SRM arrives via rail at the SRM Receiving Inspection and Storage (RIS) facility at Building 945, V31, and V33. The core arrives via air and the liquid engines via surface and are transported to the Vehicle Assembly Building/Horizontal Test Facility (VAB) at Building 8401. Payload fairing arrives via air and is trucked to the Payload Fairing Cleaning Facility at Building 8337. Each of these vehicle components are received, inspected, and tested to various degrees at these facilities. Next, the components are transported to SLC-4E where they are assembled and checked out on the launch mount.

Assembly begins with the SRMs, which mechanically interface with the launch mount, followed by Stage 1 and 2 which mechanically interface with the SRMs. Checkout of the launch vehicle is accomplished systematically by a building block approach beginning with components testing followed by subsystem and system test and culminates in a combined system test to verify the launch vehicle integrity prior to mating of the payload.

Titan II is launched from SLC-4W. Titan II conducts the same processing procedures as Titan IV except for the SRMs.

The process of converting a Titan II ICBM to a space launch vehicle begins with removal from storage at Norton AFB. The engines (first and second stages) are removed and shipped to Aerojet Technical Systems in Sacramento, where they enter a refurbishment program and are "hot-fired" to characterize the baseline engine performance.

After the engines are removed, the first and second stage core structures are shipped to Martin Marietta in Denver where they undergo the following modifications: (1) remanufacture of the Stage II oxidizer tank to provide additional capacity; (2) manufacture of adapter rings to allow mating of payload adapters/US adapters vice reentry vehicle; (3) fabrication of additional equipment truss to carry unique space launch systems (destruct systems, telemetry systems, etc.); (4) installation of electrical cabling, umbilical connectors, and other unique space flight items; and, (5) receipt and inspection of various new subsystems from vendors.

After completing the Martin Marietta factory processing, the Titan II "Kit" (except engines and guidance subsystems) is ready for shipment to VAFB where final assembly occurs. At VAFB, as at the Martin Marietta Denver facility, resources assigned to the Titan programs are shared. That is, the core facilities and manpower support both the Titan II and Titan IV programs. The common facilities consists of the VAB and the Payload Fairing Processing Facility (PLFPF). The final assembly process begins in the VAB when the Titan II "Kit" is assembled up to the state where it can either be shipped to SLC-4W for launch processing or be placed into storage pending a launch call-up.

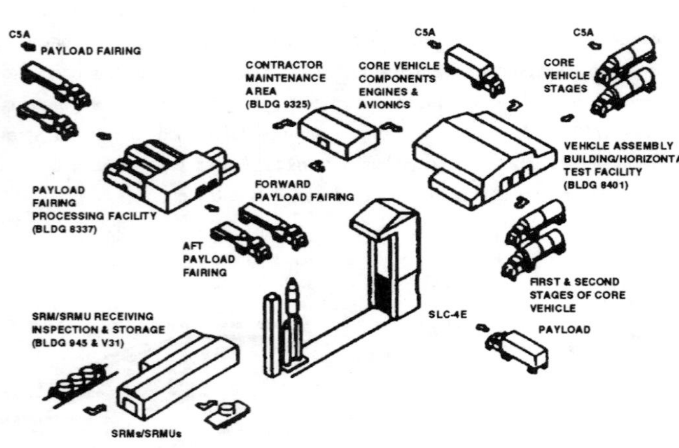

Illustration of VAFB Titan IV Processing

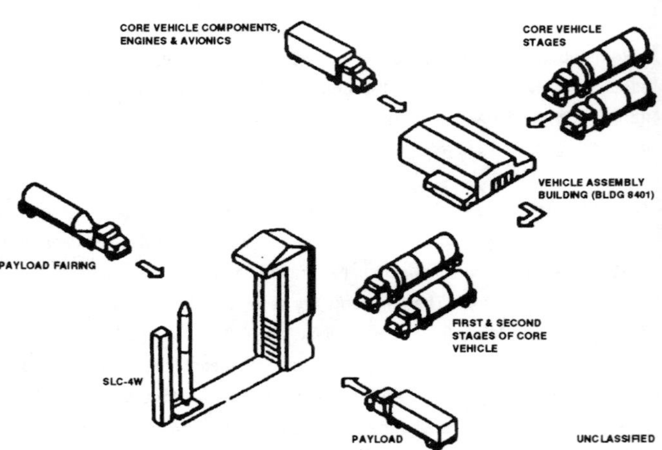

Illustration of VAFB Titan II Processing

Titan Operations

(continued)

Flight Sequence

Titan II Typical Flight Sequence

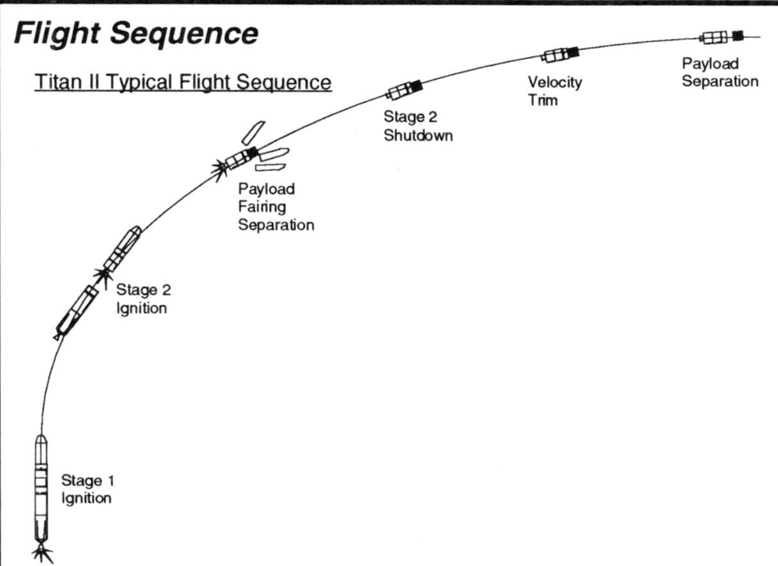

Time (min:sec)	Events
-00:05	Go Inertial
-00:02	Stage 1 Ignition Signal
00:00	Liftoff
02:38	Stage 1 Separation
03:31	Payload Fairing Separation
05:33	Stage 2 Shutdown
05:49	Enable ACS
06:18	Begin Velocity Trim
06:46	End Velocity Trim (Nominal)
07:14	Payload Separation Maneuver
08:39	Payload Separation

Reference Orbit: 100 x 100 nm (185 x 185 km), 90°

Titan III Typical Flight Sequence

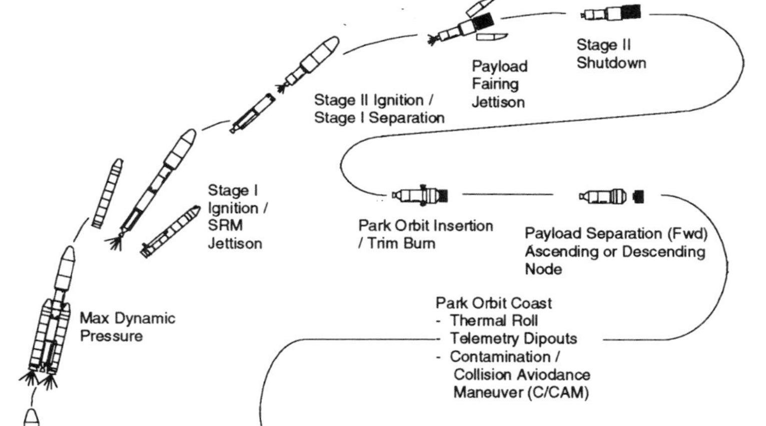

Time (min:sec)	Events
00:00	Stage 0 Ignition
00:54	Max Dynamic Pressure
01:48	Stage 1 Ignition
01:56	Stage 0 Jettison
04:29	Stage 2 Ignition
04:30	Stage 1 Separation
04:40	Payload Fairing Separation
08:14	Stage 2 Shutdown
08:30	Park Orbit Insertion
08:40	Trim Burn
67:16	Deploy Payload 1
155:50	Deploy Payload 2
201:50	Launch Vehicle Burn to Depletion (Deorbit)

Reference Orbit: 80 x 140 nm (148 x 259 km), 28.6°
Deploying Both Payloads at Ascending Node

Titan IV Typical Flight Sequence

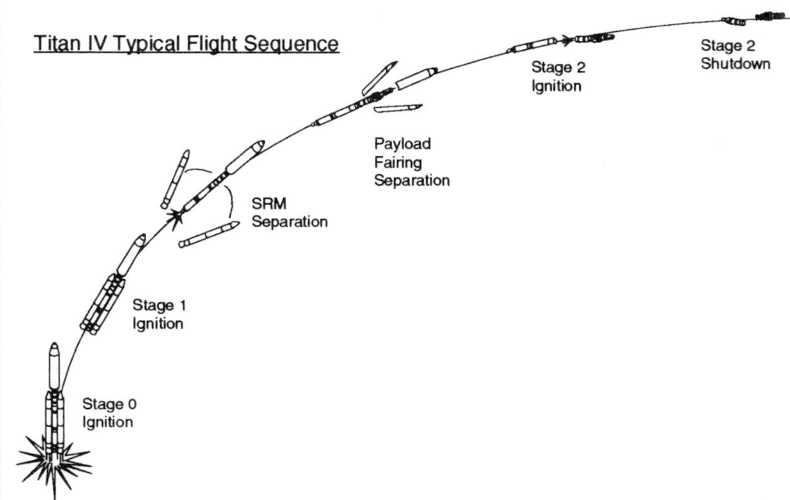

Time (min:sec)	Events	Altitude (ft)
00:00	Stage 0 Ignition	0
02:00	Stage 1 Ignition	158375
02:12	Stage 0 Separation	186398
03:50	Payload Fairing Separation	383614
05:08	Stage 2 Ignition	501535
05:09	Stage 1 Separation	502624
08:52	Stage 2 Shutdown	608391
09:18	Stage 2 Jettison	607604

Reference Orbit: 100 x 100 nm (185 x 185 km), 90°

Titan Payload Accommodations

	Titan II	Titan III	Titan IV
Payload Compartment Maximum Payload Diameter	111.5 in (2,832 mm)	143.7 in (3,650 mm)	180 in (4,570 mm)
Maximum Cylinder Length	144 in (3,658 mm) for 20 ft PLF 204 in (5,182 mm) for 25 ft PLF 264 in (6,706 mm) for 30 ft PLF	Dual Carrier - 227.5 in (5,778 mm) top compartment, 217.0 in (5,512 mm) bottom compartment Single Carrier - 317.0 in (8,049 mm)	742, 622, 502, or 382 in (18,850, 15,800, 12,750, or 9,700 mm)
Maximum Cone Length	61.75 in (1,568 mm)	169.5 in (4,305 mm) for both dual and single carrier	238 in (6,045 mm)
Payload Adapter Interface Diameter	56.15 in (1,426 mm) 36.0 in (915 mm)	45.0 in (1,143 mm) PAM-D 52.0 in (1,321 mm) ORBUS 7 63.0 in (1,600 mm) PAM-DII / III 76.0 in (1,920 mm) E-SCOTS 91.0 in (2,311 mm) ORBUS 21	??
Payload Integration Nominal Mission Schedule Begins	T-36 months	T-33 months	T-33 months
Launch Window Latest Countdown Hold Not Requiring Recycling	T-3 minutes	T-5 minutes	T-5 minutes
On-Pad Storage Capability	30 days fueled	30 days fueled	30 days fueled
Latest Access to Payload	T-15 days (approx.) or T-5 hrs through access doors	T-10 days (approx.) or T-3 hrs through access doors	??
Environment Maximum Load Factors	+4.0 to +10.0 g axial, ±2.5 g lateral	+2.5, -5.0 g axial, ±1.7 g lateral	+3.3, -6.5 g axial, ±1.5 g lateral
Minimum Lateral / Longitudinal Payload Frequency	2-10 Hz avoid <6 Hz / 12 - 24 Hz	10 Hz / 26 Hz	>2.5 Hz avoid 6-10 Hz / 17-24 Hz
Maximum Overall Acoustic Level	139.6 dB w/o blankets 134.4 dB w/ blankets 20,25 ft PLF 130.0 dB w/ blankets 30 ft PLF (1/3 Octave)	142 dB (Full Octave)	139.3 dB (Full Octave)
Maximum Flight Shock	200 g at 500 Hz	4,100 g at 1,250 Hz	2,000 g at 5,000 Hz
Maximum Dynamic Pressure on Fairing	750 lb/ft2 (35,910 N/m2)	950 lb/ft2 (45,490 N/m2)	926 lb/ft2 (44,340 N/m2)
Maximum Pressure Change in Fairing	0.5 psi/s (3.5 KPa/s). PLF could be kept attached until orbit.	1.0 psi/s (6.9 KPa/s)	0.4 psi/s (2.7 KPa/s) except 0.5 psi/s (3.5 KPa/s) for <8 sec
Cleanliness Level in Fairing (Prior to Launch)	Class 10,000	Class 10,000	Class 5,000
Payload Delivery Standard Orbit and Accuracy (3 sigma)	100 nm circular, +1, -0.5 nm (185 km, +1.8, -0.9 km) Inclination: 90.0 ± 0.15 deg	Perigee: 80 +2, -0 nm (148 +3.6, -0 km) Apogee: 140 ±6 nm (259 ±11 km) Inclination: 28.6 ± 0.01 deg	Perigee: 60 ±1.1 nm (111 ±2.0 km) Apogee: 177 ±4.4 nm (328 ±8.1 km) Inclination: 28.6 ± 0.01 deg
Attitude Accuracy (3 sigma)	All axes: ±2.0 deg; ±1.0 deg/sec	Pitch,Yaw: ±1.3 deg; ±0.2 deg/sec Roll: ±1.3 deg; ±0.3 deg/sec	??
Nominal Payload Separation Rate	2 ft/sec (0.6 m/s) w/ booster retros	2 ft/sec (0.6 m/s)	2 ft/sec (0.6 m/s)
Deployment Rotation Rate Available	0–2 rpm (more with spin table)	0–2 rpm (more with spin table)	0–2 rpm
Loiter Duration in Orbit	1.5 hrs (batteries optional)	3.5 hrs (longer is available)	??
Maneuvers (Thermal / Collision / Telemetry)	Yes	Yes	Yes for No Upper Stage vehicle

Titan Notes

Publications

User's Guide

Payload User's Guide: Titan II Space Launch Vehicle, Martin Marietta Astronautics Group, CDRL-051A2, August 1986.

User's Handbook: Titan IV (Preliminary), Martin Marietta Astronautics Group, CDRL 011A2, December 1988.

Customer Handbook: Titan III Commercial Launch Services, Martin Marietta Commercial Titan, Inc., Issue No. 1, December 1987.

Technical Publications

"Titan II SLV Secondary Payload Capability", A. Butts, M. Nance, Martin Marietta Astronautics Group, AIAA Small Satellite Conference, Aug 27-30, 1990.

"A Historical Look at United States Launch Vehicles: 1967-Present", ANSER, STDN 90-4, second edition, February 1990.

"Titan II Small Multisatellite Mission Approach," A. Butts, D. Giere, et al., Martin Marietta, AIAA Small Satellite Conference, September 1989.

"The Commercial Evolution of the Titan Program", S. Isakowitz, Martin Marietta, AIAA-88-3119, AIAA 24th Joint Propulsion Conference, Boston, MA, July 11-13, 1988.

"Commercial Titan: A Proven Derivative", R. Dutton and S. Isakowitz, Martin Marietta, AIAA-87-1796, AIAA 12th International Communication Satellite Systems Conference, Arlington, VA, March 13-17, 1988.

"The Commercial Titan Launch Vehicle", J. Troutman, S. Isakowitz, Martin Marietta, AIAA 23rd Joint Propulsion Conference, San Diego, CA, June 29-July 2, 1987.

Acronyms

A/C - air conditioner
ACS - attitude control system
CCAFS - Cape Canaveral Air Force Station
CST - Combined Systems Test
ELV - expendable launch vehicle
FPS - feet per second
GEO - geosynchronous orbit
HTPB - hydroxy terminated polybutadiene
ICBM - intercontinental ballistic missile
ITL - Integrate Transfer Launch
IUS - Inertial Upper Stage
KSC - Kennedy Space Center
LC - launch complex
LEO - low Earth orbit
MIS - Motor Inert Component Assembly Storage Building
MST - Mobile Service Tower
N2H4 - anhydrous hydrazine
N2O4 - nitrogen tetroxide
NASA - National Aeronautics and Space Administration
PAM - Payload Assist Module
PBAN - polybutadiene acrylonitrile acrylic acid
PLF - payload fairing
PLFPF - Payload Fairing Processing Facility
RIS - Segment Ready Inspection Building
SAS - Segment Arrival Storage Area
SLC - Space Launch Complex
SMAB - Solid Motor Assembly Building
SMARF - Solid Motor Assembly & Readiness Facility
SRM - solid rocket motor
SRMU - Solid Rocket Motor Upgrade
SRS - Segment Ready Storage Building
TOS - Transfer Orbit Stage
USAF - United States Air Force
VAB - Vehicle Assembly Building
VAFB - Vandenberg Air Force Base
VIB - Vertical Integration Building

Historical Launch Systems

Table of Contents

Europe: Europa

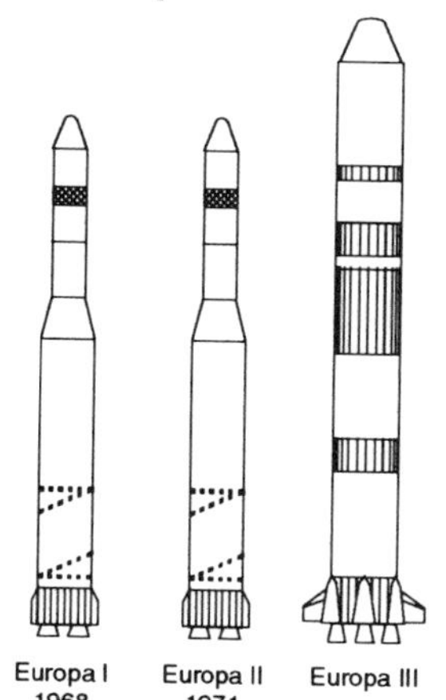

	Europa I	Europa II	Europa III
First Flight	1968	1971	—
East LEO	2,980 lb (1,350 kg)	3,180 lb (1,440 kg)	12,100 lb (5,500 kg)

Vehicle	**Europa I / II**
System Height	99.6 ft (30.35 m)
Gross Mass	247,000 lb (112,000 kg)
Stage 1	Blue Streak (Britain)
- Length	55.9 ft (17.04 m)
- Diameter	10.0 ft (3.05 m)
- Propellant	LOX / RP-1
- Thrust (SL)	2 x 150K lb (666K N)
Stage 2	Coralie (France)
- Length	18.0 ft (5.50 m)
- Diameter	6.56 ft (2.00 m)
- Propellant	UDMH / N204
- Thrust (vac)	1 x 61.7K lb (275K N)
Stage 3	Astris (Germany)
- Length	12.5 ft (3.81 m)
- Diameter	6.63 ft (2.02 m)
- Propellant	A50 / N2O4
- Thrust (vac)	1 x 5.07 lb (22.6K N)
Stage 4 (Europa II only)	Perigee System (France)
- Length	6.63 ft (2.02 m)
- Diameter	2.40 ft (0.73 m)
- Propellant	Solid
- Thrust	??
Fairing	(Italy)
- Length	13.2 ft (4.01 m)
- Diameter	6.56 ft (2.00 m)
	Europa III
System Height	120 ft (36.5 m)
Diameter	12.5 ft (3.8 m)
Stage 1 Engine	4 x Viking II
- Propellant	UDMH / N2O4
- Thrust	485K lb (2,160K N)
Stage 2 Engine	1
- Propellant	LOX/LH2
- Thrust	44K lb (195K N)
- Chamber	2,175 psia (150 bar)

Reference:
- Jane's Spaceflight Directory, 1987
- TRW Space Log 1957-87
- Missiles & Rockets, K. Gatland, 1975
- Europa III System Specifications, 1973
- Europa II Specification Sheet
- Europa I Description

The seven-nation European Launcher Development Organization (ELDO) was set up on February 29, 1964. The initiative came from the United Kingdom (UK) anxious to find a use for Blue Streak, originally developed as a strategic missile, as the first stage of an independent European launch system. The engine for the Blue Streak (RZ.2) was licensed to Rolls Royce from a U.S. firm Rocketdyne and General Dynamics's Atlas ICBM experience was shared with de Havilland.

Even before Blue Streak was abandoned as a weapon, design studies had been made at the Royal Aircraft Establishment, Farnborough, to see if it could be used as the basis of a satellite launcher in combination with a modification of the Black Knight rocket which was being developed to test reentry techniques for the IRBM. It was estimated that the two-stage concept, known as Black Prince, could place a payload of 2,750 lb (793 kg) into a circular orbit at 298 nm (552 km).

After Blue Streak cancellation the UK approached other European nations for a joint project. France and West Germany accepted only if it would not be an all-British rocket. Instead stages would be developed separately by the principal collaborators. Thus was born the highly political rocket called Europa I.

ELDO member countries were Belgium, France, West Germany, Italy, Netherlands, and the UK, plus Australia which was equally anxious to keep the Woomera Test Range, north of Adelaide, in business. With France providing the Coralie second stage, Germany the Astris third stage, Italy the satellite test vehicles, Belgium the ground guidance station, and the Netherlands the telemetry links and other equipment, Europa I was built. Under the initial agreement the UK agreed to contribute about 38% of the total budget plus Blue Streak and its test facilities.

The first stage, Blue Streak, was successfully launched ballistic five times at Woomera between 1964–1966. On both of the first two suborbital multistage launches in 1967 with the French second stage and an inert German third stage, the second stage failed to ignite. The next three launches in 1968–1970 were orbital attempts but each time the third stage failed, and on the last attempt the fairing did not jettison as planned. Thus ended the Europa I program and the ELDO venture in Woomera, Australia—still without a satellite in orbit.

Despite growing doubts about the project and the British 1970 decision to withdraw, ELDO pressed ahead with development of the more advanced Europa II, same as Europa I but with a French-built fourth-stage perigee motor able to place a satellite of up to 440 lb (200 kg) in geostationary orbit. Europa II launches were to be conducted from a new pad at Kourou, French Guiana, for both polar and easterly launch capability and about a 35% LEO payload gain due to its low latitude. The objective of Europa II was to launch two Franco-German Symphonie telecommunications satellites in 1973 and 1974. Unfortunately the first test firing of Europa II, from Kourou on November 5, 1971, with the world's space journalists observing, ended in disaster. After 2-1/2 min the launcher deviated from its trajectory, became overstressed and exploded.

YEAR	DATE	VEHICLE	SITE	PAYLOAD
1968	Nov-30	Europa I	Woomera	STV 1 (F-7)
	Failure - Premature Stg 3 shutdown			
1969	Jul-2	Europa I	Woomera	STV 2 (F-8)
	Failure - Stg 3 did not ignite			
1970	Jun-12	Europa I	Woomera	STV 3 (F-9)
	Failure - Stg 3 thrust lost and no fairing jettison			
1971	Nov-5	Europa II	Kourou	STV 4 (F-11)
	Failure - Inertial guidance ceased to function			

Launch site was Woomera, Australia (31.1°S, 136.8°E) and Kourou, French Guiana (5.2°N, 52.8°W)

Although urgent remedial action was taken after the mishap and the twelfth rocket was being made ready for launching at Kourou in October 1973, the vehicle was never flown. This was because the ministers of France and West Germany had decided to stop all further contributions to the program and at an ELDO Council Meeting on April 27, 1973, the Europa I / II program was cancelled. Up to that time Europe had spent about $745 million on ELDO.

The French, nevertheless, were still anxious to see a European launcher despite the apathy of their ELDO partners. Already, in December 1972, development of Europa III based largely on French and German rocket technology had failed to receive support from the respective parliaments. The UK government made it plain that it had no interest in launchers because satellites could be launched more cheaply in the U.S. The two-stage Europa III was designed to have four Viking II rocket engines using UDMH/N2O4 propellants. The top stage was expected to have an advanced LOX/hydrogen high-pressure engine being developed by MBB, and there were plans to produce this engine in collaboration with SEP of France by the 'Cryo-rocket' consortium. This larger launch vehicle had been regarded as a means of placing satellites of 3,400 lb (1,550 kg) into geotransfer orbit.

However, continuing budgetary difficulties within ELDO and wavering national priorities resulted in a French proposal which promised to reduce the development cost and provided a technical plan for a new launcher to take over from Europa. France guaranteed it would finance more than half of the program itself and would take responsibility for cost overruns. The new launcher was code-named L3S (the French acronym for third-generation substitution launcher). It was a three-stage design and eliminated the high-energy top stage engine. However, it retained the Europa III first stage and carried a second stage with a single Viking II engine and a new third stage, still using LOX/hydrogen, but with a low-pressure engine which owed something to French development experience. Although heavier than Europa III, with less ambitious technology, the vehicle could theoretically achieve at least the same performance. The Europa II launch pad at Kourou would be used for the L3S program. Thus was born the L3S (renamed Ariane) rocket adopted as one of the major projects of the European Space Agency (ESA) which came into existence in 1973.

Historical Launch Systems

France: Diamant

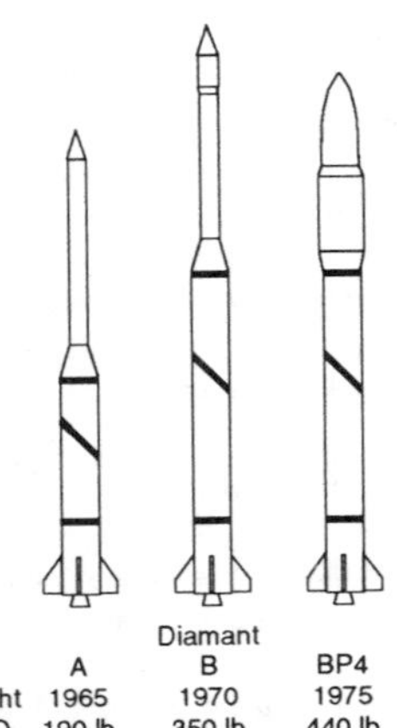

Vehicle	A	Diamant B	BP4
First Flight	1965	1970	1975
East LEO	190 lb (85 kg)	350 lb (160 kg)	440 lb (200 kg)

Diamant A, France's first satellite-launcher, was the final step in the 'precious stones' experimental series designed in 1960 as part of the national rocket program. Emeraude, the first stage, was a liquid-propellant rocket delivering thrust for 93 seconds; Topaze, the second stage, was powered by solid propellant and delivered thrust for 44 seconds. The first and second stages were tested individually several times in 1964 and 1965, and their combined form, called Saphir, was successfully fired from the Hammaguir base on October 9, 1965. The Diamant's third stage, Rubis, also solid propellant, provided thrust for 45 seconds.

Prime contractor for the construction of the Diamant rocket stages was SEREB (Company for the Study and Manufacture of Ballistic Missiles) in cooperation with the French space agency CNES (Centre National d'Etudes Spatiales) and the French aerospace industry.

France became the third nation after the Soviet Union and United States to achieve orbital capability when the three-stage Diamant A placed a satellite into orbit on November 26, 1965, eight years after Sputnik I.

All Diamant A launches were conducted at Hammaguir. This military base in Algeria's Sahara Desert about 660 miles (1,062 km) southwest of Colomb-Bechar was used as a rocket testing and development site from the mid 1950s until France was required to evacuate the base in 1967 following Algerian independence. The installations at Hammaguir included a base, annexes for the technical workshops at the Georges Leger base, missile assembly and storage hangars, and housing for 600 people. The firing range immediately south of the base had three launch pads and was used when the base at Colomb-Bechar could not be used, or was too small, or for security reasons. The southeast azimuth was used for long distance firing and the east-northeast azimuth for satellite launching. Under the terms of the Evian agreements, France vacated the Hammaguir installations on July 1, 1967, after which they were transferred to French Guiana. It had an intermediate range ballistic missile (IRBM) test corridor stretching 1,900 mi (3,000 km) southeast to Fort Lamy. About 300 rockets were fired in upper-atmosphere and missile tests. France flew the four orbital Diamant A launches there from 1965 to 1967.

It was in 1964 that France decided to transfer its national launch center from Hammaguir to Kourou, French Guiana, for the Diamant B vehicle. Though hot and wet, it provides a unique location for geosynchronous satellite launches given its proximity to the equator. Also, a launch vehicle can travel 1,900 mi (3,000 km) north or east without crossing over land.

YEAR	DATE	VEHICLE	SITE	PAYLOAD
1965	Nov-26	Diamant A	Hammaguir	A-1
1966	Feb-17	Diamant A	Hammaguir	D-1 (Diapason)
1967	Feb-8	Diamant A	Hammaguir	D-1C
	Feb-15	Diamant A	Hammaguir	D-1D
1970	Mar-10	Diamant B	Kourou	DIAL
	Dec-12	Diamant B	Kourou	Peole 1
1971	Apr-15	Diamant B	Kourou	D-2A
	Dec-5	Diamant B	Kourou	??
	Failure - Stg 2			
1973	May-21	Diamant B	Kourou	Pollux / Castor
	Failure - Stg 3 did not ignite			
1975	Feb-6	Diamant B	Kourou	Pollux / Castor
	May-17	Diamant BP4	Kourou	Beta
	Sep-27	Diamant BP4	Kourou	D-2B

Launch site was Hammaguir, Algerian Sahara (31.0°N, 8.0°W) and Kourou, French Guiana (5.2°N, 52.8°W)

The center was fully operational at the beginning of 1969. First Diamant B launch was at Kourou on March 10, 1970.

Coincident with the move to Kourou, CNES improved the capability of the national launcher by developing a new first stage to obtain Diamant B. This new version of the French national launcher had increased tankage allowing the first stage to accommodate 26,750 lb (12,130 kg) of N_2O_4 and 13,000 lb (5,900 kg.) of UDMH. Stage 2 was similar to Diamant A and Stage 3 was derived from the perigee stage of the Europa launcher; the glass fiber casing contained 1,540 lb (700 kg) of solid propellant. Thus, the payload delivered into a close Earth orbit was increased by nearly two-fold.

The last of the launcher family was Diamant BP4 of which the first stage was designated L17 with a propellant weight of 37,400 lb (17,000 kg). The second stage was a RITA I solid-propellant motor from the ballistic program, and stage three was the same as the stage employed in Diamant B. The jettisonable nose shrouds were derived from those used on the abandoned British national launcher Black Arrow. The up-rated rocket was capable of placing payloads of 440 lb (200 kg) into a 162 nm (300 km) equatorial orbit as compared with 350 lb (160 kg) for the previous Diamant B. First payloads were the backup satellites D-5A/D-5B Pollux and Castor (replacing those which failed to reach orbit in 1973).

Of 12 total Diamant launches, 10 were successful. After the Diamant campaign, CNES proposed that Europe develop a heavier launcher, the L3S, which was adopted in 1973 and was the forerunner of the Ariane launch system. Much useful technology was transferred to Ariane although the Diamant launch pad at Kourou have not been used since 1976.

Vehicle	**Diamant A**	**Diamant B**	**Diamant BP4**
System Height	62 ft (18.9 m)	77 ft (23.5 m)	??
Gross Mass	40,600 lb (18,400 kg)	55,000 lb (25,000 kg)	??
Stage 1	Emeraude	Stretched Emeraude	Stretched Emeraude
- Type / Thrust	Liquid / 66K lb (294K N)	Liquid / 77K lb (343K N)	Liquid / 88K lb (392K N)
Stage 2	Topaze	Topaze	Rita I
- Type / Thrust	Solid / 97K lb (431K N)	Solid / 97K lb (431K N)	Solid / ??
Stage 3	Rubis	Europa II Perigee System	Europa II Perigee System
- Type / Thrust	Solid / 11.6K lb (52.0K N)	Solid / ??	Solid / ??

Reference:
- Jane's Spaceflight Directory, 1987
- Missiles & Rockets, K. Gatland, 1975

Japan: H-1

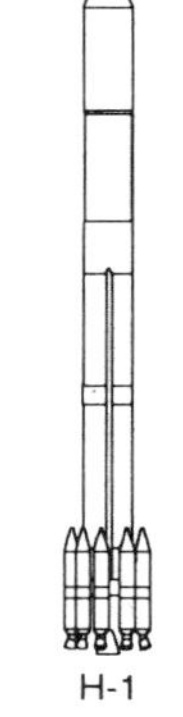

Designation	H-1
First Launch	1986
Total Length	132 ft (40.3 m)
Core Diameter	8.0 ft (2.44 m)
Total Weight	308,000 lb (140,000 kg)
LEO Payload	7,000 lb (3,200 kg)
GTO Payload	2,400 lb (1,100 kg)
GEO Payload	1,200 lb (550 kg)

Vehicle Description

Same as the N-2 vehicle except new LOX/LH2 second stage and engine, higher mass fraction third stage, and improved inertial guidance.

Historical Summary

The H-1 vehicle, developed as a successor of N-series rocket, employed a new domestically developed cryogenic second stage, inertial guidance system, and third stage solid motor, while the first stage, strap-on booster and fairing remained the same as N-2 vehicle, that is, manufactured by license. A three-stage H-1 rocket could launch a 1,200 lb (550 kg) payload into GEO. A two-stage H-1 was available for lower orbits. Two successful test flights of the H-1 were flown, a two stage version in August 13, 1986, and a three stage version on August 27, 1987. A total of nine H-1 launches were conducted through 1992.

Launch Record

	YEAR	DATE	VEHICLE	SITE	PAYLOAD
1	1986	Aug-13	H-1	OLS	Ajisai (EGS) + Fuji (JAS-1)
2	1987	Aug-27	H-1	OLS	Kiku-5 (ETS-5)
3	1988	Feb-19	H-1	OLS	Sakura-3a (CS-3a)
4		Sep-16	H-1	OLS	Sakura-3b (CS-3b)
5	1989	Sep-6	H-1	OLS	Himawari-4 (GMS-4)
6	1990	Feb-7	H-1	OLS	Momo-1b (MOS-1b)
7		Aug-28	H-1	OLS	Yuri-3a (BS-3a)
8	1991	Aug-25	H-1	OLS	Yuri-3b (BS-3b)
9	1992	Feb-11	H-1	OLS	Fuyo-1 (JERS-1)

OLS (Osaki Launch Site) at Tanegashima Space Center

General Description

Summary

The H-1 was a two or three stage launcher developed to meet the need for launching larger geostationary satellites than the N-2 launcher it was derived from. The most significant difference from the N-2 was the newly developed cryogenic second stage developed by NASDA, which will enable 1.6 times more payload to GEO. The first launch occurred 13 August 1986.

User	NASDA
Launch Service Agency	NASDA
Key Organizations	
Prime Contractor	Mitsubishi Heavy Industries (Under license to McDonnell Douglas to produce Stage 1 and fairing, Rocketdyne to produce Stage 1 engine, and Thiokol to produce Castor II. Produce Stage 2 at Mitsubishi and Stage 3 at Nissan Motor)
Vehicle	
System Height	132.2 ft (40.30 m)
Payload Fairing Size	8.0 ft (2.44 m) diameter by 26.0 ft (7.91 m) height
Gross Mass	308,000 lb (140,000 kg)
Planned Enhancements	None
Operations	
Primary Missions	LEO, GTO, GEO
Compatible Upper Stages	Stage 3 Nissan motor
First Launch	1986
Success / Flight Total	7 / 7 for H-1, 0 / 0 for H-2
Launch Site	Osaki Launch Site (30.2°N, 130.6°E)
Launch Azimuth	85°-135° (max. inclination is 31°- ??°)
Nominal Flight Rate	2 / yr
Planned Enhancements	None
Performance	
100 nm (185 km) circ, 30°	7,000 lb (3,200 kg) for two-stage
100 nm (185 km) circ, 90°	4,800 lb (2,200 kg) for two-stage
Geotransfer Orbit, 28°	2,400 lb (1,100 kg) for three-stage
Geosynchronous Orbit	1,200 lb (550 kg) for three-stage
Financial Status	
Estimated Launch Price	13 billion Yen ($90 million in 1990$) at 1/yr

Historical Launch Systems

Japan: H-1 Vehicle
(continued)

Vehicle

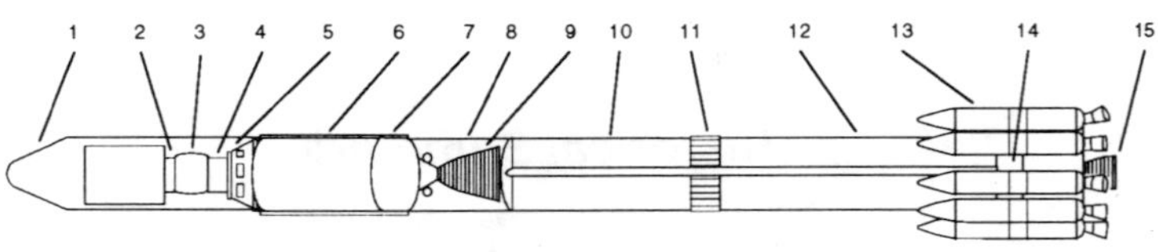

1. Payload Fairing
2. Payload Attach Fitting
3. Third Stage Solid Motor
4. Spin Table
5. Guidance Section
6. Second Stage LH2 Tank
7. Second Stage LOX Tank
8. Adapter Section
9. Second Stage Engine
10. First Stage RJ-1 Tank
11. Center Body Section
12. First Stage LOX Tank
13. Strap On Booster
14. Skirt Section
15. First Stage Main Engine

Overall

Length	132.2 ft (40.30 m)
Gross Mass	308,000 lb (140,000 kg)
Thrust at Liftoff	410,000 lb (1,800,000 N)

Stages

	Strap-On Booster	Stage 1	Stage 2	Stage 3 (Optional)
Dimension:				
Length	23.8 ft (7.25 m)	33.9 ft (10.32 m)	33.9 ft (10.32 m)	7.68 ft (2.34 m)
Diameter	2.6 ft (0.79 m)	8.0 ft (2.44 m)	8.17 ft (2.49 m)	4.40 ft (1.34 m)
Mass: (each)				
Propellant Mass	8.22K lb (3.73K kg)	180K lb (81.4K kg)	19.4K lb (8.8K kg)	4.06K lb (1.84K kg)
Gross Mass	9.87K lb (4.48K kg)	189K lb (85.8K kg)	23.4K lb (10.6K kg)	4.85K lb (2.20K kg) with spin table
Structure:				
Type	Monocoque	Isogrid	Isogrid	Monocoque
Material	Steel	Aluminum	Aluminum	Titanium
Propulsion:				
Propellant	CTPB	LOX / RJ-1	LOX / LH2	HTPB
Average Thrust (each)	49.6K lb (220K N) SL	170K lb (756K N) SL 2K lb (9K N) SL verniers	23.15K lb (103K N) vac	17K lb (77K N) vac
Engine Designation	Castor II - TX-354-5	MB-3 Block III	LE-5	UM-129A
Number of Engines	9 (1 segment ea.)	1	1	1
Isp	235 sec SL	253 sec SL 209 sec SL verniers	447.8 sec vac	291 sec vac
Feed System	—	Gas Generator	Gas Generator	—
Chamber Pressure	638 psia (44.0 bar)	569 psia (39.2 bar)	526 psia (36.3 bar)	?? psia (?? bar)
Mixture Ratio (O/F)	—	2.15	5.5	—
Throttling Capability	—	100% only	100% only ?	—
Expansion Ratio	7.45:1	8:1	140:1	53.9:1
Restart Capability	No	No	Yes	No
Tank Pressurization	—	High pressure nitrogen gas	Fuel-GH2, Ox-hot GHe	—
Control-Pitch,Yaw	Fixed 11°	Hydraulic gimbal	Hydraulic gimbal (±3.5°) RCS for coast	Spin stabilized
Roll	Controlled by stage 1	2 verniers engines	RCS	
Events:				
Nominal Burn Time	39 sec	270 sec, 276 sec verniers	370 sec	68 sec
Stage Shutdown	Burn to depletion	Burn to depletion	Predetermined velocity	Burn to depletion
Stage Separation	Spring ejection	Spring ejection	Spring ejection / RCS	Spring ejection

Remarks:

The first stage had an engine section that housed the Rocketdyne MB-3 main engine, two Rocketdyne vernier engines, and provided the aft attachments for the strap-on solid-propellant motors. The cylindrical isogrid RJ-1 fuel and liquid oxygen tanks were separated by a center body section that housed control electronics, ordnance sequencing equipment, and telemetry system. The MB-3 was a single-start, liquid-bipropellant rocket engine. Thrust augmentation was provided by nine unsegmented solid-propellant rocket motors, six ignited at liftoff and the remaining three ignited in flight.

The second stage had an integral tank with a common bulkhead, and was made from an aluminum alloy in an isogrid structure. The outer surface was heat-insulated by polyurethane foams and by aluminum-evaporated mylar. Both the guidance and engine structure were cone shaped semi-monocoque structure assembly made from aluminum alloy. The second stage used a hydrogen/oxygen propulsion system, which consisted of a cryogenic engine called LE-5, propellant feed system, and reaction control system. The restart capability of the second stage afforded mission flexibility to the mission designer without the use of a third stage.

LE-5 was a gas generator cycled, turbopumped, relatively low chamber pressured engine. The regenerative cooled combustion chamber and the dump cooled nozzle extension generated a high expansion ratio of 140. In steady state, the hydrogen and oxygen turbopumps were separated and driven in series by hot gas from the gas generator which operated with liquid hydrogen and oxygen. In the startup phase, however, a hydrogen bleed cycle was used to facilitate start/restart of engine, that is, tapping hydrogen from the main combustion chamber to drive both turbopumps. Another feature of LE-5 was to have an Engine Control Box (ECB) which was a kind of microcomputer. ECB received signals from the guidance computer such as second stage start signal, second engine lock-in signal, and second engine cut-off command, to start and stop the engine. Four bottles for restarting missions or one bottle for nonrestarting missions were installed as the ambient temperature helium system. This system was used for pressurization of the propellant tank before the second stage ignition, control of the valves, purging, etc. The cryogenic gaseous helium system had two bottles inside the liquid hydrogen tank, and was heated by the LE-5 engine and used to pressurize the liquid oxygen tank during powered flight of the second stage. The gaseous hydrogen (GH2) for pressurizing the liquid hydrogen tank during powered flight of the second stage was tapped off from the LE-5 engine.

The third stage is a spin-stabilized solid motor made of titanium with a carbon/carbon and CFRP nozzle.

Japan: H-1 Vehicle
(continued)

Payload Fairing

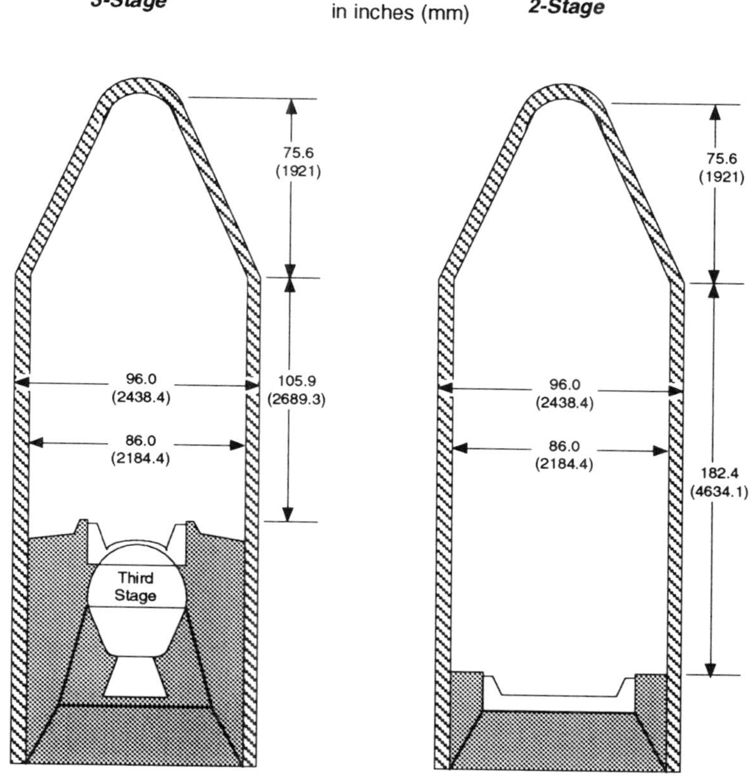

Length	26.0 ft (7.91 m)
Diameter	8.0 ft (2.44 m)
Mass	1320 lb (600 kg)
Sections	2
Structure	Isogrid
Material	Aluminum

Remarks: The fairing was attached to the second stage guidance section and was jettisoned in two sections by the separation nuts and detonating fuse. An RF transparent window could be installed on the conical or cylindrical section of the fairing for each satellite mission. An umbilical connector and access doors could be installed on the cylindrical section. The acoustical absorption blankets are mounted on the cylindrical section of the fairing.

Avionics

An inertial guidance system, called NICE (NASDA Inertial Guidance and Control Equipment) and located at the top of the second stage, consisted of an inertial measurement unit (IMU), guidance computer, rate gyro package, and electronics package. The principal functions of NICE were flight sequence control functions to control engine start and stage separation, attitude control functions to maintain a stable vehicle attitude, guidance function to correct flight path deviations, and pressurizing control function of the second stage tank pressure.

The IMU consisted of four gimbal frameworks supporting a platform on which three accelerometers and three gyros were fixed. Rate gyro packages were installed in the first and second stage to secure control stability. The guidance computer was microprogrammable with 16 bit-word length and had the main memory size of 16K words. The attitude control and guidance cycles were 40 Hz and 1.25 Hz, respectively.

Attitude Control System

The first stage engine was gimbaled to provide pitch and yaw control. The two vernier engines provided roll control during main-engine burn, and attitude control after cutoff and before second-stage separation.

The second stage attitude control was done by hydraulic engine gimbaling and a hydrazine monopropellant reaction control system. The second stage gimbaled engine provided for pitch and yaw control. The LE-5 hydraulic pump was used for the engine gimbal actuation system during powered flight of the second stage and was driven by the auxiliary turbine of the LE-5 engine. The auxiliary hydraulic pump driven by the electric motor was used to maintain the neutral position of the engine before the second stage engine ignition.

Roll control during powered flight and three axes control during coast was provided by the reaction control system (RCS). This system also provided thrust needed for settling / retention of propellant (LOX/LH2) prior to second stage ignition. The RCS consisted of two modules installed in the second stage engine section. There are two types, one was for restart missions and the other for non-restart missions. The restart missions required 119 lb (54 kg) of hydrazine propellant and the non-restart required only 26 lb (12 kg). The restart missions used helium as a pressurizing gas and the non-restart used nitrogen. Each type uses 12 thrusters that provided 4 lb (18 N) of thrust. The third stage was spin-stabilized.

Historical Launch Systems

Japan: H-1 Operations
(continued)

Launch Site

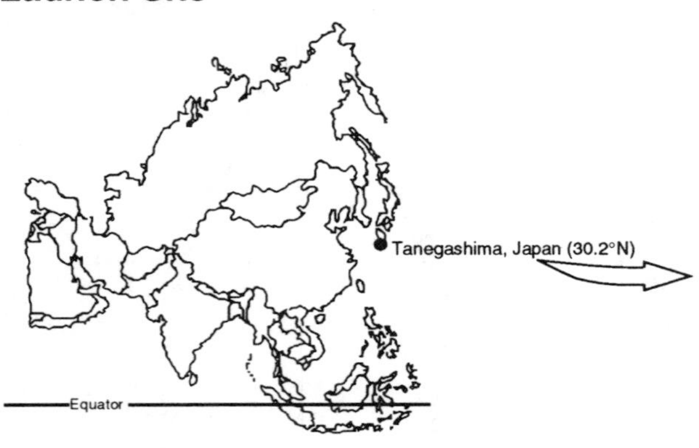

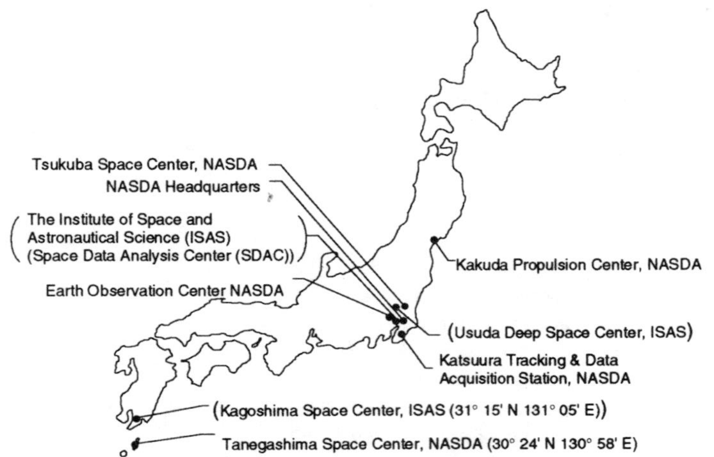

Launch Facilities

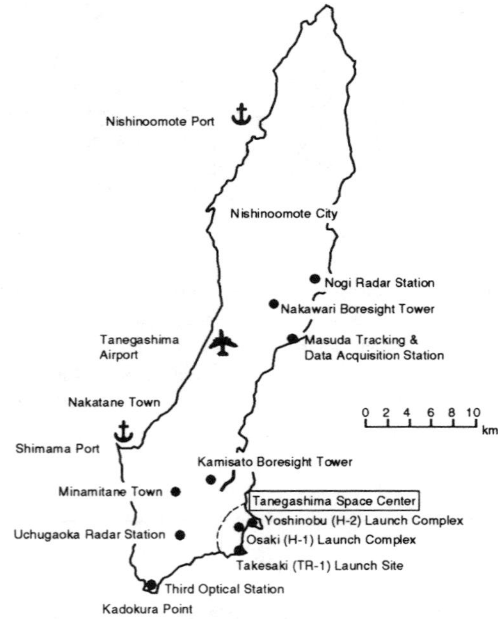

Location of NASDA Facilities on Tanegashima Island

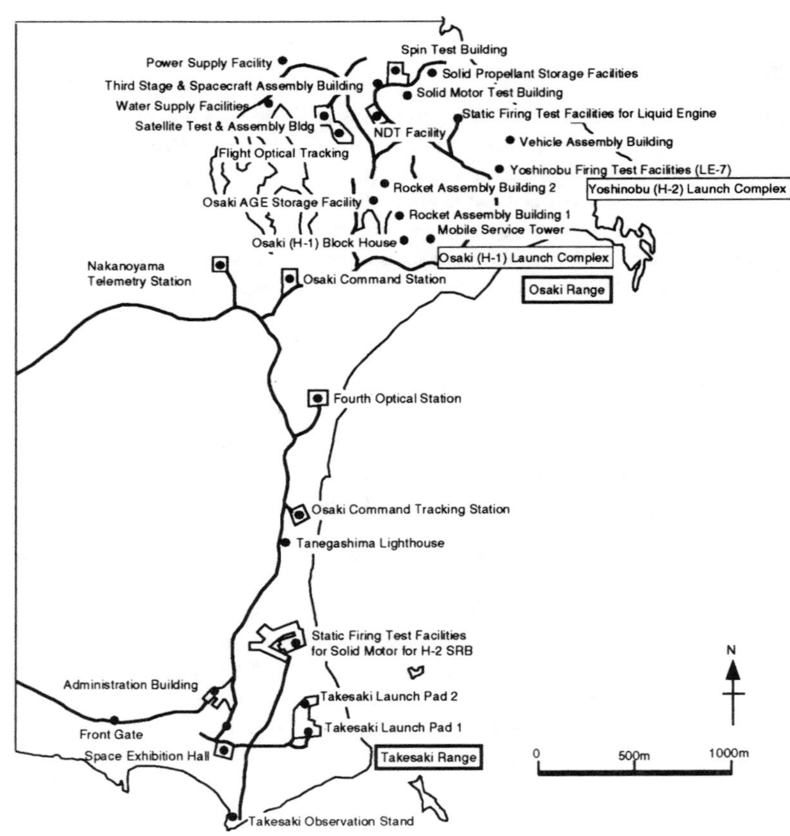

Location of Facilities at Tanegashima Space Center

Japan: H-1 Operations
(continued)

Launch Processing

Tanegashima Space Center. The Tanegashima Space Center is located on the southeast of Tanegashima Island, Kagoshima. Tanegashima Island, which is 36 mi (58 km) in diameter with a population of about 43,000, is located 50 mi (80 km) off the southern coast of Kyushu, the southern most island in the main Japanese chain. The facilities include the Takesaki Range for small rockets, and the Osaki Range for the H-1 launch vehicles and H-2 launch vehicles. The center also includes the Masuda Tracking and Data Acquisition Station, the Nogi and Uchugaoka radar stations, and three optical tracking stations. The center occupies approximately 3.3 sq-mi (8.6 sq-km) of land, where combustion test facilities on the ground for liquid and solid rocket engine are installed. It is the largest launch site in Japan. The major tasks of the center are to check, assemble and launch rockets, and to perform tracking and control after launch. It plays a major role of applications satellite launching and combustion tests for solid rocket motors and liquid rocket engines. Due to fishermen's objections to the noise and hazards associated with the launches over their fishing grounds, launches have been restricted to two launch periods of each year—January 15 to end of February and August 1 to September 15.

H-1 Operations. A number of facilities located on the Osaki Range supported the launches of the H-1.

The H-1 Rocket Launch Complex includes a launch pad and two umbilical masts and a Mobile Service Tower (MST). The launch pad (launch deck) was used to launch rockets and has a cantilever structure 21 ft (6.4 m) tall, 39 ft (12 m) wide, and weighs 375K lb (170K kg). The two masts supplied power, conditioned air, and high-pressure gases as well as the feed pipes and conduits for liquid hydrogen, liquid oxygen, and other propellants which fed the vehicle just prior to the launch. Mast No. 1 is 115 ft (35 m) tall and 11.5 ft (3.5 m) wide and mast No. 2 is 161 ft (49 m) tall and 13 ft (4 m) wide. The MST was used for assembly, checks and preparation of the launch vehicle and was retracted approximately 330 ft (100 m) on rails when all the launch preparations had been completed. The MST is 220 ft (67 m) tall, 85 ft (26 m) wide, and weighs 6.2M lb (2.8M kg). In addition to these facilities, the center has aerospace ground equipment such as propellant storage and supply facilities, ground power sources, and hydraulic sources.

The First and Second Rocket Assembly Buildings were used for accepting the first and second stages of the H-1 launch vehicle, for checking and making adjustments before installation on the launch pad, and for functional tests and adjustments of launch vehicle onboard equipment such as gyros and guidance equipment.

The Solid Motor Test Building was used for acceptance inspections and check of strap-on boosters (SOB's), third stage solid rocket motors, pyrotechnics, etc. The Non-Destructive Test Facility was used for betatron x-ray inspection of H-1 third stage motor and satellite apogee motor. The Spin Test Building is where the H-1 third stage motor and satellite were assembled, checked, and adjusted for mounting on the spin table and satellite separation section. The Satellite Test and Assembly Building contains a clean test room for overall checkouts of payload prior to mounting on the rocket, such as satellite acceptance inspections and overall functional tests. The Third-Stage and Satellite Assembly Building is where H-1 third stage rocket motors and satellites were assembled, checked, and adjusted in this building prior to mounting them on the launch vehicle.

The Osaki Block House is a semibasement explosion proof facility that contains a launch control room on the ground floor and a telemetry room on the first floor. The house was used to operate and command launches.

The Osaki Range Control Center (RCC) has the rooms for range control, ground safety, computer, and optical control. The center provided the information needed for work such as check and test of the satellite and launch vehicle in prelaunch operation, launch terminal countdown, tracking, and ground and flight safety. The center also monitored the core facilities on the Tanegashima Island and down range stations during launch. The First, Third, and Fourth Optical Tracking Stations simultaneously tracked and observed the launch vehicle using optical equipment and transmitted the obtained data to the Osaki RCC. The Nakanoyama Telemetry Station received telemetry data sent from each stage of the launched rocket such as acceleration, pressure, and temperature and also sent data for flight safety decisions to the Osaki RCC.

Launch operations for the H-1 were conducted with the basic concept unchanged from the N-2 vehicle, that is, a build-on the pad approach. After receiving inspection, the first stage was erected on the launch pad about 8 weeks before launch. Then the second stage and strap-on boosters were mated with the first stage in the MST. Vehicle system checks continued until the launch rehearsal was performed about two weeks before launch.

Payload inspection and testing were performed at the Satellite Test and Assembly Building. When necessary, the Third Stage and Satellite Assembly Building was used for apogee motor installation, propellant loading, and payload/third-stage assembly. The payload or payload/third stage was then transported to the MST using a special canister, and mated with the second stage about a week before launch. Final operations included payload fairing installation, vehicle system checks, pyrotechnic system arming, propellant loading, and final countdown.

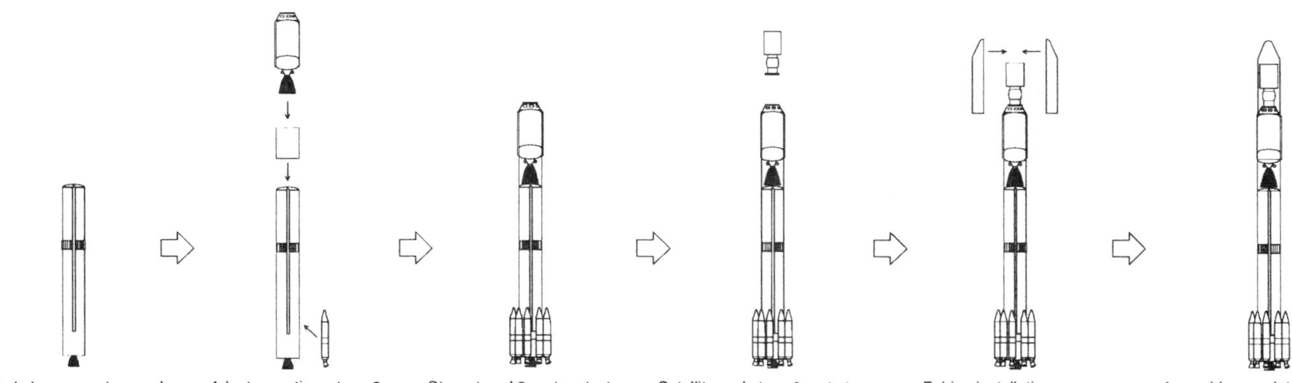

H-1 Launch Vehicle Processing Flow

Japan: H-1 Operations
(continued)

H-1 GTO Mission

Flight Time (min:sec)	Events	Flight Time (min:sec)	Events
00:00	Liftoff	05:04	Payload fairing separation
00:03	Roll program start	10:30	Pitch program finish
00:08	Roll program finish	10:37	2nd stage engine cut-off
00:08	Pitch program start	12:30	Pitch program start (coast flight phase)
00:39	6 SOBs burn-out	14:10	Pitch program finish (coast flight phase)
00:40	3 SOBs ignition	14:10	Yaw program start (coast flight phase)
01:19	3 SOBs burn-out	15:00	Yaw program finish (coast flight phase)
01:25	9 SOBs separation	24:00	3rd stage spin-up
04:16	Pitch program finish	24:05	2nd/3rd stage separation
04:26	1st stage engine cutoff	24:28	3rd stage motor ignition
04:32	Vernier engine cutoff	25:30	3rd stage motor burn-out
04:34	1st/2nd stage separation	26:30	3rd stage / satellite separation
04:39	2nd stage engine ignition	26:32	3rd stage tumble start
04:48	Pitch program start		

Payload Accommodations

Payload Compartment	
Maximum Payload Diameter	86.0 in (2184 mm)
Maximum Cylinder Length	105.9 in (2689 mm) for 3-stage vehicle 182.4 in (4634 mm) for 2-stage vehicle
Maximum Cone Length	75.6 in (1921 mm)
Payload Adapter	
Interface Diameter	37.75 in (959 mm) for 3-stage vehicle 57.1 in (1450 mm) for 2-stage vehicle
Payload Integration	
Nominal Mission Schedule Begins	T-24 months
Launch Window	
Latest Countdown Hold Not Requiring Recycling	T-7 min (next opportunity in 1.5 hours)
On-Pad Storage Capability	4.5 hrs for a fueled vehicle
Latest Access to Payload	T-11 hrs through access doors
Environment	
Maximum Load Factors	+9.3 g axial, ±2 g lateral for 3-stage +8.0g axial, ±2 g lateral for 2-stage
Minimum Lateral / Longitudinal Payload Frequency	15 Hz / 35 Hz
Maximum Overall Acoustic Level	141 dB (full octave)
Maximum Flight Shock	2,000 g from 1,500–4,000 hz for three-stage 2,000 g from 800–4,000 hz for two-stage
Maximum Dynamic Pressure on Fairing	700 lb/ft^2 (33,520 N/m^2)
Maximum Pressure Change in Fairing	0.28 psi/s (1.9 KPa/s)
Cleanliness Level in Fairing (Prior to Launch)	Class 10,000
Payload Delivery	
Standard Orbit and Accuracy (3 sigma)	GEO - ±648 nm (1,200 km), ±0.3° inclination
Attitude Accuracy (3 sigma)	All axes: ±2 deg, ±0.5 deg/sec
Nominal Payload Separation Rate	6.5 ft/s (2 m/s)
Deployment Rotation Rate Available	40 to 100 rpm for 3-stage vehicle 0 to ?? for 2-stage vehicle
Loiter Duration in Orbit	?? hrs
Maneuvers (Thermal / Collision / Telemetry)	Yes

Japan: N-Vehicle

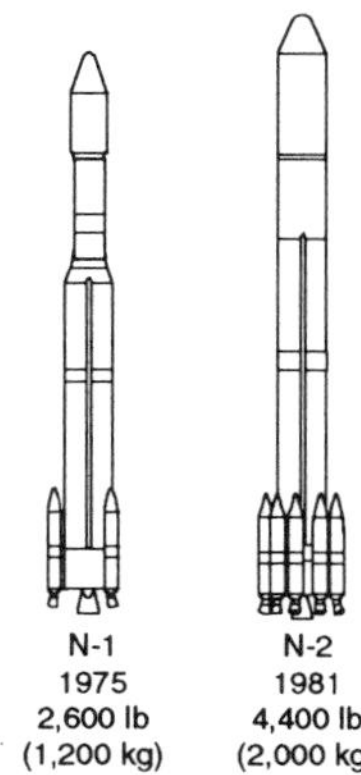

	N-1	N-2
First Flight	1975	1981
East LEO	2,600 lb (1,200 kg)	4,400 lb (2,000 kg)

The Japanese were licensed by the U.S. Department of State to build the entire N-Vehicle in Japan. However, for reasons of cost and convenience, they purchased many items in the U.S. rather than building these components themselves. McDonnell Douglas Corporation, the U.S. builder of the Delta launch vehicle, provided the bulk of the assistance on the N-Vehicle (overall design, production, and launch operations). Mitsubishi Heavy Industries was prime contractor for NASDA on the N-Vehicle program. Mitsubishi produced the Delta booster under license from McDonnell Douglas and the first-stage engine (MB-3) under license from Rocketdyne. Through the technical assistance of Rocketdyne, they developed the second stage engine (LE-3), and through McDonnell Douglas, they developed the second stage structure and nose fairing. Delta tankage (McDonnell Douglas), Castor II strap-ons (Thiokol), solid-propellant third stage (Thiokol), control systems (Honeywell), and system analysis (TRW) were purchased from the U.S.

From 1975 to 1982, seven N-1 vehicles were successfully flown. Because the N-1 performance was not sufficient for operation of commercial communication systems or other applications, the N-2 was developed. The N-2 used nine strap-ons instead of three, and improved the first, second and third stages and the inertial guidance system. From 1981 to 1987, eight N-2 vehicles were successfully launched. N-1 was capable of launching 570 lb (260 kg) to geotransfer orbit and the N-2 1575 lb (715 kg). All 15 N-Vehicle launches were flown from the Osaki launch site at the Tanegashima Space Center on the southern tip of Japan.

YEAR	DATE	VEHICLE	SITE	PAYLOAD
1975	Sep-9	N-1	Osaki	Kiku-1 (ETS-I)
1976	Feb-29	N-1	Osaki	UME-1 (ISS-1)
1977	Feb-23	N-1	Osaki	Kiku-2 (ETS-II)
1978	Feb-16	N-1	Osaki	UME-2 (ISS-2)
1979	Feb-6	N-1	Osaki	Ayame (ECS-1)
1980	Feb-22	N-1	Osaki	Ayame-2 (ECS-2)
1981	Feb-11	N-2	Osaki	Kiku-3 (ETS-IV)
	Aug-10	N-2	Osaki	Himawari-2 (GMS-2)
1982	Sep-3	N-1	Osaki	Kiku-4 (ETS-III)
1983	Feb-4	N-2	Osaki	Sakura-2A (CS-2A)
	Aug-5	N-2	Osaki	Sakura-2B (CS-2B)
1984	Jan-23	N-2	Osaki	Yuri-2A (BS-2A)
	Aug-2	N-2	Osaki	Himawari-3 (GMS-3)
1986	Feb-12	N-2	Osaki	Yuri-2B (BS-2B)
1987	Feb-18	N-2	Osaki	Momo-1 (MOS-1)

Launch site is Osaki at Tanegashima Space Center (30.4°N, 131.0°E)

Vehicle	N-1	N-2
System Height	107 ft (32.6 m)	116 ft (35.4 m)
Gross Mass	199,000 lb (90,400 kg)	297,000 lb (135,000 kg)
Max Diameter	8.0 ft (2.44 m)	8.0 ft (2.44 m)
Stage 0	3 x Castor II solid	9 x Castor II solid
Stage 1	long-tank Thor	extended long-tank Thor
- Propellant	LOX/kerosene	LOX/kerosene
- Engine	1 x MB-3	1 x MB-3
Stage 2	SSPS	SSPS
- Propellant	N2O4 / UDMH-hydrazine	N2O4 / UDMH-hydrazine
- Engine	1 x LE-3	1 x AJ10-118F
Stage 3	TE-M-364-14 solid	TE-M-364-14 solid

Reference:
- Foreign Space Launch Vehicles, Battelle, Dec 30, 1983
- TRW Space Log 1957-87

USSR: B-1

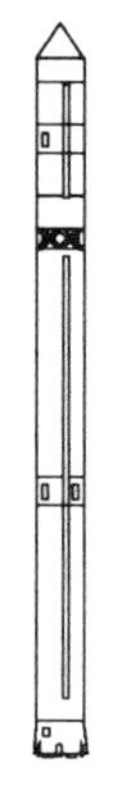

USSR Name	—
U.S. Name	SL-7
Sheldon Name	B-1
First Launch	1962
East LEO	1,300 lb (600 kg)

The B-1 (or SL-7) was used by the Soviets for small payloads. A Soviet name of the vehicle has never been announced. The first stage of the B-1, adapted from the SS-4 Sandal intermediate range ballistic missile, is a straight cylinder of about 66 ft (20 m) long by 5.4 ft (1.65 m) in diameter. It is powered by a single, four chamber engine, the RD-214, which burns kerosene and nitric acid; total thrust in vacuum is 163,000 lb (74,000 kg) and the specific impulse 264 sec. The engine was developed between 1952 and 1957.

The second stage, attached to the first stage by an open truss structure, is a cylinder almost exactly 26 ft (8 m) long and 5.4 ft (1.65 m) in diameter. It is powered by a single chamber engine and several fixed nozzles, burns LOX and asymmetrical dimethyl hydrazine, and develops 24,000 lb (11,000 kg) of thrust. Vacuum specific impulse is 352 sec. This engine, the RD-119, was developed between 1958 and 1962 and the complete vehicle was first used in 1962 to launch Kosmos 1.

For the first five years of its use, it was launched exclusively from Kapustin Yar in underground silos, probably relating to its original development as a missile. With the introduction of the Plesetsk facility in 1966, the vehicle was switched to this new site and used a conventional above ground pad, with a gradual phase-out of Kapustin Yar. Last B-1 launch was in 1977 from Plesetsk, after 144 orbital missions. It has been used to launch the small scientific Kosmos and the military ferret Kosmos satellites, and also early launches in the Interkosmos program.

YEAR	LAUNCHES (SUCCESSFUL)
1962	7
1963	4
1964	7
1965	7
1966	7
1967	13
1968	16
1969	14
1970	18
1971	12
1972	12
1973	10
1974	6
1975	5
1976	4
1977	2
Total	144

Vehicle	B-1
System Height	98 ft (30 m)
Gross Mass	??
Core Diameter	5.4 ft (1.65 m)
Stage 1	Scandal SS-4 IRBM
- Propellant	nitric acid / kerosene
- Engine	RD-214 (4 chamber)
Stage 2	
- Propellant	LOX / UDMH
- Engine	RD-119

Launch site is Plesetsk, Soviet Union (62.8°N, 40.1°E) and Kapustin Yar, Soviet Union (48.4°N, 45.8°E)

Reference:
- Jane's Spaceflight Directory, 1987
- "Soviet Launch Vehicle Designations", Ralph Gibbons, Spaceflight, pg. 54-60,80, February 19, 1977
- TRW Space Log 1957-87

Historical Launch Systems

USSR: G-1-e

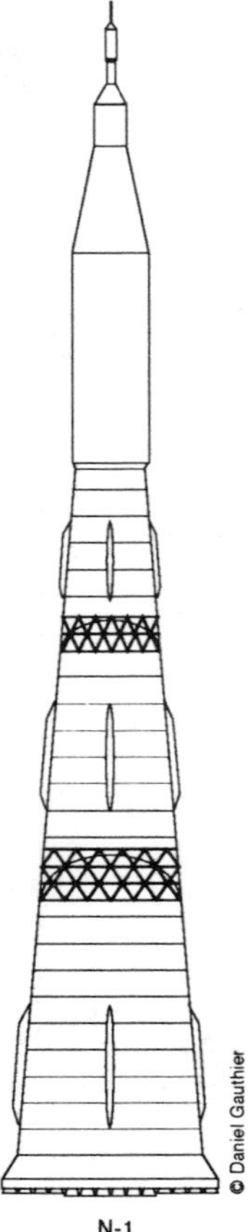

USSR Name	N-1
U.S. Name	SL-15
Sheldon Name	G-1-e
First Launch	1969
East LEO	est. 220,000 lb (100,000 kg)

The G-1-e (or SL-15) was designed to support the Soviet lunar program. Although developed in the strictest secrecy, its existence was known since the mid 1960s that the Soviets had been developing a superbooster, at least equal to the U.S. Saturn V. The vehicle had received a variety of names and designations by outside analysts ("Webbs Giant", "TT-5", "Lenin"), but only recently has it been officially designated by the Soviets as "N-1".

The vehicle was designed by Sergei Korolev. Although V. P. Glushko, the other well known Soviet designer, supported the need for a heavy-lift booster in the 1960s, he disagreed strongly with Korolev on how to design the vehicle, particularly with the selection of propellants. His opposition eventually caused Korolev to seek assistance from inexperienced rocket engine manufacturers, a factor which may have been more than a little responsible for significant development delays and four launch failures. The N-1 was 370 ft (113 m tall), 56 ft (17 m) in diameter at the base, and employed 30 engines in the first stage for a total thrust of 10.1M lb (45M N). Ironically, Glushko personally canceled the program in 1974 when he finally replaced Korolev's successor, V. P. Mishin. In its place, Glushko initiated the Energia/Buran project.

The vehicle was designed so that three stages would place it in Earth orbit. The two top stages were used for trans orbit injection. In the rush to beat the United States to the Moon, an early decision was made for a full hardware test mission in Earth orbit, similar to the Apollo 9 flight, even though the N-1 had not been man-rated nor had the 30 engine first stage been clustered fired in a ground test. In fact, the N-1 had been test flown only once, on February 21, 1969 when it suffered an explosive failure at an altitude of 40,000 ft (12,200 m). The second flight was a rehearsal mission whereby a three-man Soyuz would dock with the N-1 in orbit. However, on July 3, 1969, before the Soyuz launch, the N-1 failed. Before clearing the tower, an oxygen turbopump feeding one of the N-1's main engines disintegrated, flooding oxygen and steel shards onto a fire at the base of the booster. A series of explosions rocked the N-1, which began to keel over because of the asymmetric thrust from the vehicle's 30 engine first stage. Sensing trouble, the onboard computer shut down the remaining combustion chambers in the first stage and the N-1 fell back on the pad, touching off a fireball seen from space. The explosion that followed did severe damage to an identical pad a third of a mile away and utterly destroyed the launch complex, including its 58-story turning tower gantry. So badly was the Earth scorched that the scars are still visible in commercial satellite photos.

The accident sounded the death knell for the Soviet lunar program, even though the Soviets tried twice more to prove their N-1 in 1971 and 1972. Both flights ended in failure. In the 1971 flight, the vehicle developed an uncontrolled roll seconds after liftoff. The onboard computer shut down the engines and the N-1 stack again collapsed back, completely destroying the second pad and gantry, which had been badly damaged in 1969.

The last 1972 flight almost made it to first stage cutoff. Staging had begun, with the six inner first stage engines starting shutdown. Suddenly, longitudinal pogo oscillations caused a rupture in a propellant line. Fire was followed by an explosion and, at 107 sec into the flight, the computer shut down the 24 outer engines still burning. The N-1 was climbing through 25 mi (40 km) altitude when the range safety officer destroyed the vehicle. After four successive failures, the program was canceled.

YEAR	DATE	VEHICLE	SITE	PAYLOAD
1969	Feb-21	G-1	Tyuratam	R&D
	Failure - Explosion at 40,000 ft (12,200 m) altitude			
	Jul-3	G-1	Tyuratam	R&D
	Failure - Stg 1 engine failed at liftoff			
1971	Jun-24	G-1	Tyuratam	R&D
	Failure - Uncontrolled roll immediately after liftoff			
1972	Nov-22	G-1	Tyuratam	R&D
	Failure - Pogo oscillation at Stg 1 initial cutoff			

Launch site is Tyuratam, Soviet Union (45.6°N, 63.4°E)

Vehicle	<u>G-1-e</u>
System Height	370 ft (113 m)
Gross Mass	??
Max. Diameter	56 ft (17 m)
Propellant	LOX/RP-1 for all stages
Stage 1	
- No. of Engines	30
- Total Thrust	10.1M lb (45M N)
Stage 2	
- No. of Engines	8
- Total Thrust	3.1M lb (13.8M N)
Stage 3	
- No. of Engines	4
- Total Thrust	360.8K lb (1605K N)
Stage 4	Translunar boost
- No. of Engines	1
- Total Thrust	90.2K lb (401K N)
Stage 5	Lunar orbit insert & desc.
- No. of Engines	1
- Total Thrust	19.2K lb (85K N)

Reference:
- "Soviet Union Reveals Moon Rocket Design", Aviation Week, pp. 58–59, February 18, 1991.
- The Soviet Year in Space 1989, Nicholas Johnson, Teledyne Brown Engineering, Colorado Springs, CO.
- The Soviet Race to the Moon, C.P. Vick, R. DeMeis, Aerospace America, p. 23, November 1990.
- Jane's Spaceflight Directory, 1987.
- "Soviet Launch Vehicle Designations", Ralph Gibbons, Spaceflight, pp. 54–60,80, February 19, 1977.

United Kingdom: Black Arrow

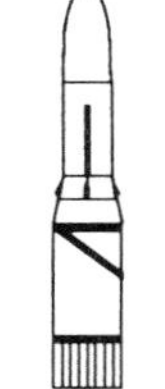

	Black Arrow
First Flight	1971
Polar LEO	240 lb (110 kg)

The United Kingdom's first nationally launched satellite was also its last. This attempt at a national launch vehicle was announced in September 1964. It was decided to develop a small three-stage vehicle from the successful Westland Black Knight research rocket. Construction and supply of three launchers, to be called Black Arrow, with the first two stages directly evolved from Black Knight, was ordered in March 1967. The purpose was to develop and test in space new components for communications satellites, and to develop a tool for space research. Black Arrow was a three stage launch vehicle. Stages one and two had liquid fueled Gamma engines using high-test peroxide (HTP) and kerosene; stage three had a solid-fueled apogee motor.

The first stage consisted of an engine bay, pressurized HTP tank, unpressurized kerosene tank and unpressurized equipment bay (housing servo system electronic components and instrumentation). There were eight combustion chambers supplied by four turbopumps arranged in four pairs, each pair being gimbaled in one axis to give control in pitch, yaw, and roll. The Black Knight hydraulic servo system was used in the new rocket without modification.

The second stage comprised engine pressurized HTP and kerosene tanks and a sealed equipment bay containing servo electronic equipment, telemetry, attitude reference unit, flight sequence program unit, tracking beacon, and command destruct receivers. The engine had two combustion chambers of Black Knight design but was fitted with kerosene-cooled extension nozzles to improve the vacuum specific impulse. Both chambers were fully gimbaled to give control in pitch, yaw, and roll and were supplied by a single turbopump of new design. The fluid pressure for servo control was provided by kerosene bled from the fuel pump delivery.

The third stage, comprising the solid-fueled apogee motor and payload, was mounted on a spin table attached to the second stage. It was spun up immediately before separation to provide axial stability up to the time of payload separation.

Black Arrow was stabilized in flight by an autopilot which used a high-precision attitude reference unit and separate rate gyros mounted in the second stage electronics bay. It was maintained in its predetermined flight path by precession of the attitude reference unit gyro in the pitch plane. Flight events were controlled by a mechanical program unit driven by a synchronous electric motor. Separation of the first stage after burnout was achieved by separation rockets attached to the engine bay skirt of stage two. After ignition of the second stage engine, the skirt was jettisoned.

After burnout of the second stage, the combined second and third stages coasted to apogee during which time they were stabilized by a gas-jet system in stage two, and by the same means rotated into the correct attitude for third stage ignition. As the vehicle achieved apogee, rockets attached to the spin table were fired, spinning up the third stage. Separation devices then operated, releasing the third stage from the spin table under the action of springs. After a time delay the apogee motor fired, accelerating the payload to orbital velocity. The payload was separated and de-spun as necessary.

Black Arrow launches were conducted at Woomera, Australia. About 270 mi (430 km) north of Adelaide and running eastwards for 1,250 mi (2,000 km) across the desert, the Woomera missile and rocket range was started jointly by the UK and Australia in 1946 and cost them well over £200 million. Australian hopes that what was originally a test range for ballistic missiles and sounding rockets, and for shooting down pilotless aircraft, would have a future as a launch center for satellites were not fulfilled. During the period when the

YEAR	DATE	VEHICLE	SITE	PAYLOAD
1970	Sep-2	Black Arrow	Woomera	Orba
	Failure - Premature stg 2 shutdown			
1971	Oct-28	Black Arrow	Woomera	Prospero

Launch site was Woomera, Australia (31.1°S, 136.8°E)

Vehicle	Black Arrow
System Height	43 ft (13.1 m)
Stage 1	
- Heritage	Black Night Stage 1
- Diameter	6.5 ft (2 m)
- Propellant	Peroxide / Kerosene
- No. of Engines	4 (1 turbopumps+2 chambers ea.)
Stage 2	
- Heritage	Black Night Stage 2
- Diameter	4.5 ft (1.37 m)
- Propellant	Peroxide / Kerosene
- No. of Engines	1 (1 turbopump + 2 chambers)
Stage 3	
- Propellant	Solid

Reference:
• Jane's Spaceflight Directory, 1987
• Missiles & Rockets, K. Gatland, 1975

British Blue Streak was being developed as the main stage of ELDO's Europa rocket, a thriving township of 4,500 with over 500 houses supplied with power and water across 100 mi (160 km) of desert sprang up. But Woomera was not suitable for equatorial launches and the French preferred to develop their own center at Kourou, French Guiana. After successive cuts in its contributions to the range, the UK announced that there would be no more work for Woomera after 1976. The small test satellite launched into polar orbit by the Australian SPARTA rocket (a modified U.S. Redstone) in 1967, and Black Arrow in 1971 are likely to be the only two satellite ever orbited from Woomera.

The first Black Arrow, being tested without a live third stage on June 28, 1969, went out of control soon after liftoff due to a minor electronic fault and had to be destroyed in flight. As a result the second rocket, originally to have been the first orbital attempt with all stages 'live', had to be restricted to a suborbital test. Launched on March 4, 1970 this second vehicle was successful, the payload and second stage impacting as planned in the Indian Ocean some 1,900 mi (3,050 km) northwest of Woomera some 25 minutes after launch.

With the next launch of Black Arrow there was a chance at last of the UK becoming the sixth nation after the Soviet Union, the United States, France, Japan, and China to achieve orbit by its own efforts. However, when Black Arrow R.2 was launched at Woomera September 2, it failed to achieve orbital velocity. Although all three stages fired, the second stage cut out 13 seconds too soon apparently due to a pressurization fault.

Political events were moving against further development. Black Arrow was canceled by the government in July 1971 although just one further launch was permitted at Woomera. On October 28, 1971, Black Arrow R.3 lifted off bearing in its nose the Prospero X-3 satellite. This spin-stabilized pumpkin-shaped 160 lb (72.5 kg) spacecraft was successfully injected into a near-polar orbit.

The entire project was managed by the then Ministry of Technology and conducted by the Royal Aircraft Establishment in conjunction with the Australian Weapons Research Establishment at Woomera.

Historical Launch Systems

United States: Jupiter C / Redstone / Juno II

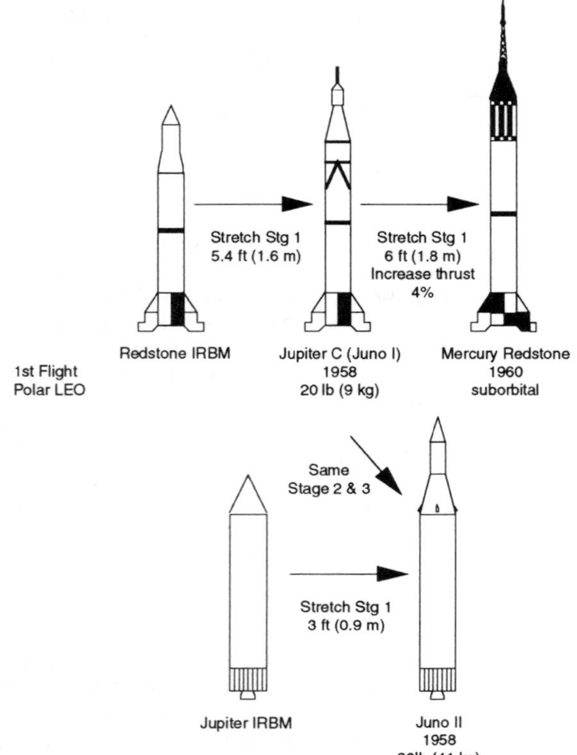

YEAR	DATE	VEHICLE	SITE	PAYLOAD
1958	Feb-1	Jupiter C	CCAFS	Explorer 1
	Mar-5	Jupiter C	CCAFS	Explorer 2
	Failure - Stg 4 failed to ignite			
	Mar-26	Jupiter C	CCAFS	Explorer 3
	Jul-26	Jupiter C	CCAFS	Explorer 4
	Aug-24	Jupiter C	CCAFS	Explorer 5
	Failure - Upper stage fired in wrong direction			
	Oct-23	Jupiter C	CCAFS	Beacon 1
	Failure - Upper stage separate prior to burnout			
	Dec-6	Juno II	CCAFS	Pioneer 3
1959	Mar-3	Juno II	CCAFS	Pioneer 4
	Jul-16	Juno II	CCAFS	Explorer S-1
	Failure -			
	Aug-14	Juno II	CCAFS	Beacon 2
	Failure - Stg 1, upper stage malfunction			
	Oct-13	Juno II	CCAFS	Explorer 7
1960	Mar-23	Juno II	CCAFS	Explorer S-46
	Failure - Upper stage malfunction			
	Nov-3	Juno II	CCAFS	Explorer 8
	Dec-19	Mercury Redstone	CCAFS	SUBORBITAL: MR-1A (unmanned capsule)
1961	Jan-31	Mercury Redstone	CCAFS	SUBORBITAL: MR-2 (chimp "Ham")
	Feb-24	Juno II	CCAFS	Explorer S-45
	Failure - Stg 3 & 4 ignition malfunction			
	Mar-24	Mercury Redstone	CCAFS	SUBORBITAL: MR-BD
	Apr-27	Juno II	CCAFS	Explorer 11
	May-5	Mercury Redstone	CCAFS	SUBORBITAL: MR-3 (A. Shepard)
	May-24	Juno II	CCAFS	Explorer S-45A
	Failure - Stg 2 ignition malfunction			
	Jul-21	Mercury Redstone	CCAFS	SUBORBITAL: MR-4 (V. Grissom)

Launch site is Cape Canaveral Air Force Station, FL (28.5°N, 81.0°W)

Ballistic Missiles: After World War II, much rivalry existed between the U.S. military services for control of ballistic missiles. When it was decided that the Air Force would be responsible for intercontinental ballistic missiles (ICBMs), the Army continued to compete with the Air Force for intermediate range ballistic missiles (IRBMs). The Von Braun German team, which was under Army control, was moved away from Fort Bliss, TX, and White Sands, NM to Redstone Arsenal, Huntsville, AL for IRBM development. The Redstone missile was based on the expansion of the original German V-2 technology. The Redstone evolved out of the Hermes design studies and was christened with its permanent name in 1952.

While the Air Force relied heavily on private industry for development and production, the Army was more wedded to use of the government arsenal system, with a somewhat lesser role for industry, coming primarily in the production phase. The Army used Chrysler as the manufacturing contractor, and Rocketdyne supplied the engine. This engine was the smaller of the two engines adapted from the V-2 and developed for the Navaho booster. The Army Redstone was deployed in the field until it was replaced by the Pershing two-stage solid-fueled rocket.

In direct competition with the Air Force Thor, Redstone IRBM was followed by the Jupiter IRBM at Redstone Arsenal. The Army Jupiter IRBM was designed to fly 1,500 mi (2,400 km). Jupiter had a larger diameter and shorter length than the Air Force Thor, but used essentially the same engine from Rocketdyne as the Thor. This engine was based on larger of the two engines developed for the Navaho booster. Chrysler also had the production contract for the Jupiter frame. The Army's hopes for a growing role in long range missile operations were seriously set back by a decision in 1956 that Jupiter operations would be conducted by the Air Force. Thirty each were deployed in Italy and Turkey and were withdrawn from use in 1963 after the Cuban missile crisis.

Jupiter C: While the Jupiter IRBM was under development, questions arose regarding its reentry characteristics. As a result, the Army developed the Jupiter C to conduct simulated reentry tests. The Jupiter C was a modified Redstone IRBM with a new combination of solid rockets as upper stages.

Jupiter C had a slightly longer tank than the regular Redstone, 37.5 ft (11.4 m) instead of 32.08 ft (9.8 m). The second stage was developed by the Jet Propulsion Laboratory, and consisted of a cluster of 11 scaled-down Sergeant missiles each 4 ft (1.2 m) long and 6 in. (15 cm) in. diameter. Each developed 1,600 lb (7,120 N) of thrust. The third stage consisted of three more of the same scaled-down Sergeants. These upper stages were mounted in an aluminum tub rotated at high speed by an electric motor in order to spin-stabilize them. The Jupiter C made the flight which led to the first recovery of an object from outer space after riding 600 mi (965 km) up and 1,200 mi (1,930 km) away, with a one-third scale Jupiter shroud. Its potential range was 3,400 mi (5,500 km).

In the mid 1950s the Army offered the Redstone with extra staging for the initial U.S. space launch attempts, but Vanguard was selected instead. However, after the launch of Sputnik 1 in October 1957, the Army came back with a fresh offer of Redstone in the Jupiter C version because it had already been prepared for Army reentry tests. By changing the weight of the payload, and choosing the right trajectory, the same rocket combination became the first successful launcher of a U.S. satellite, Explorer 1, weighing 18 lb (8.2 kg) including its expended rocket casing. Flights were carried out by Jupiter C nine times, beginning in 1956 and ending in 1958. These included three reentry tests, three successful satellite launches, and three satellite launch failures. After the fact, under a new nomenclature system, the orbital Jupiter C's were thereafter referred to as Juno I.

United States: Jupiter C / Redstone / Juno II
(continued)

Mercury Redstone: Even though the Jupiter C (Juno I) was abandoned early, the Redstone itself reappeared in the space program in connection with the Mercury program. The propellant tank was lengthened again by an additional 6 ft (1.8 m) to give a longer burning time. The thrust was raised slightly from 75,000 lb (334,000 N) to 78,000 lb (347,000 N). Instrumentation was improved, including autopilot and abort sensing system, and an adaptor section was added to receive the Mercury manned capsule. Flight tests began later in 1960.

The Mercury Redstone was flown five times (six if one counts the first attempt which shut down before lifting from the pad, but which sent the escape rocket atop the capsule off on its own). Two more flights were canceled because MR-1A (unmanned repeat), MR-2 with Ham the ape, MR-BD, MR-3 with Alan Shepard and MR-4 with Gus Grissom had supplied the data needed to support the orbital flight to come.

Juno II: The Jupiter C (Juno I) had such limited orbital capacity, that it was natural for the Army to offer the Jupiter IRBM as the first stage of a new orbital system. In this guise the new combination was called Juno II.

The Jupiter IRBM stage was 58 ft (17.7 m) long and 105 in. (2.67 m) in diameter. It weighed about 110,000 lb (50,000 kg), had a range of 1,600 mi (2,575 km). It burned LOX and RP-1 and used the Rocketdyne engine of 150,000 lb (667,000 N) thrust. It had inertial guidance and carried a nuclear warhead.

For the orbital vehicle, the Jupiter tank was lengthened by 3 ft (0.9 m) and to be expedient, the same cluster package of scaled down Sergeants which had been used on Juno I was also used on Juno II, making a very suboptimal launch vehicle for the size of the first stage. But at least it could be readied quickly and relatively cheaply. Ten Juno II vehicles were used between 1958 and 1961 when the program ended.

Juno II was used for the two Army tries at lunar flights and achieved in the case of Pioneer 4, the first successful U.S. flyby of the Moon and solar orbit with a 13 lb (5.9 kg) payload. Pioneer 3 fell back to Earth short of the Moon. Explorer 7, 8, and 11 reached Earth orbit. But Beacon 2 and Explorers S-1, S-46, S-45, and S-45a all failed to reach Earth orbit. Juno II had an Earth orbital capability of about 90 lb (41 kg).

Although never flown, there was a proposal for Juno III which would have used a larger size solid-fueled rocket in a cluster of 12 for the second stage, with an orbital capacity of about 500–600 lb (225–275 kg). The Advanced Research Projects Agency (ARPA) also asked for a proposal for a new launch vehicle with an orbital capacity of about 500 lb (225 kg). Several designs were considered, all using the Jupiter as first stage, but this new combination, which would have been called the Juno IV, did not advance beyond the design stage. The Jupiter and Redstone also figured in the Army design of the large clustered design which was called the Juno V, which later became the Saturn I first stage.

Vehicle	**Jupiter C** (Juno I)	**Mercury Redstone**	**Juno II**
Launch Site	CCAFS	CCAFS (LC-5, 6)	CCAFS
System Height	69.9 ft (21.3 m)	83.4 ft (25.4 m)	72 ft (22 m)
Stage 1			
- Heritage	Redstone IRBM	Redstone IRBM	Jupiter IRBM
- Diameter	70 in (1.8 m)	70 in (1.8 m)	105 in (2.67 m)
- Length	37.5 ft (11.4 m)	43.5 ft (13.3)	44 ft (13.4 m)
- Propellant	LOX/ethyl alcohol	LOX/ethyl alcohol	LOX/RP-1
- Thrust	75K lb (334K N)	78K lb (347K N)	150K lb (667K N)
Stage 2	(cluster of 11 missiles)	None	(cluster of 11 missiles)
- Heritage	scaled-down Sergeant missile	—	scaled-down Sergeant missile
- Diameter	4 in (0.1 m) each missile	—	4 in (0.1 m) each missile
- Length	4 ft (1.2 m) each missile	—	4 ft (1.2 m) each missile
- Propellant	Solid	—	Solid
- Thrust	1600 lb (7120 N) each missile	—	1600 lb (7120 N) each missile
Stage 3	(cluster of 3 missiles each same as above)	None	(cluster of 3 missiles each same as above)

Reference:
- U.S. Civilian Space Programs 1958-78, House Subcommittee on Space Science & Applications
- TRW Space Log 1957-87

United States: Saturn

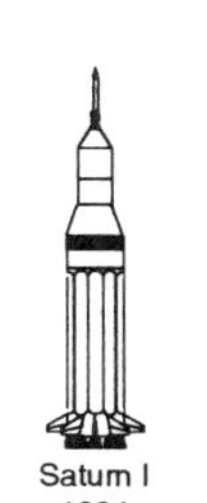

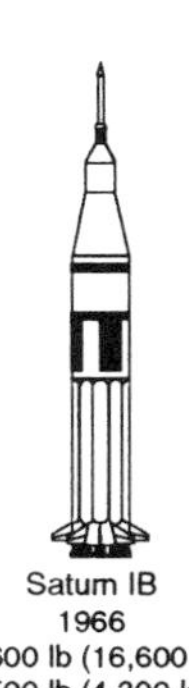

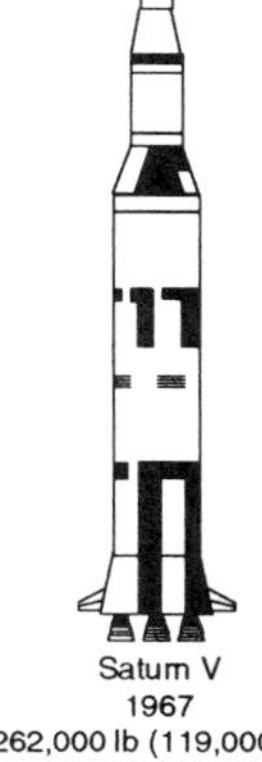

	Saturn I	Saturn IB	Saturn V
First Orbit	1964	1966	1967
East LEO	22,500 lb (10,200 kg)	36,600 lb (16,600 kg)	262,000 lb (119,000 kg)
Earth Escape	—	9,500 lb (4,300 kg)	110,000 lb (50,000 kg)

In April 1957, Dr. Werner von Braun proposed developing a launcher capable of placing payloads of 20,000 to 40,000 lb (9,100 to 18,200 kg) into a low Earth orbit (LEO). Because this project followed the successful Jupiter missile program, also under the direction of Dr. von Braun, the project was named after the next planet in line—Saturn. The Saturn program was conceived at the Army Ballistic Missile Agency (ABMA). In November 1958 the Advanced Research Projects Agency approved the development of the first stage (S-I) of the Saturn vehicle. The Saturn program was transferred to NASA on July 1, 1960.

The first eight S-I stages were designed, developed, manufactured, and tested by NASA Marshall Space Flight Center (MSFC) personnel. Later, this responsibility was transferred to the Space Division of the Chrysler Corporation. The Douglas Aircraft Company was awarded, in July 1960, a contract to develop the second stage of the Saturn I. In October 1961, the

Historical Launch Systems

United States: Saturn
(continued)

first of four Saturn I-Block I vehicles was launched, and in January 1964, the first Saturn I-Block II vehicle, having an S-IV upper stage, was launched. The enlargement and refining of the S-IV stage design and the replacement of six Pratt & Whitney RL10 engines with a single Rocketdyne J-2 engine, produced the S-IVB stage used on the Saturn IB vehicle. The S-IVB stage also served as the third stage for the Saturn V vehicle. Early Saturn I flights released water into the upper atmosphere for the physics experiment, Project High Water. The three operational Saturn I flights launched Pegasus meteoroid detector satellites. The Saturn IB was used to develop and test Apollo hardware and software in rehearsal for lunar missions with the Saturn V. In 1973 the Saturn IB served as the manned vehicle for the three Skylab visits, and in 1975, the last Saturn IB launched the Apollo Command and Docking Modules for the Apollo-Soyuz Test Project (ASTP). The Saturn I/IB Program ended in 1975 with a perfect launch record. Nine of 12 Saturn IBs built were launched successfully and three were never used.

In 1962, President Kennedy established a manned lunar landing before the end of the decade as a national goal. Shortly thereafter, NASA announced plans to develop Saturn V. Boeing, Rockwell, McDonnell Douglas, and IBM received contracts for design and manufacturing of vehicle components under direction of MSFC. The size and complexity of the Saturn V required parallel development of new ground facilities (Launch Complex 39) at Cape Canaveral for assembly and launch. The first operational Saturn V launched an unmanned Apollo spacecraft in November 1967. The first manned mission for the Saturn V occurred in December 1968 and marked the first manned voyage to orbit the moon. This launch demonstrated great confidence in the vehicle since the previous launch suffered from numerous problems: severe S-IC pogo oscillations, in-flight shutdown of two S-II engines, failed second restart of S-IVB, and buckling of the Command and Support Module support structure.

In July 1969, the Saturn V boosted Apollo 11 to the first manned lunar landing. Apollo 12 achieved its mission objectives despite being hit with lightning at launch. Only one mission, Apollo 13, was forced to return before reaching the lunar surface because of pressure loss in the oxygen tank of the service module. The launch itself was saved when the vehicle prevented a catastrophic failure when it shut down the center engine on S-II after experiencing severe pogo oscillations. Conservative Saturn design margins enabled the launch of the last three Apollo missions although 13% more performance was needed than originally required. Twelve three-stage configurations were launched as part of the Apollo program through December 1972. A two stage configuration of the Saturn V was used in 1973 to place the first manned U.S. space station, Skylab, into Earth orbit. The Saturn made it to orbit despite a payload solar array accidentally sliding down the vehicle and cutting an explosive charge that did not separate the S-IC/S-II interstage as planned. The vehicle was phased out following the Skylab mission.

YEAR	DATE	VEHICLE	SITE	PAYLOAD
1961	Oct-27	Saturn I	CCAFS	SUBORBITAL: R&D (propulsion test)
1962	Apr-25	Saturn I	CCAFS	SUBORBITAL: R&D (Project High Water I)
	Nov-16	Saturn I	CCAFS	SUBORBITAL: R&D (Project High Water II)
1963	Mar-28	Saturn I	CCAFS	SUBORBITAL: R&D (H-1 engine out test)
1964	Jan-29	Saturn I	CCAFS	R&D (1st S-IV test)
	May-28	Saturn I	CCAFS	R&D (boilerplate Apollo test)
	Sep-18	Saturn I	CCAFS	R&D (boilerplate Apollo test)
1965	Feb-16	Saturn I	CCAFS	Pegasus I (micrometeriod experiment)
	May-25	Saturn I	CCAFS	Pegasus II (micrometeriod experiment)
	Jul-30	Saturn I	CCAFS	Pegasus III (micrometeriod experiment)
1966	Feb-26	Saturn IB	CCAFS	R&D (unmanned Apollo)
	Jul-5	Saturn IB	CCAFS	R&D (S-IVB tests)
	Aug-25	Saturn IB	CCAFS	R&D (unmanned Apollo)
1967	Nov-9	Saturn V	CCAFS	Apollo 4 (test all three stages and restart)
1968	Jan-22	Saturn IB	CCAFS	Apollo 5 (1st Lunar Module test flight)
	Apr-4	Saturn V	CCAFS	Apollo 6 (emergency system detection test)
	Failure - Stg 3 failed to restart			
	Oct-11	Saturn IB	CCAFS	Apollo 7 (1st manned Apollo, lunar orbit)
	Dec-21	Saturn V	CCAFS	Apollo 8 (manned Apollo, lunar orbit)
1969	Mar-3	Saturn V	CCAFS	Apollo 9 (command and lunar module docking)
	May-18	Saturn V	CCAFS	Apollo 10 (manned Apollo, lunar orbit)
	Jul-16	Saturn V	CCAFS	Apollo 11 (manned Apollo, lunar landing)
	Nov-14	Saturn V	CCAFS	Apollo 12 (manned Apollo, lunar landing)
1970	Apr-11	Saturn V	CCAFS	Apollo 13 (manned Apollo, abort mission)
1971	Jan-31	Saturn V	CCAFS	Apollo 14 (manned Apollo, lunar landing)
	Jul-26	Saturn V	CCAFS	Apollo 15 (manned Apollo, lunar land rover)
1972	Apr-16	Saturn V	CCAFS	Apollo 16 (manned Apollo, lunar land rover)
	Dec-7	Saturn V	CCAFS	Apollo 17 (manned Apollo, lunar land rover)
1973	May-14	Saturn V	CCAFS	Skylab Workshop (Skylab module delivery)
	May-25	Saturn IB	CCAFS	Skylab 2 (manned visit to Skylab)
	Jul-28	Saturn IB	CCAFS	Skylab 3 (manned visit to Skylab)
	Nov-16	Saturn IB	CCAFS	Skylab 4 (manned visit to Skylab)
1975	Jul-15	Saturn IB	CCAFS	Apollo ASTP (manned docking with Soyuz)

Launch site is Cape Canaveral Air Force Station, Florida (28.5°N, 81.0°W)

Saturn Vehicle Design Options Considered

Name: Stage	Thrust		Engines	
C-1:				
S-I	1,500K lb	(6,670K N)	8 H-1	Max. Diameter: 21.7 ft (6.6 m)
S-IV	70K lb	(311K N)	4 RL-10B	Length (no payload): ??
S-V	35K lb	(156K N)	2 RL-10B	LEO Capability: 22.5K lb (10.2K kg)
C-1 (rev): became the Saturn I				
S-I	1,500K lb	(6,670K N)	8 H-1	Max. Diameter: 21.7 ft (6.6 m)
S-IV	90K lb	(400K N)	6 RL-10A	Length (no payload): 139 ft (42.4 m)
				LEO Capability: 22.5K lb (10.2K kg)
C-1B: became the Saturn IB				
S-I	1,500K lb	(6,670K N)	8 H-1	Max. Diameter: 21.7 ft (6.6 m)
S-IVB	200K lb	(890K N)	1 J-2	Length (no payload): 142 ft (43.3 m)
				LEO Capability: 36.6K lb (16.6K kg)
C-2:				
S-I	1,500K lb	(6,670K N)	8 H-1	Max. Diameter: 21.7 ft (6.6 m)
S-II	800K lb	(3,560K N)	4 J-2	Length (no payload): ??
S-IV	70K lb	(311K N)	4 RL-10B	LEO Capability: 45K lb (20.4K kg)
S-V	35K lb	(156K N)	2 RL-10B	
C-3:				
S-IB	3,000K lb	(13,300K N)	2 F-1	Max. Diameter: 26.6 ft (8.1 m)
S-II	800K lb	(3,560K N)	4 J-2	Length (no payload): ??
S-IV	70K lb	(311K N)	4 RL-10B	LEO Capability: 80K lb (36.3K kg)
C-4:				
S-IB	6,000K lb	(26,700K N)	4 F-1	Max. Diameter: 26.6 ft (8.1 m)
S-II	800K lb	(3,560K N)	4 J-2	Length (no payload): ??
S-IVB	200K lb	(890K N)	1 J-2	LEO Capability: ??
C-5: became the Saturn V				
S-IC	7,500K lb	(33,400K N)	5 F-1	Max. Diameter: 33.0 ft (10.1 m)
S-II	1,000K lb	(4,440K N)	5 J-2	Length (no payload): 281 ft (85.6 m)
S-IVB	200K lb	(890K N)	1 J-2	LEO Capability: 262K lb (119K kg)
C-8:				
S-ID	12,000K lb	(53,400K N)	8 F-1	Max. Diameter: ??
S-IIB	1,600K lb	(7,120K N)	8 J-2	Length (no payload): ??
S-IVB	200K lb	(890K N)	1 J-2	LEO Capability: ??
Nova:				
N-I	12,000K lb	(53,400K N)	8 F-1	Max. Diameter: 45.6 ft (13.9 m)
S-IB	3,000K lb	(13,300K N)	2 F-1	Length (no payload): 265 ft (80.8 m)
S-II	800K lb	(3,560K N)	4 J-2	LEO Capability: 350K lb (159K kg)
S-V	35K lb	(156K N)	2 RL-10B	
Nova (rev):				
N-I	12,000K lb	(53,400K N)	8 F-1	Max. Diameter: 50.0 ft (15.2 m)
N-II	4,000K lb	(17,800K N)	4 M-1	Length (no payload): 280 ft (83.3 m)
N-III	200K lb	(890K N)	1 J-2	LEO Capability: 375K lb (170K kg)

Note: S-V is a modified Centaur

United States: Saturn
(continued)

Saturn I and IB Vehicle

Launch Escape Tower
Command Module
Service Module
Lunar Module Adapter
Instrument Unit
LH2 Tank
LOX Tank
J-2 Engines (1)
S-IVB Stage
Outer LOX Tank
RP-1 Tank
S-IB Stage
H-1 Engines (8)
Center LOX Tank

Saturn IB

Overall	**Saturn I**	**Saturn IB**
System Length	120 ft (36.6 m)	224 ft (68.3 m) includes 86 ft (26.2 m) for Apollo spacecraft
Gross Mass	1.116M lb (0.506M kg)	1.297M lb (0.588M kg)

	Stage 1 (S-IB)	**Stage 2* for Saturn IB (S-IVB)**	**Stage 2* for Saturn I (S-IV)**
Dimension:			
Length	80.3 ft (24.5 m)	59.0 ft (18.0 m)	40 ft (12.2 m)
Diameter	21.4 ft (6.52 m)	21.7 ft (6.61 m)	18 ft (5.5 m)
Mass: (each)			
Propellant Mass	899K lb (408K kg)	233K lb (106K kg)	??
Gross Mass	980K lb (444K kg)	255K lb (116K kg)	96K lb (43.5K kg)
Structure:			
Type	Skin-Stringer	Isogrid	??
Material	Aluminum	Aluminum	Aluminum
Propulsion:			
Propellant	LOX / RP-1	LOX / LH2	LOX / LH2
Average Thrust (each)	205K lb (912K N) SL each or 1,640K lb (7,295K N) total	205K lb (912K N) vac	15K lb (66.7K N) vac ea. or 90K lb (400K N) total
Engine Designation	H-1	J-2	RL-10A-3
Number of Engines	8	1	6
Isp	232 sec SL	425 sec vac	444 sec SL
Feed System	Gas generator	Gas generator	Split expander
Chamber Pressure	689 psia (47.5 bar)	703 psia (48.5 bar)	465 psia (32.1 bar)
Mixture Ratio (O/F)	2.23	5.5	5.0
Throttling Capability	100% only	??	100% only
Expansion Ratio	8:1	27:1	??
Restart Capability	No	1 restart	Multiple
Tank Pressurization	Gas Generator ?	Cold helium, hot GH2	Cold helium, hot GH2
Control-Pitch, Yaw, Roll	Gimbal 4 outboard nozzles	Gimbal 1 nozzle (±7.0°) & auxiliary propulsion system	Gimbal nozzles
Events:			
Nominal Burn Time	??	??	??
Stage Shutdown	Burn to depletion	Command shutdown	Command shutdown
Stage Separation	Retro-rockets	Retro-rockets	Retro-rockets

* Does not include instrument unit by IBM (3 ft (0.9 m) length, 2,760 lb (1,252 kg))

Remarks:

The Saturn IB vehicle consisted of the S-IB first stage, S-IVB second stage, and instrument unit (IU). The Saturn I was the same except for a different second stage (S-IV).

The S-IB first stage basic design concept incorporated Jupiter and Redstone components because of their high reliability and qualification status. The S-IB stage was analogous to the Saturn I S-I stage but with lightened structure, uprated engines, a simplified propulsion system, and reduced instrumentation. The main stage body was a cluster of nine propellant tanks, consisting of four fuel and four oxidizer 70 in. (1.8 m) diameter Redstone tanks arranged alternately around a larger center oxidizer 105 in. (2.67 m) diameter Jupiter tank. Each tank had antislosh baffles to minimize propellant turbulence in flight. Stage electrical and instrumentation equipment were located in the forward and aft skirts of the fuel tanks. A tail unit assembly supported the aft tank cluster and provided a mounting surface for the engines. Eight fin assemblies supported the vehicle on the launcher and improve the aerodynamic characteristics of the vehicle. A stainless steel honeycomb heat shield enclosed the aft tail unit to protect against the engine exhaust. A firewall above the engines separated the propellant's tanks from the engine compartment. Eight Rocketdyne H-1 engines, repackaged from Thor and Jupiter weapon systems, boosted the vehicle during the first phase of powered flight. The four inboard engines were stationary, and the four outboard engines gimbaled for flight control. Two hydraulic actuators position each outboard engine on signal from the inertial guidance system.

The S-IVB used a single propellant tank with common bulkhead design and was powered by one J-2 engine. A closed loop hydraulic system gimbaled the J-2 engine for pitch and yaw control during flight. An Auxiliary Propulsion System (APS), using two APS modules on the exterior aft skirt, provided vehicle roll control during flight and three axis control during the coast mode. The exact propellant mass load needed for orbital insertion with minimum residuals at cutoff was determined prior to launch. A propellant utilization (PU) system helped load this accurate mass and controlled its utilization in flight. Instrumentation equipment was located in the forward and aft skirts. The aft interstage connected the S-IVB skirt to the S-IB spider beam unit. The aft skirt/aft interstage junction was the separation plane. The Saturn I S-IV was dimensionally smaller than S-IVB and used 6 RL-10 engines instead of one J-2 engine.

The instrument unit was an unpressurized, cylindrical, load-supporting structure of sandwich-type construction. It housed electrical equipment to guide, control, and monitor the launch vehicle and an environmental conditioning system (ECS) that provided a satisfactory operating environment for this equipment.

Historical Launch Systems

United States: Saturn
(continued)

Saturn V Vehicle

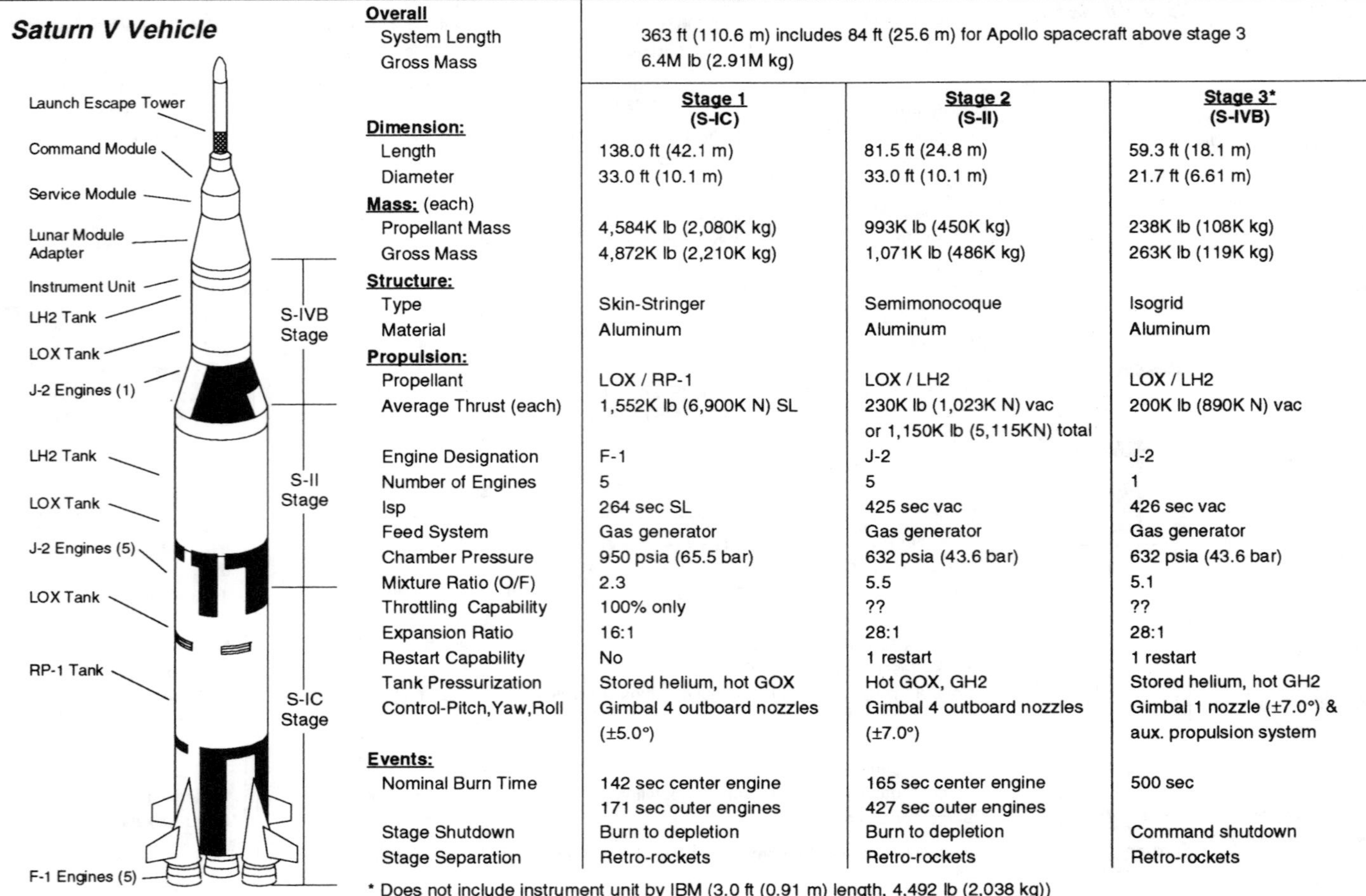

Overall	
System Length	363 ft (110.6 m) includes 84 ft (25.6 m) for Apollo spacecraft above stage 3
Gross Mass	6.4M lb (2.91M kg)

	Stage 1 (S-IC)	**Stage 2 (S-II)**	**Stage 3* (S-IVB)**
Dimension:			
Length	138.0 ft (42.1 m)	81.5 ft (24.8 m)	59.3 ft (18.1 m)
Diameter	33.0 ft (10.1 m)	33.0 ft (10.1 m)	21.7 ft (6.61 m)
Mass: (each)			
Propellant Mass	4,584K lb (2,080K kg)	993K lb (450K kg)	238K lb (108K kg)
Gross Mass	4,872K lb (2,210K kg)	1,071K lb (486K kg)	263K lb (119K kg)
Structure:			
Type	Skin-Stringer	Semimonocoque	Isogrid
Material	Aluminum	Aluminum	Aluminum
Propulsion:			
Propellant	LOX / RP-1	LOX / LH2	LOX / LH2
Average Thrust (each)	1,552K lb (6,900K N) SL	230K lb (1,023K N) vac or 1,150K lb (5,115KN) total	200K lb (890K N) vac
Engine Designation	F-1	J-2	J-2
Number of Engines	5	5	1
Isp	264 sec SL	425 sec vac	426 sec vac
Feed System	Gas generator	Gas generator	Gas generator
Chamber Pressure	950 psia (65.5 bar)	632 psia (43.6 bar)	632 psia (43.6 bar)
Mixture Ratio (O/F)	2.3	5.5	5.1
Throttling Capability	100% only	??	??
Expansion Ratio	16:1	28:1	28:1
Restart Capability	No	1 restart	1 restart
Tank Pressurization	Stored helium, hot GOX	Hot GOX, GH2	Stored helium, hot GH2
Control-Pitch,Yaw,Roll	Gimbal 4 outboard nozzles (±5.0°)	Gimbal 4 outboard nozzles (±7.0°)	Gimbal 1 nozzle (±7.0°) & aux. propulsion system
Events:			
Nominal Burn Time	142 sec center engine 171 sec outer engines	165 sec center engine 427 sec outer engines	500 sec
Stage Shutdown	Burn to depletion	Burn to depletion	Command shutdown
Stage Separation	Retro-rockets	Retro-rockets	Retro-rockets

* Does not include instrument unit by IBM (3.0 ft (0.91 m) length, 4,492 lb (2,038 kg))

Remarks:

The three stage Saturn V launch vehicle consisted of the S-IC first stage, the S-II second stage, the S-IVB third stage, and the Instrument Unit (IU).

The S-IC stage was designed and built by Boeing to provide the first stage boost for the Saturn V. It was powered by five F-1 rocket engines. The F-1 engine was then the most powerful liquid propellant engine ever developed and flown. Four engines were equally spaced in a circular pattern near the stage perimeter. These engines could be gimbaled for vehicle attitude control. The fifth engine was fixed and was mounted on stage centerline. Each outboard engine was protected from aerodynamic loading by a conically shaped engine fairing. Four fixed stabilizing fins augmented the stability of the Saturn V vehicle. The S-IC boosted the vehicle to a burnout velocity of 5,118 mph (8,236 km/hr) at an altitude of 37 mi (60 km) in 2.7 minutes. It then separated from the S-II stage and fell to Earth approximately 415 mi (670 km) downrange.

The S-II Saturn second stage was designed and built by Rockwell. The propellants for its five J-2 engines were carried in a single propellant tank with common bulkhead design. Four outer J-2 engines were equally spaced on a 17 ft (5.3 m) diameter circle and were capable of being gimbaled through a ±7.0 degree pattern for thrust vector control. The fifth engine was mounted on stage centerline and was fixed. The S-II stage contained a propellant utilization system to monitor propellant consumption. This system was capable of changing the engine mixture ratio from 5.5:1 to 4.8:1 during stage operation. At engine cutoff, the S-II stage separated from the S-IVB stage and following a suborbital path, reentered the atmosphere where it disintegrated. The S-II accelerated the vehicle to 14,775 mph (23,770 km/hr) at an altitude of 110 mi (177 km) in 6 min.

The S-IVB Saturn third stage built by McDonnell Douglas was powered by a single J-2 engine. The S-IVB used a single propellant tank with common bulkhead design. A closed loop hydraulic system gimbaled the J-2 engine for pitch and yaw control during flight. An auxiliary propulsion system (APS) using two APS modules on the aft skirt, provided vehicle roll control during flight and three-axes control during the coast mode. A propellant utilization (PU) system monitored S-IVB propellant consumption during all phases of stage operation. Propellant utilization was controlled by a two-position mixture ratio control valve. Instrumentation equipment was located in the forward and aft skirts. The S-IVB fired for 2.5 minutes to place the vehicle in an Earth-parking orbit. For lunar missions, the J-2 was refired for 5 minutes on the second orbit, injecting the vehicle into a lunar transfer trajectory with a velocity of 23,500 mph (37,800 km/hr).

The Instrument Unit was designed and manufactured by IBM to house major vehicle electronics. It was a cylindrical structure which was installed on top of the S-IVB stage. The IU contained the guidance, navigation, and control equipment which guided the vehicle through its Earth orbits and subsequently, into its mission trajectory. In addition, it contained telemetry, communications, tracking and crew safety systems, along with their supporting electrical power and environmental control systems.

United States: Saturn
(continued)

Saturn V Operations

The Vehicle Assembly Building (VAB) provided a protected environment for receipt and checkout of the propulsion stages and instrument unit, erection of launch vehicle and spacecraft in the vertical position on the mobile launcher, and integrated checkout of the assembled space vehicle. The VAB is a totally enclosed structure covering approximately 350,000 sq-ft (32,400 sq-m) of ground. It is 525 ft (160 m) high, 518 ft (158 m) wide and 715 ft (218 m) long. The principal elements of the VAB are the low bay area and the high bay area. The low bay area provided space for receiving and preparing the S-II and S-IVB stages, the instrument unit and the Apollo spacecraft. The high bay area was used for receiving and preparing the S-IC stage, vehicle assembly, and integrated vehicle checkout. Four fully assembled Saturn V vehicles could be accommodated in the high bay area.

The launch control center (LCC) served as the focal point for overall direction control, and surveillance of space vehicle checkout and launch. The LCC is located adjacent to the VAB at a distance of 3 mi (4.8 km) from the launch pad. This allowed viewing of vehicle liftoff without site hardening. The LCC is a four story building approximately 380 by 180 ft (116 m by 55 m). The first floor contained offices, laboratories, shops and communications control. The second floor housed telemetry, tracking, instrumentation, and data reduction facilities. And, the third floor was divided into four separate control areas. Each control area contained a firing room, computer room, mission control room, test conductor platform and office space. The LCC is connected by buried cableways to all key areas of the launch complex.

The mobile launcher (ML) is a transportable steel structure which, with the crawler-transporter, provided the capability to move the erected vehicle to the launch pad. The base of the mobile launcher is 25 ft (7.6 m) high, 160 ft (48.8 m) long and 135 ft (41.1 m) wide. The mobile launcher base contained test, support and servicing equipment. The umbilical tower was 380 ft (115.8 m) high and provided vehicle support and servicing access. The crawler-transporter is 131 ft (39.9 m) long and 114 ft (34.7 m) wide. It has a mass of approximately 6 million lb (2.7 million kg). The crawler-transporter was capable of lifting, transporting, and lowering the ML, with assembled vehicle, without the aid of auxiliary equipment. Crawler-transporter unloaded speed is 2 mph (3.2 km/hr), 1 mph (1.6 km/hr) with full load on level ground, and 0.5 mph (0.8 km/hr) with full load on a 5% grade. It had a minimum turning radius of 500 ft (152 m).

The launch pad provided a stable foundation for the ML during Apollo/Saturn V launch and prelaunch operations and an interface to the ML for ML and vehicle systems. Launch Complex 39 (LC-39) is composed of two launch pads located 3 mi (4.8 km) from the LCC. Each launch site is an eight-sided polygon measuring approximately 3,000 ft (914 m) across. The launch pad is cellular reinforced concrete structure with a top elevation of 48 ft (14.6 m) above sea level (42 ft (12.8 m) above grade elevation). The longitudinal axis of the pad is oriented north-south, with the crawlerway and ramp approach from the south. The ramp, with a five percent grade, provides access to the pad from the crawlerway. A flame trench 58 ft (17.7 m) wide by 450 ft (137.2 m) long, bisects the pad. This trench opens to grade at the north end. The 700,000 lb (317,500 kg) mobile wedge-type flame deflector is mounted on rails in the trench.

Reference:
- A Historical Look at US Launch Vehicles 1967-present, ANSER, 2nd edition, Feb. 1990
- US Civilian Space Programs 1958-78, House Subcom. on Space Science & Applications
- Saturn V Flight Manual - SA510, NASA MSFC, MSFC-MAN-510, Updated 25 Jun 1971
- Aeronautics & Astronautics, AIAA, pg. 75-78, Jul/Aug 1976
- US Space Launch Systems, Navy Space Systems Activity, NSSA-R-20-72-2, Rev. 1 Jul 1977

Typical Saturn V Launch Sequence

HR:MIN:SEC	EVENTS
-9:30:00.0	Switch vehicle environmental conditioning from air to GN2
-8:59:00.0	Guidance and control system checks
-7:28:00.0	S-IVB liquid oxygen loading
-7:04:00.0	S-II liquid oxygen loading
-6:02:00.0	S-IC liquid oxygen loading
-4:54:00.0	S-II liquid hydrogen loading
-4:11:00.0	S-IVB liquid hydrogen loading
-3:28:00.0	Astronauts onboard
-0:10:00.0	Prepare to launch test
-0:05:30.0	Safe and Arm devices armed
-0:05:00.0	Retract and lock Command Module access
-0:03:07.0	Start automatic sequence
-0:00:50.0	Transfer to internal power
-0:00:30.0	S-IC engine hydraulic system flight activation
-0:00:17.0	Guidance reference release
-0:00:16.2	Retract and lock S-IC forward service arm
-0:00:08.9	S-IC ignition command
-0:00:05.3	Monitor S-IC engine thrust buildup
-0:00:01.0	Holddown arm release
+0:00:00.4	Liftoff
+0:02:15.8	S-IC center engine cutoff
+0:02:38.8	S-IC outboard engine cutoff
+0:02:40.5	S-IC/S-II separation complete
+0:02:42.2	S-II ignition
+0:03:10.5	S-II aft interstage drop
+0:07:38.8	S-II center engine cutoff
+0:09:13.7	S-II outboard engine cutoff
+0:09:14.8	S-II/S-IVB separation complete
+0:09:17.9	S-IVB first burn ignition
+0:11:42.2	S-IVB first burn cutoff
+0:11:54.3	Parking orbit insertion by auxiliary propulsion system (APS)
+2:49:57.9	S-IVB second burn ignition
+2:55:53.9	S-IVB second burn cutoff
+2:56:03.9	Translunar injection
+4:15:54.1	S-IVB / spacecraft separation
+5:29.54.1	Initiate maneuver to S-IVB APS impact burn attitude

S-IC is first stage; S-II is second stage: and S-IVB is third stage

Interior of Vehicle Assembly Building (VAB)

Historical Launch Systems

United States: SCOUT History

Vehicle Description

Configuration	X-1	X-2	X-3	X-4	A-1	B-1	D-1	F-1	G-1
Year	1960	1962	1963	1964	1965	1965	1972	1974	1979
Stage 1	Algol IC	Algol ID	Algol IIA	Algol IIB	Algol IIB	Algol IIB	Algol IIIA	Algol IIIA	Algol IIIA
Stage 2	Castor I	Castor I	Castor I	Castor I	Castor II	Castor IIA	Castor IIA	Castor IIA	Castor IIA
Stage 3	Antares I	Antares II	Antares II	Antares II	Antares II	Antares IIA	Antares IIA	Antares IIB	Antares IIIA
Stage 4	Altair I	Altair I	Altair I	Altair II	Altair II	Altair IIIA	Altair IIIA	Altair IIIA	Altair IIIA
Easterly, lb (kg)	130 (59)	168 (76)	192 (87)	227 (103)	269 (122)	315 (143)	408 (185)	425 (193)	459 (208)
Polar, lb (kg)	100 (45)	130 (59)	150 (68)	176 (80)	207 (94)	256 (116)	326 (148)	344 (156)	366 (166)
Equatorial, lb (kg)	—	—	—	—	—	333 (151)	437 (198)	448 (203)	485 (220)

Note: In about 1966 it was decided that the initial version of a motor would be designated A. At this time all motors in use were so designated. A small change in a motor that affected performance would be a letter change, a major change would be a number change. The Algol II and IIA, and the Altair III and IIIA are the same motors. Both the United Technology and Thiokol motors used the Altair IIIA designation as the performance, fit, and appearance were the same.

SCOUT
(also SCOUT G-1)

Historical Summary

The SCOUT launch vehicle, the first U.S. launch vehicle to use solid fuel exclusively, was designed by Vought Astronautics as a booster for DoD, NASA, and foreign space probes and orbital and reentry research payloads. SCOUT is an acronym which stands for Solid Controlled Orbital Utility Test. Flights began in 1960 and concluded in 1994 after 118 launches.

In the earliest days of the U.S. space program, the National Advisory Committee for Aeronautics (NACA, now NASA) needed a launch vehicle to investigate the unknown realm of space flight. At the same time, solid rocket motors were being developed for missile applications, and the Air Force was interested in an advanced solid rocket motor test system. Thus, NACA and the Air Force agreed to undertake a cooperative development program. The SCOUT program, to be based at Langley Field, was approved by NASA in 1958.

Vought Astronautics was awarded a contract in 1959 for the design and development of structural elements of the SCOUT vehicle and launch tower. In 1960, NASA increased Vought's responsibilities in the program by making them the prime integrator of the SCOUT vehicle. Since that time, Loral Vought Systems has produced and launched the SCOUT for NASA, DoD, and foreign payloads.

The SCOUT launch vehicle underwent systematic upgrading during its development. The standard vehicle was a solid-propellant, four-stage booster. While keeping the basic same configuration over the years, the SCOUT evolved to the higher performance vehicle through a number of steps. Each of the four stages was replaced at least once since the original version.

The original first-stage Algol rocket was developed by Aerojet from their Jupiter and Polaris programs. The first version, the Algol I (XM-68), had a thrust of 86,000 lb (382,500 N) and a 40 in. (1.02 m) diameter steel case and a length of about 31 ft (9.4 m). Algol IIA was introduced in 1963, using the Aerojet 40 KS motor. After a flight failure, the nozzle was redesigned and became the Algol IIB. Algol IIIA, built by United Technology Corporation Chemical Systems Division, emerged in 1972. Although still the same length as the Algol IIB the Algol IIIA was 45 in. (1.14 m) in diameter and provided a thrust of 104,500 lb (464,700 N), increasing the payload lifting capacity by about 30%.

Just as the SCOUT's first stage evolved from the Polaris missile, the second stage was derived from the Sergeant missile. In its space application, the Sergeant stage, more commonly called the Castor, has been part of many different sounding and test rockets. SCOUT used an altitude straight nozzle while other programs use straight or canted sea level nozzles. The first motor used, the Castor I (TX-33), was produced by Thiokol and remained in use until 1965. At that time, the Castor II (TX-354) was developed by Thiokol with higher energy propellant and a better grain design allowing more propellant. These Castor II motors also were employed as the principal strap-on with the Thor and Delta launch vehicles. The last version used in SCOUT, the Castor IIA (TX-354-3), was very similar to the Castor II. The Castor I had a length of 19.4 ft (5.9 m) and a thrust of approximately 54,000 lb (24,500 kg). Castor II had a thrust of 52,200 lb (232,000 N) but had a longer burn time than Castor I. Castor IIA was 20.7 ft (6.31 m) long and provided a thrust of 60,100 lb (267,400 N).

The third stage of SCOUT is the Antares, manufactured until 1979 by Hercules and then by Thiokol. It has a fiberglass case. The Antares I was 11.2 ft (3.4 m) long and had a diameter of approximately 30 in. (760 mm) and a thrust of 14,500 lb (64,500 N). As the design evolved from the Antares I through the Antares II, IIA, and IIB (all designated X-259) to the Antares IIIA (TEM-762), the thrust increased to 18,200 lb (80,800 N) although fitting in the same envelope.

The fiberglass case Altair fourth stage was modified from the Vanguard program. Following Thiokol's Altair I (X-248) fourth stage was Altair II (X-258), which was used for SCOUT and Burner programs. In this improved version the Altair II had an increased thrust of 5,800 lb (25,800 N). The SCOUT-G used the Altair IIIA (TEM-640) fourth stage, which also provided 5,800 lb (25,800 N) of thrust. A fifth-stage velocity package could be added, increasing the SCOUT's hypersonic reentry performance, enabling high elliptical orbits in deep space, or extending its probe capabilities to the sun. One SCOUT flight, the E-1 vehicle used a fifth stage, the Alcyone 1A from Hercules Bacchus.

United States: SCOUT History
(continued)

Launch Record

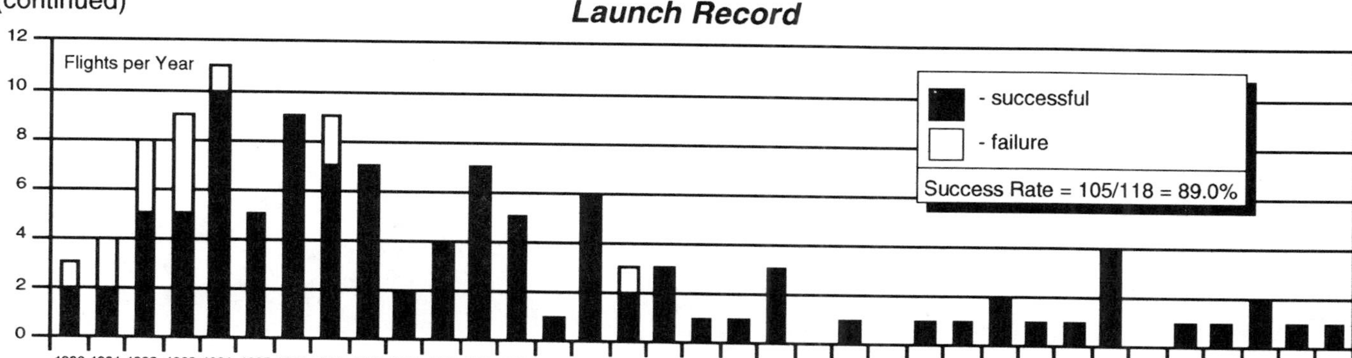

	YEAR	DATE	VEHICLE	SITE	PAYLOAD
1	1960	Jul-1	X-1 (T)	WFF	Sim. Probe
2		Oct-4	X-1 (T)	WFF	Sim. Probe
3		Dec-14	X-1 (T)	WFF	S-56
		Failure - Stage 2 ignition malfunction			
4	1961	Feb-16	X-1 (T)	WFF	S-56A (Explorer 9)
5		Jun-30	X-1 (T)	WFF	S-55
		Failure - Stage 3 ignition malfunction			
6		Aug-25	X-1 (T)	WFF	S-55A (Explorer 13)
		Failure - Inappropriate orbit			
7		Oct-19	X-1 (T)	WFF	P-21
8	1962	Mar-1	X-1 (T)	WFF	RE-1
9		Mar-29	X-1 (T)	WFF	P-21A
10		Apr-26	X-2 (T)	SLC-5	Solrad 4B
		Failure - Failed to orbit			
11		May-23	X-2 (T)	SLC-5	P-35A
		Failure - Failed to orbit			
12		Aug-23	X-2 (T)	SLC-5	P-35B
13		Aug-31	X-3 (T)	WFF	RE-2
		Failure - Late stage 3 ignition			
14		Dec-16	X-3 (T)	WFF	S-55B (Explorer 16)
15		Dec-18	X-3 (T)	SLC-5	Transit 5A
16	1963	Feb-19	X-3 (T)	SLC-5	P-35C
17		Apr-5	X-3 (T)	SLC-5	Transit 5A-2
		Failure - Failed to orbit			
18		Apr-26	X-2 (T)	SLC-5	P-35D
		Failure - Failed to orbit			
19		May-22	X-3 (T)	WFF	RFD-1
20		Jun-15	X-3 (T)	SLC-5	Transit 5A-3
21		Jun-28	X-4 (T)	WFF	CRL-1
22		Jul-20	X-3 (T)	WFF	RE-3
		Failure - Stage 1 nozzle failure			
23		Sep-27	X-4 (T)	SLC-5	P-35E
		Failure - Failed to orbit			
24		Dec-19	X-4	SLC-5	S-56B (Explorer 19)
25	1964	Mar-27	X-3	WFF	UK-2 (Ariel 2)
26		Jun-3	X-4	SLC-5	Transit 5C-1
27		Jun-26	X-4	SLC-5	CRL-2
		Failure - Stage 2 destruct			
28		Jul-20	X-4	WFF	SERT
29		Aug-18	X-4	WFF	RE-4
30		Aug-25	X-4	SLC-5	S-48 (Explorer 20)
31		Oct-9	X-4	WFF	RFD-2
32		Oct-10	X-4	SLC-5	Beacon Explorer-B (Explorer 22)
33		Nov-6	X-4	WFF	S-55C (Explorer 23)
34		Nov-21	X-4	SLC-5	Air Destiny / Injun-B (Explorers 24/25)
35		Dec-15	X-4	WFF	San Marco A
36	1965	Apr-29	X-4	WFF	Beacon Explorer-C (Explorer 27)
37		Aug-10	B-1	WFF	SECOR 5
38		Nov-18	X-4	WFF	SOLRAD-A (Explorer 30)
39		Dec-6	X-4	SLC-5	FR-1
40		Dec-21	A-1	SLC-5	Transit 5
41	1966	Jan-28	A-1	SLC-5	Transit 6
42		Feb-9	X-4	WFF	RE-E
43		Mar-25	A-1	SLC-5	Transit 7
44		Apr-22	B-1	SLC-5	OV3-1
45		May-18	A-1	SLC-5	Transit B
46		Jun-10	B-1	WFF	OV3-4
47		Aug-4	B-1	SLC-5	OV3-3
48		Aug-17	A-1	SLC-5	Transit 9
49		Oct-28	B-1	SLC-5	OV3-2
50	1967	Jan-31	B-1	SLC-5	OV3-5
		Failure - Failed to orbit			
51		Apr-13	A-1	SLC-5	Transit 10
52		Apr-26	B-1	SM	San Marco B
53		May-5	A-1	SLC-5	UK-3 (Ariel 3)
54		May-18	A-1	SLC-5	Transit 11
55		May-29	B-1	SLC-5	ESRO-2A
		Failure - Stage 3 failure			
56		Sep-25	A-1	SLC-5	Transit 12
57		Oct-19	B-1	WFF	RAM C-A
58		Dec-4	B-1	SLC-5	OV3-6
59	1968	Mar-1	A-1	SLC-5	Transit 13
60		Mar-5	B-1	WFF	SOLRAD-B (Explorer 37)
61		Apr-27	A-1	WFF	RE-F
62		May-17	B-1	SLC-5	ESRO 2B (Iris-B)
63		Aug-8	B-1	SLC-5	Air Density / Injun-C (Explorers 39/40)
64		Aug-22	B-1	WFF	RAM C-B
65		Oct-3	B-1	SLC-5	ESRO 1A (Aurora)
66	1969	Oct-1	B-1	SLC-5	ESRO 1B (Boreas)
67		Nov-7	B-1	SLC-5	GRS-A (Azur 1)
68	1970	Aug-27	A-1	SLC-5	Transit 14
69		Sep-30	B-1	WFF	RAM C-C
70		Nov-9	B-1	WFF	OFO/RMS
71		Dec-12	B-1	SM	Small Astronomy Satellite-A (Explorer 42)
72	1971	Apr-24	B-1	SM	San Marco C
73		Jun-20	B-1	WFF	PAET
74		Jul-8	B-1	WFF	SOLRAD-C (Explorer 44)
75		Aug-16	B-1	WFF	FR-2 (EOLE)
76		Sep-20	B-1	WFF	GRP-A Barium Ion Cloud
77		Nov-15	B-1	SM	Small Scientific Satellite-A (Explorer 45)
78		Dec-11	B-1	SLC-5	UK-4 (Ariel 4)
79	1972	Aug-13	D-1	WFF	Meteoroid Technology Satellite (Explorer 46)
80		Sep-2	B-1	SLC-5	TIP I
81		Nov-15	D-1	SM	Small Astronomy Satellite-B (Explorer 48)
82		Nov-21	D-1	SLC-5	ESRO IV
83		Dec-16	D-1	SLC-5	AEROS A
84	1973	Oct-29	A-1	SLC-5	Transit 15
85	1974	Feb-18	D-1	SM	San Marco C2
86		Mar-8	D-1	SLC-5	UKX-4 (Miranda)
87		Jun-3	E-1	SLC-5	Hawkeye (Explorer 52)
88		Jul-16	D-1	SLC-5	AEROS B
89		Aug-30	D-1	SLC-5	ANS-A
90		Oct-15	B-1	SM	UK-5 (Ariel 5)
91	1975	May-8	D-1	SM	Small Astronomy Satellite-C (Explorer 53)
92		Oct-11	D-1	SLC-5	TIP II
93		Dec-5	D-1	SLC-5	Dual Air Density
		Failure - Failed to orbit			
94	1976	May-22	B-1	SLC-5	P76-5
95		Jun-18	D-1	WFF	GP-A (Redshift)
96		Sep-1	D-1	SLC-5	TIP III
97	1977	Oct-27	D-1	SLC-5	Transat
98	1978	Apr-26	D-1	SLC-5	HCMM
99	1979	Feb-18	D-1	WFF	SAGE
100		Jun-2	D-1	WFF	UK-6 (Ariel 6)
101		Oct-30	G-1	SLC-5	MAGSAT
102	1981	May-14	G-1	SLC-5	Nova I
103	1983	Jun-27	G-1	SLC-5	HILAT
104	1984	Oct-11	G-1	SLC-5	Nova III
105	1985	Aug-2	G-1	SLC-5	SOOS I (A/B)
106		Dec-12	G-1	WFF	ASAT Target (1/2)
107	1986	Nov-13	G-1	SLC-5	Polar Bear
108	1987	Sep-16	G-1	SLC-5	SOOS II (A/B)
109	1988	Mar-25	G-1	SM	San Marco D
110		Apr-25	G-1	SLC-5	SOOS III (A/B)
111		Jun-15	G-1	SLC-5	Nova II
112		Aug-25	G-1	SLC-5	SOOS IV (A/B)
113	1990	May-9	G-1	SLC-5	MACSAT
114	1991	Jun-29	G-1	SLC-5	REX
115	1992	Jul-3	G-1	SLC-5	Sampex
116		Nov-21	G-1	SLC-5	MSTI-1
117	1993	Jun-25	G-1	SLC-5	Radcal
118	1994	May-8	G-1	SLC-5	MSTI II

* Department of Defense (DoD) are classified and full launch details are not disclosed.
(T) - Test Launch

WFF (Wallops Flight Facility) at Wallops, Island, VA
SLC-5 at Vandenberg AFB, CA
SM (San Marco) at Ngwana Bay off the coast of Kenya

Historical Launch Systems

United States: SCOUT General Description
(continued)

Scout-G1

Summary

The SCOUT launch vehicle was developed in the late 1950s to launch small payloads into Earth orbit and to serve as a launch vehicle for probe and reentry studies. The final SCOUT 1, the G1, was a four-stage solid propellant vehicle consisting of the following stages: Algol IIIA, Castor IIA, Antares IIIA, and Altair IIIA. SCOUT vehicles were launched from the Wallops Flight Facility, Vandenberg Air Force Base, and the San Marco Range. The San Marco range was operated by the Italian Government in cooperation with NASA. The SCOUT was used primarily to launch NASA and foreign scientific missions, Navy and Air Force satellites.

Status —

Key Organizations

User - DoD, NASA, commercial

Launch Service Agency - Loral Vought Systems

Prime Contractor - Loral Vought Systems (Integration)

Principal Subcontractors
- United Technologies CSD (Algol IIIA)
- Thiokol (Castor IIA, Antares IIA, Altair IIIA)
- Honeywell (Avionics)

Vehicle

System Height	75.1 ft (22.9 m)
Payload Fairing Size	2.8 ft (0.86 m) diameter by 10.7 ft (3.27 m) height 2.8 ft (0.86 m) diameter by 12.0 ft (3.65 m) height 3.5 ft (1.07 m) diameter by 12.4 ft (3.78 m) height
Gross Mass	47,950 lb (21,750 kg)

Operations

Primary Missions	LEO, GTO to all inclinations
Compatible Upper Stages	BE-3 fifth stage built by Hercules and STAR 17 motor built by Thiokol.
First Launch	1960 for SCOUT X-1 1979 for SCOUT G-1
Success / Flight Total	105 / 188 (all versions)
Launch Site	VAFB - SLC-5 (34.7°N, 120.6°W) WFF - Pad 3 (37.9°N, 75.4°W) San Marco (2.9°S, 40.3°E)
Launch Azimuth	SLC-5 - 164°-287° (max. inclination was 76°- 146°) WFF - 85-109°, 126-129° (max. incl 38-41°,50- 52°) San Marco - 82°-130° (max. inclination is 2.9°- 38°)
Nominal Flight Rate	10–12 / yr

Performance

100 nm (185 km) circ, 2.9°	600 lb (270 kg)
100 nm (185 km) circ, 37.7°	560 lb (255 kg)
100 nm (185 km) circ, 90°	460 lb (210 kg)
Geotransfer Orbit, 28°	120 lb (54 kg) with upper stage
Geosynchronous Orbit	—

Remarks None

United States: SCOUT Vehicle
(continued)

SCOUT Overall

Length: 75.1 ft (22.9 m)
Gross Mass: 47,950 lb (21,750 kg)
Thrust at Liftoff: 104,500 lb (464,700 N)

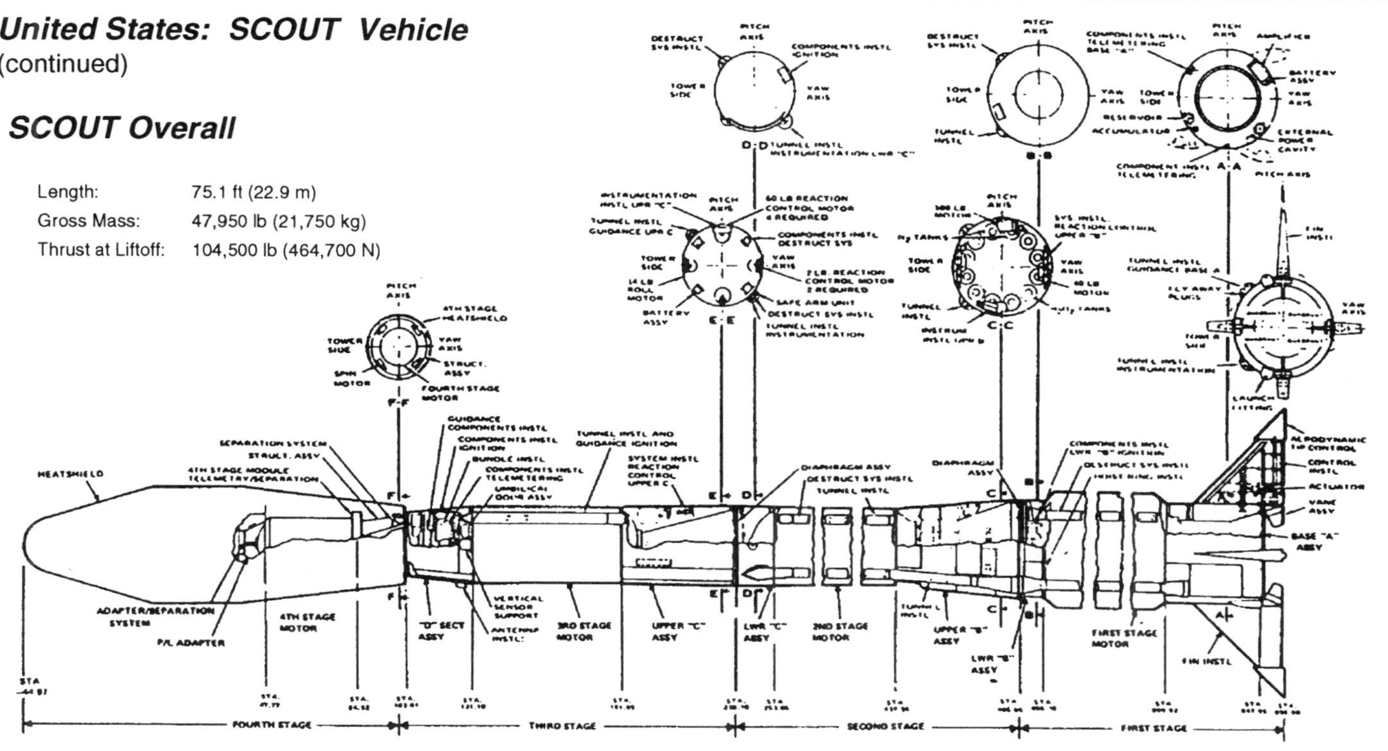

SCOUT Stages	Stage 1 (Algol IIIA)	Stage 2 (Castor IIA)	Stage 3 (Antares IIIA)	Stage 4 (Altair IIIA)
Dimension:				
Length	30.8 ft (9.40 m)	20.7 ft (6.31 m)	11.5 ft (3.51 m)	4.9 ft (1.48 m)
Diameter	3.75 ft (1.14 m)	2.6 ft (0.79 m)	2.5 ft (0.76 m)	1.7 ft (0.51 m)
Mass: (each)				
Propellant Mass	28,003 lb (12,702 kg)	8,196 lb (3,718 kg)	2,833 lb (1,285 kg)	603 lb (273 kg)
Gross Mass	32,485 lb (147,35 kg)	10,625 lb (4,819 kg)	3,610 lb (1,637 kg)	710 lb (322 kg)
Structure:				
Type	Monocoque	Monocoque	Monocoque	Monocoque
Case Material	Steel	Steel	Kevlar/Epoxy	Fiberglass
Propulsion:				
Propellant	PBAN	CTPB	HTPB	CTPB
Avg. Thrust (each)	104,500 lb (464,700 N)vac 93,250 lb (414,800 N) SL	60,100 lb (267,400 N) vac	18,200 lb (80,800 N) vac	5,800 lb (25,800 N) vac
Number of Motors	1	1	1	1
Number of Segments	1	1	1	1
Isp	259 sec vac/229 sec SL	280 sec vac	295 sec vac	288 sec vac
Chamber Pressure	450 psia (31.0 bar)	700 psia (4.82M Pa)	700 psia (48.2 bar)	670 psia (46.2 bar)
Expansion Ratio	6.5:1	21.2:1	58.8:1	50.5:1
Control-Pitch, Yaw, Roll	Aero fins and jet vanes	H2O2 RCS	H2O2 RCS	Spin-stabilized
Events:				
Nominal Burn Time	84 sec	41 sec	48 sec	34 sec
Stage Shutdown	Burn to depletion	Burn to depletion	Burn to depletion	Burn to depletion
Stage Separation	"Blow out" diaphragm	"Blow out" diaphragm	Spring ejection	Spring ejection

Remarks:
The Algol IIIA, SCOUT first stage propulsion unit, was manufactured by United Technologies Corporation/Chemical Systems Division, San Jose, CA. The Algol IIIA combined a steel motor case with a lightweight reinforced plastic nozzle. The Castor IIA motor SCOUT second stage propulsion unit, is manufactured by Thiokol Corporation, Huntsville, AL. The Antares IIIA rocket motor, SCOUT third stage propulsion unit, was manufactured by the Thiokol Corporation, Elkton Division. The motor case was Kevlar and epoxy composite. The nozzle used a 4D carbon/carbon throat insert, a carbon-phenolic exit cone and titanium housing. The Altair IIIA rocket motor, SCOUT fourth stage propulsion unit, is manufactured by Thiokol Corporation, Elkton Division. The motor is filament wound of fiberglass. The four solid propellant rocket motors of the SCOUT vehicle were joined by interstage structures referred to as "transition sections". Each transition section was divided into upper and lower portions at the stage separation plane. Frangible "blow-out" diaphragms joined the first and second, and the second and third stages. The diaphragm formed an internal clamp by the threaded periphery that engaged two structural threaded rings at the separation plane. Blast pressure of the upstage motor ruptured the diaphragm, disengaging the periphery and allowing the stage to separate. The third and fourth stages were joined by a "cold separation" arrangement of springs held compressed by a clamp retainer flange. Explosive bolt clamps released the flanges, effecting separation by spring loaded ejection force. The optional fifth stage included a similar cold separation system.

Historical Launch Systems

United States: SCOUT Vehicle
(continued)

Payload Fairing

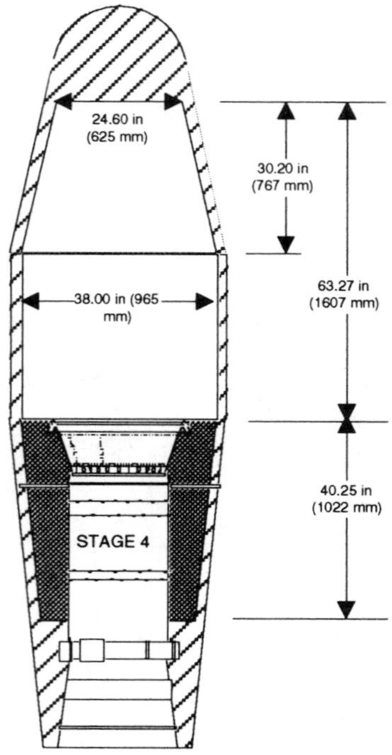

	34 in (0.86 m) Diameter (Short)	34 in (0.86 m) Diameter Long	42 in (1.07 m) Diameter
Length	10.7 ft (3.27 m)	12.0 ft (3.65 m)	12.4 ft (3.78 m)
Diameter	2.8 ft (0.86 m)	2.8 ft (0.86 m)	3.5 ft (1.07 m)
Mass	260 lb (118 kg)	296 lb (134 kg)	356 lb (161.5 kg)
Sections	2	2	2
Structure	Monocoque	Monocoque	Monocoque
Material	Honeycomb composite	Honeycomb composite	Honeycomb composite

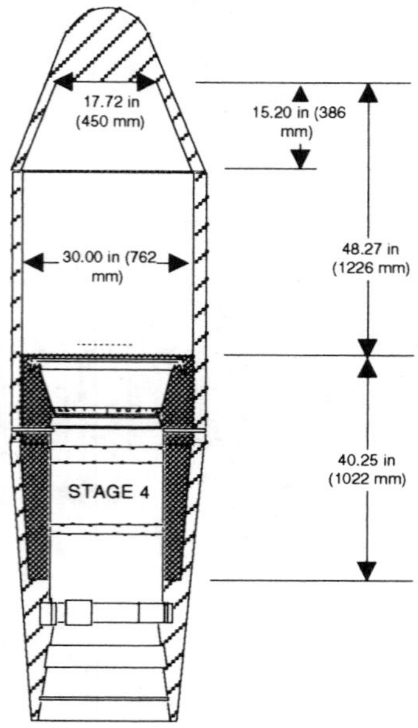

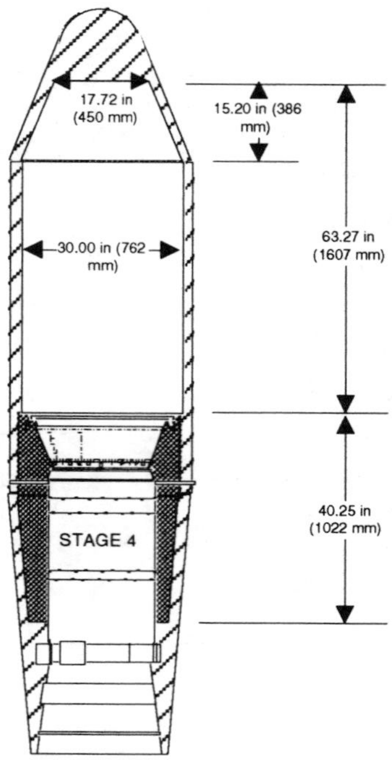

Remarks

Three payload fairings were available for the SCOUT vehicle, two 34 in. (0.86m) diameter and one 42 in. (1.07m) diameter. These payload fairings were a fiberglass laminate and honeycomb composite structure with a metal nose cap attached to the launch tower side of the shell. Steel half rings at the base of the payload fairing accepted a joining clamp. Shell restraint was provided through a series of overcenter latches along the separation plane and a joining clamp at the base. The latches and clamp were triggered and released by drawbars attached to bellcranks and a ballistic actuator near the forward end of the payload fairing. The ballistic actuator contained dual squibs, either of which provided sufficient energy to cause separation. Upon release of the latches and clamp, contained springs forced the payload fairing apart. A bumper is located in the payload fairing at the forward flange of the fourth stage motor case. Loads were distributed between payload fairing and motor through the bumper. Cutouts in the bumper were available.

Avionics

The avionics system provided an attitude reference and the resultant control signals necessary for stabilization of the vehicle in its three orthogonal axes corresponding to pitch, yaw, and roll during vertical probe, reentry, or orbital flight programs.

The yaw and roll axes were maintained at the launch reference while the pitch axis was programmed through a preselected angle corresponding to the desired vehicle zero-lift trajectory. Miniature integrating rate gyros contained within the inertial reference package detected any angular deviation about the vehicle programmed path and generate proportional error signals. These error signals were then summed with corresponding rate signals and were transmitted to the appropriate control subsystem such that the vehicle was continuously programmed to the gyro reference axes. If required, such as in the case of a dog-leg trajectory, the yaw axis could also be programmed through a preselected angle.

In addition to the "strapped down" gyro sensors, the system contained a relay unit for power and ignition switching, an intervalometer to provide precise scheduling of events during flight, a programmer to provide torquing voltages to the pitch gyro, an electronic signal conditioner to convert the gyro outputs to proper control signals, and the associated 400 cycle inverter and batteries.

Instrumentation System. Vehicle operations, through fourth stage separation, were monitored by a 21-channel system. Two of the channels were commutated allowing an additional 56 data signals at a repetition rate of 10 samples per second. The transmitter operated at 2230.5 mhz. The nominal RF power output was 6 watts delivered to three aerodynamic blade antennas located below the fairing. Fourth stage operation was monitored by a fourth stage 12-channel system. The frequency range and operation was similar to that of the primary telemetry system.

Radar Tracking System. A beacon was employed for radar tracking. Minimum peak RF power output was 500 watts, single pulse. Tracking data were generally available up to ignition of the fourth stage.

United States: SCOUT Vehicle
(continued)

Attitude Control System

First Stage Controls. In the liftoff configuration, the vehicle was aerodynamically stable. A proportional control system featuring a combination of jet vanes and aerodynamic tip control surfaces, operated by hydraulic servo actuators, was used to control the vehicle throughout the entire first stage burning period. The jet vanes provided the majority of the control force during the thrusting phase. The aerodynamic tip controls provided all the control force during the coasting phase following burnout of the first stage.

Second and Third Stage Controls. Second and third stage reaction control systems (RCS) were based on the same concept of operation as the first stage but differed in the method used to generate the control force. The control forces for these two stages were provided by hydrogen peroxide (H_2O_2) reaction jet motors which were operated as "on off" systems. The motors were so placed that moments were set up about each of the three axes: pitch, yaw, and roll. The motors were monopropellant and utilized 90% H_2O_2. Propellant pressurization was provided by compressed nitrogen (N_2) gas. Thrust motors for second stage included four 500-lb (227 kg) and four 40-lb (18 kg) motors fed by ten H_2O_2 tanks, and for the third stage included four 60-lb (27 kg) and four 14-lb (6.4 kg) motors fed by two H_2O_2 tanks.

Fourth Stage Spin Up System. The fourth stage, including the payload, received the proper spatial orientation from the control exerted by the first three stages after which it was spin-stabilized by a combination of four impulse spin motors. The miniaturized rocket spin motors were mounted tangentially in the skirt at the base of the fourth stage. Spin up began approximately 6 sec prior to fourth stage ignition.

Historical Launch Systems

United States: SCOUT Performance
(continued)

WFF (37.7°) / SCOUT Performance

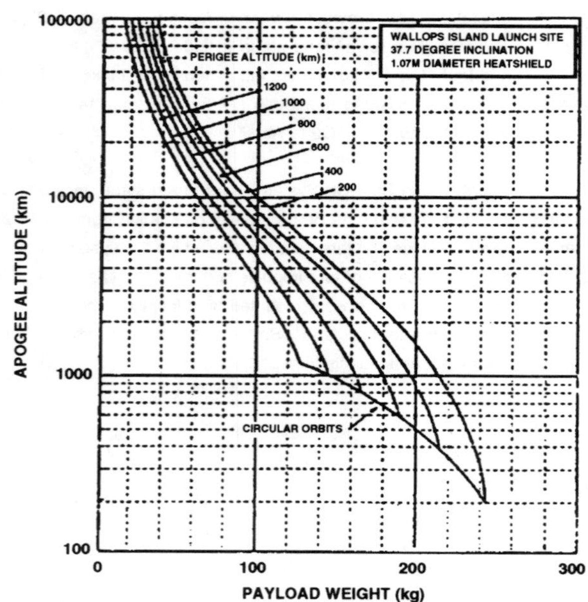

VAFB (90°) / SCOUT Performance

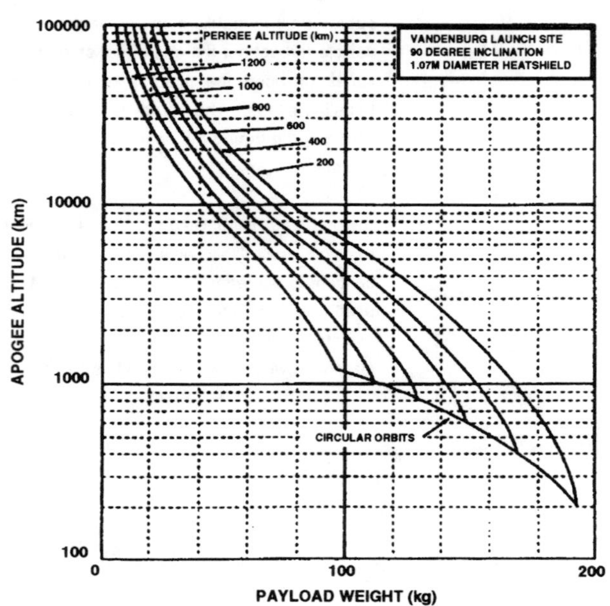

San Marco (2.9°) / SCOUT Performance

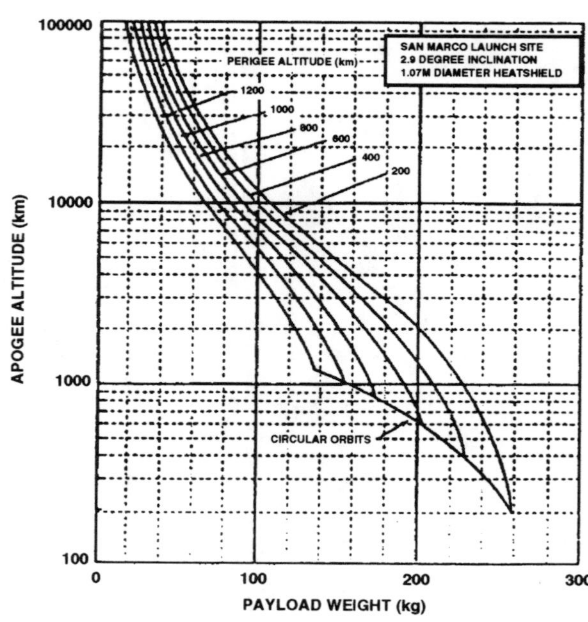

United States: SCOUT Operations
(continued)

Launch Site

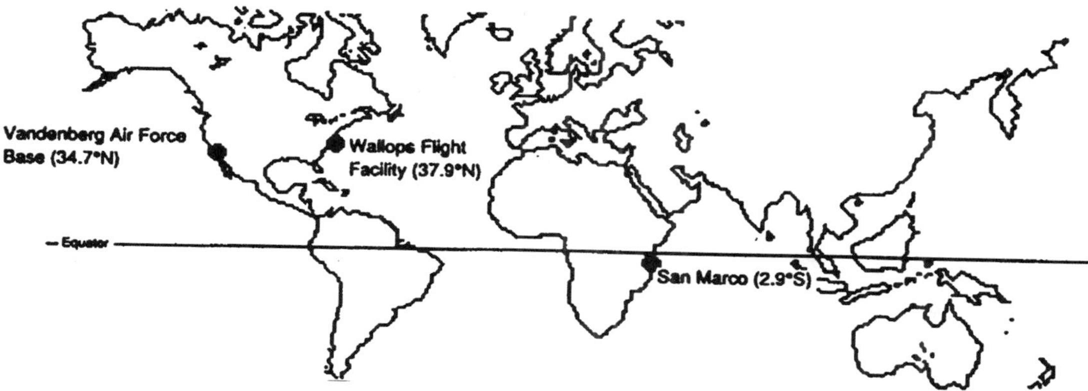

Launch Facilities

Wallops Flight Facility

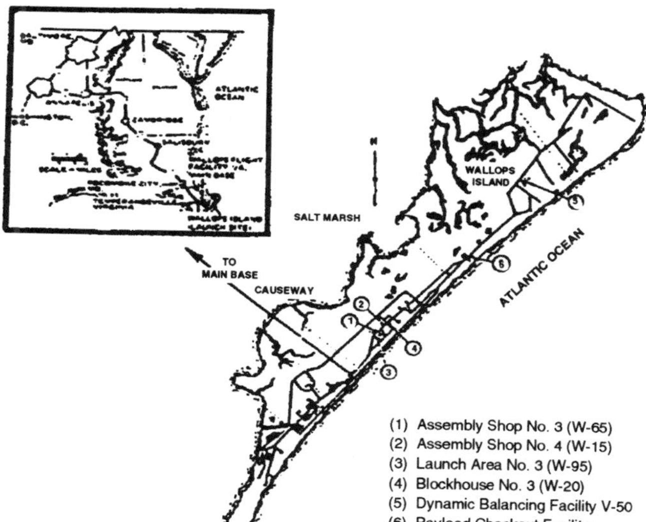

WFF Launch Facilities Layout

Vandenberg Air Force Base

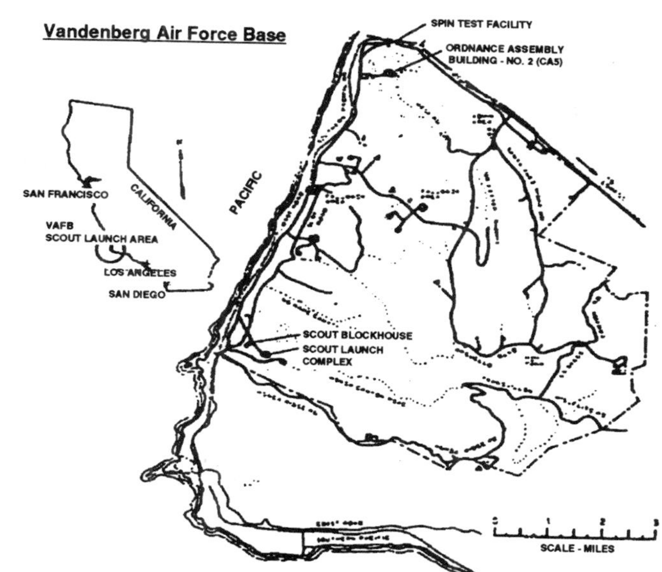

VAFB Launch Facilities Layout

San Marco

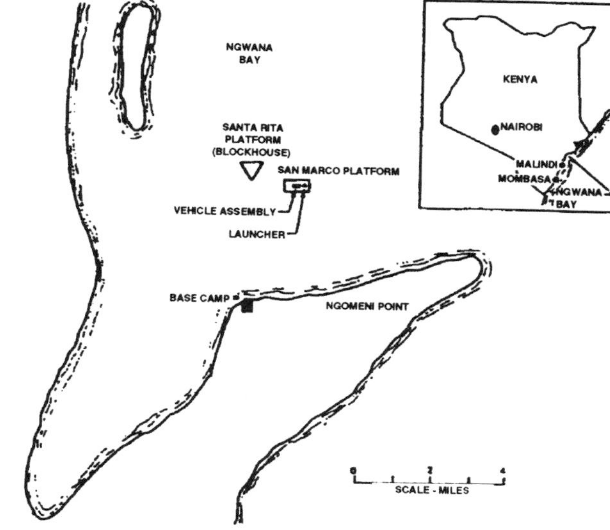

San Marco Launch Facilities Layout

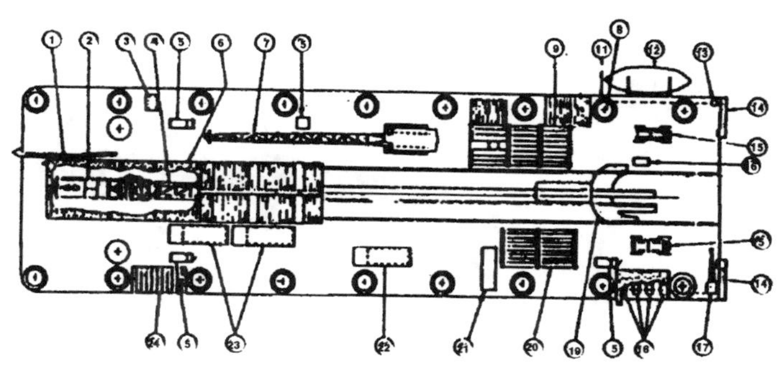

1. Submarine Cable Ramp
2. Launcher
3. Pyrotechnic Storage Room
4. Vehicle Transporter
5. Hatch
6. Shelter S-1
7. Crane, 35 Ton
8. Anemometer Pole
9. Shelter S-2
10. Kitchen / Dining Area
11. Water Reservoir
12. Life Boat
13. Test Rocket Launcher
14. Platform Access Ladders
15. Platform Air Compressors
16. Capstan Hut
17. Crane, 5 Ton
18. Power Plant
19. Nike Launcher Pad
20. Shelter S-3
21. Workshop Trailer Van
22. Electrical Trailer Van
23. Vehicle Test Set Trailer Van
24. H202 Storage Shed

San Marco Platform

Historical Launch Systems

United States: SCOUT Operations
(continued)

Launch Processing

There were three locations with the facilities capable of launching the SCOUT vehicle. The Wallops Flight Facility (WFF) was used for eastern launches, the Vandenberg Air Force Base was used for high inclination missions and the San Marco platform was used for low inclination missions.

Wallops Flight Facility. The launch site for SCOUT vehicle on the East Coast of the United States was located at the Wallops Flight Facility (WFF) on the Atlantic Coast of the Delmarva Peninsula. WFF consisted of the Wallops Island firing range and the administrative and technical service support facilities on the nearby mainland. Wallops Flight Center is located in a rural area where commercial sources of material and/or services are limited.

There were three key facilities for SCOUT at WFF. (1) The SCOUT Assembly Building No. 3 is located on the island near the launch area No. 3 site. The assembly building is a large L-shaped building used for receiving, storage, assembly, and checkout of the SCOUT vehicle. The building is divided into six bays, separated by heavy partitions of reinforced concrete. Bays 2 through 6 were used for individual step assembly buildup areas. After the individual steps had been assembled they were moved to bay 1 for vehicle buildup. A transporter trailer was the work platform on which the SCOUT assembly and checkout was performed. After the SCOUT checkout was completed the SCOUT vehicle was transported to the launcher. (2) The Dynamic Balancing Facility, located on the north end of Wallops Island, provided a means for dynamically balancing the payload, its mounting hardware, adapters, and fourth stage rocket motor. (3) The payload assembly and checkout facilities were temperature controlled masonry buildings located midway between the Scout launch area No. 3 and the Spin Test Facility.

Vandenberg Air Force Base. The launch site for SCOUT vehicles on the West coast of the United States was located at Vandenberg Air Force Base (VAFB), CA. There were three key facilities for SCOUT at VAFB. (1) The NASA Spacecraft Laboratory Building, Bldg. 836 is located in the administration area of south Vandenberg AFB. This building houses areas available for spacecraft agency offices in the west half of the building, and the NASA GSFC Spacecraft Laboratory in the east half. The Spacecraft Laboratory maintains a clean room. (2) The Ordnance Assembly Building at South Vandenberg AFB was used for the SCOUT vehicle. It is called the OAB No. 2 (Building 960). This building was used for motor checkout, SCOUT vehicle assembly buildup and systems checkout. (3) The Spin Test Facility at South Vandenberg AFB consists of a control building, spin balance building, and spin test operations support building. The vehicle was launched from Space Launch Complex (SLC) 5.

San Marco. The San Marco Equatorial Mobile Range is located in Ngwana Bay, formerly Formosa Bay, of the Indian Ocean off the east coast of Kenya, Africa. The launch complex consists of the San Marco and Santa Rita floatable platforms. Fabrication of the San Marco complex was performed by the Italian government. The San Marco platform served as the launching platform and contained the launcher and ground support equipment necessary for vehicle assembly, checkout, and launch. Vehicle assembly and checkout was performed inside the launcher shelter on the transporter. Spacecraft assembly and preinstallation checks could be performed in Shelter S-3, an environmentally controlled room. The 17 ft (5.2 m) by 25 ft (7.6 m) shelter is located on the deck of the San Marco platform. The Santa Rita platform served as a combined range control blockhouse and logistics facility. The Range Control Center is located in a compartment below deck. Vehicle launch control was conducted from a trailer-van blockhouse located on the main deck.

Logistic support was provided from a base camp located on the mainland nearby. The range is accessible by air or sea from all parts of the world. Nairobi is the port-of-entry for air passengers and cargo in Kenya. Mombasa is the port-of-entry for sea shipments. The SCOUT rocket motors and other hazardous material shipped by sea were normally delivered directly to the San Marco platform.

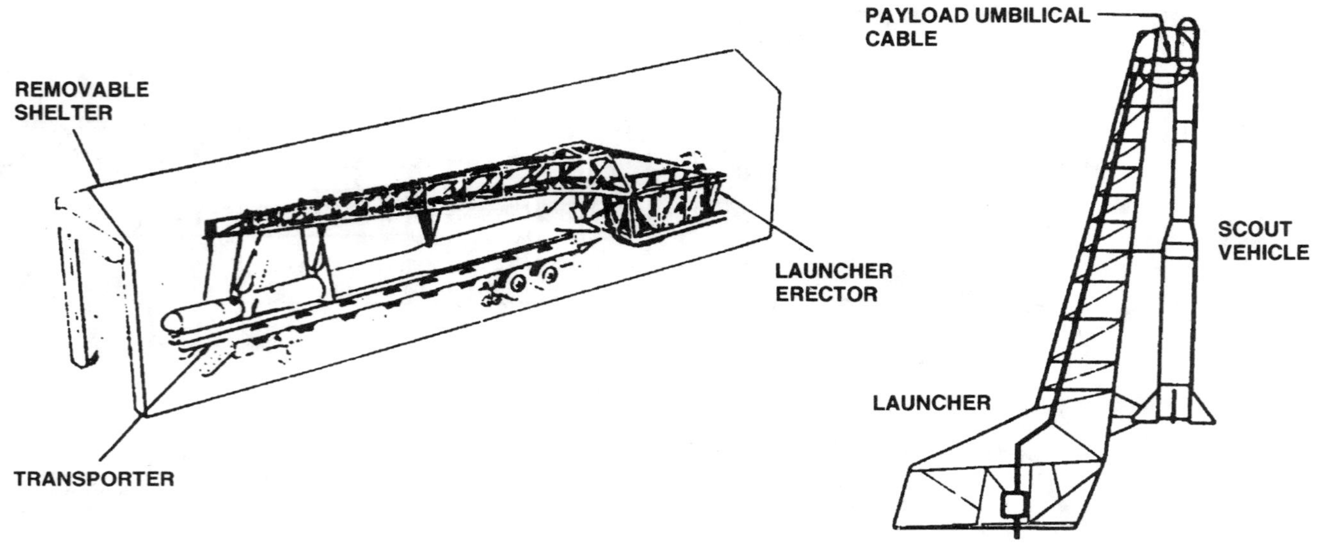

United States: SCOUT Operations
(continued)

Launch Processing
(continued)

SCOUT Standard Launch Complex. The major components of the SCOUT Standard Launch Complex were the launcher, the movable shelter, and the blockhouse. The configuration of the launch complex at each range was functionally identical but differed physically.

The SCOUT launch complex employed a dual-purpose launcher which, permitted checkout of the vehicle in the horizontal position and launching of the vehicle in the vertical position. The launcher was provided with a movable base which permitted azimuth control up to 140 degrees. This range of azimuth alignment exceeded the allowable limits established by all three ranges. A cantilevered elevating launch boom provided pitch control to the 90 degree position. The SCOUT could be launched at any angle from vertical to 20 degrees from vertical. Vehicle erection and azimuth positioning were accomplished by a blockhouse-controlled electromechanical system which included a position-indicating display of the blockhouse. All vehicle electrical connections were cabled to the blockhouse by way of a fly-away umbilical connector. The payload umbilicals could be either pull-away or fly away.

The movable shelter provided controlled environmental protection for prelaunch operation of the SCOUT vehicle. It is a galvanized steel building, mounted on rails, with power operated doors on each end and work platforms along each side. The shelter is approximately 140 ft (42.7 m) long, 24 ft (7.3 m) wide, and 31 ft (9.4 m) high.

Vehicle launch control was performed from the blockhouse by the launch team operating under the direction of a test director. The SCOUT launch control consoles were arranged in the blockhouse in two rows with the operator consoles making up the front row and the supervisor consoles making up the back row.

Launch Operations. Launch site operations began approximately 36 working days before launch (R-36) when the rocket motors were received, uncrated, and inspected. Each motor took two to five days to complete. At R-33, the interstage assemblies were received, uncrated, and inspected. This operation took approximately three days to complete. At R-30, the vehicle interstage and rocket motor assembly was started. The base A and first stage motor were assembled on the transporter. B section and second stage motor, C section, third stage motor, and D section assemblies were assembled on dollies and then transferred to the transporter for vehicle assembly. Approximately 11 days were required to assemble the vehicle. At R-19, the assembled vehicle systems checks were started, approximately 10 days were required to complete these checks. On R-11, the all systems and the preliminary RFI checks were performed. At R-9, the assembled vehicle was mated on the launcher.

Operation at the pad began with the blockhouse modifications for the payload, and preparation of the launcher. This task, depending upon complexity, took from 6 to 10 days. At R-9, the assembled vehicle was mated to the launcher followed by an electronic functional test on R-8. The balanced fourth stage/ payload assembly was installed at R-6. The ordnance items were installed at R-4. A payload umbilical retraction check was optional at R-2 when the dress rehearsal of the operation countdown was preformed. Final vehicle preparations were made the day before launch.

Operations	Begin	Delta Time
Prepare Launch Complex	R-29	10 days
Blockhouse Modification for Payload	R-25	1–4 days
Launch Complex Operational Check	R-19	2 days
Mate Vehicle to Launcher	R-9	1 day
Electronic Functional	R-8	1 day
Payload Installation and Checks	R-6	1 day
Fourth Stage Telemetry Checks	R-5	1 day
Ordnance Installation	R-4	3 days
Heatshield Installation and Final Inspection	R-3	1 day
Dress Rehearsal	R-2	1 day
Recovery (Prepare Vehicle for Launch)	R-1	1 day
Final Countdown	R-0	

Typical On-Pad Operations

Historical Launch Systems

United States: SCOUT Operations
(continued)

Flight Sequence

Countdown Time	Task Title	Delta Time
T-7 hr	Communications Checks	5 min
T-6 hr 55 min	Launch Console and Vehicle GSE System Activation	25 min
T-6 hr 30 min	Electronics Systems Checkout	165 min
T-X	Spacecraft Checkout	
T-4 hr 45 min	Reaction Control System Fueling	105 min
T-3 hr	Vehicle/Launcher Securing and Erection	75 min
T-1 hr 45 min	Ignition and Destruct Systems Checkout	45 min
T-1 hr	Countdown Evaluation	30 min
T-30 min	Terminal Countdown	30 min

T-X: Spacecraft functional checkout time may vary due to different requirements or design of each spacecraft and therefore may occur prior to T-320 min. Normally spacecraft checks are conducted concurrent with vehicle tasks and may start prior to vehicle tasks if required.

Typical Launch Countdown Sequence

SCOUT Typical Flight Sequence

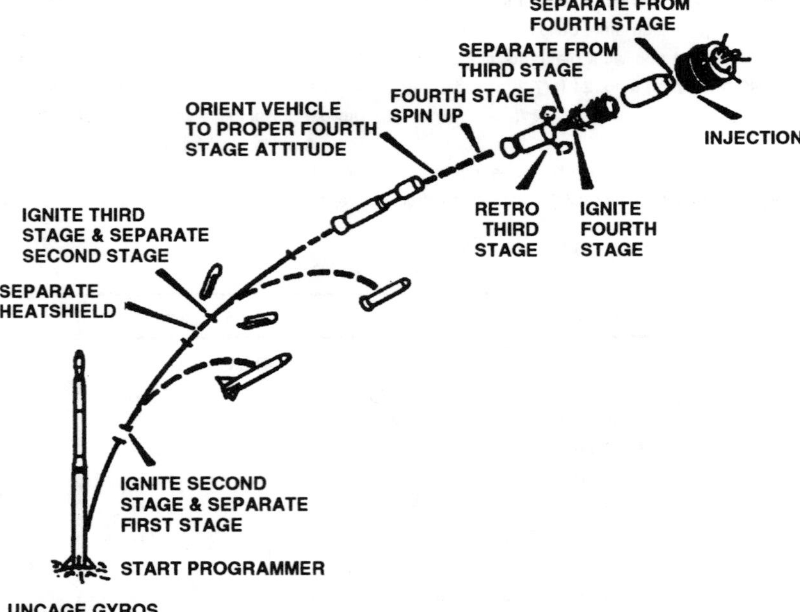

SCOUT
(270 nm (500 km) circular)

Flight Time (min:sec)	Events
00:00	Stage 1 Ignition
01:24.84	Stage 1 Burnout
01:29.30	Stage 2 Ignition
02:10.65	Stage 2 Burnout
02:13.95	Fairing Ejection
02:15.65	Stage 3 Ignition
03:04.04	Stage 3 Burnout
09:27.59	Stage 4 Ignition
10:01.45	Stage 4 Burnout

United States: SCOUT Payload Accommodations
(continued)

Payload Compartment	
Maximum Payload Diameter	30.00 in (762 mm) for 34 in fairing - opt 38.00 in (965 mm) for 42 in fairing - std
Maximum Cylinder Length	33.07 in (840 mm) for 34 in (short) - opt 48.07 in (1220 mm) for 34 in (long) - opt 33.07 in (840 mm) for 42 in fairing - std
Maximum Cone Length	15.20 in (386 mm) for 34 in fairing - opt 30.20 in (767 mm) for 42 in fairing - std
Payload Adapter	
Interface Diameter	10.20 in (259 mm) designated E-G - opt 10.25 in (260 mm) designated 200E - opt 24.25 in (616 mm) designated 25E - std
Payload Integration	
Nominal Mission Schedule Begins	T-24 months
Launch Window	
Latest Countdown Hold Not Requiring Recycling	T-2 min
On-Pad Storage Capability	Indefinite
Latest Access to Payload	T-3 hrs
Environment	
Maximum Load Factors	+9.8 g axial
Minimum Lateral / Longitudinal Payload Frequency	15–30 Hz / ? Hz
Maximum Overall Acoustic Level	137 dB (one third octave)
Maximum Flight Shock	30 g at 75–100 hz, longitudinal 60 g at 200–225 hz, lateral
Maximum Dynamic Pressure on Fairing	2,100 lb/ft^2 (100,560 N/m^2)
Maximum Pressure Change in Fairing	2 psi/s (14 KPa/s)
Cleanliness Level in Fairing (Prior to Launch)	Class 10,000
Payload Delivery	
Standard Orbit and Accuracy (3 sigma)	300 nm (550 km) circular orbit: ±59 nmi (110 km), ±0.9° inclination
Attitude Accuracy (3 sigma)	±1.23°
Nominal Payload Separation Rate	2.2–5.8 ft/s (0.67–1.76 m/s)
Deployment Rotation Rate Available	140–180 rpm flight spin rate
Loiter Duration in Orbit	None
Maneuvers (Thermal / Collision / Telemetry)	No

Historical Launch Systems

United States: Thor

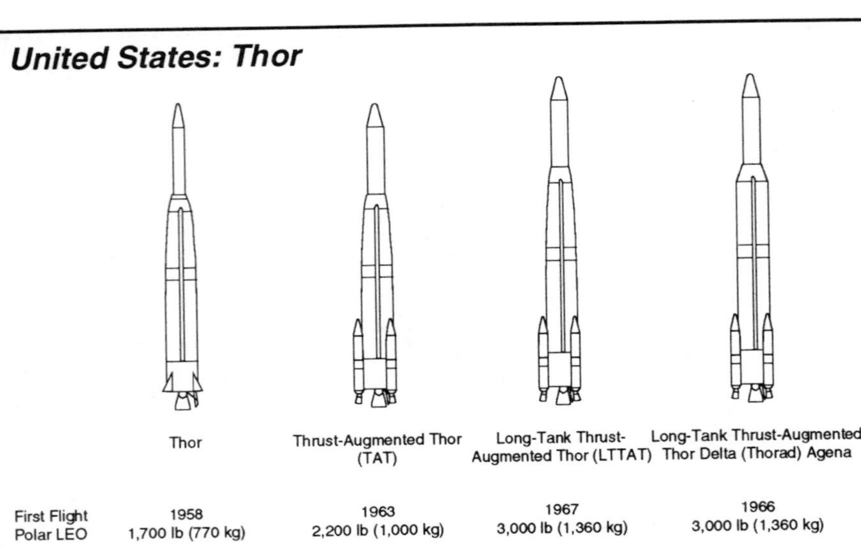

YEAR	DATE	VEHICLE	SITE	PAYLOAD
1958	Aug-17	Thor Able	CCAFS	Pioneer
	Failure - Stg 1 malfunction			
	Oct-11	Thor Able	CCAFS	Pioneer 1
	Nov-8	Thor Able	CCAFS	Pioneer 2
	Failure - Stg 3 ignition unsuccessful			
1959	Feb-28	Thor Agena A	VAFB	Discover 1
	Apr-13	Thor Agena A	VAFB	Discover 2
	Jun-3	Thor Agena A	VAFB	Discover 3
	Failure - After Agena ignition, no telemetry			
	Jun-25	Thor Agena A	VAFB	Discover 4
	Failure - Insufficient stg 2 velocity			
	Aug-7	Thor Able	CCAFS	Explorer 6
	Aug-13	Thor Agena A	VAFB	Discover 5
	Aug-19	Thor Agena A	VAFB	Discover 6
	Sep-17	Thor Able	CCAFS	Transit 1A
	Nov-7	Thor Agena A	VAFB	Discover 7
	Nov-20	Thor Agena A	VAFB	Discover 8
	Failure - Stg 1 premature cutoff			

SUMMARY OF REMAINING YEARS — **Success/Total**	Thor	TAT	LTTAT	Thorad-Agena
Pre-1967	147 / 155	54 / 57	—	1 / 1
1967	4 / 4	6 / 6	6 / 6	—
1968	2 / 2	3 / 3	7 / 7	0 / 1
1969	1 / 1	—	3 / 3	7 / 7
1970	2 / 2	1 / 1	—	6 / 6
1971	3 / 3	—	1 / 1	4 / 5
1972	2 / 2	—	—	2 / 2
1973	1 / 1	—	—	—
1974	2 / 2	—	—	—
1975	1 / 1	—	—	—
1976	1 / 2	—	—	—
1977	1 / 1	—	—	—
1978	1 / 1	—	—	—
1979	1 / 1	—	—	—
1980	0 / 1	—	—	—
Total	169 / 179	64 / 67	17 / 17	20 / 22

Launch site was Cape Canaveral Air Force Station, FL (28.5°N, 81.0°W)
Vandenberg Air Force Base, CA (34.7°N, 120.6°N)

In the early 1950s, the Air Force determined that they needed their own intermediate range ballistic missile (IRBM). A joint Army-Navy program, Jupiter, was under way when the Air Force awarded the Douglas Aircraft Company the research and development contract for the Thor IRBM weapon system on December 27, 1955.

Since the United States believed the Soviets to be two to three years ahead in ballistic missile development, the Thor program was placed on a "maximum risk" basis in which the primary objective was gross performance. The necessary scientific and engineering groundwork had to be completed within a year. On October 26, 1956, just 10 months after the contract was signed, the first Thor missile was delivered. The first launch on January 25, 1957, failed because a weld on the LOX transfer line ruptured. The next four launches also failed. By the ninth attempt, however, the missile reached a range of 2,300 mi (3,700 km).

The Thor IRBM, designated Douglas Standard Vehicle number two (DSV-2), was 60 ft in length and 8 ft (2.4 m) in diameter at its widest section. It carried one nuclear warhead and had a range of approximately 2,300 mi (3,700 km). Sixty of these ballistic missile configuration were placed in Great Britain from 1957 to 1963.

The Thor also was mated with upper stages and used as a space launch vehicle. The original booster, designated DSV-2A, was 56 ft (17 m) in length and 8 ft (2.4 m) in diameter. It employed one Rocketdyne MB-3 Blk II engine (also known as LR-79-NA11) and burned LOX/RJ-1 to produce 150,000 lb (667,000 N) of thrust at liftoff. Two small vernier engines provided roll control and pitch adjustment for the main engine. The verniers were also produced by Rocketdyne and created 1,000 lb (4,400 N) of thrust each, using the same propellants as the main engine. The verniers were designed to continue burning nine seconds after main engine cutoff. The vehicle weighed 107,230 lb (48,640 kg) at liftoff, not including the upper stage and payload.

Vehicle	**Thor**	
System Height	Up to 110 ft (33.5 m)	
Gross Mass	Up to 169K lb (51.5K kg)	
Upper Stage	Thor - Burner II/IIA, Agena A/B/D, Altair, FW-4; TAT-Agena B/D; LTTAT-Agena D; Thorad-Agena D	
Stage 0	Castor I (Thiokol TX-33-52)	
- Diameter	2.6 ft (0.8 m)	
- Length	19.75 ft (6.0 m)	
- Propellant	solid	
Stage 1	Thor	
- Diameter	8 ft (2.4 m)	
- Length	56 ft (17.1 m), long-tank 67 ft (20.4 m)	
- Propellant	LOX / RJ-1	
- Engine / Thrust	MB-3 Block 2 / 150K lb (667K N) MB-3 Block 3 / 170K lb (756K N)	
Stage 2	Burner II	Agena D
- Length	- 5.17 ft (1.6 m)	- 20.5 ft (6.25 m)
- Diameter	- 5.08 ft (1.55 m)	- 5 ft (1.5 m)
- Propellant	- solid	- storable liquid
- Thrust	- 10K lb (44K N)	- 16K lb (71K N)

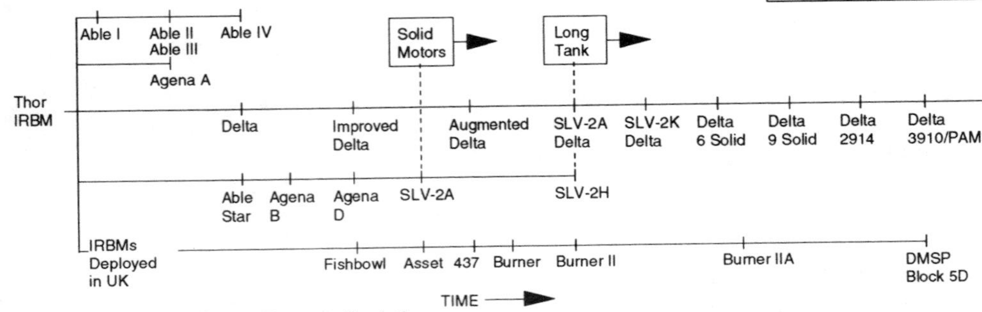

Thor IRBM / Space Booster Evolution

Reference:
- A Historical Look at US Launch Vehicles 1967-present, ANSER, 2nd edition, February 1990
- U.S. Civilian Space Programs 1958-78, House Subcommittee on Space Science & Applications
- US Space Launch Systems, Navy Space Systems Activity, NSSA-R-20-72-2, Revision 1 July 1977
- TRW Space Log 1957-87

United States: Thor
(continued)

The DSV-2A was mated with either the Lockheed Agena A, Agena B, or Agena D upper stage. The first Thor-Agena A was launched on February 28, 1959, the first Thor-Agena B on October 26, 1960, and the first Thor-Agena D on June 27, 1962.

In 1963, three Thiokol TX-33-52 (Sergeant) solid rocket boosters and an improved main engine, the MB-3 Blk III (LR-79-NA13) were added to increase liftoff thrust to 317,050 lb (1,410,300 N). The three solid rocket boosters, which were attached at the base of the Thor at 120 degree intervals, increased total booster weight at liftoff to 135,400 lb (61,420 kg). The booster's structure was modified to support the increased thrust levels. This vehicle, designated the DSV-2C became known as the Thrust-Augmented Thor (TAT) (see Figure II.A-2). Two DSV-2Cs were flown with the structural modifications but without the solids in February and March 1963. The first TAT using solids was launched on February 28, 1963. The Agena D, and later the Delta, upper stages were mated with the TAT.

Nearly all Thor-Agena combinations were flown by the Air Force. Other second stages were available and used by NASA and other agencies. Thor-Able was first flown by NASA on October 11, 1958. The Able upper stage used the second and third stage of Vanguard. The successful Thor-Delta (later to be known as the Delta) was first flown on May 13, 1969 (see the Delta section for the history of this vehicle). The United States Navy used the Thor booster mated with the Able Star and launched this combination 16 times, beginning on June 22, 1960. The Air Force mated an Altair upper stage to the Thor booster, and this combination was flown six times, the first on January 19, 1965. In addition, the Air Force used the Burner II and Burner IIA upper stages, flying this combination frequently after the first launch in September 1966.

In addition to the TAT, other modifications were made to the original Thor booster. The Long Tank Thrust-Augmented Thor (LTTAT or DSV-2L), as its name implies, was a stretched version of the TAT. The tank lengths were increased 11 ft (3.35 m), from 56 to 67 ft (17.1 to 20.4 m), which increased the burn time by 21 seconds, from 146 to 167 seconds. The LTTAT was first flown on May 9, 1967 with an Agena D upper stage. The Burner II and Delta upper stages were also used with the LTTAT. In 1966, the Long-Tank Thrust-Augmented Thor Delta (Thorad) was developed. The Thorad was launched with Agena D, Agena B, and the Delta upper stages.

As a missile or space booster, the Thor was launched over 500 times. The last launch was a Thor-Delta on October 6, 1981. The TAT Delta, LTTAT Delta, and Thorad-Delta (primarily used by NASA) are often referred to as Delta launch vehicles [see the *Delta* section for more information]. There currently are nine unassigned Thor vehicles in storage at Norton Air Force Base.

United States: Vanguard

	Vanguard
First Flight	1957
East LEO	20 lb (9 kg)

YEAR	DATE	VEHICLE	SITE	PAYLOAD
1957	Dec-6	Vanguard	CCAFS	Vanguard TV3
	Failure - Lost after 2 seconds			
1958	Feb-5	Vanguard	CCAFS	Vanguard TV3 (backup)
	Failure - Control system malfunction			
	Mar-17	Vanguard	CCAFS	Vanguard 1
	Apr-28	Vanguard	CCAFS	Vanguard TV5
	Failure - Stg 3 ignition malfunction			
	May-27	Vanguard	CCAFS	Vanguard SLV 1
	Failure - Improper stg 3 trajectory			
	Jun-26	Vanguard	CCAFS	Vanguard SLV 2
	Failure - Premature stg 2 cutoff			
	Sep-26	Vanguard	CCAFS	Vanguard SLV 3
	Failure - Insufficient stg 2 thrust			
1959	Feb-17	Vanguard	CCAFS	Vanguard 2
	Apr-13	Vanguard	CCAFS	Vanguard SLV 5
	Failure - Stg 2 damaged at separation			
	Jun-22	Vanguard	CCAFS	Vanguard SLV 6
	Failure - Stg 2 propulsion malfunction			
	Sep-18	Vanguard	CCAFS	Vanguard 3

Vehicle	Vanguard
System Height	72 ft (21.9 m)
Gross Mass	22.6K lb (10.25K kg)
Stage 1	
- Heritage	Viking sounding rocket
- Diameter	45 in (1.1 m)
- Propellant	LOX / RP-1
- Thrust	27K lb (120K N)
Stage 2	
- Heritage	Aerobee sounding rocket
- Propellant	UDMH / WF nitric acid
- Thrust	7.5K lb (33K N)
Stage 3	
- Propellant	Solid

Launch site was Cape Canaveral Air Force Station, FL (28.5°N, 81.0°W)

In 1955, on the recommendation of the U.S. National Committee for the International Geophysical Year, the United States decided to launch a series of Earth satellites to conduct a range of scientific experiments. There were several options open, and the Homer Joe Stewart Committee of the Department of Defense decided of these that a proposal from the Naval Research Laboratory (NRL) was the best choice. Behind the scenes President Eisenhower was said to be in favor of putting a maximum separation between any developing military missiles and the new satellite project.

The NRL Vanguard was to be derived from the scientific sounding rocket, the Viking, a vehicle whose design effort started in 1946 and grew out of V-2 technology. Two remaining Vikings were used to conduct suborbital flights of test payloads smaller than the full-scale Vanguard payloads to follow. Viking needed a number of changes before it could be used for orbital purposes under the Vanguard name, including a boost in its capacity plus the addition of upper stages. The original Reaction Motors Company (later merged with Thiokol) rocket motor of 20,000 lb (88,000 N) thrust was replaced by a General Electric 27,000 lb (120,000 N) thrust motor developed in the Hermes program, a whole collection of projects aimed at developing rocket technology. Fuel was switched from LOX and alcohol to LOX and RP-1. The Martin Company (later Martin Marietta) was retained for building the stage.

For the second stage, the Navy's Aerobee sounding rocket was selected for adaptation. This Aerojet improved Aerobee rocket had a thrust of 7,500 lb (33,000 N) and used storable propellants, UDMH and white fuming nitric acid (WFNA). The third stage was to be a new solid-fueled rocket with competing contracts given to Grand Central Rocket and to Allegany Ballistics Laboratory (ABL) of the Hercules Powder Company. In the end, it was the latter version which was used.

The whole assemblage was 72 ft (21.9 m) long and 45 in. (1.1 m) in diameter. The takeoff weight was 22,600 lb (10,250 kg), and the goal was to put a satellite payload of 20 lb (9 kg) in orbit.

The Vanguard received bad press because of its delay and its several spectacular failures. In fact, due to these problems the first successful U.S. launch was conducted by a Jupiter C vehicle. The second U.S. satellite, however, was put up by Vanguard, and two more followed later, after which it was dropped from use. Flights in the Vanguard program occurred between 1956 (Vikings) or 1957 (Vanguard) and 1959.

Reference:
- U.S. Civilian Space Programs 1958-78, House Subcommittee on Space Science & Applications
- TRW Space Log 1957-87